Lecture Notes in Artificial Intelligence 11051

Subseries of Lecture Notes in Computer Science

More information about this series at http://www.springer.com/series/1244

Michele Berlingerio · Francesco Bonchi
Thomas Gärtner · Neil Hurley
Georgiana Ifrim (Eds.)

Machine Learning and Knowledge Discovery in Databases

European Conference, ECML PKDD 2018
Dublin, Ireland, September 10–14, 2018
Proceedings, Part I

Springer

Editors
Michele Berlingerio
IBM Research - Ireland
Dublin, Ireland

Neil Hurley ⓘ
University College Dublin
Dublin, Ireland

Francesco Bonchi
Institute for Scientific Interchange
Turin, Italy

Georgiana Ifrim ⓘ
University College Dublin
Dublin, Ireland

Thomas Gärtner
University of Nottingham
Nottingham, UK

ISSN 0302-9743   ISSN 1611-3349 (electronic)
Lecture Notes in Artificial Intelligence
ISBN 978-3-030-10924-0   ISBN 978-3-030-10925-7 (eBook)
https://doi.org/10.1007/978-3-030-10925-7

Library of Congress Control Number: 2018965756

LNCS Sublibrary: SL7 – Artificial Intelligence

This Springer imprint is published by the registered company Springer Nature Switzerland AG
The registered company address is: Gewerbestrasse 11, 6330 Cham, Switzerland

Preface

We are delighted to introduce the proceedings of the 2018 edition of the European Conference on Machine Learning and Principles and Practice of Knowledge Discovery in Databases (ECML-PKDD 2018). The conference was held in Dublin, Ireland, during September 10–14, 2018. ECML-PKDD is an annual conference that provides an international forum for the discussion of the latest high-quality research results in all areas related to machine learning and knowledge discovery in databases, including innovative applications. This event is the premier European machine learning and data mining conference and builds upon a very successful series of ECML-PKDD conferences.

The scientific program was of high quality and consisted of technical presentations of accepted papers, plenary talks by distinguished keynote speakers, workshops, and tutorials. Accepted papers were organized in five different tracks:

- The Conference Track, which featured research contributions, presented as part of the main conference program
- The Journal Track, which featured papers that were reviewed and published separately in special issues of the Springer journals *Machine Learning* and *Data Mining and Knowledge Discovery*, and that were selected as suitable for presentation at the conference
- The Applied Data Science Track, which focused on the application of data science to practical real-world scenarios, including contributions from academia, industry, and non-governmental organizations
- The Demo Track, which presented working demonstrations of prototypes or fully operational systems that exploit data science techniques
- The Nectar Track, which presented an overviews of recent scientific advances at the frontier of machine learning and data mining in conjunction with other disciplines, as published in related conferences and journals

In addition to these tracks, the conference also included a PhD Forum in which PhD students received constructive feedback on their research progress and interacted with their peers. Co-located with the conference, this year there were 17 workshops on related research topics and six tutorial presentations.

In total, 95% of the accepted papers in the conference have accompanying software and/or data and are flagged as Reproducible Research papers in the proceedings. This speaks to the growing importance of reproducible research for the ECML-PKDD community. In the online proceedings each Reproducible Research paper has a link to the code and data made available with the paper. We believe this is a tremendous resource for the community and hope to see this trend maintained over the coming years.

We are very happy with the continued interest from the research community in our conference. We received 353 papers for the main conference track, of which 94 were accepted, yielding an acceptance rate of about 26%. This allowed us to define a very rich program with 94 presentations in the main conference track. Moreover, there was a 26% acceptance rate (143 submissions, 37 accepted) to the Applied Data Science Track and 15% to the Journal Track (151 submissions, 23 accepted). Including the Nectar Track papers and some Journal Track papers from the 2017 special issues that were held over for presentation until 2018, we had in total 166 parallel scientific talks during the three main conference days.

The program also included five plenary keynotes by invited speakers: Misha Bilenko (Head of Machine Intelligence and Research Yandex, Moscow, Russia), Corinna Cortes (Head of Google Research New York, USA), Aristides Gionis (Professor, Department of Computer Science, Aalto University, Helsinki, Finland), Cynthia Rudin (Associate Professor of Computer Science and Electrical and Computer Engineering, Duke University, Durham, North Carolina, USA), and Naftali Tishby (Professor, School of Engineering and Computer Science, Hebrew University of Jerusalem, Israel).

This year, ECML-PKDD attracted over 630 participants from 42 countries. It attracted substantial attention from industry both through sponsorship and submission/participation at the conference and workshops. Moreover, ECML-PKDD hosted a very popular Nokia Women in Science Luncheon to discuss the importance of awareness of equal opportunity and support for women in science and technology.

The Awards Committee selected research papers considered to be of exceptional quality and worthy of special recognition:

- ML Student Best Paper Award: "Hyperparameter Learning for Conditional Mean Embeddings with Rademacher Complexity Bounds," by Kelvin Hsu, Richard Nock and Fabio Ramos
- KDD Student Best Paper Award: "Anytime Subgroup Discovery in Numerical Domains with Guarantees," by Aimene Belfodil, Adnene Belfodil and Mehdi Kaytoue.

We would like to thank all participants, authors, reviewers, and organizers of the conference for their contribution to making ECML-PKDD 2018 a great scientific event.

We would also like to thank the Croke Park Conference Centre and the student volunteers. Thanks to Springer for their continuous support and Microsoft for allowing us to use their CMT software for conference management and providing support throughout. Special thanks to our many sponsors and to the ECML-PKDD Steering Committee for their support and advice. Finally, we would like to thank the organizing institutions: the Insight Centre for Data Analytics, University College Dublin, Ireland, and IBM Research, Ireland.

September 2018

Michele Berlingerio
Francesco Bonchi
Thomas Gärtner
Georgiana Ifrim
Neil Hurley

Organization

ECML PKDD 2018 Organization

General Chairs

Michele Berlingerio   IBM Research, Ireland
Neil Hurley   University College Dublin, Ireland

Program Chairs

Michele Berlingerio   IBM Research, Ireland
Francesco Bonchi   ISI Foundation, Italy
Thomas Gärtner   University of Nottingham, UK
Georgiana Ifrim   University College Dublin, Ireland

Journal Track Chairs

Björn Bringmann   McKinsey & Company, Germany
Jesse Davis   Katholieke Universiteit Leuven, Belgium
Elisa Fromont   IRISA, Rennes 1 University, France
Derek Greene   University College Dublin, Ireland

Applied Data Science Track Chairs

Edward Curry   National University of Ireland Galway, Ireland
Alice Marascu   Nokia Bell Labs, Ireland

Workshop and Tutorial Chairs

Carlos Alzate   IBM Research, Ireland
Anna Monreale   University of Pisa, Italy

Nectar Track Chairs

Ulf Brefeld   Leuphana University of Lüneburg, Germany
Fabio Pinelli   Vodafone, Italy

Demo Track Chairs

Elizabeth Daly   IBM Research, Ireland
Brian Mac Namee   University College Dublin, Ireland

PhD Forum Chairs

Bart Goethals   University of Antwerp, Belgium
Dafna Shahaf   Hebrew University of Jerusalem, Israel

Discovery Challenge Chairs

Martin Atzmüller   Tilburg University, The Netherlands
Francesco Calabrese   Vodafone, Italy

Awards Committee

Tijl De Bie   Ghent University, Belgium
Arno Siebes   Uthrecht University, The Netherlands
Bart Goethals   University of Antwerp, Belgium
Walter Daelemans   University of Antwerp, Belgium
Katharina Morik   TU Dortmund, Germany

Professional Conference Organizer

Keynote PCO   Dublin, Ireland

www.keynotepco.ie

ECML PKDD Steering Committee

Michele Sebag   Université Paris Sud, France
Francesco Bonchi   ISI Foundation, Italy
Albert Bifet   Télécom ParisTech, France
Hendrik Blockeel   KU Leuven, Belgium and Leiden University, The Netherlands
Katharina Morik   University of Dortmund, Germany
Arno Siebes   Utrecht University, The Netherlands
Siegfried Nijssen   LIACS, Leiden University, The Netherlands
Chedy Raïssi   Inria Nancy Grand-Est, France
João Gama   FCUP, University of Porto/LIAAD, INESC Porto L.A.,
                         Portugal
Annalisa Appice   University of Bari Aldo Moro, Italy
Indré Žliobaité   University of Helsinki, Finland
Andrea Passerini   University of Trento, Italy
Paolo Frasconi   University of Florence, Italy
Céline Robardet   National Institute of Applied Science Lyon, France

Jilles Vreeken	Saarland University, Max Planck Institute for Informatics, Germany
Sašo Džeroski	Jožef Stefan Institute, Slovenia
Michelangelo Ceci	University of Bari Aldo Moro, Italy
Myra Spiliopoulu	Magdeburg University, Germany
Jaakko Hollmén	Aalto University, Finland

Area Chairs

Michael Berthold	Universität Konstanz, Germany
Hendrik Blockeel	KU Leuven, Belgium and Leiden University, The Netherlands
Ulf Brefeld	Leuphana University of Lüneburg, Germany
Toon Calders	University of Antwerp, Belgium
Michelangelo Ceci	University of Bari Aldo Moro, Italy
Bruno Cremilleux	Université de Caen Normandie, France
Tapio Elomaa	Tampere University of Technology, Finland
Johannes Fürnkranz	TU Darmstadt, Germany
Peter Flach	University of Bristol, UK
Paolo Frasconi	University of Florence, Italy
João Gama	FCUP, University of Porto/LIAAD, INESC Porto L.A., Portugal
Jaakko Hollmén	Aalto University, Finland
Alipio Jorge	FCUP, University of Porto/LIAAD, INESC Porto L.A., Portugal
Stefan Kramer	Johannes Gutenberg University Mainz, Germany
Giuseppe Manco	ICAR-CNR, Italy
Siegfried Nijssen	LIACS, Leiden University, The Netherlands
Andrea Passerini	University of Trento, Italy
Arno Siebes	Utrecht University, The Netherlands
Myra Spiliopoulu	Magdeburg University, Germany
Luis Torgo	Dalhousie University, Canada
Celine Vens	KU Leuven, Belgium
Jilles Vreeken	Saarland University, Max Planck Institute for Informatics, Germany

Conference Track Program Committee

Carlos Alzate	Roberto Bayardo	Indrajit Bhattacharya
Aijun An	Martin Becker	Marenglen Biba
Fabrizio Angiulli	Srikanta Bedathur	Silvio Bicciato
Annalisa Appice	Jessa Bekker	Mario Boley
Ira Assent	Vaishak Belle	Gianluca Bontempi
Martin Atzmueller	Andras Benczur	Henrik Bostrom
Antonio Bahamonde	Daniel Bengs	Tassadit Bouadi
Jose Balcazar	Petr Berka	Pavel Brazdil

Dariusz Brzezinski
Rui Camacho
Longbing Cao
Francisco Casacuberta
Peggy Cellier
Loic Cerf
Tania Cerquitelli
Edward Chang
Keke Chen
Weiwei Cheng
Silvia Chiusano
Arthur Choi
Frans Coenen
Mário Cordeiro
Roberto Corizzo
Vitor Santos Costa
Bertrand Cuissart
Boris Cule
Tomaž Curk
James Cussens
Alfredo Cuzzocrea
Claudia d'Amato
Maria Damiani
Tijl De Bie
Martine De Cock
Juan Jose del Coz
Anne Denton
Christian Desrosiers
Nicola Di Mauro
Claudia Diamantini
Uwe Dick
Tom Diethe
Ivica Dimitrovski
Wei Ding
Ying Ding
Stephan Doerfel
Carlotta Domeniconi
Frank Dondelinger
Madalina Drugan
Wouter Duivesteijn
Inŝ Dutra
Dora Erdos
Fabio Fassetti
Ad Feelders
Stefano Ferilli
Carlos Ferreira

Cesar Ferri
Răzvan Florian
Eibe Frank
Elisa Fromont
Fabio Fumarola
Esther Galbrun
Patrick Gallinari
Dragan Gamberger
Byron Gao
Paolo Garza
Konstantinos Georgatzis
Pierre Geurts
Dorota Glowacka
Nico Goernitz
Elsa Gomes
Mehmet Gönen
James Goulding
Michael Granitzer
Caglar Gulcehre
Francesco Gullo
Stephan Günnemann
Tias Guns
Sara Hajian
Maria Halkidi
Jiawei Han
Mohammad Hasan
Xiao He
Denis Helic
Daniel Hernandez-Lobato
Jose Hernandez-Orallo
Thanh Lam Hoang
Frank Hoeppner
Arjen Hommersom
Tamas Horvath
Andreas Hotho
Yuanhua Huang
Eyke Hüllermeier
Dino Ienco
Szymon Jaroszewicz
Giuseppe Jurman
Toshihiro Kamishima
Michael Kamp
Bo Kang
Andreas Karwath
George Karypis
Mehdi Kaytoue

Latifur Khan
Frank Klawonn
Jiri Klema
Tomas Kliegr
Marius Kloft
Dragi Kocev
Levente Kocsis
Yun Sing Koh
Alek Kolcz
Irena Koprinska
Frederic Koriche
Walter Kosters
Lars Kotthoff
Danai Koutra
Georg Krempl
Tomas Krilavicius
Yamuna Krishnamurthy
Matjaz Kukar
Meelis Kull
Prashanth L. A.
Jorma Laaksonen
Nicolas Lachiche
Leo Lahti
Helge Langseth
Thomas Lansdall-Welfare
Christine Largeron
Pedro Larranaga
Silvio Lattanzi
Niklas Lavesson
Binh Le
Freddy Lecue
Florian Lemmerich
Jiuyong Li
Limin Li
Jefrey Lijffijt
Tony Lindgren
Corrado Loglisci
Peter Lucas
Brian Mac Namee
Gjorgji Madjarov
Sebastian Mair
Donato Malerba
Luca Martino
Elio Masciari
Andres Masegosa
Florent Masseglia

Ernestina Menasalvas
Corrado Mencar
Rosa Meo
Pauli Miettinen
Dunja Mladenic
Karthika Mohan
Anna Monreale
Joao Moreira
Mohamed Nadif
Ndapa Nakashole
Jinseok Nam
Mirco Nanni
Amedeo Napoli
Nicolo Navarin
Benjamin Negrevergne
Benjamin Nguyen
Xia Ning
Kjetil Norvag
Eirini Ntoutsi
Andreas Nurnberger
Barry O'Sullivan
Dino Oglic
Francesco Orsini
Nikunj Oza
Pance Panov
Apostolos Papadopoulos
Panagiotis Papapetrou
Ioannis Partalas
Gabriella Pasi
Dino Pedreschi
Jaakko Peltonen
Ruggero Pensa
Iker Perez
Nico Piatkowski
Andrea Pietracaprina
Gianvito Pio
Susanna Pirttikangas
Marc Plantevit
Pascal Poncelet
Miguel Prada
Philippe Preux
Buyue Qian
Chedy Raissi
Jan Ramon
Huzefa Rangwala
Zbigniew Ras

Chotirat Ratanamahatana
Jan Rauch
Chiara Renso
Achim Rettinger
Fabrizio Riguzzi
Matteo Riondato
Celine Robardet
Juan Rodriguez
Fabrice Rossi
Celine Rouveirol
Stefan Rueping
Salvatore Ruggieri
Yvan Saeys
Alan Said
Lorenza Saitta
Tomoya Sakai
Alessandra Sala
Ansaf Salleb-Aouissi
Claudio Sartori
Pierre Schaus
Lars Schmidt-Thieme
Christoph Schommer
Matthias Schubert
Konstantinos Sechidis
Sohan Seth
Vinay Setty
Junming Shao
Nikola Simidjievski
Sameer Singh
Andrzej Skowron
Dominik Slezak
Kevin Small
Gavin Smith
Tomislav Smuc
Yangqiu Song
Arnaud Soulet
Wesllen Sousa
Alessandro Sperduti
Jerzy Stefanowski
Giovanni Stilo
Gerd Stumme
Mahito Sugiyama
Mika Sulkava
Einoshin Suzuki
Stephen Swift
Andrea Tagarelli

Domenico Talia
Letizia Tanca
Jovan Tanevski
Nikolaj Tatti
Maguelonne Teisseire
Georgios Theocharous
Ljupco Todorovski
Roberto Trasarti
Volker Tresp
Isaac Triguero
Panayiotis Tsaparas
Vincent S. Tseng
Karl Tuyls
Niall Twomey
Nikolaos Tziortziotis
Theodoros Tzouramanis
Antti Ukkonen
Toon Van Craenendonck
Martijn Van Otterlo
Iraklis Varlamis
Julien Velcin
Shankar Vembu
Deepak Venugopal
Vassilios S. Verykios
Ricardo Vigario
Herna Viktor
Christel Vrain
Willem Waegeman
Jianyong Wang
Joerg Wicker
Marco Wiering
Martin Wistuba
Philip Yu
Bianca Zadrozny
Gerson Zaverucha
Bernard Zenko
Junping Zhang
Min-Ling Zhang
Shichao Zhang
Ying Zhao
Mingjun Zhong
Albrecht Zimmermann
Marinka Zitnik
Indré Žliobaité

Applied Data Science Track Program Committee

Oznur Alkan	Sidath Handurukande	Nikunj Oza
Carlos Alzate	Souleiman Hasan	Ioannis Partalas
Nicola Barberi	Georges Hebrail	Milan Petkovic
Gianni Barlacchi	Thanh Lam	Fabio Pinelli
Enda Barrett	Neil Hurley	Yongrui Qin
Roberto Bayardo	Hongxia Jin	Rene Quiniou
Srikanta Bedathur	Anup Kalia	Ambrish Rawat
Daniel Bengs	Pinar Karagoz	Fergal Reid
Cuissart Bertrand	Mehdi Kaytoue	Achim Rettinger
Urvesh Bhowan	Alek Kolcz	Stefan Rueping
Tassadit Bouadi	Deguang Kong	Elizeu Santos-Neto
Thomas Brovelli	Lars Kotthoff	Manali Sharma
Teodora Sandra	Nick Koudas	Alkis Simitsis
Berkant Barla	Hardy Kremer	Kevin Small
Michelangelo Ceci	Helge Langseth	Alessandro Sperduti
Edward Chang	Freddy Lecue	Siqi Sun
Soumyadeep Chatterjee	Zhenhui Li	Pal Sundsoy
Abon Chaudhuri	Lin Liu	Ingo Thon
Javier Cuenca	Jiebo Luo	Marko Tkalcic
Mahashweta Das	Arun Maiya	Luis Torgo
Viktoriya Degeler	Silviu Maniu	Radu Tudoran
Wei Ding	Elio Masciari	Umair Ul
Yuxiao Dong	Luis Matias	Jan Van
Carlos Ferreira	Dimitrios Mavroeidis	Ranga Vatsavai
Andre Freitas	Charalampos	Fei Wang
Feng Gao	Mavroforakis	Xiang Wang
Dinesh Garg	James McDermott	Wang Wei
Guillermo Garrido	Daniil Mirylenka	Martin Wistuba
Rumi Ghosh	Elena Mocanu	Erik Wittern
Martin Gleize	Raul Moreno	Milena Yankova
Slawek Goryczka	Bogdan Nicolae	Daniela Zaharie
Riccardo Guidotti	Maria-Irina Nicolae	Chongsheng Zhang
Francesco Gullo	Xia Ning	Yanchang Zhao
Thomas Guyet	Sean O'Riain	Albrecht Zimmermann
Allan Hanbury	Adegboyega Ojo	

Nectar Track Program Committee

Annalisa Appice	Peter Flach	Ernestina Menasalvas
Martin Atzmüller	Johannes Fürnkranz	Mirco Musolesi
Hendrik Blockeel	Joao Gama	Franco Maria
Ulf Brefeld	Andreas Hotho	Maryam Tavakol
Tijl De	Kristian Kersting	Gabriele Tolomei
Kurt Driessens	Sebastian Mair	Salvatore Trani

Demo Track Program Committee

Gustavo Carneiro	Brian Mac Namee	Konstantinos Skiannis
Derek Greene	Susan McKeever	Jerzy Stefanowski
Mark Last	Joao Papa	Luis Teixeira
Vincent Lemaire	Niladri Sett	Grigorios Tsoumakas

Invited Talks Abstracts

Building Production Machine Learning Systems

Misha Bilenko

Head of Machine Intelligence and Research, Yandex, Moscow, Russia

Abstract. How does one build a production ML system that can effectively incorporate corrections, while avoiding the typical risks and engineering costs of online learning methods? An effective solution requires a combination of very latest and some well-dated algorithms. Parametric machine learning methods - such as neural networks, boosted trees, factorization methods and their ensembles - yield state-of-the-art results on ML benchmarks and competitions. Real-world deployments of ML systems, however, differ dramatically from those static settings. We discuss issues that differentiate production and academic ML systems, leading to the need for combining parametric models with their non-parametric brethren, i.e, modern variants of Nearest Neighbor algorithms. The combined approach is particularly suitable for systems where incorporating corrections must be accomplished rapidly, as illustrated by some lively real-life examples from a large-scale conversational assistant.

Bio: Misha Bilenko heads the Machine Intelligence and Research (MIR) division at Yandex, which integrates research and product development in core AI areas: machine learning, dialog systems, speech recognition and synthesis, machine translation and computer vision. Before Yandex, he led the Machine Learning Algorithms team at Microsoft, which shipped ML technologies in multiple products across all Microsoft divisions. He started his career in the Machine Learning Group in Microsoft Research after receiving his Ph.D. in Computer Science from the University of Texas at Austin and stints at Google and IBM Research.

Combatting Misinformation and Building Trust in the Media

Corinna Cortes

Head of Google Research, New York, USA

Abstract. Trust in journalism, search engines and social media has been on a sharp decline over the last years. "Fake News" is a reality and internet users struggle to distinguish credible information from falsehood. In this talk we will discuss tools and methodology we as researchers at Google are building to restore online trust and combat misinformation.

Bio: Corinna Cortes is the Head of Google Research, NY, where she is working on a broad range of theoretical and applied large-scale machine learning problems. Corinna speaks about how she got into the tech world, and some work that is happening within the Research team at Google. Prior to Google, Corinna spent more than ten years at AT&T Labs – Research, formerly AT&T Bell Labs, where she held a distinguished research position. Corinna's research work is well-known in particular for her contributions to the theoretical foundations of support vector machines (SVMs), for which she jointly with Vladimir Vapnik received the 2008 Paris Kanellakis Theory and Practice Award, and her work on data-mining in very large data sets for which she was awarded the AT&T Science and Technology Medal in the year 2000. Corinna received her MS degree in Physics from University of Copenhagen and joined AT&T Bell Labs as a researcher in 1989. She received her Ph.D. in computer science from the University of Rochester in 1993. Corinna is also a competitive runner and a mother of twins.

Combating Bias and Polarization
in Online Media

Aristides Gionis

Department of Computer Science, Aalto University, Helsinki, Finland

Abstract. Online social media are a major venue of public discourse today, hosting the opinions of hundreds of millions of individuals. Social media are often credited for providing a technological means to break information barriers and promote diversity and democracy. In practice, however, the opposite effect is often observed: users tend to favor content that agrees with their existing world-view, get less exposure to conflicting viewpoints, and eventually create "echo chambers" and increased polarization. Arguably, without any kind of moderation, current social-media platforms gravitate towards a state in which net-citizens are constantly reinforcing their existing opinions. In this talk we present an ongoing line of work on analyzing and moderating online social discussions. We first consider the questions of detecting controversy using network structure and content. We then address the problem of designing algorithms to break filter bubbles, reduce polarization, and increase diversity. We discuss a number of different strategies such as user and content recommendation, as well as approaches based on information cascades.

Bio: Aristides Gionis is a professor in the department of Computer Science in Aalto University. He has been a visiting professor in the University of Rome (2016) and a senior research scientist in Yahoo! Research (2006–2012). He is currently serving as an action editor in the Data Management and Knowledge Discovery journal (DMKD), an associate editor in the ACM Transactions on Knowledge Discovery from Data (TKDD), and a managing editor in Internet Mathematics. He has contributed in several areas of data science, such as algorithmic data analysis, web mining, social-media analysis, data clustering, and privacy-preserving data mining. His current research is funded by the Academy of Finland (projects Nestor, Agra, AIDA) and the European Commission (project SoBigData).

New Algorithms for Interpretable Machine Learning

Cynthia Rudin

Associate Professor of Computer Science and Electrical
and Computer Engineering
Duke University, Durham, North Carolina, USA

Abstract. What if every black box machine learning model could be replaced with one that was equally accurate but also interpretable? If we could do this, we would identify flaws in our models and data that we could not see before. Perhaps we could prevent some of the poor decisions in criminal justice and medicine that are caused by problems with using black box models. We could also eliminate the need for "explanations" that are misleading and often wrong. I will present algorithms for (i) interpretable neural networks for computer vision, (ii) certifiably optimal decision lists, and (iii) certifiably optimal scoring systems (sparse linear models with integer coefficients). Models from these algorithms can often be used in place of black box models, while achieving the same accuracy.

Bio: Cynthia Rudin is an associate professor of computer science, electrical and computer engineering, and statistics at Duke University, and directs the Prediction Analysis Lab. Previously, Prof. Rudin held positions at MIT, Columbia, and NYU. She holds an undergraduate degree from the University at Buffalo, and a PhD in applied and computational mathematics from Princeton University. She is the recipient of the 2013 and 2016 INFORMS Innovative Applications in Analytics Awards, an NSF CAREER award, was named as one of the "Top 40 Under 40" by Poets and Quants in 2015, and was named by Businessinsider.com as one of the 12 most impressive professors at MIT in 2015. Work from her lab has won 10 best paper awards in the last 5 years. She is past chair of the INFORMS Data Mining Section, and is currently chair of the Statistical Learning and Data Science section of the American Statistical Association. She also serves on (or has served on) committees for DARPA, the National Institute of Justice, the National Academy of Sciences (for both statistics and criminology/law), and AAAI.

Information Theory of Deep Learning

Naftali Tishby

School of Engineering and Computer Science, Hebrew University of Jerusalem,
Israel

Abstract. I will present new results on the organization and interpretability of the many layers in deep learning, which stem from my information bottleneck theory of deep neural networks.

Bio: Dr. Naftali Tishby is a professor of Computer Science, and the incumbent of the Ruth and Stan Flinkman Chair for Brain Research at the Edmond and Lily Safra Center for Brain Science (ELSC) at the Hebrew University of Jerusalem. He is one of the leaders of machine learning research and computational neuroscience in Israel and his numerous ex-students serve at key academic and industrial research positions all over the world. Prof. Tishby was the founding chair of the new computer-engineering program, and a director of the Leibnitz research center in computer science, at the Hebrew university. Tishby received his PhD in theoretical physics from the Hebrew university in 1985 and was a research staff member at MIT and Bell Labs from 1985 and 1991. Prof. Tishby was also a visiting professor at Princeton NECI, University of Pennsylvania, UCSB, and IBM research. His current research is at the interface between computer science, statistical physics, and computational neuroscience. He pioneered various applications of statistical physics and information theory in computational learning theory. More recently, he has been working on the foundations of biological information processing and the connections between dynamics and information. He has introduced with his colleagues new theoretical frameworks for optimal adaptation and efficient information representation in biology, such as the Information Bottleneck method and the Minimum Information principle for neural coding.

Contents – Part I

Deep Learning

Contents – Part II

Kernel Methods

Learning Paradigms

Matrix and Tensor Analysis

Online and Active Learning

Pattern and Sequence Mining

Probabilistic Models and Statistical Methods

Recommender Systems

Transfer Learning

Contents – Part III

ADS Engineering and Design

ADS Financial/Security

ADS Health

ADS Sensing/Positioning

Nectar Track

Demo Track

Adversarial Learning

Image Anomaly Detection
with Generative Adversarial Networks

Lucas Deecke[1]($\boxtimes$), Robert Vandermeulen[2], Lukas Ruff[3], Stephan Mandt[4], and Marius Kloft[2]

[1] University of Edinburgh, Edinburgh, Scotland, UK
l.deecke@ed.ac.uk
[2] TU Kaiserslautern, Kaiserslautern, Germany
{vandermeulen,kloft}@cs.uni-kl.de
[3] Hasso Plattner Institute, Potsdam, Germany
lukas.ruff@hpi.de
[4] University of California, Irvine, CA, USA
mandt@uci.edu

Abstract. Many anomaly detection methods exist that perform well on low-dimensional problems however there is a notable lack of effective methods for high-dimensional spaces, such as images. Inspired by recent successes in deep learning we propose a novel approach to anomaly detection using generative adversarial networks. Given a sample under consideration, our method is based on searching for a good representation of that sample in the latent space of the generator; if such a representation is not found, the sample is deemed anomalous. We achieve state-of-the-art performance on standard image benchmark datasets and visual inspection of the most anomalous samples reveals that our method does indeed return anomalies.

1 Introduction

Given a collection of data it is often desirable to automatically determine which instances of it are unusual. Commonly referred to as anomaly detection, this is a fundamental machine learning task with numerous applications in fields such as astronomy [11,43], medicine [5,46,51], fault detection [18], and intrusion detection [15,19]. Traditional algorithms often focus on the low-dimensional regime and face difficulties when applied to high-dimensional data such as images or speech. Second to that, they require the manual engineering of features.

Deep learning omits manual feature engineering and has become the de-facto approach for tackling many high-dimensional machine learning tasks. This is largely a testament of its experimental performance: deep learning has helped to achieve impressive results in image classification [24], and is setting new standards in domains such as natural language processing [25,50] and speech recognition [3].

L. Deecke and R. Vandermeulen—Equal contributions.

© Springer Nature Switzerland AG 2019
M. Berlingerio et al. (Eds.): ECML PKDD 2018, LNAI 11051, pp. 3–17, 2019.
https://doi.org/10.1007/978-3-030-10925-7_1

In this paper we present a novel deep learning based approach to anomaly detection which uses generative adversarial networks (GANs) [17]. GANs have achieved state-of-the-art performance in high-dimensional generative modeling. In a GAN, two neural networks – the discriminator and the generator – are pitted against each other. In the process the generator learns to map random samples from a low-dimensional to a high-dimensional space, mimicking the target dataset. If the generator has successfully learned a good approximation of the training data's distribution it is reasonable to assume that, for a sample drawn from the data distribution, there exists some point in the GAN's latent space which, after passing it through the generator network, should closely resembles this sample. We use this correspondence to perform anomaly detection with GANs (ADGAN).

In Sect. 2 we give an overview of previous work on anomaly detection and discuss the modeling assumptions of this paper. Section 3 contains a description of our proposed algorithm. In our experiments, see Sect. 4, we both validate our method against traditional methods and showcase ADGAN's ability to detect anomalies in high-dimensional data.

2   Background

Here we briefly review previous work on anomaly detection, touch on generative models, and highlight the methodology of GANs.

2.1   Related Work

Anomaly Detection. Research on anomaly detection has a long history with early work going back as far as [12], and is concerned with finding unusual or *anomalous* samples in a corpus of data. An extensive overview over traditional anomaly detection methods as well as open challenges can be found in [6]. For a recent empirical comparison of various existing approaches, see [13].

Generative models yield a whole family of anomaly detectors through estimation of the data distribution p. Given data, we estimate $\hat{p} \approx p$ and declare those samples which are unlikely under $\hat{p}$ to be anomalous. This guideline is roughly followed by traditional non-parametric methods such as kernel density estimation (KDE) [40], which were applied to intrusion detection in [53]. Other research targeted mixtures of Gaussians for active learning of anomalies [42], hidden Markov models for registering network attacks [39], and dynamic Bayesian networks for traffic incident detection [48].

Deep Generative Models. Recently, variational autoencoders (VAEs) [22] have been proposed as a deep generative model. By optimizing over a variational lower bound on the likelihood of the data, the parameters of a neural network are tuned in such a way that samples resembling the data may be generated from a Gaussian prior. Another generative approach is to train a pair of deep convolutional neural networks in an autoencoder setup (DCAE) [33] and producing samples by decoding random points on the compression manifold. Unfortunately, none of these approaches yield a tractable way of estimating p. Our approach uses a deep generative model in the context of anomaly detection.

Deep Learning for Anomaly Detection. Non-parametric anomaly detection methods suffer from the curse of dimensionality and are thus often inadequate for the interpretation and analysis of high-dimensional data. Deep neural networks have been found to obviate many problems that arise in this context. As a hybrid between the two approaches, deep belief networks were coupled with one-class support vector machines to detect anomalies in [14]. We found that this technique did not work well for image datasets, and indeed the authors included no such experiments in their paper.

A recent work proposed an end-to-end deep learning approach, aimed specifically at the task of anomaly detection [45]. Similarly, one may employ a network that was pretrained on a different task, such as classification on ImageNet [8], and then use this network's intermediate features to extract relevant information from images. We tested this approach in our experimental section.

GANs, which we discuss in greater depth in the next section, have garnered much attention with its performance surpassing previous deep generative methods. Concurrently to this work, [46] developed an anomaly detection framework that uses GANs in a similar way as we do. We discuss the differences between our work and theirs in Sect. 3.2.

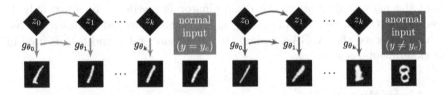

Fig. 1. An illustration of ADGAN. In this example, ones from MNIST are considered normal ($y_c = 1$). After an initial draw from p_z, the loss between the first generation $g_{\theta_0}(z_0)$ and the image x whose anomaly we are assessing is computed. This information is used to generate a consecutive image $g_{\theta_1}(z_1)$ more alike x. After k steps, samples are scored. If x is similar to the training data (red example, $y = y_c$), then a similar object should be contained in the image of g_{θ_k}. For a dissimilar x (blue example, $y \neq y_c$), no similar image is found, resulting in a large loss. (Color figure online)

2.2   Generative Adversarial Networks

GANs, which lie at the heart of ADGAN, have set a new state-of-the-art in generative image modeling. They provide a framework to generate samples that are approximately distributed to p, the distribution of the training data $\{x_i\}_{i=1}^n \triangleq \mathcal{X} \subseteq \mathbb{R}^d$. To achieve this, GANs attempt to learn the parametrization of a neural network, the so-called generator g_θ, that maps low-dimensional samples drawn from some simple noise prior p_z (e.g. a multivariate Gaussian) to samples in the image space, thereby inducing a distribution q_θ (the push-forward of p_z with respect to g_θ) that approximates p. To achieve this a second neural network, the discriminator d_ω, learns to classify the data from p and q_θ. Through an alternating training procedure the discriminator becomes better at separating samples from p and samples from q_θ, while the generator adjusts θ to fool the discriminator, thereby approximating p more closely. The objective function of the GAN framework is thus:

$$\min_\theta \max_\omega \left\{ V(\theta, \omega) = \mathbb{E}_{x \sim p}[\log d_\omega(x)] + \mathbb{E}_{z \sim p_z}[\log(1 - d_\omega(g_\theta(z)))] \right\}, \quad (1)$$

where z are vectors that reside in a latent space of dimensionality $d' \ll d$.[1] A recent work showed that this minmax optimization (1) equates to an empirical lower bound of an f-divergence [37].[2]

GAN training is difficult in practice, which has been shown to be a consequence of vanishing gradients in high-dimensional spaces [1]. These instabilities can be countered by training on integral probability metrics (IPMs) [35,49], one instance of which is the 1-Wasserstein distance.[3] This distance, informally defined, is the amount of work to pull one density onto another, and forms the basis of the Wasserstein GAN (WGAN) [2]. The objective function for WGANs is

$$\min_\theta \max_{\omega \in \Omega} \left\{ W(\theta, \omega) = \mathbb{E}_{x \sim p}[d_\omega(x)] - \mathbb{E}_{z \sim p_z}[d_\omega(g_\theta(z))] \right\}, \quad (2)$$

where the parametrization of the discriminator is restricted to allow only 1-Lipschitz functions, i.e. $\Omega = \{\omega : \|d_\omega\|_L \leq 1\}$. When compared to classic GANs, we have observed that WGAN training is much more stable and is thus used in our experiments, see Sect. 4.

3   Algorithm

Our proposed method (ADGAN, see Algorithm 1) sets in after GAN training has converged. If the generator has indeed captured the distribution of the training data then, given a new sample $x \sim p$, there should exist a point z in the latent space, such that $g_\theta(z) \approx x$. Additionally we expect points away from the support of p to have no representation in the latent space, or at least occupy a small portion of the probability mass in the latent distribution, since they are easily discerned by d_ω as not coming from p. Thus, given a test sample x, if there exists no z such that $g_\theta(z) \approx x$, or if such a z is difficult to find, then it can

---

[1] That p may be approximated via transformations from a low-dimensional space is an assumption that is implicitly motivated from the manifold hypothesis [36].

[2] This lower bound becomes tight for an optimal discriminator, making apparent that $V(\theta, \omega^*) \propto \mathrm{JS}[p|q_\theta]$.

[3] This is achieved by restricting the class over which the IPM is optimized to functions that have Lipschitz constant less than one. Note that in Wasserstein GANs, an expression corresponding to a lower bound is optimized.

be inferred that x is not distributed according to p, i.e. it is anomalous. Our algorithm hinges on this hypothesis, which we illustrate in Fig. 1.

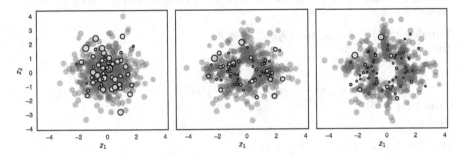

Fig. 2. The coordinates (z_1, z_2) of 500 samples from MNIST are shown, represented in a latent space with $d' = 2$. At different iterations t of ADGAN, no particular structure arises in the z-space: samples belonging to the normal class (●) and the anomalous class (●) are scattered around freely. Note that this behavior also prevents $p_z(z_t)$ from providing a sensible anomaly score. The sizes of points correspond to the reconstruction loss between generated samples and their original image $\ell(g_\theta(z_t), x)$. The normal and anomalous class differ markedly in terms of this metric. (Color figure online)

3.1 ADGAN

To find z, we initialize from $z_0 \sim p_z$, where p_z is the same noise prior also used during GAN training. For $t = 1, \ldots, k$ steps, we backpropagate the reconstruction loss ℓ between $g_\theta(z_t)$ and x, making the subsequent generation $g_\theta(z_{t+1})$ more like x. At each iteration, we also allow a small amount of flexibility to the parametrization of the generator, resulting in a series of mappings from the latent space $g_{\theta_0}(z_0), \ldots, g_{\theta_k}(z_k)$ that more and more closely resembles x. Adjusting θ gives the generator additional representative capacity, which we found to improve the algorithm's performance. Note that these adjustments to θ are not part of the GAN training procedure and θ is reset back to its original trained value for each new testing point.

To limit the risk of seeding in unsuitable regions and address the non-convex nature of the underlying optimization problem, the search is initialized from n_{seed} individual points. The key idea underlying ADGAN is that if the generator was trained on the same distribution x was drawn from, then the average over the final set of reconstruction losses $\{\ell(x, g_{\theta_{j,k}}(z_{j,k}))\}_{j=1}^{n_{\text{seed}}}$ will assume low values, and high values otherwise. In Fig. 2 we track a collection of samples through their search in a latent space of dimensionality $d' = 2$.

Our method may also be understood from the standpoint of approximate inversion of the generator. In this sense, the above backpropagation finds latent vectors z that lie close to $g_\theta^{-1}(x)$. Inversion of the generator was previously

---

Algorithm 1. Anomaly Detection using Generative Adversarial Networks (ADGAN).

Input: parameters $(\gamma, \gamma_\theta, n_{\text{seed}}, k)$, sample x, GAN generator g_θ, prior p_z, reconstruction loss ℓ

Initialize $\{z_{j,0}\}_{j=1}^{n_{\text{seed}}} \sim p_z$ and $\{\theta_{j,0}\}_{j=1}^{n_{\text{seed}}} \triangleq \theta$
for $j = 1$ **to** n_{seed} **do**
   for $t = 1$ **to** k **do**
      $z_{j,t} \leftarrow z_{j,t-1} - \gamma \cdot \nabla_{z_{j,t-1}} \ell(g_{\theta_{j,t-1}}(z_{j,t-1}), x)$
      $\theta_{j,t} \leftarrow \theta_{j,t-1} - \gamma_\theta \cdot \nabla_{\theta_{j,t-1}} \ell(g_{\theta_{j,t-1}}(z_{j,t-1}), x)$
   end for
end for

Return: $(1/n_{\text{seed}}) \sum_{j=1}^{n_{\text{seed}}} \ell(g_{\theta_{j,k}}(z_{j,k}), x)$

---

studied in [7], where it was verified experimentally that this task can be carried out with high fidelity. In addition [29] showed that generated images can be successfully recovered by backpropagating through the latent space.[4] Jointly optimizing latent vectors and the generator parametrization via backpropagation of reconstruction losses was investigated in detail by [4]. The authors found that it is possible to train the generator entirely without a discriminator, still yielding a model that incorporates many of the desirable properties of GANs, such as smooth interpolations between samples.

3.2 Alternative Approaches

Given that GAN training also gives us a discriminator for discerning between real and fake samples, one might reasonably consider directly applying the discriminator for detecting anomalies. However, once converged, the discriminator exploits checkerboard-like artifacts on the pixel level, induced by the generator architecture [31,38]. While it perfectly separates real from forged data, it is not equipped to deal with samples which are completely unlike the training data. This line of reasoning is verified in Sect. 4 experimentally.

Another approach we considered was to evaluate the likelihood of the final latent vectors $\{z_{j,k}\}_{j=1}^{n_{\text{seed}}}$ under the noise prior p_z. This approach was tested experimentally in Sect. 4, and while it showed some promise, it was consistently outperformed by ADGAN.

In [46], the authors propose a technique for anomaly detection (called Ano-GAN) which uses GANs in a way somewhat similar to our proposed algorithm. Their algorithm also begins by training a GAN. Given a test point x, their algorithm searches for a point z in the latent space such that $g_\theta(z) \approx x$ and computes the reconstruction loss. Additionally they use an intermediate

---

[4] While it was shown that any $g_\theta(z)$ may be reconstructed from some other $z_0 \in \mathbb{R}^{d'}$, this does not mean that the same holds for an x not in the image of g_θ.

Table 1. ROC-AUC of classic anomaly detection methods. For both MNIST and CIFAR-10, each model was trained on every class, as indicated by y_c, and then used to score against remaining classes. Results for KDE and OC-SVM are reported both in conjunction with PCA, and after transforming images with a pre-trained Alexnet.

DATASET	y_c	KDE		OC-SVM		IF	GMM	DCAE	ANOGAN	VAE	ADGAN
		PCA	ALEXNET	PCA	ALEXNET						
MNIST	0	0.982	0.634	0.993	0.962	0.957	0.970	0.988	0.990	0.884	**0.999**
	1	0.999	0.922	**1.000**	0.999	**1.000**	0.999	0.993	0.998	0.998	0.992
	2	0.888	0.654	0.881	0.925	0.822	0.931	0.917	0.888	0.762	**0.968**
	3	0.898	0.639	0.931	0.950	0.924	0.951	0.885	0.913	0.789	**0.953**
	4	0.943	0.676	0.962	**0.982**	0.922	0.968	0.862	0.944	0.858	0.960
	5	0.930	0.651	0.881	0.923	0.859	0.917	0.858	0.912	0.803	**0.955**
	6	0.972	0.636	0.982	0.975	0.903	**0.994**	0.954	0.925	0.913	0.980
	7	0.933	0.628	0.951	**0.968**	0.938	0.938	0.940	0.964	0.897	0.950
	8	0.924	0.617	0.958	0.926	0.814	0.889	0.823	0.883	0.751	**0.959**
	9	0.940	0.644	**0.970**	0.969	0.913	0.962	0.965	0.958	0.848	0.965
		0.941	0.670	0.951	0.958	0.905	0.952	0.919	0.937	0.850	**0.968**
CIFAR-10	0	0.705	0.559	0.653	0.594	0.630	**0.709**	0.656	0.610	0.582	0.661
	1	0.493	0.487	0.400	0.540	0.379	0.443	0.435	0.565	**0.608**	0.435
	2	**0.734**	0.582	0.617	0.588	0.630	0.697	0.381	0.648	0.485	0.636
	3	0.522	0.531	0.522	0.575	0.408	0.445	0.545	0.528	**0.667**	0.488
	4	0.691	0.651	0.715	0.753	0.764	0.761	0.288	0.670	0.344	**0.794**
	5	0.439	0.551	0.517	0.558	0.514	0.505	**0.643**	0.592	0.493	0.640
	6	**0.771**	0.613	0.727	0.692	0.666	0.766	0.509	0.625	0.391	0.685
	7	0.458	0.593	0.522	0.547	0.480	0.496	**0.690**	0.576	0.516	0.559
	8	0.595	0.600	0.719	0.630	0.651	0.646	0.698	0.723	0.522	**0.798**
	9	0.490	0.529	0.475	0.530	0.459	0.384	**0.705**	0.582	0.633	0.643
		0.590	0.570	0.587	0.601	0.558	0.585	0.583	0.612	0.524	**0.634**

discriminator layer d'_ω and compute the loss between $d'_\omega(g_\theta(z))$ and $d'_\omega(x)$. They use a convex combination of these two quantities as their anomaly score.

In ADGAN we never use the discriminator, which is discarded after training. This makes it easy to couple ADGAN with any GAN-based approach, e.g. LSGAN [32], but also any other differentiable generator network such as VAEs or moment matching networks [27]. In addition, we account for the non-convexity of the underlying optimization by seeding from multiple areas in the latent space. Lastly, during inference we update not only the latent vectors z, but jointly update the parametrization θ of the generator.

4 Experiments

Here we present experimental evidence of the efficacy of ADGAN. We compare our algorithm to competing methods on a controlled, classification-type task and show anomalous samples from popular image datasets. Our main findings are that ADGAN:

- outperforms non-parametric as well as available deep learning approaches on two controlled experiments where ground truth information is available;

– may be used on large, unsupervised data (such as LSUN bedrooms) to detect anomalous samples that coincide with what we as humans would deem unusual.

4.1 Datasets

Our experiments are carried out on three benchmark datasets with varying complexity: (i) MNIST [26] which contains grayscale scans of handwritten digits. (ii) CIFAR-10 [23] which contains color images of real world objects belonging to ten classes. (iii) LSUN [52], a dataset of images that show different scenes (such as bedrooms, bridges, or conference rooms). For all datasets the training and test splits remain as their default. All images are rescaled to assume pixel values in $[-1, 1]$.

4.2 Methods and Hyperparameters

We tested the performance of ADGAN against four traditional, non-parametric approaches commonly used for anomaly detection: (i) KDE [40] with a Gaussian kernel. The bandwidth is determined from maximum likelihood estimation over ten-fold cross validation, with $h \in \{2^0, 2^{1/2}, \ldots, 2^4\}$. (ii) One-class support vector machine (OC-SVM) [47] with a Gaussian kernel. The inverse length scale is selected with automated tuning, as proposed by [16], and we set $\nu = 0.1$. (iii) Isolation forest (IF) [30], which was largely stable to changes in its parametrization. (iv) Gaussian mixture model (GMM). We allowed the number of components to vary over $\{2, 3, \ldots, 20\}$ and selected suitable hyperparameters by evaluating the Bayesian information criterion.

For the methods above we reduced the feature dimensionality before performing anomaly detection. This was done via PCA [41], varying the dimensionality over $\{20, 40, \ldots, 100\}$; we simply report the results for which best performance on a small holdout set was attained. As an alternative to a linear projection, we evaluated the performance of both methods after applying a non-linear transformation to the image data instead via an Alexnet [24], pretrained on ImageNet. Just as on images, the anomaly detection is carried out on the representation in the final convolutional layer of Alexnet. This representation is then projected down via PCA, as otherwise the runtime of KDE and OC-SVM becomes problematic.

We also report the performance of two end-to-end deep learning approaches: VAEs and DCAEs. For the DCAE we scored according to reconstruction losses, interpreting a high loss as indicative of a new sample differing from samples seen during training. In VAEs we scored by evaluating the evidence lower bound (ELBO). We found this to perform much better than thresholding directly via the prior likelihood in the latent space or other more exotic approaches, such as scoring from the variance of the inference network.

In both DCAEs and VAEs we use a convolutional architecture similar to that of DCGAN [44], with batch normalization [20] and ReLU activations in each layer. We also report the performance of AnoGAN. To put it on equal footing,

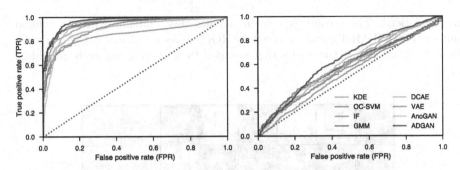

Fig. 3. ROC curves for one-versus-all prediction of competing methods on MNIST (left) and CIFAR10 (right), averaged over all classes. KDE and OC-SVM are shown in conjunction with PCA, for detailed performance statistics see Table 1.

we pair it with DCGAN [44], the same architecture also used for training in our approach.

ADGAN requires a trained generator. For this purpose, we trained on the WGAN objective (2), as this was much more stable than using GANs. The architecture was fixed to that of DCGAN [44]. Following [34] we set the dimensionality of the latent space to $d' = 256$.

For ADGAN, the searches in the latent space were initialized from the same noise prior that the GAN was trained on (in our case a normal distribution). To take into account the non-convexity of the problem, we seeded with $n_{seed} = 64$ points. For the optimization of latent vectors and the parameters of the generator we used the Adam optimizer [21].[5] When searching for a point in the latent space to match a test point, we found that more iterations helped the performance, but this gain saturates quickly. As a trade-off between execution time and accuracy we found $k = 5$ to be a good value, and used this in the results we report. Unless otherwise noted, we measured reconstruction quality with a squared L_2 loss.

4.3 One-Versus-All Classification

The first task is designed to quantify the performance of competing methods. The experimental setup closely follows the original publication on OC-SVMs [47] and we begin by training models on data from a single class from MNIST. Then we evaluate each model's performance on 5000 items randomly selected from the test set, which contains samples from all classes. In each trial, we label the classes unseen in training as anomalous.

Ideally, a method assigns images from anomalous classes (say, digits 1-9) a higher anomaly score than images belonging to the normal class (zeros). Varying the decision threshold yields the receiver operating characteristic (ROC), shown

---

[5] From a quick parameter sweep, we set the learning rate to $\gamma = 0.25$ and $(\beta_1, \beta_2) = (0.5, 0.999)$. We update the generator with $\gamma_\theta = 5 \cdot 10^{-5}$, the default learning rate recommended in [2].

in Fig. 3 (left). The second experiment follows this guideline with the colored images from CIFAR-10, and the resulting ROC curves are shown in Fig. 3 (right). In Table 1, we report the AUCs that resulted from leaving out each individual class.

Fig. 4. Starting from the top left, the first three rows show samples contained in the LSUN bedrooms validation set which, according to ADGAN, are the most anomalous (have the highest anomaly score). Again starting from the top left corner, the bottom rows contain images deemed normal (have the lowest score).

In these controlled experiments we highlight the ability of ADGAN to outperform traditional methods at the task of detecting anomalies in a collection of high-dimensional image samples. While neither table explicitly contains results from scoring the samples using the GAN discriminator, we did run these experiments for both datasets. Performance was weak, with an average AUC of 0.625 for MNIST and 0.513 for CIFAR-10. Scoring according to the prior likelihood p_z of the final latent vectors worked only slightly better, resulting in an average AUC of 0.721 for MNIST and 0.554 for CIFAR-10. Figure 2 gives an additional visual intuition as to why scoring via the prior likelihood fails to give a sensible anomaly score: anomalous samples do not get sent to low probability regions of the Gaussian distribution.

4.4   Unsupervised Anomaly Detection

In the second task we showcase the use of ADGAN in a practical setting where no ground truth information is available. For this we first trained a generator on LSUN scenes. We then used ADGAN to find the most anomalous images within the corresponding validation sets containing 300 images. The images associated with the highest and lowest anomaly scores of three different scene categories are shown in Figs. 4, 5, and 6. Note that the large training set sizes in this experiment would complicate the use of non-parametric methods such as KDE and OC-SVMs.

Fig. 5. Scenes from LSUN showing conference rooms as ranked by ADGAN. The top rows contain anomalous samples, the bottom rows scenes categorized as normal.

Fig. 6. Scenes from LSUN showing churches, ranked by ADGAN. Top rows: anomalous samples. Bottom rows: normal samples.

To additionally quantify the performance on LSUN, we build a test set from combining the 300 validation samples of each scene. After training the generator on bedrooms only we recorded whether ADGAN assigns them low anomaly scores, while assigning high scores to samples showing any of the remaining scenes. This resulted in an AUC of 0.641.

As can be seen from visually inspecting the LSUN scenes flagged as anomalous, our method has the ability to discern usual from unusual samples. We infer that ADGAN is able to incorporate many properties of an image. It does not merely look at colors, but also takes into account whether shown geometries are canonical, or whether an image contains a foreign object (like a caption). Opposed to this, samples that are assigned a low anomaly score are in line with a classes' *Ideal Form*. They show plain colors, are devoid of foreign objects, and were shot from conventional angles. In the case of bedrooms, some of the least anomalous samples are literally just a bed in a room.

5   Conclusion

We showed that searching the latent space of the generator can be leveraged for use in anomaly detection tasks. To that end, our proposed method: (i) delivers state-of-the-art performance on standard image benchmark datasets; (ii) can be used to scan large collections of unlabeled images for anomalous samples.

To the best of our knowledge we also reported the first results of using VAEs for anomaly detection. We remain optimistic that boosting its performance is possible by additional tuning of the underlying neural network architecture or an informed substitution of the latent prior.

Accounting for unsuitable initializations by jointly optimizing latent vectors and generator parameterization are key ingredients to help ADGAN achieve strong experimental performance. Nonetheless, we are confident that approaches such as initializing from an approximate inversion of the generator as in ALI [9,10], or substituting the reconstruction loss for a more elaborate variant, such as the Laplacian pyramid loss [28], can be used to improve our method further.

Acknowledgments. We kindly thank reviewers for their constructive feedback, which helped to improve this work. LD gratefully acknowledges funding from the School of Informatics, University of Edinburgh. LR acknowledges financial support from the German Federal Ministry of Transport and Digital Infrastructure (BMVI) in the project OSIMAB (FKZ: 19F2017E). MK and RV acknowledge support from the German Research Foundation (DFG) award KL 2698/2-1 and from the Federal Ministry of Science and Education (BMBF) award 031B0187B.

References

1. Arjovsky, M., Bottou, L.: Towards principled methods for training generative adversarial networks. In: International Conference on Learning Representations (2017)
2. Arjovsky, M., Chintala, S., Bottou, L.: Wasserstein generative adversarial networks. In: International Conference on Machine Learning, pp. 214–223 (2017)
3. Bahdanau, D., Cho, K., Bengio, Y.: Neural machine translation by jointly learning to align and translate. In: International Conference on Learning Representations (2015)
4. Bojanowski, P., Joulin, A., Lopez-Paz, D., Szlam, A.: Optimizing the latent space of generative networks. In: International Conference on Machine Learning (2018)
5. Campbell, C., Bennett, K.P.: A linear programming approach to novelty detection. In: Advances in Neural Information Processing Systems, pp. 395–401 (2001)
6. Chandola, V., Banerjee, A., Kumar, V.: Anomaly detection: a survey. ACM Comput. Surv. (CSUR) **41**(3), 15 (2009)
7. Creswell, A., Bharath, A.A.: Inverting the generator of a generative adversarial network. arXiv preprint arXiv:1611.05644 (2016)
8. Deng, J., Dong, W., Socher, R., Li, L.J., Li, K., Fei-Fei, L.: Imagenet: a large-scale hierarchical image database. In: Computer Vision and Pattern Recognition, pp. 248–255. IEEE (2009)
9. Donahue, J., Krähenbühl, P., Darrell, T.: Adversarial feature learning. In: International Conference on Learning Representations (2017)
10. Dumoulin, V., et al.: Adversarially learned inference. In: International Conference on Learning Representations (2017)
11. Dutta, H., Giannella, C., Borne, K., Kargupta, H.: Distributed top-k outlier detection from astronomy catalogs using the DEMAC system. In: International Conference on Data Mining, pp. 473–478. SIAM (2007)
12. Edgeworth, F.: XLI. on discordant observations. Lond. Edinb. Dublin Philos. Mag. J. Sci. **23**(143), 364–375 (1887)
13. Emmott, A.F., Das, S., Dietterich, T., Fern, A., Wong, W.K.: Systematic construction of anomaly detection benchmarks from real data. In: ACM SIGKDD Workshop on Outlier Detection and Description, pp. 16–21. ACM (2013)
14. Erfani, S.M., Rajasegarar, S., Karunasekera, S., Leckie, C.: High-dimensional and large-scale anomaly detection using a linear one-class SVM with deep learning. Pattern Recognit. **58**, 121–134 (2016)
15. Eskin, E.: Anomaly detection over noisy data using learned probability distributions. In: International Conference on Machine Learning (2000)
16. Evangelista, P.F., Embrechts, M.J., Szymanski, B.K.: Some properties of the gaussian kernel for one class learning. In: de Sá, J.M., Alexandre, L.A., Duch, W., Mandic, D. (eds.) ICANN 2007. LNCS, vol. 4668, pp. 269–278. Springer, Heidelberg (2007). https://doi.org/10.1007/978-3-540-74690-4_28
17. Goodfellow, I., et al.: Generative adversarial nets. In: Advances in Neural Information Processing Systems, pp. 2672–2680 (2014)
18. Görnitz, N., Braun, M., Kloft, M.: Hidden Markov anomaly detection. In: International Conference on Machine Learning, pp. 1833–1842 (2015)
19. Hu, W., Liao, Y., Vemuri, V.R.: Robust anomaly detection using support vector machines. In: International Conference on Machine Learning, pp. 282–289 (2003)
20. Ioffe, S., Szegedy, C.: Batch normalization: accelerating deep network training by reducing internal covariate shift. In: International Conference on Machine Learning, pp. 448–456 (2015)

21. Kingma, D., Ba, J.: Adam: A method for stochastic optimization. arXiv preprint arXiv:1412.6980 (2014)
22. Kingma, D.P., Welling, M.: Auto-encoding variational Bayes. arXiv preprint arXiv:1312.6114 (2013)
23. Krizhevsky, A., Hinton, G.: Learning multiple layers of features from tiny images. Technical report, University of Toronto (2009)
24. Krizhevsky, A., Sutskever, I., Hinton, G.E.: Imagenet classification with deep convolutional neural networks. In: Advances in Neural Information Processing Systems, pp. 1097–1105 (2012)
25. Le, Q., Mikolov, T.: Distributed representations of sentences and documents. In: International Conference on Machine Learning, pp. 1188–1196 (2014)
26. LeCun, Y.: The MNIST database of handwritten digits (1998). http://yann.lecun.com/exdb/mnist
27. Li, Y., Swersky, K., Zemel, R.: Generative moment matching networks. In: International Conference on Machine Learning, pp. 1718–1727 (2015)
28. Ling, H., Okada, K.: Diffusion distance for histogram comparison. In: Computer Vision and Pattern Recognition, pp. 246–253. IEEE (2006)
29. Lipton, Z.C., Tripathi, S.: Precise recovery of latent vectors from generative adversarial networks. In: International Conference on Learning Representations, Workshop Track (2017)
30. Liu, F.T., Ting, K.M., Zhou, Z.H.: Isolation forest. In: International Conference on Data Mining, pp. 413–422. IEEE (2008)
31. Lopez-Paz, D., Oquab, M.: Revisiting classifier two-sample tests. In: International Conference on Learning Representations (2017)
32. Mao, X., Li, Q., Xie, H., Lau, R.Y., Wang, Z., Smolley, S.P.: Least squares generative adversarial networks. In: International Conference on Computer Vision, pp. 2794–2802. IEEE (2017)
33. Masci, J., Meier, U., Cireşan, D., Schmidhuber, J.: Stacked convolutional autoencoders for hierarchical feature extraction. Artificial Neural Networks and Machine Learning (ICANN), pp. 52–59 (2011)
34. Metz, L., Poole, B., Pfau, D., Sohl-Dickstein, J.: Unrolled generative adversarial networks. In: International Conference on Learning Representations (2017)
35. Müller, A.: Integral probability metrics and their generating classes of functions. Adv. Appl. Probab. **29**(2), 429–443 (1997)
36. Narayanan, H., Mitter, S.: Sample complexity of testing the manifold hypothesis. In: Advances in Neural Information Processing Systems, pp. 1786–1794 (2010)
37. Nowozin, S., Cseke, B., Tomioka, R.: f-GAN: training generative neural samplers using variational divergence minimization. In: Advances in Neural Information Processing Systems, pp. 271–279 (2016)
38. Odena, A., Dumoulin, V., Olah, C.: Deconvolution and checkerboard artifacts. Distill **1**(10), e3 (2016)
39. Ourston, D., Matzner, S., Stump, W., Hopkins, B.: Applications of hidden Markov models to detecting multi-stage network attacks. In: Proceedings of the 36th Annual Hawaii International Conference on System Sciences. IEEE (2003)
40. Parzen, E.: On estimation of a probability density function and mode. Ann. Math. Stat. **33**(3), 1065–1076 (1962)
41. Pearson, K.: On lines and planes of closest fit to systems of points in space. Philos. Mag. **2**(11), 559–572 (1901)
42. Pelleg, D., Moore, A.W.: Active learning for anomaly and rare-category detection. In: Advances in Neural Information Processing Systems, pp. 1073–1080 (2005)

43. Protopapas, P., Giammarco, J., Faccioli, L., Struble, M., Dave, R., Alcock, C.: Finding outlier light curves in catalogues of periodic variable stars. Mon. Not. R. Astron. Soc. **369**(2), 677–696 (2006)
44. Radford, A., Metz, L., Chintala, S.: Unsupervised representation learning with deep convolutional generative adversarial networks. arXiv preprint arXiv:1511.06434 (2015)
45. Ruff, L., et al.: Deep one-class classification. In: International Conference on Machine Learning (2018)
46. Schlegl, T., Seeböck, P., Waldstein, S.M., Schmidt-Erfurth, U., Langs, G.: Unsupervised anomaly detection with generative adversarial networks to guide marker discovery. In: Niethammer, M., et al. (eds.) IPMI 2017. LNCS, vol. 10265, pp. 146–157. Springer, Cham (2017). https://doi.org/10.1007/978-3-319-59050-9_12
47. Schölkopf, B., Platt, J.C., Shawe-Taylor, J., Smola, A.J., Williamson, R.C.: Estimating the support of a high-dimensional distribution. Technical report MSR-TR-99-87, Microsoft Research (1999)
48. Singliar, T., Hauskrecht, M.: Towards a learning traffic incident detection system. In: Workshop on Machine Learning Algorithms for Surveillance and Event Detection, International Conference on Machine Learning (2006)
49. Sriperumbudur, B.K., Fukumizu, K., Gretton, A., Schölkopf, B., Lanckriet, G.R.: On integral probability metrics, ϕ-divergences and binary classification. arXiv preprint arXiv:0901.2698 (2009)
50. Sutskever, I., Vinyals, O., Le, Q.V.: Sequence to sequence learning with neural networks. In: Advances in Neural Information Processing Systems, pp. 3104–3112 (2014)
51. Wong, W.K., Moore, A.W., Cooper, G.F., Wagner, M.M.: Bayesian network anomaly pattern detection for disease outbreaks. In: International Conference on Machine Learning, pp. 808–815 (2003)
52. Xiao, J., Hays, J., Ehinger, K.A., Oliva, A., Torralba, A.: Sun database: large-scale scene recognition from abbey to zoo. In: Computer Vision and Pattern Recognition, pp. 3485–3492. IEEE (2010)
53. Yeung, D.Y., Chow, C.: Parzen-window network intrusion detectors. In: International Conference on Pattern Recognition, vol. 4, pp. 385–388. IEEE (2002)

Image-to-Markup Generation via Paired Adversarial Learning

Jin-Wen Wu[1,2(✉)], Fei Yin[1], Yan-Ming Zhang[1], Xu-Yao Zhang[1],
and Cheng-Lin Liu[1,2,3]

[1] NLPR, Institute of Automation, Chinese Academy of Sciences,
Beijing, People's Republic of China
{jinwen.wu,fyin,ymzhang,xyz,liucl}@nlpr.ia.ac.cn
[2] University of Chinese Academy of Sciences, Beijing, People's Republic of China
[3] CAS Center for Excellence of Brain Science and Intelligence Technology,
Beijing, People's Republic of China

Abstract. Motivated by the fact that humans can grasp semantic-invariant features shared by the same category while attention-based models focus mainly on discriminative features of each object, we propose a scalable paired adversarial learning (PAL) method for image-to-markup generation. PAL can incorporate the prior knowledge of standard templates to guide the attention-based model for discovering semantic-invariant features when the model pays attention to regions of interest. Furthermore, we also extend the convolutional attention mechanism to speed up the image-to-markup parsing process while achieving competitive performance compared with recurrent attention models. We evaluate the proposed method in the scenario of handwritten-image-to-LaTeX generation, i.e., converting handwritten mathematical expressions to LaTeX. Experimental results show that our method can significantly improve the generalization performance over standard attention-based encoder-decoder models.

Keywords: Paired adversarial learning · Semantic-invariant features
Convolutional attention · Handwritten-image-to-LaTeX generation

1 Introduction

The image-to-markup problem has attracted interest of researchers from 1960s [2]. The main target of the research is recognition for the printed/handwritten mathematical expressions (MEs). Different from typical sequence-to-sequence tasks such as machine translation [4] and speech recognition [8], image-to-markup generation converts the two-dimensional (2D) images into sequences of structural presentational languages. More specifically, it has to scan the two dimensional grids to generate markup of the symbols and the implicit spatial operators, such as subscript and fractions. Image-to-markup generation is also different from other image-to-sequence tasks such as image captioning [7,24] and text string

© Springer Nature Switzerland AG 2019
M. Berlingerio et al. (Eds.): ECML PKDD 2018, LNAI 11051, pp. 18–34, 2019.
https://doi.org/10.1007/978-3-030-10925-7_2

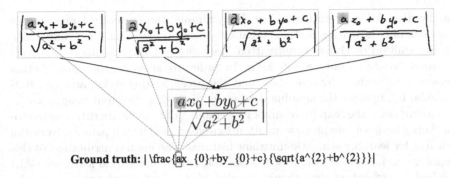

Ground truth: | \frac{ax_{0}+by_{0}+c}{\sqrt{a^{2}+b^{2}}}|

Fig. 1. MEs written by different people (top) and their standard printed template (center). These images have the same ground-truth markup (bottom). The red cells indicate the attention at same symbol a. Same symbols might be written in very different styles while share invariant features that represent the same semantic meaning. (Color figure online)

recognition in optical character recognition [30], in that input images in image-to-markup problem contain much more structural information and spatial relations than general images in computer vision.

Traditional approaches for the image-to-markup problem use handcrafted grammars to handle symbol segmentation, symbol recognition and structural analysis. Although grammar-driven approaches [1,3,23,26] can achieve high performance in practice, they require a large amount of manual work to develop grammatical rules. Furthermore, grammar-driven structural analysis is also highly computationally demanding.

Recently, methods based on deep neural networks have been proposed for image-to-markup generation and achieved great success [10,11,20,28,29]. For example, the model WYGIWYS extended the attention-based encoder-decoder architecture to image-to-markup problem [11]. It encodes printed MEs images with a multi-layer CNN and a bidirectional LSTM and employs a LSTM as the recurrent attention based decoder to generate the target LaTeX format markup. To speedup the method, authors of [10] improved the original WYGIWYS with coarse-to-fine attention and performed experiments on synthetic handwritten MEs. These studies show the data-driven attention-based models can be as effective as the grammar-based approaches while exploiting no prior knowledge of the language.

For the image-to-markup problem, it is especially important to ensure the translation of each local region of the input image. Motivated by this observation, the model WAP [29] records the history of attention at all local regions for improving the coverage of translation. An improved version of WAP uses deep gated recurrent unit (GRU) to encode the online trajectory information of handwritten MEs [28], and has achieved the state-of-the-art performance using an ensemble of five models.

Despite the progresses achieved so far, handwritten-image-to-markup generation is still a very challenging task due to the highly variable handwriting styles compared with printed images, see Fig. 1. On the other hand, well-annotated handwritten MEs are rather scarce. For example, the currently largest public database of handwritten mathematical expression recognition, the Competition on Recognition of Online Handwritten Mathematical Expressions (CROHME) database, contains only $8,836$ MEs. In order to alleviate the contradiction between the limited training data and the great writing-style variation, it is common to augment the training dataset by distorting the input images [20].

To overcome the scarcity of annotated training data in handwritten-image-to-markup generation, we propose an attention-based model with paired adversarial learning for learning semantic-invariant features. The main contributions of this paper are as follows: (1) we present a scalable paired adversarial learning (PAL) method incorporating the prior knowledge of standard templates to guide the attention-based model to learn intrinsic semantic-invariant features; (2) we use a fully convolutional attention based decoder to speed up the image-to-markup decoding without losing accuracy; (3) we introduce a novel multi-directional transition layer that can be easily extended to other deep convolutional networks for accessing 2D contextual information.

2   Background

Before describing our proposed method, we briefly review the generative adversarial network (GAN) [14] in Sect. 2.1 and the convolutional attention (Conv-Attention) model [13] proposed for sequence-to-sequence learning in Sect. 2.2.

2.1   Generative Adversarial Network

GAN is a well-known adversarial learning method originally presented for generative learning by Goodfellow et al. [14]. It generally consists of a generator G and a discriminator D, which are trained with conflicting objectives:

$$\min_G \max_D V(G, D) = E_{\mathbf{x} \sim p_{data}(\mathbf{x})}[\log D(\mathbf{x})] + E_{\mathbf{z} \sim p_{\mathbf{z}}(\mathbf{z})}[\log(1 - D(G(\mathbf{z})))] \quad (1)$$

where $\mathbf{x}$ denotes the target real sample, $\mathbf{z}$ is the input noise and $D(\mathbf{x})$ is the probability that the sample is real. G tries to forge real samples to confuse D while D tries to distinguish fake samples from real ones. Adversarial learning method does force both G and D to improve and has been proven effective in producing highly realistic samples [6,12,25].

Recently, the idea of adversarial learning in GAN has been applied to the image-to-image translation task and demonstrates very encouraging results [18,31]. It is interesting to observe that D successfully guides G to learn the style information from the two domains and realize style transfer from the source domain to the target domain. Another work related to our proposal is the domain adaptation with GAN [5], in which G is guided by D to find a *domain-invariant representation* to represent two domains with different distributions. Inspired by these works, we design an attention-based model to grasp semantic-invariant features from symbols with different writing-styles under the adversarial learning framework.

2.2   Convolutional Attention

Though recurrent attention performs well in dealing with sequential problems, it is still time consuming due to its sequential structure. In this section, we briefly introduce the fully convolutional attention (Conv-Attention) [13], which is proposed for machine translation and shows competitive performance while being more efficient.

Suppose the input sequence is $\mathbf{w} = (w_1, \cdots, w_N)$. $\mathbf{w}$ is then embedded in a distributional space as $\mathbf{x} = (x_1, \cdots, x_N), x_j \in \mathbb{R}^D$ and the absolute position of input elements is embedded as $\mathbf{p} = (p_1, \cdots, p_N), p_j \in \mathbb{R}^D$. The input of the encoder is finally represented as $\mathbf{e} = (x_1 + p_1, \cdots, x_N + p_N)$ to guarantee the model's sense of order and the output sequence of the encoder is $\mathbf{f} = (f_1, \cdots, f_N), f_j \in \mathbb{R}^D$. This process has been also applied to the output elements already generated by the decoder.

The encoder and the Conv-Attention based decoder share a simple block structure. Each block (or referred to layer) contains a one dimensional convolution and a subsequent non-linearity, and computes the output states with a fixed number of input elements. Each convolution kernel of the decoder blocks is parameterized as $W \in \mathbb{R}^{2D \times kD}$ with a base $b_w \in \mathbb{R}^{2D}$, where k is the kernel width and D is the channel dimension of the input features. This convolution kernel maps k concatenated input elements which are embedded in D dimensions to a single output $o_j \in \mathbb{R}^{2D}$. The following non-linearity of one dimensional convolution is chosen as gated linear units (GLU) [9] that implements a gating mechanism over each output element $o_j = [o_{j1} \ o_{j2}] \in \mathbb{R}^{2D}$:

$$GLU(o_j) = o_{j1} \odot \sigma(o_{j2}) \tag{2}$$

where the $\odot$ denotes the point-wise multiplication and gates $\sigma(o_{j2})$ determine which parts of o_{j1} are relevant. The output $GLU(o_j) \in \mathbb{R}^D$ is half the channel dimension of the input o_j.

Conv-Attention uses a separate attention mechanism for each decoder block. It first computes the state summary s_i^l with previous target embedding $\mathbf{t} = (t_1, \cdots, t_T), t_i \in \mathbb{R}^D$ and current hidden state of the l-th block $\mathbf{h}^l = (h_1^l, \cdots, h_T^l), h_i^l \in \mathbb{R}^D$ as:

$$s_i^l = W_s^l h_i^l + b_s^l + t_i \tag{3}$$

Next, attention score $\alpha_{i,j}^l$ of state i and source element j is calculated via a dot-product between f_j of the feature sequence $\mathbf{f}$ and state summary s_i^l:

$$\alpha_{i,j}^l = \frac{\exp(s_i^l \cdot f_j)}{\sum_{w=1}^N \exp(s_i^l \cdot f_w)} \tag{4}$$

After the weights $\alpha_{i,j}^l$ have been computed, the context vector is calculated as:

$$c_i^l = \sum_{j=1}^N \alpha_{i,j}^l (f_j + e_j) \tag{5}$$

Then, the context vector is simply added to the corresponding hidden feature h_i^l. This operation can be considered as attention with multiple *hops* [27], which improves the model's ability to access more attention history. Furthermore, to improve the information flow between blocks, residual connections are added from input to output as ResNet in [16].

3   Paired Adversarial Learning

The motivation of our work is to make the model learn semantic-invariant features of patterns to conquer the difficulties caused by the writing-style variation and the small sample size. Roughly speaking, for each handwritten image in the training set, we first generate its printed image template by compiling the LaTeX format label with a general LaTeX editor. Then, we force the attention-based encode-decoder model to extract similar features for both the handwritten image and its printed image template, which is implemented under the adversarial learning framework.

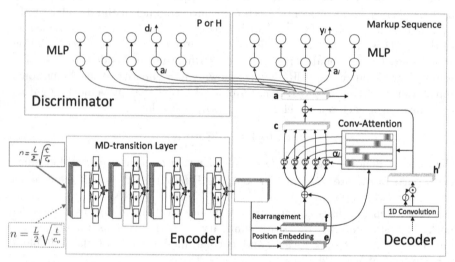

Fig. 2. Architecture of the paired adversarial learning (PAL). When training, each handwritten image is input with its paired printed template (bottom left). The encoder-decoder model and the discriminator are trained alternatively. We use Conv-Attention based decoder here to speed up the image-to-markup decoding. Theoretically, the attention model can also be substituted with any standard recurrent attention.

The proposed model consists of three parts (see Fig. 2): an encoder that extracts features from images, a decoder that parses the sequence of features outputted by the encoder and generates the markup, and a discriminator trained against the encoder-decoder model to force it to learn the semantic-invariant feature of each pattern. In the following subsections, we will sequentially introduce the encoder, the decoder, the learning objective of the proposed paired adversarial learning method, and the training algorithm.

3.1 Multi-directional Encoder

The basic architecture of the encoder in this work is adapted from the fully convolutional network (FCN) model in [29]. The difference is that we introduce a novel layer, named MD-transition layer, equipped after each convolutional block of the deep FCN model. We utilize the multi-dimensional long short-term memory (MDLSTM) [15] to improve FCN's ability to access the 2D contextual information and apply a pooling layer [21] before the MDLSTM layer to improve computation efficiency. We refer to this architecture as MD-transition layer.

MDLSTM employs LSTM layers in up, down, left and right directions. Different LSTM layers are executed in parallel to improve the computation efficiency. After the LSTM layers, we collapse the feature maps of different directions by simply summing them up. The LSTM layers in the horizontal and vertical directions are calculated as:

$$(y_{i,j}, c_{i,j}) = LSTM(x_{i,j}, y_{i\pm1,j}, y_{i,j\pm1}, c_{i\pm1,j}, c_{i,j\pm1}) \tag{6}$$

where y and c denote the output feature vector and inner state of the cell, respectively, and $x_{i,j}$ denotes the input vector of the feature map at position (i,j). The $LSTM$ denotes the mapping function of general LSTM networks which process the input sequence over space or time. With this set up, the subsequent FCN block is enabled to access more past and future contextual information in both horizontal and vertical directions.

3.2 Decoder with Convolutional Attention

We extend the decoder with Conv-Attention [13] to generate markup for the images and speed up the decoding. Different from machine translation, image-to-markup generation is a 2D-to-sequence problem. Since the outputs of the encoder are in the form of 2D feature maps rather than feature sequences, we have to propose a conversion method that preserves feature information as much as possible.

Suppose the output feature map of the multi-directional feature extractor sizes $H \times W \times D$. We split the feature map by columns and then concatenate them to get the feature sequence $\mathbf{f} = (f_1, \cdots, f_N), f_j \in \mathbb{R}^D, N = H \times W$. Then, in order to guarantee the position information during conversion, we add $\mathbf{f}$ with the embedding of the absolute position and get embedded feature sequence $\mathbf{e} = (e_1, \cdots, e_N), e_j \in \mathbb{R}^D$. Here, $\mathbf{e}$ is not the input of the encoder as the original work. After getting $\mathbf{f}$ and $\mathbf{e}$, we compute the context vectors as Sect. 2.2. With rearrangement of the feature map and position embedding, Conv-Attention can be successfully applied to image-to-markup generation. In this study, blocks number l of the Conv-Attention is set to 3. Via the multi-step attention, the model is enabled to access more attention history, thereby improving the ability of consistent tracking with its attention.

3.3 Paired Adversarial Learning

Adversarial learning for image-to-markup task are more complex than image generation, since *mismatch* between two sequences of feature vectors can easily cause the discriminator converging to irrelevant features, and thus lose the ability of guiding the encoder-decoder model to learn the semantic-invariant features. To settle this problem, first, we pair each handwritten image with its same-size printed template to ensure that the length of the two feature sequences are same. Second, since the labels of the paired images are same, the feature vectors at the same position of these two feature sequences are forced to be extracted from related regions with the attention mechanism.

Specifically, let $\mathbf{a}(x, \phi_E) = (a_1, \cdots, a_T), a_i \in \mathbb{R}^D$ denote the feature sequence at the last feature extraction layer of the decoder. Here, x is the input handwritten image x_h or its paired printed template x_p and ϕ_E denotes the parameters of the encoder-decoder model. Our model learns the semantic-invariant features with the guide of a discriminator D which judges whether a feature vector comes from the handwritten images or the printed templates. Let $D(a_i(x, \phi_E), \phi_D)$ represent the probability that feature vector a_i comes from a printed image and ϕ_D denotes the parameters of D. The objective function is defined as:

$$\mathcal{L}_D = E_{(x_h, x_p) \sim X}[E_{a_i(x_p, \phi_E) \sim \mathbf{a}}[\log D(a_i(x_p, \phi_E), \phi_D)] + \\ E_{a_i(x_h, \phi_E) \sim \mathbf{a}}[\log(1 - D(a_i(x_h, \phi_E), \phi_D)]] \tag{7}$$

where $X = \{(x_h, x_p)\}$ is the set of paired training images. D is optimized to maximize the probability of assigning correct labels to the extracted featu res by maximizing $\mathcal{L}_D$. On the contrary, the encoder-decoder model is trained to learn semantic-invariant features to confuse D by minimizing $\mathcal{L}_D$.

Moreover, the primary goal of the encoder-decoder model is to extract discriminative features and generate the correct markup. Thus, the decoder has to convert the feature sequence to the markup by a classification layer as:

$$p(y_{a_i} = y_i | x; \phi_E) = \frac{\exp(C(y_{a_i} = y_i | \mathbf{a}(x, \phi_E))}{\sum_{l=1}^{L} \exp(C(y_{a_i} = l | \mathbf{a}(x, \phi_E))} \tag{8}$$

Here, $y_i \in Y = \{1, \ldots, L\}, L$ denotes the total class number of the label set, y_{a_i} is the prediction of feature vector a_i in the feature sequence $\mathbf{a}(x, \phi_E)$.

Ideally, features extracted from both the printed and handwritten images should be classified correctly with high probabilities. The cross-entropy objective function for classifying the features learned from printed images is defined as:

$$\mathcal{L}_{C_p} = -E_{x_p \sim X_p}[\sum_{i=1}^{T} \log p(y_{a_i} = y_i | x_p; \phi_E)] \tag{9}$$

where $X_p = \{x_p\}$ is the set of printed image templates. Similarly, the loss function for classifying the features learned from handwritten images is defined as:

$$\mathcal{L}_{C_h} = -E_{x_h \sim X_h}[\sum_{i=1}^{T} \log p(y_{a_i} = y_i | x_h; \phi_E)] \tag{10}$$

where $X_h = \{x_h\}$ is the set of handwritten images.

In summary, we train the attention-based encoder-decoder model by minimizing the loss function of:

$$\mathcal{L}_E = \mathcal{L}_{C_p} + \mathcal{L}_{C_h} + \lambda\mathcal{L}_D \tag{11}$$

λ is a hyper-parameter that controls the tradeoff between the discriminative features and the semantic-invariant features. When $\lambda = 0$, the method is a general attention-based encoder-decoder model trained on the paired samples. When λ increases, the method will focus more on learning the semantic-invariant features and extract less discriminative features for the classification layer to generate the predictions.

3.4   Training Procedure

The encoder-decoder model and discriminator D are trained jointly with the paired adversarial learning algorithm. D is optimized with the objective of distinguishing the sequences of feature vectors extracted from the handwritten images and the printed templates. Contrarily, the encoder-decoder model is optimized to extract more sophisticated semantic-invariant features to fool D. Meanwhile, the encoder-decoder model is trained to maximize the probability of ground-truth markup symbols of the input images. The importance of these two objective function is balanced via the hyper-parameter λ.

See details in Algorithm 1. We sample minibatch of the paired samples to train the encoder-decoder model and D for every training cycle. The encoder-decoder model is trained one time first, and D is trained k times then. The parameters of these models are updated by adaptive moment estimation (Adam). Specifically, we update the parameters for the encode-decoder model as:

$$\phi_E \leftarrow \phi_E - Adam(\frac{\partial(\mathcal{L}_{C_p} + \mathcal{L}_{C_h} + \lambda\mathcal{L}_D)}{\partial\phi_E}, \eta_E) \tag{12}$$

And for the discriminator by:

$$\phi_D \leftarrow \phi_D + Adam(\frac{\partial\mathcal{L}_D}{\partial\phi_D}, \eta_D) \tag{13}$$

Here, the $Adam$ is the function to compute the updated value of the adaptive moment estimation with the gradient and learning rate, η_E denotes the learning rate for the encoder-decoder model and η_D denotes the learning rate of the discriminator. See more details in Algorithm 1.

Algorithm 1. The Paired Adversarial Learning Algorithm

---

1 Paired x_h with its printed template x_p by compiling its label y to get the training set $((x_h, x_p), y) \in (X, Y)$;

2 Initialize the encoder-decoder model and the discriminator randomly with parameters ϕ_E and ϕ_D;

3 **repeat**

4   //Update the encoder-decoder model

5   Sample minibatch of m pairs of samples $\{(x_h, x_p)^{(1)}, \ldots, (x_h, x_p)^{(m)}\}$ from the training set;

6   Update the encode-decoder model by:
$$\phi_E \leftarrow \phi_E - Adam(\frac{\partial(\mathcal{L}_{C_h} + \mathcal{L}_{C_p} + \lambda\mathcal{L}_D)}{\partial\phi_E}, \eta_E);$$

7   //Update the discriminator for k steps

8   **for** k *steps* **do**

9   Sample minibatch of m pairs of samples $\{(x_h, x_p)^{(1)}, \ldots, (x_h, x_p)^{(m)}\}$ from the training set;

10   Update the discriminator by: $\phi_D \leftarrow \phi_D + Adam(\frac{\partial\mathcal{L}_D}{\partial\phi_D}, \eta_D)$;

11   **end**

12 **until** $\mathcal{L}_{C_h}$ *converged*;

13 //Get the final model for the handwritten-image-to-markup generation

14 Parameterize the encoder-decoder model by: ϕ_E;

15 **return** *The encoder-decoder model*;

---

4 Experiments

4.1 Datasets

We validate our proposal on handwritten-image-to-LaTeX generation with the large public dataset available from the Competition on Recognition of Online Handwritten Mathematical Expressions (CROHME) [22]. CROHME 2013 dataset consists of 8,836 training samples and 671 test samples. The training set of CROHME 2014 is same as CROHME 2013, but the 986 handwritten samples of the test set are newly collected and labeled. We use the CROHME 2013 test set as the validation set to estimate our model during training process and test the final model on the CROHME 2014 test set. The number of symbol classes for both the CROHME 2013 and CROHME 2014 are 101. Each mathematical expression in the dataset is stored in InkML format, which contains the trajectory coordinates of the handwritten strokes and the LaTeX and MathML format markup ground truth. Models for handwritten-image-to-LaTeX generation are evaluated at expression level by the expression recognition rate (ExpRate), which is the index that ranks the participate systems in all the CROHME competitions. A markup generation of the input image is right if the markup for all the symbols and spatial operators are generated correctly with the right order. This expression level metric is useful to evaluate all the symbols and their structures are translated rightly.

In this study, we have not used the online trajectory information of the strokes, we just connect adjacent coordinate points in the same strokes to get the offline images of the handwritten MEs. Each printed image template of training data is simply gotten by compiling the LaTeX format label with a general LaTeX editor. Then, all the images are normalized to the height of 128 pixels. And images in each minibatch are all padded to the same width as the largest one with background pixels. We use the preprocessing to ensure that the features extracted from different images are the same size.

Table 1. Configurations of the PAL model

Input: $H(128) \times W \times D(1)$ binary image	
Encoder	
CNN Block	[3 × 3 conv-32, BN, ReLU] × 4
MD-transition Layer	MaxPooling, MD-LSTM-64
CNN Block	[3 × 3 conv-64, BN, ReLU] × 4
MD-transition Layer	MaxPooling, MD-LSTM-64, Dropout 0.2
CNN Block	[3 × 3 conv-64, BN, ReLU] × 4
MD-transition Layer	MaxPooling, MD-LSTM-128, Dropout 0.25
CNN Block	[3 × 3 conv-128, BN, ReLU] × 4
MD-transition Layer	MaxPooling, MD-LSTM-256, Dropout 0.35
Decoder	
Conv-Attention	[3 conv-256, GLU, Dropout 0.5] × 3
MLP Layer	256 units, Dropout 0.5
MLP Layer	L units, Dropout 0.5, Softmax
Discriminator	
MLP Layer	512 units, ReLU, Dropout 0.2
MLP Layer	1 unit, Sigmoid

4.2 Model Configurations

In this section, we briefly summarize the configurations of our proposed PAL model. See details in Table 1. The encoder model is adapted from the deep FCN of WAP [29], but equipped with a MD-transition layer after each CNN block. Each CNN block of the encoder contains four convolutional layers, and each convolutional layer is equipped with a batch normalization layer [17] and a rectified linear unit (ReLU) [19]. The filter size of the convolutional layers is 3 × 3 and convolution stride size is 1 × 1. When a feature map is input to the hidden convolutional layer, it is zero-padded by one pixel to keep the size fixed. In addition to the size, channels of the feature maps are also fixed within the CNN blocks. Every pooling layer in the MD-transition layer is set as max-pooling with 2 × 2 kernel and 2 × 2 stride to reduce the size of the feature map.

The decoder model consists of 3 Conv-Attention blocks and a subsequent multi-layer perception (MLP). CNN block in the Conv-Attention model contains a one-dimensional convolutional layer with kernel width 3 and stride size 1.

And the one-dimensional convolutional layer is equipped with a GLU nonlinear activation function introduced in the Sect. 2.2. The discriminator D is a MLP with two fully connected layers. We employ dropout for our proposal to prevent the over-fitting. L in the table denotes the total class number of the symbols in the markup ground truth. All models are implemented in Torch and trained on 4 Nvidia TITAN X GPUs.

Table 2. ExpRate (%) of different systems on CROHME 2014 test set

System	ExpRate (%)	≤1 (%)	≤2 (%)	≤3 (%)
I	37.22	44.22	47.26	50.20
II	15.01	22.31	26.57	27.69
IV	18.97	28.19	32.35	33.37
V	18.97	26.37	30.83	32.96
VI	25.66	33.16	35.90	37.32
VII	26.06	33.87	38.54	39.96
WYGIWYS*	28.70	-	-	-
End-to-end	35.19(18.97)	-	-	-
WAP*	44.42	58.40	62.20	63.10
PAL	39.66	56.80	65.11	70.49
PAL*	**47.06**	**63.49**	**72.31**	**78.60**

4.3   Validation on CROHME

We compare our proposal with the submitted systems from CROHME 2014 and some attention-based models presented recently for handwritten-image-to-LaTeX generation. The results of these systems are listed in Table 2. Systems I to VII are the participants in CROHME 2014 and the next three systems from WYGIWYS* to WAP* are attention-based models presented recently. To make fair comparison, system III are erased from Table 2 because it has used unofficial extra training data and the attention models listed in Table 2 are all trained with offline images. The ExpRate $\leq 1(\%), \leq 2(\%), \leq 3(\%)$ denote the accuracy for markup generation with one to three symbol-level error and showing the room for the models to be further improved.

Our proposed PAL model outperforms system I, which wins the first place on CROHME 2014 and named *seehat*, with more than 2% ExpRate. More importantly, it is interesting to observe that the one to three symbol-level error of our proposal has been significantly reduced due to the grasp of semantic-invariant features for each symbol. The sign * in Table 2 denotes utilizing an ensemble of 5 differently initialized models to improve the performance [29]. WYGIWYS is the first attention-based model proposed for image-to-markup generation [11]. WYGIWYS with ensemble methods finally achieves an ExpRate of 28.70%. The End-to-end indicates encoder-decoder model in [20], which has a similar architecture to WYGIWYS. It achieves a ExpRate of 18.97%, and 35.19% then by distorting the training images and bringing the number of training images to 6 times. WAP* here indicates the state-of-the-art model WAP [29] trained with ensemble methods and not uses the online trajectory information like other attention-based model here. We use the same ensemble method to get PAL*, and the result shows our proposed PAL model outperforms WAP under the same conditions. While the ensemble method can effectively improve performance, it requires much more memory space to run as fast as a single model through parallel computing.

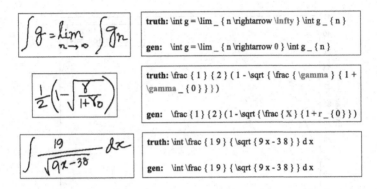

Fig. 3. Examples of the handwritten images and generated markup of our proposal

In order to make the results more intuitive, in Fig. 3 we show some handwritten MEs of the CROHME 2014 test set as well as their markup generated by our proposal. The red symbols of the *gen* indicate the incorrect markup generations, and the blue symbols of the *truth* indicate the corresponding right markup in the ground truth or the markup our proposal failed to generate. The results show our proposal are effective in dealing with the complex 2D structures and the symbols with various writing styles. It is worth noting that some symbols are too similar or written too scribbled, even humans could be confused.

4.4 Comparison of Different λ

In this section, we further analyze how the hype-parameter λ in Eq. (11) affects the performance of our proposed PAL model. By balancing the influence of the loss for markup generation and features discrimination, λ controls the trade-off between discriminative features and semantic-invariant features the encoder-decoder model learned. When λ is small, discriminative features comprise the majority loss of the encoder-decoder model and dominate the gradient backward to it. With the increasing of λ, the encoder-decoder model masters more semantic-invariant features of same symbols in the printed templates and the handwritten images. However, when λ going too large, the model will focus too much on semantic-invariant features and even try to generate same feature sequences for both printed and handwritten images to confuse the discriminator. This will lead to less discriminative features for different categories and cause the decreasing of markup generation accuracy. For an extreme case, the model may only pays attention to regions of the background and even generates $G(x)$ that equals to a constant at each step to fool the discriminator D. Therefore, an appropriate λ plays an important role in the PAL model. We explore different λ for the model while keeping the other configurations of the model fixed the and then evaluate these models with different λ on the CROHME dataset. The results is shown in Fig. 4.

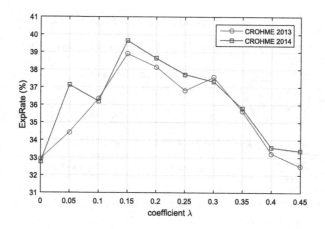

Fig. 4. Comparison of different λ on CROHME dataset

4.5   Analysis of Print Templates

It is worth noting that print templates for training are also crucial to the PAL model. Firstly, just like we cannot use printed English books to teach humans to recognize handwritten Chinese characters, the templates need to have related semantic information with the target images. Thus, the attention-based encoder-decoder model can learn semantic-invariant features for each specific symbol in the paired images. Secondly, the distribution of standard templates needs to be easier to learn. In this way, simple templates can guide the model in dealing with complex samples through paired adversarial learning.

We first validate the ability of Conv-Attention based encoder-decoder model without paired adversarial learning to generate markup for the print images. Specifically, we compile the LaTeX format markup ground truth of CROHME 2013 test set to get the printed CROHME 2013 test set (CROHME 2013 P) and train the model only on the standard printed templates, see Conv-Attention P in Table 3. Then we get the same model but trained on only the handwritten images (Conv-Attention H). Surprisingly, the accuracy of printed-image-to-markup generation is more than double of the handwritten-image-to-markup generation.

Table 3. Analysis of the influence of printed templates

ExpRate (%)	CROHME 2013	CHROHME 2014	CROHME 2013 P
Conv-Attention H	31.30	27.18	-
Conv-Attention P	0.15	-	76.15
Conv-Attention P&H	32.94	33.87	-
PAL	38.90	39.66	57.82
PAL GD	41.73	39.05	-

However, the model Conv-Attention P appears to have been over fitted in the printed images when tested on the handwritten CROHME 2013 test set.

After that, we mix these two samples to conduct experiments (Conv-Attention P&H). The experimental results show that the distributions of these two kinds of samples is relatively close, and the adding of the printed templates is helpful to the markup generation for the handwritten images. The model increased about 1% ExpRate compared with Conv-Attention H when validated on CROHME 2013 test set and the generalization is significantly enhanced when test on CROHME 2014 test set. When we train the model with paired adversarial learning (PAL) and set λ as 0.15, we find that this increase becomes even more apparent, whether it is validation or test. We also tested the PAL model on CROHME 2013 P, the result shows that the model does lose some knowledge about the distribution of the print images compared with Conv-Attention P.

In addition, we have made some global distortions for the printed templates to further explore the influence of the print templates' distribution. We rote the standard printed sample with 4 angles randomly choose from $-2°$ to $2°$ with a interval of $0.5°$ but excluding $0°$ (the minus sign here represents counterclockwise). Then we add them to the standard printed templates without distortion and re-pair each of these printed templates with the handwritten one that owned the same label. The new $8,836 * 5$ image pairs are used to train the model called PAL GD. Interestingly, we find that the accuracy of the validation has been further improved, but the accuracy of the test has slightly decreased. It is believed that if the distortions are done more elaborately, the test accuracy will also be improved, but this contradicts our original intention of training the attention-based model through easy-to-get templates with paired adversarial learning. Therefore, we haven't conducted further experiments on the distortion.

5   Conclusion

In this paper, we introduce a novel paired adversarial learning to guide the attention-based model to learn the semantic-invariant features as human when focusing attention on specific objects. Our proposal incorporates the prior knowledge of simple templates and improves the performance of an attention-based model on more complex tasks. The proposal performs much better than other systems under the same training conditions on CROHME 2014. We also extend a fully convolutional attention from machine translation to speed-up the decoding of the image-to-markup generation.

In future work, we plan to explore the language model based attention to make the neural network more like human when generating the markup for the input images. We will also apply the paired adversarial learning in more fields such as text string recognition in optical character recognition to improve the models performance.

Acknowledgements. This work has been supported by the National Natural Science Foundation of China (NSFC) Grants 61721004, 61411136002, 61773376, 61633021 and 61733007.

The authors want to thank Yi-Chao Wu for insightful comments and suggestion.

References

1. Álvaro, F., Sánchez, J.A., Benedí, J.M.: An integrated grammar-based approach for mathematical expression recognition. Pattern Recogn. **51**, 135–147 (2016)
2. Anderson, R.H.: Syntax-directed recognition of hand-printed two-dimensional mathematics. In: Symposium on Interactive Systems for Experimental Applied Mathematics: Proceedings of the Association for Computing Machinery Inc. Symposium, pp. 436–459. ACM (1967)
3. Awal, A.M., Mouchère, H., Viard-Gaudin, C.: A global learning approach for an online handwritten mathematical expression recognition system. Pattern Recogn. Lett. **35**, 68–77 (2014)
4. Bahdanau, D., Cho, K., Bengio, Y.: Neural machine translation by jointly learning to align and translate. arXiv preprint arXiv:1409.0473 (2014)
5. Bousmalis, K., Silberman, N., Dohan, D., Erhan, D., Krishnan, D.: Unsupervised pixel-level domain adaptation with generative adversarial networks. In: The IEEE Conference on Computer Vision and Pattern Recognition, pp. 95–104 (2017)
6. Chen, X., Duan, Y., Houthooft, R., Schulman, J., Sutskever, I., Abbeel, P.: InfoGAN: interpretable representation learning by information maximizing generative adversarial nets. In: Advances in Neural Information Processing Systems, pp. 2172–2180 (2016)
7. Cho, K., Courville, A., Bengio, Y.: Describing multimedia content using attention-based encoder-decoder networks. IEEE Trans. Multimedia **17**(11), 1875–1886 (2015)
8. Chorowski, J.K., Bahdanau, D., Serdyuk, D., Cho, K., Bengio, Y.: Attention-based models for speech recognition. In: Advances in Neural Information Processing Systems, pp. 577–585 (2015)
9. Dauphin, Y.N., Fan, A., Auli, M., Grangier, D.: Language modeling with gated convolutional networks. arXiv preprint arXiv:1612.08083 (2016)
10. Deng, Y., Kanervisto, A., Ling, J., Rush, A.M.: Image-to-markup generation with coarse-to-fine attention. In: International Conference on Machine Learning, pp. 980–989 (2017)
11. Deng, Y., Kanervisto, A., Rush, A.M.: What you get is what you see: a visual markup decompiler. arXiv preprint arXiv:1609.04938 (2016)

12. Denton, E.L., Chintala, S., Fergus, R., et al.: Deep generative image models using a Laplacian pyramid of adversarial networks. In: Advances in Neural Information Processing Systems, pp. 1486–1494 (2015)
13. Gehring, J., Auli, M., Grangier, D., Yarats, D., Dauphin, Y.N.: Convolutional sequence to sequence learning. arXiv preprint arXiv:1705.03122 (2017)
14. Goodfellow, I., et al.: Generative adversarial nets. In: Advances in Neural Information Processing Systems, pp. 2672–2680 (2014)
15. Graves, A., Schmidhuber, J.: Offline handwriting recognition with multidimensional recurrent neural networks. In: Advances in Neural Information Processing Systems, pp. 545–552 (2009)
16. He, K., Zhang, X., Ren, S., Sun, J.: Deep residual learning for image recognition. In: The IEEE Conference on Computer Vision and Pattern Recognition, pp. 770–778 (2016)
17. Ioffe, S., Szegedy, C.: Batch normalization: accelerating deep network training by reducing internal covariate shift. arXiv preprint arXiv:1502.03167 (2015)
18. Isola, P., Zhu, J.Y., Zhou, T., Efros, A.A.: Image-to-image translation with conditional adversarial networks. arXiv preprint (2017)
19. Krizhevsky, A., Sutskever, I., Hinton, G.E.: ImageNet classification with deep convolutional neural networks. In: Advances in Neural Information Processing Systems, pp. 1097–1105 (2012)
20. Le, A.D., Nakagawa, M.: Training an end-to-end system for handwritten mathematical expression recognition by generated patterns. In: 14th International Conference on Document Analysis and Recognition, vol. 1, pp. 1056–1061. IEEE (2017)
21. LeCun, Y., Bottou, L., Bengio, Y., Haffner, P.: Gradient-based learning applied to document recognition. Proc. IEEE **86**(11), 2278–2324 (1998)
22. Mouchere, H., Viard-Gaudin, C., Zanibbi, R., Garain, U.: ICFHR 2014 competition on recognition of on-line handwritten mathematical expressions (CROHME 2014). In: 14th International Conference on Frontiers in Handwriting Recognition, pp. 791–796. IEEE (2014)
23. Mouchere, H., Zanibbi, R., Garain, U., Viard-Gaudin, C.: Advancing the state of the art for handwritten math recognition: the CROHME competitions, 2011–2014. Int. J. Doc. Anal. Recogn. **19**(2), 173–189 (2016)
24. Qureshi, A.H., Nakamura, Y., Yoshikawa, Y., Ishiguro, H.: Show, attend and interact: perceivable human-robot social interaction through neural attention q-network. In: International Conference on Robotics and Automation, pp. 1639–1645. IEEE (2017)
25. Radford, A., Metz, L., Chintala, S.: Unsupervised representation learning with deep convolutional generative adversarial networks. arXiv preprint arXiv:1511.06434 (2015)
26. Sayre, K.M.: Machine recognition of handwritten words: a project report. Pattern Recogn. **5**(3), 213–228 (1973)
27. Sukhbaatar, S., Weston, J., Fergus, R., et al.: End-to-end memory networks. In: Advances in Neural Information Processing Systems, pp. 2440–2448 (2015)
28. Zhang, J., Du, J., Dai, L.: A gru-based encoder-decoder approach with attention for online handwritten mathematical expression recognition. arXiv preprint arXiv:1712.03991 (2017)
29. Zhang, J., et al.: Watch, attend and parse: an end-to-end neural network based approach to handwritten mathematical expression recognition. Pattern Recogn. **71**, 196–206 (2017)

30. Zhou, X.D., Wang, D.H., Tian, F., Liu, C.L., Nakagawa, M.: Handwritten Chinese/Japanese text recognition using semi-Markov conditional random fields. IEEE Trans. Pattern Anal. Mach. Intell. **35**(10), 2413–2426 (2013)
31. Zhu, J.Y., Park, T., Isola, P., Efros, A.A.: Unpaired image-to-image translation using cycle-consistent adversarial networks. arXiv preprint arXiv:1703.10593 (2017)

Toward an Understanding of Adversarial Examples in Clinical Trials

Konstantinos Papangelou[1]($\boxtimes$) (iD), Konstantinos Sechidis[1] (iD), James Weatherall[2], and Gavin Brown[1]

[1] School of Computer Science, University of Manchester, Manchester M13 9PL, UK
{konstantinos.papangelou,konstantinos.sechidis,
gavin.brown}@manchester.ac.uk
[2] Advanced Analytics Centre, Global Medicines Development,
AstraZeneca, Cambridge SG8 6EE, UK
james.weatherall@astrazeneca.com

Abstract. Deep learning systems can be fooled by small, worst-case perturbations of their inputs, known as adversarial examples. This has been almost exclusively studied in supervised learning, on vision tasks. However, adversarial examples in *counterfactual* modelling, which sits outside the traditional supervised scenario, is an overlooked challenge. We introduce the concept of *adversarial patients*, in the context of counterfactual models for clinical trials—this turns out to introduce several new dimensions to the literature. We describe how there exist multiple *types* of adversarial example—and demonstrate different consequences, e.g. ethical, when they arise. The study of adversarial examples in this area is rich in challenges for accountability and trustworthiness in ML–we highlight future directions that may be of interest to the community.

Keywords: Adversarial examples · Counterfactual modelling
Randomised clinical trials · Subgroup identification

1 Introduction

For personalised medicine, a major goal is to predict whether or not a patient will benefit from a particular treatment. The challenge here is to model the outcome of a patient under different treatment scenarios. This task sits outside traditional supervised learning, phrased as a causal inference problem, i.e. modelling the causal effect of a treatment on the outcome of a patient. Recently, Deep Neural Networks (DNNs) have been adopted in this area, achieving significant improvements on the estimation of individual-level treatment effects [1,8,17]. While DNNs have proven their merits in various domains, it has also been shown

---

Electronic supplementary material The online version of this chapter (https://doi.org/10.1007/978-3-030-10925-7_3) contains supplementary material, which is available to authorized users.

M. Berlingerio et al. (Eds.): ECML PKDD 2018, LNAI 11051, pp. 35–51, 2019.
https://doi.org/10.1007/978-3-030-10925-7_3

that they are susceptible to *adversarial examples*—small perturbations of the data, carefully designed to deceive the model. This area has received significant attention from the community, e.g. [3,6,13,18]. When DNNs are used in safety-critical applications, such as healthcare, *accountability* becomes crucial [4]. One such application, where the errors may result in personal, ethical, financial and legal consequences, is in personalised medicine.

While adversarial examples have been studied extensively in the traditional supervised setting, their properties in counterfactual models create significant new challenges. We introduce the concept of *adversarial patients*, the analogue of adversarial examples in the counterfactual (healthcare) scenario. We show that in counterfactual models there exist adversarial directions that have not been examined before. By extending well-established results from supervised learning to the counterfactual setting, we show how the derived *adversarial patients* may affect clinical decisions. We note that, in the supervised adversarial literature, a common fear is the creation of *intentional* adversarial examples, creating security risks. In contrast for healthcare, the caution should be for *unintentional* adversarial patients, with edge-case health characteristics, leading to significant ethical dilemmas. Drug development is a time consuming and expensive process spanning sometimes 10–15 years [14], where the end result may affect the lives of millions, thus we argue that the study of adversarial patients is important and worthy of investigation. In particular, we note:

- They can act as warning flags—cases where the deployed model may go wrong, increasing the level of trust.
- They are in accordance with recent regulations for interpretable and accountable procedures in ML [4].

In Sect. 2 we review the concept of adversarial examples in traditional supervised problems, and in Sect. 3 the basics of counterfactual modelling, which sets our focus. From Sect. 4 onward, we introduce the concept of *adversarial patients* in counterfactual models. We present an empirical study in Sect. 5, and conclude with a discussion of promising future directions.

2   Background: Adversarial Machine Learning

Adversarial examples are carefully crafted inputs that deceive ML models to misclassify them [18]. In vision tasks such adversarial images are indistinguishable (to the human eye) from their original counterparts, but can fool a classifier into making incorrect predictions with high confidence [6,9,12,13,18].

To formally define adversarial examples, we denote a single example as the tuple $\{\mathbf{x}, y\}$ where $\mathbf{x} \in \mathbb{R}^d$ and $y \in \{0, 1\}$, and the class prediction from some model $\hat{f}$ is given by $h(\mathbf{x}) = \mathbb{1}(\hat{f}(\mathbf{x}) > 0.5)$. Suppose now we have a correct classification, i.e. $h(\mathbf{x}) = y$. An adversarial example, $\mathbf{x}_{adv}$, can be constructed solving the following optimisation problem, where the particular l-norm is chosen per application:

$$\underset{\mathbf{x}_{adv}}{\arg\min} \quad \|\mathbf{x} - \mathbf{x}_{adv}\|_l \quad \text{s.t.} \quad h(\mathbf{x}_{adv}) \neq y \tag{1}$$

i.e. the closest input vector which the model $\hat{f}$ misclassifies. Note that this requires knowledge of a ground truth y, which, as we will see, is not the case in counterfactual models. Different choices of the distance measure and the optimisation procedure have resulted in many variations (e.g. [6,9,12,13]). Among those, a well-established approach, which we will adopt in this work, is the *fast gradient sign method* (FGSM) [6], which fixes the maximum perturbation and maximises an arbitrary, differentiable loss function $J(\cdot)$, as so:

$$\underset{\mathbf{x}_{adv}}{\arg\max} \quad J(\hat{f}(\mathbf{x}_{adv}), y) \quad \text{s.t.} \quad \|\mathbf{x} - \mathbf{x}_{adv}\|_\infty \leq \theta, \quad \text{for some} \quad \theta > 0 \qquad (2)$$

Once adversarial examples are created, they can deceive the model $\hat{f}$ in deployment. A complementary field of study has emerged, in defending models against such examples. A common method is *adversarial training*, where they are included in a new round of training, iteratively increasing the robustness of the model. Research has shown adversarial training to be an effective regulariser [6,11,18], especially in combination with others for DNNs, such as Dropout.

The study of adversarial examples is especially important in the healthcare domain, where incorrect predictions may have grave repercussions. However, the standard supervised learning paradigm does not always apply in healthcare. In the following section we introduce the basics of *counterfactual modelling*, which brings new dimensions to the adversarial literature.

3   Counterfactual Modelling in Clinical Trials

3.1   The Potential Outcomes Framework

A clinical trial is a time consuming and expensive procedure with several phases. At each phase more participants are recruited in order to evaluate the safety and efficacy of a drug before it is deployed. In phases I and II preliminary results on the safety of the drug are gathered, normally using a few hundred patients. Phases III and IV are of particular relevance to the ML community, since predictive models are usually built to estimate properties of the drug. In phase III an essential task is the identification of subgroups of patients with good response to the drug, usually requiring a model to estimate individual level treatment effects as an intermediate step [5,10]. In phase IV the drug is used in clinical practice, and a predictive model (or rules derived from it) might assist clinicians to estimate the likely outcome for a particular patient.

Inference of individual treatment effects can be formulated as a causal inference problem—for each patient we have some baseline (pre-treatment) features, the administered treatment, and the outcome. In contrast to supervised problems, we do not observe the outcome under the *alternative treatment*, i.e. we do not observe the *counterfactual* outcome. We can address this class of problems with the *potential outcomes framework* [15].

We define the treatment variable T, as a random variable[1] that indicates the treatment received by each subject; here for simplicity we assume $T \in \{0, 1\}$.

---

[1] We use upper-case for random variables and lower-case for their realisations.

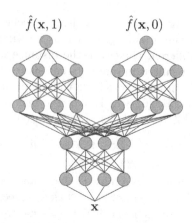

Fig. 1. Example of a multi-task counterfactual neural network.

The potential outcomes under the two treatment scenarios are functions of the subjects $\mathbf{X}$, denoted as $Y_1(\mathbf{X}), Y_0(\mathbf{X})$, while the observed outcome is $Y(\mathbf{X})$. For simplicity, we will omit the dependence of the outcomes on $\mathbf{X}$. For each subject we only observe the outcome under the actual treatment received: if $T = 0$ then $Y = Y_0$, and if $T = 1$ we observe $Y = Y_1$. The difficulty arises from the fact for any given patient, we cannot observe both Y_0 and Y_1.

The subjects that receive the drug ($T = 1$) are referred to as the *treated group*, while those who receive the baseline treatment ($T = 0$) are the *control group*. For subjects $\mathbf{X}$ the observed factual outcome can be expressed as a random variable in terms of the two potential outcomes: $Y = TY_1 + (1 - T)Y_0$. Given a set of subjects with their assigned treatments and their factual outcomes, the goal of *counterfactual modelling* is to train a model that will allow us to infer the counterfactual outcome that would have occurred if we had flipped the treatment: $Y^{cf} = (1 - T)Y_1 + TY_0$.

Building on this, the *individual treatment effect* (ITE) of a subject $\mathbf{X} = \mathbf{x}$ can be expressed as the expected difference between the two potential outcomes:

$$ITE(\mathbf{x}) := \mathbb{E}_{Y_1 \sim p(Y_1 | \mathbf{x})}[Y_1 \mid \mathbf{x}] - \mathbb{E}_{Y_0 \sim p(Y_0 | \mathbf{x})}[Y_0 \mid \mathbf{x}]$$

For treatment effects to be identifiable from the observed data, certain assumptions must hold (Sect. 1 of the supplementary material) and have been adopted for the estimation of ITE in observational studies (e.g. [17] for a recent example). We will focus on the case of randomised clinical trials, where *strongly ignorable* treatment assignment (i.e. treatment assignment depends only on the observed features—in our case is random) holds by design.

3.2 NNs for Estimation of Individual Treatment Effects

The adaptation of supervised learning models for causal inference tasks has gained much recent attention, both for estimation of individual treatment effects

[2,20] and for subgroup identification [5,10]. Multi-task Deep Neural Nets have shown significant improvement over traditional baselines [1,17], capturing the interactions between the two groups in a shared representation, while learning the response surface for each group separately. In multi-task models, learning can be achieved following an "alternating" approach during training, where at each iteration we use either a treated or a control batch [1]. At every iteration we update the weights of the shared layers and, depending on the batch, only the corresponding task-specific layers, i.e. for a treatment batch, only the left branch in Fig. 1 is updated, and the parameters of the right branch are fixed.

Note that in contrast to Sect. 2, now we will have two outputs, one for each treatment group. These can be seen as separate models. Let us denote each output as $\hat{f}(\mathbf{x}, t), t \in \{0, 1\}$. The individual treatment effect can be estimated from the outputs of the two branches as, $\hat{\text{ITE}}(\mathbf{x}) = \hat{f}(\mathbf{x}, 1) - \hat{f}(\mathbf{x}, 0)$. A common evaluation measure of ITE that has been used extensively in the literature [1,8, 17] is the *expected Precision in Estimation of Heterogeneous Effect* (PEHE):

$$\varepsilon_{PEHE} = \int ((\hat{f}(\mathbf{x}, 1) - \hat{f}(\mathbf{x}, 0)) - ITE(\mathbf{x}))^2 p(\mathbf{x}) d\mathbf{x} \tag{3}$$

The use of the true ITE assumes the knowledge of the ground truth for both potential outcomes and therefore can be estimated only in (semi-)synthetic scenarios. In real-world applications, a common approximation of ε_{PEHE} is to estimate it using the factual outcome of the nearest neighbour that received the opposite treatment [8,17].

Counterfactual modelling differs significantly from traditional supervised settings, and introduces new challenges that have not been examined in the adversarial learning literature. The key contributions of this paper are:

- We show that counterfactual models introduce new sets of adversarial directions, some of which cannot be observed, but provably still exist.
- We show how we can use FGSM to identify adversarial examples with respect to the observed (factual) outcome and discuss their implications.
- We highlight that for subgroup identification (common in clinical trials) we may not be interested in the outcomes, but the *individual treatment effect*. We show that the ITE adds a new set of adversarial directions, the implications of which we shall discuss.
- We demonstrate the effect of adversarial training on the generalisation performance of counterfactual models.
- Building on our analysis, we present several new research directions, which may be of interest to the community (Sect. 6).

4 A Theoretical Study of Adversarial Directions

In this section we introduce the concept of adversarial patients in counterfactual models. We will assume a *hypothetical* scenario, where we know both potential outcomes. This will allow us to show that in counterfactual models there can exist multiple adversarial directions, with potentially different consequences.

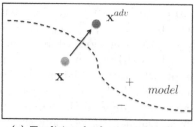
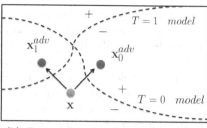

(a) Traditional adversarial setting   (b) Counterfactual adversarial setting

Fig. 2. In traditional adversarial settings there is a single model which we are trying to fool [3,6]. Counterfactual adversarial examples, or "adversarial patients" can exist in several forms. Here we illustrate how the adversarial patient that fools the treatment model may be orthogonal to that which fools the control model. Note that in practice, since we know one of the outcomes, we can identify only one of the two directions.

From now on, we will focus on the case of binary outcomes and binary treatment. We will denote a single example as $\{\mathbf{x}, y_1, y_0\}$, where $\mathbf{x} \in \mathbb{R}^d$ and $y_1, y_0 \in \{0, 1\}$ are the potential outcomes of $\mathbf{x}$. Note that in practice for a given $\mathbf{x}$ we observe only one of them. As we already mentioned, we will use $\hat{f}(\mathbf{x}, t)$ to denote the output of the counterfactual network. For binary outcomes, this is the probability of observing a positive outcome with $(t = 1)$ or without treatment $(t = 0)$, estimated by a model $\hat{f}(\cdot, t)$. The predicted outcomes of a model will be denoted as $h(\mathbf{x}, t) = 1(\hat{f}(\mathbf{x}, t) > 0.5)$.

4.1 Adversarial Patients and Potential Outcomes

Let us assume for now the hypothetical scenario, in which both potential outcomes are observable. In traditional supervised problems there will be a single model $\hat{f}$, which we are trying to deceive. On the other hand, in counterfactual modelling for two levels of treatment there will be two models, as shown in Fig. 2. The model $\hat{f}(\cdot, 1)$ defines the decision boundary between the patients who have outcome $y_1 = 1$ and $y_1 = 0$ while the model $\hat{f}(\cdot, 0)$ distinguishes between those who have $y_0 = 1$ and $y_0 = 0$. This results in the following definition of counterfactual adversarial examples, or *adversarial patients*.

Definition 1 (Adversarial Patient and Potential Outcomes). *An adversarial patient with respect to its potential outcome* y_t, $\mathbf{x}_t^{adv} \in \mathbb{R}^d$, *where* $t \in \{0, 1\}$ *is a solution to the following optimisation problem:*

$$\underset{\mathbf{x}_t^{adv}}{\arg\min} \quad \|\mathbf{x} - \mathbf{x}_t^{adv}\|_l \quad s.t. \quad h(\mathbf{x}_t^{adv}, t) \neq y_t$$

Notice that in contrast to Eq. (1), Definition 1 admits the existence of two adversarial patients, one for each treatment group (Fig. 2). Let us define the loss function on each group as the cross-entropy loss $J(\hat{f}(\mathbf{x},t), y_t) = -y_t \log \hat{f}(\mathbf{x},t) - (1 - y_t)\log(1 - \hat{f}(\mathbf{x},t)), t \in \{0,1\}$. Then, restating the optimisation problem as a maximisation of the loss results in the following modification of Eq. (2):

$$\underset{\mathbf{x}_t^{adv}}{\arg\max} \quad J(\hat{f}(\mathbf{x}_t^{adv}, t), y_t) \quad \text{s.t.} \quad \|\mathbf{x} - \mathbf{x}_t^{adv}\|_\infty \leq \theta \quad \text{for} \quad t \in \{0,1\} \quad (4)$$

Optimisation of Eq. (4) can lead to a fast way of creating adversarial patients with respect to their potential outcomes by taking the first-order Taylor expansion of $J(\cdot)$ and keeping a fixed $\|\cdot\|_\infty$ perturbation. One advantage of phrasing the adversarial problem in terms of the loss $J(\cdot)$ is that it allows us to generalise to other functions, leading to a family of adversarial directions. Consequently, we can identify a general adversarial direction that jointly maximises the loss on both groups. To show that, let us define the joint loss as the convex combination of the cross-entropy losses on each group separately:

$$J_{joint}(\hat{f}(\mathbf{x},1), y_1, \hat{f}(\mathbf{x},0), y_0) = \pi \cdot J(\hat{f}(\mathbf{x},1), y_1) + (1 - \pi) \cdot J(\hat{f}(\mathbf{x},0), y_0) \quad (5)$$

where $\pi \in [0,1]$ is a constant. We conjecture that π controls the relative importance of each loss and therefore can lead to adapting adversarial patients in a cost-sensitive framework (we will discuss this further in Sect. 6). An adversarial patient with respect to the joint loss will be derived by modifying an initial patient $\mathbf{x}$ to the direction that decreases the confidence of the model on making the correct prediction for both tasks (Fig. 3(a)). Indeed, differentiating Eq. (5) with respect to $\mathbf{x}$ will result in the following expression:

$$\nabla_{\mathbf{x}} J_{joint}(\cdot) = \pi \cdot \beta_{y_1} \cdot \nabla_{\mathbf{x}} \hat{f}(\mathbf{x},1) + (1 - \pi) \cdot \beta_{y_0} \cdot \nabla_{\mathbf{x}} \hat{f}(\mathbf{x},0),$$

$$\text{where} \quad \beta_{y_t=1} = -\frac{1}{\hat{f}(\mathbf{x},t)} \quad \text{and} \quad \beta_{y_t=0} = \frac{1}{1 - \hat{f}(\mathbf{x},t)} \quad (6)$$

Notice that $\beta_{y_t=1} < 0$ and $\beta_{y_t=0} > 0$, since $\hat{f}(\mathbf{x},t) \in (0,1)$. Suppose, for example, that $y_1 = 0$ and $y_0 = 1$. Then $\nabla_{\mathbf{x}} J_{joint}(\cdot) = \pi \cdot \beta_{y_1=0} \cdot \nabla_{\mathbf{x}} \hat{f}(\mathbf{x},1) + (1 - \pi) \cdot \beta_{y_0=1} \cdot \nabla_{\mathbf{x}} \hat{f}(\mathbf{x},0)$ and the adversarial patient will be in a direction indicated by the weighted difference between the two gradients. In the next section, we show that stating the optimisation problem in terms of maximising a loss function results in identifying a second form of adversarial patients – those that can affect ITE.

4.2   Adversarial Patients and ITE

Let us continue with the hypothetical scenario of having knowledge of both potential outcomes. We will show that in counterfactual models there exist new types of adversarial directions, that are not a direct extension of existing definitions of adversarial examples. To see that let us define first the region, in which we will search for adversarial patients as $R(\mathbf{x},\theta) = \{\mathbf{x}' : \|\mathbf{x} - \mathbf{x}'\|_l \leq \theta\}, \theta > 0$. For what follows we will consider $R(\cdot) = \|\cdot\|_\infty$, but other measures of distance can also be used. Suppose that for a patient $\mathbf{x}$ the potential outcomes are $y_1 = 1$ and $y_0 = 0$ and we have a model $\hat{f}$ that correctly assigns these outcomes with probabilities $\hat{f}(\mathbf{x}, 1) = 0.9$ and $\hat{f}(\mathbf{x}, 0) = 0.2$. Now consider a patient $\mathbf{x}'$ that maximises the joint loss on the potential outcomes within a region $R(\mathbf{x}, \theta)$ and has the same potential outcomes as $\mathbf{x}$ but the model assigns them with probabilities $\hat{f}(\mathbf{x}', 1) = 0.6$ and $\hat{f}(\mathbf{x}', 0) = 0.4$. In this case $\mathbf{x}$ will have the same potential outcomes within the region $R(\mathbf{x}, \theta)$ but in terms of the estimated $\hat{\text{ITE}}$ there is a large difference between $\mathbf{x}$ and $\mathbf{x}'$ ($\hat{\text{ITE}}(\mathbf{x}) = 0.7$ and $\hat{\text{ITE}}(\mathbf{x}') = 0.2$). In certain tasks, such as subgroup identification, where the smoothness of ITE is crucial, $\mathbf{x}'$ can be considered as adversarial.

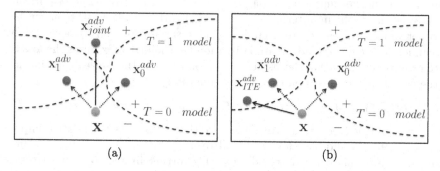

Fig. 3. Counterfactual models introduce, at least two, additional adversarial directions. An adversarial direction with respect to the joint loss on both groups will result in an adversarial patient $\mathbf{x}^{adv}_{joint}$ that would harm both models simultaneously (a). An adversarial direction with respect to ITE will result in an adversarial patient $\mathbf{x}^{adv}_{ITE}$ that maximally affects the difference between the potential outcomes (b).

Maximisation of a loss on $\hat{\text{ITE}}$ within a fixed area of interest, $R(\mathbf{x}, \theta)$, will allow us to identify edge cases that may result to significantly different estimations. To quantify the loss on $\hat{\text{ITE}}$ we can adopt the ε_{PEHE}, which was defined in Eq. (3). The empirical $\hat{\varepsilon}_{PEHE}$ on a patient $\mathbf{x}$ will be:

$$\hat{\varepsilon}_{PEHE}(\hat{f}(\mathbf{x}, 1), y_1, \hat{f}(\mathbf{x}, 0), y_0) = ((\hat{f}(\mathbf{x}, 1) - \hat{f}(\mathbf{x}, 0)) - (y_1 - y_0))^2 \quad (7)$$

If there is a treatment effect within the region $R(\mathbf{x}, \theta)$, i.e. the equality $\hat{f}(\mathbf{x}', 1) = \hat{f}(\mathbf{x}', 0), \forall \mathbf{x}' \in R(\mathbf{x}, \theta)$ does not hold, then there may be two edge cases: at least one example that maximally increases $\hat{\text{ITE}}(\mathbf{x})$ and/or at least one that maximally decreases it. Maximisation of Eq. (7) with respect to $\mathbf{x}$ results in identifying the worst case perturbations with respect to the true ITE. We can define the new set of adversarial patients as follows.

Definition 2 (Adversarial Patient and ITE). *An adversarial patient with respect to ITE, $\mathbf{x}^{adv}_{ITE} \in \mathbb{R}^d$, is a solution to the following optimisation problem:*

$$\underset{\mathbf{x}^{adv}_{ITE}}{\arg\max} \quad \hat{\varepsilon}_{PEHE}(\hat{f}(\mathbf{x}^{adv}_{ITE}, 1), y_1, \hat{f}(\mathbf{x}^{adv}_{ITE}, 0), y_0) \quad s.t. \quad \mathbf{x}^{adv}_{ITE} \in R(\mathbf{x}, \theta)$$

Differentiating Eq. (7) with respect to **x** will result in the following expression, where we have omitted the constants:

$$\nabla_{\mathbf{x}}\hat{\varepsilon}_{PEHE}(\cdot) = \gamma_{y_1-y_0} \cdot \nabla_{\mathbf{x}}\hat{f}(\mathbf{x},1) - \gamma_{y_1-y_0} \cdot \nabla_{\mathbf{x}}\hat{f}(\mathbf{x},0),$$

$$\text{where} \quad \gamma_1 = \hat{f}(\mathbf{x},1) - \hat{f}(\mathbf{x},0) - 1, \quad \gamma_{-1} = \hat{f}(\mathbf{x},1) - \hat{f}(\mathbf{x},0) + 1 \quad (8)$$

$$\text{and} \quad \gamma_0 = \hat{f}(\mathbf{x},1) - \hat{f}(\mathbf{x},0)$$

Notice that since $\hat{f}(\mathbf{x},t) \in (0,1)$, we will have $\gamma_1 < 0$, $\gamma_{-1} > 0$ and $\gamma_0 = I\hat{T}E$. Comparison of Eq. (6) and Eq. (8) reveals that when $y_1 \neq y_0$ both directions lead in decreasing the confidence of the two models on making the correct predictions. If $y_1 = y_0$ then the sign of the adversarial direction will depend on $I\hat{T}E$. For example if, for a patient **x**, $I\hat{T}E(\mathbf{x}) > 0$, then an adversarial patient will be modified to a direction that increases it. In this case the model assigns the wrong treatment effect with higher confidence (Fig. 3(b)).

To summarise, in counterfactual models there exist, at least, the following adversarial patients, with potentially different consequences:

1. Adversarial patients with respect to the treatment model result in potentially different predictions about the benefit under the administered treatment (e.g. chemotherapy) irrespectively of the control model.
2. Adversarial patients with respect to the control model result in potentially different predictions about the benefit under the baseline treatment (e.g. standard care) irrespectively of the treatment model.
3. A family of adversarial patients that maximise the joint loss on both models result in potentially different predictions about the joint outcome (e.g. simultaneously changing the predictions about the effect of chemotherapy and standard care).
4. Adversarial patients with respect to ITE maximise the loss on the difference between the potential outcomes leading to wrong estimations of ITE with higher confidence, which may affect treatment decisions and lead to undesirable consequences (e.g. financial).

In the following section we verify the existence of the different adversarial directions and describe which of those can be observed or approximated as well as their implications in practical applications.

5   Adversarial Patients in Practice

So far we considered the hypothetical scenario of observing both potential outcomes. This allowed us to identify the existence of new adversarial directions that challenge the local stability of the potential outcomes and the local smoothness of ITE. We now turn to the realistic case, where we only observe one of the outcomes. In practice, among the two directions that can deceive each model (Fig. 2), we can reliably identify only the one that corresponds to the observed outcome.

Algorithm 1. Generating Adversarial Patients

Input : Patient $\mathbf{x}$, factual outcome y_t, model $\hat{f}(\cdot, t)$, perturbation θ, step α,
            iterations m
Output: Adversarial patient $\mathbf{x}_t^{adv}$
$\mathbf{x}_t^{adv} = \mathbf{x}$
for $i := 1$ *to* m **do**
  $\quad \mathbf{x}_t^{adv} = \mathbf{x}_t^{adv} + \alpha \cdot sign(\nabla_{\mathbf{x}} J(\hat{f}(\mathbf{x}_t^{adv}, t), y_t))$
  $\quad$ **if** $\|\mathbf{x}_t^{adv} - \mathbf{x}\|_{\infty} > \theta$ **then**
  $\quad\quad\quad$ Project $\mathbf{x}_t^{adv}$ onto the boundary of the feasible region

Return $\mathbf{x}_t^{adv}$

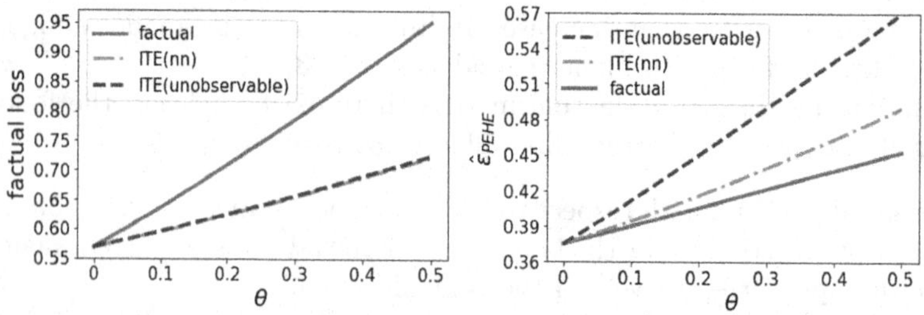

Fig. 4. In counterfactual models we can identify adversarial patients with respect to the potential outcomes and with respect to ITE. To show the existence of these different types we create adversarial patients by maximising the factual loss (factual), $\hat{\varepsilon}_{PEHE}$ (ITE) and a nearest neighbour approximation of $\hat{\varepsilon}_{PEHE}$ (ITE(nn)). We report the effect of the different types of adversarial patients on the two loss functions as the size of the search area $R(\mathbf{x}, \theta)$ increases.

Algorithm 1 describes how to construct adversarial patients with iterative FGSM [9] for a model $\hat{f}(\cdot, t)$ using projected gradient ascent.

In order to perform adversarial training using adversarial patients that require the counterfactual outcome, we will adopt a nearest neighbour model by taking into account relevant factual outcomes. Nearest neighbour approximations of $\hat{\varepsilon}_{PEHE}$ have also been adopted in recent works, either directly as part of the objective function [8], or as a measure of the generalisation performance for hyperparameter selection [17]. We will focus on two types of adversarial patients – those that affect the loss on the observed (factual) outcome and those that affect ITE, using the nearest neighbour approximation.

To verify the effect of the different types of adversarial directions we consider the following outcome function [5], which we will call Simulated Model 1 (SM1): $\text{logit}(f(\mathbf{X}, T)) = -1 + 0.5X_1 + 0.5X_2 - 0.5X_7 + 0.5X_2X_7 + \lambda T1(\mathbf{x} \in S)$, where the features X_j are drawn independently from a normal distribution, $X_j \sim N(0, 1), j = 1, ..., 15$. We choose $\lambda = 2$ and S as the region $X_0 < 0.545 \cap X_1 > -0.545$. We create a sample of 2000 examples and we average the results over 100 realisations of the outcome with 50/25/25 train/validation/test splits. To create adversarial patients we use Algorithm 1 where we also considered modifications of line 3 by substituting the factual loss with: 1. $\hat{\varepsilon}_{PEHE}$ and 2. a nearest neighbour approximation of $\hat{\varepsilon}_{PEHE}$. Figure 4 verifies the different effect of each adversarial direction. Notice that adversarial patients that maximise $\hat{\varepsilon}_{PEHE}$ can have an impact on the factual loss (and vice-versa). However their effect will be small since these directions do not always cause the examples to cross the decision boundary. In fact they can even make the model to be more confident for a correctly predicted factual outcome.

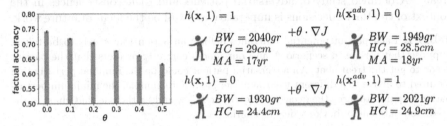

Fig. 5. Adversarial patients can deceive deep counterfactual models. Here we observe that small, worst-case perturbations (in only 2 or 3 of the 25 features) result in a different predicted outcome, that may be in contrast to clinical intuition.

5.1 Case Study: Adversarial Patients as Warning Flags

To show the effect of adversarial patients with respect to their factual outcome in a real-world scenario we consider the Infant Health and Development Program (IHDP) [7]. The dataset is from a randomised experiment evaluating the effect of specialist home visits on children's cognitive test scores. It comprises of 985 subjects and 25 features (6 continuous, 19 binary) measuring characteristics of children and their mothers. For the outcome function we used the response surface B of Hill [7] to form a binary classification task. We averaged the results over 1000 realisations of the outcome with 65/25/10 train/validation/test split and report the test set accuracy on the observed (factual) outcome (Fig. 5). We used a network with 2 shared layers and 2 group-specific with 50 nodes each[2].

To ensure *interpretable* adversarial patients, we restrict ourselves to an area $R(\mathbf{x}, \theta)$ modifying only three features: birth weight (BW), head circumference (HC) and mother's age (MA), and round each to the closest value it can take according to its domain. We observe in Fig. 5 that a small modification (we modify only 3 features, and fix the maximum perturbation to a minimal value) is enough to create adversarial patients that deceive the model. Consider the second patient in Fig. 5. We know that if the child receives the treatment ($t = 1$) she will not have a positive outcome ($y_1 = 0$). However, for a child $\mathbf{x}_1^{adv}$ with two slightly different features the model would predict the outcome will be positive

---

[2] Further details on the outcome function and evaluation protocol can be found in the supplementary material.

$(h(\mathbf{x}_1^{adv}, 1) = 1)$. Such small perturbations being responsible for potentially life-changing decisions is a significant ethical dilemma. Adversarial patients may result in different treatment decisions, a taxonomy of which is given in Table 1. In our example, for the initial patient the treatment could be either ineffective or harmful depending on the counterfactual outcome. A small perturbation results in concluding that the treatment is either essential or unnecessary. Such a small perturbation may lead to a different treatment decision, such as "treat" or "no further action required", while in reality the right decision could be "do not treat". A detailed study of adversarial patients and their consequences in the context of treatment decisions is imperative and will be the focus of future work.

Table 1. Each pair of potential outcomes defines an action that needs to be taken. Suppose that we are in scenario A and our model correctly predicts that the patient needs to get the treatment. An adversarial patient could be the minimal perturbation required to arrive to a different treatment decision. For example, the minimum perturbation required to force the decision "Do not treat" would be the one that changes the predicted outcome of both models (scenario D).

Scenario	Treatment status	Prognosis	Possible decision
A. $y_1 = 1, y_0 = 0$	Essential	Negative	Treat
B. $y_1 = 1, y_0 = 1$	Unnecessary	Positive	No further action
C. $y_1 = 0, y_0 = 0$	Ineffective	Negative	Search for alternative
D. $y_1 = 0, y_0 = 1$	Harmful	Positive	Do not treat

5.2 Case Study: Subgroup Identification

Exploratory subgroup identification is a critical task in phase III trials. Subgroups are usually defined in terms of a few features[3] associated with a trial population with some desirable property, such as improved treatment effect. As described by Sechidis et al. [16], the outcome can be realised as a function of two types of features, each one providing useful information for different tasks: 1. *prognostic* features that affect the outcome irrespectively of the treatment and 2. *predictive* features that affect the outcome through their interaction with the treatment.

Prognostic features are important for clinical trial planning as they can be considered for stratification and randomisation [14]. Predictive features are important for treatment effect estimation. A significant body of literature has focused on finding subgroups of patients with enhanced treatment effect [10]. Formally such a subgroup $\hat{S}$ includes all patients for which $\hat{\text{ITE}}(\mathbf{x}) > \delta$, $\forall \mathbf{x} \in \hat{S}$, where δ is a constant chosen based on the clinical setting.

To show how multi-task counterfactual NNs can be used for subgroup identification we again consider our simulated model SM1. In this model X_1 and X_2 have both prognostic and predictive effect, while X_7 is solely prognostic – i.e. it

---

[3] Since a subgroup needs to be interpretable for clinicians, they are defined in terms of the features with the strongest predictive effect (usually less than 3 features) [5].

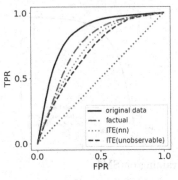 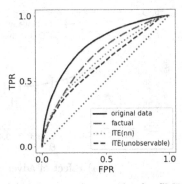

(a) Adversarial examples for SM1   (b) Adversarial examples for SM2

Fig. 6. Adversarial patients with identical predictive features can have significantly different estimated treatment effects. We created adversarial patients ($\theta = 0.4$) with respect to the factual outcome (factual) and with respect to ITE using the nearest neighbour approximation (ITE(nn)). We also report results for examples created with the true counterfactual outcome (ITE) for comparison. Different estimations of ITE may result in incorrectly removing(adding) patients from(to) the subgroup.

affects the outcome but not ITE. We also consider a modification of SM1 where the subgroup is defined as the region $X_3 < 0.545 \cap X_4 > -0.545$ (we will refer to this case as SM2). In this case X_3, X_4 are solely predictive, while X_1, X_2, X_7 will have only a prognostic effect. In this case, ITE should be influenced only by X_3 and X_4. We evaluate the performance on subgroup identification using true positive/false positive rates while varying the threshold δ (a patient $\mathbf{x}$ is included in the subgroup if $\hat{\text{ITE}}(\mathbf{x}) > \delta$) and comparing the resulting subgroup, $\hat{S}$, with the ground truth S. The datasets were created similarly to Sect. 5 and the results were averaged over 100 realisations of the outcome.

To see whether the estimated $\hat{\text{ITE}}$ can be influenced by non-predictive features, we created adversarial patients that are identical with respect to their predictive features. Figure 6 shows that adversarial patients can deceive the model to wrongly remove(add) them from(to) the subgroup, since the TPR/FPR curves are closer to the centre diagonal line than the original. Notice that adversarial patients with respect to ITE have higher effect (i.e. closer to the diagonal) than adversarial patients with respect to the factual outcome. Our results validate that patients with identical predictive features can have different $\hat{\text{ITE}}$.

We can counter this bias with an adversarial training approach, making the model robust to change in non-predictive features. We create adversarial patients with respect to their factual outcome and with respect to ITE using the nearest neighbour approximation of $\hat{\varepsilon}_{PEHE}$. We restrict the area of search $R(\mathbf{x}, \theta)$ by keeping constant the two features with the highest predictive strength. Following Goodfellow et al. [6] we use equal weight on legitimate/adversarial examples.

We observe that adversarial training acts as a regulariser improving the generalisation performance, measured by $\hat{\varepsilon}_{PEHE}$ (left side of Fig. 7). On the right

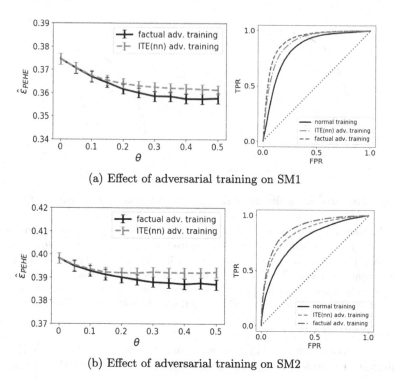

(a) Effect of adversarial training on SM1

(b) Effect of adversarial training on SM2

Fig. 7. We trained a counterfactual NN on two outcome functions and report the estimated error $\hat{\varepsilon}_{PEHE}$ (left column) and TPR/FPR curves for $\theta = 0.25$ (right column). We observe that adversarial training with respect to the factual outcome acts as a strong regulariser for ITE estimation. The effect of adversarial training is attributed to the model becoming insensitive to changes on prognostic features.

of Fig. 7 we see that adversarial training can also improve the performance on the task of subgroup identification. Therefore training the model to make similar predictions for patients differing only on their prognostic/irrelevant part leads to better estimates of ITE. Note that here we assumed knowledge of the predictive features, which can be based on suggestions from domain experts or derived algorithmically [16]. In practice it is also common to know at least a subset of the prognostic features, as they are often used for stratification (among other tasks) [14]. In the supplementary material we present additional results: e.g. modify all features, and more simulated models.

6   Discussion and Future Research

The study of adversarial examples in the context of clinical trials, raises several challenges for research. Such small perturbations being responsible for potentially life-changing decisions is a significant ethical issue. Here we highlight a "wish-list" of issues we believe could be addressed by the community.

New problems

- How can we ensure medically plausible and clinically interpretable adversarial patients?
- How can we reliably create adversarial patients with respect to population level metrics such as average treatment effect, which may have influence on issues like drug pricing?
- To the best of our knowledge, this paper is the first to tackle subgroup identification using NNs. An in-depth study of adversarial patients for robust subgroup identification is likely to be a fruitful area of work.
- Medical regulatory bodies demand strong evidence, ideally guarantees, not conjectures. What properties can we prove about adversarial patients, e.g. generalisation bounds? How should policy makers regard this?

Ethical issues

- Healthcare is rife with cost-sensitive decision making—with imbalanced ethical, personal and financial costs, for most choices that arise. How can we identify/use adversarial patients with this in mind?
- Identifying adversarial directions common to several classes of model may be suitable to influence policy—for example if we find particular health characteristics/features to be sensitive, they should be treated with care when recording from patients.

Making use of, and contributing to medical knowledge

- How can we build domain-specific medical knowledge (e.g. co-morbidities or known metabolic pathways where the drug is targeted) into the identification of adversarial patients to be used as warning flags?
- Our work has highlighted that we can have prognostic or predictive adversarial directions. How can we use this knowledge to better identity trustworthy predictive biomarkers for reliable treatment assignments?
- Outcomes of trials are often *structured*, e.g. multiple correlated and/or hierarchical outcomes. How can we adapt adversarial methods for this?

Technical issues

- A defining aspect of adversarial patients is they can be created with respect to an approximation of the ground truth label (e.g. ITE with nearest neighbours). What is the influence of this on the quality/plausibility of the created adversarial examples?
- The adversarial methodologies we use (e.g. FGSM, adversarial training) are for purely illustrative purposes. There is a significant body of literature (e.g. [11,12,19]) that could be adopted in this context.

7   Conclusions

We have studied the idea of adversarial examples in counterfactual models, for personalised medicine. The concept of "adversarial patients", can exist in many forms—we have shown that some adversarial directions cannot be observed, but still provably exist. We showed that small input perturbations (but still medically plausible) of multi-task counterfactual networks can lead to predictions that may not be in accordance with human intuition. Training a model to account for them can affect critical tasks in the clinical trial setting, such as exploratory subgroup identification. We propose that the study of adversarial patients in personalised medicine, where mispredictions can result in potentially life changing decisions, is an imperative research direction for the community to address.

Acknowledgments. K.P. was supported by the EPSRC through the Centre for Doctoral Training Grant [EP/1038099/1]. K.S. was funded by the AstraZeneca Data Science Fellowship at the University of Manchester. G.B. was supported by the EPSRC LAMBDA project [EP/N035127/1].

References

1. Alaa, A.M., Weisz, M., van der Schaar, M.: Deep counterfactual networks with propensity-dropout. In: ICML Workshop on Principled Approaches to Deep Learning (2017)
2. Athey, S., Imbens, G.: Recursive partitioning for heterogeneous causal effects. Proc. Natl. Acad. Sci. **113**(27), 7353–7360 (2016)
3. Biggio, B., et al.: Evasion attacks against machine learning at test time. In: Blockeel, H., Kersting, K., Nijssen, S., Železný, F. (eds.) ECML PKDD 2013. LNCS (LNAI), vol. 8190, pp. 387–402. Springer, Heidelberg (2013). https://doi.org/10.1007/978-3-642-40994-3_25
4. European Parliament and Council of the European Union: Regulation on the protection of natural persons with regard to the processing of personal data and on the free movement of such data, and repealing directive 95/46/EC (data protection directive). L119, pp. 1–88 (2016)
5. Foster, J.C., Taylor, J.M., Ruberg, S.J.: Subgroup identification from randomized clinical trial data. Stat. Med. **30**(24), 2867–2880 (2011)
6. Goodfellow, I.J., Shlens, J., Szegedy, C.: Explaining and harnessing adversarial examples. In: ICLR (2015)
7. Hill, J.L.: Bayesian nonparametric modelling for causal inference. J. Comput. Graph. Stat. **20**(1), 217–240 (2011)
8. Johansson, F., Shalit, U., Sontag, D.: Learning representations for counterfactual inference. In: ICML, pp. 3020–3029 (2016)
9. Kurakin, A., Goodfellow, I., Bengio, S.: Adversarial machine learning at scale. In: ICLR (2017)
10. Lipkovich, I., Dmitrienko, A., D'Agostino Sr., R.B.: Tutorial in biostatistics: data-driven subgroup identification and analysis in clinical trials. Stat. Med. **36**(1), 136–196 (2017)
11. Miyato, T., Maeda, S.I., Koyama, M., Nakae, K., Ishii, S.: Distributional smoothing with virtual adversarial training. In: ICLR (2016)
12. Moosavi-Dezfooli, S.M., Fawzi, A., Frossard, P.: DeepFool: a simple and accurate method to fool deep neural networks. In: Proceedings of the IEEE Conference on Computer Vision and Pattern Recognition, pp. 2574–2582 (2016)
13. Papernot, N., McDaniel, P., Jha, S., Fredrikson, M., Celik, Z.B., Swami, A.: The limitations of deep learning in adversarial settings. In: EuroS&P pp. 372–387. IEEE (2016)

14. Ruberg, S.J., Shen, L.: Personalized medicine: four perspectives of tailored medicine. Stat. Biopharm. Res. **7**(3), 214–229 (2015)

15. Rubin, D.B.: Estimating causal effects of treatments in randomized and nonrandomized studies. J. Educ. Psychol. **66**(5), 688 (1974)

16. Sechidis, K., Papangelou, K., Metcalfe, P.D., Svensson, D., Weatherall, J., Brown, G.: Distinguishing prognostic and predictive biomarkers: an information theoretic approach. Bioinformatics **1**, 12 (2018)

17. Shalit, U., Johansson, F., Sontag, D.: Estimating individual treatment effect: generalization bounds and algorithms. In: ICML (2017)

18. Szegedy, C., et al.: Intriguing properties of neural networks. In: ICLR (2014)

19. Tramèr, F., Kurakin, A., Papernot, N., Goodfellow, I., Boneh, D., McDaniel, P.: Ensemble adversarial training: attacks and defenses. In: ICLR (2018)

20. Wager, S., Athey, S.: Estimation and inference of heterogeneous treatment effects using random forests. J. Am. Stat. Assoc. (2017, just-accepted)

ShapeShifter: Robust Physical Adversarial Attack on Faster R-CNN Object Detector

Shang-Tse Chen[1]($\boxtimes$), Cory Cornelius[2], Jason Martin[2],
and Duen Horng (Polo) Chau[1]

[1] Georgia Institute of Technology, Atlanta, GA, USA
{schen351,polo}@gatech.edu
[2] Intel Corporation, Hillsboro, OR, USA
{cory.cornelius,jason.martin}@intel.com

Abstract. Given the ability to directly manipulate image pixels in the digital input space, an adversary can easily generate imperceptible perturbations to fool a Deep Neural Network (DNN) image classifier, as demonstrated in prior work. In this work, we propose *ShapeShifter*, an attack that tackles the more challenging problem of crafting physical adversarial perturbations to fool image-based object detectors like Faster R-CNN. Attacking an object detector is more difficult than attacking an image classifier, as it needs to mislead the classification results in multiple bounding boxes with different scales. Extending the digital attack to the physical world adds another layer of difficulty, because it requires the perturbation to be robust enough to survive real-world distortions due to different viewing distances and angles, lighting conditions, and camera limitations. We show that the *Expectation over Transformation* technique, which was originally proposed to enhance the robustness of adversarial perturbations in image classification, can be successfully adapted to the object detection setting. *ShapeShifter* can generate adversarially perturbed stop signs that are consistently mis-detected by Faster R-CNN as other objects, posing a potential threat to autonomous vehicles and other safety-critical computer vision systems. Code related to this paper is available at: https://github.com/shangtse/robust-physical-attack.

Keywords: Adversarial attack · Object detection · Faster R-CNN

1 Introduction

Adversarial examples are input instances that are intentionally designed to fool a machine learning model into producing a chosen prediction. The success of Deep Neural Network (DNN) in computer vision does not exempt it from this threat. It is possible to bring the accuracy of a state-of-the-art DNN image classifier down to near zero percent by adding imperceptible adversarial perturbations [5,22]. The existence of adversarial examples not only reveals intriguing

© Springer Nature Switzerland AG 2019
M. Berlingerio et al. (Eds.): ECML PKDD 2018, LNAI 11051, pp. 52–68, 2019.
https://doi.org/10.1007/978-3-030-10925-7_4

theoretical properties of DNN, but also raises serious practical concerns on its deployment in security and safety critical systems. Autonomous vehicle is an example application that cannot be fully trusted before guaranteeing the robustness to adversarial attacks. The imperative need to understand the vulnerabilities of DNNs attracts tremendous interest among machine learning, computer vision, and security researchers.

Although many adversarial attack algorithms have been proposed, attacking a real-world computer vision system is difficult. First of all, most of the existing attack algorithms only focus on the image classification task, yet in many real-world use cases there will be more than one object in an image. Object detection, which recognizes and localizes multiple objects in an image, is a more suitable model for many vision-based scenarios. Attacking an object detector is more difficult than attacking an image classifier, as it needs to mislead the classification results in multiple bounding boxes with different scales [14].

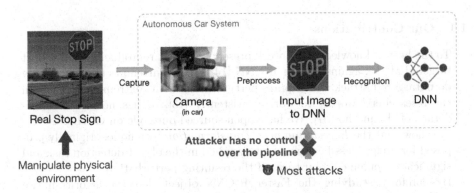

Fig. 1. Illustration motivating the need of physical adversarial attack, from attackers' perspectives, as they typically do not have full control over the computer vision system pipeline.

Further difficulty comes from the fact that DNN is usually only a component in the whole computer vision system pipeline. For many applications, attackers usually do not have the ability to directly manipulate data inside the pipeline. Instead, they can only manipulate the things outside of the system, i.e., those things in the physical environment. Figure 1 illustrates the intuition behind *physical adversarial attacks*. To be successful attacks, physical adversarial attacks must be robust enough to survive real-world distortions due to different viewing distances and angles, lighting conditions, and camera limitations.

There has been prior work that can either attack object detectors digitally [23], or attack image classifiers physically [6,10,19]. However, so far the existing attempts to physically attack object detectors remain unsatisfactory. A perturbed stop sign is shown in [13] that cannot be detected by the Faster R-CNN object detector [18]. However, the perturbation is very large and they tested it with poor texture contrast against the background, making the perturbed stop sign hard to see even by human. A recent short note [7] claims to be able to generate some adversarial stickers that, when attaching to a stop sign, can fool the YOLO object detector [17] and can be transferable to also fool Faster R-CNN. However, they did not reveal the algorithm used to create the sticker and only show a video of indoor experiment with short distance. For other threat models and adversarial attacks in computer vision, we refer the interested readers to the survey of [1].

In this work, we propose *ShapeShifter*, the first robust targeted attack that can fool a state-of-the-art Faster R-CNN object detector. To make the attack robust, we adopt the *Expectation over Transformation* technique [3,4], and adapt it from the image classification task to the object detection setting. As a case study, we generate some adversarially perturbed stop signs that can consistently be mis-detected by Faster R-CNN as the target objects in real drive-by tests. Our contributions are summarized below.

1.1 Our Contributions

- To the best of knowledge, our work presents the first reproducible and robust targeted attack against Faster R-CNN [14]. Recent attempts either can only do untargeted attack and requires perturbations with "extreme patterns" (in the researchers' words) to work consistently [13], or has not revealed the details of the method [7]. We have open-sourced our code on GitHub[1].
- We show that the *Expectation over Transformation* technique, originally proposed for image classification, can be applied in the object detection task and significantly enhance robustness of the resulting perturbation.
- By carefully studying the Faster R-CNN object detector algorithm, we overcome non-differentiability in the model, and successfully perform optimization-based attacks using gradient descent and backpropagation.
- We generate perturbed stop signs that can consistently fool Faster R-CNN in real drive-by tests (videos available on the GitHub repository), calling for imperative need to improve and fortify vision-based object detectors.

2 Background

This section provides background information of adversarial attacks and briefly describes the Faster R-CNN object detector that we try to attack in this work.

2.1 Adversarial Attack

Given a trained machine learning model C and a benign instance $x \in \mathcal{X}$ that is correctly classified by C, the goal of the untargeted adversarial attack is to find another instance $x' \in \mathcal{X}$, such that $C(x') \neq C(x)$ and $d(x, x') \leq \epsilon$ for some

---

[1] https://github.com/shangtse/robust-physical-attack.

distance metric $d(\cdot, \cdot)$ and perturbation budget $\epsilon > 0$. For targeted attack, we further require $C(x') = y'$ where $y' \neq C(x)$ is the target class. Common distance metrics $d(\cdot, \cdot)$ in the computer vision domain are ℓ_2 distance $d(x, x') = ||x - x'||_2^2$ and ℓ_∞ distance $d(x, x') = ||x - x'||_\infty$.

The work of [22] was the first to discover the existence of adversarial examples for DNNs. Several subsequent works have improved the computational cost and made the perturbation highly imperceptible to human [8,15]. Most adversarial attack algorithms against DNNs assume that the model is differentiable, and use the gradient information of the model to tweak the input instance to achieve the desired model output [5]. Sharif et al. [19] first demonstrated a physically realizable attack to fool a face recognition model by wearing an adversarially crafted pair of glasses.

2.2   Faster R-CNN

Faster R-CNN [18] is one of the state-of-the-art general object detectors. It adopts a 2-stage detection strategy. In the first state, a region proposal network is used to generate several class-agnostic bounding boxes called region proposals that may contain objects. In the second stage, a classifier and a regressor are used to output the classification results and refined bounding box coordinates for each region proposal, respectively. The computation cost is significantly reduced by sharing the convolutional layers in the two stages. Faster R-CNN is much harder to attack, as a single object can be covered by multiple region proposals of different sizes and aspect ratios, and one needs to mislead the classification results in all the region proposals to fool the detection.

3   Threat Model

Existing methods that generate adversarial examples typically yield impercep-tible perturbations that fool a given machine learning model. Our work, fol-lowing [19], generates perturbations that are perceptible but constrained such that a human would not be easily fooled by such a perturbation. We examine this kind of perturbation in the context of object detection (e.g., stop sign). We chose this use case because of object detector's possible uses in security-related and safety-related settings (e.g., autonomous vehicles). For example, attacks on traffic sign recognition could cause a car to miss a stop sign or travel faster than legally allowed.

We assume the adversary has white-box level access to the machine learning model. This means the adversary has access to the model structure and weights to the degree that the adversary can both compute outputs (i.e., the forward pass) and gradients (i.e., the backward pass). It also means that the adversary does not have to construct a perturbation in real-time. Rather, the adversary can study the model and craft an attack for that model using methods like Carlini-Wagner attack [5]. This kind of adversary is distinguished from a black-box adversary who is defined as having no such access to the model architecture or weights. While our choice of adversary is the most powerful one, existing research has shown it is possible to construct imperceptible perturbations without white-box level access [16]. However, whether our method is capable of generating perceptible perturbations with only black-box access remains an open question. Results from Liu et al. [12] suggest that iterative attacks (like ours) tend not to transfer as well as non-iterative attacks.

Unlike previous work, we restrict the adversary such that they cannot manipulate the digital values of pixels gathered from the camera that each use case uses to sense the world. This is an important distinction from existing imperceptible perturbation methods. Because those methods create imperceptible perturbations, there is a high likelihood such perturbations would not fool our use cases when physically realized. That is, when printed and then presented to the systems in our use cases, those perturbations would have to survive both the printing process and sensing pipeline in order to fool the system. This is not an insurmountable task as Kurakin et al. [10] have constructed such imperceptible yet physically realizable adversarial perturbations for image classification systems.

Finally, we also restrict our adversary by limiting the shape of the perturbation the adversary can generate. This is important distinction for our use cases because one could easily craft an odd-shaped "stop sign" that does not exist in the real world. We also do not give the adversary the latitude of modifying all pixels in an image like Kurakin et al. [10], but rather restrict them to certain pixels that we believe are both inconspicuous and physically realistic.

4 Attack Method

Our attack method, *ShapeShifter*, is inspired by the iterative, change-of-variable attack described in [5] and the *Expectation over Transformation* technique [3,4]. Both methods were originally proposed for the task of image classification. We describe these two methods in the image classification setting before showing how to extend them to attack the Faster R-CNN object detector.

4.1 Attacking an Image Classifier

Let $F : [-1, 1]^{h \times w \times 3} \to \mathbb{R}^K$ be an image classifier that takes an image of height h and width w as input, and outputs a probability distribution over K classes. The goal of the attacker is to create an image x' that looks like an object x of class y, but will be classified as another target class y'.

Change-of-variable Attack. Denote $L_F(x, y) = L(F(x), y)$ as the loss function that calculates the distance between the model output $F(x)$ and the target label y. Given an original input image x and a target class y', the change-of-variable attack [5] propose the following optimization formulation.

$$\underset{x' \in \mathbb{R}^{h \times w \times 3}}{\arg \min} \; L_F(\tanh(x'), y') + c \cdot || \tanh(x') - x ||_2^2. \tag{1}$$

The use of *tanh* ensures that each pixel is between $[-1, 1]$. The constant c controls the similarity between the modified object x' and the original image x. In practice, c can be determined by binary search [5].

Expectation over Transformation. The *Expectation over Transformation* [3,4] idea is simple: adding random distortions in each iteration of the optimization to make the resulting perturbation more robust. Given a transformation t that can be translation, rotation, and scaling, $M_t(x_b, x_o)$ is an operation that transforms an object image x_o using t and then overlays it onto a background image x_b. $M_t(x_b, x_o)$ can also include a masking operation that only keeps a certain area of x_o. This will be helpful when one wants to restrict the shape of the perturbation. After incorporating the random distortions, Eq. (1) becomes

$$\underset{x' \in \mathbb{R}^{h \times w \times 3}}{\arg \min} \ \mathbb{E}_{x \sim X, t \sim T} \left[L_F(M_t(x, \tanh(x')), y') \right] + c \cdot \| \tanh(x') - x_o \|_2^2, \quad (2)$$

where X is the training set of background images. When the model F is differentiable, this optimization problem can be solved by gradient descent and back-propagation. The expectation can be approximated by the empirical mean.

4.2 Extension to Attacking Faster R-CNN

An object detector $F : [-1, 1]^{h \times w \times 3} \rightarrow (\mathbb{R}^{N \times K}, \mathbb{R}^{N \times 4})$ takes an image as input and outputs N detected objects. Each detection includes a probability distribution over K pre-defined classification classes as well as the location of the detected object, represented by its 4 coordinates. Note that it is possible for an object detector to output more or fewer detected objects, depending on the input image, but for simplicity we select top-N detected objects ranked by confidence.

As described in Subsect. 2.2, Faster R-CNN adopts a 2-stage approach. The region proposal network in the first stage outputs several region proposals, and the second stage classifier performs classification within each of the region proposals. Let $rpn(x) = \{r_1, \ldots, r_m\}$, where each r_i is a region proposal represented as its four coordinates, and let x_r be a sub-image covered by region r. Denote $L_{F_i}(x, y) = L(F(x_{r_i}), y)$, i.e., the loss of the classification in the i-th region proposal. We can simultaneously attack all the classifications in each region proposal by doing the following optimization.

$$\underset{x' \in \mathbb{R}^{h \times w \times 3}}{\arg \min} \ \mathbb{E}_{x \sim X, t \sim T} \left[\frac{1}{m} \sum_{r_i \in rpn(M_t(x'))} L_{F_i}(M_t(x'), y') \right] + c \cdot \| \tanh(x') - x_o \|_2^2,$$

$$(3)$$

where we abuse the notation $M_t(x') = M_t(x, \tanh(x'))$ for simplicity. However, for computational issues, most models prune the region proposals by using heuristics like non-maximum suppression [18]. The pruning operations are usually non-differentiable, making it hard to optimize equation (3) end to end.

Therefore, we approximately solve this optimization problem by first run a forward pass of the region proposal network, and fixed the pruned region proposals as fixed constants to the second stage classification problem in each iteration. We empirically find this approximation sufficient to find a good solution.

5   Evaluation

We evaluate our method by fooling a pre-trained Faster R-CNN model with Inception-v2 [21] convolutional feature extraction component. The model was trained on the Microsoft Common Objects in Context (MS-COCO) dataset [11] and is publicly available in the Tensorflow Object Detection API [9] model zoo repository[2].

The MS-COCO dataset contains 80 general object classes ranging from people and animals to trucks and cars and other common objects. Although our method can potentially be used to attack any classes, we choose to focus on attacking the stop sign class due to its importance and relevance to self-driving cars, where a vision-based object detector may be used to help make decisions. An additional benefit of choosing the stop sign is its flat shape that can easily be printed on a paper. Other classes, like dogs, are less likely to be perceived as real objects by human when printed on a paper. While 3D printing adversarial examples for image recognition is possible [3], we leave 3D-printed adversarial examples against object detectors as future work.

5.1   Digitally Perturbed Stop Sign

We generate adversarial stop signs by performing the optimization process described in Eq. 3. The hyperparameter c is crucial in determining the perturbation strength. A smaller value of c will result in a more conspicuous perturbation, but the perturbation will also be more robust to real-world distortions when we do the physical attack later.

However, it is hard to choose an appropriate c when naively using the ℓ_2 distance to a real stop sign as regularization. To obtain a robust enough perturbation, a very small c needs to be used, which has the consequence of creating stop signs that are difficult for humans to recognize. The ℓ_2 distance is not a perfect metric for human perception, which tends to be more sensitive to color changes on lighter-colored objects. Due to this observation, we only allow the perturbation to change the red part of the stop sign, leaving the white text intact. This allows us to generate larger and more robust perturbation, while providing enough contrast between the lettering and red parts so that a human can easily recognize the perturbation as a stop sign. The adversarial stop sign generated in [13] does not consider this and is visually more conspicuous. Automating this procedure for other objects we leave as future work.

---

[2] http://download.tensorflow.org/models/object_detection/
faster_rcnn_inception_v2_coco_2017_11_08.tar.gz.

We performed two targeted attacks and one untargeted attack. We choose person and sports ball as the two target classes because they are relatively similar in size and shape to stop signs. Our method allows attackers to use any target classes, however the perturbation needs to achieve its means and fool the object detector. For some target classes, this may mean creating perturbations so large in deviation that they may appear radically different from the victim class. We also noticed that some classes are easier to be detected at small scales, such as *kite*, while other classes (e.g., *truck*) could not be detected when the object was too small. This may be an artifact of the MS-COCO dataset that the object detector was trained on. Nevertheless, ultimately the attacker has a choice in target class and, given ample time, can find the target class that best fools the object detector according to their means.

For each attack, we generated a high confidence perturbation and a low level perturbation. The high confidence perturbations were generated using a smaller value of c, thus making them more conspicuous but also more robust. Depending upon the target class, it may be difficult to generate an effective perturbation. We manually chose c for each target class so that the digital attack achieves high success rate while keeping the perturbation not too conspicuous, i.e., we tried to keep the color as red as possible. We used $c = 0.002$ for the high confidence perturbations and $c = 0.005$ for the low confidence perturbations in the "sports ball" targeted attack and the untargeted attack. We used $c = 0.005$ and $c = 0.01$ for the high and low confidence perturbations in the "person" targeted attack, respectively. The 6 perturbations we created are shown in Fig. 2.

5.2   Physical Attack

We performed physical attacks on the object detector by printing out the perturbed stop signs shown in Fig. 2. We then took photos from a variety of distances and angles in a controlled indoor setting. We also conducted drive-by tests by recording videos from a moving vehicle that approached the signs from a distance. The lightning conditions varied from recording to recording depending upon the weather at the time.

Equipment. We used a Canon Pixma Pro-100 photo printer to print out signs with high-confidence perturbations, and an HP DesignJet to print out those with low-confidence perturbations[3]. For static images, we used a Canon EOS Rebel T7i DSLR camera, equipped with a EF-S 18-55mm IS STM lens. The videos in our drive-by tests are shot using an iPhone 8 Plus mounted on the windshield of a car.

Indoor Experiments. Following the experimental setup of [6], we took photos of the printed adversarial stop sign, at a variety of distances (5′ to 40′) and

---

[3] We used two printers to speed up our sign production, since a sign can take more than 30 min to produce.

(a) Person (low)   (b) Sports ball (low)   (c) Untargeted (low)

(d) Person (high)   (e) Sports ball (high)   (f) Untargeted (high)

Fig. 2. Digital perturbations we created using our method. Low confidence perturbations on the top and high confidence perturbations on the bottom.

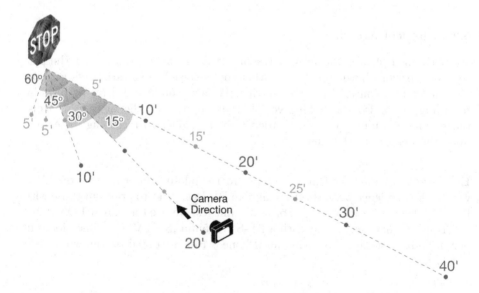

Fig. 3. Indoor experiment setup. We take photos of the printed adversarial sign, from multiple angles (0°, 15°, 30°, 45°, 60°, from the sign's tangent), and distances (5′ to 40′). The camera locations are indicated by the red dots, and the camera always points at the sign. (Color figure online)

Table 1. Our **high-confidence** perturbations succeed at attacking at a variety of distances and angles. For each distance-angle combination, we show the detected class and the confidence score. If more than one bounding boxes are detected, we report the highest-scoring one. Confidence values lower than 30% is considered undetected.

Distance	Angle	Person	(Conf.)	Sports ball	(Conf.)	Untargeted	(Conf.)
5′	0°	person	(.77)	sports ball	(.61)	clock	(.35)
5′	15°	person	(.91)	cake	(.73)	clock	(.41)
5′	30°	person	(.93)	cake	(.66)	cake	(.39)
5′	45°	person	(.69)	cake	(.61)	*stop sign*	(.62)
5′	60°	*stop sign*	(.93)	*stop sign*	(.70)	*stop sign*	(.88)
10′	0°	person	(.55)	cake	(.34)	clock	(.99)
10′	15°	person	(.63)	cake	(.33)	clock	(.99)
10′	30°	person	(.51)	cake	(.55)	clock	(.99)
15′	0°	undetected	—	cake	(.49)	clock	(.99)
15′	15°	person	(.57)	cake	(.53)	clock	(.99)
20′	0°	person	(.49)	sports ball	(.98)	clock	(.99)
20′	15°	person	(.41)	sports ball	(.96)	clock	(.99)
25′	0°	person	(.47)	sports ball	(.99)	*stop sign*	(.91)
30′	0°	person	(.49)	sports ball	(.92)	undetected	—
40′	0°	person	(.56)	sports ball	(.30)	*stop sign*	(.30)
Targeted success rate	87%		40%		N/A		
Untargeted success rate	93%		93%		73%		

Table 2. As expected, low-confidence perturbations achieve lower success rates.

Distance	Angle	Person	(Conf.)	Sports ball	(Conf.)	Untargeted	(Conf.)
5′	0°	*stop sign*	(.87)	cake	(.90)	cake	(.41)
5′	15°	*stop sign*	(.63)	cake	(.93)	cake	(.34)
5′	30°	person	(.83)	cake	(.84)	*stop sign*	(.48)
5′	45°	*stop sign*	(.97)	*stop sign*	(.94)	*stop sign*	(.82)
5′	60°	*stop sign*	(.99)	*stop sign*	(.99)	*stop sign*	(.89)
10′	0°	*stop sign*	(.83)	*stop sign*	(.99)	undetected	—
10′	15°	*stop sign*	(.79)	*stop sign*	(.94)	undetected	—
10′	30°	*stop sign*	(.60)	*stop sign*	(.98)	*stop sign*	(.78)
15′	0°	*stop sign*	(.52)	*stop sign*	(.94)	*stop sign*	(.31)
15′	15°	*stop sign*	(.33)	*stop sign*	(.93)	undetected	—
20′	0°	*stop sign*	(.42)	sports ball	(.73)	undetected	—
20′	15°	person	(.51)	sports ball	(.83)	cell phone	(.62)
25′	0°	*stop sign*	(.94)	sports ball	(.87)	undetected	—
30′	0°	*stop sign*	(.94)	sports ball	(.95)	*stop sign*	(.79)
40′	0°	*stop sign*	(.95)	undetected	—	*stop sign*	(.52)
Targeted success rate	13%		27%		N/A		
Untargeted success rate	13%		53%		53%		

angles (0°, 15°, 30°, 45°, 60°, from the sign's tangent). This setup is depicted in Fig. 3 where camera locations are indicated by red dots. The camera always pointed at the sign. We intended these distance-angle combinations to mimic a vehicle's points of view as it would approach the sign from a distance [13]. Tables 1 and 2 summarize the results for our *high-confidence* and *low-confidence* perturbations, respectively. For each distance-angle combination, we show the detected class and the detection's confidence score. If more than one bounding boxes are detected, we report the highest-scoring one. Confidence values lower than 30% were considered undetected; we decided to use the threshold of 30%, instead of the default 50% in the Tensorflow Object Detection API [9], to impose a stricter requirement on ourselves (the "attacker"). Since an object can be detected as a stop sign and the target class simultaneously, we consider our attack to be successful only when the confidence score of the target class is the highest among all of the detected classes.

Table 1 shows that our high-confidence perturbations achieve a high attack success rate at a variety of distances and angles. For example, we achieved a targeted success rate 87% in misleading the object detector into detecting the stop sign as a *person*, and an even higher untargeted success rate of 93% when our attack goal is to cause the detector to either fail to detect the stop sign (e.g., at 15′ 0°) or to detect it as a class that is *not* a stop sign. The *sports ball* targeted attack has a lower targeted success rate but achieves the same untargeted success rate. Our untargeted attack consistently misleads the detection into the *clock* class in medium distances, but is less robust for longer distances. Overall, the perturbation is less robust to very high viewing angle (60° from the sign's tangent), because we did not simulate the viewing angle distortion in the optimization.

The low-confidence perturbations (Table 2), as expected, achieve a much lower attack success rate, suggesting the need to use higher-confidence perturbations when we conduct the more challenging drive-by tests (as we shall describe in the next section). Table 3 shows some sample high-confidence perturbations from our indoor experiments.

Drive-By Tests. We performed drive-by tests at a parking lot so as not to disrupt other vehicles with our stop signs. We put a purchased real stop sign as a control and our printed perturbed stop sign side by side. Starting from about 200 feet away, we slowly drove (between 5 mph to 15 mph) towards the signs while simultaneously recording video from the vehicle's dashboard at 4 K resolution and 24 FPS using an iPhone 8 Plus. We extracted all video frames, and for each frame, we obtained the detection results from Faster R-CNN object detection model. Because our low confidence attacks showed relatively little robustness indoors, we only include the results from our high-confidence attack. Similar to our indoor experiments, we only consider detections that had a confidence score of at least 30%.

In Fig. 4, we show sample video frames (rectangular images) to give the readers a sense of the size of the signs relative to the full video frame; we also

Table 3. Sample high-confidence perturbations from indoor experiments. For complete experiment results, please refer to Table 1.

Dist.	Angle	Target: person	Target: sports ball	Untargeted
40'	0°	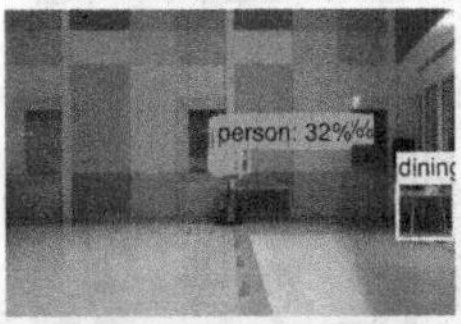		
10'	0°	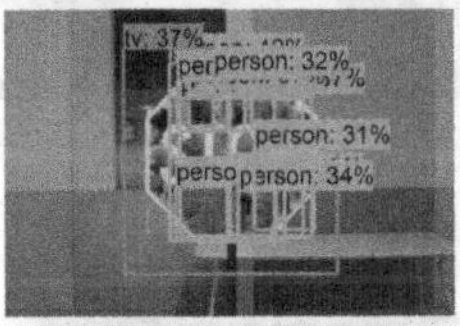		
10'	30°	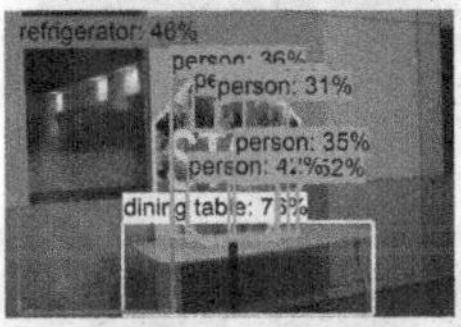		
5'	60°	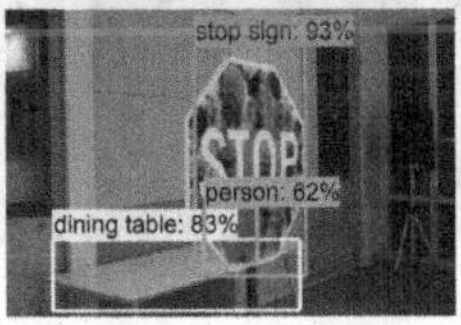		

show zoomed-in views (square images) that more clearly show the Faster R-CNN detection results.

The *person-perturbation* in Fig. 4a drive-by totaled 405 frames as partially shown in the figure. The real stop-sign in the video was correctly detected in every frame with high confidence. On the other hand, the perturbed stop sign was only correctly detected once, while 190 of the frames identified the perturbed stop sign as a person with medium confidence. For the rest of the 214 frames the object detector failed to detect anything around the perturbed stop sign.

The video we took with the *sports-ball-perturbation* shown in Fig. 4b had 445 frames. The real stop sign was correctly identified all of the time, while the perturbed stop sign was never detected as a stop sign. As the vehicle (video camera) moved closer to the perturbed stop sign, 160 of the frames were detected as a sports ball with medium confidence. One frame was detected as *apple* and *sports ball* and the remaining 284 frames had no detection around the perturbed stop sign.

Finally, the video of the untargeted perturbation (Fig. 4c) totaled 367 frames. While the unperturbed stop sign was correctly detected all of the time, the perturbed stop sign was detected as *bird* 6 times and never detected for the remaining 361 frames.

a. Target Class: Person

b. Target Class: Sports Ball

c. Untargeted

Fig. 4. Snapshots of the drive-by test results. In (a), the person perturbation was detected 47% of the frames as a person and only once as a stop sign. The perturbation in (b) was detected 36% of the time as a sports ball and never as a stop sign. The untargeted perturbation in (c) was detected as *bird* 6 times and never detected as a stop sign or anything else for the remaining frames.

Exploring Black-Box Transferability. We also sought to understand how well our high-confidence perturbations could fool other object detection models. For image recognition, it is known that high-confidence targeted attacks fail to transfer [12].

To this end, we fed our high-confidence perturbations into 8 other MS-COCO-trained models from the Tensorflow detection model zoo[4]. Table 4 shows how well our perturbation generated from the Faster R-CNN Inception-V2 transfer to other models. To better understand transferability, we examined the worse case. That is, if a model successfully detects a stop sign in the image, we say the perturbation has failed to transfer or attack that model. We report the number of images (of the 15 angle-distance images in our indoor experiments) where a model successfully detected a stop sign with at least 30% confidence. We also report the maximum confidence of all of those detected stop sign.

Table 4. Black-box transferability of our 3 perturbations. We report the number of images (of the 15 angle-distance images) that failed to transfer to the specified model. We consider the detection of any stop sign a "failure to transfer." Our perturbations fail to transfer for most models, most likely due to the iterative nature of our attack.

Model	Person	(Conf.)	Sports ball	(Conf.)	Untargeted	(Conf.)
Faster R-CNN Inception-V2	3	(.93)	1	(.70)	5	(0.91)
SSD MobileNet-V2	2	(.69)	8	(.96)	15	(1.00)
SSD Inception-V2	11	(1.00)	14	(.99)	15	(1.00)
R-FCN ResNet-101	4	(.82)	10	(.85)	15	(1.00)
Faster R-CNN ResNet-50	13	(.00)	15	(1.00)	15	(1.00)
Faster R-CNN ResNet-101	15	(.99)	13	(.97)	15	(1.00)
Faster R-CNN Inc-Res-V2	1	(.70)	0	(.00)	12	(1.00)
Faster R-CNN NASNet	14	(1.00)	15	(1.00)	15	(1.00)

Table 4 shows the lack of transferability of our generated perturbations. The untargeted perturbation fails to transfer most of the time, followed by the sports ball perturbation, and finally the person perturbation. The models most susceptible to transferability were the Faster R-CNN Inception-ResNet-V2 model, followed by the SSD MobileNet-V2 model. Iterative attacks on image recognition also usually fail to transfer [12], so it is not surprising that our attacks fail to transfer as well. We leave the thorough exploration of transferability as future work.

Fig. 5. Example stop signs from the MS-COCO dataset. Stop signs can vary by language, by degree of occlusion by stickers or modification by graffiti, or just elements of the weather. Each stop sign in the images is correctly detected by the object detector with high confidence (99%, 99%, 99%, and 64%, respectively).

---

[4] https://github.com/tensorflow/models/blob/master/research/object_detection/g3doc/detection_model_zoo.md.

6 Discussion and Future Work

There is considerable variation in the physical world that real systems will have to deal with. Figure 5 shows a curated set of non-standard examples of stop signs from the MS-COCO dataset[5]. The examples show stop signs in a different language, or that have graffiti or stickers applied to them, or that have been occluded by the elements. In each of these cases, it is very unlikely a human would misinterpret the sign as anything else but a stop sign. They each have the characteristic octagonal shape and are predominantly red in color. Yet, the object detector sees something else.

Unlike previous work on adversarial examples for image recognition, our adversarial perturbations are overt. They, like the examples in Fig. 5, exhibit large deviations from the standard stop sign. A human would probably notice these large deviations, and a trained human might even guess they were constructed to be adversarial. But they probably would not be fooled by our perturbations. However an automated-system using an off-the-shelf object detector would be fooled, as our results show. Our digital perturbation shown in Fig. 2e does look like a baseball or tennis ball has been painted on the upper right hand corner. Figure 4b shows how the object detector detects this part of the image as a sports ball with high confidence. This might seem unfair, but attackers have much latitude when these kind of models are deployed in automated systems. Even in non-automated systems a human might not think anything of Fig. 2d because it does not exhibit any recognizable person-like features.

Attackers might also generate perturbations without restricting the shape and color, and attach them to some arbitrary objects, like a street light or a trash bin. An untrained eye might see these perturbations as some kind of artwork, but the autonomous system might see something completely different. This attack, as described in [20], could be extended to object detectors using our method.

Defending against these adversarial examples has proven difficult. Many defenses fall prey to the so-called "gradient masking" or "gradient obfuscating" problem [2]. The most promising defense, adversarial training, has yet to scale up to models with good performance on the ImageNet dataset. Whether adversarial training can mitigate our style of overt, large-deviation (e.g., large ℓ_p distance) perturbations is also unclear.

7 Conclusion

We show that the state-of-the-art Faster R-CNN object detector, while previously considered more robust to physical adversarial attacks, can actually be

---

[5] Full resolution images of the examples in Fig. 5 can be found at: http://cocodataset.org/#explore?id=315605, http://cocodataset.org/#explore?id=214450, http://cocodataset.org/#explore?id=547465, and http://cocodataset.org/#explore?id=559484.

attacked with high confidence. Our work demonstrates vulnerability in MS-COCO-learned object detectors and posits that security and safety critical systems need to account for the potential threat of adversarial inputs to object detection systems.

Many real-world systems probably do not use an off-the-shelf pre-trained object detector as in our work. Why would a system with safety or security implications care to detecting sports balls? Most probably do not. Although it remains to be shown whether our style of attack can be applied to safety or security critical systems that leverage object detectors, our attack provides the means to test for this new class of vulnerability.

Acknowledgements. This work is supported in part by NSF grants CNS-1704701, TWC-1526254, and a gift from Intel.

References

1. Akhtar, N., Mian, A.S.: Threat of adversarial attacks on deep learning in computer vision: a survey. IEEE Access **6**, 14410–14430 (2018)
2. Athalye, A., Carlini, N., Wagner, D.: Obfuscated gradients give a false sense of security: circumventing defenses to adversarial examples. In: Proceedings of the 35th International Conference on Machine Learning (2018)
3. Athalye, A., Sutskever, I.: Synthesizing robust adversarial examples. In: Proceedings of the 35th International Conference on Machine Learning (2018)
4. Brown, T.B., Mané, D., Roy, A., Abadi, M., Gilmer, J.: Adversarial patch. arXiv preprint arXiv:1712.09665 (2017)
5. Carlini, N., Wagner, D.: Towards evaluating the robustness of neural networks. In: Proceedings of the 38th IEEE Symposium on Security and Privacy, pp. 39–57 (2017)
6. Evtimov, I., et al.: Robust physical-world attacks on machine learning models. In: Proceedings of the IEEE Conference on Computer Vision and Pattern Recognition (2018)
7. Eykholt, K., et al.: Note on attacking object detectors with adversarial stickers. arXiv preprint arXiv:1712.08062 (2017)
8. Goodfellow, I.J., Shlens, J., Szegedy, C.: Explaining and harnessing adversarial examples. In: International Conference on Learning Representations (2015). https://openreview.net/forum?id=BJm4T4Kgx
9. Huang, J., et al.: Speed/accuracy trade-offs for modern convolutional object detectors. In: Proceedings of the IEEE Conference on Computer Vision and Pattern Recognition, pp. 3296–3297 (2017)
10. Kurakin, A., Goodfellow, I., Bengio, S.: Adversarial examples in the physical world. In: International Conference on Learning Representations (Workshop) (2017). https://openreview.net/forum?id=HJGU3Rodl
11. Lin, T.-Y., et al.: Microsoft COCO: common objects in context. In: Fleet, D., Pajdla, T., Schiele, B., Tuytelaars, T. (eds.) ECCV 2014. LNCS, vol. 8693, pp. 740–755. Springer, Cham (2014). https://doi.org/10.1007/978-3-319-10602-1_48
12. Liu, Y., Chen, X., Liu, C., Song, D.: Delving into transferable adversarial examples and black-box attacks. In: International Conference on Learning Representations (2017). https://openreview.net/forum?id=Sys6GJqxl

13. Lu, J., Sibai, H., Fabry, E.: Adversarial examples that fool detectors. arXiv preprint arXiv:1712.02494 (2017)
14. Lu, J., Sibai, H., Fabry, E., Forsyth, D.: No need to worry about adversarial examples in object detection in autonomous vehicles. arXiv preprint arXiv:1707.03501 (2017)
15. Moosavi-Dezfooli, S.M., Fawzi, A., Frossard, P.: Deepfool: a simple and accurate method to fool deep neural networks. In: Proceedings of the IEEE Conference on Computer Vision and Pattern Recognition, pp. 2574–2582 (2016)
16. Papernot, N., McDaniel, P., Goodfellow, I., Jha, S., Celik, Z.B., Swami, A.: Practical black-box attacks against machine learning. In: Proceedings of the 12th ACM on Asia Conference on Computer and Communications Security, pp. 506–519 (2017)
17. Redmon, J., Farhadi, A.: YOLO9000: better, faster, stronger. In: Proceedings of the IEEE Conference on Computer Vision and Pattern Recognition, pp. 6517–6525 (2017)
18. Ren, S., He, K., Girshick, R., Sun, J.: Faster R-CNN: towards real-time object detection with region proposal networks. In: Advances in Neural Information Processing Systems, pp. 91–99 (2015)
19. Sharif, M., Bhagavatula, S., Bauer, L., Reiter, M.K.: Accessorize to a crime: Real and stealthy attacks on state-of-the-art face recognition. In: Proceedings of the 23rd ACM SIGSAC Conference on Computer and Communications Security, pp. 1528–1540 (2016)
20. Sitawarin, C., Bhagoji, A.N., Mosenia, A., Chiang, M., Mittal, P.: Darts: Deceiving autonomous cars with toxic signs. arXiv preprint arXiv:1802.06430 (2018)
21. Szegedy, C., Vanhoucke, V., Ioffe, S., Shlens, J., Wojna, Z.: Rethinking the inception architecture for computer vision. In: Proceedings of the IEEE Conference on Computer Vision and Pattern Recognition, pp. 2818–2826 (2016)
22. Szegedy, C., et al.: Intriguing properties of neural networks. In: International Conference on Learning Representations (2014). https://openreview.net/forum?id=kklr_MTHMRQjG
23. Xie, C., Wang, J., Zhang, Z., Zhou, Y., Xie, L., Yuille, A.L.: Adversarial examples for semantic segmentation and object detection. In: Proceedings of the IEEE International Conference on Computer Vision, pp. 1378–1387 (2017)

Anomaly and Outlier Detection

GRIDWATCH: Sensor Placement and Anomaly Detection in the Electrical Grid

Bryan Hooi[1,2]([✉]), Dhivya Eswaran[1], Hyun Ah Song[1], Amritanshu Pandey[3],
Marko Jereminov[3], Larry Pileggi[3], and Christos Faloutsos[1]

[1] School of Computer Science, Carnegie Mellon University, Pittsburgh, USA
{bhooi,deswaran,hyunahs,christos}@cs.cmu.edu
[2] Department of Statistics, Carnegie Mellon University, Pittsburgh, USA
[3] Department of Electrical and Computer Engineering,
Carnegie Mellon University, Pittsburgh, USA
{amritanp,mjeremin,pileggi}@andrew.cmu.edu

Abstract. Given sensor readings over time from a power grid consisting of nodes (e.g. generators) and edges (e.g. power lines), how can we most accurately detect when an electrical component has failed? More challengingly, given a limited budget of sensors to place, how can we determine where to place them to have the highest chance of detecting such a failure? Maintaining the reliability of the electrical grid is a major challenge. An important part of achieving this is to place sensors in the grid, and use them to detect anomalies, in order to quickly respond to a problem. Our contributions are: **(1) Online anomaly detection:** we propose a novel, online anomaly detection algorithm that outperforms existing approaches. **(2) Sensor placement:** we construct an optimization objective for sensor placement, with the goal of maximizing the probability of detecting an anomaly. We show that this objective has the property of submodularity, which we exploit in our sensor placement algorithm. **(3) Effectiveness:** Our sensor placement algorithm is provably near-optimal, and both our algorithms outperform existing approaches in accuracy by 59% or more (F-measure) in experiments. **(4) Scalability:** our algorithms scale **linearly**, and our detection algorithm is **online**, requiring bounded space and constant time per update. Code related to this paper is available at: https://github.com/bhooi/gridwatch.

1 Introduction

Improving the efficiency and security of power delivery is a critically important goal, in the face of disturbances arising from severe weather, human error, equipment failure, or even intentional intrusion. Estimates [5] suggest that reducing outages in the U.S. grid could save $49 billion per year, reduce emissions by 12 to 18%, while improving efficiency could save an additional $20.4 billion per year.

© Springer Nature Switzerland AG 2019
M. Berlingerio et al. (Eds.): ECML PKDD 2018, LNAI 11051, pp. 71–86, 2019.
https://doi.org/10.1007/978-3-030-10925-7_5

A key part of achieving this goal is to use sensor monitoring data to quickly identify when parts of the grid fail, so as to quickly respond to the problem.

A major challenge is scalability - power systems data can be both high-volume and received in real time, since the data comes from sensors which are continuously monitoring the grid. This motivates us to develop fast methods that work in this online (or streaming) setting. When each new data point is received, the algorithm should update itself efficiently.

Hence, our goal is an online anomaly detection algorithm:

Informal Problem 1 (Online Anomaly Detection)

- *Given: A graph $\mathcal{G} = (\mathcal{V}, \mathcal{E})$, and a subset S of nodes which contain sensors. For each sensor, we have a continuous stream of values of real and imaginary voltage $V(t)$ and current $I(t)$ measured by these sensors.*
- *Find: At each time t, compute an anomalousness score $A(t)$, indicating our confidence level that an anomaly occurred (i.e. a transmission line failed).*

For cost reasons, it is generally infeasible to place sensors at every node. Hence, an important follow-up question is where to place sensors so as to maximize the probability of detecting an anomaly.

Informal Problem 2 (Sensor Placement)

- *Given: A budget k of the number of sensors we can afford, a graph $\mathcal{G} = (\mathcal{V}, \mathcal{E})$, and a simulator that allows us to simulate sensor readings at each node.*
- *Find: A set of nodes $S \subseteq \mathcal{V}$, which are the locations we should place our sensors, such that $|S| = k$.*

In contrast to most approaches, our anomaly detection algorithm, GRIDWATCH-D, uses a domain-dependent approach which exploits the fact that electrical sensors consist of a voltage reading at a node as well as the current along each adjacent edge. This allows us to detect anomalies more accurately, even when using an online approach. Next, we propose GRIDWATCH-S, a sensor placement algorithm. The main idea is to define an objective which estimates our probability of successfully detecting an anomaly, then show that this objective has the submodularity property, allowing us to optimize it with approximation guarantees using an efficient greedy algorithm.

Figure 1a shows the sensors selected by GRIDWATCH-S: red circles indicate positions chosen. Figure 1b shows the anomaly scores (black line) output by GRIDWATCH-D, which accurately match the ground truth. Figure 1c shows that GRIDWATCH-S outperforms baselines on the CASE2869 data.

Our contributions are as follows:

1. **Online anomaly detection:** we propose a novel, online anomaly detection algorithm, GRIDWATCH-D, that outperforms existing approaches.
2. **Sensor placement:** we construct an optimization objective for sensor placement, with the goal of maximizing the probability of detecting an anomaly. We show that this objective has the property of 'submodularity,' which we exploit to propose our sensor placement algorithm.

3. **Effectiveness:** Our sensor placement algorithm, GRIDWATCH-S, is provably near-optimal. In addition, both our algorithms outperform existing approaches in accuracy by 59% or more (F-measure) in experiments.
4. **Scalability:** Our algorithms scale linearly, and GRIDWATCH-D is online, requiring bounded space and constant time per update.

Reproducibility: Our code and data are publicly available at github.com/bhooi/gridwatch.

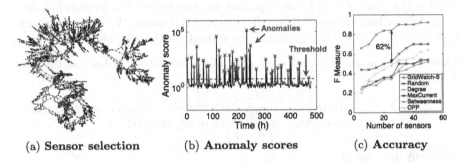

(a) **Sensor selection**   (b) **Anomaly scores**   (c) **Accuracy**

Fig. 1. (a) GRIDWATCH-S provably selects near-optimal sensor locations. Red circles indicate positions chosen for sensors, in the CASE2869 graph. (b) GRIDWATCH-D computes anomaly scores (black line) on CASE2869. Red crosses indicate ground truth - notice 100% true alarms (all black spikes above blue line are true alarms) and only 4 false dismissals (red crosses below blue line). (c) F-measure of GRIDWATCH-S compared to baselines on CASE2869. (Color figure online)

2   Background and Related Work

Time Series Anomaly Detection. Numerous algorithms exist for anomaly detection in univariate time series [17]. For multivariate time series, LOF [8] uses a local density approach. Isolation Forests [20] partition the data using a set of trees for anomaly detection. Other approaches use neural networks [34], distance-based [28], and exemplars [15]. However, none of these consider sensor selection.

Anomaly Detection in Temporal Graphs. [4] finds anomalous changes in graphs using an egonet (i.e. neighborhood) based approach, while [10,22] uses community-based approaches. [3] finds change points in dynamic graphs, while other partition-based [2] and sketch-based [29] exist for anomaly detection. However, these methods require fully observed edge weights (i.e. all sensors present), and do not consider sensor selection.

Power Grid Monitoring. A number of works consider the Optimal PMU Placement (OPP) problem [9], of optimally placing sensors in power grids, typically to make as many nodes as possible fully observable, or minimizing mean-squared error. Greedy [19], convex relaxation [16], integer program [12], simulated annealing [7] have been proposed. However, these do not perform anomaly detection.

[21,27,36] consider OPP in the presence of branch outages, but not anomalies in general, and due to their use of integer programming, only use small graphs of size at most 60.

Epidemic and Outbreak Detection. [18] proposed CELF, for outbreak detection in networks, such as water distribution networks and blog data, also using a submodular objective function. Their setting is a series of cascades spreading over the graph, while our input data is time-series data from sensors at various edges of the graph. For epidemics, [11,25] consider targeted immunization, such as identifying high-degree [25] or well-connected [11] nodes. We show experimentally that our sensor selection algorithm outperforms both approaches.

Table 1. Comparison of related approaches: only GRIDWATCH satisfies all the listed properties.

	Temporal [8,17], etc.	Graph-based [4,22,30]	OPP [9,16,19], etc.	Immunize [11,25]	GridWatch
Anomaly detection	✓	✓			✓
Online algorithm	✓				✓
Using graph data		✓	✓	✓	✓
Sensor selection			✓	✓	✓
Approx guarantee					✓

Table 1 summarizes related work. GRIDWATCH differs from existing methods in that it performs anomaly detection using an online algorithm, and it selects sensor locations with a provable approximation guarantee.

2.1 Background: Submodular Functions

A function f defined on subsets of $\mathcal{V}$ is submodular if whenever $\mathcal{T} \subseteq \mathcal{S}$ and $i \notin \mathcal{S}$:

$$f(\mathcal{S} \cup \{i\}) - f(\mathcal{S}) \leq f(\mathcal{T} \cup \{i\}) - f(\mathcal{T}) \tag{1}$$

Intuitively, this can be interpreted as *diminishing returns*: the left side is the gain in f from adding i to $\mathcal{S}$, and the right side is the gain from adding i to $\mathcal{T}$. Since $\mathcal{T} \subseteq \mathcal{S}$, this says that as $\mathcal{T}$ 'grows' to $\mathcal{S}$, the gains from adding i can only diminish.

[23] showed that nondecreasing submodular functions can be optimized by a greedy algorithm with a constant-factor approximation guarantee of $(1 - 1/e)$. These were extended by [32] to the non-constant sensor cost setting.

Table 2. Symbols and definitions

Symbol	Interpretation
$\mathcal{G} = (\mathcal{V}, \mathcal{E})$	Input graph
$\mathcal{S}$	Subset of nodes to place sensors on
n	Number of nodes
s	Number of scenarios
$\mathcal{N}_i$	Set of edges adjacent to node i
$V_i(t)$	Voltage at node i at time t
$I_e(t)$	Current at edge e at time t
$S_{ie}(t)$	Power w.r.t. node i and edge e at time t
$\Delta S_{ie}(t)$	Power change: $\Delta S_{ie}(t) = S_{ie}(t) - S_{ie}(t-1)$
$X_i(t)$	Sensor vector for scenario i at time t
c	Anomalousness threshold parameter
$\tilde{\mu}_i(t)$	Median of sensor i at time t
$\tilde{\sigma}_i(t)$	Inter-quartile range of sensor i at time t
$a_i(t)$	Sensor-level anomalousness for sensor i at time t
$A(t)$	Total anomalousness at time t

3   GRIDWATCH-D Anomaly Detection Algorithm

Preliminaries. Table 2 shows the symbols used in this paper.

In this section, we are given a graph $\mathcal{G} = (\mathcal{V}, \mathcal{E})$ and a fixed set of sensors $\mathcal{S} \subseteq \mathcal{V}$. Each sensor consists of a central node i on which voltage $V_i(t) \in \mathbb{C}$ is measured, at each time t. Note that complex voltages and currents are used to take phase into account, following standard practice in circuit analysis (this paper will not presume familiarity with this). Additionally, for sensor i, letting $\mathcal{N}_i$ be the set of edges adjacent to i, we are given the current $I_e \in \mathbb{C}$ along each edge $e \in \mathcal{N}_i$.

For sensor i and edge $e \in \mathcal{N}_i$, define the power w.r.t. i along edge e as $S_{ie}(t) = V_i(t) \cdot I_e(t)^*$, where $*$ is the complex conjugate. We find that using power (rather than current) provides better anomaly detection in practice. However, when considering the edges around a single sensor i, variations in current result in similar variations in power, so they perform the same role.

3.1   Types of Anomalies

Our goal is to detect single edge deletions, i.e. a transmission line failure. Single edge deletions affect the voltage and current in the graph in a complex, nonlinear way, and can manifest themselves in multiple ways. Consider the illustrative power grid shown by the graphs in Fig. 2. The power grid consists of a single generator, a single load, and power lines of uniform resistance. When the edge marked in the black cross fails, current is diverted from some edges to others, causing some edges to have increased current flow (blue edges), and thus

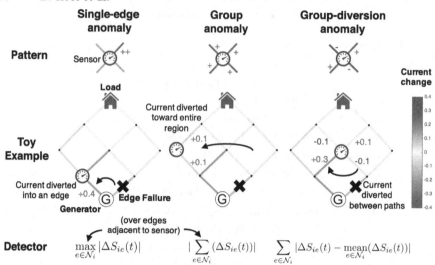

Fig. 2. Domain-aware model for anomalies: edge failures form 3 patterns. Edge color indicates change in current due to the edge failure: blue is an increase; red is a decrease. *Left:* diversion into a single edge. *Center:* diversion from the right to the left side of the graph, forming a group anomaly. *Right:* diversion between paths, forming a group diversion anomaly. (Color figure online)

increased power, and others to have decreased current flow (red edges). Current flows are computed using a standard power grid simulator, Matpower [37].

In the leftmost plot, the edge deletion diverts a large amount of current into a single edge, resulting in a highly anomalous value (+0.4) along a single edge. To detect single-edge anomalies, we consider the largest absolute change in power in the edges adjacent to this sensor. Formally, letting $\Delta S_{ie}(t) = S_{ie}(t) - S_{ie}(t-1)$.

Definition 1 (Single-Edge Detector). *The detector at sensor i is:*

$$x_{SE,i}(t) = \max_{e \in \mathcal{N}_i} |\Delta S_{ie}(t)| \tag{2}$$

In the middle plot, the edge deletion cuts off a large amount of current that would have gone from the generator toward the right side of the graph, diverting it into the left side of the graph. This results in some nodes in the left region with all their neighboring edges having positive changes (blue), such as the leftmost node. Individually, these changes may be too small to appear anomalous, but in aggregate, they provide stronger evidence of an anomaly. Hence, the group anomaly detector computes the sum of power changes around sensor i, then takes the absolute value:

Definition 2 (Group Anomaly Detector). *The detector at sensor i is:*

$$x_{GA,i}(t) = |\sum_{e \in \mathcal{N}_i} (\Delta S_{ie}(t))| \tag{3}$$

In the right plot, the edge deletion diverts current between nearby edges. In particular, current diversions around the central node cause it to have neighbors which greatly differ from each other: 2 positive edges and 2 negative edges. If this diversion is large enough, this provides stronger evidence of an anomaly than simply looking at each edge individually. Hence, the group diversion detector measures the 'spread' around sensor i by looking at the total absolute deviation of power changes about sensor i:

Definition 3 (Group Diversion Detector). *The detector at sensor i is:*

$$x_{GD,i}(t) = \sum_{e \in \mathcal{N}_i} |\Delta S_{ie}(t) - \operatorname*{mean}_{e \in \mathcal{N}_i}(\Delta S_{ie}(t))| \qquad (4)$$

3.2 Proposed Anomaly Score

Having computed our detectors, we now define our anomaly score. For each sensor i, concatenate its detectors into a vector:

$$X_i(t) = [x_{SE,i}(t) \quad x_{GA,i}(t) \quad x_{GD,i}(t)] \qquad (5)$$

Sensor i should label time t as an anomaly if any of the detectors greatly deviate from their historical values. Hence, let $\tilde{\mu}_i(t)$ and $\tilde{\sigma}_i(t)$ be the historical median and inter-quartile range (IQR)[1] [35] of $X_i(t)$ respectively: i.e. the median and IQR of $X_i(1), \cdots, X_i(t-1)$. We use median and IQR generally instead of mean and standard deviation as they are robust against anomalies, since our goal is to detect anomalies.

Thus, define the sensor-level anomalousness as the maximum number of IQRs that any detector is away from its historical median. The infinity-norm $\| \cdot \|_\infty$ denotes the maximum absolute value of a vector.

Definition 4 (Sensor-level anomalousness). *Sensor-level anomalousness is:*

$$a_i(t) = \left\| \frac{X_i(t) - \tilde{\mu}_i(t)}{\tilde{\sigma}_i(t)} \right\|_\infty \qquad (6)$$

Finally, the **overall anomalousness** at time t is the maximum of $a_i(t)$ over all sensors. Taking maximums allows us to determine the *location* (not just time) of an anomaly, by looking at which sensor contributed toward the maximum.

Definition 5 (Overall anomalousness). *Overall anomalousness at time t is:*

$$A(t) = \max_{i \in \mathcal{S}} a_i(t) \qquad (7)$$

Algorithm 1 summarizes our GRIDWATCH-D anomaly detection algorithm. Note that we can maintain the median and IQR of a set of numbers in a streaming manner using reservoir sampling [33]. Hence, the NORMALIZE operation in Line 5 takes a value of $\Delta S_{ie}(t)$, subtracts its historical median and divides by the historical IQR for that sensor. This ensures that sensors with large averages or spread do not dominate.

---

[1] IQR is a robust measure of spread, equal to the difference between the 75% and 25% quantiles.

Algorithm 1. GRIDWATCH-D online anomaly detection algorithm

Input : Graph $\mathcal{G}$, voltage $V_i(t)$, current $I_i(t)$
Output: Anomalousness score $A(t)$ for each t, where higher $A(t)$ indicates
greater certainty of an anomaly

1 **for** t *received as a stream:* **do**
2   **for** $i \in \mathcal{S}$ **do**
3        $S_{ie}(t) \leftarrow V_i(t) \cdot I_e^*(t) \; \forall \, e \in \mathcal{N}_i$ ▷Power
4        $\Delta S_{ie}(t) \leftarrow S_{ie}(t) - S_{ie}(t-1)$ ▷Power differences
5        $\Delta S_{i\cdot}(t) \leftarrow$ NORMALIZE$(\Delta S_{i\cdot})$
6   Compute detectors $x_{SE,i}(t)$, $x_{GA,i}(t)$ and $x_{GD,i}(t)$ using Eq. (2) to (4)
7   Concatenate detectors: $X_i(t) = [x_{SE,i}(t) \;\; x_{GA,i}(t) \;\; x_{GD,i}(t)]$
8        $\tilde{\mu}_i(t) \leftarrow$ UPDATEMEDIAN$(\tilde{\mu}_i(t-1), X_i(t))$ ▷Historical median
9        $\tilde{\sigma}_i(t) \leftarrow$ UPDATEIQR$(\tilde{\sigma}_i(t-1), X_i(t))$ ▷Historical IQR
10        $a_i(t) \leftarrow \| \frac{X_i(t)-\tilde{\mu}_i(t)}{\tilde{\sigma}_i(t)} \|_\infty$ ▷Sensor-level anomalousness
11     $A(t) = \max_{i \in \mathcal{S}} a_i(t)$ ▷Overall anomalousness

---

Lemma 1. GRIDWATCH-D *is online, and requires bounded memory and time.*

Proof. We verify from Algorithm 1 that GRIDWATCH-D's memory consumption is $O(|\mathcal{S}|)$, and updates in $O(|\mathcal{S}|)$ time per iteration, which are bounded (regardless of the length of the stream). ■

4   Sensor Placement: GRIDWATCH-S

So far, we have detected anomalies using a fixed set of sensors. We now consider how to select locations for sensors to place given a fixed budget of k sensors to place. Our main idea will be to construct an optimization objective for the anomaly detection performance of a subset $\mathcal{S}$ of sensor locations, and show that this objective has the 'submodularity' property, showing that a greedy approach gives approximation guarantees.

Note the change in problem setting: we are no longer monitoring for anomalies online in time series data, since we are now assuming that the sensors have not even been installed yet. Instead, we are an offline planner deciding where to place the sensors. To do this, we use a model of the system in the form of its graph $\mathcal{G}$, plugging it into a simulator such as Matpower [37] to generate a dataset of ground truth anomalies and normal scenarios, where the former contain a randomly chosen edge deletion, and the latter do not.

4.1   Proposed Optimization Objective

Intuitively, we should select sensors $\mathcal{S}$ to maximize the **probability of detecting an anomaly**. This probability can be estimated as the fraction of ground truth anomalies that we successfully detect. Hence, our optimization objective, $f(\mathcal{S})$, will be the fraction of anomalies that we successfully detect when using

GRIDWATCH-D, with sensor set $\mathcal{S}$. We will now formalize this and show that it is submodular.

Specifically, define $X_i(r)$ as the value of sensor i on the rth anomaly, analogous to (5). Also define $\tilde{\mu}_i$ and $\tilde{\sigma}_i$ as the median and IQR of sensor i on the full set of normal scenarios. Also let $a_i(r)$ be the sensor-level anomalousness of the rth anomaly, which can be computed as in Definition 4 plugging in $\tilde{\mu}_i$ and $\tilde{\sigma}_i$:

$$a_i(r) = \left\| \frac{X_i(r) - \tilde{\mu}_i}{\tilde{\sigma}_i} \right\|_\infty \tag{8}$$

Define overall anomalousness w.r.t. $\mathcal{S}$, $A(r, \mathcal{S})$, analogously to Definition 5:

$$A(r, \mathcal{S}) = \max_{i \in \mathcal{S}} a_i(r) \tag{9}$$

Given threshold c, anomaly r will be detected by sensor set $\mathcal{S}$ if and only if $A(r, \mathcal{S}) > c$. Hence, our optimization objective is to maximize the fraction of detected anomalies:

$$\underset{\mathcal{S} \subseteq \mathcal{V}, |\mathcal{S}| = k}{\text{maximize}} \; f(\mathcal{S}), \text{ where } f(\mathcal{S}) = \frac{1}{s} \sum_{r=1}^{s} \mathbf{1}\{A(r, \mathcal{S}) > c\} \tag{10}$$

4.2 Properties of Objective

Our optimization objective $f(\mathcal{S})$ is submodular: informally, it exhibits diminishing returns. The more sensors we add, the smaller the marginal gain in detection probability.

Theorem 1. *Detection probability $f(\mathcal{S})$ is submodular, i.e. for all subsets $\mathcal{T} \subseteq \mathcal{S}$ and nodes $i \in \mathcal{V} \setminus \mathcal{S}$:*

$$f(\mathcal{S} \cup \{i\}) - f(\mathcal{S}) \leq f(\mathcal{T} \cup \{i\}) - f(\mathcal{T}) \tag{11}$$

Proof

$$f(\mathcal{S} \cup \{i\}) - f(\mathcal{S}) = \frac{1}{s} \sum_{r=1}^{s} \left(\mathbf{1}\{A(r, \mathcal{S} \cup \{i\}) > c\} - \mathbf{1}\{A(r, \mathcal{S}) > c\} \right)$$

$$= \frac{1}{s} \sum_{r=1}^{s} \left(\mathbf{1}\{ \max_{j \in \mathcal{S} \cup \{i\}} a_j(r) > c \} - \mathbf{1}\{ \max_{j \in \mathcal{S}} a_j(r) > c \} \right)$$

$$= \frac{1}{s} \sum_{r=1}^{s} \left(\mathbf{1}\{ a_i(r) > c \wedge \max_{j \in \mathcal{S}} a_j(r) \leq c \} \right)$$

$$\leq \frac{1}{s} \sum_{r=1}^{s} \left(\mathbf{1}\{ a_i(r) > c \wedge \max_{j \in \mathcal{T}} a_j(r) \leq c \} \right)$$

$$= f(\mathcal{T} \cup \{i\}) - f(\mathcal{T})$$

∎

Theorem 2. $f(S)$ *is nondecreasing, i.e.* $f(T) \leq f(S)$ *for all subsets* $T \subseteq S$.

Proof

$$f(S) = \frac{1}{s}\sum_{r=1}^{s} A(r, S) = \frac{1}{s}\sum_{r=1}^{s} \max_{j \in S} a_j(r) \geq \frac{1}{s}\sum_{r=1}^{s} \max_{j \in T} a_j(r) = f(T)$$

∎

4.3 Proposed GRIDWATCH-S Algorithm

We exploit this submodularity using an efficient greedy algorithm that starts from S as the empty set, and iteratively adds the best sensor to maximize $f(S)$, until the budget constraint $|S| = k$ is reached. Algorithm 2 describes our GRIDWATCH-S algorithm.

Algorithm 2. GRIDWATCH-S sensor selection algorithm

Input : Graph $\mathcal{G}$, voltage $V_i(t)$, current $I_i(t)$, budget k, sensor scores $a_i(r)$
 from (8)
Output: Chosen sensor set S
1 $S = \{\}$
2 Initialize $A(r) = 0 \; \forall \, r \in S$ ▷Overall anomalousness is all zero since $S = \{\}$
3 **while** $|S| < k$ **do**
4 | **for** $i \notin S$ **do**
5 | | $\delta_i \leftarrow \frac{1}{s}\sum_{r=1}^{s} \mathbf{1}\{\max(A(r), a_i(r)) > c\}$ ▷Objective value if we added i to
   | | S
6 | $i^* \leftarrow \underset{i \notin S}{\arg\max} \; \delta_i$ ▷Greedily add the sensor that maximizes objective
7 | $S \leftarrow S \cup \{i^*\}$
8 | $A(r) = \max(A(r), a_{i^*}(r)) \; \forall \, r \in S$

4.4 Approximation Bound

The nondecreasing and submodularity properties of f imply that Algorithm 2 achieves at least $1 - 1/e$ ($\approx 63\%$) of the value of the optimal sensor placement. Letting $\hat{S}$ be the set returned by Algorithm 2, and S^* be the optimal set:

Theorem 3

$$f(\hat{S}) \geq (1 - 1/e)f(S^*) \tag{12}$$

Proof. This follows from [23] since f is nondecreasing and submodular. ∎

5   Experiments

We design experiments to answer the following questions:

- **Q1. Anomaly Detection Accuracy:** on a fixed set of sensors, how accurate are the anomalies detected by GRIDWATCH-S compared to baselines?
- **Q2. Sensor Selection:** how much does sensor selection using GRIDWATCH-S improve the anomaly detection performance compared to baselines?
- **Q3. Scalability:** how do our algorithms scale with the graph size?

Our code and data are publicly available at github.com/bhooi/gridwatch. Experiments were done on a 2.4 GHz Intel Core i5 Macbook Pro, 16 GB RAM running OS X 10.11.2.

Data. We use 2 graphs, CASE2869 and CASE9241, which accurately represent different parts of the European high voltage network [37]. CASE2869 contains 2869 nodes (generators or buses) and 2896 edges (power lines or transformers). CASE9241 contains 9241 nodes and 16049 edges.

5.1   Q1. Anomaly Detection Accuracy

In this section, we compare GRIDWATCH-D against baseline anomaly detection approaches, given a fixed set of sensors.

Experimental Settings. For each graph, the sensor set for all algorithms is chosen as a uniformly random set of nodes of various sizes (the sizes are plotted in the x-axis of Fig. 3). Then, out of 480 time ticks, we first sample 50 random time ticks as the times when anomalies occur. In each such time tick, we deactivate a randomly chosen edge (i.e. no current can flow over that edge).

Using MatPower [37], we then generate voltage and current readings at each sensor. This requires an input time series of loads (i.e. real and reactive power at each node): we use load patterns estimated from real data [31] recorded from the Carnegie Mellon University (CMU) campus for 20 days from July 29 to August 17, 2016, scaled to a standard deviation of $0.3 \cdot \sigma$, with added Gaussian noise of $0.2 \cdot \sigma$, where σ is the standard deviation of the original time series [31].

This results in a time series of 480 time ticks (hourly data from 20 days), at each time recording the voltage at each sensor and the current at each edge adjacent to one of the sensors. Given this input, each algorithm then returns a ranking of the anomalies. We evaluate this using standard metrics, AUC (area under the ROC curve) and F-measure ($\frac{2 \cdot \text{precision} \cdot \text{recall}}{\text{precision} + \text{recall}}$), the latter computed on the top 50 anomalies output by each algorithm.

Baselines. Dynamic graph anomaly detection approaches [4,6,10,22,30] cannot be used as they require graphs with fully observed edge weights. Moreover, detecting failed power lines with all sensors present can be done by simply checking if any edge has current equal to 0, which is trivial. Hence, instead, we compare GRIDWATCH-D to the following multidimensional anomaly detection methods: Isolation Forests [20], Vector Autoregression (VAR) [14], Local Outlier Factor

(LOF) [8], and Parzen Window [24]. Each uses the currents and voltages at the given sensors as features. For VAR the norms of the residuals are used as anomaly scores; the remaining methods return anomaly scores directly. For Isolation Forests, we use 100 trees (following the scikit-learn defaults [26]). For VAR we select the order by maximizing AIC, following standard practice. For LOF we use 20 neighbors, and 20 neighbors for Parzen Window.

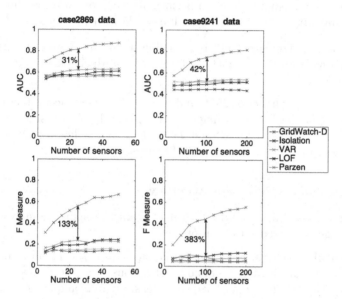

Fig. 3. Accurate anomaly detection: GRIDWATCH-D outperforms the baselines. Left plots are for CASE2869; right plots are for CASE9241.

Figure 3 shows that GRIDWATCH-D outperforms the baselines, by 31% to 42% Area under the Curve (AUC) and 133% to 383% F-Measure. The gains in performance likely come from the use of the 3 domain-knowledge based detectors, which combine information from the currents surrounding each sensor in a way that makes it clearer when an anomaly occurs.

Further testing shows that GRIDWATCH-D's 3 detectors all play a role: e.g. on CASE2869, for 50 sensors, GRIDWATCH-D has F-measure 0.67, but only using single detectors 1, 2 or 3 (where detector 1 refers to the detector in Definition 1, and so on) gives F-measures of 0.51, 0.6 or 0.56 respectively.

5.2   Q2. Sensor Selection Quality

We now evaluate GRIDWATCH-S. We use the same settings as in the previous sub-section, except that the sensors are now chosen using either GRIDWATCH-S, or one of the following baselines. We then compute the anomaly detection performance of GRIDWATCH-D as before on each choice of sensors. For GRIDWATCH-S we use $c = 15$. For our simulated data sizes, we assume 2000 anomalies and 480 normal scenarios.

Baselines: randomly selected nodes (*Random*); highest degree nodes (*Degree*); nodes with highest total current in their adjacent edges (*MaxCurrent*); highest betweenness centrality [13] nodes, i.e. nodes with the most shortest paths passing through them, thus being the most 'central' (*Betweenness*); a power-grid based Optimal PMU Placement algorithm using depth-first search (*OPP* [7]).

Figure 4 shows that GRIDWATCH-S outperforms the baselines, by 18 to 19% Area under the Curve (AUC) and 59 to 62% F-Measure.

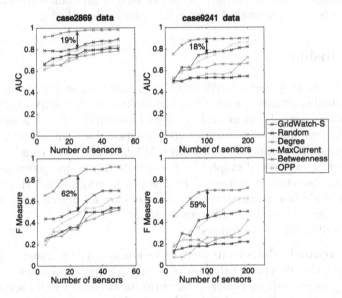

Fig. 4. GRIDWATCH-S provides effective sensor selection: sensor selection using GRIDWATCH-S results in higher detection accuracy than baselines.

Figure 1b shows the GRIDWATCH-S scores on CASE2869 over time, when using the maximum 200 sensors, with red crosses where true anomalies exist. Spikes in anomaly score match very closely with the true anomalies.

5.3   Q3. Scalability

Finally, we evaluate the scalability of GRIDWATCH-D and GRIDWATCH-S. To generate graphs of different sizes, we start with the IEEE 118-bus network [1], which represents a portion of the US power grid in 1962, and duplicate it $2, 4, \cdots, 20$ times. To keep our power grid connected, after each duplication, we add edges from each node to its counterpart in the last duplication; the parameters of each such edge are randomly sampled from those of the actual edges. We then run GRIDWATCH-D and GRIDWATCH-S using the same settings as the previous sub-section. Figure 5b shows that GRIDWATCH-D and GRIDWATCH-S scale linearly. The blue line is the best-fit regression line.

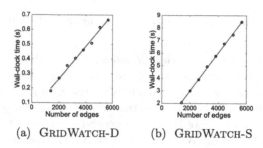

(a) GRIDWATCH-D   (b) GRIDWATCH-S

Fig. 5. Our algorithms scale linearly: wall-clock time of (a) GRIDWATCH-D and (b) GRIDWATCH-S against number of edges in $\mathcal{G}$. (Color figure online)

6   Conclusion

In this paper, we proposed GRIDWATCH-D, an online algorithm that accurately detects anomalies in power grid data. The main idea of GRIDWATCH-D is to design domain-aware detectors that combine information at each sensor appropriately. We then proposed GRIDWATCH-S, a sensor placement algorithm, which uses a submodular optimization objective. While our method could be technically applied to any type of graph-based sensor data (not just power grids), the choice of our detectors is motivated by our power grid setting. Hence, future work could study how sensitive various detectors are for detecting anomalies in graph-based sensor data from different domains.

Our contributions are as follows:

1. **Online anomaly detection:** we propose a novel, online anomaly detection algorithm, GRIDWATCH-D that outperforms existing approaches.
2. **Sensor placement:** we construct an optimization objective for sensor placement, with the goal of maximizing the probability of detecting an anomaly. We show that this objective is submodular, which we exploit in our sensor placement algorithm.
3. **Effectiveness:** Due to submodularity, GRIDWATCH-S, our sensor placement algorithm is provably near-optimal. In addition, both our algorithms outperform existing approaches in accuracy by 59% or more (F-measure) in experiments.
4. **Scalability:** Our algorithms scale linearly, and GRIDWATCH-D is online, requiring bounded space and constant time per update.

Reproducibility: Our code and data are publicly available at github.com/bhooi/gridwatch.

Acknowledgment. This material is based upon work supported by the National Science Foundation under Grant No. CNS-1314632, IIS-1408924, and by the Army Research Laboratory under Cooperative Agreement Number W911NF-09-2-0053, and in part by the Defense Advanced Research Projects Agency (DARPA) under award no. FA8750-17-1-0059 for the RADICS program. Any opinions, findings, and conclusions or recommendations expressed in this material are those of the author(s) and do not necessarily reflect the views of the National Science Foundation, or other funding parties. The U.S. Government is authorized to reproduce and distribute reprints for Government purposes notwithstanding any copyright notation here on.

References

1. IEEE power systems test case archive. http://www2.ee.washington.edu/research/pstca/. Accessed 15 Mar 2017
2. Aggarwal, C.C., Zhao, Y., Philip, S.Y.: Outlier detection in graph streams. In: 2011 IEEE 27th International Conference on Data Engineering (ICDE), pp. 399–409. IEEE (2011)
3. Akoglu, L., Faloutsos, C.: Event detection in time series of mobile communication graphs. In: Army Science Conference, pp. 77–79 (2010)
4. Akoglu, L., McGlohon, M., Faloutsos, C.: Oddball: spotting anomalies in weighted graphs. In: Zaki, M.J., Yu, J.X., Ravindran, B., Pudi, V. (eds.) PAKDD 2010. LNCS (LNAI), vol. 6119, pp. 410–421. Springer, Heidelberg (2010). https://doi.org/10.1007/978-3-642-13672-6_40
5. Amin, S.M.: US grid gets less reliable [the data]. IEEE Spectr. 48(1), 80–80 (2011)
6. Araujo, M., et al.: Com2: fast automatic discovery of temporal ('comet') communities. In: Tseng, V.S., Ho, T.B., Zhou, Z.-H., Chen, A.L.P., Kao, H.-Y. (eds.) PAKDD 2014. LNCS (LNAI), vol. 8444, pp. 271–283. Springer, Cham (2014). https://doi.org/10.1007/978-3-319-06605-9_23
7. Baldwin, T., Mili, L., Boisen, M., Adapa, R.: Power system observability with minimal phasor measurement placement. IEEE Trans. Power Syst. 8(2), 707–715 (1993)
8. Breunig, M.M., Kriegel, H.P., Ng, R.T., Sander, J.: LOF: identifying density-based local outliers. In: ACM Sigmod Record, vol. 29, pp. 93–104. ACM (2000)
9. Brueni, D.J., Heath, L.S.: The PMU placement problem. SIAM J. Discret. Math. 19(3), 744–761 (2005)
10. Chen, Z., Hendrix, W., Samatova, N.F.: Community-based anomaly detection in evolutionary networks. J. Intell. Inf. Syst. 39(1), 59–85 (2012)
11. Cohen, R., Havlin, S., Ben-Avraham, D.: Efficient immunization strategies for computer networks and populations. Phys. Rev. Lett. 91(24), 247901 (2003)
12. Dua, D., Dambhare, S., Gajbhiye, R.K., Soman, S.: Optimal multistage scheduling of PMU placement: an ILP approach. IEEE Trans. Power Deliv. 23(4), 1812–1820 (2008)
13. Freeman, L.C.: Centrality in social networks conceptual clarification. Soc. Netw. 1(3), 215–239 (1978)
14. Hamilton, J.D.: Time Series Analysis, vol. 2. Princeton University Press, Princeton (1994)
15. Jones, M., Nikovski, D., Imamura, M., Hirata, T.: Anomaly detection in real-valued multidimensional time series. In: International Conference on Bigdata/Socialcom/Cybersecurity. Stanford University, ASE. Citeseer (2014)
16. Kekatos, V., Giannakis, G.B., Wollenberg, B.: Optimal placement of phasor measurement units via convex relaxation. IEEE Trans. Power Syst. 27(3), 1521–1530 (2012)
17. Keogh, E., Lin, J., Lee, S.H., Van Herle, H.: Finding the most unusual time series subsequence: algorithms and applications. Knowl. Inf. Syst. 11(1), 1–27 (2007)
18. Leskovec, J., Krause, A., Guestrin, C., Faloutsos, C., VanBriesen, J., Glance, N.: Cost-effective outbreak detection in networks. In: KDD, pp. 420–429. ACM (2007)

19. Li, Q., Negi, R., Ilić, M.D.: Phasor measurement units placement for power system state estimation: a greedy approach. In: 2011 IEEE Power and Energy Society General Meeting, pp. 1–8. IEEE (2011)
20. Liu, F.T., Ting, K.M., Zhou, Z.H.: Isolation forest. In: ICDM, pp. 413–422. IEEE (2008)
21. Magnago, F.H., Abur, A.: A unified approach to robust meter placement against loss of measurements and branch outages. In: Proceedings of the 21st 1999 IEEE International Conference Power on Industry Computer Applications, PICA 1999, pp. 3–8. IEEE (1999)
22. Mongiovi, M., Bogdanov, P., Ranca, R., Papalexakis, E.E., Faloutsos, C., Singh, A.K.: Netspot: spotting significant anomalous regions on dynamic networks. In: SDM, pp. 28–36. SIAM (2013)
23. Nemhauser, G.L., Wolsey, L.A., Fisher, M.L.: An analysis of approximations for maximizing submodular set functions-I. Math. Program. 14(1), 265–294 (1978)
24. Parzen, E.: On estimation of a probability density function and mode. Ann. Math. Stat. 33(3), 1065–1076 (1962)
25. Pastor-Satorras, R., Vespignani, A.: Immunization of complex networks. Phys. Rev. E 65(3), 036104 (2002)
26. Pedregosa, F., et al.: Scikit-learn: machine learning in Python. J. Mach. Learn. Res. 12, 2825–2830 (2011)
27. Rakpenthai, C., Premrudeepreechacharn, S., Uatrongjit, S., Watson, N.R.: An optimal PMU placement method against measurement loss and branch outage. IEEE Trans. Power Deliv. 22(1), 101–107 (2007)
28. Ramaswamy, S., Rastogi, R., Shim, K.: Efficient algorithms for mining outliers from large data sets. In: ACM Sigmod Record, vol. 29, pp. 427–438. ACM (2000)
29. Ranshous, S., Harenberg, S., Sharma, K., Samatova, N.F.: A scalable approach for outlier detection in edge streams using sketch-based approximations. In: SDM, pp. 189–197. SIAM (2016)
30. Shah, N., Koutra, D., Zou, T., Gallagher, B., Faloutsos, C.: TimeCrunch: interpretable dynamic graph summarization. In: KDD, pp. 1055–1064. ACM (2015)
31. Song, H.A., Hooi, B., Jereminov, M., Pandey, A., Pileggi, L., Faloutsos, C.: PowerCast: mining and forecasting power grid sequences. In: Ceci, M., Hollmén, J., Todorovski, L., Vens, C., Džeroski, S. (eds.) ECML PKDD 2017. LNCS (LNAI), vol. 10535, pp. 606–621. Springer, Cham (2017). https://doi.org/10.1007/978-3-319-71246-8_37
32. Sviridenko, M.: A note on maximizing a submodular set function subject to a knapsack constraint. Oper. Res. Lett. 32(1), 41–43 (2004)
33. Vitter, J.S.: Random sampling with a reservoir. ACM Trans. Math. Softw. (TOMS) 11(1), 37–57 (1985)
34. Yi, S., Ju, J., Yoon, M.K., Choi, J.: Grouped convolutional neural networks for multivariate time series. arXiv preprint arXiv:1703.09938 (2017)
35. Yule, G.U.: An Introduction to the Theory of Statistics. C. Griffin, limited, London (1919)
36. Zhao, Y., Goldsmith, A., Poor, H.V.: On PMU location selection for line outage detection in wide-area transmission networks. In: 2012 IEEE Power and Energy Society General Meeting, pp. 1–8. IEEE (2012)
37. Zimmerman, R.D., Murillo-Sánchez, C.E., Thomas, R.J.: Matpower: steady-state operations, planning, and analysis tools for power systems research and education. IEEE Trans. Power Syst. 26(1), 12–19 (2011)

Incorporating Privileged Information to Unsupervised Anomaly Detection

Shubhranshu Shekhar$^{(\boxtimes)}$ and Leman Akoglu

Heinz College of Information Systems and Public Policy,
Carnegie Mellon University, Pittsburgh, USA
{shubhras,lakoglu}@andrew.cmu.edu

Abstract. We introduce a new unsupervised anomaly detection ensemble called SPI which can harness *privileged* information—data available only for training examples but not for (future) test examples. Our ideas build on the Learning Using Privileged Information (LUPI) paradigm pioneered by Vapnik et al. [17,19], which we extend to unsupervised learning and in particular to anomaly detection. SPI (for Spotting anomalies with Privileged Information) constructs a number of frames/fragments of knowledge (i.e., density estimates) in the privileged space and *transfers* them to the anomaly scoring space through "imitation" functions that use only the partial information available for test examples. Our generalization of the LUPI paradigm to unsupervised anomaly detection shepherds the field in several key directions, including (i) *domain-knowledge-augmented detection* using expert annotations as PI, (ii) *fast detection* using computationally-demanding data as PI, and (iii) *early detection* using "historical future" data as PI. Through extensive experiments on simulated and real datasets, we show that augmenting privileged information to anomaly detection significantly improves detection performance. We also demonstrate the promise of SPI under all three settings (*i–iii*); with PI capturing expert knowledge, computationally-expensive features, and future data on three real world detection tasks. Code related to this paper is available at: http://www.andrew.cmu.edu/user/shubhras/SPI.

1 Introduction

Outlier detection in point-cloud data has been studied extensively [1]. In this work we consider a unique setting with a much sparser literature: the problem of augmenting privileged information into unsupervised anomaly detection. Simply put, privileged information (PI) is *additional* data/knowledge/information that is available only at the learning/model building phase for (subset of) training examples, which however is unavailable for (future) test examples.

Electronic supplementary material The online version of this chapter (https://doi.org/10.1007/978-3-030-10925-7_6) contains supplementary material, which is available to authorized users.

© Springer Nature Switzerland AG 2019
M. Berlingerio et al. (Eds.): ECML PKDD 2018, LNAI 11051, pp. 87–104, 2019.
https://doi.org/10.1007/978-3-030-10925-7_6

The LUPI Framework. Learning Using Privileged Information (LUPI) has been pioneered by Vapnik et al. first in the context of SVMs [17,19] (PI-incorporated SVM is named SVM+), later generalized to neural networks [18]. The setup involves an Intelligent (or non-trivial) Teacher at learning phase, who provides the Student with *privileged* information (like explanations, metaphors, etc.), denoted x_i^*, about each training example x_i, $i = 1 \ldots n$. The key point in this paradigm is that *privileged information is not available at the test phase* (when Student operates without guidance of Teacher). Therefore, the goal is to build models (in our case, detectors) that can leverage/incorporate such additional information but *yet, not depend on the availability of PI at test time.*

Example: The additional information x_i^*'s belong to space X^* which is, generally speaking, different from space X. In other words, the feature spaces of vectors x_i^*'s and x_i's do not overlap. As an example, consider the task of identifying cancerous biopsy images. Here the images are in pixel space X. Suppose that there is an Intelligent Teacher that can recognize patterns in such images relevant to cancer. Looking at a biopsy image, Teacher can provide a description like "Aggressive proliferation of A-cells into B-cells" or "Absence of any dynamic". Note that such descriptions are in a specialized language space X^*, different from pixel space X. Further, they would be available only for a set of examples and not when the model is to operate autonomously in the future.

LUPI's Advantages: LUPI has been shown to (*i*) improve rate of convergence for learning, i.e., require asymptotically fewer examples to learn [19], as well as (*ii*) improve accuracy, when one can learn a model in space X^* that is not much worse than the best model in space X (i.e., PI is intelligent/non-trivial) [18]. Motivated by these advantages, LUPI has been applied to a number of problems from action recognition [13] to risk modeling [14] (expanded in Sect. 5). However, the focus of all such work has mainly been on *supervised* learning.

LUPI for Anomaly Detection. The only (perhaps straightforward) extension of LUPI to unsupervised anomaly detection has been introduced recently, generalizing SVM+ to the One-Class SVM (namely OC-SVM+) [2] for malware and bot detection. The issue is that OC-SVM is not a reliable detector since it assumes that normal points can be separated from origin in a single hyperball—experiments on numerous benchmark datasets with ground truth by Emmott et al. that compared popular anomaly detection algorithms find that OC-SVM ranks at the bottom (Table 1, pg. 4 [6]; also see our results in Sect. 4). We note that the top performer in [6] is the Isolation Forest (iForest) algorithm [11], an ensemble of randomized trees.

Our Contributions: Motivated by LUPI's potential value to learning and the scarcity in the literature of its generalization to anomaly detection, we propose a new technique called SPI (pronounced 'spy'), for Spotting anomalies with Privileged Information. Our work *bridges the gap (for the first time) between LUPI and unsupervised ensemble based anomaly detection* that is considered state-of-the-art [6]. We summarize our main contributions as follows.

- **Study of LUPI for anomaly detection:** We analyze how LUPI can benefit anomaly detection, not only when PI is truly unavailable at test time (as in traditional setup) but also when PI is strategically and willingly avoided at test time. We argue that data/information that incurs overhead on resources ($$$/storage/battery/etc.), timeliness, or vulnerability, if designated as PI, can enable resource-frugal, early, and preventive detection (expanded in Sect. 2).
- **PI-incorporated detection algorithm:** We show how to incorporate PI into ensemble based detectors and propose SPI, which constructs frames/fragments of knowledge (specifically, density estimates) in the privileged space (X^*) and *transfers* them to the anomaly scoring space (X) through "imitation" functions that use only the partial information available for test examples. To the best of our knowledge, ours is the first attempt to leveraging PI for improving the state-of-the-art *ensemble methods* for anomaly detection within an *unsupervised* LUPI framework. Moreover, while SPI augments PI within the tree-ensemble detector iForest [11], our solution can easily be applied to any other ensemble based detector (Sect. 3).
- **Applications:** Besides extensive simulation experiments, we employ SPI on three real-world case studies where PI respectively captures (*i*) expert knowledge, (*ii*) computationally-expensive features, and (*iii*) "historical future" data, which demonstrate the benefits that PI can unlock for anomaly detection in terms of accuracy, speed, and detection latency (Sect. 4).

Reproducibility: Implementation of SPI and real world datasets used in experiments are open-sourced at http://www.andrew.cmu.edu/user/shubhras/SPI.

2 Motivation: How Can LUPI Benefit Anomaly Detection?

The implications of the LUPI paradigm for anomaly detection is particularly exciting. Here, we discuss a number of detection scenarios and demonstrate that LUPI unlocks advantages for anomaly detection problems in multiple aspects.

In the original LUPI framework [19], privileged information (hereafter PI) is defined as data that is available *only* at training stage for training examples but *unavailable at test time* for test examples. Several anomaly detection scenarios admit this definition directly. Interestingly, PI can also be specified as *strategically "unavailable"* for anomaly detection. That is, one can willingly avoid using certain data at test time (while incorporating such data into detection models at train phase[1]) in order to achieve resource efficiency, speed, and robustness. We organize detection scenarios into two with PI as (truly) Unavailable vs. Strategic, and elaborate with examples below. Table 1 gives a summary.

---

[1] Note that training phase in anomaly detection does not involve the use of any labels.

Table 1. Types of data used in anomaly detection with various overhead on resources ($$$, storage, battery, etc.), timeliness, and/or risk, *if used as privileged information can enable resource-frugal, early, as well as preventive detection.*

Properties vs. type of privileged info	Unavailable vs. Strategic	Need Resources	Cause Delay	Incur Risk
1. "historical future" data	U	n/a	n/a	n/a
2. after-the-fact data	U	n/a	n/a	n/a
3. advanced technical data	U	n/a	n/a	n/a
4. restricted-access data	U, S	✓		
5. expert knowledge	U, S	✓	✓	
6. compute-heavy data	S	✓	✓	
7. unsafe-to-collect data	S		✓	✓
8. easy-target-to-tamper data	S			✓

Unavailable PI: This setting includes typical scenarios, where PI is (truly) unknown for test examples.

1. *"historical future" data*: When training an anomaly detection model with offline/historical data that is over time (e.g., temporal features), one may use values both before *and after* time t while creating an example for each t. Such data is PI; not available when the model is deployed to operate in real-time.

2. *after-the-fact data*: In malware detection, the goal is to detect before it gets hold of and harms the system. One may have historical data for some (training) examples from past exposures, including measurements of system variables (number of disk/port read/writes, CPU usage, etc.). Such after-the-exposure measurements can be incorporated as PI.

3. *advanced technical data*: This includes scenarios where some (training) examples are well-understood but those to be detected are simply unknown. For example, the expected behavior of various types of apps on a system may be common domain knowledge that can be converted to PI, but such knowledge may not (yet) be available for new-coming apps.

Strategic PI: Strategic scenarios involve PI that can in principle be acquired but is willingly avoided at test time to achieve gains in resources, time, or risk.

4. *restricted-access data*: One may want to build models that do not assume access to private data or intellectual property at test time, such as source code (for apps or executables), *even if* they could be acquired through resources. Such information can also be truly unavailable, e.g. encrypted within the software.

5. *expert knowledge*: Annotations about some training examples may be available from experts, which are truly unavailable at test time. One could also strategically choose to avoid expert involvement at test time, which (a) may be costly to obtain and/or (b) cause significant delay, especially for real-time detection.

6. *compute-heavy data*: One may strategically choose not to rely on features that are computationally expensive to obtain, especially in real-time detection, but rather use such data as PI (which can be extracted offline at training phase). Such features not only cause delay but also require compute resources (which e.g., may drain batteries in detecting malware apps on cellphones).

7. *unsafe-to-collect data*: This involves cases where collecting PI at test time is unsafe/dangerous. For example, the slower a drone moves to capture high-resolution (privileged) images for surveillance, not only it causes delay but more importantly, the more susceptible it becomes to be taken down.

8. *easy-target-to-tamper data*: Finally, one may want to avoid relying on features that are easy for adversaries to tamper with. Examples to those features include self-reported data (like age, location, etc.). Such data may be available reliably for some training examples and can be used as PI.

In short, by strategically designating PI one can achieve resource, timeliness, and robustness gains for various anomaly detection tasks. Designating features that need resources as PI → allow resource-frugal ("lazy") detection; features that cause delay as PI → allow early/speedy detection; and designating features that incur vulnerability as PI → allow preventive and more robust detection.

In this subsection, we laid out a long list of scenarios that make LUPI-based learning particularly attractive for anomaly detection. In our experiments (Sect. 4) we demonstrate its premise for scenarios 1., 5. and 6. above using three real world datasets, while leaving others as what we believe interesting future investigations.

3 Privileged Info-Augmented Anomaly Detection

The Learning Setting. Formally, the input for the anomaly detection model at learning phase are tuples of the form

$$\mathcal{D} = \{(\boldsymbol{x}_1, \boldsymbol{x}_1^*), (\boldsymbol{x}_2, \boldsymbol{x}_2^*), \ldots, (\boldsymbol{x}_n, \boldsymbol{x}_n^*)\},$$

where $\boldsymbol{x}_i = (x_i^1, \ldots, x_i^d) \in X$ and $\boldsymbol{x}_i^* = (x_i^{*1}, \ldots, x_i^{*p}) \in X^*$. Note that this is an unsupervised learning setting where label information, i.e., y_i's are not available. The privileged information is represented as a feature vector $\boldsymbol{x}^* \in \mathbb{R}^p$ that is in space X^*, which is *additional to and different from* the feature space X in which the primary information is represented as a feature vector $\boldsymbol{x} \in \mathbb{R}^d$.

The important distinction from the traditional anomaly detection setting is that the input to the (trained) detector at testing phase are feature vectors

$$\{\boldsymbol{x}_{n+1}, \boldsymbol{x}_{n+2}, \ldots, \boldsymbol{x}_{n+m}\}.$$

That is, the (future) test examples do not carry any privileged information. The anomaly detection model is to score the incoming/test examples and make decisions solely based on the primary features $\boldsymbol{x} \in X$.

In this text, we refer to space X^* as the *privileged space* and to X as the *decision space*. Here, a key assumption is that the information in the privileged space is intelligent/nontrivial, that is, it allows to create models $f^*(\boldsymbol{x}^*)$ that detect anomalies with vectors $\boldsymbol{x}^*$ corresponding to vectors $\boldsymbol{x}$ with higher accuracy than models $f(\boldsymbol{x})$. As a result, the main question that arises which we address in this work is: "how can one use the knowledge of the information in space X^* to improve the performance of the desired model $f(\boldsymbol{x})$ in space X?".

In what follows, we present a first-cut attempt to the problem that is a natural knowledge transfer between the two feature spaces (called FT for feature transfer). We then lay out the shortcomings of such an attempt, and present our proposed solution SPI. We compare to FT (and other baselines) in experiments.

3.1 First Attempt: Incorporating PI by Transfer of Features

A natural attempt to learning under privileged information that is unavailable for test examples is to treat the task as a *missing data problem*. Then, typical techniques for data imputation can be employed where missing (privileged) features are replaced with their predictions from the available (primary) features.

In this scheme, one simply maps vectors $\boldsymbol{x} \in X$ into vectors $\boldsymbol{x}^* \in X^*$ and then builds a detector model in the transformed space. The goal is to find the transformation of vectors $\boldsymbol{x} = (x^1, \ldots, x^d)$ into vectors $\boldsymbol{\phi}(\boldsymbol{x}) = (\phi_1(\boldsymbol{x}), \ldots, \phi_p(\boldsymbol{x}))$ that minimizes the expected risk given as

$$R(\boldsymbol{\phi}) = \sum_{j=1}^{p} \min_{\phi_j} \int (x^{*j} - \phi_j(\boldsymbol{x}))^2 p(x^{*j}, \boldsymbol{x}) dx^{*j} d\boldsymbol{x}, \qquad (1)$$

where $p(x^{*j}, \boldsymbol{x})$ is the joint probability of coordinate x^{*j} and vector $\boldsymbol{x}$, and functions $\phi_j(\boldsymbol{x})$ are defined by p regressors.

Here, one could construct approximations to functions $\phi_j(\boldsymbol{x})$, $j = \{1, \ldots, p\}$ by solving p regression estimation problems based on the training examples

$$(\boldsymbol{x}_1, x_1^{*j}), \ldots, (\boldsymbol{x}_n, x_n^{*j}), \quad j = 1, \ldots, p,$$

where $\boldsymbol{x}_i$'s are input to each regression ϕ_j and the jth coordinate of the corresponding vector $\boldsymbol{x}_i^*$, i.e. x_i^{*j}'s are treated as the output, by minimizing the regularized empirical loss functional

$$R(\phi_j) = \min_{\phi_j} \sum_{i=1}^{n} (x_i^{*j} - \phi_j(\boldsymbol{x}_i))^2 + \lambda_j \text{penalty}(\phi_j), \quad j = 1, \ldots, p. \qquad (2)$$

Having estimated the transfer functions $\hat{\phi}_j$'s (using linear or non-linear regression techniques), one can then learn any desired anomaly detector $f(\hat{\boldsymbol{\phi}}(\boldsymbol{x}))$ using the training examples, which concludes the learning phase. Note that the detector does not require access to privileged features $\boldsymbol{x}^*$ and can be employed solely on primary features $\boldsymbol{x}$ of the test examples $i = n+1, \ldots, m$.

3.2   Proposed SPI: Incorporating PI by Transfer of Decisions

Treating PI as missing data and predicting x^* from x could be a difficult task, when privileged features are complex and high dimensional (i.e., p is large). Provided $f^*(x^*)$ is an accurate detection model, a more direct goal would be to *mimic its decisions*—the scores that f^* assigns to the training examples. Mapping *data* between two spaces, as compared to *decisions*, would be attempting to solve a more general problem, that is likely harder and unnecessarily wasteful.

The general idea behind transferring decisions/knowledge (instead of data) is to identify a small number of elements in the privileged space X^* that well-approximate the function $f^*(x^*)$, and then try to transfer them to the decision space—through the approximation of those elements in space X. This is the knowledge transfer mechanism in LUPI by Vapnik and Izmailov [17]. They illustrated this mechanism for the (supervised) SVM classifier. We generalize this concept to unsupervised anomaly detection.

The knowledge transfer mechanism uses three building blocks of knowledge representation in AI, as listed in Table 2. We first review this concept for SVMs, followed by our proposed SPI. While SPI is clearly different in terms of the task it is addressing as well as in its approach, as we will show, it is inspired by and builds on the same fundamental mechanism.

Table 2. Three building blocks of knowledge representation in artificial intelligence, in context of SVM-LUPI for classification [17] and SPI for anomaly detection [this paper].

	SVM-LUPI	SPI (Proposed)
1. Fundamental elements of knowledge	Support vectors	Isolation trees
2. Frames (fragments) of the knowledge	Kernel functions	Tree anomaly scores
3. Structural connections of the frames	Weighted sum	Weighted sum (by L2R)

Knowledge Transfer for SVM: The *fundamental elements* of knowledge in the SVM classifier are the support vectors. In this scheme, one constructs two SVMs; one in X space and another in X^* space. Without loss of generality, let $x_1, \ldots, x_t$ be the support vectors of SVM solution in space X and $x_1^*, \ldots, x_{t^*}^*$ be the support vectors of SVM solution in space X^*, where t and t^* are the respective number of support vectors.

The decision rule f^* in space X^* (which one aims to mimic) has the form

$$f^*(x^*) = \sum_{k=1}^{t^*} y_k \alpha_k^* K^*(x_k^*, x^*) + b^*, \tag{3}$$

where $K^*(x_k^*, x^*)$ is the kernel function of similarity between support vector x_k^* and vector $x^* \in X^*$, also referred as the *frames* (or *fragments*) of knowledge. Equation (3) depicts the *structural connection* of these fragments, which is a weighted sum with learned weights α_k^*'s.

The goal is to approximate each fragment of knowledge $K^*(x_k^*, x^*)$, $k = 1, \ldots, t^*$ in X^* using the fragments of knowledge in X; i.e., the t kernel functions $K(x_1, x), \ldots, K(x_t, x)$ of the SVM trained in X. To this end, one maps t-dimensional vectors $z = (K(x_1, x), \ldots, K(x_t, x)) \in Z$ into t^*-dimensional vectors $z^* = (K^*(x_1^*, x^*), \ldots, K^*(x_{t^*}^*, x^*)) \in Z^*$ through t^* regression estimation problems. That is, the goal is to find regressors $\phi_1(z), \ldots, \phi_{t^*}(z)$ in X such that

$$\phi_k(z_i) \approx K^*(x_k^*, x_i^*), \quad k = 1, \ldots, t^* \tag{4}$$

for all training examples $i = 1, \ldots, n$. For each $k = 1, \ldots, t^*$, one can construct the approximation to function ϕ_k by training a regression on the data

$$\{(z_1, K^*(x_k^*, x_1^*)), \ldots, (z_n, K^*(x_k^*, x_n^*))\}, \ k = 1, \ldots, t^*,$$

where we regress vectors z_i's onto scalar output $K^*(x_k^*, x_i^*)$'s to obtain $\hat{\phi}_k$.

For the prediction of a test example x, one can then replace each $K^*(x_k^*, x^*)$ in Eq. (3) (which requires privileged features x^*) with $\hat{\phi}_k(z)$ (which mimics it, using only the primary features x—to be exact, by first transforming x into z through the frames $K(x_j, x), j = 1, \ldots, t$ in the X space).

Knowledge Transfer for SPI: In contrast to mapping of features from space X to space X^*, knowledge transfer of decisions maps space Z to Z^* in which fragments of knowledge are represented. Next, we show how to generalize these ideas to anomaly detection with no label supervision. Figure 1 shows an overview.

To this end, we utilize a state-of-the-art ensemble technique for anomaly detection, called Isolation Forest [11] (hereafters IF, for short), which builds a set of extremely randomized trees. In essence, each tree approximates density in a random feature subspace and anomalousness of a point is quantified by the sum of such partial estimates across all trees.

In this setting, one can think of the individual trees in the ensemble to constitute the *fundamental elements* and the partial density estimates (i.e., individual anomaly scores from trees) to constitute the *fragments* of knowledge, where the structural connection of the fragments is achieved by an unweighted sum.

Similar to the scheme with SVMs, we construct two IFs; one in X space and another in X^* space. Let $\mathcal{T} = T_1, \ldots, T_t$ denote the trees in the ensemble in X and $\mathcal{T}^* = T_1^*, \ldots, T_{t^*}^*$ the trees in the ensemble in X^*, where t and t^* are the respective number of trees (prespecified by the user, typically a few 100s). Further, let $S^*(T_k^*, x^*)$ denote the anomaly score estimated by tree T_k^* for a given x^* (the lower the more anomalous; refer to [11] for details of the scoring). $S(T_k, x)$ is defined similarly. Then, the anomaly score s^* for a point x^* in space X^* (which we aim to mimic) is written as

$$s^*(x^*) = \sum_{k=1}^{t^*} S^*(T_k^*, x^*), \tag{5}$$

which is analogous to Eq. (3). To mimic/approximate each fragment of knowledge $S^*(T_k^*, x^*)$, $k = 1, \ldots, t^*$ in X^* using the fragments of knowledge in X; i.e., the

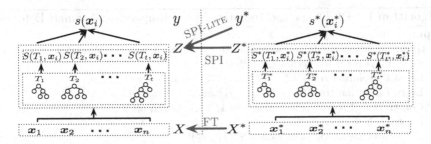

Fig. 1. Anomaly detection with PI illustrated. FT maps *data* between spaces (Sect. 3.1) whereas SPI (and "light" version SPI-LITE) mimic *decisions* (Sect. 3.2).

t scores for $\boldsymbol{x}$: $S(T_1, \boldsymbol{x}), \ldots, S(T_t, \boldsymbol{x})$ of the IF trained in X, we estimate t^* regressors $\phi_1(\boldsymbol{z}), \ldots, \phi_{t^*}(\boldsymbol{z})$ in X such that

$$\phi_k(\boldsymbol{z}_i) \approx S^*(T_k^*, \boldsymbol{x}_i^*), \quad k = 1, \ldots, t^* \tag{6}$$

for all training examples $i = 1, \ldots, n$, where $\boldsymbol{z}_i = (S(T_1, \boldsymbol{x}_i), \ldots, S(T_t, \boldsymbol{x}_i))$. Simply put, each $\hat{\phi}_k$ is an approximate mapping of all the t scores from the ensemble $\mathcal{T}$ in X to an individual score (fragment of knowledge) by tree T_k^* of the ensemble $\mathcal{T}^*$ in X^*. In practice, we learn a mapping from the leaves rather than the trees of $\mathcal{T}$ for a more granular mapping. Specifically, we construct vectors $\boldsymbol{z}_i = (\boldsymbol{z}_{i1}', \ldots, \boldsymbol{z}_{it}')$ where each $\boldsymbol{z}_{ik}'$ is a size ℓ_k vector in which the value at index leaf$(T_k, \boldsymbol{x}_i)$ is set to $S(T_k, \boldsymbol{x}_i)$ and other entries to zero. Here, ℓ_k denotes the number of leaves in tree T_k and leaf$(\cdot)$ returns the index of the leaf that $\boldsymbol{x}_i$ falls into in the corresponding tree (note that $\boldsymbol{x}_i$ belongs to exactly one leaf of any tree, since the trees partition the feature space).

SPI-lite: A "light" version. We note that instead of mimicking each individual fragment of knowledge $S^*(T_k^*, \boldsymbol{x}^*)$'s, one could also directly mimic the "final decision" $s^*(\boldsymbol{x}^*)$. To this end, we also introduce SPI-LITE, which estimates a *single* regressor $\phi(\boldsymbol{z}_i) \approx s^*(\boldsymbol{x}_i^*)$ for $i = 1, \ldots, n$ (also see Fig. 1). We compare SPI and SPI-LITE empirically in Sect. 4.

Learning to Rank (L2R) Like in X^*: An important challenge in learning to accurately mimic the scores s^*'s in Eq. (5) is to make sure that the regressors ϕ_k's are very accurate in their approximations in Eq. (6). Even then, it is hard to guarantee that the final ranking of points by $\sum_{k=1}^{t^*} \hat{\phi}_k(\boldsymbol{z}_i)$ would reflect their ranking by $s^*(\boldsymbol{x}_i^*)$. Our ultimate goal, after all, is to *mimic the ranking* of the ensemble in X^* space since anomaly detection is a ranking problem at its heart.

To this end, we set up an additional pairwise learning to rank objective as follows. Let us denote by $\boldsymbol{\phi}_i = (\hat{\phi}_1(\boldsymbol{z}_i), \ldots, \hat{\phi}_{t^*}(\boldsymbol{z}_i))$ the t^*-dimensional vector of estimated knowledge fragments for each training example i. For each pair of training examples, we create a tuple of the form $((\boldsymbol{\phi}_i, \boldsymbol{\phi}_j), p_{ij}^*)$ where

$$p_{ij}^* = P(s_i^* < s_j^*) = \sigma(-(s_i^* - s_j^*)), \tag{7}$$

Algorithm 1. SPI-TRAIN: Incorporating PI to Unsupervised Anomaly Detector

---
Input: training examples $\{(\boldsymbol{x}_1, \boldsymbol{x}_1^*), \ldots, (\boldsymbol{x}_n, \boldsymbol{x}_n^*)\}$
Output: detection model (ensemble-of-trees) $\mathcal{T}$ in X space; regressors $\hat{\phi}_k$'s, $k = 1, \ldots, t^*$; $\boldsymbol{\beta}$ (or $\boldsymbol{\gamma}$ for kernelized L2R)
1: Learn t^* isolation trees $\mathcal{T}^* = \{T_1^*, \ldots, T_{t^*}^*\}$ on $\boldsymbol{x}_i^*$'s $i = 1, \ldots, n$
2: Learn t isolation trees $\mathcal{T} = \{T_1, \ldots, T_t\}$ on $\boldsymbol{x}_i$'s $i = 1, \ldots, n$
3: Construct leaf score vectors $\boldsymbol{z}_i$'s, $i = 1, \ldots, n$, based on $\mathcal{T}$
4: **for each** $k = 1, \ldots, t^*$ **do**
5:    Learn regressor $\hat{\phi}_k$ of $\boldsymbol{z}_i$'s onto $S^*(T_k^*, \boldsymbol{x}_i^*)$'s
6:    Obtain $\boldsymbol{\beta}$ by optimizing C in (9) (or $\boldsymbol{\gamma}$ for kernelized C_ψ)
7: **end for**

Algorithm 2. SPI-TEST: PI-Augmented Unsupervised Anomaly Detection

---
Input: test examples $\{\boldsymbol{x}_{n+1}, \ldots, \boldsymbol{x}_{n+m}\}$; $\mathcal{T}$, $\hat{\phi}_k$'s $k = 1, \ldots, t^*$, $\boldsymbol{\beta}$ (or $\boldsymbol{\gamma}$ if kernelized)
Output: estimated anomaly scores $\{s_{n+1}, \ldots, s_{n+m}\}$ for all test examples
1: **for each** test example $\boldsymbol{x}_e$, $e = n+1, \ldots, n+m$ **do**
2:    Construct leaf score vector $\boldsymbol{z}_e = (z_{e1}', \ldots, z_{et}')$ where entry in each z_{ek}' for index leaf$(T_k, \boldsymbol{x}_e)$ is set to $S(T_k, \boldsymbol{x}_e)$ and to 0 o.w., for $k = 1, \ldots, t$
3:    Construct $\boldsymbol{\phi}_e = (\hat{\phi}_1(\boldsymbol{z}_e), \ldots, \hat{\phi}_{t^*}(\boldsymbol{z}_e))$
4:    Estimate anomaly score as $s_e = \boldsymbol{\beta}\boldsymbol{\phi}_e^T$ (or $s_e = \sum_{l=1}^n \gamma_l K(\boldsymbol{\phi}_l, \boldsymbol{\phi}_e)$ if kernelized)
5: **end for**

---

which is the probability that i is ranked ahead of j by anomalousness in X^* space (recall that lower s^* is more anomalous), where $\sigma(v) = 1/(1 + e^{-v})$ is the sigmoid function. Notice that the larger the gap between the anomaly scores of i and j, the larger this probability gets (i.e., more surely i ranks above j).

Given the training pair tuples above, our goal of learning-to-rank is to estimate $\boldsymbol{\beta} \in \mathbb{R}^{t^*}$, such that

$$p_{ij} = \sigma(\Delta_{ij}) = \sigma(\boldsymbol{\beta}\boldsymbol{\phi}_i^T - \boldsymbol{\beta}\boldsymbol{\phi}_j^T) = \sigma(-\hat{s}_i^* + \hat{s}_j^*)) \approx p_{ij}^*, \quad \forall i,j \in \{1, \ldots, n\}. \quad (8)$$

We then utilize the cross entropy as our cost function over all (i,j) pairs, as

$$\min_{\boldsymbol{\beta}} \ C = \sum_{(i,j)} -p_{ij}^* \log(p_{ij}) - (1 - p_{ij}^*)\log(1 - p_{ij}) = \sum_{(i,j)} -p_{ij}^* \Delta_{ij} + \log(1 + e^{\Delta_{ij}}) \quad (9)$$

where p_{ij}^*'s are given as input to the learning as specified in Eq. (7) and p_{ij} is denoted in Eq. (8) and is parameterized by $\boldsymbol{\beta}$ that is to be estimated.

The objective function in (9) is convex and can be solved via a gradient-based optimization, where $\frac{dC}{d\boldsymbol{\beta}} = \sum_{(i,j)} (p_{ij} - p_{ij}^*)(\boldsymbol{\phi}_i - \boldsymbol{\phi}_j)$ (details omitted for brevity). More importantly, in case the linear mapping $s_i^* \approx \boldsymbol{\beta}\boldsymbol{\phi}_i^T$ is not sufficiently accurate to capture the desired pairwise rankings, the objective can be *kernelized* to learn a *non-linear* mapping that is likely more accurate. The idea is to write $\boldsymbol{\beta}_\psi = \sum_{l=1}^n \gamma_l \psi(\boldsymbol{\phi}_l)$ (in the transformed space) as a weighted linear combination of (transformed) training examples, for feature transformation function $\psi(\cdot)$ and parameter vector $\boldsymbol{\gamma} \in \mathbb{R}^n$ to be estimated. Then, Δ_{ij} in objective (9) in the transformed space can be written as

$$\Delta_{ij} = \sum_{l=1}^{n} \gamma_l [\psi(\boldsymbol{\phi}_l)\psi(\boldsymbol{\phi}_i)^T - \psi(\boldsymbol{\phi}_l)\psi(\boldsymbol{\phi}_j)^T] = \sum_{l=1}^{n} \gamma_l [K(\boldsymbol{\phi}_l, \boldsymbol{\phi}_i) - K(\boldsymbol{\phi}_l, \boldsymbol{\phi}_j)]. \quad (10)$$

The kernelized objective, denoted C_ψ, can also be solved through gradient-based optimization where we can show partial derivatives (w.r.t. each γ_l) to be equal to $\frac{\partial C_\psi}{\partial \gamma_l} = \sum_{(i,j)}(p_{ij} - p_{ij}^*)[K(\boldsymbol{\phi}_l, \boldsymbol{\phi}_i) - K(\boldsymbol{\phi}_l, \boldsymbol{\phi}_j)]$. Given the estimated γ_l's, prediction of score is done by $\sum_{l=1}^{n} \gamma_l K(\boldsymbol{\phi}_l, \boldsymbol{\phi}_e)$ for any (test) example e.

The SPI Algorithm: We outline the steps of SPI for both training and testing (i.e., detection) in Algorithms 1 and 2, respectively. Note that the test-time detection no longer relies on the availability of privileged features for the test examples, but yet be able to leverage/incorporate them through its training.

4 Experiments

We design experiments to evaluate our methods in two different settings:

1. **Benchmark Evaluation:** We show the effectiveness of augmenting PI (see Table 3) on 17 publicly available benchmark datasets.[2]
2. **Real-world Use Cases:** We conduct experiments on LingSpam[3] and BotOrNot[4] datasets to show that (i) domain-expert knowledge as PI improves spam detection, (ii) compute-expensive PI enables fast detection at test time, and (iii) "historical future" PI allows early detection of bots.

Baselines. We compare both SPI and SPI-LITE to the following baselines:

1. IF(X-only): Isolation Forest [11] serves as a simple baseline that operates solely in decision space X. PI is not used neither for modeling nor detection.
2. OC-SVM+ (PI-incorporated): OC+ for short, is an extension of (unsupervised) One-Class SVM that incorporates PI as introduced in [2].
3. FT(PI-incorporated): This is the direct feature transfer method that incorporates PI by learning a mapping $X \rightarrow X^*$ as we introduced in Sect. 3.1.

* IF* (X^*-only): IF that operates in X^* space. We report performance by IF* only for reference, since PI is unavailable at test time.

4.1 Benchmark Evaluation

The benchmark datasets do not have an explicit PI representation. Therefore, in our experiments we introduce PI as explained below.

[2] http://agents.fel.cvut.cz/stegodata/Loda.zip.
[3] http://csmining.org/index.php/ling-spam-datasets.html.
[4] https://botometer.iuni.iu.edu/bot-repository/datasets/caverlee-2011/caverlee-2011.zip.

Table 3. Mean Average Precision (MAP) on benchmark datasets (avg'ed over 5 runs) for $\gamma = 0.7$. Numbers in parentheses indicate rank of each algorithm on each dataset. IF* (for reference only) reports MAP in the X^* space.

Datasets	$p + d$	n	IF	OC+	FT	SPI-LITE	SPI	IF*
breast-cancer	30	357	0.1279 (4)	0.0935 (6)	0.0974 (5)	0.4574 (3)	**0.5746** (2)	0.6773 (1)
ionosphere	33	225	0.0519 (4)	**0.2914** (1)	0.0590 (3)	0.0512 (5)	0.0470 (6)	0.0905 (2)
letter-recognition	617	4197	0.0889 (6)	0.1473 (4)	0.0908 (5)	0.3799 (3)	**0.6413** (2)	0.9662 (1)
multiple-features	649	1200	0.1609 (5)	0.1271 (6)	0.2044 (4)	0.6589 (3)	**0.8548** (2)	1.0000 (1)
wall-following-robot	24	2923	0.1946 (5)	0.2172 (4)	0.1848 (6)	0.4331 (3)	**0.5987** (2)	0.7538 (1)
cardiotocography	27	1831	0.2669 (5)	0.6107 (4)	0.2552 (6)	0.6609 (3)	**0.6946** (2)	0.8081 (1)
isolet	617	4497	0.1533 (5)	0.1561 (4)	0.1303 (6)	0.5084 (3)	**0.7124** (2)	0.9691 (1)
libras	90	216	0.1368 (5)	0.4479 (4)	0.0585 (6)	0.5175 (3)	**0.6806** (2)	1.0000 (1)
parkinsons	22	147	0.0701 (6)	0.0964 (4)	0.0714 (5)	0.1556 (3)	**0.1976** (1)	0.1778 (2)
statlog-satimage	36	3594	0.2108 (6)	0.5347 (5)	0.5804 (4)	0.9167 (3)	**0.9480** (2)	0.9942 (1)
gisette	4971	3500	0.1231 (4)	0.0814 (6)	0.0977 (5)	0.5593 (3)	**0.8769** (2)	0.9997 (1)
waveform-1	21	3304	0.1322 (4)	0.1481 (3)	0.0841 (6)	0.1234 (5)	**0.1556** (2)	0.4877 (1)
madelon	500	1300	0.7562 (5)	0.1167 (6)	**0.9973** (2)	0.9233 (4)	0.9925 (3)	1.0000 (1)
synthetic-control	60	400	0.3207 (6)	0.7889 (4)	0.6870 (5)	0.8103 (3)	**0.8539** (2)	0.9889 (1)
waveform-2	21	3304	0.1271 (5)	**0.2828** (2)	0.1014 (6)	0.1778 (3)	0.1772 (4)	0.2944 (1)
statlog-vehicle	18	629	0.1137 (6)	0.3146 (5)	0.6326 (4)	0.6561 (3)	**0.7336** (2)	1.0000 (1)
statlog-segment	18	1320	0.1250 (6)	0.2323 (4)	0.1868 (5)	0.3304 (3)	**0.3875** (2)	0.7399 (1)
(Average Rank)			(5.11)	(4.23)	(4.88)	(3.29)	(2.35)	(1.11)

Generating Privileged Representation. For each dataset, we introduce PI by perturbing normal observations. We designate a small random fraction ($= 0.1$) of n normal data points as anomalies. Then, we randomly select a subset of p attributes and add zero-mean Gaussian noise to the designated anomalies along the selected subset of attributes with matching variances of the selected features. The p selected features represent PI since anomalies stand-out in this subspace due to added noise, while the rest of the d attributes represent X space. Using normal observations allows us to control for features that could be used as PI. Thus we discard the actual anomalies from these datasets where PI is unknown.

We construct 4 versions per dataset with varying fraction γ of perturbed features (PI) retained in X^* space. In particular, each set has γp features in X^*, and $(1 - \gamma)p + d$ features in X for $\gamma \in \{0.9, 0.7, 0.5, 0.3\}$.

Results. We report the results on perturbed datasets with $\gamma = 0.7$[5] as fraction of features retained in space X^*. Table 3 reports mean Average Precision (area under the precision-recall curve) against 17 datasets for different methods. The results are averaged across 5 independent runs on stratified train-test splits.

---

[5] The results with $\gamma \in \{0.9, 0.5, 0.3\}$ are similar and reported in the supplementary material available at http://www.andrew.cmu.edu/user/shubhras/SPI.

Our SPI outperforms competition in detection performance in most of the datasets. To compare the methods statistically, we use the non-parametric Friedman test [5] based on the average ranks. Table 3 reports the ranks (in parentheses) on each dataset as well as the average ranks. With p-value $= 2.16 \times 10^{-11}$, we reject the null hypothesis that all the methods are equivalent using Friedman test. We proceed with Nemenyi post-hoc test to compare the algorithms pairwise and to find out the ones that differ significantly. The test identifies performance of two algorithms to be significantly dif-

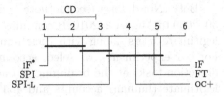

Fig. 2. Average rank of algorithms (w.r.t. MAP) and comparison by the Nemenyi test. Groups of methods not significantly different (at p-val $= 0.05$) are connected with horizontal lines. CD depicts critical distance required to reject equivalence. Note that SPI is significantly better than the baselines.

ferent if their average ranks differ by at least the "critical difference" (CD). In our case, comparing 6 methods on 17 datasets at significance level $\alpha = 0.05$, CD $= 1.82$.

Results of the post-hoc test are summarized through a graphical representation in Fig. 2. We find that SPI is significantly better than all the baselines. We also notice that SPI has no significant difference from IF^* which uses PI at test time, demonstrating its effectiveness in augmenting PI. While all the baselines are comparable to SPI-LITE, its average rank is better (also see last row in Table 3), followed by other PI-incorporated detectors, and lastly IF with no PI.

Average Precision (AP) is a widely-accepted metric to quantify overall performance of ranking methods like anomaly detectors. We also report average rank of the algorithms against other popular metrics including AUC of ROC curve, NDCG@10 and PRECISION@10 in Fig. 3. Notice that the results are consistent across measures, SPI and SPI-LITE performing among the best.

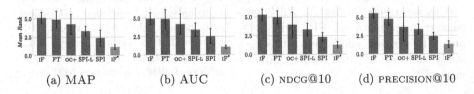

(a) MAP   (b) AUC   (c) NDCG@10   (d) PRECISION@10

Fig. 3. SPI and SPI-LITE outperform competition w.r.t. different evaluation metrics. Average rank (bars) across benchmark datasets. IF^* shown for reference.

4.2 Real-World Use Cases

Data Description. LingSpam dataset (see footnote 3) consists of 2412 non-spam and 481 spam email messages from a linguistics mailing-list. We evaluate two use cases (1) domain-expert knowledge as PI and (2) compute-expensive PI on LingSpam.

BotOrNot dataset (see footnote 4) is collected from Twitter during December 30, 2009 to August 2, 2010. It contains 22,223 content polluters (bots) and 19,276 legitimate users, along with their number of followings over time and tweets. For our experiments, we select accounts with age less than 10 days (for early detection task) at the beginning of dataset collection. The subset contains 901 legitimate (human) accounts and 4535 bots. We create 10 sets containing all the legitimate and a random 10% sample of the bots. We evaluate use case (3) "historical future" as PI and report the results averaged over these sets.

Case 1: Domain-Expert Knowledge as PI for Email Spam Detection.
X^* **space:** The Linguistic Inquiry and Word Count (LIWC) software[6] is a widely used text analysis tool in social sciences. It uses a manually-curated keyword dictionary to categorize text into 90 psycholinguistic classes. Construction of LIWC dictionary relies exclusively on human experts which is a slow and evolving process. For the LingSpam dataset, we use the percentage of word counts in each class (assigned by LIWC software) as the privileged features.

X **space:** The bag-of-word model is widely used as feature representation in text analysis. As such, we use the term frequencies for our email corpus as the primary features.

Figure 4 shows the detection performance[7] of algorithms in ROC curves (averaged over 15 independent runs on stratified train-test splits). We find that IF, which does not leverage PI but operates solely in X space, is significantly worse than most PI-incorporated methods. OC-SVM+ is nearly as poor as IF despite using PI—this is potentially due to OC-SVM being a poor anomaly detector in the first place, as shown in [6] and as we argued in Sect. 1. All knowledge transfer methods, SPI, SPI-LITE, and FT, perform similarly on this case study, and are as good as IF*, directly using X^*.

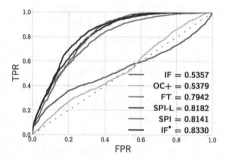

Fig. 4. Detection performance on Case 1: using expert knowledge as PI. Legend depicts the AUC values. PI-incorporated detectors (except OC-SVM+) outperform non-PI IF and achieve similar performance to IF*.

Case 2: Compute-Expensive Features as PI for Email Spam Detection.
X^* **space:** Beyond bag-of-words, one can use syntactic features to capture *stylistic* differences between spam and non-spam emails. To this end, we extract features from the parse trees of emails using the StanfordParser[8]. The parser

---

[6] https://liwc.wpengine.com/.
[7] See supplementary material quantifying the performance of methods against other ranking metrics.
[8] https://nlp.stanford.edu/software/lex-parser.shtml.

provides the taxonomy (tree) of Part-of-Speech (PoS) tags for each sentence, based on which we construct (*i*) PoS bi-gram frequencies, and (*ii*) quantitative features (width, height, and horizontal/vertical imbalance) of the parse tree.

On average, StanfordParser requires 66 s[9] to parse and extract features from a single raw email in LingSpam. Since the features are computationally demanding, we incorporate those as PI to facilitate faster detection at test time.

X space: We use the term frequencies as the primary features as in Case 1.

Figure 5(a) shows the detection performance (see footnote 7) of methods in terms of AUC under ROC. We find that IF^* using (privileged) syntactic features achieves lower AUC of ~0.65 as compared to ~0.83 using (privileged) LIWC features in Case 1. Accordingly, all methods perform relatively lower, suggesting that the syntactic features are less informative of spam than psycholinguistic ones. Nonetheless, we observe that the performance ordering remains consistent, where IF ranks at the bottom and SPI and SPI-LITE get closest to IF^*.

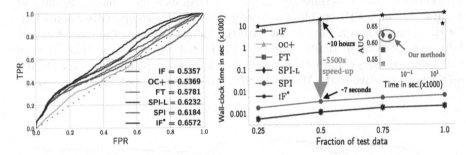

Fig. 5. Comparison of detectors on Case 2: using computationally-expensive features as PI. (a) detection performance, legend depicts AUC values; and (b) wall-clock time required (in seconds, note the logarithmic scale) vs. test data size [inset plot on top right: AUC vs. time (methods depicted with symbols)].

Figure 5(b) shows the comparison of wall-clock time required by each detector to compute the anomaly scores at test time for varying fraction of test data. On average, SPI achieves 5500× speed-up over IF^* that employs the parser at test time. This is a considerable improvement of response time for comparable accuracy. Also notice the inset plot showing the AUC vs. total test time, where our proposed SPI and SPI-LITE are closest to the ideal point at the top left.

Case 3: "Historical Future" as PI for Twitter Bot Detection.
We use temporal data from the activity and network evolution of an account to capture behavioral differences between a human and a bot. We construct temporal features including volume, rate-of-change, and lag-autocorrelations of the number of followings. We also extract temporal features from text such as count of tweets, links, hash-tags and mentions.

---

[9] Using a single thread on 2.2 GHz Intel Core i7 CPU with 8 cores and 16 GB RAM.

X^* **space:** All the temporal features within f_t days in the future (relative to detection at time t) constitute privileged features. Such future values would not be available at any test time point but can be found in historical data.

X **space:** Temporal features within h_t days in the past as well as static user features (from screen name and profile description) constitute primary features.

Figure 6(a) reports the detection performance of algorithms in terms of ROC curves (averaged over 10 sets) at time $t = 2$ days after the data collection started; for $h_t = 2$, $f_t = 7$.[10] The findings are similar to other cases: SPI and SPI-LITE outperform the competing methods in terms of AUC and OC-SVM+ performs similar to non-PI IF; demonstrating that knowledge transfer based methods are more suitable for real-world use cases.

Figure 6(b) compares the detection performance of SPI and IF over time; for detection at $t = \{0, 1, 2, 3, 4\}$. As time passes, historical data grows as $h_t = \{0, 1, 2, 3, 4\}$ where "historical future" data is fixed at $f_t = 7$ for PI-incorporated methods. Notice that at time $t = 1$, SPI achieves similar detection performance to IF's performance at $t = 2$ that uses more historical data of 2 days. As such, SPI enables 24 h early detection as compared to non-PI IF for the same accuracy. Notice that with the increase in historical data, the performances of both methods improve, as expected. At the same time, that of SPI improves faster, ultimately reaching a higher saturation level, specifically $\sim$7% higher relative to IF. Moreover, SPI gets close to IF*'s level in just around 3 days.

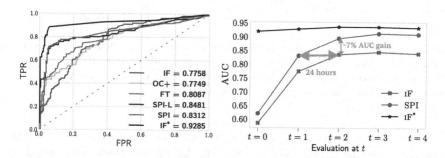

Fig. 6. Comparison of detectors on Case 3: using "historical future" data as PI. (a) SPI outperforms competition in performance and is closest to IF*'s; (b) SPI achieves same detection performance as IF 24 h earlier, and gets close to IF* in 3 days of history.

5   Related Work

We review the history of LUPI, follow up and related work on learning with side/hidden information, as well as LUPI-based anomaly detection.

---

[10] Same conclusions can be drawn for $f_t \in \{1, 3, 5, 7\}$ (see supplementary material).

Learning Under Privileged Information: The LUPI paradigm is introduced by Vapnik and Vashist [19] as the SVM+ method, where, Teacher provides Student not only with (training) examples but also explanations, comparisons, metaphors, etc. which accelerate the learning process. Roughly speaking, PI adjusts Student's concept of similarity between training examples and reduces the amount of data required for learning. Lapin et al. [10] showed that learning with PI is a particular instance of importance weighting in SVMs. Another such mechanism was introduced more recently by Vapnik and Izmailov [17], where knowledge is transferred from the space of PI to the space where the decision function is built. The general idea is to specify a small number of fundamental concepts of knowledge in the privileged space and then try to transfer them; i.e., construct additional features in decision space via e.g., regression techniques in decision space. Importantly, the knowledge transfer mechanism is not restricted to SVMs, but generalizes, e.g. to neural networks [18].

LUPI has been applied to a number of different settings including clustering [7,12], metric learning [8], learning to rank [15], malware and bot detection [2,3], risk modeling [14], as well as recognizing objects [16], actions and events [13].

Learning with Side/Hidden Information: Several other work, particularly in computer vision [4,20], propose methods to learn with data that is unavailable at test time referred as side and hidden information (e.g., text descriptions or tags for general images, facial expression annotations for face images, etc.). In addition, Jonschkowski et al. [9] describe various patterns of learning with side information. All of these work focus on supervised learning problems.

LUPI-Based Anomaly Detection: With the exception of One-Class SVM (OC-SVM+) [2], which is a direct extension of Vapnik's (supervised) SVM+, the LUPI framework has been utilized only for supervised learning problems. While anomaly detection has been studied extensively [1], we are unaware of any work other than [2] leveraging privileged information for unsupervised anomaly detection. Motivated by this along with the premises of the LUPI paradigm, we are the first to design a new technique that ties LUPI with unsupervised tree-based ensemble methods, which are considered state-of-the-art for anomaly detection.

6   Conclusion

We introduced SPI, a new ensemble approach that leverages privileged information (data available only for training examples) for unsupervised anomaly detection. Our work builds on the LUPI paradigm, and to the best of our knowledge, is the first attempt to incorporating PI to improve the state-of-the-art ensemble detectors. We validated the effectiveness of our method on both benchmark datasets as well as three real-world case studies. We showed that SPI and SPI-LITE consistently outperform the baselines. Our case studies leveraged a variety of privileged information—"historical future", complex features, expert knowledge—and verified that SPI can unlock multiple benefits for anomaly detection in terms of detection latency, speed, as well as accuracy.

Acknowledgements. This research is sponsored by NSF CAREER 1452425 and IIS 1408287. Any conclusions expressed in this material are of the authors and do not necessarily reflect the views, expressed or implied, of the funding parties.

References

1. Aggarwal, C.C.: Outlier analysis. Data Mining, pp. 237–263. Springer, Cham (2015). https://doi.org/10.1007/978-3-319-14142-8_8
2. Burnaev, E., Smolyakov, D.: One-class SVM with privileged information and its application to malware detection. In: ICDM Workshops, pp. 273–280 (2016)
3. Celik, Z.B., McDaniel, P., Izmailov, R., Papernot, N., Swami, A.: Extending detection with forensic information. arXiv:1603.09638 (2016)
4. Chen, J., Liu, X., Lyu, S.: Boosting with side information. In: Lee, K.M., Matsushita, Y., Rehg, J.M., Hu, Z. (eds.) ACCV 2012. LNCS, vol. 7724, pp. 563–577. Springer, Heidelberg (2013). https://doi.org/10.1007/978-3-642-37331-2_43
5. Demšar, J.: Statistical comparisons of classifiers over multiple data sets. J. Mach. Learn. Res. **7**, 1–30 (2006)
6. Emmott, A., Das, S., Dietterich, T., Fern, A., Wong, W.-K.: Systematic construction of anomaly detection benchmarks from real data. In: KDD ODD (2013)
7. Feyereisl, J., Aickelin, U.: Privileged information for data clustering. Inf. Sci. **194**, 4–23 (2012)
8. Fouad, S., Tino, P., Raychaudhury, S., Schneider, P.: Incorporating privileged information through metric learning. IEEE Neural Netw. Learn. Syst. **24**(7), 1086–1098 (2013)
9. Jonschkowski, R., Höfer, S., Brock, O.: Patterns for learning with side information. arXiv:1511.06429 (2015)
10. Lapin, M., Hein, M., Schiele, B.: Learning using privileged information: SVM+ and weighted SVM. Neural Netw. **53**, 95–108 (2014)
11. Liu, F.T., Ting, K.M., Zhou, Z.-H.: Isolation forest. In: ICDM (2008)
12. Marcacini, R.M., Domingues, M.A., Hruschka, E.R., Rezende, S.O.: Privileged information for hierarchical document clustering: a metric learning approach. In: ICPR, pp. 3636–3641 (2014)
13. Niu, L., Li, W., Xu, D.: Exploiting privileged information from web data for action and event recognition. Int. J. Comput. Vis. **118**(2), 130–150 (2016)
14. Ribeiro, B., Silva, C., Chen, N., Vieira, A., das Neves, J.C.: Enhanced default risk models with SVM+. Expert Syst. Appl. **39**(11), 10140–10152 (2012)
15. Sharmanska, V., Quadrianto, N., Lampert, C.H.: Learning to rank using privileged information. In: ICCV, pp. 825–832 (2013)
16. Sharmanska, V., Quadrianto, N., Lampert, C.H.: Learning to transfer privileged information. arXiv:1410.0389 (2014)
17. Vapnik, V., Izmailov, R.: Learning with intelligent teacher: similarity control and knowledge transfer. In: Gammerman, A., Vovk, V., Papadopoulos, H. (eds.) SLDS 2015. LNCS (LNAI), vol. 9047, pp. 3–32. Springer, Cham (2015). https://doi.org/10.1007/978-3-319-17091-6_1
18. Vapnik, V., Izmailov, R.: Knowledge transfer in SVM and neural networks. Ann. Math. Artif. Intell. **81**(1–2), 3–19 (2017)
19. Vapnik, V., Vashist, A.: A new learning paradigm: learning using privileged information. Neural Netw. **22**(5–6), 544–557 (2009)
20. Wang, Z., Ji, Q.: Classifier learning with hidden information. In: CVPR (2015)

L1-Depth Revisited: A Robust Angle-Based Outlier Factor in High-Dimensional Space

Ninh Pham[(✉)]

Department of Computer Science,
University of Copenhagen, Copenhagen, Denmark
pham@di.ku.dk

Abstract. Angle-based outlier detection (ABOD) has been recently emerged as an effective method to detect outliers in high dimensions. Instead of examining neighborhoods as proximity-based concepts, ABOD assesses the broadness of angle spectrum of a point as an outlier factor. Despite being a *parameter-free* and robust measure in high-dimensional space, the exact solution of ABOD suffers from the cubic cost $O(n^3)$ regarding the data size n, hence cannot be used on large-scale data sets.

In this work we present a *conceptual* relationship between the ABOD intuition and the L1-depth concept in statistics, one of the earliest methods used for detecting outliers. Deriving from this relationship, we propose to use L1-depth as a variant of angle-based outlier factors, since it only requires a quadratic computational time as proximity-based outlier factors. Empirically, L1-depth is competitive (often superior) to proximity-based and other proposed angle-based outlier factors on detecting high-dimensional outliers regarding both efficiency and accuracy.

In order to avoid the quadratic computational time, we introduce a simple but efficient sampling method named *SamDepth* for estimating L1-depth measure. We also present theoretical analysis to guarantee the reliability of SamDepth. The empirical experiments on many real-world high-dimensional data sets demonstrate that SamDepth with $\sqrt{n}$ samples often achieves very competitive accuracy and runs several orders of magnitude faster than other proximity-based and ABOD competitors. Data related to this paper are available at: https://www.dropbox.com/s/nk7nqmwmdsatizs/Datasets.zip. Code related to this paper is available at: https://github.com/NinhPham/Outlier.

1 Introduction

Outlier detection is the process of detecting anomalous patterns that do not conform to an expected behavior. According to Hawkins [8], an outlier would be

Research supported by the Innovation Fund Denmark through the DABAI project.

Electronic supplementary material The online version of this chapter (https://doi.org/10.1007/978-3-030-10925-7_7) contains supplementary material, which is available to authorized users.

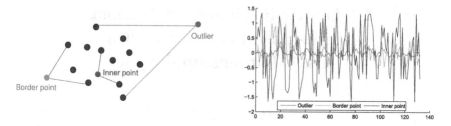

Fig. 1. Variance of angles of different types of points. Outliers have small variances whereas inliers have large variances.

"an observation which deviates so much from other observations as to arouse suspicions that it was generated by a different mechanism". Detecting such outlier patterns is a fundamental and well studied data mining task due to its several application domains, such as fraud detection in finance, author verification for forensic investigation, and detecting anomalous patterns for medical diagnosis.

One of the earliest methods to detect outliers is based the concept of depth in statistics [11,18] due to the natural correlation between depth-based measure and outlierness measure. The depth-based methods organize the data points in many layers, with the expectation that the "deepest" layers with large depth values contain points close to the center of the point cloud and outliers are likely to appear in the shallow layers with small depth values. However, since most outlier detection applications often arise in high-dimensional domains and most of depth-based methods do not scale up with data dimensionality [9], depth-based approaches suffer from a computational bottleneck for searching for high-dimensional outliers.

Since then, several outlierness measures based on the notion of proximity have been proposed to detect high-dimensional outliers. Due to the phenomenon "curse of dimensionality", proximity-based approaches in the literature which are implicitly or explicitly based on the concept of proximity in Euclidean distance metric between points in full-dimensional space do not work efficiently and effectively. Traditional solutions to detect distance-based global outliers [4,13,19] and density-based local outliers [5,16] often suffer from the high computational cost due to their core operation, near(est) neighbor search in high dimensions. Moreover, the high-dimensional data is often very sparse and therefore the measures like distances or nearest neighbors may not be qualitatively meaningful [1,3].

In order to alleviate the effects of the "curse of dimensionality", Kriegel et al. [14] proposed a novel outlier ranking approach based on the broadness of angle spectrum of data. The approach named *Angle-based Outlier Detection (ABOD)* evaluates the degree of outlierness on the *variance of the angles (VOA)* between a point and all other pairs of points in the data set. The intuition of ABOD, as shown in Fig. 1, is that the smaller the angle variance of the point has, the more likely it is an outlier. Since angles are more stable than distances, the ABOD approach does not substantially deteriorate in high-dimensional data. It is worth noting that the proposed outlierness measure in [14], called *ABOF*, does not deal directly with the intuition of variance of angles. Indeed, ABOF assesses the weighted variance of weighted cosine of angles where the both weights are the corresponding distances between the assessed point and other pairs of points. The variant notion ABOF with weight of distances is more robust than the original intuition VOA in low-dimensional space since it allows distance affects the outlierness measure.

Despite many advantages of alleviating the effects of the "curse of dimensionality" and being a *parameter-free* measure, there are two intrinsic drawbacks with the ABOD approaches.

- There is no theoretical foundation connecting to the ABOD observation so it is difficult to understand and explain the outlierness behaviors detected by ABOD.
- The cubic time complexity taken to compute angle-based measures is very significant and problematic for many applications with large-scale data sets.

To avoid the cubic time complexity, Kriegel et al. [14] also proposed a heuristic approximation variant of ABOF, called *approxABOF*, for efficient computations. Instead of computing ABOF over all other pairs in the point set, approx-ABOF computes ABOF value over all pairs in the k-nearest neighbor (kNN) set. Hence, approxABOF requires a quadratic time complexity used in sequential search for kNN. Moreover, there is no analysis on the approximation error between approxABOF and ABOF and hence the reliability of detecting outliers using approxABOF is not guaranteed.

Recently, Pham and Pagh [17] investigated the variance of angles (VOA) as an ABOD outlierness measure and proposed an efficient algorithm for the ABOD approach. They proved that VOA is well preserved under random projections and introduced *FastVOA*, a near-linear time algorithm for estimating VOA for the point set. Despite many advantages of the fast running time and the quality of approximation guarantee, FastVOA might introduce large approximation errors. The large approximation errors result in detection performance degradation when the VOA gap between outliers and inliers is rather small. Furthermore, large approximation error of estimation can be problematic when combining VOA with other outlier factors to build outlier detection ensembles [2,24].

In this work, we investigate both mentioned drawbacks of the ABOD method. We examine the first drawback via a well-established concept of *data depth* in statistics [18,22]. In particular, we consider the *L1-depth* notion [21,23], which intuitively measures how much additional probability mass needed for moving a point in a set to the multivariate median of its point set. We study the notion of L1-depth in details and provide a strong conceptual relationship between L1-depth and the ABOD observation. Deriving from this relationship, we propose to use L1-depth as a variant of angle-based outlier factor since it requires a *quadratic* time computation as proximity-based outlier factors. Empirically, ABOD using L1-depth is superior to using VOA and ABOF, i.e. the computational cost is much smaller and the outlier detection accuracy is much higher.

To overcome the drawback of quadratic computational time, we introduce a simple but efficient sampling method named *SamDepth* for estimating

L1-depth measure. The empirical experiments on many real-world high-dimensional data sets demonstrate that SamDepth often runs much faster, provide smaller approximation errors and therefore more accurate outlier rankings than other ABOD competitors. Especially, SamDepth with $\sqrt{n}$ samples where n is the data size achieves very competitive accuracy and runs several orders of magnitude faster than other proximity-based and ABOD methods on several large-scale data sets.

2   Notation and Background

Given a point set $\mathbb{S} \subseteq \mathcal{R}^d$ of size n and a point $\boldsymbol{p} \in \mathbb{S}$, we denote by $P = \mathbb{S} \setminus \{\boldsymbol{p}\}$ since most of outlier factors of $\boldsymbol{p}$ are evaluated on the set P. We also denote by $(\boldsymbol{a}, \boldsymbol{b})$ a pair of any two *different* points in P. As we will elaborate later, ABOD algorithms compute outlier factors of $\boldsymbol{p}$ by values dependent on $\boldsymbol{p}$ and *each* pair $(\boldsymbol{a}, \boldsymbol{b})$. Hence, we will use the notation $\sum_{a,b}$ for short to represent for the summation on $\boldsymbol{a}, \boldsymbol{b} \in P$.

For a given pair $(\boldsymbol{a}, \boldsymbol{b})$, we denote by Θ_{apb} the angle between the difference vectors $\boldsymbol{p} - \boldsymbol{a}$ and $\boldsymbol{p} - \boldsymbol{b}$. Since we will show a conceptual relationship between L1-depth measure and the variance of angles (VOA), an ABOD outlier factor, we will describe VOA and discuss about the time complexity of a naïve algorithm to compute exactly this measure.

Definition 1. *The variance of angle of a point $\boldsymbol{p}$ is computed via the first moment MOA1 and the second moment MOA2 of the angle Θ_{apb} between $\boldsymbol{p}$ and each pair $(\boldsymbol{a}, \boldsymbol{b})$. That is*

$$VOA(\boldsymbol{p}) = \boldsymbol{Var}\,[\Theta_{apb}] = MOA2(\boldsymbol{p}) - (MOA1(\boldsymbol{p}))^2$$

where $MOA2(\boldsymbol{p})$ and $MOA1(\boldsymbol{p})$ are defined as follows:

$$MOA2(\boldsymbol{p}) = \frac{\sum_{a,b} \Theta_{apb}^2}{(n-1)(n-2)}; MOA1(\boldsymbol{p}) = \frac{\sum_{a,b} \Theta_{apb}}{(n-1)(n-2)}.$$

Note that the VOA definition is identical to the definition in [17] since we take into account both Θ_{apb} and Θ_{bpa}. A naïve algorithm computing VOA for n points takes $O(n^3)$ time since computing $VOA(\boldsymbol{p})$ for each $\boldsymbol{p}$ takes $O(n^2)$ time. Note that VOA values are often very small and this challenges approximation methods to have good approximation errors in order to preserve the ABOD ranking.

3   L1-Depth and Its Conceptual Relationship with the ABOD Intuition

This section will study the L1-depth concept in statistics and provide a conceptual relationship between L1-depth concept and variance of angles of the ABOD intuition. We also discuss some benefits derived from this relationship.

3.1 L1-Depth as an ABOD Measure

Vardi and Zhang [23] studied the multivariate L1-median point (i.e. the point that minimizes the weighted sum of the Euclidean distances to all points in a high-dimensional cloud). Associating to the multivariate L1-median concept, they also proposed a simple close-form formula for the data depth called L1-depth function. The L1-depth function shares the same spirit with other proposed data depth [18], that is deeper points with larger depth are relatively closer to the center of the cloud (i.e. the L1-median).

Of the various depth notions, L1-depth is computationally efficient, i.e. $O(n)$ for each point. It has been used in clustering and classification tasks for microarray gene expression data [12] and novelty detection in taxonomic applications [7]. The definition of L1-depth (L1D) is as follows:

Definition 2 ([7, Eq. 3], [23, Eq. 4.3]). *L1-depth*

$$L1D(\boldsymbol{p}) = 1 - \frac{1}{n-1} \left\| \sum_{a \in P} \frac{\boldsymbol{p} - \boldsymbol{a}}{\|\boldsymbol{p} - \boldsymbol{a}\|} \right\|$$

It is clear that L1-depth shares the same spirit as VOA and ABOF on dealing with the angle spectrum and being a parameter-free measure but has more efficient computation, i.e. $O(n)$ time for each point. In particular, the intuition of L1-depth concept is very similar to the ABOD idea since it assesses the broadness of *directions* of distance vectors, and the smaller $L1D(\boldsymbol{p})$ is, the more likely $\boldsymbol{p}$ is an outlier. Considering again Fig. 1, for inliers within the cluster, their L1D values will be close to 1. However, the L1D value of the outlier tends to be close to 0 since most the other points locate in some particular direction.

The following lemma shows that L1-depth can be derived from the sum of cosine of angles between a point $\boldsymbol{p}$ and all other pairs of points. This lemma also sheds the light on the conceptual relationship between L1-depth and variance of angles in the ABOD methods.

Lemma 3.

$$(1 - L1D(\boldsymbol{p}))^2 = \frac{1}{n-1} + \frac{1}{(n-1)^2} \sum_{a,b} \cos \Theta_{apb}$$

Proof. Using the extension

$$\left\| \sum_i \boldsymbol{x}_i \right\|^2 = \sum_i \|\boldsymbol{x}_i\|^2 + \sum_{i \neq j} \langle \boldsymbol{x}_i, \boldsymbol{x}_j \rangle,$$

we have:

$$(1 - L1D(\boldsymbol{p}))^2 = \frac{1}{(n-1)^2} \left\| \sum_{a \in P} \frac{\boldsymbol{p} - \boldsymbol{a}}{\|\boldsymbol{p} - \boldsymbol{a}\|} \right\|^2$$

$$= \frac{1}{(n-1)^2} \left(\sum_{a \in P} \left\| \frac{\boldsymbol{p} - \boldsymbol{a}}{\|\boldsymbol{p} - \boldsymbol{a}\|} \right\|^2 + \sum_{a,b} \left\langle \frac{\boldsymbol{p} - \boldsymbol{a}}{\|\boldsymbol{p} - \boldsymbol{a}\|}, \frac{\boldsymbol{p} - \boldsymbol{b}}{\|\boldsymbol{p} - \boldsymbol{b}\|} \right\rangle \right)$$

$$= \frac{1}{(n-1)^2} \left((n-1) + \sum_{a,b} \cos \Theta_{apb} \right) = \frac{1}{n-1} + \frac{1}{(n-1)^2} \sum_{a,b} \cos \Theta_{apb}$$

□

Intuitively, since $\cos(x)$ is a strictly monotonically decreasing function in the range $[0, \pi]$, L1D will be correlated to the first moment of angles MOA1. Hence, the outlier ranking produced by L1D is highly positively correlated to MOA1's one. Mathematically, we can exploit the Taylor series approximation $\cos(x) \approx 1 - x^2/2$ on Lemma 3 to show the relationship between L1D and the second moment MOA2 as follows.

$$(1 - L1D(\boldsymbol{p}))^2 = \frac{1}{(n-1)^2} \left((n-1) + \sum_{a,b} \cos \Theta_{apb} \right)$$

$$\approx \frac{1}{(n-1)^2} \left((n-1) + \sum_{a,b} \left(1 - \frac{\Theta_{apb}^2}{2} \right) \right)$$

$$= \frac{1}{(n-1)^2} \left((n-1)^2 - \sum_{a,b} \frac{\Theta_{apb}^2}{2} \right)$$

$$= 1 - \frac{1}{(n-1)^2} \sum_{a,b} \frac{\Theta_{apb}^2}{2}$$

$$= 1 - \frac{(n-1)(n-2)MOA2(p)}{2(n-1)^2}$$

$$= 1 - \frac{n-2}{2(n-1)} MOA2(p). \tag{1}$$

When the Taylor series approximation $\cos(x) \approx 1 - x^2/2$ provides a small error, the outlier factor L1D is highly proportional to MOA2. Therefore, we can use both VOA, the *central* second moment and MOA2, the second moment of angles as angle-based outlier factors, and the smaller $MOA2(\boldsymbol{p})$ or $VOA(\boldsymbol{p})$ is, the more likely $\boldsymbol{p}$ is an outlier.

3.2 An Empirical Study and Benefits from the Conceptual Relationship

In order to confirm our theoretical finding, we compute the exact VOA, MOA1, MOA2, and L1D values on a synthetic data set generated by a Gaussian mixture

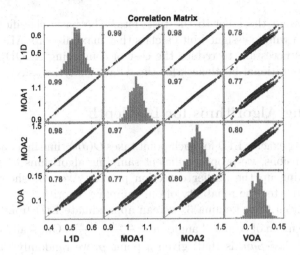

Fig. 2. Kendall's rank correlation between L1D, MOA1, MOA2, VOA measures on the synthetic data set.

as provided in the original ABOD paper [14]. The 100-dimensional synthetic data set contains 1000 inliers generated by independent Gaussian distributions and 10 outliers generated by a uniform distribution. Figure 2 shows very high Kendall's rank correlation coefficients between L1D, MOA1, MOA2 and VOA on the data set. We note that the Spearman's rank correlation coefficients among them are even higher and not reported here.

It is very clear that the outlier ranking based on L1D is almost identical to that of MOA1, MOA2 and very highly correlated to that of VOA. This means that instead of using VOA, the *central* second moment of angle distribution, we can use L1D, an approximation of the second moment with less computational resource for outlier ranking in high-dimensional data sets.

Next we discuss about the two benefits from this observed relationship, including an algorithmic benefit for approximating VOA and an application benefit on using ABOD measures for other data analytics tasks.

Algorithmic Benefit: Note that the time complexity of computing $L1D(\mathbf{p})$ is $O(n)$. Hence we can use $L1D(\mathbf{p})$ to derive an approximation of the second moment $MOA2(\mathbf{p})$ (see Eq. 1), which can replace the main computational resource in FastVOA [17]. Also note that FastVOA approximates the first moment $MOA1(\mathbf{p})$ in $O(n \log n)$ time. Combining these two approximations, we can estimate VOA of all n points with high accuracy in quadratic time without utilizing the AMS Sketches. In small data sets, this combination runs faster and provides better outlier detection performance than FastVOA [17].

Application Benefit: Since L1-depth is a variant of ABOD measures and since depth notions have been used in clustering and classification [10,12], we can use other ABOD measures, including VOA and ABOF on these settings. For example in classification tasks, instead of using kNN relationship, we can use ABOD measures to assign the label to the test data. The basic classification rule is that a test data will be assigned into a class that maximizes its ABOD measure. Therefore, it is necessary to reduce the cost of computing ABOD measures to avoid the computational bottleneck for these data analytic tasks.

4   Sampling Algorithms for L1-Depth

Since computing exactly L1D for each point takes $O(dn)$ time in a data set of size n with d dimensions, we propose efficient sampling algorithms to approximate L1D for speeding up the outlier detection process. We also show theoretical analysis to guarantee the reliability of our sampling algorithms.

As can be seen from Lemma 3, we can approximate L1D from an accurate estimate of the mean of cosine of angles $\mu = \frac{\sum_{a,b} \cos \Theta_{apb}}{(n-1)(n-2)}$. Our standard sampling method called *BasicSam* is that, given a point p, we randomly sample a pair (a,b) in P and define a random variable

$$X = \begin{cases} \cos \Theta_{apb} & \text{if } a \text{ and } b \text{ are chosen;} \\ 0 & \text{otherwise.} \end{cases}$$

Then, we have $\mathbf{E}[X] = \mu$. Using Hoeffding's inequality with $t = O(\frac{1}{\epsilon^2} \log \frac{1}{\delta})$ independent random sample *pairs*, we can guarantee an absolute approximation error at most ϵ with probability at least $1 - \delta$. We note that for t random pairs BasicSam takes $O(dt)$ time since it needs to compute $2t$ difference vectors $p - a$ and $p - b$.

As can be seen from Definition 2, computing directly $L1D(p)$ takes $O(dn)$ time. We now exploit this property to avoid sampling random pairs and propose *SamDepth*, a more efficient sampling method for L1D, as shown in Algorithm 1.

Algorithm 1. SamDepth (n)

Input: A data set $\mathbb{S}$ of size n and a point $p \in \mathbb{S}$
Output: An estimate of $L1D(p)$
1 Sample without replacement a subset $S \subset \mathbb{S} \backslash \{p\}$ of $t = \sqrt{n}$ points ;
2 Compute the norm $m = \left\| \sum_{a \in S} \frac{p-a}{\|p-a\|} \right\|^2$;
3 Output $1 - \sqrt{\frac{1}{n-1} + \frac{n-2}{n-1} \left(\frac{m}{t(t-1)} - \frac{1}{t-1} \right)}$ as an estimate of $L1D(p)$;

Instead of sampling a random pair, we sample without replacement a subset $S \subset P$ of t points, and define a random variable $Z = \sum_{a,b} Z_{ab}$ where

$$Z_{ab} = \begin{cases} \cos \Theta_{apb} / t(t-1) & \text{if } a \text{ and } b \text{ are in } S; \\ 0 & \text{otherwise.} \end{cases}$$

Since $\mathbf{Pr}\,[a,b \in S] = \mathbf{Pr}\,[a \in S | b \in S]\,\mathbf{Pr}\,[b \in S] = \frac{t(t-1)}{(n-1)(n-2)}$, we have $\mathbf{E}\,[Z] = \mu$. Note that we can evaluate Z in $O(dt)$ time due to Lemma 3 as follows

$$Z = \frac{\sum_{a,b \in S} \cos \Theta_{apb}}{t(t-1)} = \frac{1}{t(t-1)} \left\| \sum_{a \in S} \frac{p-a}{\|p-a\|} \right\|^2 - \frac{1}{t-1}.$$

Hence, we can estimate

$$L1D(p) \approx 1 - \sqrt{\frac{1}{n-1} + \frac{n-2}{n-1} \left(\frac{1}{t(t-1)} \left\| \sum_{a \in S} \frac{p-a}{\|p-a\|} \right\|^2 - \frac{1}{t-1} \right)}.$$

Theoretical Analysis: For notational simplicity, let σ^2 be the variance of cosine of angles. Hence we have $\frac{\sum_{a,b} \cos^2 \Theta_{apb}}{(n-1)(n-2)} = \sigma^2 + \mu^2$. For simplicity, we also assume that $\cos \Theta_{apb} \geq 0$ for *any* $a,b,p \in \mathbb{S}$. The following theorem shows the upper bound of variance of estimator provided by SamDepth.

Theorem 4.

$$\mathbf{Var}\,[Z] \leq \left(\sigma^2 + \mu^2\right) \left(\frac{1}{t-1} - \frac{1}{n-2} \right)$$

Proof. Due to limited space, we just sketch the proof. In order to bound $\mathbf{Var}\,[Z] = \mathbf{E}\,[Z^2] - \mu^2$ where $Z = \sum_{a,b} Z_{ab}$, we decompose it into three terms corresponding to $\sum_{a,b} \cos^2 \Theta_{apb}$, $\sum_{a \neq a', b \neq b'} \cos \Theta_{apb} \cos \Theta_{a'pb'}$, and $\sum_{a,b \neq b'} \cos \Theta_{apb} \cos \Theta_{apb'}$. Since $\sum_{a,b \neq b'} \cos \Theta_{apb} \cos \Theta_{apb'} \leq (n-3) \sum_{a,b} \cos^2 \Theta_{apb}$ by Cauchy–Schwarz inequality and the contribution of the second term is negative, we can bound $\mathbf{Var}\,[Z]$ using only the first term, which leads to the result. $\square$

Discussion: It is worth noting that SamDepth with $t = n-1$ provides an exact $L1D(p)$ while the basic sampling method can only give an estimate. When t is large, SamDepth gives sufficiently small variance of estimator, and hence results in negligible loss on outlier detection using L1D. For the general case when any $\cos \Theta_{apb}$ might be negative, we will consider random variables $Y_{ab} = \frac{1}{t(t-1)} \frac{1+\cos \Theta_{apb}}{2}$ instead. Applying Theorem 4, we can also bound the variance of estimator provided by SamDepth. Due to limited space, we leave the detail of the proof in the supplementary material[1].

Parameter Setting and Reproducibility: In order to make SamDepth completely parameter-free and efficient, we simply set $t = \sqrt{n}$. Hence SamDepth computes an unbiased estimate of $L1D(p)$ in $O(d\sqrt{n})$ time, which is significantly faster than $O(dn)$ time required by standard proximity-based outlier detectors. For reproducibility, we have released a C++ source code of SamDepth[2]. Our

---

[1] https://www.dropbox.com/s/yzbam4heruglj4i/Supplementary.pdf.
[2] https://github.com/NinhPham/Outlier.

empirical experiments on 14 real-world high-dimensional data sets demonstrate that SamDepth with $\sqrt{n}$ samples often achieves very competitive accuracy and runs several orders of magnitude faster than other proximity-based and ABOD competitors.

5   Experiment

We implemented SamDepth and other competitors in C++ and conducted experiments on a 3.40 GHz core i7 Windows platform with 32 GB of RAM. We compared the performance of SamDepth with ABOD detectors using L1D, VOA and ABOF and other proximity-based detectors on real-world high-dimensional data sets. We used the area under the ROC curve (AUC) to evaluate the accuracy of our *unsupervised* outlier detection methods since they deliver outlier rankings. For measuring efficiency, we computed the total running time in seconds for each detector. All results are over 5 runs of the algorithms.

Table 1. Data set properties: short names, number of points n, dimensionality d, and number of outliers o.

	Mam	Shuttle	Cover	Cardio	KDD	Spam	Opt	Mnist	Musk	Arr	Speech	Isolet	Mfeat	Ads
n	11183	49097	286048	2126	60839	4207	5216	7603	3062	452	3686	945	440	1966
d	6	9	10	21	41	57	64	100	166	274	400	617	649	1555
o	260	3511	2747	471	246	1679	150	700	97	66	60	45	40	368

5.1   Experiment Setup

Due to the cubic time complexity of VOA and ABOF, we will compute these exact values in some small data sets for comparison. For large-scale data sets, we used FastVOA [17] and FastABOF [14] to approximate VOA and ABOF, respectively. Below is the list of all implemented algorithms used in our experiment.

- **L1D**: L1D (exact), BasicSam (basic sampling), SamDepth.
- **VOA**: VOA (exact), FastVOA [17].
- **ABOF**: ABOF (exact), FastABOF ($k = \lceil 0.1 \cdot n \rceil$) [14].
- **Proximity-based factors**: kNN [19], kNNW [4] with fixed parameter $k = 10$ and LOF [5] with fixed parameter $k = 40$.

For FastVOA, we used 100 random projections and fixed the size of AMS Sketches $s_1 = 3200, s_2 = 5$ in all experiments. We used $k = \lceil 0.1 \cdot n \rceil$ for FastABOF as suggested in [14] since if k is small, approxABOF is simply the local outlier factor of ABOF and cannot reflect well the original ABOF idea. For consistency, we used $k = 10$ for kNN and kNNW as used in [19] and [4]. For LOF, we used $k = 40$ as suggested in [5]. We note that finding the best parameter k for proximity-based methods is an extremely time-consuming task, especially for large-scale data sets. The brute force procedure to find the best parameter k requires $O(\kappa n \log n)$ time for evaluating κ AUC scores where κ is the largest possible value of k. With the KDDCup99 data set of size $n = 60839$, this process will need approximately 100 h to finish for $\kappa = 1000$ on our machine.

It is worth noting that in the realistic cases where we do not actually know the outlier labels, incorrect settings in parameter-laden measures may cause the algorithms to fail in finding the true anomaly patterns. Parameter-free outlier factors including L1D, VOA and ABOF would limit our ability to impose our prejudices or presumptions on the problem, and "let the data speak for themselves".

5.2 Data Sets

We conducted experiments on real-world high-dimensional data sets, including widely used data sets in literature and semantically meaningful data sets with interpretation for outliers from popular resources, as shown in Table 1.

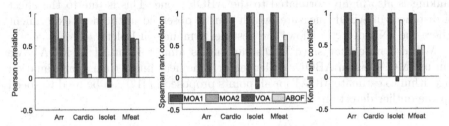

Fig. 3. Correlation coefficients between L1D and MOA1, MOA2, VOA and ABOF on 4 data sets: Arr, Cardio, Isolet and Mfeat.

- [20][3]: Shuttle, Optdigits (Opt for short), Mnist, Musk, Arrhythmia (Arr for short), Speech, Mammography (Mam for short), ForestCover (Cover for short).
- [6][4]: Cardiotocography (Cardio for short, 22% of outliers, not normalized, duplicates), KDDCup99 (KDD for short, normalized, duplicates, idf weighted), SpamBase (Spam for short, 40% of outliers, not normalized, without duplicates), InternetAds (Ads for short, 19%, normalized, w.o. duplicates)
- [15][5]: Isolet (classes C, D, E as inliers and random points in class Y as outliers) and Multiple Features (Mfeat for short, classes 6 and 9 as inliers and random points in class 0 as outliers).

5.3 Relationship Between L1D and ABOD Measures

This subsection conducted experiments on evaluating the correlation between the proposed outlier factor L1D with other ABOD measures. We computed the exact values of L1D, MOA1, MOA2, VOA and ABOF and evaluated the statistical relationships using correlation coefficients, including Pearson correlation

---

[3] http://odds.cs.stonybrook.edu/.
[4] http://www.dbs.ifi.lmu.de/research/outlier-evaluation/DAMI/.
[5] https://archive.ics.uci.edu/ml/datasets.html.

coefficients, Spearman's rank correlation coefficients and Kendall's rank correlation coefficients. We computed these coefficients over 4 small data sets, including Arr, Cardio, Isolet and Mfeat, and demonstrates the results in Fig. 3.

It is clear that the outlier rankings based on L1D are almost identical to that of MOA1 and MOA2. L1D is also highly correlated to ABOF, except the low-dimensional data set Cardio. This is due to the fact that the distance's effect is significant in ABOF for low-dimensional data.

The L1D's rankings are also highly correlated to the VOA's ones, except the Isolet data set. In fact, on Isolet, the inlier classes C, D, and E have very similar MOA1 values, whereas the average MOA1 of the outlier class Y significantly deviates from the rest. Considering variance as a central second moment, we know that the VOA's ranking of outliers will change significantly compared to the MOA1's ranking, which leads to the situation where there is a negative correlation between L1D (or MOA1, MOA2) and VOA. Besides, on Isolet, the L1D's ranking is also highly correlated to the ABOF's one. This is due to the effect of distances in ABOF measure, which can be observed in the next experiment where the kNN detector shows the best performance on Isolet.

We also note that since L1D's ranking is almost identical to MOA1, we can use MOA1 as an ABOD outlier factor. The near-linear time approximation algorithm to estimate MOA1 for all points proposed in [17] can be used to speed up the outlier detection process.

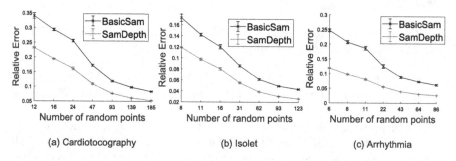

(a) Cardiotocography   (b) Isolet   (c) Arrhythmia

Fig. 4. Relative approximation errors provided by SamDepth and BasicSam on the Cardio, Isolet and Arr data sets when increasing the sample size.

5.4   Relative Approximation Errors

This subsection presents experiments to measure *relative* approximation errors of SamDepth and BasicSam on previous data sets, including Cardio, Isolet and Arr. For the sake of comparison, we used t random *points* for SamDepth and $\lceil t/2 \rceil$ random *pairs* for BasicSam due to the fact that BasicSam needs two random points for each random pair.

Figure 4 displays the *average* relative approximation errors provided by SamDepth and BasicSam when varying the number of sample *points* t in range $\lceil \frac{\sqrt{n}}{i} \rceil$ where $i = \{\frac{1}{4}, \frac{1}{3}, \cdots, 3, 4\}$. It is clear that the average relative errors and its variances of both sampling methods decrease dramatically when increasing the sample size. Particularly, SamDepth with $\sqrt{n}$ samples provides average relative errors less than $\epsilon = 0.1$ on the three data sets. Since the errors of SamDepth are significantly smaller than that of BasicSam, SamDepth will achieve higher accuracy than BasicSam on detecting outliers using L1-depth.

5.5   Outlier Detection Performance

In this subsection, we compare the outlier detection performance of sampling methods using the AUC value (i.e. the area under the ROC curve). The AUC value for an ideal outlier ranking is 1 when all outliers are top-ranked points. The AUC value of a "less than perfect" outlier detection algorithm is typically less than 1.

We again used 3 data sets, Arr and Isolet as high-dimensional data sets and Cardio as a low-dimensional data set to measure AUC values of L1D measure provided by the exact and sampling solutions. We studied the performance of sampling methods where we varied the sample size as described in the previous subsections. Figure 5 reveals the AUC values of BasicSam and SamDepth compared to the exact solution (L1D) on the data sets.

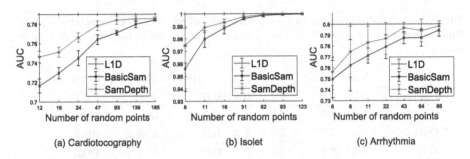

Fig. 5. Comparison of AUC values of SamDepth, BasicSam and L1D when varying number of samples on 3 data sets.

It is clear that the AUC values of sampling methods converge to the AUC value of the exact solution and the deviations of sampling methods are significantly reduced when increasing the sample size. SamDepth provides superior performance compared to BasicSam regarding both AUC values and its deviation on the 3 data sets. Since SamDepth outperforms BasicSam on detecting outliers, we will use SamDepth with L1D to compare the outlier detection performance with other proximity-based and ABOD-based outlier factors.

In order to quantify ABOD measures, we compare L1D, VOA, and ABOF with proximity-based measures, including kNN, kNNW and LOF. For L1D, we study the exact method and SamDepth with sample size $\sqrt{n}$. Since VOA and ABOF require $O(n^3)$ time for exact values, we computed exact values for small data sets and used approximation methods, including FastVOA and FastABOF, for large data sets. Table 2 shows the AUC values and Table 3 depicts the running time in seconds on all used data set, except the 4 large-scale data sets Mam, Shuttle, Cover and KDD.

Table 2. Comparison of AUC values of several outlier detectors on high-dimensional data sets. The top-2 AUCs are in boldface on each data set.

	kNN	kNNW	LOF	L1D	SamDepth	VOA	FastVOA	ABOF	FastABOF
Cardio	0.62	0.61	0.63	**0.79**	**0.78**	**0.78**	0.77	0.57	0.55
SpamBase	**0.71**	**0.68**	0.51	0.49	0.49	–	0.44	–	0.54
OptDigits	0.41	0.40	0.54	**0.56**	0.55	–	**0.62**	–	0.47
Mnist	0.82	0.80	0.72	**0.84**	0.82	–	0.57	–	**0.86**
Musk	0.64	0.24	0.41	**0.91**	**0.89**	0.79	0.79	0.1	0.06
Arrhythmia	**0.81**	**0.80**	**0.81**	**0.80**	0.79	0.68	0.56	**0.81**	0.79
Speech	0.48	**0.52**	0.49	0.47	0.47	0.40	0.50	0.47	**0.51**
Isolet	**0.96**	0.94	0.26	1	1	0.44	0.32	1	0.83
Mfeat	0.41	0.41	0.37	**0.95**	**0.92**	0.90	0.79	0.49	0.45
InternetAds	**0.70**	**0.72**	0.67	0.69	0.68	0.44	0.57	0.68	0.69
Avg AUC	0.66	0.61	0.54	**0.75**	**0.74**	–	0.59	–	0.58

Table 3. Comparison of running time (in seconds) of several outlier detectors on high-dimensional data sets. The smallest running time values are in boldface on each data set.

	kNN	kNNW	LOF	L1D	SamDepth	FastVOA	FastABOF
Cardio	0.3	0.3	0.4	0.4	**0.03**	17.6	3.1
SpamBase	1.6	1.6	2.1	3.4	**0.2**	36.1	47.3
OptDigits	2.9	2.7	3.6	5.3	**0.2**	45.6	102.2
Mnist	8.2	8.3	10.0	17.4	**0.6**	66.0	460.0
Musk	2.1	2.1	2.4	4.7	**0.2**	26.9	51.2
Arrhythmia	0.06	0.08	0.09	0.16	**0.02**	3.6	0.3
Speech	6.8	6.8	7.0	15.9	**0.8**	31.5	196.1
Isolet	0.6	0.6	0.7	1.6	**0.1**	8.2	5.8
Mfeat	0.14	0.14	0.17	0.34	**0.05**	3.6	0.7
InternetAds	6.9	7.1	7.3	17.0	**1.0**	16.7	125.6
Avg Time	3.0	3.0	3.4	6.6	**0.3**	25.6	99.2

In general, L1D provides superior performance compared to other outlier factors regarding the AUC with the highest average value of 75%. In particular, L1D significantly outperforms VOA and ABOF for all 7 small data sets regarding both accuracy and efficiency. Its AUC values are in the top-2 of 7 over 10 used data sets. While AUC scores of SamDepth, kNN and kNNW are in the top-2 of 4 data sets, FastVOA and FastABOF show slightly less detection performance due to the approximation errors. Among the proximity-based factors, kNN shows the

Table 4. Comparison of AUC values and running time in seconds of representative outlier detectors on 4 large-scale data sets.

Methods	AUC				Time (s)			
	Mam	Shuttle	Cover	KDDCup99	Mam	Shuttle	Cover	KDDCup99
kNN	**0.85**	0.76	**0.85**	0.85	8	138	4837	387
SamDepth	0.84	**0.99**	**0.85**	**0.99**	**0.2**	3	**265**	**10**
FastVOA	0.79	0.71	0.74	**0.99**	123	625	15413	1637
FastABOF	0.62	0.66	0.81	0.57	30	330	12956	3557

superior performance and LOF shows the inferior performance in average. Hence we used KNN as a representative algorithm for the proximity-based methods on large-scale experiments.

Regarding both effectiveness and efficiency, SamDepth illustrates substantial advantages with the second highest average AUC 74% but runs up to several orders of magnitude faster than other methods. In average, SamDepth runs approximately 10× faster than proximity-based methods, 22× faster than exact L1D, 85× faster than FastVOA, and 330× faster than FastABOF.

We conclude the empirical evaluation by depicting the performance of detectors on 4 large-scale data sets, including Mam, Shuttle, Cover and KDD. For each type of outlier factors, we used its representative algorithm, including kNN, SamDepth, FastVOA and FastABOF. Since the data set's size is very large, FastABOF with $k = \lceil 0.1 \cdot n \rceil$ would not finish after 10 h. Hence we set $k = \lceil \sqrt{n} \rceil$ for FastABOF. Table 4 reveals the AUC values and running time in seconds on 4 large-scale data sets. Again, SamDepth provides superior performance compared to the other methods. It almost obtains the highest AUC values and runs several orders of magnitude faster than other competitors.

6 Conclusions

The paper investigates the *parameter-free* angle-based outlier detection (ABOD) in high-dimensional data. Exploiting the conceptual relationship between the ABOD intuition and the L1-depth notion (L1D), we propose to use L1D as a robust variant of ABOD measures, which only requires a quadratic computational time. Empirical experiments on many real-world high-dimensional data sets show that L1D is superior to other ABOD measures, such as ABOF and VOA, and very competitive to other proximity-based measures, including kNN, kNNW and LOF on detecting high-dimensional outliers regarding ROC AUC scores.

In order to avoid the high computational complexity of L1D measures, we propose SamDepth, a simple but efficient sampling algorithm which often runs faster and achieves very comparable outlier detection performance compared to the exact method. Especially, SamDepth with $\sqrt{n}$ samples shows the superior performance compared to widely used detectors regarding both effectiveness and efficiency on many real-world high-dimensional data sets.

Acknowledgments. We would like to thank Rasmus Pagh for useful discussion and comments in the early stage of this work. We thank members of the DABAI project and anonymous reviewers for their constructive comments and suggestions.

References

1. Aggarwal, C.C., Hinneburg, A., Keim, D.A.: On the surprising behavior of distance metrics in high dimensional space. In: Van den Bussche, J., Vianu, V. (eds.) ICDT 2001. LNCS, vol. 1973, pp. 420–434. Springer, Heidelberg (2001). https://doi.org/10.1007/3-540-44503-X_27
2. Aggarwal, C.C., Sathe, S.: Outlier Ensembles: An Introduction. Springer, Cham (2017). https://doi.org/10.1007/978-3-319-54765-7
3. Aggarwal, C.C., Yu, P.S.: Outlier detection for high dimensional data. In: Proceedings of SIGMOD 2001, pp. 37–46 (2001)
4. Angiulli, F., Pizzuti, C.: Outlier mining in large high-dimensional data sets. IEEE Trans. Knowl. Data Eng. **17**(2), 203–215 (2005)
5. Breunig, M.M., Kriegel, H.-P., Ng, R.T., Sander, J.: LOF: identifying density-based local outliers. In: Proceedings of SIGMOD 2000, pp. 93–104 (2000)
6. Campos, G.O., et al.: On the evaluation of unsupervised outlier detection: measures, datasets, and an empirical study. Data Min. Knowl. Discov. **30**(4), 891–927 (2016)
7. Chen, Y., Bart Jr., H.L., Dang, X., Peng, H.: Depth-based novelty detection and its application to taxonomic research. In: Proceedings of ICDM 2007, pp. 113–122 (2007)
8. Hawkins, D.: Identification of Outliers. Chapman and Hall, London (1980)
9. Hugg, J., Rafalin, E., Seyboth, K., Souvaine, K.: An experimental study of old and new depth measures. In: Proceedings of ALENEX 2006, pp. 51–64 (2006)
10. Jeong, M., Cai, Y., Sullivan, C.J., Wang, S.: Data depth based clustering analysis. In: Proceedings of SIGSPATIAL 2016, pp. 29:1–29:10 (2016)
11. Johnson, T., Kwok, I., Ng, R.T.: Fast computation of 2-dimensional depth contours. In: Proceedings of KDD 1998, pp. 224–228 (1998)
12. Jörnsten, R.: Clustering and classification based on the L1 data depth. J. Multivar. Anal. **90**(1), 67–89 (2004)
13. Knorr, E.M., Ng, R.T.: Algorithms for mining distance-based outliers in large datasets. In: Proceedings of VLDB 1998, pp. 392–403 (1998)
14. Kriegel, H.-P., Schubert, M., Zimek, A.: Angle-based outlier detection in high-dimensional data. In: Proceedings of KDD 2008, pp. 444–452 (2008)
15. Lichman, M.: UCI machine learning repository (2013)
16. Papadimitriou, S., Kitagawa, H., Gibbons, P.B., Faloutsos, C.: LOCI: fast outlier detection using the local correlation integral. In: Proceedings of ICDE 2003, pp. 315–326 (2003)
17. Pham, N., Pagh, R.: A near-linear time approximation algorithm for angle-based outlier detection in high-dimensional data. In: Proceedings of KDD 2012, pp. 877–885 (2012)
18. Preparata, F.P., Shamos, M.: Computational Geometry: An Introduction. Springer, New York (1985). https://doi.org/10.1007/978-1-4612-1098-6
19. Ramaswamy, S., Rastogi, R., Shim, K.: Efficient algorithms for mining outliers from large data sets. In: Proceedings of SIGMOD 2000, pp. 427–438 (2000)
20. Rayana, S.: ODDS library (2016)

21. Serfling, R.: A depth function and a scale curve based on spatial quantiles. In: Dodge, Y. (ed.) Statistical Data Analysis Based on the L_1-Norm and Related Methods. SIT, pp. 25–38. Birkhäuser Basel, Basel (2002). https://doi.org/10.1007/978-3-0348-8201-9_3

22. Tukey, J.W.: Mathematics and picturing data. In: Proceedings of the International Congress of Mathematicians Vancouver, pp. 523–531 (1974)

23. Vardi, Y., Zhang, C.-H.: The multivariate L1-median and associated data depth. Proc. Natl. Acad. Sci. U. S. A. **97**(4), 1423–1426 (2000)

24. Zimek, A., Campello, R.J.G.B., Sander, J.: Ensembles for unsupervised outlier detection: challenges and research questions a position paper. SIGKDD Explor. **15**(1), 11–22 (2013)

Beyond Outlier Detection: LookOut for Pictorial Explanation

Nikhil Gupta[1(✉)], Dhivya Eswaran[2], Neil Shah[3], Leman Akoglu[2], and Christos Faloutsos[2]

[1] IIT Delhi, New Delhi, India
Nikhil.Gupta.cs514@cse.iitd.ac.in
[2] CMU, Pittsburgh, USA
{deswaran,lakoglu,christos}@cs.cmu.edu
[3] Snap Inc., Santa Monica, USA
nshah@snap.com

Abstract. Why is a given point in a dataset marked as an outlier by an off-the-shelf detection algorithm? Which feature(s) explain it the best? What is the best way to convince a human analyst that the point is indeed an outlier? We provide succinct, interpretable, and simple pictorial explanations of outlying behavior in multi-dimensional real-valued datasets while respecting the limited attention of human analysts. Specifically, we propose to output a few focus-plots, i.e., pairwise feature plots, from a few, carefully chosen feature sub-spaces. The proposed LookOut makes four contributions: (a) **problem formulation:** we introduce an "analyst-centered" problem formulation for explaining outliers via focus-plots, (b) **explanation algorithm:** we propose a plot-selection objective and the LookOut algorithm to approximate it with optimality guarantees, (c) **generality:** our explanation algorithm is *both* domain- and detector-agnostic, and (d) **scalability:** LookOut scales linearly with the size of input outliers to explain and the explanation budget. Our experiments show that LookOut performs near-ideally in terms of maximizing explanation objective on several real datasets, while producing visually interpretable and intuitive results in explaining groundtruth outliers. Code related to this paper is available at: https://github.com/NikhilGupta1997/Lookout.

Keywords: Outlier detection · Pictorial explanation · Interpretability

1 Introduction

Given a multi-dimensional dataset of real-valued features, e.g., sensor measurements, and a list of outliers (identified by an off-the-shelf "black-box" detector or any other external mechanism), how can we *explain* the outliers to a human analyst in a succinct, effective, and interpretable fashion?

Outlier detection is a widely studied problem. Numerous detectors exist for point data [1,5,18], time series [12], as well as graphs [2,3]. However, the literature on outlier explanation is surprisingly sparse. Given that the outcomes

© Springer Nature Switzerland AG 2019
M. Berlingerio et al. (Eds.): ECML PKDD 2018, LNAI 11051, pp. 122–138, 2019.
https://doi.org/10.1007/978-3-030-10925-7_8

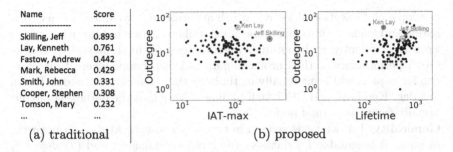

Name	Score
Skilling, Jeff	0.893
Lay, Kenneth	0.761
Fastow, Andrew	0.442
Mark, Rebecca	0.429
Smith, John	0.331
Cooper, Stephen	0.308
Tomson, Mary	0.232
...	...

(a) traditional   (b) proposed

Fig. 1. Compared to traditional ranked list output (a: *wordy, lengthy, no explanation*), LOOKOUT produces simple, interpretable explanations (b: *visual, succinct, interpretable*) (Color figure online)

(alerts) of a detector often go through a "vetting" procedure by human analysts, it is beneficial to provide explanations for such alerts which can empower analysts in sense-making and reduce their efforts in troubleshooting. Moreover, such explanations should justify the outliers succinctly in order to save analyst time.

Our work sets out to address the above outlier explanation problem. Consider the following **example situation:** Given performance metrics from hundreds of machines within a large company, an analyst could face two relevant scenarios.

- *Detected outliers:* For monitoring, s/he could use any "black-box" outlier detector to spot machines with suspicious values of metric(s). Here, we are oblivious to the specific detector, knowing only that it flags outliers, but does not produce any interpretable explanation.
- *Dictated outliers:* Alternatively, outlying machines may get reported to the analyst externally (e.g., they crash or get compromised).

In both scenarios, the analyst would be interested in understanding *in what ways* the pre-identified outlying machines (detected or dictated) differ from the rest.

In this work, we propose a new approach called LOOKOUT, for explaining a given set of outliers, and apply it to various, relational and non-relational, settings. At its heart, LOOKOUT provides interpretable *pictorial* explanations through simple, easy-to-grasp focus-plots (Definition 1), which "incriminate" the given outliers the most. We summarize our contributions as follows.

- **Outlier Explanation Problem Formulation:** We introduce a new formulation that explains outliers through "focus-plots". In a nutshell, given the list of outliers from a dataset with real-valued features, we aim to find a few 2D plots on which the total outlier "blame" is maximized. Our emphasis is on two key aspects: (a) *interpretability:* our plots visually incriminate the outliers, and (b) *succinctness:* we show only a few plots to respect the analyst's attention; the analysts can then quickly interpret the plots, spot the outliers, and verify their abnormality given the discovered feature pairs.
- **Succinct Quantifiable Explanation Algorithm** LOOKOUT: We propose the LOOKOUT algorithm to solve our explanation problem. Specifically, we

develop a plot selection objective, which quantifies the 'goodness' of an explanation and lends itself to monotone submodular function optimization, which we solve efficiently with optimality guarantees. Figure 1 illustrates LOOK-OUT's performance on the Enron communications network, where it discovers two focus-plots which maximally incriminate the given outlying nodes: Enron founder "Ken Lay" and CEO "Jeff Skilling." Note that the outliers stand out visually from the normal nodes.

- **Generality:** LOOKOUT is general in two respects: it is (a) *domain-agnostic*, meaning it is suitable for datasets from various domains, and (2) *detector-agnostic*, meaning it can be employed to explain outliers produced by any detector or identified through any other mechanism (e.g., crash reports).
- **Scalability:** We show that LOOKOUT requires time linear on (i) the number of plots to choose explanations from, (ii) the number of outliers to explain and (iii) the user-specified budget for explanations (see Lemma 7 and Fig. 5).

We experiment with several real-world datasets from diverse domains including e-mail communications and astronomy, which demonstrate the effectiveness, interpretability, succinctness and generality of our approach.

Reproducibility: Our datasets are publicly available (See Sect. 5.1) and LOOK-OUT is open-sourced at https://github.com/NikhilGupta1997/Lookout.

2   Related Work

While there is considerable prior work on outlier detection [3,6,12], literature on outlier description is comparably sparse. Several works aim to find an optimal feature subspace which distinguishes outliers from normal points. [14] aims to find a subspace which maximizes differences in outlier score distributions of all points across subspaces. [17] instead takes a constraint programming approach which aims to maximize differences between neighborhood densities of known outliers and normal points. An associated problem focuses on finding minimal, or optimal feature subspaces for each outlier. [15] aims to give "intensional knowledge" for each outlier by finding minimal subspaces in which the outliers deviate sufficiently from normal points using pruning rules. [7,8] use spectral embeddings to discover subspaces which promote high outlier scores, while aiming to preserve

Properties vs. Methods	Knorr et al. [15]	Dang et al.[7]	Angiulli et al. [4]	Micenkova et al. [19]	Kopp et al. [16]	Keller et al. [14]	LOOKOUT
Quantifiable explanations	✔	✔	✔		✔		✔
Budget-conscious					✔		✔
Visually interpretable							✔
Scalable	✔				✔	✔	✔

Fig. 2. Comparison with other outlier description approaches, in terms of four desirable properties

distances of normal points. [19] instead employs sparse classification of an inlier class against a synthetically-created outlier class for each outlier in order to discover small feature spaces which discern it. [16] proposes combining decision rules produced by an ensemble of short decision trees to explain outliers. [4] augments the per-outlier problem to include outlier groups by searching for single features which differentiate many outliers.

All in all, none of these works meet several key desiderata for outlier description: (a) quantifiable explanation quality, (b) budget-consciousness towards analysts, (c) visual interpretability, and (d) a scalable descriptor, which is subquadratic on the number of nodes and at worst polynomial on (low) dimensionality. Figure 2 shows that unlike existing approaches, our LOOKOUT approach is designed to give quantifiable explanations which aim to maximize *incrimination*, respect human attention-budget and visual interpretability constraints, and scale linearly on the number of outliers.

3 Preliminaries and Problem Statement

3.1 Notation

Let $\mathcal{V}$ be the set of input data points, where each point $v \in \mathcal{V}$ originates from $\mathbb{R}^d$ and $n = |\mathcal{V}|$ is the total number of points. Here, $d = |\mathcal{F}|$ is the dimensionality of the dataset and $\mathcal{F} = \{f_1, f_2, \ldots, f_d\}$ is the set of real-valued features (either directly given, or extracted, e.g., from a relational dataset). The set of outlying points given as input is denoted by $\mathcal{A} \subseteq \mathcal{V}$, $|\mathcal{A}| = k$. Typically, $k \ll n$. Table 1 summarizes the frequently used notation.

3.2 Intuition and Proposed Problem

The explanations we seek to generate should be simple, interpretable, and easy to illustrate to humans who will ultimately leverage the explanations. To this end, we decide to use *focus-plots* (Definition 1) for outlier justification, due to their visual appeal and interpretability. A formal definition is given below.

Definition 1 (Focus-plot). *Given a dataset of points $\mathcal{V}$, a pair of features $f_x, f_y \in \mathcal{F}$ (where $\mathcal{F}$ is the set of real-valued features) and an input set of outliers $\mathcal{A}$, focus-plot $p \in \mathcal{P}$ is a 2-d scatter plot of all points, with f_x on x-axis, f_y on y-axis, 'drawing attention' to the set of "maxplained" (maximally explained) outliers $\mathcal{A}_p \subseteq \mathcal{A}$ best explained by this feature pair.*

Intuitively, our pictorial outlier explanation is a set of *focus-plots*, each of which "blames" or "explains away" a subset of the input outliers, whose outlierness is best showcased by the corresponding pair of features. That is, we consider $\binom{d}{2} = \frac{d(d-1)}{2}$ 2-d spaces by generating all pairwise feature combinations. Within each 2-d space, we then score the points in $\mathcal{A}$ by their outlierness (Sect. 4.1).

Let us denote the set of all $\binom{d}{2}$ focus-plots by $\mathcal{P}$. Even for small values of d, showing all the focus-plots would be too overwhelming for the analyst. Moreover, some outliers could redundantly show up in multiple plots. Ideally, we would identify only a few focus-plots, which could "blame" or "explain away" the outliers to the largest possible extent. In other words, our goal would be to output a small subset S of $\mathcal{P}$, on which points in $\mathcal{A}$ receive high outlier scores (Sect. 4.2). Given this intuition, we formulate our problem below.

Table 1. Symbols and definitions

Symbol	Definition
$\mathcal{V}$	Set of data points, $|\mathcal{V}| = n$
$\mathcal{A}$	Input set of outliers, $|\mathcal{A}| = k$
$\mathcal{F}$	Set of features, $|\mathcal{F}| = d$
$\mathcal{P}$	Set of focus-plots, $|\mathcal{P}| = d(d-1)/2 = l$
$s_{i,j}$	Outlier score of $a_i \in \mathcal{A}$ in plot $p_j \in \mathcal{P}$
S	Subset of selected focus-plots
$f(S)$	Explanation objective function
$\Delta_f(p \mid S)$	Marginal gain of plot p w.r.t S
b	Budget, i.e., maximum cardinality of S

Problem 2 (Outlier Explanation).

- **Given** (a) a dataset on points $\mathcal{V}$ consisting of real-valued features $\mathcal{F}$, (b) a list of outliers $\mathcal{A} \subseteq \mathcal{V}$, either (1) detected by an off-the-shelf detector or (2) dictated by external information, and (c) a fixed budget of b focus-plots,
- **find** the *best* such focus-plots $S \subseteq \mathcal{P}$, $|S| = b$, so as to *to maximize* the *total maximum outlier score of outliers* that we can "blame" through the b plots.

4   Proposed Algorithm LOOKOUT

In this section, we detail our approach for scoring the input outliers by focus-plots, the overall complexity analysis of LOOKOUT, and conclude with discussion.

4.1   Scoring by Feature Pairs

Given all the points $\mathcal{V}$, with marked outliers $\mathcal{A} \subseteq \mathcal{V}$, and their given (or extracted) features $\mathcal{F} \in \mathbb{R}^d$, our first step is to quantify how much "blame" we can attribute to each input outlier in $\mathbb{R}^2$. As previously mentioned, 2-d spaces are easy to illustrate visually with focus-plots. Moreover, outliers in 2-d are easy to interpret: e.g., "point a has too many/too few $y =$ dollars for its $x =$ number of accounts". Given a focus-plot, an analyst can easily discern the outliers visually and come up with such explanations without any further supervision.

We construct 2-d spaces (f_x, f_y) by pairing the features $\forall x, y = \{1, \ldots, d\}, x \neq y$ (order does not matter). Each focus-plot $p_j \in \mathcal{P}$ corresponds to such a pair of features, $j = \{1, \ldots, \binom{d}{2}\}$. For scoring, we consider two different scenarios, depending on how the input outliers were obtained.

If the outliers are *detected* by some "black-box" detector available to the analyst, we can employ the same detector on all the nodes (this time in 2-d) and thus obtain the scores for the nodes in $\mathcal{A}$.

If the outliers are *dictated*, i.e. reported externally, then the analyst could use any off-the-shelf detector, such as LOF [5], DB-outlier [15], etc. In this work, we use the Isolation Forest (iForest) detector [18] for two main reasons: (a) it boasts *constant* training time and space complexity (i.e., independent of n) due to its sampling strategy, and (b) it has been shown empirically to outperform alternatives [9] and is thus state-of-the-art. However, note that none of these existing detectors has the ability to *explain* the outliers, especially iForest, as it is an ensemble approach.

By the end of the scoring process, each outlier receives $|\mathcal{P}| = l = \binom{d}{2}$ scores.

4.2   Plot Selection Objective

While scoring in small, 2-d spaces is easy and can be trivially parallelized, presenting all such focus-plots to the analyst would not be productive given their limited attention budget. As such, our next step is to carefully select a short list of plots that best blame all the outliers collectively, where the plot budget can be specified by the user.

Objective Function. While selecting plots, we aim to incorporate the following criteria:

- **incrimination power;** such that the outliers are scored as highly as possible,
- **high expressiveness;** where each plot incriminates multiple outliers, so that the explanation is *sublinear* in the number of outliers, and
- **low redundancy;** such that the plots do not explain similar sets of outliers.

We next introduce our objective criterion which satisfies the above requirements.

At this step of the process, we can conceptually think of a complete, weighted bipartite graph between the k input outliers $\mathcal{A} = \{a_1, \ldots, a_k\}$ and l focus-plots $\mathcal{P} = \{p_1, \ldots, p_l\}$, in which edge weight $s_{i,j}$ depicts the outlier score that a_i received from p_j, as illustrated in Fig. 3.

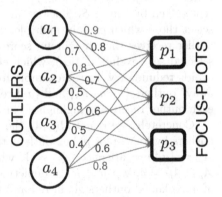

Fig. 3. LookOut with $k = 4$ outliers, $l = 3$ focus-plots, and budget $b = 2$: p_1 is picked first due to maximum *total incrimination* (sum of edge weights = 2.9); next p_3 is chosen over p_2, due to its higher marginal gain (0.4 vs 0.2) (Color figure online)

Algorithm 1. LOOKOUT

Data: dataset of points $\mathcal{V}$, outliers $\mathcal{A}$, set of all focus-plots $\mathcal{P}$, budget b

Result: pictorial outlier explanation $\mathcal{S}$, which is a set of focus-plots

1 **for** $p_j \in \mathcal{P}$ **do**
2      $D_j \leftarrow$ iForest constructed using $\mathcal{V}$ and the two features in plot p_j;
3   **for** $a_i \in \mathcal{A}$ **do**
4          $\mid$ $s_{i,j} \leftarrow$ anomaly score given by detector D_j to point a_i;
5   **end**
6 **end**
7 initialize $\mathcal{S} \leftarrow \emptyset$;
8 **while** $|\mathcal{S}| < b$ **do**
9   recompute marginal gain $\Delta_f(p \mid \mathcal{S}) \; \forall \, p \in \mathcal{P} \setminus \mathcal{S}$;   // using Eq. 4
10     $p^* \leftarrow \arg\max_{p \in \mathcal{P} \setminus \mathcal{S}} \Delta_f(p \mid \mathcal{S})$;
11     $\mathcal{S} \leftarrow \mathcal{S} \cup \{p^*\}$;
12 **end**
13 **return** $\mathcal{S}$;

We formulate our objective to *maximize* the total *maximum outlier score* of each outlier amongst the selected plots:

$$\underset{\mathcal{S} \subseteq \mathcal{P}, |\mathcal{S}| = b}{\text{maximize}} \quad f(\mathcal{S}) = \sum_{a_i \in \mathcal{A}} \max_{p_j \in \mathcal{S}} s_{i,j} \tag{1}$$

Here, our objective function, $f(\mathcal{S})$, can be considered the *total incrimination score* given by subset $\mathcal{S}$. Since we are limited with a budget of plots, we aim to select those which explain multiple outliers to the best extent. Note that each outlier receives their maximum score from exactly one of the plots among the selected set (excluding ties), which effectively *partitions* the explanations and avoids redundancy. In the example from Fig. 3, focus-plots p_1 and p_3 "explain away" outliers $\{a_1, a_2, a_3\}$ and $\{a_4\}$ respectively, where the maximum score that each outlier receives is highlighted in red font.

Concretely, we denote by $\mathcal{A}_p$ the set of *maxplained* (maximally explained) outliers by focus-plot p, i.e., outliers that receive their highest score from p, i.e. $\mathcal{A}_p = \{a_i | p = \arg\max_{p_j \in \mathcal{S}} s_{i,j}\}$, where we break ties at random. Note that $\mathcal{A}_p \cap \mathcal{A}_{p'} = \emptyset, \; \forall \, p, p' \in \mathcal{P}$. In depicting a plot p to the analyst, we mark the set of maxplained outliers $\mathcal{A}_p$ in red and the rest in $\mathcal{A} \setminus \mathcal{A}_p$ in blue – see Fig. 1.

4.3 Approximation Algorithm LOOKOUT

Having defined our plot selection objective, we need to devise a subset selection algorithm to optimize Eq. (1), for a budget b. Notice that the optimal subset selection is a combinatorial task which we can show to be **NP-hard**.

Lemma 3. *The focus-plot selection problem in Eq. (1) is NP-hard.*

Proof. We sketch the proof by a reduction from the Maximum Coverage (Max-Cover) problem, which is known to be NP-hard [10]. An instance of MaxCover involves an integer k and a collection of sets $\{S_1, \ldots, S_l\}$ each containing a list of elements, where the goal is to find k sets such that the total number of covered elements is maximized. The MaxCover problem instance maps to an instance of our problem, where each set S_j corresponds to a focus-plot p_j, each element e_i maps to an outlier a_i, and the elements (outliers) inside each set has the same unit score ($s_{i,j} = 1$ for $e_i \in S_j$) while the others outside the set has score zero ($s_{i,j} = 0$ for $e_i \notin S_j$) on the corresponding focus-plot. Since MaxCover is equivalent to a special case of our problem, we conclude that Eq. (1) is at least as hard as MaxCover. □

Therefore, our aim is to find an approximation algorithm to optimize Eq. (1).

Properties of Our Objective. Fortunately, our objective $f(\cdot)$ exhibits three key properties that enable us to use a greedy algorithm with an approximation guarantee. Specifically, our objective $f : 2^{|\mathcal{P}|} \to \mathbb{R}^+ \cup \{0\}$ is (i) *non-negative*, since the outlier scores take non-negative values, often in $[0, 1]$, e.g., using iForest [18], (ii) *non-decreasing* (see Lemma 4) and (iii) *submodular* (see Lemma 5).

Lemma 4 (Monotonicity). f *is non-decreasing, i.e., for any* $\mathcal{S} \subseteq \mathcal{T}$, $f(\mathcal{S}) \leq f(\mathcal{T})$.

Proof. $f(\mathcal{S}) = \sum_{a_i \in \mathcal{A}} \max_{p_j \in \mathcal{S}} s_{i,j} \leq \sum_{a_i \in \mathcal{A}} \max_{p_j \in \mathcal{T}} s_{i,j} = f(\mathcal{T})$ □

Lemma 5 (Submodularity). f *is submodular, i.e., for any two sets* $\mathcal{S} \subseteq \mathcal{T}$ *and a focus-plot* $p_{j^*} \in \mathcal{P} \setminus \mathcal{T}$, $f(\mathcal{S} \cup \{p_{j^*}\}) - f(\mathcal{S}) \geq f(\mathcal{T} \cup \{p_{j^*}\}) - f(\mathcal{T})$.

Proof

$$f(\mathcal{S} \cup \{p_{j^*}\}) - f(\mathcal{S}) = \sum_{a_i \in \mathcal{A}} \left[\max_{p_j \in \mathcal{S} \cup \{p_{j^*}\}} s_{i,j} - \max_{p_j \in \mathcal{S}} s_{i,j} \right]$$

$$= \sum_{a_i \in \mathcal{A}} \left(s_{i,j^*} - \max_{p_j \in \mathcal{S}} s_{i,j} \right) \cdot \mathbb{I}\left[s_{i,j^*} > \max_{p_j \in \mathcal{S}} s_{i,j} \right]$$

$$\geq \sum_{a_i \in \mathcal{A}} \left(s_{i,j^*} - \max_{p_j \in \mathcal{T}} s_{i,j} \right) \cdot \mathbb{I}\left[s_{i,j^*} > \max_{p_j \in \mathcal{S}} s_{i,j} \right] \quad (2)$$

$$\geq \sum_{a_i \in \mathcal{A}} \left(s_{i,j^*} - \max_{p_j \in \mathcal{T}} s_{i,j} \right) \cdot \mathbb{I}\left[s_{i,j^*} > \max_{p_j \in \mathcal{T}} s_{i,j} \right] \quad (3)$$

$$= f(\mathcal{T} \cup \{p_{j^*}\}) - f(\mathcal{T})$$

where $\mathbb{I}[\cdot]$ is the indicator function and Eqs. (2) and (3) follow from the fact that $\max_{p_j \in \mathcal{S}} s_{i,j} \leq \max_{p_j \in \mathcal{T}} s_{i,j}$ whenever $\mathcal{S} \subseteq \mathcal{T}$. □

Proposed LOOKOUT **Algorithm.** Submodular functions which are non-negative and non-decreasing admit approximation guarantees under a greedy approach identified by Nemhauser et al. [21]. The greedy algorithm starts with the empty set S_0. In iteration t, it adds the element (in our case, focus-plot) that maximizes the *marginal gain* Δ_f in function value, defined as

$$\Delta_f(p|S_{t-1}) = f(S_{t-1} \cup \{p\}) - f(S_{t-1}) \tag{4}$$

That is,

$$S_t := S_{t-1} \cup \{ \arg\max_{p \in \mathcal{P} \setminus S_{t-1}} \Delta_f(p|S_{t-1}) \} .$$

This leads to LOOKOUT explanation algorithm, given in Algorithm 1. Its approximation guarantee is given in Lemma 6.

Lemma 6 (63% **approximation guarantee**). *Given $\mathcal{A}, \mathcal{P}$ and budget b, let $\hat{S}$ be the output of* LOOKOUT *(Algorithm 1). Suppose $S^* = \arg\max_{S \subseteq \mathcal{P}, |S|=b} f(S)$ is an optimal set of focus-plots. Then:*

$$f(\hat{S}) \geq \left(1 - \frac{1}{e}\right) f(S^*) \tag{5}$$

Proof. This follows from [21] since by design, our plot selection objective f is non-negative, non-decreasing and submodular. □

4.4 Computational Complexity Analysis

Lemma 7. LOOKOUT *total time complexity is $O(l \log n'(k+n') + klb)$, for sample size $n' < n$, and is sub-linear in total number of input points n.*

Proof. We study complexity in two parts: (1) scoring the given outliers (Sect. 4.1) and (2) selecting focus-plots to present to the user (Sect. 4.2).

(1) For each focus-plot, we train an iForest model [18] in 2-d. Following their recommended setup, we sub-sample n' points and train t (100 in [18]) randomized isolation trees. The depth of each tree is $O(\log n')$, where each point is evaluated at each level for the threshold/split conditions. Therefore, training iForest with t trees takes $O(tn' \log n')$. Then, scoring $|\mathcal{A}| = k$ outliers takes $O(tk \log n')$. Total complexity of training and scoring on all plots is $O(lt \log n'(k + n'))$. Note that this can also be done per plot independently in parallel to reduce time.

(2) At each iteration of the greedy selection algorithm, we compute the marginal gain for each yet-unselected plot of being added to our select-set in $O(kl)$. Marginal gain per plot can also be computed independently in parallel. Among the remaining plots, we pick the one with the largest marginal gain. Finding the maximum among all gains takes $O(l)$ via a linear scan. We repeat this process b times until the budget is exhausted. Total selection complexity is thus $O(klb)$.

The overall complexity of both parts is effectively $O(l \log n'(k + n') + klb)$, since t is a constant.   □

Notice that the total number of focus-plots, $l = d^2$, is quadratic in number of features. Typically, d is small (<100). In high dimensions, we could either use parallelism (multi-core machines are commodity), or drop features with low kurtosis as done earlier [18] or other feature selection criteria [13].

4.5 Discussion

Here we answer some questions that may be in the reader's mind.

1. *How do we define "outlier?"* We defer this question to the off-the-shelf outlier detection algorithm (iForest [18], LOF [5], etc.). Our focus here is to succinctly and interpretably *explain* what makes the selected items stand out from the rest.
2. *Why focus-plots?* Using focus-plots for justification is an essential, conscious choice we make for several reasons: (a) scatter plots are easy to look at and quickly interpret (b) they are universal and non-verbal, in that we need not use language to convey the outlierness of points – even people unfamiliar with the context of Enron will agree that the point "Jeff Skilling" in Fig. 1 is far away from the rest, and (c) they show where the outliers lie *relative to the normal points* – the contrastive visualization of points is more convincing than stand-alone rules.
3. *How do we choose the budget b?* We designed our objective function to be budget-conscious, and let the budget be specified by the analyst (user). If not specified, we use $b = 7$, since humans have a working memory of size "seven, plus or minus two" [20].
4. *Why not decision trees to separate outliers from the rest?* While arguably interpretable, decisions trees are not easy to visualize the points when higher than depth 3. Moreover, they try to find balanced splits which would try to cluster the outliers – which is unlikely for outliers. Also, decision trees are not budget-conscious, i.e. they would not necessarily produce the minimum description. Finally, they do not provide any quantifiable explanations, i.e. incrimination per outlier like our $s_{i,j}$ scores – the splits are binary.

5 Experiments

In this section, we empirically evaluate LOOKOUT on three, diverse datasets. Our experiments were designed to answer the following questions:

[Q1] **Quality of Explanation:** How well can LOOKOUT "explain" or "blame" the given outliers?

[Q2] **Scalability:** How does LOOKOUT scale with the input graph size and the number of outliers?

[Q3] **Discoveries:** Does LOOKOUT lead to interesting and intuitive explanations on real world data?

These are addressed in Sects. 5.3, 5.4 and 5.5 respectively. Before detailing our empirical findings, we describe the datasets used and our experimental setup.

5.1 Dataset Description

To illustrate the generality of our proposed domain-agnostic pictorial outlier explanations algorithm LOOKOUT, we select our datasets from diverse domains: - e-mail communication (ENRON), co-authorship (DBLP), pulsar identification (HTRU), and glass composition (GLASS). All datasets are publicly available and the first two are unipartite, directed and undirected resp., time-evolving graph datasets. The latter two are multi-feature datasets consisting of continuous values. A brief description is given below and a summary is provided in Table 2.

Table 2. Datasets with labeled outliers (that we explain) studied in this work

Dataset	Type	# points	# features	Description
ENRON[a]	Graph	151	12	e-mail communications
DBLP[b]	Graph	1.3M	12	Co-authorship
HTRU[c]	Feature	17.9K	8	Pulsar identification
GLASS[d]	Feature	213	9	Glass composition

[a]http://networkdata.ics.uci.edu/netdata/html/EnronMailUSC1.html
[b]http://konect.uni-koblenz.de/networks/dblp_coauthor
[c]https://archive.ics.uci.edu/ml/datasets/Glass+Identification
[d]https://archive.ics.uci.edu/ml/datasets/HTRU2

ENRON: This dataset consists of 19K emails exchanged between 151 ENRON employees during the period surrounding the scandal[5] (May 1999–June 2002).
DBLP: This dataset contains the co-authorship network of 1.3M authors over 25 years from 1990 to 2014. The networks are collected at yearly granularity.
HTRU: This dataset describes a sample of 17.9K pulsar (rapidly rotating neutron star) candidates collected during the High Time Resolution Universe Survey. Radio emissions have been binary classified as either background noise or as pulsar radiation. Features are extracted from the radio emission pattern curves.
GLASS: This dataset consists of a multi-class classification of 213 glass samples with element-wise compositions of each sample as features. There are a total of seven classes which are clustered into two distinct types: (Classes 1–4) window glass and (Classes 5–7) non-window glass.

---

[5] https://en.wikipedia.org/wiki/Enron_scandal.

5.2 Experimental Setup

Graph Feature Extraction: We extract the following intuitive and easy-to-understand features from our graph datasets (ENRON, DBLP) in order to generate pictorial explanations: (1) indegree and (2) outdegree for the number of unique in- and out- neighbors of every node, (3) inweight-v and (4) outweight-v for the total weight of in- and out- edges incident on each node, (5) inweight-r and (6) outweight-r for the count of in- and out- edges (including repetitions) incident on each node, (7) average-IAT, (8) IAT-variance, (9–11) minimum-IAT, median-IAT, and maximum-IAT to capture various statistics of inter-arrival time (IAT) between edges and finally, (12) lifetime- for the time gap between the first and the last edge [2,11,22].

Groundtruth: To obtain "ground-truth" outliers for LOOKOUT input, we use the iForest [18] algorithm on given or extracted features. This yields a ranked list of points with scores in $[0,1]$ (higher value suggests higher abnormality), from which we pick the desired top k. Analogously, we use iForest for computing the outlier score in each focus-plot. We note that the analyst is free to choose any outlier detector(s) for both/either of the above stages, making LOOKOUT *detector-agnostic*. However, it is recommended that the same methods be used for both stages to ensure ranking similarities.

Evaluation Metric: We quantify the quality of explanation provided by a set of plots $\mathcal{S}$ using its *incrimination* score which is a normalized form of our objective:

$$incrimination(\mathcal{S}) = \frac{1}{C} \cdot f(\mathcal{S}) \qquad (6)$$

where C is the normalization constant equal to the maximum achievable objective (see Eq. (1)) when all plots are selected, i.e., $C = f(\mathcal{P})$.

Baselines: Due to the lack of comparable prior works, we use a naïve version of our approach, called LOOKOUT-NAÏVE which ignores the submodularity of our objective. Instead, LOOKOUT-NAÏVE assigns a score to each plot by summing up scores for all given outliers and chooses the top b plots for a given budget b. For the sake of comparison, we compare both LOOKOUT and LOOKOUT-NAÏVE with a RANDOM baseline in which random b plots are chosen for the given budget b.

All experiments were performed on an OSX computer with 16 GB memory. RANDOM baseline incriminations and runtimes were averaged over 10 trials.

5.3 Quality of Explanation

Figure 4 compares the *incrimination* scores of both LOOKOUT-NAÏVE and LOOKOUT on the ENRON, HTRU and GLASS datasets for several choices of k and b. The red dotted line indicates the ideal value, *incrimination*$(\mathcal{P}) = 1$, i.e., the highest achievable incrimination (by selecting all plots). Figure 4 shows that LOOKOUT consistently outperforms the baselines and rapidly converges to the ideal *incrimination* with increasing budget.

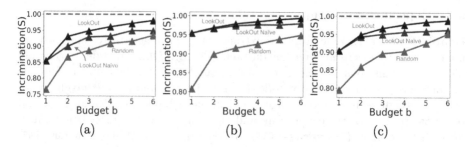

Fig. 4. LookOut vs. baselines: (a) Enron with $k = 10$ outliers, (b) Htru with $k = 32$ outliers and (c) Glass with $k = 28$ outliers (Color figure online)

5.4 Scalability

We empirically studied how LookOut runtime varies with (i) number of focus-plots l and (ii) the number of outliers k.

To study the variation of runtime with the number of focus-plots, we vary the number of features which are taken into consideration. Figure 5 (left) illustrates linear scaling with respect to number of focus-plots for the Glass dataset.

We also study the variation of runtime with the number of outliers, as feature extraction incurs a constant overhead on each dataset. Figure 5 (right) shows linear scaling with the number of outliers for a Dblp subgraph with 10K edges.

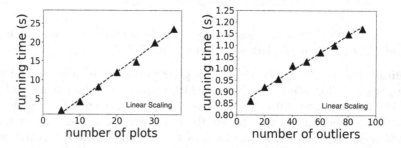

Fig. 5. LookOut **scales linearly** with (left) number of focus-plots to consider and right) number of outliers

5.5 Discoveries

In this section, we present our discoveries using LookOut on all four real world datasets. Scoring in 2-d was performed using iForest with $t = 100$ trees and sample size $\psi = 64$ (Enron, Htru, Glass) and $\psi = 256$ (Dblp). We use dictated outliers for Enron, Htru and Glass, and detected outliers for Dblp dataset to demonstrate performance in both settings.

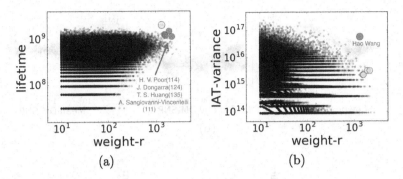

Fig. 6. Discoveries using LOOKOUT **on detected outliers:** LOOKOUT partitions and explains outlier detection results from iForest on DBLP (a–b)

ENRON *(CEO & CFO explained by large out-degree)*
We used two top actors in the ENRON scandal, *Kenneth Lay* (CEO) and *Jeff Skilling* (CFO) as dictated outliers for LOOKOUT and sought explanations for their abnormality based on internal e-mail communications. With $b = 2$, LOOK-OUT produced the focus-plots shown in Fig. 1 (right). Explanations indicate that *Jeff Skilling* had an unusually large IAT-max for the number of employees he communicated with (outdegree). On the other hand, *Kenneth Lay* sent emails to an abnormally large number of employees (outdegree) given the time range during which he emailed anyone (lifetime).

DBLP *(high h-index authors explained by large lifetime and high co-authorships)*
We obtained ground truth outliers by running iForest on the high-dimensional space spanned by the extracted graph features. With $k = 5$, the detected outlying authors were *Jack Dongarra, Thomas S. Huang, Alberto L. Sangiovanni-Vincentelli, H. Vincent Poor*, and *Hao Wang*. The explanations provided by LOOKOUT with $b = 2$ are shown in Fig. 6a–b. Thus, the outlying authors users are partitioned into two groups. The members of the first group, *Jack Dongarra, Thomas S. Huang, Alberto L. Sangiovanni-Vincentelli*, and *H. Vincent Poor* are outlying because they had unusually high duration during which they published papers (lifetime) and total number of co-authorships (inweight-r). This is consistent with their high h-indices obtained from their respective Google scholar pages (see brackets in Fig. 6a). The second group consists of only *Hao Wang*, who was also outlying in the first focus-plot, but is best explained by very high IAT-variance for his inweight-r value, shown in Fig. 6b.

HTRU *(pulsars correlated with skewness and extra kurtosis of integrated profile)*
The given radio emission samples were classified as either random noise or pulsar generated. We subsampled datapoints from the pulsar class and considered them as our set of dictated outliers with $k = 32$. We ran LOOKOUT on this subsampled dataset with $b = 3$ and obtained the focus-plots shown in Fig. 7a–c. The explanations infer high values of skewness and excess kurtosis strongly

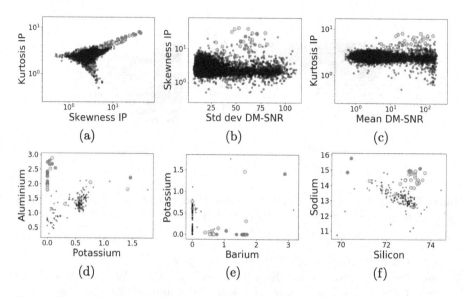

Fig. 7. **Discoveries using** LOOKOUT **on dictated outliers:** LOOKOUT explains outlier characteristics on HTRU (a–c) and GLASS (d–f) for budget $b = 3$.

indicate a pulsar emission. We observe that only the first focus-plot, Fig. 7a, succeeds to provide a suitable explanation for our set of detected anomalies. This is quantitatively explained in Fig. 4c where budget $b = 1$ has high incrimination and on further increasing the budget only a small marginal gain is observed.

GLASS (*headlamp glass explained by high Aluminium and Barium content*)
The dataset contains seven classes pertaining to different types of glass. Broadly these seven classes are split into two categories: *window based glass* (class 1–4) and *non-window based glass* (class 5–7). To compare glass composition between these two categories we stitched together a subset of the original dataset by including only classes 1, 2, 3, 4 & 7. Here class 7 (*headlamps*) is considered the set of dictated outliers with $k = 28$. The explanations provided by LOOKOUT, on the newly constructed dataset, with $b = 3$ are shown in Fig. 7d–f. The first two focus-plots Fig. 7d–e reflect higher `aluminium` and `barium` concentrations in *headlamps* as compared to *window glass*. `Aluminium` is used as a reflective coating and the presence of `barium`, in the form of oxides (borosilicate glass), helps induce heat resistant properties – both properties we expect to find in headlamps. Concurrently, we observe a very low or nearly zero concentration of `potassium` in headlamp glass. `Potassium` is to used to toughen glass and is found in windows which need to be resistant to adverse weather conditions.

Note that on all datasets, outlying points are visually distinguishable, and often complementary between focus-plots. This is in line with our desired explanation task, and achieved as a result of our LOOKOUT subset selection objective.

6 Conclusions

In this work, we formulated and tackled the problem of succinctly and interpretably explaining outliers to human analysts. We made the following contributions: (a) **problem formulation:** we formulate our goal for explaining outliers using a budget of visually interpretable focus-plots, (b) **explanation algorithm:** we propose a submodular objective to quantify explanation quality and propose the LookOut method for solving it approximately with guarantees, (c) **generality:** we show that LookOut can work with diverse domains and any detection algorithm, and (d) **scalability:** we show theoretically and empirically that LookOut scales linearly in the number of input outliers as well as the total number of focus-plots to chose from. We conduct experiments on real-world datasets: e-mail communication, co-authorship, pulsar identification, and glass composition and demonstrate that LookOut produces qualitatively interpretable explanations for "ground-truth" outliers and achieves strong quantitative performance in maximizing our proposed objective.

Acknowledgments. This material is based upon work supported by the National Science Foundation (NSF) under Grants No. CNS-1314632, IIS-1408924 and IIS 1408287, by NSF CAREER 1452425 and the PwC Risk and Regulatory Services Innovation Center at Carnegie Mellon University. Any opinions, findings, and conclusions or recommendations expressed in this material are those of the author(s) and do not necessarily reflect the views of the National Science Foundation, or other funding parties. The U.S. Government is authorized to reproduce and distribute reprints for Government purposes notwithstanding any copyright notation here on.

References

1. Aggarwal, C.C.: Outlier Analysis. Springer, New York (2013). https://doi.org/10.1007/978-1-4614-6396-2
2. Akoglu, L., McGlohon, M., Faloutsos, C.: oddball: spotting anomalies in weighted graphs. In: Zaki, M.J., Yu, J.X., Ravindran, B., Pudi, V. (eds.) PAKDD 2010. LNCS, vol. 6119, pp. 410–421. Springer, Heidelberg (2010). https://doi.org/10.1007/978-3-642-13672-6_40
3. Akoglu, L., Tong, H., Koutra, D.: Graph based anomaly detection and description: a survey. Data Min. Knowl. Discov. **29**(3), 626–688 (2015)
4. Angiulli, F., Fassetti, F., Palopoli, L.: Discovering characterizations of the behavior of anomalous subpopulations. IEEE TKDE **25**(6), 1280–1292 (2013)
5. Breunig, M.M., Kriegel, H.P., Ng, R.T., Sander, J.: LOF: identifying density-based local outliers. SIGMOD Rec. **29**(2), 93–104 (2000)
6. Chandola, V., Banerjee, A., Kumar, V.: Anomaly detection: a survey. ACM Comput. Surv. (CSUR) **41**(3), 15 (2009)
7. Dang, X.H., Assent, I., Ng, R.T., Zimek, A., Schubert, E.: Discriminative features for identifying and interpreting outliers. In: ICDE, pp. 88–99 (2014)
8. Dang, X.H., Micenková, B., Assent, I., Ng, R.T.: Local outlier detection with interpretation. In: Blockeel, H., Kersting, K., Nijssen, S., Železný, F. (eds.) ECML PKDD 2013. LNCS, vol. 8190, pp. 304–320. Springer, Heidelberg (2013). https://doi.org/10.1007/978-3-642-40994-3_20

9. Emmott, A.F., Das, S., Dietterich, T., Fern, A., Wong, W.K.: Systematic construction of anomaly detection benchmarks from real data. In: KDD Workshop on Outlier Detection and Description (2013)

10. Garey, M.R., Johnson, D.S.: Computers and Intractability: A Guide to the Theory of NP-Completeness. Freeman, New York (1979)

11. Giatsoglou, M., Chatzakou, D., Shah, N., Beutel, A., Faloutsos, C., Vakali, A.: ND-SYNC: detecting synchronized fraud activities. In: Cao, T., Lim, E.-P., Zhou, Z.-H., Ho, T.-B., Cheung, D., Motoda, H. (eds.) PAKDD 2015. LNCS, vol. 9078, pp. 201–214. Springer, Cham (2015). https://doi.org/10.1007/978-3-319-18032-8_16

12. Gupta, M., Gao, J., Aggarwal, C.C., Han, J.: Outlier detection for temporal data: a survey. IEEE Trans. Knowl. Data Eng. **26**(9), 2250–2267 (2014)

13. Traina Jr., C., Traina, A.J.M., Wu, L., Faloutsos, C.: Fast feature selection using fractal dimension. JIDM **1**(1), 3–16 (2010)

14. Keller, F., Müller, E., Wixler, A., Böhm, K.: Flexible and adaptive subspace search for outlier analysis. In: CIKM, pp. 1381–1390. ACM (2013)

15. Knorr, E.M., Ng, R.T.: Finding intensional knowledge of distance-based outliers. In: VLDB, pp. 211–222 (1999)

16. Kopp, M., Pevný, T., Holena, M.: Interpreting and clustering outliers with sapling random forests. In: ITAT (2014)

17. Kuo, C.T., Davidson, I.: A framework for outlier description using constraint programming. In: AAAI, pp. 1237–1243 (2016)

18. Liu, F.T., Ting, K.M., Zhou, Z.H.: Isolation forest. In: ICDM, pp. 413–422 (2008)

19. Micenková, B., Ng, R.T., Dang, X.H., Assent, I.: Explaining outliers by subspace separability. In: ICDM, pp. 518–527 (2013)

20. Miller, G.: The magic number seven plus or minus two: some limits on our automatization of cognitive skills. Psychol. Rev. **63**, 81–97 (1956)

21. Nemhauser, G.L., Wolsey, L.A.: Best algorithms for approximating the maximum of a submodular set function. Math. Oper. Res. **3**(3), 177–188 (1978)

22. Shah, N., et al.: EdgeCentric: anomaly detection in edge-attributed networks. In: ICDM Workshops, pp. 327–334. IEEE (2016)

ConOut: Contextual Outlier Detection with Multiple Contexts: Application to Ad Fraud

M. Y. Meghanath[1($\boxtimes$)], Deepak Pai[2], and Leman Akoglu[1]

[1] Heinz College of Information Systems and Public Policy,
Carnegie Mellon University, Pittsburgh, USA
{meghanam,lakoglu}@andrew.cmu.edu
[2] Adobe, Bangalore, India
dpai@adobe.com

Abstract. Outlier detection has numerous applications in different domains. A family of techniques, called contextual outlier detectors, are based on a *single, user-specified* demarcation of data attributes into indicators and contexts. In this work, we propose ConOut, a new contextual outlier detection technique that leverages *multiple* contexts that are *automatically* identified. Importantly, ConOut is a *one-click* algorithm—it does not require any user-specified (hyper)parameters. Through experiments on various real-world data sets, we show that ConOut outperforms existing baselines in detection accuracy. Further, we motivate and apply ConOut to the advertisement domain to identify fraudulent publishers, where ConOut not only improves detection but also provides statistically significant revenue gains to advertisers: a minimum of 57% compared to a naïve fraud detector; and ~20% in revenue gains as well as ~34% in mean average precision compared to its nearest competitor. Code related to this paper is available at: https://github.com/meghanathmacha/ConOut, https://cmuconout.github.io/.

1 Introduction

Outlier detection is a fundamental data mining task and has important applications in medicine, finance and advertisement industry [1]. A family of methods in the outlier detection literature focuses on context based detection. A contextual outlier is defined as an instance whose behavior deviates markedly from instances that share similar contexts. The contextual outlier detection (COD) techniques are aimed to incorporate two main ideas. First, they avoid assigning a higher outlier score to instances that stand out in attributes that are not directly indicative of outlierness, called *contextual attributes*. Second, they aim to tease out the instances, whose behavior, defined by *indicator attributes*, deviates markedly only in sub-populations identified by similar contexts. This demarcation between the two types of attributes, namely contextual and indicator, has been recently shown to improve the detection performance of outlier detection techniques in various domains [9,13,17], and is an active area of interest.

© Springer Nature Switzerland AG 2019
M. Berlingerio et al. (Eds.): ECML PKDD 2018, LNAI 11051, pp. 139–156, 2019.
https://doi.org/10.1007/978-3-030-10925-7_9

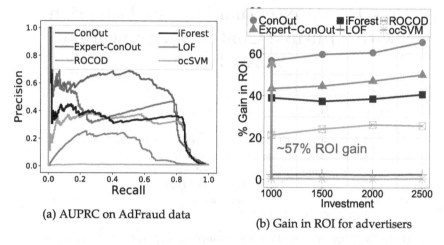

(a) AUPRC on AdFraud data

(b) Gain in ROI for advertisers

Fig. 1. Proposed CONOUT achieves significant improvements in *both* (a) detection & (b) revenue gains compared to existing techniques. (See Sect. 3.1 for details)

Motivating Application: To exemplify the key ideas of COD, consider the domain of advertisement fraud where one of the primary goals is to identify publishers[1] that illicitly generate fake eyeballs to increase their revenue by a variety of schemes. Advertisement fraud is well documented to have multiple mechanisms that generate these fake eyeballs [4]. Outlier detection has been used to detect fraudulent publishers in the advertisement industry (refer [14] for a survey). Publishers can be characterized by a long list of features[2] derived from ad request data as well as external sources (See e.g., Table 2). For ease of illustration, let us consider three attributes - average clicks on ads (`clicks`), average number of ad impressions served (`impressions`) and host country (`country`) of the publisher to identify fraudulent publishers.

Why COD? Which `country` a publisher belongs to can not be a direct indication of fraudulent activity. However, `country` could be used to infer `clicks` and `impressions`, hence could be used as a contextual attribute. Traditional detection techniques that consider `country` equally important to the other two variables would be adding additional noise to the model. On the other hand, if one were to not consider `country` altogether, the auxiliary information provided by considering the sub-populations created by `country` would not be incorporated. Hence, one would expect a model similar to [9,13], which treats `country` as a context, and `clicks` and `impressions` as indicators to perform better. However, in practice, given a long list of publisher features and the complex generating mechanism of ad fraud, the assumption of a *single, user-specified* demarcation of attributes in the existing COD techniques is questionable.

---

[1] A publisher is an entity that provides real estate on their website to host ads.
[2] Throughout the paper, attributes and features are used interchangeably.

Why Multiple Contexts? Viewing `clicks` as an indicator attribute is justified since fraudulent publishers are expected to have higher clicks compared to other publishers. However, to avoid detection, illegitimate publishers often employ schemes to mimic legitimate publishers and camouflage their clicks, while feeding off the long tail impression revenue [4]. In such a scheme, `clicks` may no longer be indicative of fraudulent behavior directly but would serve as a context to assess the deviation in `impressions` served. This ambiguity of viewing `clicks` as a context or an indicator attribute further debates the assumption of a *single, user-specified* demarcation and motivates the need for *multiple* contexts that current COD approaches do not explore. To better detect fraudulent publishers, an approach that incorporates both the contexts is required. Also, given the uncertainty of the role of different attributes, a data-driven identification of contexts and indicator attributes would be needed.

To address current limitations, we introduce a novel COD technique called CONOUT that does not rely on a pre-specified context, rather automatically identifies and incorporates multiple contexts. Our work is motivated and applied to the publisher fraud detection problem in the ad domain. It outperforms a list of existing techniques, including one with domain expert-specified context, in identifying fraudulent publishers (Fig. 1a). CONOUT also achieves statistically significant revenue gains when compared to its competitors. In particular, CONOUT provides more than 57% gains (Fig. 1b) in terms of return on investment (ROI) to the advertiser when compared to a naïve fraud detector and ~20% gains when compared to its nearest competitor. We summarize our notable contributions as follows.

- **Automatic context formation:** To identify contexts, we develop a unified measure grounded with concepts of statistical hypothesis tests to capture dependence between the attributes. The measure can handle mixed (type) attributes and quantifies the similarity of sub-populations generated by attributes which we leverage to automate the context formation.
- **Context-incorporated detection algorithm:** CONOUT quantifies outlierness of an instance with reference to its sub-population specified by a given context. Rather than training a separate detector for each sub-population in a given context, we train a *single* density-based outlier detector and introduce a scheme for re-weighing neighbors of a given test instance (for density estimation) by their distance in the contextual space.
- **Incorporating multiple contexts:** CONOUT searches through multiple contexts to spot one in which a point deviates the most in its indicator attributes. In essence, CONOUT is an ensemble over contexts.
- **Parameter-free nature:** CONOUT relies on one parameter (a kernel bandwidth) to be specified. We introduce an unsupervised model selection procedure for tuning the parameter, as such, CONOUT requires no user input.
- **Application to ad fraud:** We provide an in depth case study of CONOUT in the ad domain. To showcase the advantage of deploying CONOUT versus competing detectors in making ad-placement decisions, we develop a cost

benefit framework to assess the ROI (a metric more relevant to this domain than detection accuracy) gained by an advertiser.

Reproducibility: Implementation of CONOUT and public data used in experiments are open-sourced at https://cmuconout.github.io/.

2   CONOUT for Outlier Detection with Multiple Contexts

Notation. Consider an input data set $\mathcal{D} = \{\mathbf{x}_1, \mathbf{x}_2, \ldots, \mathbf{x}_n\}$ containing n points in d dimensions, where $\mathcal{F}$ denotes the feature set. A context $\mathcal{C}_p$ with p attributes, referred to as *contextual attributes*, is a subset of $\mathcal{F}$ and the corresponding *indicator attributes* are denoted by $\mathcal{I}_p = \mathcal{F} \setminus \mathcal{C}_p$, where $|\mathcal{C}_p| = p$, $|\mathcal{I}_p| = d - p$.

Definition 1 (Sub-population). *Given a context $\mathcal{C}_{p_k}$ and a point $\mathbf{x}_i$, its sub-population consists of objects similar to $\mathbf{x}_i$ in the context space, i.e. w.r.t. features in $\mathcal{C}_{p_k}$. If $\mathcal{C}_{p_k}$ is categorical, all points with the same categorical value as $\mathbf{x}_i$ belong to its sub-population. For a numerical context, a distance measure capturing the similarity in the context space is used to specify a sub-population.*

Our aim is to find $\boldsymbol{s} = \{s_1, s_2, \ldots, s_n\}$, where s_i is the outlier score of instance $\mathbf{x}_i$ in $\mathcal{D}$ by automatically finding and incorporating multiple contexts. We break down the task into two sub problems.

Problem 1 (Automatic Context Formation). **Given** a dataset $\mathcal{D} \in \mathbb{R}^{n \times d}$, with feature set $\mathcal{F}$; **Find** a set of contexts $\mathcal{C} = \{\mathcal{C}_{p_1}, \mathcal{C}_{p_2}, \ldots, \mathcal{C}_{p_K}\}$ such that each $\mathcal{C}_{p_k}$ would act as a suitable frame of reference for set of indicator attributes $\mathcal{I} = \{\mathcal{I}_{p_1}, \mathcal{I}_{p_2}, \ldots, \mathcal{I}_{p_K}\}$, where $\mathcal{I}_{p_k} = \mathcal{F} \setminus \mathcal{C}_{p_k}$ (Details in Sect. 2.1).

Problem 2 (Context-incorporated Outlier Detection (COD)). **Given** a set of contexts $\mathcal{C}$ and corresponding indicator attributes $\mathcal{I}$; **Find** the outlier scores $\boldsymbol{s} = \{s_1, s_2, \ldots, s_n\}$ that incorporate the K contexts in $\mathcal{C}$ such that s_i is representative of the deviation of $\mathbf{x}_i$ in each $\mathcal{I}_{p_k}$ that share similar $\mathcal{C}_{p_k}$ (Details in Sect. 2.2).

Next we introduce CONOUT that addresses both problems stated above.

2.1   Automatic Context Formation

The underlying assumption of contextual detection is that objects sharing similar contextual attributes are expected to have similar indicator attributes [9]. The objects that share similar contextual attributes can be viewed as sub-populations of the whole data. For instance, earlier we used `country` (context) to create these sub-populations to assess `clicks` and `impressions` (indicators). These sub-populations are expected to have similar behavior and deviation from this behavior would indicate a contextual outlier. Assuming that the contexts are unknown and multiple, a naive way of forming contexts would be to consider all the subsets of the feature set d resulting in $(2^d - 1)$ contexts.[3] However, this is computationally infeasible in high dimensions.

Alternatively, since the aim of contexts is to identify sub-populations, one could group the attributes which would result in similar sub-populations. Intuitively, a pair of highly dependent attributes would result in similar sub-populations. For example, two numerical features that have a similar rank ordering of instances would produce similar sub-populations when binned. As such, a measure of rank correlation can be used to capture dependence. However, many practical datasets often consist of *both* numerical and categorical attributes. To effectively handle *mixed* attributes, i.e. capture dependence between attribute pairs of mixed type, we develop a *unified* measure by leveraging statistical tests to quantify dependence between two samples. We then use the measure to group the attributes into context groups that result in similar sub-populations. In particular, we set up hypothesis tests to handle combinations of categorical and numerical attributes, where the p value (p-val) of the test would signify the dependence between a given pair of features. Depending on the types of the attribute pair, we calculate the dependence using the following tests.

Numerical-Numerical: For a pair of numerical attributes, we use the nonparametric Spearman's rank correlation statistic that operates on the rank orderings of the two numerical features. Let us denote by $\mathbf{v}_f$ and $\mathbf{v}_{f'}$ two vectors with values for n points in $\mathcal{D}$ of two arbitrary numerical features f and f' in $\mathcal{F}$. The test statistic ρ is given by,

$$\rho = \frac{cov(\mathbf{r}_f, \mathbf{r}_{f'})}{\sigma_{\mathbf{r}_f} \sigma_{\mathbf{r}_{f'}}} \tag{1}$$

where $\mathbf{r}_f$ and $\mathbf{r}_{f'}$ correspond to vectors that hold the indices of points when ranked by values in $\mathbf{v}_f$ and $\mathbf{v}_{f'}$. Spearman's rank correlation assesses how well the relationship between two features can be described using a monotonic function. The test is widely used to quantify the dependence between both ordinal and continuous features. To determine the significance of the test, we employ a permutation test [8]. The permutation test works by randomly reshuffling the observed data to generate multiple samples. For each such permuted sample, the test statistic is computed. If one were to generate B such permuted samples, the (null) distribution of the corresponding test statistics under no dependence could be used to obtain the p-val of the observed sample.[4]

Numerical-Categorical: For a pair of numerical and categorical features, we use the Kruskal-Wallis non-parameteric test statistic that operates on the rank orderings of the numerical feature within each group of the categorical feature. The test statistic H is given by,

$$H = \frac{\sum_{c=1}^{a} n_c (\overline{r_c} - \bar{r})^2}{\sum_{c=1}^{a} \sum_{i=1}^{n_c} (r_{ic} - \bar{r})^2} \tag{2}$$

---

[3] We would need to omit the set which contains all the d features since there would be no indicator attributes left to assess the deviation.

[4] Typically a small set of random permutations of the observed data, $B = 400$ [8] is sufficient to generate a reliable significance value.

where a is the arity of the categorical feature, n_c is the number of data points of category c, r_{ic} is the rank of observation i of category c with respect to the numerical feature, $\bar{r}_c$ is the average rank of all observations in category c, $\bar{r}$ is the average rank of all the observations. Under the null hypothesis of independence of the two samples, the Kruskal-Wallis statistic asymptotically follows a Chi-squared distribution with $(a-1)$ degrees of freedom, which is used to determine the significance of the test statistic. To account for the possibility of ties, we perform a permutation test as before.

Categorical-Categorical: For a pair of categorical features, we use the Chi-square statistic which quantifies the differences in the observed and expected frequency distribution of the two features. The statistic χ^2 is given by,

$$\chi^2 = \sum_{c=1}^{a} \sum_{c'=1}^{a'} \frac{(O_{cc'} - E_{cc'})^2}{E_{cc'}} \tag{3}$$

where a, a' are the arities of the two categorical features, $O_{cc'}$ and $E_{cc'}$ denote the observed and expected number of instances of type c, c'. Chi-square tests are widely used as a test of independence to assess whether unpaired observations on two samples, expressed in a contingency table, are independent of each other. The test statistic follows a Chi-squared distribution with $(a-1)(a'-1)$ degrees of freedom and is used to assess the significance of the test. To address the possibility of very few observations in a cell $O_{cc'}$, which could lead to inaccurate inference, we employ a permutation test to compute the p-val.

Unified Measure: The p-val of each of the tests listed above signifies the dependence between the pair of features. If a pair of features have a low p-val, one could reliably reject the null hypothesis of the two features being independent. Hence, to capture the dependence of any two features, we use $1 - p\text{-val} \in [0, 1]$ as our unified measure of dependence.

Forming Context Groups: Next, we perform a clustering of the features into context groups based on the unified (dependence) measure, using an algorithm that automatically decides the number of context groups, denoted by G, $G \le d$. This is achieved either by employing X-means [11] or by performing a hierarchical clustering using the gap statistic [7] to automatically estimate the number of clusters. By construction, features in each context group are highly dependent and would result in similar sub-populations. Therefore, considering all $(2^G - 1)$ combinations of the context groups would provide a sufficient and computationally efficient proxy to the naïve way of forming contexts based on all possible $(2^d - 1)$ combinations of the original features.

This completes the formation of set of contexts $\mathcal{C} = \{\mathcal{C}_{p_1}, \mathcal{C}_{p_2}, \ldots, \mathcal{C}_{p_K}\}$, $K = (2^G - 1)$, as stated in Problem 1. Next, we introduce our proposed detection algorithm that incorporates all these K contexts.

2.2 Context-Incorporated Outlier Detection

After generating multiple contexts, we aim to use them in assigning an outlier score s_i to each instance $\mathbf{x}_i \in \mathcal{D}$. Given $(\mathcal{C}_{p_k}, \mathcal{I}_{p_k})$, a single (context, indicators) tuple, one could cluster the instances in the context space, to find sub-populations in the context and use a traditional outlier detection technique to assign scores to each sub-population based on the corresponding indicator attributes. The same heuristic could be extended to multiple contexts. There are two potential problems with this approach. First, clustering the data instances would involve non-trivial choices of the clustering algorithm (depending on the distribution of the data) and number of clusters.[5] Second, to assign the outlier scores, one would need to learn multiple models (corresponding to each sub-population in a context) which may be computationally expensive.

We address these issues by learning only a *single* model per context using the indicator attributes and weigh the scores of the model based on the similarity in the corresponding contextual attributes, completely avoiding the need to cluster the data instances into sub-populations. To achieve this, we propose a modification to the outlier scoring mechanism of isolation forest [10] (iForest), a popular tree based outlier detection technique, which we briefly review next.

iForest Overview: The core idea of iForest is to isolate a data point by recursively partitioning the feature space into random intervals. The recursive partitioning can be visualized as a binary search tree where the path of a tree from root to leaf node is a conjunction of multiple random feature splits. Broadly, iForest comprises of two phases - training and testing. In the training phase, multiple trees (denoted by t) are built by sampling a set of points (denoted by ψ) for building each tree. In the testing phase, an object traverses through trees built earlier until it reaches a leaf node. Intuitively, training instances in the same leaf with a point are its near-neighbors in the subspace specified by split-features from root to leaf (see Fig. 2a for e.g. leaf node in an iTree), where the count of such neighbors serves as a crude estimate of density (the lower the count of such near-neighbors, the more likely that the point is an outlier). iForest outlier score for an instance is given by,

$$o(\mathbf{x}_i, \psi) = 2^{-\frac{E(h(\mathbf{x}_i))}{c(\psi)}} \tag{4}$$

where $h(\mathbf{x}_i)$ is a function of number of training instances in the leaf node to which $\mathbf{x}_i$ has traversed to, $E(h(\mathbf{x}_i))$ is the average of $h(\mathbf{x}_i)$ over multiple trees and $c(\psi)$ is a normalizing constant dependent on sample size ψ. A higher value of $o(\mathbf{x}_i, \psi)$ would indicate a higher chance of the instance being an outlier.

Next, we present how we modify Eq. 4 to quantify contextual outlierness of a point in CONOUT. To avoid cluttered notations, we consider a single context and later explain how to combine the scores from all K contexts.

---

[5] Note that, earlier we perform clustering in features rather than on data instances with a carefully constructed unified measure and are less prone to the issues mentioned.

Contextual Weighing: To incorporate a context into Eq. 4, we should account for the similarity of test instance and the tree neighbors in the contextual space. Intuitively, a neighbor that is dissimilar to the test instance in the contextual space should contribute less (to density) compared to a similar neighbor that belongs to the same sub-population. Formally, since $h(\mathbf{x}_i)$ is determined based on the leaf/neighbor counts, for a test instance $\mathbf{x}_i$ and a neighbor $\mathbf{x}_j$ that shares the same leaf in a tree, we need a smooth function $\phi(\mathbf{x}_i, \mathbf{x}_j)$ that returns 1 when $\mathbf{x}_i$ and $\mathbf{x}_j$ are identical and approaches 0 for the more dissimilar $\mathbf{x}_i$ and $\mathbf{x}_j$. We note that the radial basis function suits this purpose, given by,

$$\phi_\gamma(\mathbf{x}_i, \mathbf{x}_j) = \exp(-\gamma \|\mathbf{x}_i - \mathbf{x}_j\|) \in [0, 1] , \tag{5}$$

where $\|\mathbf{x}_i - \mathbf{x}_j\|$ denotes the distance between $\mathbf{x}_i$ and $\mathbf{x}_j$ in the contextual space, and γ is the kernel bandwidth; a free parameter that controls the radius of influence. A higher γ would impose a higher penalty to the contribution of the points farther in the context space, while a lower γ would impose a relatively lesser penalty. γ can also be seen as a parameter that controls the influence of the context on the indicators. When $\gamma = 0$, the influence of context vanishes and the modified outlier score would be equivalent to the score of an iForest solely based on the indicators. Then, the modified score of a point is written as

$$o(\mathbf{x}_i, \psi, \phi_\gamma) = 2^{-\frac{E(h(\mathbf{x}_i, \phi_\gamma))}{c(\psi)}} \tag{6}$$

where this time $h(\mathbf{x}_i, \phi_\gamma)$ is a function of the *weighted* number of training instances in the same leaf as $\mathbf{x}_i$, where weights are obtained by Eq. 5.

In short, for a context $\mathcal{C}_{p_k}$, we train an iForest using the indicator attributes $\mathcal{I}_{p_k} = \mathcal{F} \setminus \mathcal{C}_{p_k}$. During the testing phase, for a point $\mathbf{x}_i$, we find the leaf it traverses to on each tree, use Eq. 5 to reweigh the points in each such leaf to estimate a weighted average count of neighbors $h(\mathbf{x}_i, \phi_\gamma)$. Then, we use Eq. 6 to assign an outlier score. We show an illustration of the idea in Fig. 2.

We remark that the proposed modification could be integrated into other outlier detection techniques that are based on near neighbors of a test instance in arbitrary subspaces. For example, the outlier score of Half Space Trees [15] could also be modified in the stated fashion.

Distance in the Context Space: The distance $\|\mathbf{x}_i - \mathbf{x}_j\|$, input to the Eq. 5 should capture the similarity in the context space of the instance $\mathbf{x}_i$ to its neighbors in a given tree. Recall from Sect. 2.1 that a context could consist of attributes of mixed type. Here, we specify the distance computation in such a space.

For the attributes that are categorical in the context space, if $\mathbf{x}_i$ and $\mathbf{x}_j$ do not share the same value in *any* categorical attribute, then they do not belong to the same sub-population (for instance if they do not belong to the same country) and hence $\mathbf{x}_j$ should have no contribution to the density estimation at $\mathbf{x}_i$, that is, $\phi_\gamma(\mathbf{x}_i, \mathbf{x}_j) = 0$, irrespective of the numerical attributes. If $\mathbf{x}_i$ and $\mathbf{x}_j$ share the same value in *all* the categorical attributes, then we use the normalized Euclidean distance on the numerical attributes.

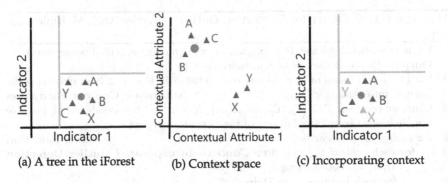

Fig. 2. Toy Example: To exemplify our *contextual weighing* scheme, consider the above example setup. In (a), we visualize the neighbors (in red) of the test instance (in green) for a single tree of iForest created by random splits on the indicator attributes. All the neighbors (A, B, C, X, Y) in the same leaf contribute equally to the outlier score of test instance irrespective of the alignment of the points in the context space (b). For the test instance to be a contextual outlier, points closer to it in context space (A, B, C) should contribute more, while points farther (X, Y) should contribute less (to density estimation). This is achieved using Eq. 5; weighing down the points farther from the test instance in context space (b) (X, Y greyed out in (c)) more than the points closer (A, B, C). (Color figure online)

Combining Scores from Multiple Contexts: In the testing phase, we compute the outlier score using Eq. 6, for all the K contexts in $\mathcal{C}$ with their corresponding indicator attributes $\mathcal{I}$. This results in K scores for each instance $\mathbf{x}_i$. Since a larger score indicates a higher chance of being an outlier, we use the maximum of the scores across all contexts as our assembling scheme, i.e., $s_i = \max_{k \in \{1,...,K\}} o_k(\mathbf{x}_i, \psi, \phi_\gamma)$. Taking the maximum achieves the purpose of teasing out a potential outlier which stands out in a particular context but is hidden in the rest of the contexts.

Choosing γ: Given the importance of γ as a parameter to control the influence of a context, we vary γ on a logarithmic grid between 10^{-3} to 10^3 and employ an unsupervised model selection approach leveraging [6]. The idea is to convert the outlier scores for a certain γ into calibrated probabilities assuming the posterior probabilities follow a logistic sigmoid function, $\sigma(s_i) = 1/(1 + \exp(-(w_0 + w_1 s_i)))$. The parameters of the sigmoid function are estimated from the scores. This allows us to use the goodness of fit criterion across different models (corresponding to various γ) to choose the γ that corresponds to the most calibrated probabilities. Formally, given $\mathbf{s}$, the outlier scores of the n instances obtained using a certain γ, let us denote by $\boldsymbol{\ell}$ a binary *latent* vector corresponding to the unobserved labels of the n instances, which takes the value $\ell_i = 1$ if the instance is an outlier and 0 otherwise. Then, the negative log likelihood function can be written as

---

Algorithm 1. CONOUT for Contextual Outlier Detection with Multiple Contexts

---

    Input: unlabeled dataset $\mathcal{D} = \{\mathbf{x}_1, \mathbf{x}_2, \ldots, \mathbf{x}_n\}$ (no user-specified parameters)
    Output: the outlier scores of n instances $s = \{s_1, s_2, \ldots, s_n\}$

1: Use $(1 - p\text{-val})$ based on appropriate test statistics in Eqs. 1, 2, 3 for clustering the features into context groups   ▷ **Automatic Context Formation**
2: Generate contexts $\mathcal{C} = \{\mathcal{C}_{p_1}, \mathcal{C}_{p_2}, \ldots, \mathcal{C}_{p_K}\}$, $K = 2^G - 1$ by considering all possible combinations (except the full set) of the G context groups
3: **for** each γ in the logarithmic grid 10^{-3} to 10^3 **do**   ▷ **Model Selection**
4:     **for** each context $\mathcal{C}_{p_k}$ in $\mathcal{C}$ **do** ▷ **Context- incorporated Outlier Detection**
5:         Train an iForest using $\mathcal{I}_{p_k} = \mathcal{F} \setminus \mathcal{C}_{p_k}$
6:         **for** each instance $\mathbf{x}_i$ in $\mathcal{D}$ **do**
7:             Compute score $o_k(\mathbf{x}_i, \psi, \phi_\gamma)$ using the modified outlier score in Eq. 6
8:     **for** each instance $\mathbf{x}_i$ in $\mathcal{D}$ **do**
9:         Assign $s_i^\gamma = \max_{k \in \{1, \ldots, K\}} o_k(\mathbf{x}_i, \psi, \phi_\gamma)$, to form s^γ
10:    Estimate the parameters in Eq. 7 via EM using s^γ
11: **Return** the scores s^γ based on the γ corresponding to the maximum loglikelihood

---

$$LL(\boldsymbol{\ell}|\boldsymbol{s}) = \sum_{i=1}^{n} \left[\log(1 + \exp(-w_0 - w_1 s_i)) + (1 - \ell_i)(w_0 + w_1 s_i) \right] \qquad (7)$$

EM Initialization: The model parameters could be estimated by employing the EM algorithm [5] that simultaneously estimates the unobserved labels ℓ_i's, and parameters w_0, w_1 of the sigmoid function. However, EM provides only a locally optimum solution for latent variable functions like Eq. 7, the quality of which strongly depends on the initialization. Thanks to the ranking provided by the scores s, we can heuristically initialize our latent variable vector $\boldsymbol{\ell}$ where for a given threshold, the instances having a score higher than the threshold are initially labeled as outliers and the rest as inliers. This is a more informed initialization compared to a random one. Since we do not know the exact threshold, we try a few random thresholds in decreasing order of the scores, run EM multiple times and pick the solution that yields the highest likelihood.

For each γ, the maximum likelihood based on the fitted parameters is noted. We then choose the γ that corresponds to the maximum likelihood of the different models. Since the number of parameters learnt in all the models is the same, this is equivalent to using a goodness of fit measure such as AIC/BIC. Note that the proposed unsupervised model selection criterion could be used for other outlier detection techniques that require user-specified (hyper)parameters. For example, the number of nearest neighbors k required by most distance or density based methods (like LOF [3]) could be tuned in a similar fashion.

This completes the estimation of outlier scores s, as stated in Problem 2, that incorporates the K contexts generated from Sect. 2.1.

Summary: We conclude by outlining the detailed steps of CONOUT in Algorithm 1. A high level abstraction of the CONOUT can be described as follows:

1. **Context Groups:** Use unified dependence measure $(1 - p\text{-val})$ to cluster the features into G context groups, where G is automatically chosen [11].
2. **Context Formation:** Form $K = 2^G - 1$ contexts by considering all possible combinations of the G context groups.
3. **Training Detectors:** For each context $\mathcal{C}_{p_k}$, $k = 1, \ldots, K$, train an iForest on corresponding indicator attributes $\mathcal{I}_{p_k} = \mathcal{F} \setminus \mathcal{C}_{p_k}$.
4. **Outlier Scoring:** For each point $\mathbf{x}_i$ in $\mathcal{D}$, assign the outlier score to be the maximum across the K contexts to form $\mathbf{s}$.

One-Click Algorithm: As evident from Algorithm 1, CONOUT does not require the user to specify any input values other than feeding in the data set $\mathcal{D}$, which makes CONOUT a one-click algorithm that runs *parameter free*. We carefully choose γ, the only hyper parameter (kernel bandwidth) within CONOUT using the probabilistic unsupervised model selection scheme.

Complexity Analysis. To conclude, we analyze the time complexity of each of the above steps. In Step 1, we cluster the features based on dependence, quantified by the unified measure $(1 - p\text{-val})$ which takes $O(n)$ for a given pair.[6] We use the scalable X-means algorithm [11], which extends k-means to automatically find the number of clusters. Initially, we randomly pick G features as the centroids. In the assignment step, we compute dependence of each feature on the G designated centers, which is $O(ndG)$. In re-centering, we need to identify, within each current cluster, the feature with the largest total dependence to others in the cluster to be designated as the new center (note that we can *not* re-center by *averaging* the columns, as we work with mixed attributes). To this end, we randomly sample a few points and pick among those the one with the largest total dependence. This takes $O(nd)$ for all clusters. Overall complexity of clustering is $O(ndG)$ where G is small. Having identified G constant number of clusters, Step 2 takes $O(K)$ for constructing $K = 2^G - 1$ contexts. In step 3, training an iForest depends on the number of trees t and sample size ψ which is $O(t\psi \log \psi)$, where depth of each tree is $O(\log \psi)$. We train an iForest for each context, resulting in $O(Kt\psi \log \psi)$. Finally, assigning an outlier score to a point involves finding its leaf (in $O(\log \psi)$) in each tree, and calculating its distance (in the given context space, in $O(d)$) to the points sharing the same leaf (smaller than ψ points in each leaf). This takes $O(Knt(\log \psi + d\psi))$ for n points and K contexts across t trees. Overall complexity of steps 1–4 is $O(Kt[(n + \psi) \log \psi + nd\psi])$. Since t and ψ are constants, the complexity is linear in data size n and d as well as the number of generated contexts K.

---

[6] Since the number of simulations B is constant, we omit this in the complexity.

3 Experiments

Datasets. In this section, we empirically evaluate the efficacy of CONOUT. To this end, we provide an in depth case study of the practical applicability of CONOUT in ad domain by comparing it to multiple baselines. In addition, we also run CONOUT on multiple publicly available data sets from ODDS[7] and UCI repositories. A summary of the data sets ordered wrt outlier % is reported in Table 1. We consider both mixed and numeric-only attribute data sets to show the efficacy of CONOUT in handling different types of attributes.

Table 1. Data set summary

Name	Size n	dim. d	Outliers (%)	Type
AdFraud	18,959	14	208 (1.09%)	Mixed
SatImage	5,803	36	71 (1.20%)	Numeric
Pens	6,870	16	156 (2.27%)	Numeric
Mammography	11,183	6	260 (2.32%)	Numeric
Seismic	2,584	18	170 (6.50%)	Mixed
Shuttle	49,097	9	3,511 (7.21%)	Numeric
Income (Adult)	48,842	14	7,841 (24.08%)	Mixed
Satellite	6,435	36	2,036 (32.21%)	Numeric

Baselines. We compare CONOUT to the following state-of-the-art approaches in both traditional outlier and contextual detection literature.

- **ROCOD:** Robust contextual outlier detection [9] (ROCOD) combines both local and global effects in outlier detection. ROCOD requires a single context to be pre-specified. Such a context is not available for any of our data sets, as such, we assign all the categorical attributes in the mixed type data sets in Table 1 as contexts and use numerical ones as indicators. However, picking contexts in numeric data sets is non-trivial and requires domain expertise. Hence, we omit ROCOD from comparison on those data sets.
- **iForest:** iForest [10] is the isolation based tree ensemble detector discussed in Sect. 2.2. We set number of trees $t = 100$ and $\psi = 256$ as suggested in the paper.
- **LOF:** Local Outlier Factor (LOF) [3] compares the local density of each point to its neighbors. We vary k (number of nearest neighbors) between 10 to 100 as suggested in [3] and pick the maximum outlier score.
- **ocSVM:** One class SVM (ocSVM) [12] is a popular outlier detection technique based on the principles of support vectors. We use the default ($\gamma = \frac{1}{n}$) radial kernel and $\nu = 0.5$ for our experiments as suggested in [12].
- **Expert-CONOUT:** Instead of generating multiple contexts in CONOUT, we use a *single* context that our industry collaborators hand-created for the AdFraud data set to investigate the efficacy of our context generation. For a fair comparison, we use the same context for ROCOD in the AdFraud data set.

3.1 Case Study: Ad Fraud Domain

CONOUT is motivated by the ad fraud problem, briefly, of identifying publishers that make revenue through a variety of illegitimate schemes. First, we provide a primer about the advertisement ecosystem in Fig. 3.

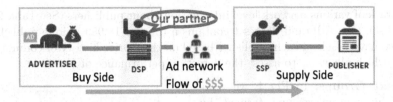

Fig. 3. Primer on advertisement ecosystem: An advertiser (buyer) is an entity that manages ad campaigns of multiple brands, a publisher (seller) provides the real-estate for hosting ads. To establish the buyer/seller relationship, three key intermediaries are involved - Display Side Platforms (DSP), Supply Side Platforms (SSP) and an ad-exchange. Publishers earn revenue based on the number of views, clicks, or actions on ads, and have incentives to commit fraud. To mitigate this, DSPs maintain an up-to-date blacklist of publishers to avoid, while bidding on ad requests. The limitation of such a list is that a fraudulent publisher could switch between multiple websites to get through them. Hence, a robust data driven approach to identify fraudulent publishers is required.

Table 2. Publisher features extracted for AdFraud. N,O,C indicate numerical, ordinal and categorical features respectively. Fields with P are protected based on the agreement with the DSP. The mean features are aggregated over multiple ad requests of a publisher.

Feature	Description	Type	Mean/Arity
Features from the DSP (Display Side Platform) collaborator			
total_visitors	# of users DSP has served ads	N	2122.87
mean_revenue	Mean revenue generated on ads	N	P
mean_bid	Mean bid amount placed by the DSP	N	P
mean_cost	Mean cost paid by the DSP	N	P
mean_conversions	Mean conversions from ads	N	0.002
mean_clicks	Mean clicks on the ads placed	N	1.247
unique_users	#of unique users DSP has served ads	N	1895.42
Features from public repositories (who.is and myip.ms)			
websites_before	# of websites earlier hosted on publisher's IP	N	90.77
websites_notworking	# of websites hosted on IP that are not working	N	52.072
websites_working	# of websites hosted on IP that are working	N	458.91
popularity	Average # of visitors per day	O	2,338,150
alexa_rank	Alexa rank of the publisher	O	570,403
host_country	Hosting country of the publisher	C	5
category	Root IAB category of the publisher	C	5

Application. To detect fraudulent publishers, we partner with a large DSP. We build the publisher features from a snapshot of ad requests served by the DSP. Additionally, we collect data from publicly available repositories[8] which

---

[7] ODDS (Outlier Detection DataSets): http://odds.cs.stonybrook.edu.

[8] https://www.whois.com/whois/, https://myip.ms/.

keep track of various network level information about publishers (See Table 2 for a full list). Our AdFraud data set contains a total of 18,959 unique publishers. We also have labels of each publisher based on their prior history of committing fraud, which allows us to assess the detection performance of CONOUT.

Context Groups: In Fig. 4, we visualize the context groups formed by performing clustering using our proposed unified measure developed in Sect. 2.1. We note that the `popularity` of a publisher is grouped together with `alexa_rank`. This is intuitive since both these attributes would be highly dependent - a higher `popularity` would mean a lower `alexa_rank`. All the features related to monetary gains are grouped together. Ads with higher bid requests cost more to the advertiser and in return

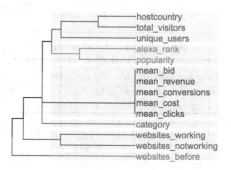

Fig. 4. Context groups of AdFraud

one could expect higher revenue, clicks and conversion. The `total_visitors` of a publisher and `unique_visitors` would be highly correlated and are grouped together. In total, we have $G = 6$ context groups forming $K = 63$ contexts.

Detection Performance: In Fig. 1a, we compare AUPRC (Area Under the Precision-Recall Curve) of CONOUT with the baselines listed earlier averaged across five independent runs. We notice that CONOUT significantly outperforms all the baselines in detecting fraudulent publishers. Interestingly, Expert-CONOUT (CONOUT with a context specified by experts from our partner DSP) is inferior to CONOUT with automatically generated contexts, demonstrating the utility of multiple contexts and the sub-optimality of a hand-made context. ROCOD with the same pre-specified context also performs poorly in comparison to CONOUT further supporting the incorporation of multiple contexts. The non-contextual techniques ocSVM and LOF perform poorly, followed by iForest indicating the importance of contexts in the ad fraud domain.

Cost-Benefit Analysis: AUPRC measures performance assuming all the fraudulent publishers are equally important to detect. However, we note that as different publishers employ different schemes, detecting the *right* set of fraudulent publishers becomes more important. For instance, a fraudulent publisher that targets costly bids would deplete the advertiser's budget quicker, returning higher gains when caught. Therefore, we use a second evaluation metric more relevant to the ad domain, which compares the benefits an advertiser obtains by employing a given detector in the bidding mechanism.

Specifically, we perform a cost benefit analysis, computing the advertiser's return on investment (ROI). This is done using a simulation mirroring the ad buying ecosystem outlined in Algorithm 2, where the advertiser decides whether or not to bid on an ad request (denoted by *bidbit*) from a given publisher based on its score by the employed detector (the higher the score/risk s_i, the more

---

Algorithm 2. Advertiser Cost-Benefit Analysis

Input: $\{s_1^1, s_2^1, \ldots, s_n^1\}, \ldots \{s_1^M, s_2^M, \ldots, s_n^M\}$, fraudulent scores of M competing detectors,
Investment (Budgets) $\boldsymbol{b} = \{b^1, b^2, \ldots, b^M\}$, initially all equal to bgt.
Output: $ROI = \{roi^1, roi^2, \ldots, roi^M\}$, Return on Investments, all set to zero initially.
1: **while** any of $b^l \in \boldsymbol{b} > 0$ **do**   ▷ *Until all the budgets are depleted*
2:    Simulate an ad request by choosing a publisher i based on $total\_visitors_i$
3:    **for** each detector l **do**
4:        Based on s_i^l, decide $bidbit^l$,   ▷ *Decide whether or not to bid on the request*
5:        **if** $bidbit^l = 1$ and i's true label is benign **then**   ▷ *Budget spent, revenue gained*
6:            $b^l = b^l - \text{mean\_cost}_i$, $roi^l = roi^l + \text{mean\_revenue}_i/\text{bgt}$
7:        **else if** $bidbit^l = 1$ and i's true label is fraudulent **then**
8:            $b^l = b^l - \text{mean\_cost}_i$ ▷ *Budget spent lost due to fraudulent scheme*
9:        **else**   ▷ $bidbit^l = 0$, *i.e., DSP does not bid due to high risk (outlier score)*
10:           b^l, roi^l stay the same   ▷ *No budget spent, no revenue received*
11: **Return** $ROI \leftarrow \{roi^1, roi^2, \ldots, roi^M\}$

---

likely the advertiser *not* to bid. The simulation incorporates the importance of a publisher based on the number of requests made (`total_visitors`), cost to the advertiser (`mean_cost`) and revenue (`mean_revenue`). The more s_i reflects the true labels of publishers, the more revenue the advertiser makes due to decisions on correct estimates, and the more they lose (to fake eyeballs) otherwise.

We perform the simulation for 10,000 times with varying initial bgt values. In Table 3, we report the relative % gain—the difference of mean ROI of a method with that of a naïve detector (that bids 98.91 times out of 100—based on the outlier % in AdFraud dataset) divided by mean ROI of naïve detector. We observe that CONOUT outperforms all the baselines and achieves a minimum of ~57% relative gain against a naïve detector. NA's at budget 500 are due to zero ROI obtained using a naïve detector. We also report the relative gain of CONOUT over its closest competitor, Expert-CONOUT (in brackets), where we use a single context provided by our industry collaborators. CONOUT provides a minimum of ~20% relative gain in ROI when compared to Expert-CONOUT.

Additionally, to assess the significance of the differences in CONOUT's ROI over the naïve detector as well as Expert-CONOUT (its closest competitor), we employ a one tailed paired two sample t-test and report the corresponding p-values in Table 3. The lower p-values indicate that CONOUT achieves

Table 3. Comparison of relative % gains in ROI with varying budgets.

Budget	CONOUT	p-value wrt naïve (wrt Exp-CONOUT)	Expert-CONOUT	ROCOD	iForest	LOF	ocSVM
500	**NA (24.58)**	2e−5 (2e−5)	NA	NA	NA	NA	NA
1000	**56.71 (19.55)**	1e−7 (4e−5)	43.38	21.26	38.98	2.40	0.40
1500	**59.78 (19.63)**	2e−6 (1e−4)	44.62	24.15	37.36	2.47	0.47
2000	**60.63 (25.65)**	2e−8 (2e−5)	47.08	26.45	38.54	2.31	0.31
2500	**65.62 (21.07)**	1e−6 (4e−4)	50.11	25.78	40.69	2.53	0.53

statistically significant gains in both cases. Moreover, the estimated ROI of CONOUT is consistent, where returns increase as investment/bgt increases. We remark that LOF, ocSVM and the naïve detector yielded no return with a bgt of 500. Given the low returns in ad domain, this is expected. On the contrary, this further highlights the improved gains obtained by CONOUT even at low budget.

3.2 CONOUT on Public Datasets

Next, we compare the detection performance of CONOUT on various data sets with ground truth outliers listed in Table 1. In Tables 4 and 5, we report the mean average precision (averaged over five independent runs) of competing methods on mixed attribute and numeric attribute data sets respectively. CONOUT consistently outperforms the baselines. The importance of incorporating multiple contexts is evident in the Seismic and Income data sets, where baselines considering single context (ROCOD[9]) or no context (iForest, LOF, ocSVM) perform worse compared to CONOUT. Results are similar on numeric data sets. Here, we do not show ROCOD as it requires a pre-specified context, which is not available nor easy to set. CONOUT outperforms the no-context baselines highlighting the potential benefits of incorporating multiple contexts.

Table 4. AUPRC on mixed datasets

Method/Data	CONOUT	ROCOD	iForest	LOF	ocSVM
AdFraud	**0.5138**	0.3011	0.3270	0.1344	0.0108
Seismic	**0.9180**	0.8083	0.8970	0.9007	0.9011
Income (Adult)	**0.5812**	0.5604	0.3128	0.2760	0.2377

Table 5. AUPRC on numeric datasets

Method/Data	CONOUT	iForest	LOF	ocSVM
Pens	**0.3574**	0.3193	0.0499	0.1413
SatImage	**0.9442**	0.9011	0.3525	0.1101
Mammography	0.1902	**0.1925**	0.1425	0.1806
Satellite	**0.6862**	0.6557	0.4037	0.3819
Shuttle	**0.9911**	0.9793	0.2481	0.4331

4 Related Work

Outlier detection has been extensively studied in the literature [1]. Contextual outlier detection (COD) is notably different and is the focus of our work. COD has been studied in [9,13,17]. The method developed in [17] applies to spatio-temporal data where the contexts comprise of spatial, temporal or spatio-temporal attributes. Direct applicability of such techniques to other types of data is not obvious.

Song et al. [13] takes a generative approach to model the relation between context and indicators. Both are modeled separately as a mixture of multiple Gaussian components. Next, a mapping function between the Gaussian components is learned using EM to incorporate the intuition of similar contexts generating similar indicators. Liang et al. [9] tackles the problem of sparsity in

---

[9] We assign the categorical attributes as contexts in Seismic and Income data sets.

the context space by proposing an ensemble of local and global estimation of the indicators. The local estimates are obtained using a kNN regression where context is used to find the neighbors. The global estimates are obtained via a linear or a non-linear regression using both indicators and contexts. All the above techniques assume that there is a *single, user-given* context.

Wang et al. [16] propose a graph based method to find contextual neighbors without the need of a demarcation. However, the contexts here are defined as a set of instances rather than a set of attributes which is different from our problem. Angiulli et al. [2] considers the problem of characterizing outliers in a *labeled* data set by automatically finding context attributes and a *single* indicator attribute to explain a group of outliers—which is not a detection technique.

5 Conclusion

We introduced CONOUT for contextual outlier detection addressing the problem of automatically finding and incorporating multiple contexts while handling mixed type attributes. In summary, we make the following contributions.

- **Automatic context formation**, by developing a unified measure that can handle mixed type attributes;
- **Leveraging multiple contexts**, by proposing a context-incorporated detection algorithm that is assembled over multiple contexts;
- **Parameter-free nature**, by tuning its (one) hyperparameter via an unsupervised model selection criterion, that makes CONOUT a *one-click* algorithm.

Through experiments on real-world data sets, we showed the effectiveness of CONOUT over existing techniques in detection performance. We motivated and applied CONOUT to the ad domain where CONOUT not only improves detection but also provides statistically significant revenue gains to advertisers.

Acknowledgments. This research is sponsored by Adobe University Marketing Research Award, NSF CAREER 1452425 and IIS 1408287. Any conclusions expressed in this material do not necessarily reflect the views expressed by the funding parties.

References

1. Aggarwal, C.C.: Outlier Analysis. Springer, New York (2013). https://doi.org/10.1007/978-1-4614-6396-2
2. Angiulli, F., Fassetti, F., Palopoli, L.: Discovering characterizations of the behavior of anomalous subpopulations. IEEE TKDE **25**(6), 1280–1292 (2013)
3. Breunig, M.M., Kriegel, H.-P., Ng, R.T., Sander, J.: LOF: identifying density-based local outliers. In: SIGMOD, vol. 29, pp. 93–104. ACM (2000)
4. Dave, V., Guha, S., Zhang, Y.: ViceROI: catching click-spam in search ad networks. In: SIGSAC, pp. 765–776. ACM (2013)
5. Dempster, A., Laird, N., Rubin, D.: Maximum likelihood from incomplete data via the EM algorithm. J. Royal Stat. Soc. **39**, 1–38 (1977)

6. Gao, J., Tan, P.-N.: Converting output scores from outlier detection algorithms into probability estimates. In: ICDM, pp. 212–221. IEEE (2006)
7. Hastie, T., Tibshirani, R., Walther, G.: Estimating the number of data clusters via the gap statistic. J. Royal Stat. Soc. B **63**, 411–423 (2001)
8. Lehmann, E.L., Romano, J.P.: Testing Statistical Hypotheses. Springer, New York (2006). https://doi.org/10.1007/0-387-27605-X
9. Liang, J., Parthasarathy, S.: Robust contextual outlier detection: where context meets sparsity. In: CIKM, pp. 2167–2172. ACM (2016)
10. Liu, F.T., Ting, K.M., Zhou, Z.-H.: Isolation forest. In: ICDM. IEEE (2008)
11. Pelleg, D., Moore, A.W., et al.: X-means: extending k-means with efficient estimation of the number of clusters. In: ICML, pp. 727–734 (2000)
12. Schölkopf, B., Williamson, R.C., Smola, A.J., Shawe-Taylor, J., Platt, J.C.: Support vector method for novelty detection. In: NIPS, pp. 582–588 (2000)
13. Song, X., Wu, M., Jermaine, C., Ranka, S.: Conditional anomaly detection. IEEE TKDE **19**(5), 631–645 (2007)
14. Spirin, N., Han, J.: Survey on web spam detection: principles and algorithms. ACM SIGKDD Explor. Newslett. **13**(2), 50–64 (2012)
15. Tan, S.C., Ting, K.M., Liu, T.F.: Fast anomaly detection for streaming data. In: IJCAI, vol. 22, p. 1511 (2011)
16. Wang, X., Davidson, I.: Discovering contexts and contextual outliers using random walks in graphs. In: ICDM, pp. 1034–1039. IEEE (2009)
17. Zheng, G., Brantley, S.L., Lauvaux, T., Li, Z.: Contextual spatial outlier detection with metric learning. In: KDD, pp. 2161–2170. ACM (2017)

Scalable and Interpretable One-Class SVMs with Deep Learning and Random Fourier Features

Minh-Nghia Nguyen and Ngo Anh Vien$^{(\boxtimes)}$

School of Electronics, Electrical Engineering and Computer Science,
Queen's University Belfast, Belfast, UK
{mnguyen04,v.ngo}@qub.ac.uk

Abstract. One-class support vector machine (OC-SVM) for a long time has been one of the most effective anomaly detection methods and extensively adopted in both research as well as industrial applications. The biggest issue for OC-SVM is yet the capability to operate with large and high-dimensional datasets due to optimization complexity. Those problems might be mitigated via dimensionality reduction techniques such as manifold learning or autoencoder. However, previous work often treats representation learning and anomaly prediction separately. In this paper, we propose autoencoder based one-class support vector machine (AE-1SVM) that brings OC-SVM, with the aid of random Fourier features to approximate the radial basis kernel, into deep learning context by combining it with a representation learning architecture and jointly exploit stochastic gradient descent to obtain end-to-end training. Interestingly, this also opens up the possible use of gradient-based attribution methods to explain the decision making for anomaly detection, which has ever been challenging as a result of the implicit mappings between the input space and the kernel space. To the best of our knowledge, this is the first work to study the interpretability of deep learning in anomaly detection. We evaluate our method on a wide range of unsupervised anomaly detection tasks in which our end-to-end training architecture achieves a performance significantly better than the previous work using separate training. Code related to this paper is available at: https://github.com/minh-nghia/AE-1SVM.

1 Introduction

Anomaly detection (AD), also known as outlier detection, is a unique class of machine learning that has a wide range of important applications, including intrusion detection in networks and control systems, fault detection in industrial manufacturing procedures, diagnosis of certain diseases in medical areas by identifying outlying patterns in medical images or other health records, cybersecurity, etc. AD algorithms are identification processes that are able to single out items or events that are different from an expected pattern, or those that have significantly lower frequencies compared to others in a dataset [8,14].

© Springer Nature Switzerland AG 2019
M. Berlingerio et al. (Eds.): ECML PKDD 2018, LNAI 11051, pp. 157–172, 2019.
https://doi.org/10.1007/978-3-030-10925-7_10

In the past, there has been substantial effort in using traditional machine learning techniques for both supervised and unsupervised AD such as principal component analysis (PCA) [6,7], one-class support vector machine (OC-SVM) [12,22,29], isolation forests [18], clustering based methods such as k-means, and Gaussian mixture model (GMM) [4,16,35], etc. Notwithstanding, they often become inefficient when being used in high-dimensional problems because of high complexity and the absence of an integrated efficient dimensionality reduction approach. There is recently a growing interest in using deep learning techniques to tackle this issue. Nonetheless, most previous work still relies on two-staged or separate training in which a low-dimensional space is firstly learned via an autoencoder. For example, the work in [13] simply proposes a hybrid architecture with a deep belief network to reduce the dimensionality of the input space and separately applies the learned feature space to a conventional OC-SVM. Robust deep autoencoder (RDA) [34] uses a structure that combines robust PCA and dimensionality reduction by autoencoder. However, this two-stage method is not able to learn efficient features for AD problems, especially when the dimensionality grows higher because of decoupled learning stages. More similar to our approach, deep clustering embedding (DEC) [31] is a state-of-the-art algorithm that integrates unsupervised autoencoding network with clustering. Even though clustering is often considered as a possible solution to AD tasks, DEC is designed to jointly optimize the latent feature space and clustering, thus would learn a latent feature space that is more efficient to clustering rather than AD.

End-to-end training of dimensionality reduction and AD has recently received much interest, such as the frameworks using deep energy-based model [33], autoencoder combined with Gaussian mixture model [36], generative adversarial networks (GAN) [21,32]. Nonetheless, these methods are based on density estimation techniques to detect anomalies as a by-product of unsupervised learning, therefore might not be efficient for AD. They might assign high density if there are many proximate anomalies (a new cluster or mixture might be established for them), resulting in false negative cases.

One-class support vector machine is one of the most popular techniques for unsupervised AD. OC-SVM is known to be insensitive to noise and outliers in the training data. Still, the performance of OC-SVM in general is susceptible to the dimensionality and complexity of the data [5], while their training speed is also heavily affected by the size of the datasets. As a result, conventional OC-SVM may not be desirable in big data and high-dimensional AD applications. To tackle these issues, previous work has only performed dimensionality reduction via deep learning and OC-SVM based AD separately. Notwithstanding, separate dimensionality reduction might have a negative effect on the performance of the consequential AD, since important information useful for identifying outliers can be interpreted differently in the latent space. On the other hand, to the best of our knowledge, studies on the application of kernel approximation and stochastic gradient descent (SGD) on OC-SVM have been lacking: most of the existing works only apply random Fourier features (RFF) [20] to the input space and treat the problem as a linear support vector machine (SVM); meanwhile,

[5,23] have showcased the prospect of using SGD to optimize SVM, but without the application of kernel approximation.

Another major issue in joint training with dimensionality reduction and AD is the interpretability of the trained models, that is, the capability to explain the reasoning for why they detect the samples as outliers, with respect to the input features. Very recently, explanation for black-box deep learning models has been brought about and attracted a respectable amount of attention from the machine learning research community. Especially, gradient-based explanation (attribution) methods [2,3,26] are widely studied as protocols to address this challenge. The aim of the approach is to analyse the contribution of each neuron in the input space of a neural network to the neurons in its latent space by calculating the corresponding gradients. As we will demonstrate, this same concept can be applied to kernel-approximated SVMs to score the importance of each input feature to the margin that separates the decision hyperplane.

Driven by those reasoning, in this paper we propose AE-1SVM that is an end-to-end autoencoder based OC-SVM model combining dimensionality reduction and OC-SVM for large-scale AD. RFFs are applied to approximate the RBF kernel, while the input of OC-SVM is fed directly from a deep autoencoder that shares the objective function with OC-SVM such that dimensionality reduction is forced to learn essential pattern assisting the anomaly detecting task. On top of that, we also extend gradient-based attribution methods on the proposed kernel-approximate OC-SVM as well as the whole end-to-end architecture to analyse the contribution of the input features on the decision making of the OC-SVM.

The remainder of the paper is organised as follows. Section 2 reviews the background on OC-SVM, kernel approximation, and gradient-based attribution methods. Section 3 introduces the combined architecture that we have mentioned. In Sect. 4, we derive expressions and methods to obtain the end-to-end gradient of the OC-SVM's decision function with respect to the input features of the deep learning model. Experimental setups, results, and analyses are presented in Sect. 5. Finally, Sect. 6 draws the conclusions for the paper.

2   Background

In this section, we briefly describe the preliminary background knowledge that is referred to in the rest of the paper.

2.1   One-Class Support Vector Machine

OC-SVM [22] for unsupervised anomaly detection extends the idea of support vector method that is regularly applied in classification. While classic SVM aims to find the hyperplane to maximize the margin separating the data points, in OC-SVM the hyperplane is learned to best separate the data points from the origin. SVMs in general have the ability to capture non-linearity thanks to the use of kernels. The kernel method maps the data points from the input feature space in $\mathcal{R}^d$ to a higher-dimensional space in $\mathcal{R}^D$ (where D is potentially infinite), where the data is linearly separable, by a transformation $\mathcal{R}^d \rightarrow \mathcal{R}^D$. The most commonly used kernel is the radial basis function (RBF) kernel defined by a similarity mapping between any two points x and x' in the input feature space, formulated by $K(x, x') = \exp(-\frac{\|x-x'\|^2}{2\sigma^2})$, with σ being a kernel bandwidth.

Let w and ρ denote the vectors indicating the weights of all dimensions in the kernel space and the offset parameter determining the distance from the origin to the hyperplane, respectively. The objective of OC-SVM is to separate all data points from the origin by a maximum margin with respect to some constraint relaxation, and is written as a quadratic program as follows:

$$\min_{w,\xi,\rho} \frac{1}{2}\|w\|^2 - \rho + \frac{1}{\nu n}\sum_{i=1}^{n}\xi_i, \tag{1}$$

$$\text{subject to } w^T\phi(x_i) \geq \rho - \xi_i, \xi_i \geq 0.$$

where ξ_i is a slack variable and ν is the regularization parameter. Theoretically, ν is the upper bound of the fraction of anomalies in the data, and also the main tuning parameter for OC-SVM. Additionally, by replacing ξ_i with the hinge loss, we have the unconstrained objective function as

$$\min_{w,\rho} \frac{1}{2}\|w\|^2 - \rho + \frac{1}{\nu n}\sum_{i=1}^{n}\max(0, \rho - w^T\phi(x_i)). \tag{2}$$

Let $g(x) = w.\phi(x_i) - \rho$, the decision function of OC-SVM is

$$f(x) = \text{sign}(g(x)) = \begin{cases} 1 & \text{if } g(x) \geq 0 \\ -1 & \text{if } g(x) < 0 \end{cases}. \tag{3}$$

The optimization problem of SVM in (2) is usually solved as a convex optimization problem in the dual space with the use of Lagrangian multipliers to reduce complexity while increasing solving feasibility. LIBSVM [9] is the most popular library that provides efficient optimization algorithms to train SVMs, and has been widely adopted in the research community. Nevertheless, solving SVMs in the dual space can be susceptible to the data size, since the function K between each pair of points in the dataset has to be calculated and stored in a matrix, resulting in an $O(n^2)$ complexity, where n is the size of the dataset.

2.2   Kernel Approximation with Random Fourier Features

To address the scalability problem of kernel machines, approximation algorithms have been introduced and widely applied, with the most two dominant being Nyströem [30] and random Fourier features (RFF) [20]. In this paper, we focus on RFF since it has lower complexity and does not require pre-training. The method is based on the Fourier transform of the kernel function, given by a Gaussian distribution $p(\omega) = \mathcal{N}(0, \sigma^{-2}\mathbb{I})$, where $\mathbb{I}$ is the identity matrix and σ is an adjustable parameter representing the standard deviation of the Gaussian process.

From the distribution p, D independent and identically distributed weights $\omega_1, \omega_2, ..., \omega_D$ are drawn. In the original work [20], two mappings are introduced, namely the combined *cosine* and *sine* mapping as $z_\omega(x) = [cos(\omega^T x) \ sin(\omega^T x)]^T$ and the offset *cosine* mapping as $z_\omega(x) = \sqrt{2}cos(\omega^T x + b)$, where the offset parameter $b \sim U(0, 2\pi)$. It has been proven in [28] that the former mapping outperforms the latter one in approximating RBF kernels due to the fact that no phase shift is introduced as a result of the offset variable. Therefore, in this paper, we only consider the combined *sine* and *cosine* mapping. As such, the complete mapping is defined as follows:

$$z(x) = \sqrt{\frac{1}{D}} \left[cos(\omega_1^T x) \ ... \ cos(\omega_D^T x) \ sin(\omega_1^T x) \ ... \ sin(\omega_D^T x) \right]^T, \qquad (4)$$

Applying the kernel approximation mappings to (2), the hinge loss can be replaced by $\max(0, \rho - \mathbf{w}^T z(x_i))$. The objective function itself is then equivalent to a OC-SVM in the approximated kernel space $\mathcal{R}^D$, and thus the optimization problem is more trivial, despite the dimensionality of $\mathcal{R}^D$ being higher than that of $\mathcal{R}^d$.

2.3   Gradient-Based Explanation Methods

Gradient-based methods exploit the gradient of the latent nodes in a neural network with respect to the input features to rate the attribution of each input to the output of the network. In the recent years, many research studies [2, 19, 26, 27] have applied this approach to explain the classification decision and sensitivity of input features in deep neural networks and especially convolutional neural networks. Intuitively, an input dimension $\mathbf{x}_i$ has larger contribution to a latent node $\mathbf{y}$ if the gradient of $\mathbf{y}$ with respect to $\mathbf{x}_i$ is higher, and vice versa.

Instead of using purely gradient as a quantitative factor, various extensions of the method has been developed, including Gradient*Input [25], Integrated gradients [27], or DeepLIFT [24]. The most recent work [2] showed that these methods are strongly related and proved conditions of equivalence or approximation between them. In addition, other non gradient-based can be re-formulated to be implemented easily like gradient-based.

3   Deep Autoencoding One-Class SVM

In this section, we present our combined model, namely Deep autoencoding One-class SVM (AE-1SVM), based on OC-SVM for anomaly detecting tasks in high-dimensional and big datasets. The model consists of two main components, as illustrated in Fig. 1 (Left). The first component is a deep autoencoder network for dimensionality reduction and feature representation of the input space. The second one is an OC-SVM for anomaly prediction based on support vectors and margin. The RBF kernel is approximated using random Fourier features. The bottleneck layer of the deep autoencoder network is forwarded directly into the Random features mapper as the input of the OC-SVM. By doing this, the autoencoder network is pressed to optimize its variables to represent the input features in the direction that supports the OC-SVM in separating the anomalies from the normal class.

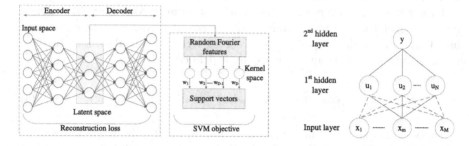

Fig. 1. (Left) Illustration of the Deep autoencoding One-class SVM architecture. (Right) Connections between input layer and hidden layers of a neural network

Let us denote $\mathbf{x}$ as the input of the deep autoencoder, $\mathbf{x}'$ as the reconstructed value of $\mathbf{x}$, and x as the latent space of the autoencoder. In addition, θ is the set of parameters of the autoencoder. As such, the joint objective function of the model regarding the autoencoder parameters, the OC-SVM's weights, and its offset is as follows:

$$Q(\theta, \mathbf{w}, \rho) = \alpha L(\mathbf{x}, \mathbf{x}') + \frac{1}{2}\|\mathbf{w}\|^2 - \rho + \frac{1}{\nu n}\sum_{i=1}^{n}\max(0, \rho - \mathbf{w}^T z(x_i)) \quad (5)$$

The components and parameters in (5) are described below

- $L(\mathbf{x}, \mathbf{x}')$ is the reconstruction loss of the autoencoder, which is normally chosen to be the L2-norm loss $L(\mathbf{x}, \mathbf{x}') = \|\mathbf{x} - \mathbf{x}'\|_2^2$.
- Since SGD is applied, the variable n, which is formerly the number of training samples, becomes the batch size since the hinge loss is calculated using the data points in the batch.
- z is the random Fourier mappings as defined in (4). Due to the random features being data-independent, the standard deviation σ of the Gaussian distribution has to be fine-tuned correlatively with the parameter ν.
- α is a hyperparameter controlling the trade-off between feature compression and SVM margin optimization.

Overall, the objective function is optimized in conjunction using SGD with backpropagation. Furthermore, the autoencoder network can also be extended to a convolutional autoencoder, which is showcased in the experiment section.

4   Interpretable Autoencoding One-Class SVM

In this section, we outline the method for interpreting the results of AE-1SVM using gradients and present illustrative example to verify its validity.

4.1 Derivations of End-to-End Gradients

Considering an input x of an RFF kernel-approximated OC-SVM with dimensionality R^d. In our model, x is the bottleneck representation of the latent space in the deep autoencoder. The expression of the margin $g(x)$ with respect to the input x is as follows:

$$g(x) = \sum_{j=1}^{D} \mathsf{w}_j z_{\omega_j}(x) - \rho = \sqrt{\frac{1}{D}} \sum_{j=1}^{D} \left[\mathsf{w}_j cos(\sum_{k=1}^{d} \omega_{jk} x_k) + \mathsf{w}_{D+j} sin(\sum_{k=1}^{d} \omega_{jk} x_k) \right] - \rho.$$

As a result, the gradient of the margin function on each input dimension $k = 1, 2, ..., d$ can be calculated as

$$\frac{\partial g}{\partial x_k} = \sqrt{\frac{1}{D}} \sum_{j=1}^{D} \omega_{jk} \left[-\mathsf{w}_j sin(\sum_{k=1}^{d} \omega_{jk} x_k) + \mathsf{w}_{j+D} cos(\sum_{k=1}^{d} \omega_{jk} x_k) \right]. \quad (6)$$

Next, we can derive the gradient of the latent space nodes with respect to the deep autoencoder's input layer (extension to convolutional autoencoder is straightforward). In general, considering a neural network with M input neurons $x_m, m = 1, 2, ..., M$, and the first hidden layer having N neurons $u_n, n = 1, 2, ..., N$, as depicted in Fig. 1 (Right). The gradient of u_n with respect to x_m can be derived as

$$G(x_m, u_n) = \frac{\partial u_n}{\partial x_m} = w_{mn} \sigma'(x_m w_{mn} + b_{mn}) \sigma(x_m w_{mn} + b_{mn}), \quad (7)$$

where $\sigma(x_m w_{mn} + b_{mn}) = u_n$, $\sigma(.)$ is the activation function, w_{mn} and b_{mn} are the weight and bias connecting x_m and u_n. The derivative of σ is different for each activation function. For instance, with a sigmoid activation σ, the gradient $G(x_m, u_n)$ is computed as $w_{mn} u_n (1 - u_n)$, while $G(x_m, u_n)$ is $w_{mn}(1 - u_n^2)$ for $tanh$ activation function.

To calculate the gradient of neuron y_l in the second hidden layer with respect to x_m, we simply apply the chain rule and sum rule as follows:

$$G(x_m, y_l) = \frac{\partial y_l}{\partial x_m} = \sum_{n=1}^{N} \frac{\partial y_l}{\partial u_n} \frac{\partial u_n}{\partial x_m} = \sum_{n=1}^{N} G(u_n, y_l) G(x_m, u_n). \quad (8)$$

The gradient $G(u_n, y_l)$ can be obtained in a similar manner to (7). By maintaining the values of G at each hidden layer, the gradient of any hidden or output layer with respect to the input layer can be calculated. Finally, combining this and (6), we can get the end-to-end gradient of the OC-SVM margin with respect to all input features. Besides, state-of-the-art machine learning frameworks like TensorFlow also implements automatic differentiation [1] that simplifies the procedures for computing those gradient values.

Using the obtained values, the decision making of the AD model can be interpreted as follows. For an outlying sample, the dimension which has higher gradient indicates a higher contribution to the decision making of the ML model. In other words, the sample is further to the boundary in that particular dimension. For each mentioned dimension, if the gradient is positive, the value of the feature in that dimension is lesser than the lower limit of the boundary. In contrast, if the gradient holds a negative value, the feature exceeds the level of the normal class.

4.2   Illustrative Example

Figure 2 presents an illustrative example of interpreting anomaly detecting results using gradients. We generate 1950 four-dimensional samples as normal instances, where the first two features are uniformly generated such that they are inside a circle with center $C(0.5, 0.5)$. The third and fourth dimensions are drawn uniformly in the range $[-0.2, 0.2]$ so that the contribution of them are significantly less than the other two dimensions. In contrast, 50 anomalies are created which have the first two dimensions being far from the mentioned circle, while the last two dimensions has a higher range of $[-2, 2]$. The whole dataset including both the normal and anomalous classes are trained with the proposed AE-1SVM model with a bottleneck layer of size 2 and sigmoid activation.

The figure on the left shows the representation of the 4D dataset on a 2-dimensional space. Expectedly, it captures most of the variability from only the first two dimensions. Furthermore, we plot the gradients of 9 different anomalous samples, with the two latter dimensions being randomized, and overall, the results have proven the aforementioned interpreting rules. It can easily be observed that the contribution of the third and fourth dimensions to the decision making of the model is always negligible. Among the first two dimensions, the ones having the value of 0.1 or 0.9 has the corresponding gradients perceptibly higher than those being 0.5, as they are further from the boundary and the sample can be considered "more anomalous" in that dimension. Besides, the gradient

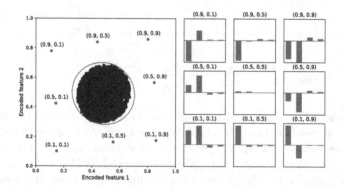

Fig. 2. Illustrative example of gradient-based explanation methods. (Left) The encoded 2D feature space from a 4D dataset. (Right) The gradient of the margin function with respect to the four original features for each testing point. Only the coordinates of first two dimensions are annotated.

of the input 0.1 is always positive due to the fact that it is lower than the normal level. In contrast, the gradient of the input 0.9 is consistently negative.

5   Experimental Results

We present qualitative empirical analysis to justify the effectiveness of the AE-1SVM model in terms of accuracy and improved training/testing time. The objective is to compare the proposed model with conventional and state-of-the-art AD methods over synthetic and well-known real world data[1].

5.1   Datasets

We conduct experiments on one generated datasets and five real-world datasets (we assume all tasks are unsupervised AD) as listed below in Table 1. The descriptions of each individual dataset is as follows:

- **Gaussian**: This dataset is taken into account to showcase the performance of the methods on high-dimensional and large data. The normal samples are drawn from a normal distribution with zero mean and standard deviation $\sigma = 1$, while $\sigma = 5$ for the anomalous instances. Theoretically, since the two groups have different distributional attributes, the AD model should be able to separate them.
- **ForestCover**: From the ForestCover/Covertype dataset [11], class 2 is extracted as the normal class, and class 4 is chosen as the anomaly class.
- **Shuttle**: From the Shuttle dataset [11], we select the normal samples from classes 2, 3, 5, 6, 7, while the outlier group is made of class 1.
- **KDDCup99**: The popular KDDCup99 dataset [11] has approximately 80% proportion as anomalies. Therefore, from the 10-percent subset, we randomly select 5120 samples from the outlier classes to form the anomaly set such that the contamination ratio is 5%. The categorical features are extracted using one-hot encoding, making 118 features in the raw input space.
- **USPS**: We select from the U.S Postal Service handwritten digits dataset [15] 950 samples from digit 1 as normal data, and 50 samples from digit 7 as anomalous data, as the appearance of the two digits are similar. The size of each image is 16×16, resulting in each sample being a flatten vector of 256 features.
- **MNIST**: From the MNIST dataset [17], 5842 samples of digit '4' are chosen as normal class. On the other hand, the set of outliers contains 100 digits from classes '0', '7', and '9'. This task is challenging due to the fact that many digits '9' are remarkably similar to digit '4'. Each input sample is a flatten vector with 784 dimensions.

---

[1] All code for reproducibility is available at https://github.com/minh-nghia/AE-1SVM.

Table 1. Summary of the datasets used for comparison in the experiments.

Dataset	Dimensions	Normal instances	Anomalies rate (%)
Gaussian	512	950	5.0
ForestCover	54	581012	0.9
Shuttle	9	49097	7.2
KDDCup99	118	97278	5.0
USPS	256	950	5.0
MNIST	784	5842	1.7

5.2 Baseline Methods

Variants of OC-SVM and several state-of-the-art methods are selected as baselines to compare the performance with the AE-1SVM model. Different modifications of the conventional OC-SVM are considered. First, we take into account the version where OC-SVM with RBF kernel is trained directly on the raw input. Additionally, to give more impartial justifications, a version where an autoencoding network exactly identical to that of the AE-1SVM model is considered. We use the same number of training epochs to AE-1SVM to investigate the ability of AE-1SVM to force the dimensionality reduction network to learn better representation of the data. The OC-SVM is then trained on the encoded feature space, and this variant is also similar to the approach given in [13].

The following methods are also considered as baselines to examine the anomaly detecting performance of the proposed model:

- **Isolation Forest** [18]: This ensemble method revolves around the idea that the anomalies in the data have significantly lower frequencies and are different from the normal points.
- **Robust Deep Autoencoder (RDA)** [34]: In this algorithm, a deep autoencoder is constructed and trained such that it can decompose the data into two components. The first component contains the latent space representation of the input, while the second one is comprised of the noise and outliers that are difficult to reconstruct.
- **Deep Clustering Embeddings (DEC)** [31]: This algorithm combines unsupervised autoencoding network with clustering. As outliers often locate in sparser clusters or are far from their centroids, we apply this method into AD and calculate the anomaly score of each sample as a product of its distance to the centroid and the density of the cluster it belongs to.

5.3 Evaluation Metrics

In all experiments, the area under receiver operating characteristic (AUROC) and area under the Precision-Recall curve (AUPRC) are applied as metrics to evaluate and compare the performance of AD methods. Having a high AUROC is necessary for a competent model, whereas AUPRC often highlights the difference between the methods regarding imbalance datasets [10]. The testing procedure follows the unsupervised setup, where each dataset is split with 1:1 ratio, and the entire training set including the anomalies is used for training the model. The output of the models on the test set is measured against the ground truth using the mentioned scoring metrics, with the average scores and approximal training and testing time of each algorithm after 20 runs being reported.

5.4   Model Configurations

In all experiments, we employ the sigmoid activation function and implement the architecture using TensorFlow [1]. We discover that for the random Fourier features, a standard deviation $\sigma = 3.0$ produces satisfactory results for all datasets. For other parameters, the network configurations of AE-1SVM for each individual dataset are as in Table 2 below.

Table 2. Summary of network configurations and training parameters of AE-1SVM used in the experiments.

Dataset	Encoding layers	ν	α	RFF	Batch size	Learning rate
Gaussian	{128, 32}	0.40	1000	500	32	0.01
ForestCover	{32, 16}	0.30	1000	200	1024	0.01
Shuttle	{6, 2}	0.40	1000	50	16	0.001
KDDCup99	{80, 40, 20}	0.30	10000	400	128	0.001
USPS	{128, 64, 32}	0.28	1000	500	16	0.005
MNIST	{256, 128}	0.40	1000	1000	32	0.001

For the MNIST dataset, we additionally implement a convolutional autoencoder with pooling and unpooling layers: conv1($5 \times 5 \times 16$), pool1(2×2), conv2($5 \times 5 \times 9$), pool2(2×2) and a feed-forward layer afterward to continue compressing into 49 dimensions; the decoder: a feed-forward layer afterward of 49×9 dimensions, then deconv1($5 \times 5 \times 9$), unpool1(2×2), deconv2($5 \times 5 \times 16$), unpool2(2×2), then a feed-forward layer of 784 dimensions. The dropout rate is set to 0.5 in this convolutional autoencoder network.

For each baseline methods, the best set of parameters is selected. In particular, for different variants of OC-SVM, the optimal values for parameter ν and the RBF kernel width are exhaustively searched. Likewise, for Isolation forest, the fraction ratio is tuned around the anomalies rate for each dataset. For RDA, DEC, as well as OC-SVM variants that involves auto-encoding network for dimensionality reduction, the autoencoder structures exactly identical to AE-1SVM are used, while the λ hyperparameter in RDA is also adjusted as it is the most important factor of the algorithm.

5.5 Results

Firstly, for the Gaussian dataset, the histograms of the decision scores obtained by different methods are presented in Fig. 3. It can clearly be seen that AE-1SVM is able to single out all anomalous samples, while giving the best separation between the two classes.

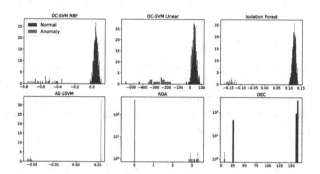

Fig. 3. Histograms of decision scores of AE-1SVM and other baseline methods.

For other datasets, the comprehensive results are given in Table 3. It is obvious that AE-1SVM outperforms conventional OC-SVM as well as the two-staged structure with decoupled autoencoder and OC-SVM in terms of accuracy in all scenarios, and is always among the top performers. As we restrict the number of training epochs for the detached autoencoder to be same as that for AE-1SVM, its performance declines significantly and in some cases its representation is even worse than the raw input. This proves that AE-1SVM can attain more efficient features to support AD task given the similar time.

Other observations can also be made from the results. For ForestCover, only the AUROC score of Isolation Forest is close, but the AUPRC is significantly lower, with three time less than that of AE-1SVM, suggesting that it has to compensate a higher false alarm rate to identify anomalies correctly. Similarly, Isolation Forest slightly surpasses AE-1SVM in AUROC for Shuttle dataset, but is subpar in terms of AUPRC, thus can be considered less optimal choice. Analogous patterns can as well be noticed for other datasets. Especially, for MNIST, it is shown that the proposed method AE-1SVM can also operate under a convolutional autoencoder network in image processing contexts.

Regarding training time, AE-1SVM outperforms other methods for Forest-Cover, which is the largest dataset. For other datasets that have high sample size, namely KDDCup99 and Shuttle, it is still one of the fastest candidates. Furthermore, we also extend the KDDCup99 experiment and train AE-1SVM model on a full dataset, and acquire promising results in only about 200 s. This verifies the effectiveness and potential application of the model in big-data circumstances. On top of that, the testing time of AE-1SVM is a notable improvement over other methods, especially Isolation Forest and conventional OC-SVM, suggesting its feasibility in real-time environments.

Table 3. Average AUROC, AUPRC, approximal train time and test time of the baseline methods and proposed method. Best results are displayed in boldface.

Dataset	Method	AUROC	AUPRC	Train	Test
Forest Cover	OC-SVM raw input	0.9295	0.0553	6×10^2	2×10^2
	OC-SVM encoded	0.7895	0.0689	2.5×10^2	8×10^1
	Isolation Forest	0.9396	0.0705	3×10^1	1×10^1
	RDA	0.8683	0.0353	1×10^2	2×10^0
	DEC	0.9181	0.0421	$\mathbf{2 \times 10^1}$	4×10^0
	AE-1SVM	**0.9485**	**0.1976**	2×10^1	$\mathbf{7 \times 10^{-1}}$
Shuttle	OC-SVM raw input	0.9338	0.4383	2×10^1	5×10^1
	OC-SVM encoded	0.8501	0.4151	2×10^1	2.5×10^0
	Isolation Forest	**0.9816**	0.7694	2.5×10^1	1.5×10^1
	RDA	0.8306	0.1872	3×10^2	2×10^{-1}
	DEC	0.9010	0.3184	$\mathbf{6 \times 10^0}$	1×10^0
	AE-1SVM	0.9747	**0.9483**	1×10^1	$\mathbf{1 \times 10^{-1}}$
KDDCup	OC-SVM raw input	0.8881	0.3400	6×10^1	2×10^1
	OC-SVM encoded	0.9518	0.3876	5×10^1	1×10^1
	Isolation Forest	0.9572	0.4148	2×10^1	5×10^0
	RDA	0.6320	0.4347	1×10^2	5×10^{-1}
	DEC	0.9496	0.3688	$\mathbf{1 \times 10^1}$	2×10^0
	AE-1SVM	0.9663	**0.5115**	3×10^1	$\mathbf{4.5 \times 10^{-1}}$
	AE-1SVM (Full dataset)	**0.9701**	0.4793	2×10^2	4×10^0
USPS	OC-SVM raw input	0.9747	0.5102	$\mathbf{2 \times 10^{-2}}$	1.5×10^{-2}
	OC-SVM encoded	0.9536	0.4722	6×10^0	$\mathbf{5 \times 10^{-3}}$
	Isolation Forest	0.9863	0.6250	2.5×10^{-1}	6×10^{-2}
	RDA	0.9799	0.5681	1.5×10^0	1.5×10^{-2}
	DEC	0.9263	0.7506	4×10^0	2.5×10^{-2}
	AE-1SVM	**0.9926**	**0.8024**	1×10^1	$\mathbf{5 \times 10^{-3}}$
MNIST	OC-SVM raw input	0.8302	0.0819	$\mathbf{2 \times 10^0}$	1×10^0
	OC-SVM encoded	0.7956	0.0584	1×10^2	$\mathbf{1 \times 10^{-1}}$
	Isolation Forest	0.7574	0.0533	4.5×10^0	1.5×10^0
	RDA	0.8464	0.0855	1×10^2	2.5×10^{-1}
	DEC	0.5522	0.0289	3.5×10^1	1.5×10^{-1}
	AE-1SVM	0.8119	0.0864	1.5×10^2	7×10^{-1}
	CAE-1SVM	**0.8564**	**0.0885**	3.5×10^3	1.5×10^0

5.6 Gradient-Based Explanation in Image Datasets

We also investigate the use of gradient-based explanation methods on the image datasets. Figure 4 illustrates the unsigned gradient maps of several anomalous digits in the USPS and MNIST datasets. The MNIST results are given by the version with convolutional autoencoder. Interesting patterns proving the correctness of gradient-based explanation approach can be observed from Fig. 4 (Left). The positive gradient maps revolve around the middle part of the images where the pixels in the normal class of digits '1' are normally bright (higher values), indicating the absence of those pixels contributes significantly to the reasoning that the samples '7' are detected as outliers. Likewise, the negative gradient maps are more intense on the pixels matching the bright pixels outside the center area of its corresponding image, meaning that the values of those pixels in the original image exceeds the range of the normal class, which is around the zero (black) level. Similar perception can be acquired from Fig. 4 (Right), as it shows the difference between each samples of digits '0', '7', and '9', to digit '4'.

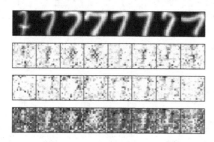 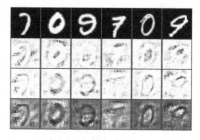

Fig. 4. (Left) The USPS experiment. (Right) The MNIST experiment. From top to bottom rows: original image, positive gradient map, negative gradient map, and full gradient map.

6 Conclusion

In this paper, we propose the end-to-end autoencoding One-class Support Vector Machine (AE-1SVM) model comprising of a deep autoencoder for dimensionality reduction and a variant structure of OC-SVM using random Fourier features for anomaly detection. The model is jointly trained using SGD with a combined loss function to both lessen the complexity of solving support vector problems and force dimensionality reduction to learn better representation that is beneficial for the anomaly detecting task. We also investigate the application of applying gradient-based explanation methods to interpret the decision making of the proposed model, which is not feasible for most of the other anomaly detection algorithms. Extensive experiments have been conducted to verify the strengths of our approach. The results have demonstrated that AE-1SVM can be effective in detecting anomalies, while significantly enhance both training and response time for high-dimensional and large-scale data. Empirical evidence of interpreting the predictions of AE-1SVM using gradient-based methods has also been presented using illustrative examples and handwritten image datasets.

References

1. Abadi, M., et al.: TensorFlow: large-scale machine learning on heterogeneous systems (2015). https://www.tensorflow.org/. Software available from tensorflow.org
2. Ancona, M., Ceolini, E., Öztireli, C., Gross, M.: Towards better understanding of gradient-based attribution methods for deep neural networks. In: International Conference on Learning Representations (2018). https://openreview.net/forum?id=Sy21R9JAW
3. Baehrens, D., Schroeter, T., Harmeling, S., Kawanabe, M., Hansen, K., Müller, K.R.: How to explain individual classification decisions. J. Mach. Learn. Res. **11**, 1803–1831 (2010)
4. Barnett, V., Lewis, T.: Outliers in Statistical Data. Wiley, New York (1974)
5. Bengio, Y., Lecun, Y.: Scaling Learning Algorithms Towards AI. MIT Press, Cambridge (2007)
6. Candès, E.J., Li, X., Ma, Y., Wright, J.: Robust principal component analysis? J. ACM (JACM) **58**(3), 11 (2011)
7. Chalapathy, R., Menon, A.K., Chawla, S.: Robust, deep and inductive anomaly detection. In: Ceci, M., Hollmén, J., Todorovski, L., Vens, C., Džeroski, S. (eds.) ECML PKDD 2017, Part I. LNCS (LNAI), vol. 10534, pp. 36–51. Springer, Cham (2017). https://doi.org/10.1007/978-3-319-71249-9_3
8. Chandola, V., Banerjee, A., Kumar, V.: Anomaly detection: a survey. ACM Comput. Surv. (CSUR) **41**(3), 15 (2009)
9. Chang, C.C., Lin, C.J.: LIBSVM: a library for support vector machines. ACM Trans. Intell. Syst. Technol. **2**(3), 27:1–27:27 (2011)
10. Davis, J., Goadrich, M.: The relationship between precision-recall and ROC curves. In: Proceedings of the 23rd International Conference on Machine Learning, ICML 2006, pp. 233–240. ACM, New York (2006)
11. Dheeru, D., Karra Taniskidou, E.: UCI machine learning repository (2017). http://archive.ics.uci.edu/ml
12. Erfani, S.M., Baktashmotlagh, M., Rajasegarar, S., Karunasekera, S., Leckie, C.: R1SVM: a randomised nonlinear approach to large-scale anomaly detection. In: Proceedings of the Twenty-Ninth AAAI Conference on Artificial Intelligence, AAAI 2015, pp. 432–438. AAAI Press (2015)
13. Erfani, S.M., Rajasegarar, S., Karunasekera, S., Leckie, C.: High-dimensional and large-scale anomaly detection using a linear one-class SVM with deep learning. Pattern Recognit. **58**(C), 121–134 (2016)
14. Grubbs, F.E.: Procedures for detecting outlying observations in samples. Technometrics **11**(1), 1–21 (1969)
15. Hull, J.J.: A database for handwritten text recognition research. IEEE Trans. Pattern Anal. Mach. Intell. **16**(5), 550–554 (1994)
16. Kim, J., Scott, C.D.: Robust kernel density estimation. J. Mach. Learn. Res. **13**, 2529–2565 (2012)
17. LeCun, Y., Cortes, C.: MNIST handwritten digit database (2010). http://yann.lecun.com/exdb/mnist/
18. Liu, F.T., Ting, K.M., Zhou, Z.H.: Isolation forest. In: Eighth IEEE International Conference on Data Mining, pp. 413–422 (2008)
19. Montavon, G., Bach, S., Binder, A., Samek, W., Müller, K.: Explaining nonlinear classification decisions with deep Taylor decomposition. Pattern Recognit. **65**, 211–222 (2017)

20. Rahimi, A., Recht, B.: Random features for large-scale kernel machines. In: Advances in Neural Information Processing Systems, vol. 20, pp. 1177–1184. Curran Associates Inc. (2008)
21. Schlegl, T., Seeböck, P., Waldstein, S.M., Schmidt-Erfurth, U., Langs, G.: Unsupervised anomaly detection with generative adversarial networks to guide marker discovery. In: Niethammer, M., et al. (eds.) IPMI 2017. LNCS, vol. 10265, pp. 146–157. Springer, Cham (2017). https://doi.org/10.1007/978-3-319-59050-9_12
22. Schölkopf, B., Williamson, R., Smola, A., Shawe-Taylor, J., Platt, J.: Support vector method for novelty detection. In: Proceedings of the 12th International Conference on Neural Information Processing Systems, NIPS 1999, pp. 582–588. MIT Press, Cambridge (1999)
23. Shalev-Shwartz, S., Singer, Y., Srebro, N., Cotter, A.: Pegasos: primal estimated sub-gradient solver for SVM. Math. Program. **127**(1), 3–30 (2011)
24. Shrikumar, A., Greenside, P., Kundaje, A.: Learning important features through propagating activation differences. In: ICML (2017)
25. Shrikumar, A., Greenside, P., Shcherbina, A., Kundaje, A.: Not just a black box: learning important features through propagating activation differences. In: ICML (2017)
26. Simonyan, K., Vedaldi, A., Zisserman, A.: Deep inside convolutional networks: visualising image classification models and saliency maps. In: Workshop at International Conference on Learning Representations (2014)
27. Sundararajan, M., Taly, A., Yan, Q.: Axiomatic attribution for deep networks. In: ICML (2017)
28. Sutherland, D.J., Schneider, J.G.: On the error of random Fourier features. In: UAI (2015)
29. Tax, D.M., Duin, R.P.: Support vector data description. Mach. Learn. **54**(1), 45–66 (2004)
30. Williams, C.K.I., Seeger, M.: Using the Nyström method to speed up kernel machines. In: Proceedings of the 13th International Conference on Neural Information Processing Systems, NIPS 2000, pp. 661–667. MIT Press, Cambridge (2000)
31. Xie, J., Girshick, R., Farhadi, A.: Unsupervised deep embedding for clustering analysis. In: Proceedings of the 33rd International Conference on International Conference on Machine Learning, ICML 2016, vol. 48, pp. 478–487. JMLR.org (2016). http://dl.acm.org/citation.cfm?id=3045390.3045442
32. Zenati, H., Foo, C.S., Lecouat, B., Manek, G., Chandrasekhar, V.R.: Efficient GAN-based anomaly detection. In: ICLR Workshop (2018)
33. Zhai, S., Cheng, Y., Lu, W., Zhang, Z.: Deep structured energy based models for anomaly detection. In: Proceedings of the 33rd International Conference on Machine Learning, ICML 2016, New York City, 19–24 June 2016, pp. 1100–1109 (2016)
34. Zhou, C., Paffenroth, R.C.: Anomaly detection with robust deep autoencoders. In: Proceedings of the 23rd ACM SIGKDD International Conference on Knowledge Discovery and Data Mining, KDD 2017, pp. 665–674. ACM, New York (2017)
35. Zimek, A., Schubert, E., Kriegel, H.P.: A survey on unsupervised outlier detection in high-dimensional numerical data. Stat. Anal. Data Min. ASA Data Sci. J. **5**(5), 363–387 (2012)
36. Zong, B., et al.: Deep autoencoding Gaussian mixture model for unsupervised anomaly detection. In: International Conference on Learning Representations (2018). https://openreview.net/forum?id=BJJLHbb0-

Group Anomaly Detection Using Deep Generative Models

Raghavendra Chalapathy[1], Edward Toth[2(✉)], and Sanjay Chawla[3]

[1] The University of Sydney and Capital Markets CRC, Sydney, Australia
[2] School of Information Technologies, The University of Sydney, Sydney, Australia
etot5316@uni.sydney.edu.au
[3] Qatar Computing Research Institute, HBKU, Doha, Qatar

Abstract. Unlike conventional anomaly detection research that focuses on point anomalies, our goal is to detect anomalous collections of individual data points. In particular, we perform group anomaly detection (GAD) with an emphasis on irregular group distributions (e.g. irregular mixtures of image pixels). GAD is an important task in detecting unusual and anomalous phenomena in real-world applications such as high energy particle physics, social media and medical imaging. In this paper, we take a generative approach by proposing deep generative models: Adversarial autoencoder (AAE) and variational autoencoder (VAE) for group anomaly detection. Both AAE and VAE detect group anomalies using point-wise input data where group memberships are known a priori. We conduct extensive experiments to evaluate our models on real world datasets. The empirical results demonstrate that our approach is effective and robust in detecting group anomalies. Code related to this paper is available at: https://github.com/raghavchalapathy/gad, https://www.cs.cmu.edu/~lxiong/gad/gad.html, https://github.com/jorjasso/SMDD-group-anomaly-detection, https://github.com/cjlin1/libsvm.

Keywords: Group anomaly detection · Adversarial · Variational Auto-encoders

1   Anomaly Detection: Motivation and Challenges

Group anomaly detection (GAD) is an important part of data analysis for many interesting group applications. Pointwise anomaly detection focuses on the study of individual data instances that do not conform with the expected pattern in a dataset. With the increasing availability of multifaceted information, GAD research has recently explored datasets involving groups or collections of observations. Many pointwise anomaly detection methods cannot detect a variety of different deviations that are evident in group datasets. For example, Muandet et al. [20] possibly discover Higgs bosons as a group of collision events in high energy particle physics whereas pointwise methods are unable to distinguish this

---

R. Chalapathy and E. Toth—Equal contribution.

© Springer Nature Switzerland AG 2019
M. Berlingerio et al. (Eds.): ECML PKDD 2018, LNAI 11051, pp. 173–189, 2019.
https://doi.org/10.1007/978-3-030-10925-7_11

anomalous behavior. Detecting group anomalies require more specialized techniques for robustly differentiating group behaviors.

GAD aims to identify groups that deviate from the regular group pattern. Generally, a group consists of a collection of two or more points and group behaviors are more adequately described by a greater number of observations. A point-based anomalous group is a collection of individual pointwise anomalies that deviate from the expected pattern. It is more difficult to detect distribution-based group anomalies where points are seemingly regular however their collective behavior is anomalous. It is also possible to characterize group anomalies by certain properties and subsequently apply pointwise anomaly detection methods. In image applications, a distribution-based anomalous group has an irregular mixture of visual features compared to the expected group pattern.

GAD is a difficult problem for many real-world applications especially involving more complicated group behaviors such as in image datasets. Xiong et al. [29] propose a novel method for detecting group anomalies however an improvement in their detection results is possible for image applications. Images are modeled as group of pixels or visual features and it may be difficult to accurately characterize anomalous images by deviating properties. For example, it is difficult to distinguish regular groups (cat images) from anomalous groups (tiger images) that possess cat whiskers but also irregular features of tiger stripes. The problem of GAD in image datasets is useful and applicable to similar challenging real-world applications where group distributions are more complex and difficult to characterize.

Figure 1 illustrates examples of point-based and distribution-based group anomalies where the innermost circle contains images exhibiting regular behaviors whereas the outer circle conveys group anomalies. Plot (A) displays tiger images as point-based group anomalies as well as rotated cat images as distribution-based group anomalies (180° rotation). In plot (B), distribution-based group anomalies are irregular mixtures of cats and dogs in a single image while plot (C) depicts anomalous images stitched from different scene categories of cities, mountains or coastlines. Our image data experiments will mainly focus on detecting group anomalies in these scenarios.

Even though the GAD problem may seem like a straightforward comparison of group observations, many complications and challenges arise. As there is a dependency between the location of pixels in a high-dimensional space, appropriate features in an image may be difficult to extract. For effective detection of anomalous images, an adequate description of images is required for model training. Complications in images potentially arise such as low resolution, poor illumination intensity, different viewing angles, scaling and rotations of images. Like other anomaly detection applications, ground truth labels are also usually unavailable for training or evaluation purposes. A number of pre-processing and extraction techniques can be applied as solutions to different aspects of these challenges.

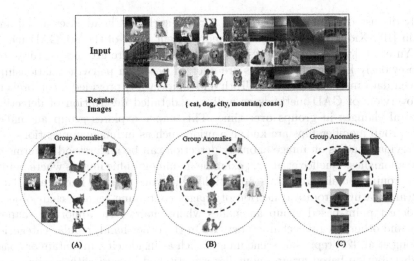

Fig. 1. Examples of point-based and distribution-based group anomalies in various image applications. The expected group behavior represents images in the inner concentric circle while the outer circle contains images that are group anomalies.

In order to detect distribution-based group anomalies in various image applications, we propose using deep generative models (DGMs). The main contributions of this paper are:

- We formulate DGMs for the problem of detecting group anomalies using a group reference function.
- Although deep generative models have been applied in various image applications, they have not been applied to the GAD problem.
- A variety of experiments are performed on both synthetic and real-world datasets to demonstrate the effectiveness of deep generative models for detecting group anomalies as compared to other GAD techniques.

The rest of the paper is organized as follows. An overview of related work is provided (Sect. 2) and preliminaries for understanding approaches for detecting group anomalies are also described (Sect. 3). We formulate our problem and then proceed to elaborate on our proposed solution that involves deep generative models (Sect. 4). Our experimental setup and key results are presented in Sect. 5 and Sect. 6 respectively. Finally, Sect. 7 provides a summary of our findings as well as recommends future directions for GAD research.

2   Background and Related Work on Group Anomaly Detection

GAD applications are emerging areas of research where most state-of-the-art techniques have been more recently developed. While group anomalies are briefly discussed in anomaly detection surveys such as Chandola et al. [4] and Austin [10], Xiong [28] elaborates on more recent state-of-the-art GAD methods. Yu et al. [33] further reviews GAD techniques where group structures are not previously known and clusters are inferred based on pairwise relationships between data instances. Recently Toth and Chawla [27] provided a comprehensive overview of GAD methods as well as a detailed description of detecting temporal changes in groups over time. This paper explores group anomalies where group memberships are known a priori such as in image applications.

Previous studies on image anomaly detection can be understood in terms of group anomalies. Quellec et al. [22] examine mammographic images where point-based group anomalies represent potentially cancerous regions. Perera and Patel [21] learn features from a collection of images containing regular chair objects and detect point-based group anomalies where chairs have abnormal shapes, colors and other irregular characteristics. On the other hand, regular categories in Xiong et al. [29] represent scene images such as inside city, mountain or coast and distribution-based group anomalies are stitched images with a mixture of different scene categories. At a pixel level, Xiong et al. [30] apply GAD methods to detect anomalous galaxy clusters with irregular proportions of RGB pixels. We emphasize detecting distribution-based group anomalies rather than point-based anomalies in our subsequent image applications.

The discovery of group anomalies is of interest to a number of diverse domains. Muandet et al. [20] investigate GAD for physical phenomena in high energy particle physics where Higgs bosons are observed as slight excesses in a collection of collision events rather than individual events. Xiong et al. [29] analyze a fluid dynamics application where a group anomaly represents unusual vorticity and turbulence in fluid motion. In topic modeling, Soleimani and Miller [25] characterize documents by topics and anomalous clusters of documents are discovered by their irregular topic mixtures. By incorporating additional information from pairwise connection data, Yu et al. [34] find potentially irregular communities of co-authors in various research communities. Thus there are many GAD application other than image anomaly detection.

A related discipline to image anomaly detection is video anomaly detection where many deep learning architectures have been applied. Sultani et al. [26] detect real-world anomalies such as burglary, fighting, vandalism and so on from CCTV footage using deep learning methods. In a review, Kiran et al. [15] compare DGMs with different convolution architectures for video anomaly detection applications. Recent work [3,23,32] illustrate the effectiveness of generative models for high-dimensional anomaly detection. Although, there are existing works that have applied deep generative models in image related applications, they have not been formulated as a GAD problem. We leverage autoencoders for DGMs to detect group anomalies in a variety of data experiments.

3 Preliminaries

In this section, a summary of state-of-the-art techniques for detecting group anomalies is provided. We also assess strengths and weaknesses of existing models, compared with the proposed deep generative models.

3.1 Mixture of Gaussian Mixture (MGM) Models

A hierarchical generative approach MGM is proposed by Xiong et al. [30] for detecting group anomalies. The data generating process in MGM assumes that each group follow a Gaussian mixture where more than one regular mixture proportion is possible. For example, an image is a distribution over visual features such as paws and whiskers from a cat image and each image is categorized into possible regular behaviors or genres (e.g. dogs or cats). An anomalous group is then characterized by an irregular mixture of visual features such as a cat and dog in a single image. MGM is useful for distinguishing multiple types of group behaviors however poor results are obtained when group observations do not appropriately follow the assumed generative process.

3.2 One-Class Support Measure Machines (OCSMM)

Muandet et al. [20] propose OCSMM to maximize the margin that separates regular class of group behaviors from anomalous groups. Each group is firstly characterized by a mean embedding function then group representations are separated by a parameterized hyperplane. OCSMM is able to classify groups as regular or anomalous behaviors however careful parameter selection is required in order to effectively detect group anomalies.

3.3 One-Class Support Vector Machines (OCSVM)

If group distributions are reduced and characterized by a single value then OCSVM from Schölkopf et al. [24] can be applied to the GAD problem. OCSVM separates data points using a parametrized hyperplane similar to OCSMM. OCSVM requires additional pre-processing to convert groups of visual features into pointwise observations. We follow a bag of features approach in Azhar et al. [1], where k-means is applied to visual image features and centroids are clustered into histogram intervals before implementing OCSVM. OCSVM is a popular pointwise anomaly detection method however it may not accurately capture group anomalies if the initial group characterizations are inadequate.

3.4 Deep Generative Models for Anomaly Detection

This section describes the mathematical background of deep generative models that will be applied for detecting group anomalies. The following notation considers data involving M groups where the mth group is denoted by G_m.

Autoencoders: An autoencoder is trained to learn reconstructions that are close to its original input. The autoencoder consists of encoder f_ϕ to embed the input to latent or hidden representation and decoder g_ψ which reconstructs the input from hidden representation. The reconstruction loss of an autoencoder is defined as the squared error between the input G_m and output $\hat{G}_m$ given by

$$L_r(G_m, \hat{G}_m) = ||G_m - \hat{G}_m||^2 \tag{1}$$

Autoencoders leverage reconstruction error as an anomaly score where data points with significantly high errors are considered to be anomalies.

Variational Autoencoders (VAE): Variational autoencoder (VAE) [14] are generative analogues to the standard deterministic autoencoder. VAE impose constraint while inferring latent variable z. The hidden latent codes produced by encoder f_ϕ is constrained to follow prior data distribution $P(G_m)$. The core idea of VAE is to infer $P(z)$ from $P(z|G_m)$ using Variational Inference (VI) technique given by

$$L(G_m, \hat{G}_m) = L_r(G_m, \hat{G}_m) + KL(f_\phi(z|x) \,||\, g_\psi(z)) \tag{2}$$

In order to optimize the Kullback–Leibler (KL) divergence, a simple reparameterization trick is applied; instead of the encoder embedding a real-valued vector, it creates a vector of means $\boldsymbol{\mu}$ and a vector of standard deviations $\boldsymbol{\sigma}$. Now a new sample that replicates the data distribution $P(G_m)$ can be generated from learned parameters $(\boldsymbol{\mu}, \boldsymbol{\sigma})$ and input this latent representation z through the decoder g_ψ to reconstruct the original group observations. VAE utilizes reconstruction probabilities [3] or reconstruction error to compute anomaly scores.

Adversarial Autoencoders (AAE): One of the main limitations of VAE is lack of closed form analytical solution for integral of the KL divergence term except for few distributions. Adversarial autoencoders (AAE) [19] avoid using the KL divergence by adopting adversarial learning, to learn broader set of distributions as priors for the latent code. The training procedure for this architecture is performed using an adversarial autoencoder consisting of encoder f_ϕ and decoder g_ψ. Firstly a latent representation z is created according to generator network $f_\phi(z|G_m)$ and the decoder reconstructs the input $\hat{G}_m$ from z. The weights of encoder f_ϕ and decoder g_ψ are updated by backpropagating the reconstruction loss between $\hat{G}_m$ and G_m. Secondly the discriminator receives z distributed as $f_\phi(z|G_m)$ and z' sampled from the true prior $P(z)$ to compute the score assigned to each ($D(z)$ and $D(z')$). The loss incurred is minimized by backpropagating through the discriminator to update its weights. The loss function for autoencoder (or generator) L_G is composed of the reconstruction error along with the loss for discriminator L_D where

$$L_G = \frac{1}{M'} \sum_{m=1}^{M'} \log D(z_m) \quad \text{and} \quad L_D = -\frac{1}{M'} \sum_{m=1}^{M'} \left[\log D(z'_m) + \log(1 - D(z_m)) \right]$$

$$(3)$$

where M' is the minibatch size while z represents the latent code generated by encoder and z' is a sample from the true prior $P(z)$.

4   Problem and Model Formulation

Problem Definition: The following formulation follows the problem definition introduced in Toth and Chawla [27]. Suppose groups $\mathcal{G} = \{\mathbf{G}_m\}_{m=1}^{M}$ are observed where M is the number of groups and the mth group has group size N_m with V-dimensional observations, that is $\mathbf{G}_m \in \mathbb{R}^{N_m \times V}$. In GAD, the behavior or properties of the mth group is captured by a characterization function denoted by $f : \mathbb{R}^{N_m \times V} \to \mathbb{R}^D$ where D is the dimensionality on a transformed feature space. After a characterization function is applied to a training dataset, group information is combined using an aggregation function $g : \mathbb{R}^{M \times D} \to \mathbb{R}^D$. A group reference is composed of characterization and aggregation functions on input groups with

$$\mathcal{G}^{(ref)} = g\left[\{f(\mathbf{G}_m)\}_{m=1}^{M} \right] \tag{4}$$

Then a distance metric $d(\cdot, \cdot) \geq 0$ is applied to measure the deviation of a particular group from the group reference function. The distance score $d\left(\mathcal{G}^{(ref)}, \mathbf{G}_m\right)$ quantifies the deviance of the mth group from the expected group pattern where larger values are associated with more anomalous groups. Group anomalies are effectively detected when characterization function f and aggregation function g respectively capture properties of group distributions and appropriately combine information into a group reference. For example in an variational autoencoder setting, an encoder function f characterizes mean and standard deviation of group distributions whereas decoder function g reconstructs the original sample. Further descriptions of functions f and g for VAE and AAE are provided in Algorithm 1.

Algorithm 1. Group anomaly detection using deep generative models

Input : Groups $\{\mathbf{G}_1, \mathbf{G}_2, \ldots, \mathbf{G}_M\}$ where $\mathbf{G}_m = (X_{ij}) \in \mathbb{R}^{N_m \times V}$

Output: Group anomaly scores **S**

1 Train AAE and VAE to obtain encoder f_ϕ and decoder g_ψ

2 **begin**

3 $\quad$ **switch** C **do**

4 $\quad\quad$ **case** *(VAE)*

5 $\quad\quad\quad$ $(\mu_m, \sigma_m) = f_\phi(z|\mathbf{G}_m)$ for $m = 1, 2, \ldots, M$

6 $\quad\quad\quad$ $(\mu, \sigma) = \frac{1}{M} \sum_{m=1}^{M} (\mu_m, \sigma_m)$

7 $\quad\quad\quad$ draw a sample from $z \sim \mathcal{N}(\mu, \sigma)$

8 $\quad\quad$ **endsw**

9 $\quad\quad$ **case** *(AAE)*

10 $\quad\quad\quad$ draw a random latent representation $z \sim f_\phi(z|\mathbf{G}_m)$

11 $\quad\quad\quad$ for $m = 1, 2, \ldots, M$

12 $\quad\quad$ **endsw**

13 $\quad$ **endsw**

14 $\quad$ **for** *(m = 1 to M)* **do**

15 $\quad\quad$ reconstruct sample using decoder $\mathcal{G}^{(ref)} = g_\psi(\mathbf{G}_m|z)$

16 $\quad\quad$ compute the score $s_m = d\left(\mathcal{G}^{(ref)}, \mathbf{G}_m\right)$

17 $\quad$ **end**

18 $\quad$ sort scores in descending order $\mathbf{S} = \{s_{(M)} > \cdots > s_{(1)}\}$

19 $\quad$ groups that are furthest from $\mathcal{G}^{(ref)}$ are more anomalous.

20 $\quad$ **return S**

21 **end**

4.1 Training the Model

The variational and adversarial autoencoder are trained according to the objective function given in Eqs. (2) and (3) respectively. The objective functions of DGMs are optimized using the standard backpropagation algorithm. Given known group memberships, AAE is fully trained on input groups to obtain a representative group reference $\mathcal{G}^{(ref)}$ described in Eq. 4. While in case of VAE, $\mathcal{G}^{(ref)}$ is obtained by drawing samples using mean and standard deviation parameters that are inferred using VAE as illustrated in Algorithm 1.

4.2 Predicting with the Model

In order to identify group anomalies, the distance of a group from the group reference $\mathcal{G}^{(ref)}$ is computed. The output scores are sorted according to descending order where groups that are furthest from $\mathcal{G}^{(ref)}$ are considered most anomalous. One convenient property of DGMs is that the anomaly detector will be inductive, i.e. it can generalize to unseen data points. One can interpret the model as learning a robust representation of group distributions. An appropriate characterization of groups results in more accurate detection where any unseen observations either lie within the reference group manifold or deviate from the expected group pattern.

5 Experimental Setup

In this section we show the empirical effectiveness of deep generative models over the state-of-the-art methods on real-world data. Our primary focus is on non-trivial image datasets, although our method is applicable in any context where autoencoders are useful e.g. speech, text.

5.1 Methods Compared

We compare our proposed technique using deep generative models (DGMs) with the following state-of-the art methods for detecting group anomalies:

- **Mixture of Gaussian Mixture (MGM) Model**, as per [30].
- **One-Class Support Measure Machines (OCSMM)**, as per [20].
- **One-Class Support Vector Machines (OCSVM)**, as per [24].
- **Variational Autoencoder (VAE)** [9], as per Eq. (2).
- **Adversarial Autoencoder (AAE)** [19], as per Eq. (3).

We used Keras [5], TensorFlow [2] for the implementation of AAE and VAE[1]. MGM[2], OCSMM[3] and OCSVM[4] are applied using publicly available code.

5.2 Datasets

We compare all methods on the following datasets:

- `synthetic` data follows Muandet et al. [20] where regular groups are generated by bivariate Gaussian distributions while anomalous groups have rotated covariance matrices.
- `cifar-10` [16] consists of 32×32 color images over 10 classes with 6000 images per class.
- `scene` image data following Xiong et al. [31] where anomalous images are stitched from different scene categories.
- `Pixabay` [11] is used to obtain tiger images as well as images of cats and dogs together. These images are rescaled to match dimensions of cat images in `cifar-10` dataset.

The real-world data experiments are previously illustrated in Fig. 1.

---

[1] https://github.com/raghavchalapathy/gad.
[2] https://www.cs.cmu.edu/~lxiong/gad/gad.html.
[3] https://github.com/jorjasso/SMDD-group-anomaly-detection.
[4] https://github.com/cjlin1/libsvm.

5.3 Parameter Selection

We now briefly discuss the model and parameter selection for applying techniques in GAD applications. A pre-processing stage is required for state-of-the-art GAD methods when dealing with images where feature extraction methods such as SIFT [18] or HOG [7] represent images as a collection of visual features. In MGM, the number of regular group behaviors T and number of Gaussian mixtures L are selected using information criteria. The kernel bandwidth smoothing parameter in OCSMM [20] is chosen as median$\{||\mathbf{G}_{m,i} - \mathbf{G}_{l,j}||^2\}$ for all $i, j \in \{1, 2, \ldots, N_m\}$ and $m, l \in 1, 2, \ldots, M$ where $\mathbf{G}_{m,i}$ represents the ith random vector in the mth group. In addition, the parameter for expected proportions of anomalies in OCSMM and OCSVM is set to the true value in the respective datasets.

When applying VAE and AAE, there are four existing network parameters that require careful selection; (a) number of convolutional filters, (b) filter size, (c) strides of convolution operation and (d) activation function. We tuned via grid search of additional hyper-parameters including the number of hidden-layer nodes $H \in \{3, 64, 128\}$ and regularization λ within range $[0, 100]$. The learning drop-out rates and regularization parameter μ were sampled from a uniform distribution in the range $[0.05, 0.1]$. The embedding and initial weight matrices are all sampled from uniform distribution within range $[-1, 1]$.

6 Experimental Results

In this section, we explore a variety of GAD experiments. As anomaly detection is an unsupervised learning problem, model evaluation is highly challenging. We employ anomaly injection where known group anomalies are injected into real-world image datasets. The performances of DGMs are evaluated against state-of-the-art GAD methods using area under precision-recall curve (AUPRC) and area under receiver operating characteristic curve (AUROC). AUPRC is more appropriate than AUROC for binary classification under class imbalanced datasets such as in GAD applications [8]. However in our experiments, a high AUPRC score indicates the effectiveness of accurately identifying regular groups while AUROC accounts for the false positive rate of detection methods.

6.1 Synthetic Data: Rotated Gaussian Distribution

Firstly we generate synthetic data where regular behavior consists of bivariate Gaussian samples while anomalous groups have rotated covariance structures. More specifically, $M = 500$ regular group distributions have correlation $\rho = 0.7$ while 50 anomalous groups are generated with correlation $\rho = -0.7$. The mean vectors are randomly sampled from uniform distributions while covariances of group distributions are given by

$$\Sigma_m = \begin{cases} \begin{pmatrix} 0.2 & 0.14 \\ 0.14 & 0.2 \end{pmatrix}, & m = 1, 2, \ldots, 500 \\ \begin{pmatrix} 0.2 & -0.14 \\ -0.14 & 0.2 \end{pmatrix}, & m = 501, 502, \ldots, 550 \end{cases} \tag{5}$$

with each group having $N_m = 1536$ observations. Since we configured the proposed DGMs with an architecture suitable for 32×32 pixels for 3 dimensions (red, green, blue), our dataset is constructed such that each group has bivariate observations with a total of 3072 values.

Parameter Settings: GAD methods are applied on the raw data with various parameter settings. MGM is trained with $T = 1$ regular scene types and $L = 3$ as the number of Gaussian mixtures. The expected proportion of group anomalies as true proportion in OCSMM and OCSVM is set to $\nu = 50/M$ where $M = 550$ or $M = 5050$. In addition, OCSVM is applied by treating each Gaussian distribution as a single high-dimensional observation.

Results: Table 1 illustrates the results of detecting distribution-based group anomalies for different number of groups. For smaller number of groups $M = 550$, state-of-the-art GAD methods achieve a higher performance than DGMs however for a larger training set with $M = 5050$, deep generative models achieve the highest performance. AAE and VAE attain similar results for both synthetic datasets. This conveys that DGMs require larger number of group observations in order to train an appropriate model.

Table 1. Task results for detecting rotated Gaussian distributions in synthetic datasets where AAE and VAE attain poor detection results for smaller datasets while they achieve the highest performances (as highlighted in gray) given a larger number of groups.

Methods	M=550		M=5050	
	AUPRC	AUROC	AUPRC	AUROC
AAE	0.9060	0.5000	1.0000	1.0000
VAE	0.9001	0.5003	1.0000	1.0000
MGM	0.9781	0.8180	0.9978	0.8221
OCSMM	0.9426	0.6097	0.9943	0.6295
OCSVM	0.9211	0.5008	0.9898	0.5310

6.2 Detecting Tigers Within Cat Images

Firstly we explore the detection of point-based group anomalies (or image anomalies) by injecting 50 anomalous images of tigers among 5000 cat images. From Fig. 1, the visual features of cats are considered as regular behavior while characteristics of tigers are anomalous. The goal is to correctly detect images of tigers (point-based group anomalies) in an unsupervised manner.

Parameter Settings: In this experiment, HOG extracts visual features as inputs for GAD methods. MGM is trained with $T = 1$ regular cat type and $L = 3$ as the number of mixtures. Parameters in OCSMM and OCSVM are set to $\nu = 50/5050$ and OCSVM is applied with k-means ($k = 40$). Following the success of the Batch Normalization architecture [12] and Exponential Linear Units (elu) [6], we have found that convolutional+batch-normalization+elu layers for DGMs provide a better representation of convolutional filters. Hence, in this experiment the autoencoder of both AAE and VAE adopts four layers of (conv-batch-normalization-elu) in the encoder part and as well as in the decoder portion of the network. AAE network parameters such as (number of filter, filter size, strides) are chosen to be (16, 3, 1) for first and second layers while (32, 3, 1) for third and fourth layers of both encoder and decoder layers. The middle hidden layer size is set to be same as rank $K = 64$ and the model is trained using Adam [13]. The decoding layer uses sigmoid function in order to capture the nonlinearity characteristics from latent representations produced by the hidden layer. Similar parameter settings are selected for DGMs in subsequent experiments.

Results: From Table 2, AAE attains the highest AUROC value of 0.9906 while OCSMM achieves a AUPRC of 0.9941. MGM, OCSMM, OCSVM are associated with high AUPRC as regular groups are correctly identified but their low AUROC scores indicate poor detection of group anomalies. Figure 2(a) further investigates the top 10 anomalous images detected by these methods and finds that AAE correctly detects all images of tigers while OCSMM erroneously captures regular cat images.

6.3   Detecting Cats and Dogs

We further investigate GAD detection where images of a single cat and dog are considered as regular groups while images with both cats and dogs are distributed-based group anomalies. The constructed dataset consists of 5050 images; 2500 single cats, 2500 single dogs and 50 images of cats and dogs together. As previously illustrated in Fig. 1(B), our goal is to detect all images with irregular mixtures of cats and dogs in an unsupervised manner.

Parameter Settings: In this experiment, HOG extracts visual features as inputs for GAD methods. MGM is trained with $T = 2$ regular cat type and $L = 3$ as the number of mixtures while OCSVM is applied with k-means ($k = 30$).

Results: Table 2 highlights (in gray) that AEE achieves the highest AUPRC and AUROC values. Other state-of-the-art GAD methods attain high AUPRC however AUROC values are relatively low. From Fig. 2(a), the top 10 anomalous images with both cats and dogs are correctly detected by AAE while OCSMM erroneously captures regular cat images. In fact, OCSMM incorrectly but consistently detects regular cats with similar results to Subsect. 6.2.

6.4   Discovering Rotated Entities

We now explore the detection of distribution-based group anomalies with 5000 regular cat images and 50 images of rotated cats. As illustrated in Fig. 1(A), images of rotated cats are anomalous compared to regular images of cats. Our goal is to detect all rotated cats in an unsupervised manner.

Parameter Settings: In this experiment involving rotated entities, HOG extracts visual features because SIFT is rotation invariant. MGM is trained with $T = 1$ regular cat type and $L = 3$ mixtures while OCSVM is applied with k-means ($k = 40$).

Results: In Table 2, AAE and VAE achieve the highest AUROC with AAE having slightly better detection results. MGM, OCSMM and OCSVM achieve a high AUPRC but low AUROC. Figure 3 illustrates the top 10 most anomalous groups where AAE correctly detects images containing rotated cats while MGM incorrectly identifies regular cats as anomalous.

(a) Tigers within cat images from `cifar-10` dataset.

(b) Images of cats and dogs within single cat and dog images using `cifar-10` dataset.

Fig. 2. Top 10 anomalous groups are presented for AAE and the best GAD method respectively where red boxes outlining images represent true group anomalies. AAE has an accurate detection of anomalous tigers injected into the `cifar-10` dataset as well as for anomalous images of both cats and dogs. On the other hand, OCSMM consistently but erroneously identifies similar cat images as the most anomalous images. (Color figure online)

6.5 Detecting Stitched Scene Images

A scene image dataset is also explored where 100 images originated from each category "inside city", "mountain" and "coast". 66 group anomalies are injected where images are stitched from two scene categories. Illustrations are provided in Fig. 1(C) where a stitched image may contain half coast and half city street view. These anomalies are challenging to detect since they have the same local features as regular images however as a collection, they are anomalous. Our objective is detect stitched scene images in an unsupervised manner.

Parameter Settings: State-of-the-art GAD methods utilize SIFT feature extraction in this experiment. MGM is trained with $T = 3$ regular scene types and $L = 4$ Gaussian mixtures while OCSVM is applied with k-means ($k = 10$). The scene image dimensions are rescaled to enable the application of an identical architecture for DGMs as implemented in previous experiments. The parameter settings for both AAE and VAE follows setup as described in Sect. 6.2.

Results: In Table 2, OCSMM achieves the highest AUROC score while DGMs are less effective in detecting distribution-based group anomalies in this experiment. We suppose that this is because only $M = 366$ groups are available for training in the scene dataset as compared to $M = 5050$ groups in previous experiments. Figure 3(b) displays the top 10 most anomalous images where OCSMM achieves a better detection results than AAE.

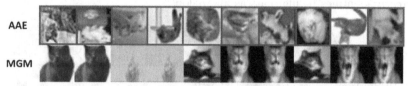

(a) Rotated cats amongst regular cats in the cifar-10 dataset.

(b) Stitched Images amongst the scene dataset.

Fig. 3. Top 10 anomalous groups are presented where red boxes outlining images represent true group anomalies in the given datasets. AAE performs well in (a) with number of groups $M = 5050$ however does not effectively detect group anomalies in (b) where number of groups is $M = 366$. MGM is unable to correctly detect any rotated cats while OSCMM is able to group anomalies in the scene dataset. (Color figure online)

6.6   Results Summary and Discussion

Table 2 summarizes the performance of detection methods in our experiments. AAE usually achieves better results than VAE as AAE has the advantage of the embedding coverage in the latent space [19]. AAE enforces a better mapping of input variables to embedding space and hence captures more robust input features. Thus AAE achieves the highest detection performance in most experiments however poor results are obtained for scene image data due to the limited number of groups. As demonstrated in our synthetic data and scene images, DGMs have a significantly worse performance on a dataset with a smaller number of groups. Thus given sufficient number of group observations for training, DGMs are effective in detecting group anomalies however poor detection occurs for a small number of groups.

Comparison of Training Times: We add a final remark about applying the proposed DGMs on GAD problems in terms of computational time and training efficiency. For example, including the time taken to calculate SIFT features on

Table 2. Summary of results for various data experiments where first two rows contains deep generative models and the later techniques are state-of-the-art GAD methods. The highest values of performance metrics are shaded in gray.

Methods	Tigers		Cats and Dogs		Rotated Cats		Scene	
	AUPRC	AUROC	AUPRC	AUROC	AUPRC	AUROC	AUPRC	AUROC
AAE	0.9449	0.9906	1.0000	1.0000	1.0000	1.0000	0.9449	0.5906
VAE	0.9786	0.9092	0.9998	0.9999	0.9999	0.9999	0.8786	0.3092
MGM	0.9881	0.5740	0.9906	0.5377	0.9919	0.6240	0.8835	0.6639
OCSMM	0.9941	0.6461	0.9930	0.5876	0.9917	0.6128	0.9140	0.7162
OCSVM	0.9909	0.5474	0.9916	0.5549	0.9894	0.5568	0.8650	0.5733

the small-scale `scene` dataset, MGM takes 42.8 s for training, 3.74 min to train OCSMM and 27.9 s for OCSVM. In comparison, the computational times for our AAE and VAE are 6.5 min and 8.5 min respectively. All the experiments involving DGMs were conducted on a MacBook Pro equipped with an Intel Core i7 at 2.2 GHz, 16 GB of RAM (DDR3 1600 MHz). The ability to leverage recent advances in deep learning as part of our optimization (e.g. training models on a GPU) is a salient feature of our approach. We also note that while MGM and OCSMM are faster to train on small-scale datasets, they suffer from at least $O(N^2)$ complexity for the total number of observations N. It is plausible that one could leverage recent advances in fast approximations of kernel methods [17] for OCSMM and studying these would be of interest in future work.

7   Conclusion

Group anomaly detection is a challenging area of research especially when dealing with complex group distributions such as image data. In order to detect group anomalies in various image applications, we clearly formulate deep generative models (DGMs) for detecting distribution-based group anomalies. DGMs outperform state-of-the-art GAD techniques in many experiments involving both synthetic and real-world image datasets however DGMs require a large number of group observations for model training. To the best of our knowledge, this is the first paper to formulate and apply DGMs to the problem of detecting group anomalies. A future direction for research involves using recurrent neural networks to detect temporal changes in a group of time series.

References

1. Azhar, R., Tuwohingide, D., Kamudi, D., Suciati, N., Sarimuddin: Batik image classification using SIFT feature extraction, bag of features and support vector machine. Procedia Comput. Sci. **72**, 24–30 (2015). The Third Information Systems International Conference 2015
2. Abadi, M., et al.: Tensorflow: large-scale machine learning on heterogeneous distributed systems. arXiv preprint arXiv:1603.04467 (2016)

3. An, J., Cho, S.: Variational autoencoder based anomaly detection using reconstruction probability. SNU Data Mining Center, Technical report (2015)
4. Chandola, V., Banerjee, A., Kumar, V.: Anomaly detection: a survey. ACM Comput. Surv. **41**(3), 15:1–15:58 (2009)
5. Chollet, F., et al.: Keras (2015). https://keras.io
6. Clevert, D.A., Unterthiner, T., Hochreiter, S.: Fast and accurate deep network learning by exponential linear units (ELUs). arXiv preprint arXiv:1511.07289 (2015)
7. Dalal, N., Triggs, B.: Histograms of oriented gradients for human detection. In: IEEE Computer Society Conference on Computer Vision and Pattern Recognition. CVPR 2005, vol. 1, pp. 886–893. IEEE (2005)
8. Davis, J., Goadrich, M.: The relationship between precision-recall and ROC curves. In: International Conference on Machine Learning (ICML) (2006)
9. Doersch, C.: Tutorial on variational autoencoders. arXiv preprint arXiv:1606.05908 (2016)
10. Hodge, V.J., Austin, J.: A survey of outlier detection methodologies. Artif. Intell. Rev. **22**, 2004 (2004)
11. Image source license: CC public domain (2018). https://pixabay.com/en/photos/tiger/
12. Ioffe, S., Szegedy, C.: Batch normalization: accelerating deep network training by reducing internal covariate shift. arXiv preprint arXiv:1502.03167 (2015)
13. Kingma, D., Ba, J.: Adam: a method for stochastic optimization. arXiv preprint arXiv:1412.6980 (2014)
14. Kingma, D.P., Welling, M.: Auto-encoding variational Bayes (Ml), pp. 1–14 (2013)
15. Kiran, B., Thomas, D.M., Parakkal, R.: An overview of deep learning based methods for unsupervised and semi-supervised anomaly detection in videos. ArXiv e-prints arXiv:1801.03149v2, January 2018
16. Krizhevsky, A., Hinton, G.: Learning multiple layers of features from tiny images. Technical report (2009)
17. Lopez-Paz, D., Sra, S., Smola, A.J., Ghahramani, Z., Schölkopf, B.: Randomized nonlinear component analysis. In: International Conference on Machine Learning (ICML) (2014)
18. Lowe, D.G.: Object recognition from local scale-invariant features. In: The Proceedings of the Seventh IEEE International Conference on Computer vision, vol. 2, pp. 1150–1157. IEEE (1999)
19. Makhzani, A., Shlens, J., Jaitly, N., Goodfellow, I., Frey, B.: Adversarial autoencoders. arXiv preprint arXiv:1511.05644 (2015)
20. Muandet, K., Schölkopf, B.: One-class support measure machines for group anomaly detection. In: Conference on Uncertainty in Artificial Intelligence (2013)
21. Perera, P., Patel, V.M.: Learning deep features for one-class classification. ArXiv e-prints arXiv:1801.05365v1, January 2018
22. Quellec, G., Lamard, M., Cozic, M., Coatrieux, G., Cazuguel, G.: Multiple-instance learning for anomaly detection in digital mammography. IEEE Trans. Med. Imaging **35**(7), 1604–1614 (2016)
23. Schlegl, T., Seeböck, P., Waldstein, S.M., Schmidt-Erfurth, U., Langs, G.: Unsupervised anomaly detection with generative adversarial networks to guide marker discovery. In: Niethammer, M., Styner, M., Aylward, S., Zhu, H., Oguz, I., Yap, P.-T., Shen, D. (eds.) IPMI 2017. LNCS, vol. 10265, pp. 146–157. Springer, Cham (2017). https://doi.org/10.1007/978-3-319-59050-9_12

24. Schölkopf, B., Platt, J.C., Shawe-Taylor, J., Smola, A.J., Williamson, R.C.: Estimating the support of a high-dimensional distribution. Neural Comput. **13**(7), 1443–1471 (2001)
25. Soleimani, H., Miller, D.J.: ATD: anomalous topic discovery in high dimensional discrete data. IEEE Trans. Knowl. Data Eng. **28**(9), 2267–2280 (2016)
26. Sultani, W., Chen, C., Shah, M.: Real-World Anomaly Detection in Surveillance Videos. ArXiv e-prints arXiv:1801.04264, January 2018
27. Toth, E., Chawla, S.: Group deviation detection: a survey. ACM Comput. Surv. (2018). (Forthcoming)
28. Xiong, L.: On Learning from Collective Data. In: Dissertations, p. 560 (2013)
29. Xiong, L., Póczos, B., Schneider, J.: Group anomaly detection using flexible genre models. In: Advances in Neural Information Processing Systems, vol. 24, pp. 1071–1079. Curran Associates Inc. (2011)
30. Xiong, L., Póczos, B., Schneider, J., Connolly, A., VanderPlas, J.: Hierarchical probabilistic models for group anomaly detection. In: AISTATS 2011 (2011)
31. Xiong, L., Póczos, B., Schneider, J.G.: Group anomaly detection using flexible genre models. In: Advances in Neural Information Processing Systems, pp. 1071–1079 (2011)
32. Xu, H., et al.: Unsupervised anomaly detection via variational auto-encoder for seasonal KPIs in web applications. arXiv preprint arXiv:1802.03903 (2018)
33. Yu, R., Qiu, H., Wen, Z., Lin, C.Y., Liu, Y.: A Survey on Social Media Anomaly Detection. ArXiv e-prints arXiv:1601.01102, January 2016
34. Yu, R., He, X., Liu, Y.: GLAD: group anomaly detection in social media analysis. In: Proceedings of the 20th ACM SIGKDD International Conference on Knowledge Discovery and Data Mining, KDD 2014, pp. 372–381. ACM, New York (2014)

Applications

Detecting Autism by Analyzing
a Simulated Social Interaction

Hanna Drimalla[1,2,3]($\boxtimes$), Niels Landwehr[1,5], Irina Baskow[2], Behnoush Behnia[4],
Stefan Roepke[4], Isabel Dziobek[2,3], and Tobias Scheffer[1]

[1] Department of Computer Science, University of Potsdam, Potsdam, Germany
tobias.scheffer@uni-potsdam.de
[2] Department of Psychology, Humboldt-Universität zu Berlin, Berlin, Germany
{hanna.drimalla,irina.baskow,isabel.dziobek}@hu-berlin.de
[3] Berlin School of Mind and Brain, Humboldt-Universität zu Berlin, Berlin, Germany
[4] Department of Psychiatry and Psychotherapy, Campus Benjamin Franklin,
Charité-Universitätsmedizin Berlin, Berlin, Germany
{behnoush.behnia,stefan.roepke}@charite.de
[5] Leibniz Institute for Agricultural Engineering and Bioeconomy, Potsdam, Germany
NLandwehr@atb-potsdam.de

Abstract. Diagnosing autism spectrum conditions takes several hours by well-trained practitioners; therefore, standardized questionnaires are widely used for first-level screening. Questionnaires as a diagnostic tool, however, rely on self-reflection—which is typically impaired in individuals with autism spectrum condition. We develop an alternative screening mechanism in which subjects engage in a simulated social interaction. During this interaction, the subjects' voice, eye gaze, and facial expression are tracked, and features are extracted that serve as input to a predictive model. We find that a random-forest classifier on these features can detect autism spectrum condition accurately and functionally independently of diagnostic questionnaires. We also find that a regression model estimates the severity of the condition more accurately than the reference screening method.

1 Introduction

Autism spectrum conditions (ASC) encompass a range of neurodevelopmental conditions that affect how an individual perceives the world and interacts with others. Around 1 in every 100 individuals has some form of autism [1] and shows the characteristic impairments in social communication and interaction as well as restricted interests and repetitive behaviors [2].

Autism is nowadays seen as a spectrum of conditions: its severity and impact on the individual's life vary [3]. Mild autism conditions with normal intelligence levels have been described as *high-functioning autism* and *Asperger syndrome.*

Charité-Universitätsmedizin Berlin—Corporate Member of Freie Universität Berlin, Humboldt-Universität zu Berlin, and Berlin Institute of Health.

M. Berlingerio et al. (Eds.): ECML PKDD 2018, LNAI 11051, pp. 193–208, 2019.
https://doi.org/10.1007/978-3-030-10925-7_12

Although symptoms of autism are already occurring early in life, individuals with high-functioning autism are often diagnosed later [4], due to compensation strategies [5] or the subtlety of the autistic symptoms. A study with college students points to a substantial proportion of individuals with normal intelligence and autism that are undiagnosed [6]. Despite the higher social-functional level, their lifetime rate of psychiatric consultations is high, reflecting a need of earlier support and diagnosis [7].

Existing diagnostic tools for adults with autism examine the altered social communication and interaction patterns in semi-structured activities with the individual (*Autism Diagnostic Observation Schedule, ADOS* [8]) as well as in diagnostic interviews with the parents (*Autism Diagnostic Interview-Revised, ADI-R* [9]). These diagnostic processes are considered to establish a "gold-standard" diagnosis and concentrate on diagnostic criteria defined by DSM-5 [2] and ICD-10 [10]. But they take several hours of time and have to be carried out by well-trained practitioners. Thus, they cannot be administered to any significant share of the population to screen for high-functioning autism.

Therefore, standardized questionnaires are often used for screening. One of the most widely applied ones is the *Autism-Spectrum Quotient (AQ)* [11], a brief self-administered questionnaire that measures traits associated with the autistic spectrum in adults with normal intelligence. In a clinical study in which both an AQ screening and a diagnosis by a medical practitioner have been observed for adults who sought out a diagnostic clinic, the AQ screening has shown an AUC of 0.78 [12]. One general concern about the AQ screening and other self-reports is the universal bias towards giving socially desirable answers [13]. In autism diagnosis, another aspect may affect the results of self-reports even more: individuals with autism spectrum condition often have an impaired introspection [14] and problems in abstract reasoning [15]. Therefore, an easy but still precise screening mechanism for autism that does not rely on self-reflection would be beneficial.

Deficits in social interaction that are part of the diagnostic criteria [2,10] include a lack of social-emotional reciprocity, lack of facial expressions, and abnormalities in eye contact and voice modulation. In this paper, we develop and evaluate a screening approach for high-functioning adults of both genders that automatically analyzes these criteria in a simulated social interaction. Section 2 presents related work to autism detection via machine learning. Section 3 describes the Simulated Interaction screening method. Section 4 presents empirical results and Sect. 5 concludes.

2   Related Work

To date, there have been only a few studies using automatic behavioral analysis to detect autism. Crippa et al. [16] monitor upper-limb movements during a specifically designed manual task with an optoelectronic system, and use features extracted from these kinematics data to detect children with autism spectrum condition. They observe a maximum accuracy of 96.7% (with a maximum precision of 93.8% and a maximum recall of 100%) on a small sample of 15 children with ASC and 15 neurotypical children. However, as the method is developed for this very specific task and is applicable only for optoelectronic systems, it provides no scalable screening mechanism. Furthermore, it focuses on a behavioral feature that is not very tightly linked to ASC.

Hashemi et al. [17] use a computer vision approach to analyze activities that are assessed by the *Autism Observation Scale for Infants* [18] in video recordings. In a small sample of three infants with and three without indication of ASC risk they find differences between the groups in head motion and gait. However, the paper makes no attempt to classify the children based on the video analysis.

Liu et al. [19] observe eye movements during a face-recognition task; such face scanning patterns have been reported to differ in autism [20]. Based on the gaze patterns of all participants, they are able to discriminate between autistic and neurotypical children with accuracy of 88.5%. Since this process involves a high-acuity eye tracker, it does not suggest itself as a scalable screening process. Another study points at the potential of eye gaze as a feature for detection of autism: Gliga et al. [21] found that eye movements of nine-month-old infants significantly predict a higher level of autism symptoms at two years of age.

A recent study [22] used machine-learning-based voice analysis to classify word utterances of children with autism spectrum condition and children with typical development. The study focuses on the classification at the level of word utterances and not individuals. Therefore, the value of this approach for diagnostic purposes remains unclear.

Beyond autism, there has been remarkable progress in using machine learning technologies to infer underlying medical and psychological conditions from behavior or appearance. Some promising results could be achieved in learning to detect depression [23], predict suicidal ideation [24] or recognize schizophrenia [25]. Using audio and video recordings of the participants, some studies focused on speech and vocal cues [26]. Other studies investigated the predictive value of facial expressions [27], gaze direction or head pose [28].

3   Simulated Interaction

In this section, we develop the *Simulated Interaction* screening method. The aim of this procedure is to detect autism via a simulated social interaction, using only a screen, a webcam, and a microphone. This problem can be divided into two subproblems: The first is to predict whether a practitioner will diagnose the individual as on the autism spectrum condition (binary classification). The second is an assessment of the degree of autism, conceptualized as the individual's value of the ADI-R, a diagnostic clinical interview with their parents (regression). Specifically, we focus on the score for the reciprocal social interaction subdomain of the ADI-R (social subscale), as we expect this score to be most sensitive to high-functioning individuals and closely corresponding to the naturalistic setting of a simulated social interaction.

The core symptoms of autism are deficits in social communication and interactions [2]. These deficits manifest themselves in a number of nuances that guide

Table 1. Simulated Interaction schedule

Speaker	Topic	Time (s)
Actress	Introduces herself, asks "what is your name?"	4
Participant	Answers	2
Actress	Describes her way to the institute, asks "how did you get to the institute today?"	90
Participant	Answers	20
Actress	Thanks the participant, switches the topic towards dining and describes how she sets the table for dinner; asks "How do you prepare a table for dinner?"	40
Participant	Answers	25
Actress	Describes her favorite food, asks "what is your favorite food?"	25
Participant	Answers	25
Actress	Describes her least favorite food, asks "which food do you dislike?"	25
Participant	Answers	25
Actress	Thanks the participant and concludes the conversation	8

the design of the screening method. First, in a social interaction, individuals tend to involuntarily mimic facial expressions of their conversation partner [29]. Individuals with autism spectrum condition are less likely to mimic the facial expressions of others spontaneously [30]. Similarly, a reduced intensity [31] or at least qualitative differences [32] of facial expressions in natural conversations have been observed for autism conditions.

Secondly, autism spectrum condition typically manifests itself in altered gaze patterns in complex environments [33] or in emotion recognition paradigms [34]. Madipakkam et al. [35] observe that patients have a tendency to avoid eye contact and instead let their gaze stray over a wider range of angles than neurotypical subjects. Thirdly, effects of autism spectrum condition on individual's voice have been reported; *e.g.,* for prosody [36] or pitch [37]—not in all studies, but especially in naturalistic settings [38].

Motivated by these findings, we design Simulated Interaction as a "dialog" between the recording of an actress and the participant about positive and negative food experiences. The actress addresses the participant directly and asks simple questions. The participant listens to the actress and answers her question while the actress nods and appears to listen. The first two parts are emotionally neutral. The third part is about the participant's favorite food and addresses a joyful topic. The final part about the participant's least favorite food raises an emotional response of disgust. The exact schedule is described in Table 1.

3.1   Feature Extraction

Based on the described phenomenology of autism condition, we concentrate our feature on gaze, voice, and facial expressions. Facial expressions can be broken down into *facial action units* according to the facial action coding system developed by [39]—each action unit is comprised of visually detectable muscle movements in the face. A major advantage of this sign-based over message-based approaches is its objectivity as it does not require any interpretation [40]. Moreover, it does not reduce the facial expression to a small set of prototypical emotional expressions [41].

We employ the OpenFace library 1.0.0 [42] to extract the occurrence and intensity of 18 facial action units as well as gaze angles for both eyes from each frame. OpenFace is an open-source tool that is capable of facial-landmark detection, head-pose estimation, facial-action-unit recognition and eye-gaze estimation. OpenFace detects and tracks facial landmarks with conditional local neural fields [43], and aligns the face.

In order to detect emotions, OpenFace extracts HOG features, and reduces the HOG features space by principal component analysis. To correct for person-specific neutral expressions, OpenFace subtracts the median value of each of the remaining features over the entire observation sequence. Finally, the activation of each action unit is determined by a support-vector classifier and its intensity by support-vector regression. OpenFace has been tested on the SEMAINE [44], DISFA [45] and BP4D-Spontaneous [46] datasets and demonstrated state-of-the-art results [47] as well as outperformed the baselines of the FERA 2015 challenge [48].

OpenFace performs appearance-based gaze estimation. On the MPIIGaze data set [49], it achieves a mean absolute error of under 10 degrees which exceeds the performance of other tools (*e.g.*, EyeTab [50]). OpenFace detects eye-region landmarks including the eyelids, iris, and pupil with conditional local neural fields, trained on the SynthesEyes dataset [51]. Based on the location of the eye and pupil, it estimates the center of the eyeball and infers the gaze vector from this center through the pupil.

To extract features of the audio recording of the participant's voice, we use the librosa library [52]. For each frame, we extract prosodic (root-mean-square energy) as well as spectral features (forty mel-frequency cepstral coefficients, MMCC). Both are standard features in speech recognition [53] and have shown to be altered under autism condition [54].

From these primary features, we extract secondary features that aggregate the values for each feature. For action units' intensity and gaze angle, we calculate arithmetic mean, standard deviation, skewness, kurtosis, the maximum and the time point of the maximum. For action unit's occurrence and the voice features, we calculate the mean values. All aggregated values are calculated for seven parts of the conversation.

3.2 Machine Learning Methods

We use SVMs and random forests as base machine-learning methods. For the SVM we use a radial basial function kernel and tune the regularization parameters in a nested cross-validation with grid search. For the random forest we use an ensemble of 1,000 different trees on different subsets of data and input variables. We tune the maximal depth of the trees and the minimum number of samples per leaf with a nested grid-search.

Additionally, we explore the use of two different convolutional neural networks all employing the 1-dimensional convolution operation. The convolution uses a stride of one and zero padding. We tune the number of filters, size of convolution and pooling, dropout rate and number of units of the dense layer via a nested grid search; Table 2 shows the search space for all hyperparameters. Both CNNs use the rectified linear activation function for the hidden units, and have one output unit with a logistic activation function. We use the cross-entropy loss function and train the networks with the gradient descent algorithm Adam. To avoid over-fitting, training is stopped when the validation-loss does not improve for three epochs.

The *StackedCNN* contains four learned layers: Two 1D-convolutional layers, one dense layer and one fully-connected output layer. Each convolutional layer is followed by a max-pooling layer. One additional drop-out layer is followed by a dense layer. The structure of the network can be seen in Fig. 1. We tune the hyperparameters via a full grid search, leading to a minimal number of 135 and a maximal number of 938,769 parameters.

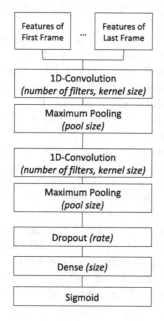

Fig. 1. Architecture of the StackedCNN (tuned hyperparameters are printed in italics).

PooledCNN is a CNN with multiple inputs that incorporates some domain knowledge into its structure. It contains seven learned convolutional layers—one for each conversation part. The input is split into the conversation parts and distributed accordingly into the seven convolutional layer. Thus, each convolutional layer receives only a part of the original input. Every convolutional layer is followed by a pooling layer. The size of the pooling equals the number of units of the previous convolutional layer leading to seven units as input for the following dense layer. The next and last layer is the fully-connected output layer. The structure of the network can be seen in Fig. 2. The hyperparameters are tuned via a random search with 20 iterations, leading to a minimal number of 618 parameters and a maximal number of 110,343 parameters.

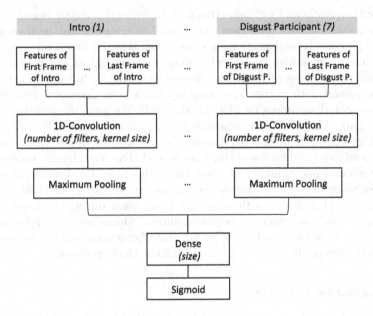

Fig. 2. Architecture of the PooledCNN (tuned hyperparameters are printed in italics).

Table 2. Hyperparameter space

Hyperparameter	Stacked	Pooled
Number of filters	{1, 2, 4, 8}	{1, 2, 4, 8}
Size of kernel	{2, 4, 8, 16, 32}	{2, 4, 8, 16, 32}
Size of pooling	{5, 25, 50}	-
Rate of dropout	{0.25, 0.5}	-
Size of dense layer	{4, 8, 16, 32, 64, 128, 256}	{8, 16, 32, 64, 128}

4   Empirical Study

This section explores the effectiveness of the Simulated Interaction screening method on a sample of patients that have been diagnosed with autism spectrum condition and a neurotypical control group. The AQ questionnaire will serve as reference screening method.

4.1   Data Collection

We record the audio and webcam stream of 44 neurotypical participants and 37 participants with autism spectrum condition. The sample is balanced regarding gender. The neurotypical participants have been selected based on a questionnaire that asks for a history of a wide range of neurological and psychological conditions; participants with any such history have been excluded from the study. Each participant with autism spectrum condition been diagnosed by a practitioner and scored according to ADOS and ADI-R. We measure autistic traits of all participants via a German version of the AQ questionnaire [11].

We film the faces of the participants in experiment rooms with constant lighting conditions and no disturbing background. The participants are recorded with the internal microphone and a webcam with a rate of 30 frames per second and a resolution of 640×480 pixels, leading to a total number of 11,340 frames. These video and audio recordings of each participant during the conversation are the raw input to detect participant's autism. OpenFace provides a success value $(0/1)$ for the face tracking of each frame. Participants with a success rate of less than 90% of the frames are excluded from the experiment.

4.2   Evaluation Protocol

To validate the results, we use a *nested cross-validation* strategy with an outer leave-one-out cross-validation loop and an inner 3-fold cross-validation loop in which we tune all hyperparameters. The hyperparameters of the neural networks are listed in Table 1.

4.3   Prediction of the Clinical Diagnosis

We will first study the ability of Simulated Interaction to predict the clinical diagnosis of an autism condition.

Comparison of Base Machine-Learning Methods. We first compare the different machine learning methods under investigation. Figure 3 compares ROC curves for the base learning methods using the full set of features. The random forest and SVM achieve the best detection using all features. The random forest achieves an area under the curve of 0.84 and the SVM an AUC of 0.81. The SVM $(p < 0.01$ according to a sign test) and the random forest $(p < 0.01$ according to a sign test) perform significantly better than the majority baseline. Their predictions correlate strongly with the autism diagnosis of the participants: the class probabilities predicted by the random forest reach a point-biserial correlation with the diagnosis of $r = 0.53$ ($p < 0.0001$).

Both neural network models perform worse than the random forest and SVM. The CNN without knowledge about the interaction parts, performs close to chance level with an AUC of 0.53 (stacked). The PooledCNN produces a better result with an AUC of 0.64. However, with a set threshold of 0.5, none of the CNNs achieves a higher accuracy than the naive baseline, which always predicts the majority class. The class probabilities predicted by the PooledCNN correlate positively with the autism diagnosis at trend-level ($r = 0.21$, $p < 0.1$).

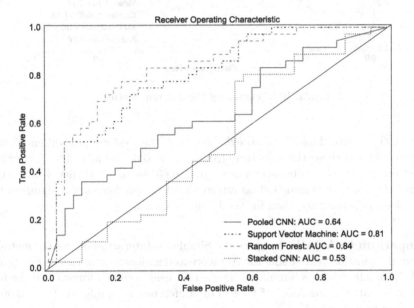

Fig. 3. ROC curves for the different classifiers

Comparison of Feature Subsets. We now explore the relative contributions of the different types of features; Fig. 4 compares the ROC curves of various feature subsets. As on the complete feature set the random forest performs best, we concentrate on this classifier for the next steps.

We compare the performance of different groups of features: the occurrence of action units, the intensity of action units, the gaze angles, all video features and the vocal features. The best prediction of the autism diagnosis is achieved with the combination of all features, as Fig. 4 shows. Statistical testing reveals that a significantly better detection is possible with the vocal features, the intensity of action units, all video features, or all features together than with the baseline

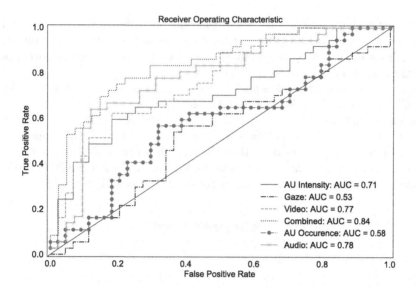

Fig. 4. ROC curves for the different features

($p < 0.05$ for all). There was no evidence that the performance differs significantly between these three feature groups ($p > 0.05$ for all). The prediction based on gaze was significantly worse ($p < 0.05$ for all of them). Comparing the facial features revealed that on a trend-level the prediction was significantly worse for the occurrence than for the intensity ($p < 0.05$).

Comparison to AQ. We compare the Simulated Interaction screening method to the AQ questionnaire. Using the AQ score as decision-function value we obtain an AUC of 0.99 whereas Simulated Interaction using random forests and the full feature set attains an AUC of 0.84. This difference is significant ($p < 0.001$) according to a sign test.

In previous studies using a clinical sample of patients, the AQ has shown an AUC of 0.78 [12] which starkly contrasts our observation of 0.99. This difference can be explained by the selection criteria for our neurotypical sample. Subjects with any history of psychological, psychiatric, and neurological treatments have been excluded from the neurotypical sample. This results in a much easier classification problem than can be expected in any clinical or broad screening setting.

Patients in a clinical environment usually seek out medical treatment because of an impaired quality of life. The AQ questionnaire has a known propensity to higher scores for individuals with obsessive-compulsive disorder, social-anxiety [55] or schizophrenia [56]. Furthermore, individuals with suspected ASD typically enter the diagnostic process at a specialized center as in the study mentioned above [12]. Despite signs of autism and self-diagnosis, only around 73% [12,57] receive an official diagnosis. In such a difficult setting, the AQ's low AUC of 0.78 and high false-positive rate of 64% [57] are unsatisfactory.

Thus, it is natural to ask whether a combination of the AQ questionnaire and the Simulated Interaction screening method is more accurate than just the AQ. This would be the case if the error cases of AQ and Simulated Interaction were independent of one another. Unfortunately, our limited sample does not allow us to answer this question. The AQ questionnaire misclassifies only one single patient in our sample, which makes it impossible to draw any conclusions about the independence of error cases or the accuracy of a combined detection model. However, since the AQ and Simulated Interaction are based on fundamentally different functional principles, our results motivate a follow-up study in with a larger clinical sample from a clinical distribution of patients.

4.4   Estimation of the Autism Degree

We will now evaluate Simulated Interactions' ability to estimate the severity of the autism condition, measured by the score of the clinical interview of the parents, ADI-R (social subscale). We use a random forest regression model; in order to further reduce the dimensionality of the feature space, we aggregate each of the secondary features over the seven parts of the interview into a single value. Figure 5 plots the gold-standard ADI-R social subscale score over the score of the regression model. For the individuals with ASC, it is possible to estimate the ADI-R social subscale with a random forest better than a mean baseline, according to a paired t-test ($p < 0.05$).

Comparison to AQ. The root mean squared error of the tree is 5.40, while the baseline produces an error of 6.42. The ADI-R social subscale values predicted

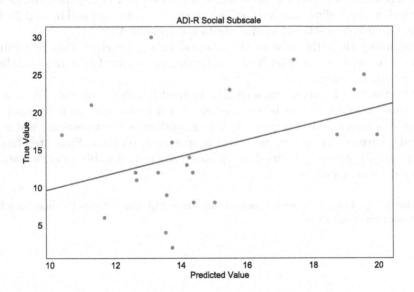

Fig. 5. Regression for the ADI-R social

by the random forest correlate positively with the true values at trend-level ($p < 0.1$). By contrast, the values of the AQ and the ADI-R social subscale are not significantly positively correlated (trend towards a negative correlation: $r = -0.42$, $p < 0.1$) and a prediction with a random forest regressor on the autism questionnaire score lead to a mean absolute error of 8.02, which is significantly worse than the prediction with the Simulated Interaction ($p < 0.05$).

5 Conclusion

Diagnosing autism spectrum condition and quantifying the severity of the condition require time and well-trained practitioners. Quantifying the severity accurately also requires access to the individual's parents. In this paper, we presented and evaluated the Simulated Interaction screening method for autism spectrum condition of high-functioning individuals. We find that it is possible to predict the binary diagnosis of autism spectrum condition with high accuracy from the facial expressions, vocal features, and gaze patterns of the individuals.

The intensity of the facial expressions and the vocal features turned out to be more informative than the occurrence of facial expressions and the gaze patterns. The webcam might not be capable to track the gaze behavior sensitively enough to detect individual differences. The results concerning the facial expressions fit the literature about qualitative differences in facial expressions.

For the prediction of the binary diagnosis, the Simulated Interaction screening method did not outperform the AQ questionnaire. However, Simulated Interaction is functionally independent of a self-assessment because it is based on fundamentally different features. While our limited sample does not allow us to draw any conclusions on the accuracy of a combination of AQ and Simulated Interaction, our findings motivate a follow-up study on a larger and more difficult clinical sample consisting of subjects with a suspected ASD.

Regarding the estimation of the severity, we can conclude that Simulated Interaction estimates the ADI-R values significantly better than a mean baseline and the AQ questionnaire.

The Simulated Interaction screening method has two principal advantages over questionnaires: first, it is not as easily biased by social desirability tendencies as a questionnaire. Secondly, it is independent of introspection—which is typically impaired in subjects with autism spectrum condition. Simulated Interaction could potentially be used as an online self-test, possibly in combination with a questionnaire.

Acknowledgment. This work was partially funded by the German Science Foundation under grant LA3270/1-1.

References

1. Mattila, M.L., et al.: Autism spectrum disorders according to DSM-IV-TR and comparison with DSM-5 draft criteria: an epidemiological study. J. Am. Acad. Child Adolesc. Psychiatry **50**(6), 583–592.e11 (2011)
2. American Psychiatric Association: Diagnostic and Statistical Manual of Mental Disorders: DSM-5, 5th edn. American Psychiatric Association, Arlington (2013)
3. Newschaffer, C.J., et al.: The epidemiology of autism spectrum disorders. Annu. Rev. Public Health **28**, 235–258 (2007)
4. Barnard, J., Harvey, V., Potter, D.: Ignored or Ineligible? The reality for adults with autism spectrum disorders. National Autistic Society (2001)
5. Harms, M.B., Martin, A., Wallace, G.L.: Facial emotion recognition in autism spectrum disorders: a review of behavioral and neuroimaging studies. Neuropsychol. Rev. **20**(3), 290–322 (2010)
6. White, S.W., Ollendick, T.H., Bray, B.C.: College students on the autism spectrum: prevalence and associated problems. Autism Int. J. Res. Pract. **15**(6), 683–701 (2011)
7. Lehnhardt, F.G., Gawronski, A., Volpert, K., Schilbach, L., Tepest, R., Vogeley, K.: Das psychosoziale funktionsniveau spätdiagnostizierter patienten mit autismus-spektrum-störungen–eine retrospektive untersuchung im erwachsenenalter. Fortschr. Neurol. Psychiatr. **80**(2), 88–97 (2012)
8. Lord, C., et al.: The autism diagnostic observation schedule-generic: a standard measure of social and communication deficits associated with the spectrum of autism. J. Autism Dev. Disord. **30**(3), 205–223 (2000)
9. Lord, C., Rutter, M., Le Couteur, A.: Autism diagnostic interview-revised: a revised version of a diagnostic interview for caregivers of individuals with possible pervasive developmental disorders. J. Autism Dev. Disord. **24**(5), 659–685 (1994)
10. World Health Organization: The ICD-10 Classification of Mental and Behavioural Disorders: Clinical Descriptions and Diagnostic Guidelines. ICD-10 Classification of Mental and Behavioural Disorders/World Health Organization. World Health Organization (1992)
11. Baron-Cohen, S., Wheelwright, S., Skinner, R., Martin, J., Clubley, E.: The autism-spectrum quotient (AQ): evidence from asperger syndrome/high-functioning autism, malesand females, scientists and mathematicians. J. Autism Dev. Disord. **31**(1), 5–17 (2001)
12. Woodbury-Smith, M.R., Robinson, J., Wheelwright, S., Baron-Cohen, S.: Screening adults for asperger syndrome using the AQ: a preliminary study of its diagnostic validity in clinical practice. J. Autism Dev. Disord. **35**(3), 331–335 (2005)
13. Van de Mortel, T.F., et al.: Faking it: social desirability response bias in self-report research. Aust. J. Adv. Nurs. **25**(4), 40 (2008)
14. Happé, F.: Theory of mind and the self. Ann. N. Y. Acad. Sci. **1001**(1), 134–144 (2003)
15. Minshew, N.J., Meyer, J., Goldstein, G.: Abstract reasoning in autism: a disassociation between concept formation and concept identification. Neuropsychology **16**(3), 327 (2002)
16. Crippa, A., et al.: Use of machine learning to identify children with autism and their motor abnormalities. J. Autism Dev. Disord. **45**(7), 2146–2156 (2015)
17. Hashemi, J., et al.: A computer vision approach for the assessment of autism-related behavioral markers. In: 2012 IEEE International Conference on Development and Learning and Epigenetic Robotics (ICDL), pp. 1–7. IEEE (2012)

18. Bryson, S.E., Zwaigenbaum, L.: Autism observation scale for infants. In: Patel, V., Preedy, V., Martin, C. (eds.) Comprehensive Guide to Autism, pp. 299–310. Springer, New York (2014). https://doi.org/10.1007/978-1-4614-4788-7_12
19. Liu, W., Li, M., Yi, L.: Identifying children with autism spectrum disorder based on their face processing abnormality: a machine learning framework. Autism Res. Off. J. Int. Soc. Autism Res. 9(8), 888–898 (2016)
20. Pelphrey, K.A., Sasson, N.J., Reznick, J.S., Paul, G., Goldman, B.D., Piven, J.: Visual scanning of faces in autism. J. Autism Dev. Disord. 32(4), 249–261 (2002)
21. Gliga, T., Bedford, R., Charman, T., Johnson, M.H.: Enhanced visual search in infancy predicts emerging autism symptoms. Curr. Biol. CB 25(13), 1727–1730 (2015)
22. Nakai, Y., Takiguchi, T., Matsui, G., Yamaoka, N., Takada, S.: Detecting abnormal word utterances in children with autism spectrum disorders: machine-learning-based voice analysis versus speech therapists. Percept. Mot. Ski. 124(5), 961–973 (2017)
23. Nasir, M., Jati, A., Shivakumar, P.G., Nallan Chakravarthula, S., Georgiou, P.: Multimodal and multiresolution depression detection from speech and facial landmark features. In: Proceedings of the 6th International Workshop on Audio/Visual Emotion Challenge, pp. 43–50. ACM (2016)
24. Laksana, E., Baltrušaitis, T., Morency, L.P., Pestian, J.P.: Investigating facial behavior indicators of suicidal ideation. In: 2017 12th IEEE International Conference on Automatic Face & Gesture Recognition (FG 2017), pp. 770–777. IEEE (2017)
25. Tron, T., Peled, A., Grinsphoon, A., Weinshall, D.: Automated facial expressions analysis in schizophrenia: a continuous dynamic approach. In: Serino, S., Matic, A., Giakoumis, D., Lopez, G., Cipresso, P. (eds.) MindCare 2015. CCIS, vol. 604, pp. 72–81. Springer, Cham (2016). https://doi.org/10.1007/978-3-319-32270-4_8
26. Moore, E., Clements, M.A., Peifer, J.W., Weisser, L.: Critical analysis of the impact of glottal features in the classification of clinical depression in speech. IEEE Trans. Bio-Med. Eng. 55(1), 96–107 (2008)
27. Cohn, J.F., et al.: Detecting depression from facial actions and vocal prosody. In: Staff, I. (ed.) 2009 3rd International Conference on Affective Computing and Intelligent Interaction, pp. 1–7. IEEE (2009)
28. Alghowinem, S., Goecke, R., Cohn, J.F., Wagner, M., Parker, G., Breakspear, M.: Cross-cultural detection of depression from nonverbal behaviour. In: IEEE International Conference on Automatic Face & Gesture Recognition and Workshops, vol. 1 (2015)
29. Seibt, B., Mühlberger, A., Likowski, K., Weyers, P.: Facial mimicry in its social setting. Front. Psychol. 6, 1122 (2015)
30. McIntosh, D.N., Reichmann-Decker, A., Winkielman, P., Wilbarger, J.L.: When the social mirror breaks: deficits in automatic, but not voluntary, mimicry of emotional facial expressions in autism. Dev. Sci. 9(3), 295–302 (2006)
31. Stagg, S.D., Slavny, R., Hand, C., Cardoso, A., Smith, P.: Does facial expressivity count? How typically developing children respond initially to children with autism. Autism 18(6), 704–711 (2014)
32. Grossman, R.B., Edelson, L.R., Tager-Flusberg, H.: Emotional facial and vocal expressions during story retelling by children and adolescents with high-functioning autism. J. Speech Lang. Hear. Res. 56(3), 1035–1044 (2013)
33. Zhao, S., Uono, S., Yoshimura, S., Kubota, Y., Toichi, M.: Atypical gaze cueing pattern in a complex environment in individuals with ASD. J. Autism Dev. Disord. 47(7), 1978–1986 (2017)

34. Wieckowski, A.T., White, S.W.: Eye-gaze analysis of facial emotion recognition and expression in adolescents with ASD. J. Clin. Child Adolesc. Psychol. **46**(1), 110–124 (2017). The official journal for the Society of Clinical Child and Adolescent Psychology, American Psychological Association, Division 53

35. Madipakkam, A.R., Rothkirch, M., Dziobek, I., Sterzer, P.: Unconscious avoidance of eye contact in autism spectrum disorder. Sci. Rep. **7**(1), 13378 (2017)

36. Shriberg, L.D., Paul, R., McSweeny, J.L., Klin, A., Cohen, D.J., Volkmar, F.R.: Speech and prosody characteristics of adolescents and adults with high-functioning autism and asperger syndrome. J. Speech Lang. Hear. Res. **44**(5), 1097–1115 (2001)

37. Sharda, M., et al.: Sounds of melodypitch patterns of speech in autism. Neurosci. Lett. **478**(1), 42–45 (2010)

38. Diehl, J.J., Watson, D., Bennetto, L., McDonough, J., Gunlogson, C.: An acoustic analysis of prosody in high-functioning autism. Appl. Psycholinguist. **30**(3), 385–404 (2009)

39. Ekman, P., Friesen, W.V.: Facial Action Coding System. Consulting Psychologists Press, Palo Alto (1978)

40. Fasel, B., Luettin, J.: Automatic facial expression analysis: a survey. Pattern Recognit. **36**(1), 259–275 (2003)

41. Tian, Y.L., Kanade, T., Cohn, J.F.: Recognizing action units for facial expression analysis. IEEE Trans. Pattern Anal. Mach. Intell. **23**(2), 97–115 (2001)

42. Baltrusaitis, T., Robinson, P., Morency, L.P.: Openface: an open source facial behavior analysis toolkit. In: 2016 IEEE Winter Conference on Applications of Computer Vision (WACV), pp. 1–10. IEEE (2016)

43. Baltrusaitis, T., Robinson, P., Morency, L.P.: Constrained local neural fields for robust facial landmark detection in the wild. In: Proceedings of the IEEE International Conference on Computer Vision Workshops, pp. 354–361 (2013)

44. McKeown, G., Valstar, M.F., Cowie, R., Pantic, M.: The semaine corpus of emotionally coloured character interactions. In: 2010 IEEE International Conference on Multimedia and Expo, pp. 1079–1084. IEEE (2010)

45. Mavadati, S.M., Mahoor, M.H., Bartlett, K., Trinh, P., Cohn, J.F.: DISFA: a spontaneous facial action intensity database. IEEE Trans. Affect. Comput. **4**(2), 151–160 (2013)

46. Zhang, X., et al.: BP4D-spontaneous: a high-resolution spontaneous 3D dynamic facial expression database. Image Vis. Comput. **32**(10), 692–706 (2014)

47. Baltrusaitis, T., Mahmoud, M., Robinson, P.: Cross-dataset learning and person-specific normalisation for automatic action unit detection. In: 2015 11th IEEE International Conference and Workshops on Automatic Face and Gesture Recognition (FG), pp. 1–6. IEEE (2015)

48. Valstar, M.F., et al.: FERA 2015 - second facial expression recognition and analysis challenge. In: 2015 11th IEEE International Conference and Workshops on Automatic Face and Gesture Recognition (FG), pp. 1–8. IEEE (2015)

49. Zhang, X., Sugano, Y., Fritz, M., Bulling, A.: Mpiigaze: Real-world dataset and deep appearance-based gaze estimation. IEEE Trans. Pattern Anal. Mach. Intell. **41**(1), 162–175 (2019)

50. Wood, E., Bulling, A.: EyeTab: model-based gaze estimation on unmodified tablet computers. In: Proceedings of the Symposium on Eye Tracking Research and Applications, pp. 207–210. ACM (2014)

51. Wood, E., Baltrusaitis, T., Zhang, X., Sugano, Y., Robinson, P., Bulling, A.: Rendering of eyes for eye-shape registration and gaze estimation. In: Proceedings of the IEEE International Conference on Computer Vision, pp. 3756–3764 (2015)

52. McFee, B., et al.: librosa: Audio and music signal analysis in python. In: Proceedings of the 14th Python in Science Conference, pp. 18–25 (2015)
53. Ittichaichareon, C., Suksri, S., Yingthawornsuk, T.: Speech recognition using MFCC. In: International Conference on Computer Graphics, Simulation and Modeling (ICGSM 2012), pp. 28–29, July 2012
54. Marchi, E., Schuller, B., Batliner, A., Fridenzon, S., Tal, S., Golan, O.: Emotion in the speech of children with autism spectrum conditions: prosody and everything else. In: Proceedings 3rd Workshop on Child, Computer and Interaction (WOCCI 2012), Satellite Event of INTERSPEECH 2012 (2012)
55. Hoekstra, R.A., Bartels, M., Cath, D.C., Boomsma, D.I.: Factor structure, reliability and criterion validity of the autism-spectrum quotient (AQ): a study in dutch population and patient groups. J. Autism Dev. Disord. **38**(8), 1555–1566 (2008)
56. Zhang, L., et al.: Psychometric properties of the autism-spectrum quotient in both clinical and non-clinical samples: Chinese version for mainland China. BMC Psychiatry **16**(1), 213 (2016)
57. Ashwood, K., et al.: Predicting the diagnosis of autism in adults using the autism-spectrum quotient (AQ) questionnaire. Psychol. Med. **46**(12), 2595–2604 (2016)

A Discriminative Model for Identifying Readers and Assessing Text Comprehension from Eye Movements

Silvia Makowski[1(✉)], Lena A. Jäger[1,2,3], Ahmed Abdelwahab[1,4], Niels Landwehr[1,4], and Tobias Scheffer[1]

[1] Department of Computer Science, University of Potsdam, August-Bebel-Straße 89, 14482 Potsdam, Germany
{silvia.makowski,lena.jaeger,tobias.scheffer}@uni-potsdam.de
[2] Department of Linguistics, University of Potsdam, Karl-Liebknecht-Straße 24–25, 14476 Potsdam, Germany
[3] Weizenbaum Institute for the Networked Society, Hardenbergstraße 32, 10623 Berlin, Germany
[4] Leibniz Institute for Agricultural Engineering and Bioeconomy, Max-Eyth-Allee 100, 14469 Potsdam, Germany
{AAbdelwahab,NLandwehr}@atb-potsdam.de

Abstract. We study the problem of inferring readers' identities and estimating their level of text comprehension from observations of their eye movements during reading. We develop a generative model of individual gaze patterns (*scanpaths*) that makes use of lexical features of the fixated words. Using this generative model, we derive a Fisher-score representation of eye-movement sequences. We study whether a Fisher-SVM with this Fisher kernel and several reference methods are able to identify readers and estimate their level of text comprehension based on eye-tracking data. While none of the methods are able to estimate text comprehension accurately, we find that the SVM with Fisher kernel excels at identifying readers.

1 Introduction

During reading, the eye proceeds in a series of rapid movements, called saccades, instead of smoothly wandering over the text. Between two saccades, the eye remains almost still for about 250 ms on average, fixating a certain position in text to obtain visual input. Saccades serve as a relocation mechanism of the eye moving the focus on average seven to nine characters wide from one fixation position to the next. Eye movements during reading are driven by complex cognitive processes involving vision, attention, language and oculomotor control [1,2]. Since a reader's eye movement behavior is precisely observable and reflects the interplay of internal processes and external stimuli for the generation of complex action [1], it is a popular research subject in cognitive psychology.

S. Makowski and L. A. Jäger—Joint first authorship.

© Springer Nature Switzerland AG 2019
M. Berlingerio et al. (Eds.): ECML PKDD 2018, LNAI 11051, pp. 209–225, 2019.
https://doi.org/10.1007/978-3-030-10925-7_13

Eye movements have long been known to vary significantly between individuals [3–5]. This property makes them interesting for biometrics. Indeed, identification based on eye movements during reading may offer several advantages in many application areas. Users can be identified unobtrusively while having access to a document they would read anyway, which saves time and attention. For biometric identification during reading, nearest-neighbor [6] and generative probabilistic models [7,8] of eye-gaze patterns have been explored.

Eye movements are believed to mirror different levels of comprehension processes involved in reading [9]. Experimental work has shown that readers' fixations are influenced by syntactic comprehension [10], semantic plausibility [11], background knowledge [12], text difficulty, and inconsistencies [13]. This motivates our goal of estimating readers' levels of text comprehension from their eye-gaze.

Gaze patterns, also referred to as *scanpaths*, that occur during reading are sequences of fixations and saccades. One can easily extract vectors of aggregated distributional features that standard learning algorithms can process [6], *e.g.,* the average fixation duration and saccade amplitude, albeit at a great loss of information. Generative graphical models [7] allow to infer the likelihood of a scanpath under reader-specific model parameters. However, since both identification and assessing text comprehension are discriminative tasks, it appears plausible that discriminatively trained models would be better suited to this task. Classifying sequences by a discriminative model involves engineering a suitable sequence kernel or another form of data representation. Recurrent neural networks tend to work well for problems for which large data sets are available to train high-capacity models. By contrast, eye movement data cannot be collected at a large scale, because this requires laboratory equipment and test subjects. We therefore focus on the development of a suitable sequence kernel. We will follow the approach of Fisher kernels because it allows us to use background knowledge in the form of a plausible generative model as the representation of scanpaths.

Based on an existing generative model by [7], we develop a model that takes into account lexical features of the fixated words to generate a scanpath. This model is then used to map scanpaths into Fisher score vectors. We classify with an SVM and the Fisher kernel function such that we exploit both the advantages of generative modeling and the strengths of discriminative classification.

The rest of this paper is organized as follows. Section 2 reviews related work. Section 3 introduces the problem setting and notation. In Sect. 4, we develop a generative model of scanpaths that takes into account the lexical features of the fixated word, and derive the corresponding Fisher kernel in Sect. 5. In Sect. 6, we show the empirical evaluation; Sect. 7 concludes.

2   Related Work

Eye movements are assumed to mirror cognitive processes involved in reading [9]. A large body of psycholinguistic evidence shows that language comprehension processes at the syntactic, semantic and pragmatic level are significant predictors for a reader's fixation durations and saccadic behavior [2,10,13–15].

For our purposes, effects of higher level text comprehension (*i.e.,* on the level of the discourse) on a reader's eye movements are most relevant. For example, it has been shown that conceptual difficulty of a text leads to a larger proportion of regressions, an increase in fixation durations, and a decrease in saccade amplitudes [16,17]. Additionally, higher global and local discourse difficulty of a text increases the number and average duration of fixations as well as the proportion of regressive saccades [13]. Semantically impossible or implausible words increase the first-pass reading time and the total reading time of a word, respectively [18,19]. Moreover, background knowledge decreases both the sum of all fixation durations on a word when reading it for the first time and the proportion of skipped words [12].

Existing attempts to exploit this eye-mind connection and actually use a reader's eye movements to predict text comprehension have crucial limitations. Copeland et al. [20–22] use the saccades between a comprehension question and the text as a feature to predict the response accuracy on this very question. Hence, these models are not trained to infer reading comprehension from the eye movements while reading a text, as claimed by the authors, but rather predict response accuracy on a question from the answer-seeking eye movements. Indeed, the practical relevance of predicting text comprehension from reading is that no questions would be needed anymore to assess a reader's comprehension of a text. Underwood et al. [23] also claim to predict text comprehension from fixation durations. However, they use the same data for training and testing their model.

Compared to the usually rather small size of the effects reflecting cognitive processes, individual variability of eye movements in reading is very large. This has been observed consistently in the psychological literature [4,24,25]. The idea behind eye movements as a biometric feature is to exploit this individual variability. Some biometric studies are based on eye movements observed in response to an artificial visual stimulus, such as a moving [26–28] or fixed [29] dot on a computer screen, or a specific image stimulus [30]. Other studies, like our paper, focus on the problem of identifying subjects while they process an arbitrary stimulus, which has the advantage that the identity can be inferred unobtrusively during routine access to a device or document. Holland and Komogortsev study identification of subjects from eye movements on arbitrary text, based on aggregated statistical features such as the average fixation duration and average saccade amplitude [6]. Rigas et al. extend this approach with additional dynamic saccadic features [31]. However, by reducing observations to a small set of real-valued features, much of the information in eye movements is lost. Landwehr et al. [7] show that by fitting subject-specific generative probabilistic models to eye movements, much higher identification accuracies can be achieved. They develop a parametric generative model [7]; as this model serves as a starting point for our method, details are given in Sect. 4.1. Abdelwahab et al. [8] extend this model to a fully Bayesian approach, in which distributions are defined by nonparametric densities inferred under a Gaussian process prior that is centered at the gamma family of distributions [8]. Both methods serve as reference methods in our experiments.

3   Problem Setting

When reading a text $\mathbf{X}$, a reader generates a scanpath that is given by a sequence $\mathbf{S} = ((q_1, d_1), \ldots, (q_T, d_T))$ of fixation positions q_t (position in text that was fixated, measured in characters) and fixation durations d_t (measured in milliseconds). This scanpath can be observed with an eye-tracking system.

Each word fixated at time t possesses lexical features that can be aggregated into a vector $\mathbf{w}_t$. Some of the models that we will study will allow the distributions of saccade amplitudes and durations to depend on such lexical features. Lexical features—for instance, word frequency or part of speech—are derived from the text $\mathbf{X}$ itself.

We study the problems of reader identification and assessing text comprehension. In reader identification, the model output y is the conjectured identity of the reader that generates scanpath $\mathbf{S}$ for text $\mathbf{X}$, from a set of individuals that are known at training time. In assessing text comprehension, the model output is the conjectured level of the reader's comprehension y of text $\mathbf{X}$. In order to annotate training and evaluation data, the ground-truth level of text comprehension can be determined, for instance, by a question-answering protocol carried out after reading. In an actual application setting, no comprehension questions are asked.

In both settings, training data consists of a set $\mathcal{D} = \{(\mathbf{S}_1, \mathbf{X}_1, y_1), \ldots, (\mathbf{S}_n, \mathbf{X}_n, y_n)\}$ of scanpaths $\mathbf{S}_1, \ldots, \mathbf{S}_n$ that have been obtained from subjects reading texts $\mathbf{X}_1, \ldots, \mathbf{X}_n$, annotated with labels $y_1, \ldots, y_n$.

4   Generative Models of Scan Paths

Landwehr et al. [7] define a parametric model $p(\mathbf{S}|\mathbf{X}, \boldsymbol{\theta})$ of scanpaths given a text $\mathbf{X}$. Fitting this model to the subset of scanpaths and texts $\mathcal{D}_y = \{(\mathbf{S}_i, \mathbf{X}_i)|(\mathbf{S}_i, \mathbf{X}_i, y_i) \in \mathcal{D}, y_i = y\}$ in the training data generated by reader y yields reader-specific models $p(\mathbf{S}|\mathbf{X}, \boldsymbol{\theta}_y)$. At application time, the prediction for a scanpath $\mathbf{S}$ on a novel text $\mathbf{X}$ can be obtained as $y^* = \text{argmax}_y p(\mathbf{S}|\mathbf{X}, \boldsymbol{\theta}_y)$. We first review this generative model [7], and then develop it into a generative model of scanpaths that takes into account the lexical features of the fixated words in Sect. 4.2. In Sect. 5, we derive the Fisher kernel and arrive at a discriminative model.

4.1   The Model of Landwehr et al. 2014

This section presents a slightly simplified version of the generative model of scanpaths $p(\mathbf{S}|\mathbf{X}, \boldsymbol{\theta})$ [7]. It reflects how readers generate fixations while reading a text and models the type and amplitude of saccadic movements and fixation durations. The joint distribution over all fixation positions and durations is assumed to factorize as

$$p(q_1, \ldots, q_T, d_1, \ldots, d_T | \mathbf{X}, \boldsymbol{\theta}) = p(q_1, d_1 | \mathbf{X}, \boldsymbol{\theta}) \prod_{t=1}^{T-1} p(q_{t+1}, d_{t+1} | q_t, \mathbf{X}, \boldsymbol{\theta}). \quad (1)$$

To model the conditional distribution $p(q_{t+1}, d_{t+1} | q_t, \mathbf{X}, \boldsymbol{\theta})$ of the next fixation position and duration given the current fixation position, the model distinguishes five *saccade types* u: a reader can refixate the current word at a character position before the current position ($u = 1$), refixate the current word at a position after the current position ($u = 2$), fixate the next word in the text ($u = 3$), move the fixation to a word after the next word ($u = 4$), or regress to fixate a word occurring earlier in the text ($u = 5$). At each time t, the model first draws a saccade type

$$u_{t+1} \sim p(u|\boldsymbol{\pi}) = \text{Mult}(u|\boldsymbol{\pi}) \quad (2)$$

from a multinomial distribution. It then draws a (signed) saccade amplitude[1] $a_{t+1} \sim p(a|u_{t+1}, \boldsymbol{\alpha}, \boldsymbol{\beta})$ from type-specific gamma distributions; that is,

$$p(a|u_{t+1} = u, \boldsymbol{\alpha}, \boldsymbol{\beta}) = \mathcal{G}(a|\alpha_u, \beta_u) \text{ for } u \in \{2, 3, 4\} \quad (3)$$

$$p(a|u_{t+1} = u, \boldsymbol{\alpha}, \boldsymbol{\beta}) = \mathcal{G}(-a|\alpha_u, \beta_u) \text{ for } u \in \{1, 5\} \quad (4)$$

where $\boldsymbol{\alpha} = \{\alpha_u | u \in \{1, ..., 5\}\}$, $\boldsymbol{\beta} = \{\beta_u | u \in \{1, ..., 5\}\}$ and $\mathcal{G}(\cdot|\alpha, \beta)$ is the gamma distribution parameterized by shape α and scale β. The current fixation position is then updated as $q_{t+1} = q_t + a_{t+1}$. The model finally draws the fixation duration $d_{t+1} \sim p(d|u_{t+1}, \boldsymbol{\gamma}, \boldsymbol{\delta})$, also from type-specific gamma distributions

$$p(d|u_{t+1} = u, \boldsymbol{\gamma}, \boldsymbol{\delta}) = \mathcal{G}(d|\gamma_u, \delta_u) \text{ for } u \in \{1, 2, 3, 4, 5\} \quad (5)$$

where $\boldsymbol{\gamma} = \{\gamma_u | u \in \{1, ..., 5\}\}$, $\boldsymbol{\delta} = \{\delta_u | u \in \{1, ..., 5\}\}$. All parameters of the model are aggregated into a parameter vector $\boldsymbol{\theta}$.

The difference between this simplified variant and the original model [7] is that the original model truncates the gamma distributions in order to fit within the limits of the text interval defined by the saccade type; for instance, to the currently fixated word for refixations. Since this truncation causes unsteadiness of the Fisher scores, we instead let the amplitudes be governed by regular gamma distributions with scale parameter α_u or γ_u and shape parameter β_u or δ_u. Furthermore, Landwehr et al. distinguish the same five saccade types for modeling saccade amplitude, but only four saccade types for modeling fixation durations, while we distinguish five saccade types for both distributions.

---

[1] Throughout our work, saccade amplitude is measured in number of characters as this metric is relatively insensitive to differences in the eye-to-screen distance, which might become relevant for practical applications of the model [32].

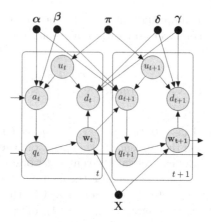

Fig. 1. Plate notation of the generative model for scanpaths with lexical features.

4.2   Generative Model with Lexical Features

We extend the model presented in Sect. 4.1 by allowing the distributions of fixation durations and saccade amplitudes to depend on lexical features $\mathbf{w}_t$ of each fixated word.

Let the random variable $\mathbf{w}_t$ denote a vector of features of the word that is fixated at time step t, such as word frequency or length (Section 6 gives more details on the features under study). We allow these features to influence the scale and shape of the gamma distributions from which the saccade amplitudes and fixation durations are generated. Hence, we model the scale and shape parameters $\alpha_u, \beta_u, \gamma_u, \delta_u$ in Eqs. 3, 4 and 5 as linear regressions on the word features $\mathbf{w}_t$ with an exponential link to ensure positivity of the gamma parameters. That is, we replace Eqs. 3, 4 and 5 by

$$p(a|u_{t+1} = u, \mathbf{w}_t, \boldsymbol{\alpha}, \boldsymbol{\beta}) = \mathcal{G}(a| \exp(\boldsymbol{\alpha}_u^T \mathbf{w}_t), \exp(\boldsymbol{\beta}_u^T \mathbf{w}_t)) \text{ for } u \in \{2,3,4\}, \quad (6)$$

$$p(a|u_{t+1} = u, \mathbf{w}_t, \boldsymbol{\alpha}, \boldsymbol{\beta}) = \mathcal{G}(-a| \exp(\boldsymbol{\alpha}_u^T \mathbf{w}_t), \exp(\boldsymbol{\beta}_u^T \mathbf{w}_t)) \text{ for } u \in \{1,5\} \quad (7)$$

and

$$p(d|u_{t+1} = u, \mathbf{w}_t, \boldsymbol{\gamma}, \boldsymbol{\delta}) = \mathcal{G}(d| \exp(\boldsymbol{\gamma}_u^T \mathbf{w}_{t+1}), \exp(\boldsymbol{\delta}_u^T \mathbf{w}_{t+1})) \text{ for } 1 \leq u \leq 5. \quad (8)$$

Note that $\boldsymbol{\alpha}_u, \boldsymbol{\beta}_u, \boldsymbol{\gamma}_u, \boldsymbol{\delta}_u$ are now vectors of regression weights from which the respective gamma parameters are computed, which are aggregated into the parameterizations $\boldsymbol{\alpha} = \{\boldsymbol{\alpha}_u | u \in \{1, ..., 5\}\}$, $\boldsymbol{\beta} = \{\boldsymbol{\beta}_u | u \in \{1, ..., 5\}\}$, $\boldsymbol{\gamma} = \{\boldsymbol{\gamma}_u | u \in \{1, ..., 5\}\}$, and $\boldsymbol{\delta} = \{\boldsymbol{\delta}_u | u \in \{1, ..., 5\}\}$. Figure 1 shows a graphical model representation.

4.3   Parameter Estimation

Given a set of scanpaths and texts $\bar{\mathcal{D}} = \{(\mathbf{S}_i, \mathbf{X}_i)\}$, model parameters can be estimated by maximum likelihood. In a generative setting, models for a specific

reader y or a specific discrete competence level y can be be estimated on a data
subset $\bar{\mathcal{D}}_y = \{(\mathbf{S}_i, \mathbf{X}_i)|(\mathbf{S}_i, \mathbf{X}_i, y_i) \in \mathcal{D}, y_i = y\}$. For the discriminative setting
we develop in Sect. 5, generative parameters are estimated on all training data
$\bar{\mathcal{D}} = \{(\mathbf{S}_i, \mathbf{X}_i)|(\mathbf{S}_i, \mathbf{X}_i, y_i) \in \mathcal{D}\}$, and a Fisher score representation is derived from
this generative model. We optimize a regularized maximum likelihood criterion

$$\boldsymbol{\theta}^* = \operatorname*{argmax}_{\boldsymbol{\theta}} \sum_{i=1}^{k} \ln p(\bar{\mathbf{S}}_i|\bar{\mathbf{X}}_i, \boldsymbol{\theta}) - \lambda \Omega(\boldsymbol{\theta}). \tag{9}$$

Given $\bar{\mathcal{D}}$, all fixation positions q_t, saccade types u_t and word features $\mathbf{w}_t$ are
known. Equation 9 thus factorizes into separate likelihood terms depending on
saccade type, amplitude, and duration parameters:

$$\begin{aligned}
\boldsymbol{\theta}^* = \operatorname*{argmax}_{\boldsymbol{\pi}, \boldsymbol{\alpha}, \boldsymbol{\beta}, \boldsymbol{\gamma}, \boldsymbol{\delta}} \bigg(& \sum_{i=1}^{k} \sum_{t=1}^{T_i} \ln \operatorname{Mult}(u_t^{(i)}|\boldsymbol{\pi}) \\
& + \sum_{i=1}^{k} \sum_{t=1}^{T_i} \ln p(a_t^{(i)}|u_t^{(i)}, \mathbf{w}_t^{(i)}, \boldsymbol{\alpha}, \boldsymbol{\beta}) - \lambda \Omega(\boldsymbol{\alpha}, \boldsymbol{\beta}) \\
& + \sum_{i=1}^{k} \sum_{t=1}^{T_i} \ln p(d_t^{(i)}|u_t^{(i)}, \mathbf{w}_t^{(i)}, \boldsymbol{\gamma}, \boldsymbol{\delta}) - \lambda \Omega(\boldsymbol{\gamma}, \boldsymbol{\delta}) \bigg) \tag{10}
\end{aligned}$$

where $u_t^{(i)}$, $a_t^{(i)}$, $d_t^{(i)}$, and $\mathbf{w}_k^{(i)}$ denote saccade types, amplitudes, fixation durations, and word features in $(\bar{\mathbf{S}}_i, \bar{\mathbf{X}}_i)$, the number of fixations in sequence $\bar{\mathbf{S}}_i$ is
written as T_i, and we have split up the regularizer into separate regularizers
$\Omega(\boldsymbol{\alpha}, \boldsymbol{\beta})$ and $\Omega(\boldsymbol{\gamma}, \boldsymbol{\delta})$ (parameter $\boldsymbol{\pi}$ is not regularized). Equation 10 can be optimized independently in saccade type parameters $\boldsymbol{\pi}$, amplitude parameters $\boldsymbol{\alpha}, \boldsymbol{\beta}$,
and duration parameters $\boldsymbol{\gamma}, \boldsymbol{\delta}$. Optimization in $\boldsymbol{\pi}$ is straightforward. Because
given $\bar{\mathcal{D}}$, saccades types are known, amplitude parameters can be optimized
independently for each saccade type; that is, optimization is independent for
each $\boldsymbol{\alpha}_u, \boldsymbol{\beta}_u$. Let $u \in \{1, ..., 5\}$, then

$$\begin{aligned}
(\boldsymbol{\alpha}_u^*, \boldsymbol{\beta}_u^*) &= \operatorname*{argmax}_{\boldsymbol{\alpha}_u, \boldsymbol{\beta}_u} \sum_{i=1}^{k} \sum_{1 \le t \le T_i : u_t^{(i)} = u} \ln p(a_t^{(i)}|u_t^{(i)} = u, \mathbf{w}_t^{(i)}, \boldsymbol{\alpha}_u, \boldsymbol{\beta}_u) - \lambda \Omega(\boldsymbol{\alpha}_u, \boldsymbol{\beta}_u) \\
&= \operatorname*{argmax}_{\boldsymbol{\alpha}_u, \boldsymbol{\beta}_u} \sum_{i=1}^{k} \sum_{1 \le t \le T_i : u_t^{(i)} = u} \ln \mathcal{G}(|a_t^{(i)}|| \exp(\boldsymbol{\alpha}_u^\top \mathbf{w}_t^{(i)}), \exp(\boldsymbol{\beta}_u^\top \mathbf{w}_t^{(i)})) \\
&\qquad - \lambda \sum_{m=0}^{M-1} \exp(\alpha_{u,m}) + \exp(\beta_{u,m}) \tag{11}
\end{aligned}$$

where M is the number of lexical features used to predict the gamma parameters
(including a bias), and $\alpha_{u,m}$, $\beta_{u,m}$ denote the m-th element of parameter vectors
$\boldsymbol{\alpha}_u$, $\boldsymbol{\beta}_u$ respectively. Note that as the linear regression on the word features
is scaled using an exponential function (Eqs. 6 and 7), we use an exponential
regularizer $\Omega(\boldsymbol{\alpha}, \boldsymbol{\beta})$. Analogously, for fixation durations,

$$(\gamma_u^*, \delta_u^*) = \underset{\gamma_u, \delta_u}{\operatorname{argmax}} \sum_{i=1}^{k} \sum_{1 \le t \le T_i : u_t^{(i)} = u} \ln p(d_t^{(i)} | u_t^{(i)} = u, \mathbf{w}_t^{(i)}, \gamma_u, \delta_u) - \lambda \Omega(\gamma_u, \delta_u)$$

$$= \underset{\gamma_u, \delta_u}{\operatorname{argmax}} \sum_{i=1}^{k} \sum_{1 \le t \le T_i : u_t^{(i)} = u} \ln \mathcal{G}(d_t^{(i)} | \exp(\gamma_u^\top \mathbf{w}_t^{(i)}), \exp(\delta_u^\top \mathbf{w}_t^{(i)}))$$

$$- \lambda \sum_{m=0}^{M-1} \exp(\gamma_{u,m}) + \exp(\delta_{u,m}). \qquad (12)$$

(α_u, β_u) and (δ_u, γ_u) are optimized using a truncated Newton method [33].

5   Discriminative Classification with Fisher Kernels

Fisher kernels [34] provide a commonly used framework that exploits generative probabilistic models as a representation of instances within discriminative classifiers. Specifically, the Fisher kernel approach involves a feature mapping of structured input—for instance, sequential input—by a projection into the gradient space of a generative probabilistic model that is previously fit on the training data via maximum likelihood. We use the generative probabilistic model developed in Sect. 4.2 to map scanpaths and lexical features into feature vectors $\mathbf{g}$. The Fisher score representation $\mathbf{g}$ for a scanpath $\mathbf{S}$ is the gradient of the log likelihood of $\mathbf{S}$ with respect to the model parameters, evaluated at the maximum likelihood estimate.

5.1   Fisher Kernel Function

The Fisher kernel function K calculates the similarity of two scanpaths $\mathbf{S}_i$, $\mathbf{S}_j$ as the inner product of their Fisher score representations $\mathbf{g}_i$ and $\mathbf{g}_j$, relative to the Riemannian metric, given by the inverse of the Fisher information matrix $\mathbf{I}$.

Definition 1 (Fisher kernel function of model with lexical features). *Let θ^* be the maximum likelihood estimate of the model defined in Sect. 4.2 on all training data. Let $\mathbf{S}_i$, $\mathbf{S}_j$ denote scanpaths on texts $\mathbf{X}_i$, $\mathbf{X}_j$. The fisher kernel between $\mathbf{S}_i$, $\mathbf{S}_j$ is*

$$K((\mathbf{S}_i, \mathbf{X}_i), (\mathbf{S}_j, \mathbf{X}_j)) = \mathbf{g}_i^\top \mathbf{I}^{-1} \mathbf{g}_j$$

where $\mathbf{g}_i = \nabla_\theta p(\mathbf{S}_i | \mathbf{X}_i, \theta)|_{\theta = \theta^*}$ and we employ the empirical version of the Fisher information matrix given by $\mathbf{I} = \frac{1}{N} \sum_{i=1}^{N} \mathbf{g}_i \mathbf{g}_i^\top$. The gradient of the log-likelihood function is derived in Proposition 2.

Proposition 2 (Gradient of log-likelihood of generative model with lexical features). *Let $\mathbf{S} = ((q_1, d_1), \ldots, (q_T, d_T))$ denote a scanpath obtained on text $\mathbf{X}$. Let $a_1, \ldots, a_T$ denote the saccade amplitudes, and $u_1, \ldots, u_T$ denote the saccade types in $\mathbf{S}$. Define for $u \in \{1, 2, 3, 4, 5\}$ the set $\{i_1^{(u)}, \ldots, i_{K_u}^{(u)}\} = \{i \in$*

$\{1, ..., T\}|u_i = u\}$. Let $\mathbf{a}_u = (|a_{i_1^{(u)}}|, ..., |a_{i_{K_u}^{(u)}}|)^\top$, $\mathbf{d}_u = (d_{i_1^{(u)}}, ..., d_{i_{K_u}^{(u)}})^\top$, and $\mathbf{W}_u$ the $K_u \times M$ matrix with row vectors $\mathbf{w}_{i_k^{(u)}}^\top$ for $1 \leq k \leq K_u$. Then the gradient of the logarithmic likelihood of the model defined in Sect. 4.2 is

$$\mathbf{g} = \nabla_\theta \ln p(\mathbf{S}|\mathbf{X}, \boldsymbol{\theta}) = (\bar{\mathbf{g}}_1^\top, \bar{\mathbf{g}}_2^\top, \bar{\mathbf{g}}_3^\top, \bar{\mathbf{g}}_4^\top, \bar{\mathbf{g}}_5^\top)^\top$$

where for $u \in \{1, 2, 3, 4, 5\}$ (and $\odot$ denotes the Hadamard product)

$$\bar{\mathbf{g}}_u = \begin{pmatrix} \pi_u^{-1} K_u \\ \mathbf{W}_u^\top \Big(\exp(\mathbf{W}_u \boldsymbol{\alpha}_u) \odot \big(\ln(\mathbf{a}_u) - \psi(\exp(\mathbf{W}_u \boldsymbol{\alpha}_u)) - \mathbf{W}_u \boldsymbol{\beta}_u \big) \Big) \\ \mathbf{W}_u^\top \Big(\mathbf{a}_u \exp(-\mathbf{W}_u \boldsymbol{\beta}_u) - \exp(\mathbf{W}_u \boldsymbol{\alpha}_u) \Big) \\ \mathbf{W}_u^\top \Big(\exp(\mathbf{W}_u \boldsymbol{\delta}_u) \odot \big(\ln(\mathbf{d}_u) - \psi(\exp(\mathbf{W}_u \boldsymbol{\delta}_u)) - \mathbf{W}_u \boldsymbol{\gamma}_u \big) \Big) \\ \mathbf{W}_u^\top \Big(\mathbf{d}_u \exp(-\mathbf{W}_u \boldsymbol{\gamma}_u) - \exp(\mathbf{W}_u \boldsymbol{\delta}_u) \Big) \end{pmatrix}$$

A proof of Proposition 2 is given in the Appendix A.

5.2 Applying the Fisher Kernel to Identification and Text Comprehension

Applying the Fisher Kernel to both prediction problems first requires to estimate the parameters of the generative model parameters on the training data. Note that we fit a global model, instead of class-specific models. In both prediction problems, we treat the scanpaths of each single line of text as an instance, and train a dual SVM with the resulting Fisher kernel. At application time, the scanpath of a text that is comprised of multiple lines is processed as multiple instances by the Fisher SVM. In order to obtain one decision-function value for the entire text, we average the decision-function values of all individual lines.

6 Empirical Study

6.1 Data Collection

Experimental Design and Materials. We let a group of 62 advanced and first-semester students read a total of 12 scientific texts on biology (6 texts) and on physics adopted from various German language textbooks. All students are native speakers of German with normal or corrected-to-normal vision and are majoring in either physics or biology. We determine each reader's comprehension of each text by presenting three comprehension questions after each text. All questions are multiple-choice questions with always one out of four options being correct. Texts have 158 words on average (minimally 126 and maximally 180).

Technical Set-Up and Procedure. Participants' eye movements are recorded with an SR Research Eyelink 1000 eyetracker (right eye monocular tracking) at a sampling rate of 1000 Hz using a desktop mounted camera system with head stabilization. After setting up the camera and familiarizing the participant with the procedure, the twelve texts are presented in randomized order. Each text fits onto a single screen. We impose no restrictions regarding the time spent on reading one text. After each text, three comprehension questions are presented on separate screens together with 4 multiple choice options. Participants cannot backtrack to the text or previous questions, or undo an answer. The total duration of the experiment is approximately 90 min; participants were paid for participating.

Lexical Features. Lexical frequency and word length are well known to affect a reader's fixation durations and saccadic behavior, such as whether a word is skipped or a regressive saccade is initiated [1, 35–37]. Hence, for each word of the stimuli, we extract different kinds of word frequency and word length measures using dlexDB [38]. Specifically, we extract type frequency (*i.e.*, the number of occurrences of a type in the corpus per million tokens), annotated type frequency (*i.e.*, the number of occurrences of a unique combination of a type, its part-of-speech, and its lemma in the corpus per million tokens), lemma frequency (*i.e.*, the total number of occurrences of types associated with this lemma in the corpus per million tokens), document frequency (*i.e.*, the number of documents with at least one occurrence of this type per 10,000 documents), type length in number of characters, type length in number of syllables, and lemma length in number of characters. All corpus-based features are log-transformed and z-score normalized. Moreover, we tag each word with the following binary lexical features: whether the word is a technical term, a technical term from physics, a technical term from biology, an abbreviation, the first word of a sentence.

6.2 Reference Methods

We compare the *Fisher SVM with lexical features* to several reference methods. The first natural baseline is the *generative model with lexical features* developed in Sect. 4.2; this comparison allows us to measure the merit of the discriminative Fisher kernel compared to the underlying generative model. The next baseline is the *Fisher SVM without lexical features*—that is, an SVM with the Fisher kernel derived from the generative model described in Sect. 4.1. We compare this discriminative model to the full *generative model (Landwehr et al. 2014)* [7] without lexical features and without the simplification introduced in Sect. 4.1.

The current gold-standard model for reader identification is the model of *Abdelwahab et al. 2016* [8]. Note that no Fisher kernel can be derived from this non-parametric generative model for lack of explicit model parameters. Since this model has been shown to outperform all previous approaches [6, 7], we exclude [6] from our comparison.

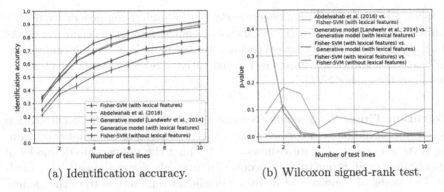

(a) Identification accuracy.   (b) Wilcoxon signed-rank test.

Fig. 2. Identification accuracy as a function of lines read at test time, where error bars show the standard error (a), and p-values of a Wilcoxon signed-rank test for comparison of model pairs (b).

6.3 Experimental Setting

For reader identification, data are split along texts, so that the same text does not appear in training and test data. We conduct a leave-one-text-out cross-validation protocol: The models are trained on 11 texts per reader and a reader is identified on the left-out text. Identification accuracy is averaged across the resulting 12 training- and test-splits and is studied as a function of the number of text lines read at test time.

For text comprehension, data are split (50/50) across readers and texts, so that neither the same reader nor the same text appears in both training and test data. This setup leads to four train-test splits, across which we average the classification accuracy.

For both problem settings, we execute another nested cross-validation inside the top-level cross-validation in which we tune the hyperparameters of all learning methods (*e.g.*, regularization parameters of the SVM and the linear model for lexical features and parameter α of the non-parametric method of Abdelwahab et al.) by grid search. We also perform feature subset selection on vector $\mathbf{w}_t$ by backward elimination in this inner cross-validation step. The nested cross-validation protocol ensures that all hyperparameters are tuned on the training part of the data.

6.4 Reader Identification

We measure the percentage of correctly identified readers from the set of 62 readers. The Fisher-SVM achieves an identification accuracy of up to 91.53% and outperforms the other evaluated models, see Fig. 2a. Figure 2b shows the p-value of a Wilcoxon signed-rank test for a comparison of several pairs of methods. We conclude that the Fisher-SVM with lexical features outperforms the model of Abdelwahab et al. significantly ($p < 0.05$) for 4 and 8 lines read, the Fisher-SVM with lexical features always outperforms the Fisher-SVM without lexical features, the Fisher-SVM always outperforms the underlying generative model, and the generative model with lexical features outperforms the generative model of Landwehr et al. without lexical features for 3 or more lines read. Including lexical features significantly improves the generative model by Landwehr et al. [7], as well as the Fisher-SVM.

Execution Time. We compare the time required to train reader-identification models for all methods under investigation as a function of the number of training texts per reader. Figure 3 shows that training the nonparametric model of Abdelwahab et al. is one to three orders of magnitude slower than all other models. The model of Landwehr et al. uses a quasi-Newton method to fit the gamma distributions, the generative model with lexical features additionally fits several linear models. Generative models are fit for each reader. By contrast, the Fisher kernel requires fitting one single model to all data and training a linear model; this turns out to be faster in some cases.

Text Comprehension. After reading a text, each subject answers three text comprehension questions. We study a binary classification problem where one class corresponds to zero or one correct answers and the other class to two or three correct answers. Table 1 shows the classification accuracies of the evaluated models[2]. No methods exceeds the classification accuracy of a model that always predicts the majority class. The discriminative models minimize the hinge loss—which is an upper bound of the zero-one loss—and reach the minimal loss by almost always predicting the majority class. The generative models are not trained to minimize any classification loss at all. They fall far short of the accuracy of the majority class but attain an AUC that is marginally above

Table 1. Accuracy and AUC ± standard error for text comprehension.

Method	Classification accuracy	AUC
Fisher-SVM (with lexical features)	0.6866 ± 0.0615	0.5071 ± 0.0282
Fisher-SVM (without lexical features)	0.6529 ± 0.0708	0.5181 ± 0.0339
Abdelwahab et al. (2016)	0.5954 ± 0.0232	0.5403 ± 0.0272
Generative model (with lexical features)	0.5273 ± 0.0261	0.5500 ± 0.0293
Generative model [Landwehr et al. 2014]	0.5206 ± 0.0207	0.5555 ± 0.0120
Majority class	0.7014 ± 0.0547	0.5

---

[2] The main memory requirement of the model of Abdelwahab et al. is quadratic in the number of instances per class; we had to discard 80% of the data at random for this problem.

random guessing. The AUC of all three models is significantly higher than 0.5 ($p < 0.05$, paired t-test). In order to validate this interpretation, we additionally train the Fisher SVM on a class-balanced data subset; with balanced classes, the Fisher SVM cannot minimize the loss without also increasing the AUC. Here, the Fisher SVM achieves an AUC of 0.54 ± 0.03 which is significantly higher than 0.5. We conclude that estimating the level of text comprehension is a difficult problem that cannot be solved at any useful level by any of the models under investigation.

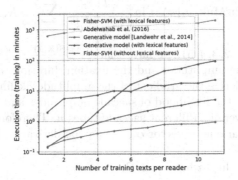

Fig. 3. Elapsed execution time of training the models for reader identification, as a function of the number of used training texts per reader on a single ten-core CPU (Intel Xeon E5-2640, 2.40 GHz).

7   Conclusions

We developed a discriminative model for the classification of scanpaths in reading. The aim was to (i) predict the readers' identity, and (ii) their level of text comprehension. To this end, we built on the work of [7] and developed a generative graphical model of scanpaths that takes into account lexical features of the fixated word, derived a Fisher representation of scanpaths from this model, and subsequently used this Fisher kernel to classify the data using an SVM. We collected eye-tracking data of 62 readers who read 12 scientific texts and answered comprehension questions for each text.

We can conclude that the inclusion of lexical features leads to a significant improvement compared to the original generative model [7], and that a discriminative model using a Fisher kernel gives an additional considerable improvement over the generative model. We conclude that this model significantly outperforms the semiparametric model of [8] in some cases, which, to the best of our knowledge, is the best published biometric model that is based on eye movements. None of the considered models was able to reliably predict reading comprehension from a reader's eye movements.

Acknowledgments. This work was partially funded by the German Science Foundation under grants SFB1294, SFB1287, and LA3270/1-1, and by the German Federal Ministry of Research and Education under grant 16DII116-DII.

A   Appendix

Proof (Proposition 2). As discussed in Sect. 4.3, the likelihood factorizes as

$$\ln p(\mathbf{S}|\mathbf{X},\boldsymbol{\theta}) = \sum_{t=1}^{T} \ln \mathrm{Mult}(u_t|\boldsymbol{\pi}) + \sum_{t=1}^{T} \ln p(a_t|u_t,\mathbf{w}_t,\boldsymbol{\alpha},\boldsymbol{\beta}) + \sum_{t=1}^{T} \ln p(d_t|u_t,\mathbf{w}_t,\boldsymbol{\gamma},\boldsymbol{\delta}).$$

For the multinomial distribution,

$$\sum_{t=1}^{T} \ln \mathrm{Mult}(u_t|\boldsymbol{\pi}) = \ln \frac{T!}{\prod_{u=1}^{5} K_u!} + \sum_{u=1}^{5} K_u \ln \pi_u$$

and thus for $u \in \{1,2,3,4,5\}$, we have that $\frac{\partial \ln p(\mathbf{S}|\mathbf{X},\boldsymbol{\theta})}{\partial \pi_u} = \frac{K_u}{\pi_u}$. Since the likelihoods of the saccade amplitudes and the fixation durations are analogous (see Eqs. 6–8), we only derive the gradient of the amplitude likelihood. As discussed in Sect. 4.3 (Eq. 11), the likelihood of saccade amplitudes and fixation durations further factorizes over the different saccade types u. Therefore, if $\alpha_{u,m}$ denotes the m-th entry of parameter vector $\boldsymbol{\alpha}_u$, its partial derivative is

$$\frac{\partial}{\partial \alpha_{u,m}} \sum_{1 \leq t \leq T: u_t = u} \ln p(a_t|u_t,\mathbf{w}_t,\boldsymbol{\alpha},\boldsymbol{\beta})$$

$$= \frac{\partial}{\partial \alpha_{u,m}} \sum_{1 \leq t \leq T: u_t = u} \ln \mathcal{G}(|a_t|\,|\exp(\boldsymbol{\alpha}_u^\top \mathbf{w}_t^{(i)}), \exp(\boldsymbol{\beta}_u^\top \mathbf{w}_t^{(i)}))$$

$$= \frac{\partial}{\partial \alpha_{u,m}} \sum_{1 \leq t \leq T: u_t = u} \ln \left(|a_t|^{\exp(\boldsymbol{\alpha}_u^\top \mathbf{w}_t)-1} \exp\left(-\frac{|a_t|}{\exp(\boldsymbol{\beta}_u^\top \mathbf{w}_t)}\right) \right)$$

$$\qquad - \ln(\Gamma(\exp(\boldsymbol{\alpha}_u^\top \mathbf{w}_t))\exp(\boldsymbol{\beta}_u^\top \mathbf{w}_t)^{\exp(\boldsymbol{\alpha}_u^\top \mathbf{w}_t)})$$

$$= \sum_{1 \leq t \leq T: u_t = u} w_{t,m} \exp(\boldsymbol{\alpha}_u^\top \mathbf{w}_t) \Big(\ln(|a_t|) - \psi(\exp(\boldsymbol{\alpha}_u^\top \mathbf{w}_t)) - \boldsymbol{\beta}_u^\top \mathbf{w}_t \Big).$$

$$(13)$$

Moreover, if $\beta_{u,m}$ denotes the m-th entry of parameter vector $\boldsymbol{\beta}_u$,

$$\frac{\partial}{\partial \beta_{u,m}} \sum_{1 \leq t \leq T: u_t = u} \ln p(|a_t|\,|u_t,\mathbf{w}_t,\boldsymbol{\alpha},\boldsymbol{\beta})$$

$$= \frac{\partial}{\partial \beta_{u,m}} \sum_{1 \leq t \leq T: u_t = u} \ln \mathcal{G}(|a_t|\,|\exp(\boldsymbol{\alpha}_u^\top \mathbf{w}_t^{(i)}), \exp(\boldsymbol{\beta}_u^\top \mathbf{w}_t^{(i)}))$$

$$= \frac{\partial}{\partial \beta_{u,m}} \sum_{1 \leq t \leq T: u_t = u} \ln \left(|a_t|^{\exp(\boldsymbol{\alpha}_u^\top \mathbf{w}_t)-1} \exp\left(-\frac{|a_t|}{\exp(\boldsymbol{\beta}_u^\top \mathbf{w}_t)}\right) \right)$$

$$\qquad - \ln(\Gamma(\exp(\boldsymbol{\alpha}_u^\top \mathbf{w}_t))\exp(\boldsymbol{\beta}_u^\top \mathbf{w}_t)^{\exp(\boldsymbol{\alpha}_u^\top \mathbf{w}_t)})$$

$$= \sum_{1 \leq t \leq T: u_t = u} w_{t,m} \Big(|a_t|\exp(-\boldsymbol{\beta}_u^\top \mathbf{w}_t) - \exp(\boldsymbol{\alpha}_u^\top \mathbf{w}_t) \Big). \qquad (14)$$

In Eqs. 13 and 14 we have exploited that the derivative of the log-gamma function is given by the digamma function ψ—i.e., $\frac{d}{dx} \ln \Gamma(x) = \psi(x)$. The claim now follows from straightforward calculation.

References

1. Kliegl, R., Nuthmann, A., Engbert, R.: Tracking the mind during reading: the influence of past, present, and future words on fixation durations. J. Exp. Psychol. Gen. **135**(1), 12–35 (2006)
2. Rayner, K.: Eye movements in reading and information processing: 20 years of research. Psychol. Bull. **124**(3), 372–422 (1998)
3. Erdmann, B., Dodge, R.: Psychologische Untersuchungen über das Lesen auf experimenteller Grundlage. Niemeyer, Halle (1898)
4. Huey, E.B.: The Psychology and Pedagogy of Reading. The Macmillan Company, New York (1908)
5. Afflerbach, P. (ed.): Handbook of Individual Differences in Reading. Routledge, New York (2015)
6. Holland, C., Komogortsev, O.V.: Biometric identification via eye movement scanpaths in reading. In: Proceedings of the 2011 International Joint Conference on Biometrics, pp. 1–8 (2011)
7. Landwehr, N., Arzt, S., Scheffer, T., Kliegl, R.: A model of individual differences in gaze control during reading. In: Proceedings of the 2014 Conference on Empirical Methods in Natural Language Processing, pp. 1810–1815 (2014)
8. Abdelwahab, A., Kliegl, R., Landwehr, N.: A semiparametric model for Bayesian reader identification. In: Proceedings of the 2016 Conference on Empirical Methods in Natural Language Processing (2016)
9. Just, M.A., Carpenter, P.A.: A theory of reading: from eye fixations to comprehension. Psychol. Rev. **87**(4), 329–354 (1980)
10. Frazier, L., Rayner, K.: Making and correcting errors during sentence comprehension: eye movements in the analysis of structurally ambiguous sentences. Cogn. Psychol. **14**(2), 178–210 (1982)
11. Staub, A., Rayner, K., Pollatsek, A., Hyönä, J., Majewski, H.: The time course of plausibility effects on eye movements in reading: evidence from noun-noun compounds. J. Exp. Psychol. Learn. Mem. Cogn. **33**(6), 1162–1169 (2007)
12. Kaakinen, J.K., Hyönä, J.: Perspective effects in repeated reading: an eye movement study. Mem. Cogn. **35**, 1323–1336 (2007)
13. Rayner, K., Chace, K.H., Slattery, T.J., Ashby, J.: Eye movements as reflections of comprehension processes in reading. Sci. Stud. Read. **10**(3), 241–255 (2006)
14. Rayner, K., Sereno, S.C.: Eye movements in reading: psycholinguistic studies. In: Gernsbacher, M.A. (ed.) Handbook of Psycholinguistics, pp. 57–81. Academic Press, San Diego (1994)
15. Clifton, C., Staub, A., Rayner, K.: Eye movements in reading words and sentences. In: Van Gompel, R.P., Fischer, M.H., Murray, W.S., Hill, R.L. (eds.) Eye Movements: A Window on Mind and Brain, pp. 341–372. Elsevier, Oxford (2007)
16. Jacobson, J.Z., Dodwell, P.C.: Saccadic eye movements during reading. Brain Lang. **8**, 303–314 (1979)
17. Rayner, K., Pollatsek, A.: The Psychology of Reading. Prentice Hall, Englewood Cliffs (1989)

18. Rayner, K., Warren, T., Juhasz, B.J., Liversedge, S.P.: The effect of plausibility on eye movements in reading. J. Exp. Psychol.: Learn. Mem. Cogn. **30**, 1290–1301 (2004)
19. Warren, T., McConnell, K., Rayner, K.: Effects of context on eye movements when reading about possible and impossible events. J. Exp. Psychol. Learn. Mem. Cogn. **34**, 1001–1007 (2008)
20. Copeland, L., Gedeon, T.: Measuring reading comprehension using eye movements. In: Proceedings of the 4th IEEE International Conference on Cognitive Infocommunication, pp. 791–796 (2013)
21. Copeland, L., Gedeon, T., Mendis, S.: Predicting reading comprehension scores from eye movements using artificial neural networks and fuzzy output error. Artif. Intell. Res. **3**(3), 35–48 (2014)
22. Copeland, L., Gedeon, T., Mendis, S.: Fuzzy output error as the performance function for training artificial neural networks to predict reading comprehension from eye gaze. In: Loo, C.K., Yap, K.S., Wong, K.W., Teoh, A., Huang, K. (eds.) ICONIP 2014, Part I. LNCS, vol. 8834, pp. 586–593. Springer, Cham (2014). https://doi.org/10.1007/978-3-319-12637-1_73
23. Underwood, G., Hubbard, A., Wilkinson, H.: Eye fixations predict reading comprehension: the relationships between reading skill, reading speed, and visual inspection. Lang. Speech **33**(1), 69–81 (1990)
24. Dixon, W.R.: Studies in the psychology of reading. In: Morse, W.S., Ballantine, P.A., Dixon, W.R. (eds.) University of Michigan Monographs in Education No. 4. The University of Michigan Press, Ann Arbor (1951)
25. Rayner, K., Pollatsek, A., Ashby, J., Clifton Jr., C.: Psychology of Reading. Psychology Press, New York (2012)
26. Komogortsev, O.V., Jayarathna, S., Aragon, C.R., Mahmoud, M.: Biometric identification via an oculomotor plant mathematical model. In: Proceedings of the 2010 Symposium on Eye-Tracking Research and Applications, pp. 1–4 (2010)
27. Rigas, I., Economou, G., Fotopoulos, S.: Human eye movements as a trait for biometrical identification. In: Proceedings of the IEEE 5th International Conference on Biometrics: Theory, Applications and Systems (2012)
28. Zhang, Y., Juhola, M.: On biometric verification of a user by means of eye movement data mining. In: Proceedings of the 2nd International Conference on Advances in Information Mining and Management, pp. 85–90 (2012)
29. Bednarik, R., Kinnunen, T., Mihaila, A., Fränti, P.: Eye-movements as a biometric. In: Kalviainen, H., Parkkinen, J., Kaarna, A. (eds.) SCIA 2005. LNCS, vol. 3540, pp. 780–789. Springer, Heidelberg (2005). https://doi.org/10.1007/11499145_79
30. Rigas, I., Economou, G., Fotopoulos, S.: Biometric identification based on the eye movements and graph matching techniques. Pattern Recogn. Lett. **33**(6), 786–792 (2012)
31. Rigas, I., Komogortsev, O., Shadmehr, R.: Biometric recognition via eye movements: saccadic vigor and acceleration cues. ACM Trans. Appl. Percept. **13**(2), 1–21 (2016)
32. Morrison, R.E., Rayner, K.: Saccade size in reading depends upon character spaces and not visual angle. Percept. Psychophys. **30**, 395–396 (1981)
33. Nocedal, J., Wright, S.J.: Numerical Optimization. Springer, New York (2006). https://doi.org/10.1007/978-3-540-35447-5
34. Jaakkola, T., Haussler, D.: Exploiting generative models in discriminative classifiers. In: Advances in Neural Information Processing Systems, pp. 487–493 (1999)
35. Rayner, K., McConkie, G.W.: What guides a reader's eye movements? Vis. Res. **16**, 829–837 (1976)

36. Rayner, K., Duffy, S.A.: Parafoveal word processing during eye fixations in reading: Effects of word frequency. Percept. Psychophys. **40**, 431–440 (1986)

37. Kliegl, R., Grabner, E., Rolfs, M., Engbert, R.: Length, frequency, and predictability effects of words on eye movements in reading. Eur. J. Cogn. Psychol. **16**(1–2), 262–284 (2004)

38. Berlin-Brandenburg Academy of Science, University of Potsdam (2011). http:// dlexdb.de

Face-Cap: Image Captioning Using Facial Expression Analysis

Omid Mohamad Nezami[1(✉)], Mark Dras[1], Peter Anderson[1,2], and Len Hamey[1]

[1] Department of Computing, Macquarie University, Sydney, Australia
omid.mohamad-nezami@hdr.mq.edu.au, {mark.dras,len.hamey}@mq.edu.au
[2] The Australian National University, Canberra, Australia
peter.anderson@anu.edu.au

Abstract. Image captioning is the process of generating a natural language description of an image. Most current image captioning models, however, do not take into account the emotional aspect of an image, which is very relevant to activities and interpersonal relationships represented therein. Towards developing a model that can produce human-like captions incorporating these, we use facial expression features extracted from images including human faces, with the aim of improving the descriptive ability of the model. In this work, we present two variants of our Face-Cap model, which embed facial expression features in different ways, to generate image captions. Using all standard evaluation metrics, our Face-Cap models outperform a state-of-the-art baseline model for generating image captions when applied to an image caption dataset extracted from the standard Flickr 30 K dataset, consisting of around 11 K images containing faces. An analysis of the captions finds that, perhaps surprisingly, the improvement in caption quality appears to come not from the addition of adjectives linked to emotional aspects of the images, but from more variety in the actions described in the captions. Code related to this paper is available at: https://github.com/omidmn/Face-Cap.

Keywords: Image captioning · Facial expression recognition
Sentiment analysis · Deep learning

1 Introduction

Image captioning systems aim to describe the content of an image using computer vision and natural language processing. This is a challenging task in computer vision because we have to capture not only the objects but also their relations and the activities displayed in the image in order to generate a meaningful description. Most of the state-of-the-art methods, including deep neural networks, generate captions that reflect the factual aspects of an image [3,8,12,16,20,35,37]; the emotional aspects which can provide richer and attractive image captions are usually ignored in this process. Emotional properties, including recognizing and

© Springer Nature Switzerland AG 2019
M. Berlingerio et al. (Eds.): ECML PKDD 2018, LNAI 11051, pp. 226–240, 2019.
https://doi.org/10.1007/978-3-030-10925-7_14

expressing emotions, are required in designing intelligent systems to produce intelligent, adaptive, and effective results [22]. Designing an image captioning system, which can recognize emotions and apply them to describe images, is still a challenge.

A few models have incorporated sentiment or other non-factual information into image captions [10,23,38]; they typically require the collection of a supplementary dataset, with a sentiment vocabulary derived from that, drawing from work in Natural Language Processing [25] where sentiment is usually characterized as one of positive, neutral or negative. Mathews et al. [23], for instance, constructed a sentiment image-caption dataset via crowdsourcing, where annotators were asked to include either positive sentiment (e.g. *a cuddly cat*) or negative sentiment (e.g. *a sinister cat*) using a fixed vocabulary; their model was trained on both this and a standard set of factual captions. Gan et al. [10] proposed a captioning model called StyleNet to add styles, which could include sentiments, to factual captions; they specified a predefined set of styles, such as humorous or romantic.

These kinds of models typically embody descriptions of an image that represent an *observer's* sentiment towards the image (e.g. *a cuddly cat* for a positive view of an image, versus *a sinister cat* for a negative one); they do not aim to capture the emotional content of the image, as in Fig. 1. This distinction has been recognized in the sentiment analysis literature: the early work of [24], for instance, proposed a graph-theoretical method for predicting sentiment expressed by a text's author by first removing text snippets that are positive or negative in terms of the actual content of the text (e.g. "The protagonist tries to protect her good name" as part of the description of a movie plot, where *good* has positive sentiment) and leaving only the sentiment-bearing text that reflects the writer's subjective view (e.g. "bold, imaginative, and impossible to resist"). We are interested in precisely this notion of content-related sentiment, in the context of an image.

In this paper, therefore, we introduce an image captioning model we term Face-Cap to incorporate emotional content from the images themselves: we automatically detect emotions from human faces, and apply the derived facial expression features in generating image captions. We introduce two variants of Face-Cap, which employ the features in different ways to generate the captions. The contributions of our work are:

1. Face-Cap models that generate captions incorporating facial expression features and emotional content, using neither sentiment image-caption paired data nor sentiment caption data, which is difficult to collect. To the authors' knowledge, this is the first study to apply facial expression analysis in image captioning tasks.
2. A set of experiments that demonstrate that these Face-Cap models outperform baseline, a state-of-the-art model, on all standard evaluation metrics. An analysis of the generated captions suggests that they improve over baseline models by better describing the actions performed in the image.

3. An image caption dataset that includes human faces which we have extracted from Flickr 30 K dataset [39], which we term FlickrFace11K. It is publicly available[1] for facilitating future research in this domain.

The rest of the paper is organized as follows. In Sect. 2, related work in image captioning and facial expression recognition is described. In Sect. 3, we explain our models to caption an image using facial expression analysis. To generate sentimentally human-like captions, we show how facial expression features are detected and applied in our image captioning models. Section 4 presents our experimental setup and the evaluation results. The paper concludes in Sect. 5.

2 Related Work

In the following subsections, we review image captioning and facial expression recognition models as they are the key parts of our work.

2.1 Image Captioning

Recent image captioning models apply a CNN model to learn the image contents (encoding), followed by a LSTM to generate the image caption (decoding). This follows the paradigm employed in neural machine translation, using deep neural networks [31] to translate an image into a caption. In terms of encoding, they are divided into two categories: global encoding and fragment-level encoding [15]. The global approach encodes an image into a single feature vector, while the fragment-level one encodes the image fragments into separate feature vectors.

As a global encoding technique, Kiros et al. [20] applied a CNN and a LSTM to capture the image and the caption information, separately. They made a joint multi-modal space to encode the information and a multi-modal log-bilinear model (in the form of a language model) to generate new captions. In comparison, Vinyals et al. [35] encoded image contents using a CNN and applied a LSTM to generate a caption for the image in an end-to-end neural network model. In general, the global encoding approaches generate captions according to the detected objects in an image; however, when the test samples are significantly different from the training ones in terms of the object locations and interactions, they often cannot generalize to the test samples in terms of appropriate captions.

With respect to fragment-level encoding, Fang et al. [8] detected words from visual regions and used a maximum entropy language model to generate candidate captions. Instead of using LSTMs, they utilized a re-ranking method called deep multi-modal similarity to select the captions. Karpathy and Fei-Fei [16] applied a region-based image captioning model consisting of two separate models to detect an image region and generate its corresponding caption. Johnson et al. [12], based on the work of Ren et al. [28] on detecting image regions, incorporated the detection and generation tasks in an end-to-end training task. Attention mechanisms (either hard or soft) were applied by Xu et al. [37] to

---

[1] https://github.com/omidmn/Face-Cap.

detect salient regions and generate their related words. In each time step, the model dynamically used the regional features as inputs to the LSTM model. The fragment-level encoding methods detect objects and their corresponding regions in an image. However, they usually neglect encoding fine and significant fragments of data such as emotions. The work that we describe next has recognised this: human captions, such as those in Fig. 1, do include sentiment, and image captioning systems should therefore also aim to do this.

There are a few models that have incorporated sentiment into image captions [10,23,38]. However, this has typically required the construction of a new dataset, and the notion of sentiment is realized via a sentiment lexicon. Mathews et al. [23] applied a model to describe images using predefined positive and negative sentiments called SentiCap. The model used a full switching method including two parallel systems, each of which includes a Convolutional Neural Network (CNN) and a Long Short-Term Memory (LSTM). The first system was used to generate factual image captions and the second one to add word-level sentiments. The latter required a specifically constructed dataset, where crowdsourced workers rewrote thousands of factual captions to incorporate terms from a list of sentiment-bearing adjective-noun pairs. You et al. [38] presented two optimum schemes to employ the predefined sentiments to generate image descriptions. Their approach is still focused on subjective descriptions of images using a given sentiment vocabulary, rather than representing the emotional content of the image.

Gan et al. [10] StyleNet system that we noted in Sect. 1 adds styles, including sentiment values, to factual captions; these styles, such as humorous or romantic. Once more, these reflect the attitude of the viewer to the image, and it is in principle possible to generate captions that do not accord with the content of the image: for instance, while happy faces of babies can be properly described using positive sentiment, it is difficult to apply negative sentiment in this context.

Fig. 1. The examples of Flickr 30 K dataset [39] including sentiments. A man in a suit and tie with a **sad** look on his face (left) and a man on a sidewalk is playing the accordion while **happy** people pass by (right).

In contrast to this work, we focus on images including human faces and recognize relevant emotions, using facial expression analyses, to generate image captions.

Furthermore, we do not use any specific sentiment vocabulary or dataset to train our models: our goal is to see whether, given the existing vocabulary, incorporating facial emotion can produce better captions.

2.2   Facial Expression Recognition

Facial expression is a form of non-verbal communication which conveys attitude, affects, and intentions of individuals. Facial features and muscles changes during time lead to facial expression [9]. Darwin started research leading to facial expressions more than one century ago [7]. Now, there is a large body of work in recognizing basic facial expressions [9, 29] most often using the framework of six purportedly universal emotions [6] of happiness, sadness, fear, surprise, anger, and disgust plus neutral expressions. Recently, to find effective representations, deep learning based methods have been successfully applied to facial expression recognition (FER) tasks. They are able to capture hierarchical structures from low- to high-level data representations thanks to their complex architectures including multiple layers. Among deep models, Convolutional Neural Networks (CNNs) have achieved state-of-the-art performances in this domain. Kahou et al. [14], as a winning submission to the 2013 Emotion Recognition in the Wild Challenge, used CNNs to recognize facial expressions. CNNs and linear support vector machines were trained to detect basic facial expressions by Tang [32], who won the 2013 FER challenge [11]. In FER tasks, CNNs can be also used for transfer learning and feature extraction. Yu and Zhang [40] used CNNs, in addition to a face detection approach, to recognize facial expressions using transfer learning. The face detection approach was applied to detect faces areas and remove irrelevant noises in the target samples. Kahou et al. [13] also used CNNs for extracting visual features together with audio features in a multi-modal framework.

As is apparent, these models usually employ CNNs with a fairly standard deep architecture to produce good results on the FER-2013 dataset [11], which is a large dataset collected 'in the wild'. Pramerdorfer et al. [27], instead, applied a combination of modern deep architectures including VGGnet [30] on the dataset. They succeeded in generating the state-of-the-art result in this domain. We similarly aim to train a facial expression recognition model that can recognize facial expressions in the wild and produce state-of-the-art performance on FER-2013 dataset. In the next step, we then use the model as a feature extractor on the images of FlickrFace11K, our extracted dataset from Flickr 30K [39]. The features will be applied as a part of our image captioning models in this work.

3   Describing an Image Using Facial Expression Analysis

In this paper, we describe our image captioning models to generate image captions using facial expression analysis, which we term Face-Cap. We use a facial expression recognition model to extract the facial expression features from an image; the Face-Cap models in turn apply the features to generate image descriptions. In the following subsections, we first describe the datasets used in this work. Second, the face pre-processing step is explained to detect faces from our image caption data, and make them exactly similar to our facial expression recognition data. Third, the faces are fed into our facial expression recognition model to extract facial expression features. Finally, we elucidate Face-Cap models, which are image captioning systems trained by leveraging additional facial expression features and image-caption paired data.

3.1 Datasets

To train our facial expression recognition model, we use the facial expression recognition 2013 (FER-2013) dataset [11]. It includes in-the-wild samples labeled *happiness, sadness, fear, surprise, anger, disgust*, and *neutral*. It consists of 35,887 examples (28,709 for training, 3589 for public and 3589 for private test), collected by means of the Google search API. The examples are in grayscale at the size of 48-by-48 pixels. We split the training set of FER-2013 into two sections after removing 11 completely black examples: 25,109 for training and 3589 for validating the model. Similar to other work in this domain [17,27,40], we use the private test set of FER-2013 for the performance evaluation of the model after the training phase. To compare with the related work, we do not apply the public test set either for training or for validating the model.

To train our image captioning models, we have extracted a subset of the Flickr 30 K dataset with image captions [39], which we term FlickrFace11K. It contains 11,696 examples including human faces, which are detected using a CNN-based face detection algorithm [18].[2] We observe that the Flickr 30 K dataset is a good source for our dataset, because it has a larger portion of samples that include human faces, in comparison with other image caption datasets such as the COCO dataset [4]. We split the FlickrFace11K samples into 8696 for training, 2000 for validation and 1000 for testing, and make them publicly available.[3] To extract the facial features of the samples, we use a face pre-processing step and a facial expression recognition model as follows.

3.2 Face Pre-processing

3.3 Facial Expression Recognition Model

In this section, using the FER-2013 dataset, we train a VGGnet model [30] to recognize facial expressions. The model's architecture is similar to recent work [27] that is state-of-the-art in this domain, and our replication gives similar performance. The classification accuracy, which is a popular performance metric on the FER-2013 dataset, on the test set of FER-2013 is 72.7%. It is around 7% better than the human performance (65 ± 5%) on the test set [11]. The output layer of the model, generated using a softmax function, includes seven neurons, corresponding to the categorical distribution probabilities over the emotion classes

---

[2] The new version (2018) of Dlib library is applied.
[3] https://github.com/omidmn/Face-Cap.

in FER-2013 including *happiness, sadness, fear, surprise, anger, disgust,* and *neutral*; we refer to this by the vector $a = (a_1, \ldots, a_7)$.

We use the network to extract the probabilities of each emotion from all faces, as detected in the pre-processing step of Sect. 3.2, in each FlickrFace11K sample.

For each image, we construct a vector of facial emotion features $s = (s_1, \ldots, s_7)$ used in the Face-Cap models as in Eq. 1.

$$s_k = \begin{cases} 1 & \text{for } k = \arg\max \sum_{1 \leq i \leq n} a_i, \\ 0 & \text{otherwise} \end{cases} \tag{1}$$

where n is the number of faces in the sample. That is, s is a one-hot encoding of the aggregate facial emotion features of the image.

3.4 Training Face-Cap

Face-Cap$_F$. In order to train the Face-Cap models, we apply a long short-term memory (LSTM) network as our caption generator, adapted from Xu et al. [37]. The LSTM is informed about the emotional content of the image using the facial features, defined in Eq. 1. It also takes the image features which are extracted by Oxford VGGnet [30], learned on the ImageNet dataset, and weighted using the attention mechanism [37]. In the mechanism, the attention-based features, including the factual content of the image, are chosen for each generated word in the LSTM. Using Eq. 2, in each time step (t), the LSTM uses the previously embedded word (x_{t-1}), the previous hidden state (h_{t-1}), the image features (z_t), and the facial features (s) to generate input gate (i_t), forget gate (f_t), output gate (o_t), input modulation gate (g_t), memory cell (c_t), and hidden state (h_t).

$$\begin{aligned}
i_t &= \sigma(W_i x_{t-1} + U_i h_{t-1} + Z_i z_t + S_i s + b_i) \\
f_t &= \sigma(W_f x_{t-1} + U_f h_{t-1} + Z_f z_t + S_f s + b_f) \\
o_t &= \sigma(W_o x_{t-1} + U_o h_{t-1} + Z_o z_t + S_o s + b_o) \\
g_t &= \tanh(W_c x_{t-1} + U_c h_{t-1} + Z_c z_t + S_c s + b_c) \\
c_t &= f_t c_{t-1} + i_t g_t \\
h_t &= o_t \tanh(c_t)
\end{aligned} \tag{2}$$

where W, U, Z, S, and b are learned weights and biases and σ is the logistic sigmoid activation function. According to Eq. 2, the facial features of each image are fixed in all time steps and the LSTM automatically learns to condition, at the appropriate time, the next generated word by applying the features. To initialize the LSTM's memory state (c_0) and hidden state (h_0), we feed the facial features through two typical multilayer perceptrons, shown in Eq. 3.

$$c_0 = \tanh_{init,c}(s), \quad h_0 = \tanh_{init,h}(s) \tag{3}$$

We use the current hidden state (h_t), to calculate the negative log-likelihood of s in each time step (Eq. 4), named the face loss function. Using this method, h_t will be able to record a combination of s, x_{t-1} and z_t in each time step.

$$L(h_t, s) = - \sum_{1 \leq i \leq 7} 1_{(i=s)} \log(p(i|h_t)) \tag{4}$$

where a multilayer perceptron generates $p(i|h_t)$, which is the categorical probability distribution of the current state across the emotion classes. In this we adapt You et al. [38], who use this loss function for injecting ternary sentiment (positive, neutral, negative) into captions. This loss is estimated and averaged, over all steps, during the training phase.

Face-Cap$_L$. The above Face-Cap$_F$ model feeds in the facial features at the initial step (Eq. 3) and at each time step (Eq. 2), shown in Fig. 2 (top). In Face-Cap$_F$, the LSTM uses the facial features for generating every word because the features are fed at each time step. Since a few words, in the ground truth captions (e.g. Fig. 1), are related to the features, this mechanism can sometimes lead to less effective results.

Our second variant of the model, Face-Cap$_L$, is as above except that the s term is removed from Eq. 2: we do not apply the facial feature information at each time step (Fig. 2 (bottom)), eliminating it from Eq. 2. Using this mechanism, the LSTM can effectively take the facial features in generating image captions and ignore the features when they are irrelevant. To handle this issue, You et al. [38] implemented the sentiment cell, working similar to the memory cell in the LSTM, initialized by the ternary sentiment. They fed the image features to initialize the memory cell and hidden state of the LSTM. In comparison, Face-Cap$_L$ uses the facial features to initialize the memory cell and hidden state rather than the sentiment cell which requires more time and memory to compute. Using the attention mechanism, our model applies the image features in generating every caption word.

4 Experiments

4.1 Evaluation Metrics and Testing

To evaluate Face-Cap$_F$ and Face-Cap$_L$, we use standard evaluation metrics including BLEU [26], ROUGE-L [21], METEOR [5], CIDEr [34], and SPICE [2]. All five metrics with larger values mean better results.

We train and evaluate all models on the same splits of FlickrFace11K.

4.2 Models for Comparison

The model of Xu et al. [37] is the starting point of Face-Cap$_F$ and Face-Cap$_L$, which is selectively attending to a visual section at each time step. We train Xu's model using the FlickrFace11K dataset.

We also look at two additional models to investigate the impact of the face loss function in using the facial features in different schemes. We train the Face-Cap$_F$ model, which uses the facial features in every time step, without calculating the face loss function (Eq. 4); we refer to this as the Face-Step model. The Face-Cap$_L$ model, which applies the facial features in the initial time step, is also modified in the same way; we refer to this as the Face-Init model.

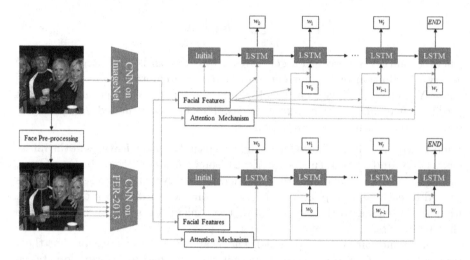

Fig. 2. The frameworks of Face-Cap$_F$ (top), and Face-Cap$_L$ (bottom). The face pre-processing and the feature extraction from the faces and the image are illustrated. The Face-Cap models are trained using the caption data plus its corresponding image features, selected using the attention mechanism, and facial features.

4.3   Implementation Details

In our implementation, the memory cell and the hidden state of the LSTM each have 512 dimensions.[4] We set the size of the word embedding layer to 300, which is initialized using a uniform distribution. The mini-batch size is 100 and the epoch limit is 20. We train the models using the Adam optimization algorithm [19]. The learning rate is initialized to 0.001, while its minimum is set to 0.0001. If there is no improvement of METEOR for two successive epochs, the learning rate is divided by two and the prior network that has the best METEOR is reloaded. This approach leads to effective results in this work. For Adam, tuning the learning rate decay, similar to our work, is supported by Wilson et al. [36]. METEOR on the validation set is used for model selection. We apply METEOR for the learning rate decay and the model selection because it shows reasonable correlation with human judgments but calculates more quickly than SPICE (as it does not require dependency parsing) [2].

Exactly the same visual feature size and vocabulary are used for all five models. As the encoder of images, in this work as for Xu et al., we use Oxford VGGnet [30] trained on ImageNet, and take its fourth convolutional layer (after ReLU), which gives $14 \times 14 \times 512$ features. For all five models, the negative log likelihood of the generated word is calculated, as the general loss function, at each time step.

---

[4] We use TensorFlow to implement the models [1].

Table 1. Comparisons of image caption results (%) on the test split of FlickrFace11K dataset. B-1, ... SPICE are standard evaluation metrics, where B-N is BLEU-N metric.

Model	B-1	B-2	B-3	B-4	METEOR	ROUGE-L	CIDEr	SPICE
Xu's model	55.95	35.43	23.06	15.69	16.96	43.71	21.94	9.30
Face-Step	58.43	37.56	24.78	16.96	**17.45**	45.04	22.83	9.90
Face-Init	56.63	36.49	24.30	16.86	17.17	44.84	23.13	9.80
Face-Cap$_F$	57.13	36.51	24.07	16.52	17.19	44.76	23.04	9.70
Face-Cap$_L$	**58.90**	**37.89**	**25.07**	**17.19**	17.44	**45.47**	**24.72**	**10.00**

4.4 Results

Overall Metrics. The experimental results are summarized in Table 1. All Face models outperform Xu's model using all standard evaluation metrics. This shows that the facial features are effective in image captioning tasks. As predicted, Face-Cap$_L$ has a better performance in comparison with other models using all the metrics except METEOR, where it is only very marginally (0.01) lower. Under most metrics, Face-Step performs second best, with the notable exception of CIDEr, suggesting that its strength on other metrics might be from use of popular words (which are discounted under CIDEr). Comparing the mechanics of the top two approaches, Face-Cap$_L$ uses the face loss function to keep the facial features and apply them at the appropriate time; however, Face-Step does not apply the face loss function. Face-Cap$_L$ only applies the facial features in the initial time step, while Face-Step uses the features in each time step, in generating an image caption. In this way, Face-Step can keep the features without applying the face loss function. This yields comparable results between Face-Cap$_L$ and Face-Step; however, the results show that applying the face loss function is more effective than the facial features in each time step. This relationship can also be seen in the results of Face-Init, which is Face-Cap$_L$ without the face loss function. The results of Face-Cap$_F$ show that a combination of applying the face loss function and the facial features in each time step is problematic.

Caption Analysis. To analyze what it is about the captions themselves that differs under the various models, with respect to our aim of injecting information about emotional states of the faces in images, we first extracted all generated adjectives, which are tagged using the Stanford part-of-speech tagger software [33]. Perhaps surprisingly, emotions do not manifest themselves in the adjectives in Face-Cap models: the adjectives used by all systems are essentially the same. This may be because adjectives with weak sentiment values (e.g. *long, small*) predominate in the training captions, relative to the adjectives with strong sentiment values (e.g. *happy, surprised*).

We therefore also investigated the difference in distributions of the generated verbs under the models. Entropy (in the information-theoretic sense) can indicate

Table 2. Comparisons of distributions of verbs in generated captions: entropies, and probability mass of the top 4 frequent verbs (*is, sitting, are, standing*)

Model	Entropy	Top 4
Xu's model	2.7864	77.05%
Face-Step	2.9059	74.80%
Face-Init	2.6792	78.78%
Face-Cap$_F$	2.7592	77.68%
Face-Cap$_L$	**2.9306**	**73.65%**

Table 3. The ranks of sample generated verbs under each model.

Model	Smiling	Looking	Singing	Reading	Eating	Laughing
Xu's model	19	n/a	15	n/a	24	n/a
Face-Step	11	**18**	10	n/a	15	n/a
Face-Init	10	21	12	n/a	14	n/a
Face-Cap$_F$	12	20	**9**	n/a	14	n/a
Face-Cap$_L$	**9**	**18**	15	**22**	**13**	**27**

which distributions are closer to deterministic and which are more spread out (with a higher score indicating more spread out) calculated using Eq. 5.

$$E = - \sum_{1 \leq i \leq n} p(x_i) \times \log_2(p(x_i)) \tag{5}$$

where E is the entropy score and n is the number of the generated unique verbs under each model. $P(x_i)$ is the probability of each generated unique verb (x_i), estimated as the Maximum Likelihood Estimate from the sample. From Table 2, Face-Cap$_L$ has the highest entropies, or the one with the greatest variability of expression. Relatedly, we look at the four most frequent verbs, which are the same for all models (*is, sitting, are, standing*) — these are verbs with relatively little semantic content, and for the most part act as syntactic props for the content words of the sentence. Table 2 also shows that Face-Cap$_L$ has the lowest proportion of the probability mass taken up by these, leaving more for other verbs.

The ranks of the generated verbs under the models, which are calculated using the numerical values of their frequency, are also interesting. Table 3 includes some example verbs; of these, *smiling, singing,* and *eating* are higher ranked under the Face-Cap models, and *reading* and *laughing* only appear under the Face-Cap$_L$ model. *Looking* is also generated only using the models including the facial features. These kinds of verbs are relevant to the facial features and show the effectiveness of applying the features in generating image captions.

Samples. Figure 3 includes a number of generated captions, for six sample images, under all models in this work. In example 1, the models that include facial features properly describe the emotional content of the image using *smiling*. The Face-Cap$_L$ model also generates *laughing* according to the emotional content of example 4. In example 3, the Face-Init and the Face-Cap$_L$ models generate *playing* which is connected to the emotional content of the example. It is perhaps because the child in the example is happy that the models generate *playing*, which

1
X: two women and a man are posing for a picture .
S: two men and a woman are smiling .
I: two men and a woman are smiling at the camera .
F: two men and a woman are smiling .
L: two men and a woman are smiling at a camera .

2
X: a man in a black shirt and a woman in a black shirt playing a guitar .
S: a group of people are playing instruments in a band .
I: a group of people are playing a drum .
F: a group of people are playing a guitar and singing .
L: a group of people are singing and playing instruments .

3
X: a boy in a red shirt is standing on a red slide .
S: a young boy in a red shirt is sitting on a red slide .
I: a young boy in a red shirt is playing with a toy car .
F: a young boy in a red shirt and blue shorts is sitting on a red slide .
L: a young boy in a red shirt is playing with a toy truck .

4
X: two women and a man are sitting at a table .
S: a man and woman are dancing .
I: a man and a woman are sitting at a table with a drink .
F: a man in a black shirt is holding a woman in a black shirt .
L: a man and a woman are laughing at a bar .

5
X: a young boy is holding a baby in a yellow shirt .
S: a young boy in a yellow shirt is holding a yellow object .
I: a woman in a yellow shirt is holding a baby .
F: a young boy in a yellow shirt is holding a yellow cup .
L: a young boy in a yellow shirt is playing with a toy .

6
X: a man in a white shirt is working on a machine .
S: a woman in a white shirt is sitting at a table with a large glass of wine in her hand
I: a woman in a white shirt is sitting at a table with a computer .
F: a woman in a white shirt is sitting at a table with a computer .
L: a woman in a white shirt is looking at a computer screen .

Fig. 3. Examples of different image captioning models including X (Xu's model), S (Face-Step), I (Face-Init), F (Face-Cap$_F$), and L (Face-Cap$_L$).

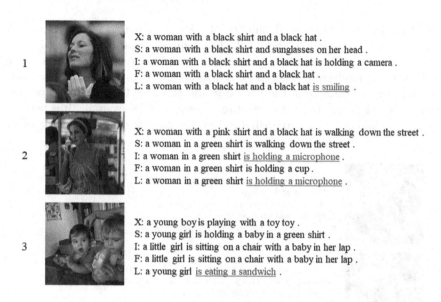

1   X: a woman with a black shirt and a black hat .
S: a woman with a black shirt and sunglasses on her head .
I: a woman with a black shirt and a black hat is holding a camera .
F: a woman with a black shirt and a black hat .
L: a woman with a black hat and a black hat is smiling .

2   X: a woman with a pink shirt and a black hat is walking down the street .
S: a woman in a green shirt is walking down the street .
I: a woman in a green shirt is holding a microphone .
F: a woman in a green shirt is holding a cup .
L: a woman in a green shirt is holding a microphone .

3   X: a young boy is playing with a toy toy .
S: a young girl is holding a baby in a green shirt .
I: a little girl is sitting on a chair with a baby in her lap .
F: a little girl is sitting on a chair with a baby in her lap .
L: a young girl is eating a sandwich .

Fig. 4. Examples of the models including various amounts of error.

has a positive sentiment connotation. In example 5, Face-Cap$_L$ also uses *playing* in a similar way. Example 2 shows that the Face-Cap models apply *singing* at the appropriate time. Similarly, *looking* is used, by Face-Cap$_L$, in example 6. *Singing* and *looking* are generated because of the facial features of people in the examples, which are related to some emotional states such as *surprised* and *neutral*. Figure 3 shows that our models can effectively apply the facial features to describe images in different ways. In Fig. 4, three examples are shown, which our models inappropriately use the facial features. *Smiling* is used to describe the emotional content of the example 1; however, the girl in the example is not happy. The results of the example 2 and 3 wrongly contain *holding a microphone* and *eating*, which are detected from the facial features, due to visual likeness.

5   Conclusion and Future Work

In this paper, we have proposed two variants of an image captioning model, Face-Cap, which employ facial features to describe images. To this end, a facial expression recognition model has been applied to extract the features from images including human faces. Using the features, our models are informed about the emotional content of the images to automatically condition the generating of image captions. We have shown the effectiveness of the models using standard evaluation metrics compared to the state-of-the-art baseline model. The generated captions demonstrate that the Face-Cap models succeed in generating image captions, incorporating the facial features at the appropriate time. Linguistic analyses of the captions suggest that the improved effectiveness in describing image content comes through greater variability of expression.

Future work can involve designing new facial expression recognition models, which can cover a richer set of emotions including *confusion* and *curiousity*; and effectively apply their corresponding facial features to generate image captions. In addition, we would like to explore alternative architectures for injecting facial emotions, like the soft injection approach of [37].

References

1. Abadi, M., et al.: TensorFlow: large-scale machine learning on heterogeneous distributed systems. arXiv preprint arXiv:1603.04467 (2016)
2. Anderson, P., Fernando, B., Johnson, M., Gould, S.: SPICE: semantic propositional image caption evaluation. In: Leibe, B., Matas, J., Sebe, N., Welling, M. (eds.) ECCV 2016. LNCS, vol. 9909, pp. 382–398. Springer, Cham (2016). https://doi.org/10.1007/978-3-319-46454-1_24
3. Anderson, P., et al.: Bottom-up and top-down attention for image captioning and VQA. arXiv preprint arXiv:1707.07998 (2017)
4. Chen, X., et al.: Microsoft coco captions: data collection and evaluation server. arXiv preprint arXiv:1504.00325 (2015)
5. Denkowski, M., Lavie, A.: Meteor universal: language specific translation evaluation for any target language. In: WMT, pp. 376–380 (2014)
6. Ekman, P.: Basic emotions. In: Dalgleish, T., Power, T. (eds.) The Handbook of Cognition and Emotion, pp. 45–60. Wiley, Sussex (1999)
7. Ekman, P.: Darwin and facial expression: a century of research in review. In: ISHK (2006)
8. Fang, H., et al.: From captions to visual concepts and back. In: CVPR. IEEE (2015)
9. Fasel, B., Luettin, J.: Automatic facial expression analysis: a survey. Pattern Recogn. **36**(1), 259–275 (2003)
10. Gan, C., Gan, Z., He, X., Gao, J., Deng, L.: StyleNet: generating attractive visual captions with styles. In: CVPR. IEEE (2017)
11. Goodfellow, I.J., et al.: Challenges in representation learning: a report on three machine learning contests. In: Lee, M., Hirose, A., Hou, Z.-G., Kil, R.M. (eds.) ICONIP 2013. LNCS, vol. 8228, pp. 117–124. Springer, Heidelberg (2013). https://doi.org/10.1007/978-3-642-42051-1_16
12. Johnson, J., Karpathy, A., Fei-Fei, L.: DenseCap: fully convolutional localization networks for dense captioning. In: CVPR, pp. 4565–4574. IEEE (2016)
13. Kahou, S.E., et al.: EmoNets: multimodal deep learning approaches for emotion recognition in video. J. Multimodal User Interfaces **10**(2), 99–111 (2016)
14. Kahou, S.E., et al.: Combining modality specific deep neural networks for emotion recognition in video. In: ICMI, pp. 543–550. ACM (2013)
15. Karpathy, A.: Connecting Images and Natural Language. Ph.D. thesis, Stanford University (2016)
16. Karpathy, A., Fei-Fei, L.: Deep visual-semantic alignments for generating image descriptions. In: CVPR, pp. 3128–3137. IEEE (2015)
17. Kim, B.K., Dong, S.Y., Roh, J., Kim, G., Lee, S.Y.: Fusing aligned and non-aligned face information for automatic affect recognition in the wild: a deep learning approach. In: CVPR Workshops, pp. 48–57. IEEE (2016)
18. King, D.E.: Dlib-ml: a machine learning toolkit. J. Mach. Learn. Res. **10**, 1755–1758 (2009)

19. Kingma, D.P., Ba, J.: Adam: a method for stochastic optimization. arXiv preprint arXiv:1412.6980 (2014)
20. Kiros, R., Salakhutdinov, R., Zemel, R.S.: Unifying visual-semantic embeddings with multimodal neural language models. arXiv preprint arXiv:1411.2539 (2014)
21. Lin, C.Y.: ROUGE: a package for automatic evaluation of summaries. In: Text Summarization Branches Out (2004)
22. Lisetti, C.: Affective computing (1998)
23. Mathews, A.P., Xie, L., He, X.: SentiCap: generating image descriptions with sentiments. In: AAAI, pp. 3574–3580 (2016)
24. Pang, B., Lee, L.: A sentimental education: sentiment analysis using subjectivity summarization based on minimum cuts. In: ACL, Barcelona, Spain, pp. 271–278, July 2004. https://doi.org/10.3115/1218955.1218990, http://www.aclweb.org/anthology/P04-1035
25. Pang, B., Lee, L.: Opinion mining and sentiment analysis. Found. Trends Inf. Retr. 2(1–2), 1–135 (2008). https://doi.org/10.1561/1500000011
26. Papineni, K., Roukos, S., Ward, T., Zhu, W.J.: BLEU: a method for automatic evaluation of machine translation. In: ACL, pp. 311–318. Association for Computational Linguistics (2002)
27. Pramerdorfer, C., Kampel, M.: Facial expression recognition using convolutional neural networks: state of the art. arXiv preprint arXiv:1612.02903 (2016)
28. Ren, S., He, K., Girshick, R., Sun, J.: Faster R-CNN: towards real-time object detection with region proposal networks. In: NIPS, pp. 91–99 (2015)
29. Sariyanidi, E., Gunes, H., Cavallaro, A.: Automatic analysis of facial affect: a survey of registration, representation, and recognition. IEEE Trans. Pattern Anal. Mach. Intell. 37(6), 1113–1133 (2015)
30. Simonyan, K., Zisserman, A.: Very deep convolutional networks for large-scale image recognition. arXiv preprint arXiv:1409.1556 (2014)
31. Sutskever, I., Vinyals, O., Le, Q.V.: Sequence to sequence learning with neural networks. In: NIPS, pp. 3104–3112 (2014)
32. Tang, Y.: Deep learning using linear support vector machines. arXiv preprint arXiv:1306.0239 (2013)
33. Toutanova, K., Klein, D., Manning, C.D., Singer, Y.: Feature-rich part-of-speech tagging with a cyclic dependency network. In: NAACL HLT, pp. 173–180. Association for Computational Linguistics (2003)
34. Vedantam, R., Lawrence Zitnick, C., Parikh, D.: CIDEr: consensus-based image description evaluation. In: CVPR, pp. 4566–4575. IEEE (2015)
35. Vinyals, O., Toshev, A., Bengio, S., Erhan, D.: Show and tell: a neural image caption generator. In: CVPR, pp. 3156–3164. IEEE (2015)
36. Wilson, A.C., Roelofs, R., Stern, M., Srebro, N., Recht, B.: The marginal value of adaptive gradient methods in machine learning. In: NIPS, pp. 4151–4161 (2017)
37. Xu, K., et al.: Show, attend and tell: neural image caption generation with visual attention. In: ICML, pp. 2048–2057 (2015)
38. You, Q., Jin, H., Luo, J.: Image captioning at will: a versatile scheme for effectively injecting sentiments into image descriptions. arXiv preprint arXiv:1801.10121 (2018)
39. Young, P., Lai, A., Hodosh, M., Hockenmaier, J.: From image descriptions to visual denotations: new similarity metrics for semantic inference over event descriptions. Trans. Assoc. Comput. Linguist. 2, 67–78 (2014)
40. Yu, Z., Zhang, C.: Image based static facial expression recognition with multiple deep network learning. In: ICMI, pp. 435–442. ACM (2015)

Pedestrian Trajectory Prediction
with Structured Memory Hierarchies

Tharindu Fernando$^{(\boxtimes)}$, Simon Denman, Sridha Sridharan, and Clinton Fookes

Image and Video Research Laboratory, SAIVT Research Program,
Queensland University of Technology, Brisbane, Australia
{t.warnakulasuriya,s.denman,s.sridharan,c.fookes}@qut.edu.au

Abstract. This paper presents a novel framework for human trajectory prediction based on multimodal data (video and radar). Motivated by recent neuroscience discoveries, we propose incorporating a structured memory component in the human trajectory prediction pipeline to capture historical information to improve performance. We introduce structured LSTM cells for modelling the memory content hierarchically, preserving the spatiotemporal structure of the information and enabling us to capture both short-term and long-term context. We demonstrate how this architecture can be extended to integrate salient information from multiple modalities to automatically store and retrieve important information for decision making without any supervision. We evaluate the effectiveness of the proposed models on a novel multimodal dataset that we introduce, consisting of 40,000 pedestrian trajectories, acquired jointly from a radar system and a CCTV camera system installed in a public place. The performance is also evaluated on the publicly available New York Grand Central pedestrian database. In both settings, the proposed models demonstrate their capability to better anticipate future pedestrian motion compared to existing state of the art. Data related to this paper are available at: https://github.com/qutsaivt/SAIVTMultiSpectralTrajectoryDataset.

Keywords: Human trajectory prediction
Structured memory networks · Multimodal information fusion
Long-term planing

1 Introduction

Understanding and predicting crowd behaviour is an important topic due to its myriad applications (surveillance, event detection, traffic flow, etc). However this remains a challenging problem due to the complex nature of human behaviour and the lack of attention that researchers pay to human navigational patterns when developing machine learning models.

Electronic supplementary material The online version of this chapter (https://doi.org/10.1007/978-3-030-10925-7_15) contains supplementary material, which is available to authorized users.

© Springer Nature Switzerland AG 2019
M. Berlingerio et al. (Eds.): ECML PKDD 2018, LNAI 11051, pp. 241–256, 2019.
https://doi.org/10.1007/978-3-030-10925-7_15

Recent neuroscience studies have revealed that humans utilise map and grid like structures for navigation [12,26]. The human brain builds a unified representation of the spatial environment, which is stored in the hippocampus [13] and guides the decision making process. Further studies [6] provide strong evidence towards a hierarchical spatial representation of these maps. Additionally in [11,20] authors have observed multiple representations of structured maps instead of one single map in the long-term memory. This idea was explored in [28] using structured memory for Deep Reinforcement Learning. To generate an output at a particular time step, the system passes the memory content through a series of convolution layers to summarise the content. We argue this is inefficient and could lead to a loss of information when modelling large spatial areas.

Motivated by recent neuroscience [12,26] and deep reinforcement leaning [28] studies, we utilise a structured memory to predict human navigational behaviour. In particular such a memory structure allows a machine learning algorithm to exploit historical knowledge about the spatial structure of the environment, and reason and plan ahead, instead of generating reflexive behaviour based on the current context. Novel contributions of this paper are summarised as follows:

- We introduce a novel neural memory architecture which effectively captures the spatiotemporal structure of the environment.
- We propose structured LSTM (St-LSTM) cells, which model the structured memory hierarchically, preserving the memories' spatiotemporal structure.
- We incorporate the neural memory network into a human trajectory prediction pipeline where it learns to automatically store and retrieve important information for decision making without any supervision.
- We introduce a novel multimodal dataset for human trajectory prediction, containing more than 40,000 trajectories from Radar and CCTV streams.
- We demonstrate how the semantic information from multiple input streams can be captured through multiple memory components and propose an effective fusion scheme that preserves the spatiotemporal structure.
- We provide extensive evaluations of the proposed method using multiple public benchmarks, where the proposed method is capable of imitating human navigation behaviour and outperforms state-of-the-art methods.

2   Related Work

The related literature can be broadly categorised into human behaviour prediction approaches, introduced in Sect. 2.1; neural memory architectures, presented in Sect. 2.2; and multimodal information fusion which we review in Sect. 2.3.

2.1   Human Behaviour Prediction

Before the dawn of deep learning, Social Force models [33,34] had been extensively applied for modelling human navigational behaviour. They rely on the attractive and repulsive forces between pedestrians to predict motion. However as shown in [1,14,16] these methods ill represent the structure of human decision making by modelling the behaviour with just a handful of parameters.

One of the most popular deep learning methods for predicting human behaviour is the Social LSTM model of [1], which removed the need for hand-crafted features by using LSTMs to encode and decode trajectory information. This method is further augmented in [16] where the authors incorporate the entire trajectory of the pedestrian of interest as well as the neighbouring pedestrians and extract salient information from these embeddings through a combination of soft and hardwired attention. Similar to [16] the works in [3,32,36] also highlight the importance of fully capturing context information. However these methods all consider short-term temporal context in the given neighbourhood, completely discarding scene structure and the longterm scene context.

2.2 Neural Memory Architectures

External memory modules are used to store historic information, and learn to automatically store and retrieve important facts to aid future predictions. Many approaches across numerous domains [14,15,17,18,22,27] have utilised memory modules to aid prediction, highlighting the importance of stored knowledge for decision making. However existing memory structures are one dimensional modules which completely ignore the environmental spatial structure. This causes a significant hindrance when modelling human navigation, since they are unable to capture the map-like structures humans use when navigating [12,26].

The work of Parisotto et al. [28] proposes an interesting extension to memory architectures where they structure the memory as a 3D block, preserving spatial relationships. However, when generating memory output they rely on a static convolution kernel to summarise the content, failing to generate dynamic responses and propagate salient information from spatial locations to the trajectory prediction module, where multiple humans can interact in the environment.

Motivated by the hierarchical sub-map structure humans use to navigate [11,20], we model our spatiotemporal memory with gated St-LSTM cells, which are arranged hierarchically in a grid structure.

2.3 Multimodal Information Fusion

Multimodal information fusion addresses the task of integrating inputs from various modalities and has shown superior performance compared to unimodal approaches [4,10] in variety of applications [2,21,35]. The simplest approach is to concatenate features to obtain a single vector representation [23,29]. However it ignores the relative correlation between the modalities [2].

More complex fusion strategies include concatenating higher level representations from individual modalities separately and then combining them together, enabling the model to learn the salient aspects of individual streams. In this direction, attempts were made using Deep Boltzmann Machines [31] and neural network architectures [9,25].

In [15] the authors explore the importance of capturing both short and longterm temporal context when performing feature fusion, utilising separate neural memory units for individual feature streams and aggregating the temporal representation during fusion. Yet this fails to preserve the spatial structure, restricting its applicability when modelling human navigation.

3   Architecture

In this section we introduce the encoding scheme utilised to embed the trajectory information of the pedestrian of interest and their neighbours; the structure and the operations of the proposed hierarchical memory; how to utilise memory output to enhance the future trajectory prediction; and an architecture for effectively coupling multimodal information streams through structured memories.

3.1   Embedding Local Neighbourhood Context

In order to embed the current short-term context of the pedestrian of interest and the local neighbourhood, we utilise the trajectory prediction framework proposed in [16]. Let the observed trajectory of pedestrian k from frame 1 to frame T_{obs} be given by,

$$X^k = [(x_1, y_1), (x_2, y_2), \ldots, (x_{T_{obs}}, y_{T_{obs}})], \tag{1}$$

where the trajectory is composed of points in a 2D Cartesian grid. Similar to [16] we utilise the soft attention operation [8] to embed the trajectory information from the pedestrian of interest (k) and generate a vector embedding $C_t^{s,k}$. To embed neighbouring trajectories the authors in [16,19] have shown that distance based hardwired attention is efficient and effective. We denote the hardwired context vector as $C_t^{h,k}$.

Now we define the combined context vector, $C_t^{*,k}$, representing the short-term context of the local neighbourhood of the kth pedestrian as,

$$C_t^{*,k} = \tanh([C_t^{s,k}, C_t^{h,k}]), \tag{2}$$

where $[.,.]$ denotes a concatenation operation. Please see [16,19] for details.

3.2   Structured Memory Network (SMN)

Let the structured memory, M, be a $l \times W \times H$ block where l is the embedding dimension of p_t^k. W is the vertical extent of the map and H is the horizontal extent. We define a function $\psi(x, y)$ which maps spatial coordinates (x, y) with $x \in \mathbb{R}$ and $y \in \mathbb{R}$ to a map grid (x', y') where $x' \in 0, \ldots, W$ and $y' \in 0, \ldots, H$. The works of [16,19] have shown that the context embeddings $C_t^{*,k}$ capture the short-term context of the pedestrian of interest and the local neighbourhood. Hence we store these embeddings in our structured memory as it represents the temporal context in that grid cell. The operations of the proposed structured memory network (SMN) can be summarised as follows,

$$h_t = \text{read}(M_t), \tag{3}$$

$$\beta_{t+1}^{(x',y')} = \text{write}(C_t^{*,k}, M_t^{(x',y')}), \tag{4}$$

$$M_{t+1} = \text{update}(M_t, w_{t+1}^{(x',y')}). \tag{5}$$

The following subsections explain these three operations.

Hierarchical Read Operation. The read operation outputs output a vector, h_t, capturing the most salient information from the entire memory for decision making in the present state. We define a hierarchical read operation which passes the current memory representation, M_t, through a series of gated, structured LSTM (St-LSTM) cells arranged in a grid like structure. Figure 1 depicts the operations of the proposed St-LSTM cells.

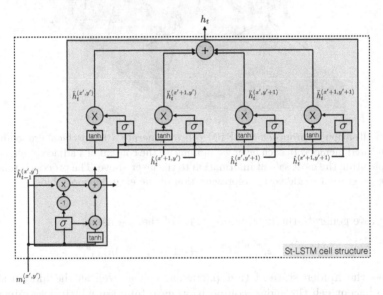

Fig. 1. The operations of the proposed St-LSTM cell. It considers the current representation of the respective memory cell and the 3 adjacent neighbours as well as the previous time step outputs and utilises gated operations to render the output in the present time step.

Let the content of (x', y') memory cell at time t be represented by $m_t^{(x',y')}$ and the three adjacent cells be represented by $m_t^{(x'+1,y')}, m_t^{(x',y'+1)}$ and $m_t^{(x'+!,y'+1)}$. As shown in Fig. 2, we first pass the current state of the memory cell through an input gate to decide how much information to pass through the gate and how much information to gather from the previous hidden state of the that particular cell, $\hat{h}_{t-1}^{(x',y')}$. This operation is given by,

$$z_t^{(x',y')} = \sigma(w_z^{(x',y')}[m_t^{(x',y')}, \hat{h}_{t-1}^{(x',y')}]),$$
$$\hat{o}_t^{(x',y')} = \tanh([m_t^{(x',y')}, \hat{h}_{t-1}^{(x',y')}]). \tag{6}$$

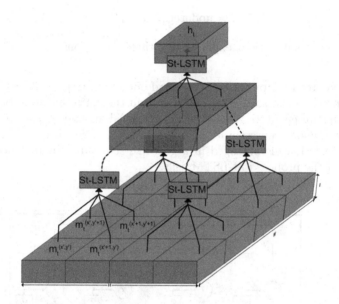

Fig. 2. Utilisation of proposed St-LSTM cell to generate a hierarchical embedding of the structured memory. In each layer we summarise the content of 4 adjacent neighbours via propagating the most salient information to the layer above. The process is repeated until we generate a single vector representation of the entire memory block.

Then we generate the new hidden state of the cell using,

$$\hat{h}_t^{(x',y')} = z_t^{(x',y')}\hat{o}_t^{(x',y')} + (1 - z_t^{(x',y')})\hat{h}_{t-1}^{(x',y')}, \tag{7}$$

and pass the hidden state of that particular cell as well as the hidden states of the adjacent cell through a composition gate function which determines the amount of information to be gathered from each of the cells as,

$$q_t^{(x',y')} = \sigma(w_q^{(x',y')}[\hat{h}_t^{(x',y')}, \hat{h}_t^{(x'+1,y')}, \hat{h}_t^{(x',y'+1)}, \hat{h}_t^{(x'+1,y'+1)}]). \tag{8}$$

Now we can generate the augmented state of the cell (x', y') as,

$$\ddot{h}_t^{(x',y')} = \tanh(\hat{h}_t^{(x',y')})q_t^{(x',y')}. \tag{9}$$

We perform the above operations to the rest of the group of 3 cells: $(x' + 1, y'), (x', y' + 1)$ and $(x' + 1, y' + 1)$; and generate the representations

$\ddot{h}_t^{(x'+1,y')}, \ddot{h}_t^{(x',y'+1)}$ and $\ddot{h}_t^{(x'+1,y'+1)}$ respectively. Then the feature embedding representation, h_t, of the merged 4 cells in the next layer of the memory is given by,

$$h_t = \ddot{h}_t^{(x',y')} + \ddot{h}_t^{(x'+1,y')} + \ddot{h}_t^{(x',y'+1)} + \ddot{h}_t^{(x'+1,y'+1)}. \tag{10}$$

We repeat this process for all cells (x', y') where $x' \in 0, \ldots, W$ and $y' \in 0, \ldots, H$. Note that as we are merging four adjacent cells we have $W/2$ and $H/2$ St-LSTM cells in the immediate next layer of the memory block. We continue merging cells until we are left with one cell summarising the entire memory block. We denote the hidden state of this cell as h_t.

Write Operation. Given the current position of the pedestrian of interest at (x, y), we first evaluate the associated location in the map grid by passing (x, y) through function, ψ, such that,

$$(x', y') = \psi(x, y). \tag{11}$$

Then we retrieve the current memory state of (x', y') as,

$$m_t^{(x',y')} = M_t^{(x',y')}. \tag{12}$$

Then by utilising the above stated vector and the short-term context of the pedestrian of interest, C_t^*, we define a write function which generates a write vector for memory update,

$$\beta_{t+1}^{(x',y')} = \text{LSTM}_w(c_t^*, m_t^{(x',y')}). \tag{13}$$

Update Operation. We update the memory map for the next time step by,

$$M_{t+1}^{(a,b)} = \begin{cases} \beta_{t+1}^{(x',y')} & \text{for } (a, b) = (x', y') \\ M_t^{(a,b)} & \text{for } (a, b) \neq (x', y') \end{cases} \tag{14}$$

The new memory map is equal to the memory map at the previous time step except for the current location of the pedestrian where we completely update the content with the generated write vector.

3.3 Trajectory Prediction with Structured Memory Hierarchies

We utilise the combined context vector $C_t^{*,k}$ representing the short-term context of the local neighbourhood of the kth pedestrian, and the generated memory output h_t to generate an augmented vector for the context representation,

$$\bar{c}_t^{(k)} = \tanh([C_t^{*,k}, h_t]), \tag{15}$$

which is used to predict the future trajectory of the pedestrian k,

$$Y_t = \text{LSTM}(p_{t-1}^k, \bar{c}_t^{(k)}, Y_{t-1}). \tag{16}$$

3.4   Coupling Multimodal Information to Improve Prediction

Using multimodal input streams allows us to capture different semantics that are present in the same scene, and compliment individual streams. For instance, in a surveillance setting, radar and video fusion is widely utilised [5, 30] as radar offers better coverage in the absence of visual light, however has a lower frame rate (<5 fps) compared to video (~25 fps) which records more fine grained motion.

As pointed out in [15], simply concatenating data from both streams leads to information loss as they contain information at different granularities. Hence it is vital to jointly back propagate among the information streams to learn the important aspects of each. Therefore, we capture streams through separate memory modules, and perform gated coupling of memory hierarchies.

We denote the two synchronised input modalities as I and R, where the trajectory of pedestrian k observed in stream I is denoted as X_I^k and the same pedestrian trajectory observed in stream R is given as X_R^k. We pass each stream separately through the local neighbourhood embedding mechanism proposed in Sect. 3.1 and generate vector embeddings $C_{t,I}^{*,k}$ and $C_{t,R}^{*,k}$ respectively. We embed these through individual memory blocks denoted as $M_{t,I}$ and $M_{t,R}$. In the absences of such trajectories (i.e due to poor coverage, occlusions, ...), we evaluate only the neighbourhood embeddings $C_{t,R}^{h,k}$ and use them as $C_{t,R}^{*,k}$.

After the hierarchical gated operations, let the memory output generated using Eq. 10 from memory $M_{t,I}$ at time instance t be denoted by $h_{t,I}$ and memory $M_{t,R}$ be denoted as $h_{t,R}$. For simplicity, in Fig. 3 we consider 2 input streams, however the proposed coupling mechanism is flexible and is able to handle any number of modalities.

Motivated by [2, 24] we perform gated modality fusion such that,

$$\begin{aligned}
\bar{h}_{t,I} &= \tanh(W_I h_{t,I}), \\
\bar{h}_{t,R} &= \tanh(W_R h_{t,R}), \\
\nu &= \sigma(W_\nu [\bar{h}_{t,I}, \bar{h}_{t,R}]),
\end{aligned} \tag{17}$$

where W_I and W_R are the weights for the respective memories and W_ν is the weight of the fusion gate. This can be seen as performing attention from one modality over the other where each modality determines the amount of information to flow from the other. We then obtain the combined feature vector,

$$h_t = \nu \bar{h}_{t,I} + (1 - \nu)\bar{h}_{t,R}, \tag{18}$$

and augment Eq. 15 to utilise information from both streams,

$$\bar{C}_t^{(k)} = \tanh([C_{t,I}^{*,k}, C_{t,R}^{*,k}, h_t]), \tag{19}$$

and predict the future trajectory using Eq. 16. In contrast to [5, 30] where simple concatenation of multimodal data is used, the proposed multi-memory architecture allows the model to store salient information of individual streams separately and propagate it effectively to the decision making process. We denote this model as $SMN(I + R)$ as it couples I and R streams to the SMN model.

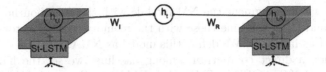

Fig. 3. Coupling multimodal information through multiple memory modules. The information from each modality is stored separately. Note that the figure shows only the top most layer in each memory.

4 Evaluation and Discussion

4.1 Datasets

We present the experimental results for the single modal framework on the publicly available New York Grand Central (GC) [34] dataset. The Grand Central dataset consist of 12,600 trajectories. For training, testing and validation we use the same splits defined in [16]. Due to the unavailability of public multimodal pedestrian trajectory data, we introduce a new large scale dataset[1]. Pedestrian trajectories from a CCTV surveillance feed (I) and Radar (R) streams, for 32 h, were collected and synchronised. Please refer to the supplementary material for statistics, calibration and synchronisation details of the dataset.

4.2 Evaluation Metrics

Following [1,16] we evaluate the performance with the following 3 error metrics: Average displacement error (ADE), Final displacement error (FDE) and Average non-linear displacement error (n-ADE). Please refer to [1,16] for details.

4.3 Evaluation of Trajectory Prediction with Single Modal Data

The evaluation of single modal trajectories is conducted on the GC dataset [34]. We compare our model against 6 state of the art baselines. The first baseline is the Social Force (SF) model of [33]. It requires the destination of the pedestrian as input, and a linear SVM is trained with ground truth destination areas for this task. The next baseline is the Social LSTM (So-LSTM) model of [1]. It requires the neighbourhood size as a hyper-parameter and is set to 32px. The soft + hardwired attention model from [16] (SHA) does not posses any memory and computes the trajectory prediction by modelling the local neighbourhood of the pedestrian of interest. We also consider the Tree Memory Network (TMN) [14] which models the memory as a tree structure. This model uses the hyper parameter δ, which defines the length of the memory as it structures a flat memory vector as a tree. We also evaluate the Neural Map (NM) model introduced in [28]. The pedestrian of interest's trajectory is embedded using a soft attention mechanism as defined in Sect. 3.1 and is stored in the memory. To provide a fair

---

[1] Available at https://github.com/qutsaivt/SAIVTMultiSpectralTrajectoryDataset.

comparison, we also augment the NM module with the neighbourhood embeddings, $C_t^{s,k}$ and $C_t^{h,k}$, combine these with the memory output vector generated from the NM as in Eq. 15. We define this model as NMA.

To provide a direct comparison among baselines we set the hidden state dimensions of So-LSTM, SHA, TMN, NM, NMA and the proposed SMN model to be 30 units. As the models NM, NMA and SMN have a map width (W) and map height (H) as hyper-parameters, we evaluate different memory sizes. Similarly, for TMN we evaluate different memory lengths δ. Please refer to the supplementary material for those evaluations. Best results are shown in Table. 1. To evaluate the relative performance of each model, we observe the trajectory for 20 frames and predict the future trajectory for the next 20 frames.

Table 1. Quantitative results with the GC dataset [34] for Social Force (SF) [33], Social LSTM (So-LSTM) [1], Soft + Hardwired Attention (SHA) [16], Tree Memory Network (TMN) [14], Neural Map (NM) [28], Neural Map Augmented (NMA) and the proposed Structured Memory Network (SMN) models. In all the methods forecast trajectories are of length 20 frames. The measured error metrics are as in Sect. 4.2.

Method	Metric		
	ADE	FDE	n-ADE
SF [33]	3.364	5.808	3.983
So-LSTM [1]	1.990	4.519	1.781
SHA [16]	1.096	3.011	0.985
TMN ($\delta = 64$) [14]	2.982	4.989	2.780
NM (W = H = 64) [28]	2.505	4.151	2.432
NMA (W = H = 64)	1.466	3.811	1.445
SMN (W = H = 128)	**0.891**	**2.899**	**0.814**

From the results tabulated in Table 1 we observe poor performance in the SF model due to its lack of capacity to model long-term history. Models So-LSTM and SHA utilise short-term history from the pedestrian of interest and the local neighbourhood and generate improved predictions accordingly.

The lack of spatial structure and context modelling in the TMN module leads to it's poor performance despite it's long-term history modelling capacity. Comparing the NM and NMA models, the performance increase from NM and NMA is due to the addition of local context, highlighting the importance of capturing both long and short-term context. The NMA model attains improved performance due to the improved modelling of the local neighbourhood, and the structured memory; however when compared to the SHA model it fails to propagate salient spatiotemporal information from the structured memory to aid the decision making. This is due to the static kernel used when generating the memory output. In contrast, we map the memory output hierarchically using the proposed St-LSTM cells and propagate salient information to the upper layer, enabling efficient information transfer to the prediction model. The proposed gated architecture considers the evolution of memory over time, where multiple humans can interact with the environment, changing the state of multiple spatial locations. Hence we are able to generate dynamic responses instead of passing the information through a static convolution kernel as in NM and NMA; enabling superior performance even with large memory sizes.

We present a qualitative evaluation of the proposed SMN model with the SHA and NMA baselines in Fig. 4. We selected these baselines as they provide the highest comparative results. The trajectories are shown in the first column where the observed part of the trajectory is denoted in green, the ground truth observations in blue, neighbouring trajectories are in purple and the predicted trajectories are shown in red (SMN), yellow (SHA) and orange (NMA).

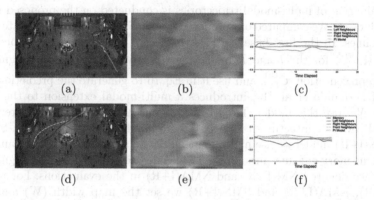

Fig. 4. Qualitative results for the GC dataset [34]: Given (in green), Ground Truth (in blue), Neighbouring (in purple) and Predicted trajectories from SMN model (in red), from SHA model (in yellow), from NMA model (in orange) along with the respective structured memory activations and relative activation contribution of each component in the prediction module. Please note that in the structured memory activations the intensity of the colour represents the degree of the activation and has been manually aligned with the figure in the first column for the clarity of visualisation. (Color figure online)

When observing the qualitative results it can be clearly seen that the proposed SMN model generates better trajectory predictions compared to the state-of-the-arts. For instance in Fig. 4 (a) and (d) we observe significant deviation of the predictions of SHA and NMA models from the ground truth. However the proposed SMN model has been able to anticipate the pedestrian motion more accurately with the improved context modelling.

From the memory activation visualisations, it is evident more attention is given to cells surrounding the trajectory of the pedestrian of interest and the neighbours. Varying levels of attention are given to the cells occupied by the neighbours. However by passing this information through the proposed gated St-LSTM cells the proposed model is able to learn salient information among the passed activations from the layer below. This can be verified by observing the relative activation plots presented in the 3rd column of Fig. 4. While in general more attention is given to the encoded trajectory information from the pedestrian of interest (PI model), in cases such as Fig. 4 (c) more attention is given to the historic neighbourhood embeddings present in memory, where as in Fig. 4 (f) the model gives more attention to the neighbours. This verifies our hypothesis that both current context information encoded within the motion of pedestrian of interest and the neighbouring trajectories as well as the information from the long-term history that preserves the structural integrity, is vital for prediction. Refer to the supplementary material for more qualitative evaluations.

4.4   Evaluation of Trajectory Prediction with Multimodal Data

The evaluation of multi-modal trajectories is conducted on the proposed multi-modal dataset. We compared our proposed model, SMN(I+R), with 4 state of the art baselines. In the first baseline, SHA(I+R), we concatenate the embeddings $C_{I,t}^{*,k}$ and $C_{R,t}^{*,k}$ for the I and R modalities directly to generate the augmented vector representation, $\bar{C}_t^{(k)}$, and use it in Eq. 16 to generate the prediction. The work of Fernando et al. [15] introduces a multi-modal extension to the TMN module. We use this model, TMN(I+R), as our next baseline. We extend the NM and NMA architectures (see Sect. 4.3) to handle multi-modal data. Similar to TMN(I+R) model we use multiple memories to store each input streams and pass the memory outputs through Eq. 17 to generate predictions. The augmented models are denoted NM(I+R) and NMA(I+R) in the evaluations. For models NM(I+R), NMA(I+R) and SMN(I+R) we set the map width (W) and map height (H) to be 128 and for TMN(I+R) we set the memory length $\delta = 64$, as this provided the best accuracies in Sect. 4.3.

Following the previous experiment, we observe the trajectory for 20 frames and predict the trajectory for the next 20 frames. After filtering out short and fragmented trajectories we are left with 40,800 trajectories. We randomly selected 28,560 trajectories for training, 10,200 for testing and 2,040 for validation (Table 2).

Similar to the evaluations in Sect. 4.3, we observe poor performance from TMN(I+R) and NM(I+R) due to their inability to capture local neighbourhood information. However we observe a significant reduction in the performance gap between SHA(I+R) and NMA(I+R), compared to the that in Table 1, which is a result of the naive fusion method used in the former model. SHA(I+R) simply concatenates the two modes together, and as such the model lacks the capacity to capture salient information from individual modes. In contrast, by capturing long-term temporal dependencies of the two modalities, the memory based coupling mechanism yields better predictions. We further augment this process in SMN(I+R) by utilising the St-LSTM cells to hierarchically capture salient information from each mode. This enables the model to jointly back propagate through the two modalities and learn the strengths and weaknesses of each, effectively complimenting the prediction module with the additional information

Table 2. Quantitative results with the proposed multimodal dataset for, Soft + Hard-wired Attention (SHA(I+R)) [16], Tree Memory Network (TMN(I+R)) [14], Neural Map (NM(I+R)) [28], Neural Map Augmented (NMA(I+R)) and the proposed Structured Memory Network (SMN(I+R)) models. In all the methods forecast trajectories are of length 20 frames. Error metrics are defined in Sect. 4.2.

Method	Metric		
	ADE	FDE	n-ADE
SHA(I+R) [16]	1.245	1.654	1.454
TMN(I+R) [14]	2.901	3.169	3.001
NM(I+R) [28]	2.015	2.741	2.344
NMA(I+R)	1.325	1.814	1.558
SMN(I+R)	**0.979**	**0.998**	**1.036**

stream. Please refer to supplementary material for qualitative evaluations of the proposed SMN(I+R) model with the SHA(I+R) and NMA(I+R) baselines.

4.5 Ablation Experiments

To further demonstrate the effectiveness of our proposed fusion approach, we conduct a series of ablation experiments, identifying the crucial components of the proposed architecture. In the same settings as the experiment in Sect. 4.4, we compare the SMN(I+R) (proposed) method to a series of counterparts constructed by removing components of the model as follows:

- **SA(I):** Uses only the soft attention context vector, $C_t^{s,k}$, and data from the image stream (I) for trajectory prediction.
- **SHA(I):** Uses both soft ($C_t^{s,k}$) and hardwired ($C_t^{h,k}$) attention vectors and data from image stream (I) for trajectory prediction
- **SMN(I):** Uses the proposed SMN model and data from Image (I) stream.
- **SA(R):** Similar to SA-I but uses data from the Radar (R) stream.
- **SHA(R):** Similar to SHA-I but uses data from the Radar (R) stream.
- **SMN(R):** Similar to SMN-I but uses data from the Radar (R) stream.
- **SA(I+R):** SA model that directly concatenates $C_{t,I}^{s,k}$ and $C_{t,R}^{s,k}$ and generates a vector embedding for Eq. 16.
- **SHA(I+R):** SHA model that directly concatenates $C_{t,I}^{*,k}$ and $C_{t,R}^{*,k}$ and generates a vector embedding for Eq. 16.
- **SMN(I+R):** Uses the model proposed in Sect. 3.4.

Note that for all *SMN* models we used $W = H = 128$.

The results of our ablation study are presented in Table 3. Models SA(I) and SA(R) perform poorly due to their inability to oversee the neighbourhood context. We observe improved performance in SHA(I) and SHA(R) with the introduction of information from neighbouring pedestrians. The combined information from both modalities contributes to the performance gain we observe in SHA(I+R) over the unimodal counterparts, verifying the observations in [4,10].

Table 3. Ablation experiment evaluations

Method	Metric		
	ADE	FDE	n-ADE
SA(I)	2.012	3.011	2.190
SHA(I)	1.235	2.731	1.442
SMN(I)	1.029	1.104	1.092
SA(R)	2.259	3.312	2.261
SHA(R)	1.613	3.070	1.892
SMN(R)	1.198	1.330	1.288
SA(I+R)	1.334	1.813	1.579
SHA(I+R)	1.245	1.654	1.454
SMN(I+R)	**0.979**	**0.998**	**1.036**

Comparing the unimodal SMN(I) and SMN(R) models with the multimodal SHA(I+R) model, the former outperforms the latter by a significant margin, emphasising the importance of capturing long-term spatial context, and propagating the information effectively to the prediction model. The introduction of a secondary modality in SMN(I+R) further improves the prediction accuracy.

We would like to further compare the results obtained from the individual models in the I and R streams. We observe a performance boost in modality I, due to the finer granularity present in the CCTV stream due to the higher frame rate, compared to the radar stream. Hence extracted trajectories are smoother compared to the trajectories from modality R, making it easier to model.

4.6 Implementation Details

We use Keras [7] for our implementation. The SMN and SMN(I+R) modules do not require any special hardware (i.e. GPUs) to run. The SMN(W = H = 128) model has 152 K trainable parameters, and SMN(I+R) (W = H = 128) has 358 K. We ran the test set in Sect. 4.3 on a single core of an Intel Xeon E5-2680 2.50 GHz CPU and the SMN algorithm was able to generate 1000 predicted trajectories with 40, 2 dimensional data points (i.e. using 20 observations to predict the next 20 data points) in 2.791 s. In a similar experiment with the test set in Sect. 4.4 we were able to generate 1000 predicted trajectories in 11.722 s.

5 Conclusions

In this paper we propose a method to anticipate complex human motion by analysing structural and temporal accordance. We extend the standard pedestrian trajectory prediction framework by introducing a novel model, Structured Memory Network (SMN), which is able to oversee the long-term history, preserving the structural integrity and improving prediction of pedestrian motion.

As an extension to the proposed SMN model, we contribute a novel data driven method to capture salient information from multiple modalities and demonstrate how to incorporate this to enhance prediction. Additionally, we introduce a novel multi-modal pedestrian trajectory dataset, collected from synchronised CCTV and Radar streams, and consisting of 40,000 pedestrian trajectories. Our evaluations on both single and multi-modal datasets demonstrate the capacity of the proposed SMN method to learn complex real world human navigation behaviour.

Acknowledgement. This research was supported in part by the Defence Science and Technology (DST) Group under the Defence Science Partnership Program. The authors acknowledge the contribution to the paper by Dr. Jason Williams, Senior Research Scientist, National Security, Intelligence, Surveillance and Reconnaissance Division of DST.

References

1. Alahi, A., Goel, K., Ramanathan, V., Robicquet, A., Fei-Fei, L., Savarese, S.: Social LSTM: human trajectory prediction in crowded spaces. In: CVPR, pp. 961–971 (2016)
2. Arevalo, J., Solorio, T., Montes-y Gómez, M., González, F.A.: Gated multimodal units for information fusion. In: ICLR (2017)
3. Bartoli, F., Lisanti, G., Ballan, L., Del Bimbo, A.: Context-aware trajectory prediction. arXiv preprint arXiv:1705.02503 (2017)
4. Bhatt, C.A., Kankanhalli, M.S.: Multimedia data mining: state of the art and challenges. Multimedia Tools Appl. **51**(1), 35–76 (2011)
5. Boström, M., Claesson, T.: Reducing false triggers in surveillance systems using sensor fusion. Master's theses in Mathematical Sciences (2017)
6. Brun, V.H., et al.: Progressive increase in grid scale from dorsal to ventral medial entorhinal cortex. Hippocampus **18**(12), 1200–1212 (2008)
7. Chollet, F.: Keras (2017). http://keras.io
8. Chorowski, J.K., Bahdanau, D., Serdyuk, D., Cho, K., Bengio, Y.: Attention-based models for speech recognition. In: NIPS, pp. 577–585 (2015)
9. Coates, A., Ng, A.Y.: The importance of encoding versus training with sparse coding and vector quantization. In: ICML, pp. 921–928 (2011)
10. Deng, L., Yu, D., et al.: Deep learning: methods and applications. Found. Trends® Sig. Process. **7**(3–4), 197–387 (2014)
11. Derdikman, D., Moser, E.I.: A manifold of spatial maps in the brain. Trends Cogn. Sci. **14**(12), 561–569 (2010)
12. Epstein, R.A., Patai, E.Z., Julian, J.B., Spiers, H.J.: The cognitive map in humans: spatial navigation and beyond. Nature Neurosci. **20**(11), 1504 (2017)
13. Fanselow, M.S., Dong, H.W.: Are the dorsal and ventral hippocampus functionally distinct structures? Neuron **65**(1), 7–19 (2010)
14. Fernando, T., Denman, S., McFadyen, A., Sridharan, S., Fookes, C.: Tree memory networks for modelling long-term temporal dependencies. Neurocomputing **304**, 64–81 (2018)
15. Fernando, T., Denman, S., Sridharan, S., Fookes, C.: Going deeper: autonomous steering with neural memory networks. In: ICCV, pp. 214–221 (2017)

16. Fernando, T., Denman, S., Sridharan, S., Fookes, C.: Soft+ hardwired attention: an LSTM framework for human trajectory prediction and abnormal event detection. arXiv preprint arXiv:1702.05552 (2017)
17. Fernando, T., Denman, S., Sridharan, S., Fookes, C.: Learning temporal strategic relationships using generative adversarial imitation learning. In: IFAAMAS (2018)
18. Fernando, T., Denman, S., Sridharan, S., Fookes, C.: Task specific visual saliency prediction with memory augmented conditional generative adversarial networks. In: WACV, pp. 1539–1548. IEEE (2018)
19. Fernando, T., Denman, S., Sridharan, S., Fookes, C.: Tracking by prediction: a deep generative model for multi-person localisation and tracking. In: WACV (2018)
20. Gobet, F., et al.: Chunking mechanisms in human learning. Trends Cogn. Sci. **5**(6), 236–243 (2001)
21. Huang, Y., Wu, Q., Wang, L.: Learning semantic concepts and order for image and sentence matching. arXiv preprint arXiv:1712.02036 (2017)
22. Kaiser, Ł., Sutskever, I.: Neural GPUs learn algorithms. In: ICLR (2016)
23. Kiela, D., Bottou, L.: Learning image embeddings using convolutional neural networks for improved multi-modal semantics. In: EMNLP, pp. 36–45 (2014)
24. Kiela, D., Grave, E., Joulin, A., Mikolov, T.: Efficient large-scale multi-modal classification. arXiv preprint arXiv:1802.02892 (2018)
25. Kiros, R., Salakhutdinov, R., Zemel, R.: Multimodal neural language models. In: ICML, pp. 595–603 (2014)
26. Madl, T., Franklin, S., Chen, K., Trappl, R., Montaldi, D.: Exploring the structure of spatial representations. PloS one **11**(6), e0157343 (2016)
27. Malinowski, M., Fritz, M.: A multi-world approach to question answering about real-world scenes based on uncertain input. In: NIPS, pp. 1682–1690 (2014)
28. Parisotto, E., Salakhutdinov, R.: Neural map: structured memory for deep reinforcement learning. In: ICLR (2018)
29. Pei, D., Liu, H., Liu, Y., Sun, F.: Unsupervised multimodal feature learning for semantic image segmentation. In: IJCNN, pp. 1–6. IEEE (2013)
30. Roy, A., Gale, N., Hong, L.: Automated traffic surveillance using fusion of doppler radar and video information. Mathe. Comput. Model. **54**(1–2), 531–543 (2011)
31. Srivastava, N., Salakhutdinov, R.R.: Multimodal learning with deep boltzmann machines. In: NIPS, pp. 2222–2230 (2012)
32. Varshneya, D., Srinivasaraghavan, G.: Human trajectory prediction using spatially aware deep attention models. arXiv preprint arXiv:1705.09436 (2017)
33. Yamaguchi, K., Berg, A.C., Ortiz, L.E., Berg, T.L.: Who are you with and where are you going? In: CVPR, pp. 1345–1352. IEEE (2011)
34. Yi, S., Li, H., Wang, X.: Understanding pedestrian behaviors from stationary crowd groups. In: CVPR, pp. 3488–3496 (2015)
35. Yuan, A., Li, X., Lu, X.: FFGS: feature fusion with gating structure for image caption generation. In: Yang, J., et al. (eds.) CCCV 2017. CCIS, vol. 771, pp. 638–649. Springer, Singapore (2017). https://doi.org/10.1007/978-981-10-7299-4_53
36. Zou, H., Su, H., Song, S., Zhu, J.: Understanding human behaviors in crowds by imitating the decision-making process. arXiv preprint arXiv:1801.08391 (2018)

Classification

Multiple Instance Learning
with Bag-Level Randomized Trees

Tomáš Komárek[1]([⊠]) and Petr Somol[2,3]

[1] Faculty of Electrical Engineering, Czech Technical University in Prague,
Technická 2, Prague 6, Czech Republic
komartom@fel.cvut.cz
[2] Institute of Information Theory and Automation, Czech Academy of Sciences,
Pod Vodárenskou věží 4, Prague 8, Czech Republic
somol@utia.cas.cz
[3] Faculty of Management, University of Economics, Prague, Czech Republic

Abstract. Knowledge discovery in databases with a flexible structure poses a great challenge to machine learning community. Multiple Instance Learning (MIL) aims at learning from samples (called bags) represented by multiple feature vectors (called instances) as opposed to single feature vectors characteristic for the traditional data representation. This relaxation turns out to be useful in formulating many machine learning problems including classification of molecules, cancer detection from tissue images or identification of malicious network communications. However, despite the recent progress in this area, the current set of MIL tools still seems to be very application specific and/or burdened with many tuning parameters or processing steps. In this paper, we propose a simple, yet effective tree-based algorithm for solving MIL classification problems. Empirical evaluation against 28 classifiers on 29 publicly available benchmark datasets shows a high level performance of the proposed solution even with its default parameter settings. Data related to this paper are available at: https://github.com/komartom/MIDatasets.jl. Code related to this paper is available at: https://github.com/komartom/BLRT.jl.

Keywords: Multiple Instance Learning · Randomized trees
Classification

1 Introduction

Multiple Instance Learning (MIL) relaxes conditions for data representation. In MIL formalism, objects of interest are described by sets $\mathcal{B}_1, \mathcal{B}_2, \ldots$ (called bags) consisting of multiple feature vectors of an identical size $\mathcal{B} = \{\mathbf{x}_1, \mathbf{x}_2, \ldots\}$, $\mathbf{x} \in \mathcal{X}$ (called instances). Each bag is attributed output variable y (e.g. label $y \in \{0, 1\}$ in a binary classification problem). The goal is to infer function $\mathcal{F}$ from training dataset $\mathcal{D} = \{(\mathcal{B}, y)_1, (\mathcal{B}, y)_2, \ldots\}$ that can predict output variables for previously unseen bags $\mathcal{F}(\mathcal{B}) = y$.

© Springer Nature Switzerland AG 2019
M. Berlingerio et al. (Eds.): ECML PKDD 2018, LNAI 11051, pp. 259–272, 2019.
https://doi.org/10.1007/978-3-030-10925-7_16

This relaxed formalism has received a lot of attention since its first formulation [11]. One of the recent application domains, where MIL formalism seems to fit better than traditional Single Instance Learning (SIL), is network security [20]. For example, in the work [16] dealing with a binary classification of network users as either infected or clean based on monitoring their web requests, the MIL approach enabled to (i) *describe the problem more naturally* – each user is represented by a bag with instances corresponding to individual requests; representing users with single feature vectors would be too restrictive, since the number of requests can vary from user to user, (ii) *save ground truth acquisition costs* – security analysts preparing a training dataset did not have to label individual requests as infection-related or benign; it was enough to provide labels on the (higher) user/bag level, resulting in a significantly lower number of items that needed to be annotated, (iii) *improve classification performance* – a MIL classifier modeling the global contextual information across multiple requests was able to achieve a higher classification precision than a SIL classifier analyzing individual requests one by one.

Although many MIL classifiers have been developed over the years, most of them work well only within their specific application domains and are less effective over a wider range of problems [8] (demonstrated in Sect. 4). A recent survey of MIL classifiers [1] has pointed out that approaches extracting global *bag-level* information achieve a higher performance in general than their counterparts operating on *instance-level* trying to infer instance labels from the bag ones (discussed in Sect. 2). Following this observation, we propose a novel bag-level MIL algorithm (Sect. 3) as an extension of traditional single-instance randomized trees [15] to the multiple-instance setting. Ensembles of tree-based learners (e.g. Extremely randomized trees [15] or Breiman's Random Forests [5]) are known to perform very well across many domains even without a careful hyperparameter tuning [12], which makes a good case for research of a general MIL tree-based model formalism.

2   Related Work

A taxonomy proposed in the review [1] categorizes MIL classifiers into two groups according to the level at which they extract the discriminative information. The first *instance-level* category considers the discriminative information to lie at the level of individual instances. It is assumed that each instance $\mathbf{x}$ can be attributed binary label $y_{\mathbf{x}} \in \{0, 1\}$ and that positive bags contain at least one positive instance $y_{\mathbf{x}} = 1$. Negative bags are supposed to not contain any positive instance. The instance labels are unknown (even in the training dataset), but are inferred during the learning process. Most algorithms based on this assumption build instance-level classifier $f(\mathbf{x}) = y_{\mathbf{x}}$ and produce bag labels by a simple aggregation of instance level scores $\mathcal{F}(\mathcal{B}) = \max_{\mathbf{x} \in \mathcal{B}} f(\mathbf{x})$. This category is mostly represented by earlier works, starting with the APR algorithm [11] proposed in 1997. APR (Axis-Parallel Rectangle) algorithm considers all instances in positive bags to be positive and expands/shrinks an hyper-rectangle in the feature space to maximize the number of positive instances falling inside, while minimizing the number of negative ones. MI-SVM [2] initially also assumes that all instances in positive bags are positive, but then maximizes a margin between the positive and negative bags by selecting a single instance to represent each bag. EM-DD [25] uses Expectation-Maximization (EM) algorithm to search for the maximum of diverse density (DD) measure. A location in the feature space has a high DD measure if the number of nearest instances from different positives bags is high and from negative bags low. Other approaches are based e.g. on boosting (MILBoost [23]) or deterministic annealing [14,17].

The second *bag-level* category (mostly represented by later works) considers the discriminative information to lie at the level of bags. Representatives of this category do not use the notion of instance labels, which does not reflect the reality in many applications [1], but rather treat bags as whole entities. That is, a bag-level classifier $\mathcal{F}(\mathcal{B})$ extracts the information from the whole bag to make a decision about the class of $\mathcal{B}$, instead of aggregating individual instance-level decisions. Bag-level methods are further categorized as either *bag-space* or *embedded-space*. Since bags are non-vectorial objects, the *bag-space* methods define a distance function (or kernel) that can compare any two bags, e.g. $\mathcal{B}_i$ and $\mathcal{B}_j$, and plug this function into a distance-based classifier such as k-NN or SVM. Examples of such functions are the minimal Hausdorff distance $d(\mathcal{B}_i, \mathcal{B}_j) = \text{argmin}_{\mathbf{x}^i \in \mathcal{B}_i, \mathbf{x}^j \in \mathcal{B}_j} \|\mathbf{x}^i - \mathbf{x}^j\|$ measuring distance between the closest instances or the Earth Movers Distance (EMD) $d(\mathcal{B}_i, \mathcal{B}_j) = \frac{\sum_i \sum_j w_{ij} \|\mathbf{x}^i - \mathbf{x}^j\|}{\sum_i \sum_j w_{ij}}$, where weights w_{ij} are obtained through an optimization process that globally minimizes the distance subject to some constrains, see [24] for details. On the other hand, the *embedded-space* methods define a transformation mapping bags into single fixed-size vectors, which effectively converts the MIL problem into the standard SIL one. In the work of MI-Kernel [13], each bag is transformed by calculating simple statistics like the maximum, minimum or mean across all feature dimensions and concatenating the results into a single vector. MILES [7] maps each bag into a feature space defined by similarities to instances in the training bags. A sparse 1-norm SVM is then applied to select only the discriminative features (i.e. instances) and construct the classifier. Bag dissimilarity [9], on the other hand, measures similarities to the training bags rather than instances.

Most prior tree-based works fall into the *instance-level* category. MIForest [17] uses the deterministic annealing approach to uncover the instance labels during the tree growing. MITI [4] introduces a weighted Gini impurity measure and modifies the tree induction procedure to prioritize expansion of nodes with positive instances. MIOForest [21] extends MITI by implementing non-linear splitting rules instead of the traditional axis-orthogonal ones and by optimal combining of individual tree outputs within the forest. To our knowledge, the only tree-based MIL classifier that operates on the level of bags is RELIC [19]. Since we consider RELIC as the closest prior work, we discuss the differences with respect to the proposed solution in detail in Sect. 3.

3   Bag-Level Randomized Trees

The bag-level randomized trees (BLRT) are trained according to the classical top-down greedy procedure for building ensembles of unpruned decision trees. Individual tree learners recursively partition a training dataset by choosing binary splitting rules until pure sample sets are obtained.

The key difference, however, lies in the conditions that are evaluated inside the splitting nodes. While nodes of standard single-instance decision trees (Eq. 1) test only whether feature f of a given sample is greater than certain value v, nodes of the proposed MIL trees also count the number of instances within the sample (i.e. bag) that accomplish the condition. This absolute count is then normalized by bag size $|\mathcal{B}|$ and compared to value $r \in [0, 1)$ (Eq. 2).

$$
\mathcal{N}_{\mathrm{SIL}}(\mathbf{x}; f, v) = \begin{cases} \text{left,} & \text{if } x_f > v, \\ \text{right,} & \text{otherwise.} \end{cases} \tag{1}
$$

$$
\mathcal{N}_{\mathrm{MIL}}(\mathcal{B}; \underbrace{f, v, r}_{\Phi}) = \begin{cases} \text{left,} & \text{if } \left[\dfrac{1}{|\mathcal{B}|} \sum_{\mathbf{x} \in \mathcal{B}} \mathbb{1}\left[x_f > v\right] \right] > r, \\ \text{right,} & \text{otherwise.} \end{cases} \tag{2}
$$

Parameter r denotes a relative count of instances $\mathbf{x}$ inside bag $\mathcal{B}$ that must satisfy the inner condition $x_f > v$ to be the whole bag passed to the left branch. It is the only additional parameter that needs to be learned from the training data together with f and v. Symbol $\mathbb{1}$ stands for an indicator function that equals one if its argument is true and zero otherwise.

Note that if bags are of size one, nodes $\mathcal{N}_{\mathrm{MIL}}$ behave like the traditional $\mathcal{N}_{\mathrm{SIL}}$ regardless the value of r parameters. The next special case is when the relative count takes extreme values, i.e. $r \in \{0, 0.\bar{9}\}$[1]. The proposed algorithm then becomes equivalent to the prior art solution known as RELIC [19]. Under this condition, the splitting rules act as either the universal or the existential quantifier. In particular, bags are tested in two possible ways: if there exists at least one instance that fulfills the inner condition or if the condition is satisfied by all instances. An experiment in Sect. 4 (Fig. 2), however, shows that the ability of the proposed algorithm to test situations also between these two extreme cases is highly beneficial on many datasets.

Search of optimal splitting parameters $\Phi^* = (f, v, r)$ during the tree growth is implemented in a randomized manner. At each node construction, a set of candidate splitting rules $\mathcal{R} = \{\Phi_1, \ldots\}$ is generated (based on local training subset $\mathcal{S} \subseteq \mathcal{D}$) among which the best one Φ^* is selected according to a score obtained by an impurity measure such as Information gain [18] or Gini impurity [6]. Specifically, for each feature f out of K randomly selected ones, T values of parameter v are drawn uniformly from interval $[x_f^{\min}, x_f^{\max})$, where $x_f^{\min}$ and $x_f^{\max}$ denote

---

[1] Technically, the value of $0.\bar{9}$ should be 1 minus the smallest representable value.

the minimum and the maximum value of feature f across all bags within the local sample set. For each such pair (f, v), other T values of parameter r are generated uniformly from interval $[0, 1)$. In total, there are $K \times T \times T$ candidate splitting rules at maximum[2]. A detail description of the tree induction procedure is given in Algorithm 1 in the form of pseudo code.

The above randomized approach is adopted from Extremely[3] randomized trees [15] and generalized to MIL setting by adding the third parameter r (i.e. the relative count). Unlike CART algorithm, used e.g. in Breiman's Random Forests [5], the randomized search does not require to go over all possible splitting points on selected features, which could be prohibitively expensive in this MIL variant of trees. Furthermore, the explicit randomization in combination with ensemble averaging makes the decision boundary more smooth, resulting in models with better or equal accuracy than that of Random Forests [15].

Algorithm 1 builds M fully grown decision trees. Each tree is trained on the whole sample set rather than a bootstrap replica as realized e.g. in Random Forests. Training on the full original sample set minimizes bias. Variance is reduced by the strong randomization in the splitting parameters combined with the output aggregation across multiple trees. From the computational point of view, the time complexity of the learning procedure is, assuming balanced trees, $\Theta(MKT^2 N_I \log N_B)$, where N_B and N_I denote the number of bags and the number of instances within the bags, respectively[4].

In the testing mode, assuming a binary classification problem (i.e. $y \in \{0, 1\}$), predictions of individual trees are aggregated by a simple arithmetic average to produce final prediction score $\hat{y} \in [0, 1]$.

4   Experiments

The proposed algorithm is evaluated on 29 real-life datasets that are publicly available e.g. on https://doi.org/10.6084/m9.figshare.6633983.v1. The datasets with meta descriptions are listed in Table 1. These classification problems are well known and cover a wide range of conditions in terms of application domains (molecule, scene, image, text, audio spectrogram, etc.), ratios of positive and negative samples (e.g. imbalanced Corel datasets), feature dimensions (from 9 to 6519) and average numbers of bag instances (from 4 to 185). For more details about the datasets we refer the reader to a recent study of MIL datasets [8].

The same collection of datasets was also used in the evaluation of 28 MIL classifiers (including their variants) implemented in the MIL matlab toolbox [22]. The last two columns of Table 1 summarize the results from the evaluation available also through http://homepage.tudelft.nl/n9d04/milweb/. We report only those classifiers that achieved the highest performance by means of AUC

---

[2] If $x_f^{\min}$ equals to $x_f^{\max}$, no splitting rules are generated on feature f.

[3] Term *extremely* corresponds to setting $T = 1$.

[4] When bags are of size one (i.e. $N_I = N_B = N$) and $T = 1$, the complexity is equivalent to the complexity of Extremely randomized trees $\Theta(MKN \log N)$.

Algorithm 1. Induction algorithm for bag-level randomized trees (binary classification problem $y \in \{0,1\}$ and numerical features are assumed).

Function Train($\mathcal{D}; M, K, T$)

   Input : A training set $\mathcal{D} = \{(\mathcal{B}, y)_1, \ldots\}$,
           a number of trees to grow M,
           a number of randomly selected features K,
           a number of generated thresholds T.
   Output: An ensemble of bag-level randomized trees $\mathcal{E}$.

   $\mathcal{E} = \emptyset$
   foreach tree in $1 \ldots M$ **do**
       $\lfloor$ $\mathcal{E} = \mathcal{E} \cup \{\text{BuildTree}(\mathcal{D}; K, T)\}$
   return $\mathcal{E}$

Function BuildTree($\mathcal{S}; K, T$)

   Input : A local training subset $\mathcal{S} \subseteq \mathcal{D}$.
   Output: A node with left and right followers or a leaf.

   if all y in $\mathcal{S}$ are equal **then**
       $\lfloor$ **return** leaf(y)
   $\mathcal{R} = \text{GenerateCandidateSplittingRules}(\mathcal{S}; K, T)$
   $\Phi^* = \text{argmax}_{\Phi \in \mathcal{R}} Score(\mathcal{S}, \Phi)$
   $\mathcal{S}_{\text{left}} = \{\mathcal{B} \in \mathcal{S} | \mathcal{N}_{\text{MIL}}(\mathcal{B}; \Phi^*) = \text{left}\}$
   $\mathcal{S}_{\text{right}} = \mathcal{S} \setminus \mathcal{S}_{\text{left}}$
   if $\mathcal{S}_{left} = \emptyset$ *or* $\mathcal{S}_{right} = \emptyset$ **then**
       $\lfloor$ **return** leaf($\frac{1}{|\mathcal{S}|} \sum_{y \in \mathcal{S}} y$)
   return node($\mathcal{N}_{\text{MIL}}(\,\cdot\,; \Phi^*)$, BuildTree($\mathcal{S}_{\text{left}}$), BuildTree($\mathcal{S}_{\text{right}}$))

Function GenerateCandidateSplittingRules($\mathcal{S}; K, T$)

   Output: A set of candidate splitting rules $\mathcal{R} = \{\Phi_1, \ldots\}$.

   $\mathcal{R} = \emptyset$
   foreach feature f in K randomly selected ones (without replacement) **do**
       Find extremes x_f^{min} and x_f^{max} on given feature f across all bags $\mathcal{B} \in \mathcal{S}$
       if $x_f^{min} \neq x_f^{max}$ **then**
           foreach value v in T uniformly drawn values from $[x_f^{min}, x_f^{max})$ **do**
               foreach value r in T uniformly drawn values from $[0, 1)$ **do**
                  $\lfloor$ $\mathcal{R} = \mathcal{R} \cup \{\Phi\}$, $\Phi = (f, v, r)$
   return $\mathcal{R}$

---

metric[5] at least on one problem. This selection yields to 13 classifiers that are listed in Table 2 together with references to their original papers.

---

[5] Area Under a ROC Curve showing the true positive rate as a function of the false positive rate. AUC is agnostic to class imbalance and classifier's threshold setting.

Table 1. Metadata about 29 used datasets together with classification scores and standard deviations presented in percent (AUC × 100). Best results are in bold face. Stars denote statistically significant ($\alpha = 0.05$) differences according to Welch's t-test.

Dataset				BLRT		Best prior art	
Name	Bags +/−	Feat.	Avg.inst.	AUC		AUC	Algorithm
Musk1	47/45	166	5	**96.8** (1.6)	*	92.9 (1.3)	MI-SVM g
Musk2	39/63	166	65	91.2 (1.8)	*	**95.3** (1.5)	MILES g
C. African	100/1900	9	4	**96.2** (0.2)		95.7 (0.4)	minmin
C. Beach	100/1900	9	4	**98.9** (0.2)	*	90.7 (0.9)	RELIC
C. Historical	100/1900	9	4	**99.2** (0.1)	*	92.9 (0.5)	EM-DD
C. Buses	100/1900	9	4	97.4 (0.2)	*	**99.5** (0.1)	minmin
C. Dinosaurs	100/1900	9	4	96.4 (0.1)	*	**99.9** (0.0)	MILES p
C. Elephants	100/1900	9	4	**97.1** (0.1)		96.9 (0.2)	minmin
C. Food	100/1900	9	4	**99.4** (0.1)	*	97.2 (0.2)	minmin
Fox	100/100	230	7	**73.3** (1.4)	*	69.8 (1.7)	MILES g
Tiger	100/100	230	6	**92.6** (1.0)	*	87.2 (1.7)	MILES g
Elephant	100/100	230	7	**95.8** (0.9)	*	91.1 (1.2)	MI-SVM g
Protein	25/168	9	138	74.9 (2.3)	*	**89.5** (1.4)	minmin
Harddrive1	191/178	61	185	**99.6** (0.2)	*	98.6 (0.1)	MILES g
Harddrive2	178/191	61	185	**99.5** (0.1)	*	98.6 (0.2)	RELIC
Mutagenesis1	125/63	7	56	**92.1** (1.3)		91.0 (0.5)	cov-coef
Mutagenesis2	13/29	7	51	**86.0** (3.5)		84.0 (3.4)	EMD
B. BrownCreeper	197/351	38	19	**99.5** (0.0)	*	96.5 (0.5)	RELIC
B. WinterWren	109/439	38	19	**99.8** (0.1)	*	99.3 (0.1)	summin
B. Pacifics.	165/383	38	19	**96.1** (0.2)	*	95.7 (0.3)	MILES g
B. Red-breasted.	82/466	38	19	**99.2** (0.2)	*	98.7 (0.4)	MILBoost
UCSBBreast.	26/32	708	35	84.5 (2.5)	*	**92.2** (3.1)	cov-coef
Newsgroups1	50/50	200	54	78.8 (2.6)	*	**89.8** (1.6)	meanmin
Newsgroups2	50/50	200	31	63.0 (4.0)	*	**78.1** (1.4)	meanmean
Newsgroups3	50/50	200	52	76.3 (4.1)		**77.4** (1.5)	meanmean
Web1	21/54	5863	29	86.5 (2.6)	*	**91.9** (0.0)	MI-SVM g
Web2	18/57	6519	30	50.7 (7.8)	*	**90.1** (0.5)	MI-SVM g
Web3	14/61	6306	34	73.4 (6.7)	*	**91.8** (0.4)	MI-SVM g
Web4	55/20	6059	31	80.0 (4.2)	*	**99.4** (0.0)	mean-inst

Since an exact experimental protocol is provided as a part of the referenced evaluation, we followed that protocol precisely. For each dataset, the protocol provides indexes of all splits in 5-times repeated 10-fold cross-validation. The material, however, does not specify any approach for hyperparameter optimization. Therefore, we evaluated the proposed model using default parameter settings. We set the number of trees to grow to $M = 500$ that should ensure convergence of the ensemble, the number of randomly selected features at each split to square root of the feature dimension $K = \sqrt{D}$, which is the default value for tree-based models, and the number of uniformly drawn values of v and r to $T = 8$.

Table 1 summarizes results from the evaluation in terms of average scores and standard deviations. Although among the prior art (28 MIL classifiers) there is no single winning solution and almost each problem is associated with a different classifier, which demonstrates the difficulty and diversity of MIL problems, the proposed model was able to outperform the best prior art algorithm for a given dataset in 17 out of 29 cases. The most significant improvement with respect to the prior art is on the group of image classification problems (Fox, Tiger and Elephant) and on some scene classification problems (Corel Beach and Corel Historical). On the other hand, the proposal is less accurate on text classification problems (Newsgroups[6] and Web), Protein and Breast datasets.

From Table 2 showing ranking of algorithms in the evaluation, it can be observed that the second best classifier with the lowest average rank (MI-SVM [2] with Gaussian kernel) ranked first only three times. Overall, the proposed algorithm works very reliably even without any hyperparamter tuning. Indeed, the proposal never ended on any of the last three positions, which is unique among all classifiers. It should be stressed though that not all prior art classifiers were evaluated on all 29 datasets. Column N/A of Table 2 indicates the number of missing evaluations.

Table 2. Number of times that each algorithm obtained each rank in the evaluation.

Rank position														Avg. rank	N/A	Algorithm
1.	2.	3.	4.	5.	6.	7.	8.	9.	10.	11.	12.	13.	14.			
17	1	0	3	2	2	1	1	0	0	2	0	0	0	3.1	0	BLRT ours
0	2	1	4	2	1	0	5	3	0	3	0	1	2	7.5	5	EM-DD [25]
0	1	1	6	4	1	1	1	3	2	2	3	1	1	7.5	2	MILBoost [23]
3	3	3	5	1	1	0	2	2	1	0	3	1	1	6.0	3	MI-SVM g [2]
1	0	2	2	5	2	6	6	1	2	0	1	0	0	6.5	1	MILES p [7]
2	5	6	1	0	2	1	1	2	0	1	3	0	4	6.5	1	MILES g [7]
1	5	3	1	2	2	2	1	4	2	2	1	1	2	6.9	0	mean-inst [13]
1	2	1	2	0	0	8	4	1	5	0	3	2	0	7.8	0	cov-coef [13]
0	3	2	0	0	3	0	1	3	6	3	4	4	0	8.9	0	RELIC [19]
2	3	2	1	3	4	4	0	0	2	5	0	1	1	6.7	1	minmin [9]
0	2	2	2	0	2	3	1	2	2	5	4	2	1	8.6	1	summin [9]
0	2	3	4	5	3	0	1	1	2	2	2	1	1	6.7	2	meanmin [9]
3	0	0	1	2	1	0	3	3	4	1	3	5	1	8.9	2	meanmean [9]
0	1	2	1	2	1	2	0	2	0	2	1	2	0	7.5	13	EMD [24]

---

[6] Except for Newsgroup3 where the proposal is competitive with the best prior art.

The non-parametric Wilcoxon signed ranks test [10] (testing whether two classifiers have equal performance) confirmed at significance level $\alpha = 0.05$ that the proposed bag-level randomized trees are superior to any other involved method. The test compared pair-wisely the proposal with every prior art method, each time using an intersection of their available datasets. The two most similarly performing methods are mean-inst [13] (p-value 0.037) and MI-SVM [2] with Gaussian kernel (p-value 0.022).

Besides the above evaluation, we also provide comparison to other tree-based MIL algorithms in Table 3, namely RELIC [19], MIOForest [21], MIForest [17], MITI [4] and RF [5]. Except for RELIC, all of them operate on *instance-level*; labels are assigned to instances and a bag is positive if it contains at least one positive instance. RF represents a naive approach where standard single-instance Random Forests are trained directly on instances that inherited bag labels. Reported classification accuracies in Table 3 are taken from the work of MIOForest [21]. Unfortunately, the classifiers were evaluated only on five pioneering datasets (i.e. Musk1-2 and the image classification problems) and their implementations are not publicly available. As can be seen from Table 3, the proposal clearly outperforms all the prior tree-based MIL solutions on these datasets.

Table 3. Comparison with other tree-based MIL classifiers. Scores refer to accuracy in percent (ACC $\times$ 100). The prior art results are taken from the work of MIOForest [21].

	BLRT[ours]	RELIC	MIOForest	MIForest	MITI	RF
Musk1	**96**	83	89	85	84	85
Musk2	**91**	81	87	82	88	78
Fox	**75**	66	68	64	N/A	60
Tiger	**90**	78	83	82	N/A	77
Elephant	**93**	80	86	84	N/A	74

In Fig. 1, we assess various variants of the proposed algorithm. Dots in each subplot represent the 29 datasets. Their (x, y) coordinates are given by AUC scores obtained by the tested variants. If a dot lies on the diagonal (i.e. $x = y$ line), there is no difference between the two tested variants from that particular dataset perspective. The first two **subplots (a-b)** illustrate the influence of the ensemble size. It can be observed that it is significantly better to use 100 trees than 10 trees, but building 500 trees usually does not bring any additional performance. Also, according to **subplot (c)**, there is almost no difference between Information gain [18] and Gini impurity measure [6] scoring functions for selecting splitting rules. The next **subplot (d)** indicates that using higher values (e.g. 16 instead of the default 8) for parameter T (i.e. the number of randomly generated values for parameters v and r at each split) might lead to over-fitting on some datasets. In **subplot (e)** we tested a variant with an absolute count[7] instead of the relative one used in Eq. 2. The variant with the absolute count, however, performed significantly worse on the majority of datasets. The last **subplot (f)** compares the proposed algorithm with its simplified alternative, where traditional Random Forests are trained on a non-optimized bag representation. To do so, all bags $\{\mathcal{B}_1, \mathcal{B}_2, \ldots\}$ are transformed into single feature vectors $\{\mathbf{b}_1, \mathbf{b}_2, \ldots\}$ of values $b_{\mathcal{B}}^{(f,v)} = \frac{1}{|\mathcal{B}|} \sum_{\mathbf{x} \in \mathcal{B}} \mathbb{1} [x_f > v]$, where for each feature f eight equally-spaced values v are generated from interval $[x_f^{\min}, x_f^{\max})$ that is estimated beforehand on the whole training sample set. As a result, the non-optimized bag representation is eight times longer than the dimensionality of instances. As can be seen from subplot (f), the Random Forests trained on the non-optimized bag representation are far inferior to the proposed algorithm on all datasets except one. This result highlights the importance to simultaneously optimize the representation parameters with the classification ones as proposed in Sect. 3.

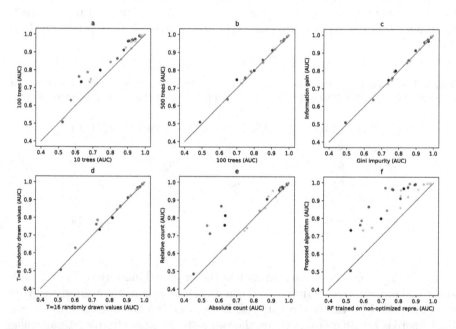

Fig. 1. Pair-wise comparisons of various configurations of the proposed algorithm on the 29 datasets. Subplots (a-b) illustrate the influence of the ensemble size, subplot (c) the impact of selected impurity measure, subplot (d) the effect of parameter T, subplot (c) the performance of the variant with the absolute count and subplot (d) compares the proposed algorithm with RF trained on the non-optimized bag representation.

---

[7] The sum in Eq. 2 is not normalized by bag size $|\mathcal{B}|$ and parameter r can take values from interval $[1, \operatorname{argmax}_{\mathcal{B} \in \mathcal{S}} |\mathcal{B}|)$.

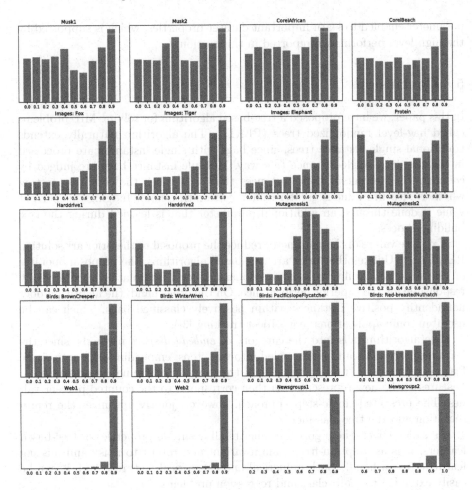

Fig. 2. Histograms of learned values of r parameters (Eq. 2). Datasets from the same source (e.g. Musk1-2, Harddrive1-2 and so forth) usually have very similar distributions that differ from the others.

Finally, Fig. 2 shows histograms of learned values of r parameters for some datasets. The first observation is that datasets from the same source (e.g. Fox, Tiger and Elephant) have very similar distributions. This demonstrates that the learned knowledge of randomized trees is not totally random as it might appear to be from the algorithm description. The next observation is that in almost all histograms (except for Mutagenesis problems) one or both extreme values of the parameter (i.e. $r \in \{0, 0.\bar{9}\}$) are the most frequent ones. As discussed in Sect. 3, the behavior of splitting rules (Eq. 2) with extreme values is approaching to the behavior of the universal or existential quantifier. On Web and Newsgroup datasets, this behavior is even dominant, meaning that the algorithm reduces to the prior art solution RELIC [19]. In the rest cases, however, the added parameter enabled to learn important dataset properties, which is supported by the high level performance reported in this section.

5   Conclusion

In this paper, we have proposed a tree-based algorithm for solving MIL problems called bag-level randomized trees (BLRT)[8]. The algorithm naturally extends traditional single-instance trees, since bags with single instances are processed in the standard single-instance tree way. Multiple instance bags are judged by counting the percent of their instances that accomplish the condition testing whether a feature value is greater than a certain threshold. Judging this percent value is done through an additional parameter that is learned during the tree building process.

Extreme values of the parameter reduce the proposal to the prior art solution RELIC [19]. Unlike other prior art tree-based algorithms, the proposal operates on the *bag-level*. Ability to analyze global bag-level information is most likely responsible for the superior performance. On the other hand, the algorithm does not identify positive instances within positively classified bags, which can be useful in some applications (e.g. object tracking [3]).

The algorithm falls into the category of *embedded-space* methods, since the learning procedure can be decoupled into two steps: embedding bags into single feature vectors and training traditional trees on top of the new representation. Features of the new representation then correspond to the counted percent values. The presented single-step approach, however, jointly optimizes the representation and the tree classifier.

As a side effect, the algorithm inherits all desirable properties of tree-based learners. It is assumption-free, scale invariant and robust to noisy and missing features. It can handle both numerical and categorical features. And, it can be easily extended to multi-class and regression problems.

Acknowledgments. This research has been supported by the Grant Agency of the Czech Technical University in Prague, grant No. SGS16/235/OHK3/3T/13.

References

1. Amores, J.: Multiple instance classification: review, taxonomy and comparative study. Artif. Intell. **201**, 81–105 (2013). https://doi.org/10.1016/j.artint.2013.06.003
2. Andrews, S., Tsochantaridis, I., Hofmann, T.: Support vector machines for multiple-instance learning. In: Proceedings of the 15th International Conference on Neural Information Processing Systems, NIPS 2002, pp. 577–584. MIT Press, Cambridge (2002). http://dl.acm.org/citation.cfm?id=2968618.2968690

---

[8] Source codes are accessible at https://github.com/komartom/BLRT.jl.

3. Babenko, B., Yang, M.H., Belongie, S.: Visual tracking with online multiple instance learning. In: 2009 IEEE Conference on Computer Vision and Pattern Recognition, pp. 983–990, June 2009. https://doi.org/10.1109/CVPR.2009.5206737
4. Blockeel, H., Page, D., Srinivasan, A.: Multi-instance tree learning. In: Proceedings of the 22nd International Conference on Machine Learning, ICML 2005, pp. 57–64. ACM, New York (2005). http://doi.acm.org/10.1145/1102351.1102359
5. Breiman, L.: Random forests. Mach. Learn. **45**(1), 5–32 (2001). https://doi.org/10.1145/1102351.1102359
6. Breiman, L., Friedman, J., Stone, C.J., Olshen, R.A.: Classification and Regression Trees. CRC Press, Boca Raton (1984)
7. Chen, Y., Bi, J., Wang, J.Z.: Miles: multiple-instance learning via embeddedinstance selection. IEEE Trans. Pattern Anal. Mach. Intell. **28**(12), 1931–1947 (2006). https://doi.org/10.1109/TPAMI.2006.248
8. Cheplygina, V., Tax, D.M.J.: Characterizing multiple instance datasets. In: Feragen, A., Pelillo, M., Loog, M. (eds.) SIMBAD 2015. LNCS, vol. 9370, pp. 15–27. Springer, Cham (2015). https://doi.org/10.1007/978-3-319-24261-3_2
9. Cheplygina, V., Tax, D.M., Loog, M.: Multiple instance learning with bag dissimilarities. Pattern Recogn. **48**(1), 264–275 (2015). https://doi.org/10.1016/j.patcog.2014.07.022. http://www.sciencedirect.com/science/article/pii/S0031320314002817
10. Demšar, J.: Statistical comparisons of classifiers over multiple data sets. J. Mach. Learn. Res. **7**, 1–30 (2006). http://dl.acm.org/citation.cfm?id=1248547.1248548
11. Dietterich, T.G., Lathrop, R.H., Lozano-Prez, T.: Solving the multiple instance problem with axis-parallel rectangles. Artif. Intell. **89**(1), 31–71 (1997). https://doi.org/10.1016/S0004-3702(96)00034-3. http://www.sciencedirect.com/science/article/pii/S0004370296000343
12. Fernández-Delgado, M., Cernadas, E., Barro, S., Amorim, D.: Do we need hundreds of classifiers to solve real world classification problems? J. Mach. Learn. Res. **15**(1), 3133–3181 (2014). http://dl.acm.org/citation.cfm?id=2627435.2697065
13. Gärtner, T., Flach, P.A., Kowalczyk, A., Smola, A.J.: Multi-instance kernels. In: Proceedings of the Nineteenth International Conference on Machine Learning, ICML 2002, pp. 179–186. Morgan Kaufmann Publishers Inc., San Francisco (2002). http://dl.acm.org/citation.cfm?id=645531.656014
14. Gehler, P.V., Chapelle, O.: Deterministic annealing for multiple-instance learning. In: Artificial Intelligence and Statistics, pp. 123–130 (2007)
15. Geurts, P., Ernst, D., Wehenkel, L.: Extremely randomized trees. Mach. Learn. **63**(1), 3–42 (2006)
16. Kohout, J., Komárek, T., Čech, P., Bodnár, J., Lokoč, J.: Learning communication patterns for malware discovery in https data. Expert Systems with Applications **101**, 129–142 (2018). https://doi.org/10.1016/j.eswa.2018.02.010. http://www.sciencedirect.com/science/article/pii/S0957417418300794
17. Leistner, C., Saffari, A., Bischof, H.: MIForests: multiple-instance learning with randomized trees. In: Daniilidis, K., Maragos, P., Paragios, N. (eds.) ECCV 2010. LNCS, vol. 6316, pp. 29–42. Springer, Heidelberg (2010). https://doi.org/10.1007/978-3-642-15567-3_3
18. Quinlan, J.R.: Induction of decision trees. Mach. Learn. **1**(1), 81–106 (1986). https://doi.org/10.1023/A:1022643204877
19. Ruffo, G.: Learning single and multiple instance decision trees for computer security applications. University of Turin, Torino (2000)

20. Stiborek, J., Pevný, T., Rehák, M.: Multiple instance learning for malware classification. Expert Syst. Appl. **93**, 346–357 (2018). https://doi.org/10.1016/ j.eswa.2017.10.036. http://www.sciencedirect.com/science/article/pii/S095741741 7307170

21. Straehle, C., Kandemir, M., Koethe, U., Hamprecht, F.A.: Multiple instance learning with response-optimized random forests. In: 2014 22nd International Conference on Pattern Recognition, pp. 3768–3773, August 2014. https://doi.org/10. 1109/ICPR.2014.647

22. Tax, D.M.J.: A matlab toolbox for multiple-instance learning, version 1.2.2, Faculty EWI, Delft University of Technology, The Netherlands, April 2017

23. Zhang, C., Platt, J.C., Viola, P.A.: Multiple instance boosting for object detection. In: Weiss, Y., Schölkopf, B., Platt, J.C. (eds.) Advances in Neural Information Processing Systems, vol. 18, pp. 1417–1424. MIT Press (2006). http://papers.nips. cc/paper/2926-multiple-instance-boosting-for-object-detection.pdf

24. Zhang, J., Marszałek, M., Lazebnik, S., Schmid, C.: Local features and kernels for classification of texture and object categories: a comprehensive study. Int. J. Comput. Vision **73**(2), 213–238 (2007)

25. Zhang, Q., Goldman, S.A.: Em-dd: an improved multiple-instance learning technique. In: Advances in Neural Information Processing Systems, pp. 1073–1080. MIT Press (2001)

One-Class Quantification

Denis dos Reis[✉], André Maletzke[✉], Everton Cherman[✉],
and Gustavo Batista[✉]

Universidade of São Paulo, São Carlos, Brazil
{denismr,andregustavo,echerman}@usp.br, gbatista@icmc.usp.br

Abstract. This paper proposes one-class quantification, a new Machine Learning task. Quantification estimates the class distribution of an unlabeled sample of instances. Similarly to one-class classification, we assume that only a sample of examples of a single class is available for learning, and we are interested in counting the cases of such class in a test set. We formulate, for the first time, one-class quantification methods and assess them in a comprehensible open-set evaluation. In an open-set problem, several "subclasses" represent the negative class, and we cannot assume to have enough observations for all of them at training time. Therefore, new classes may appear after deployment, making this a challenging setup for existing quantification methods. We show that our proposals are simple and more accurate than the state-of-the-art in quantification. Finally, the approaches are very efficient, fitting batch and stream applications. Code related to this paper is available at: https://github.com/denismr/One-class-Quantification.

Keywords: One-class quantification · Counting · Open-set recognition

1 Introduction

Quantification is the Machine Learning task that estimates the class distribution of an unlabeled sample of instances. It has numerous applications. In social sciences, quantification predicts election results by analyzing different data sources supporting a candidate [13]. In natural language processing, quantification estimates the prior probability of different senses for a given word [2]. In entomology, it infers the local density of mosquitoes in a specific area covered by an insect sensor [4], among many other applications.

In several quantification applications, we are mostly interested in counting a single or a small set of class labels. In literature, a typical approach is to model these applications as binary quantification problems and use the positive class to designate the group of interest. For example, the positive class can indicate the *Aedes* or *Anopheles* mosquito genus, vectors of terrible diseases such as Zika fever and malaria, respectively, or a specific defect regarding battery duration in mobile phones.

In many of these applications, the negative class is the universe, *i.e.*, a broad set of all categories, except the positive one. For example, the negative label

© Springer Nature Switzerland AG 2019
M. Berlingerio et al. (Eds.): ECML PKDD 2018, LNAI 11051, pp. 273–289, 2019.
https://doi.org/10.1007/978-3-030-10925-7_17

would be all possible defects reported by customers, tweets posted by users, word senses in a large corpus and insect species in an area. However, fully characterizing the universe is not a trivial task. Apart from being typically represented by a large number of categories, we can expect that examples from previously unseen classes may appear in deployment, a problem denominated as *open-set recognition*. Taking the insect sensor as an example, the number of insect species is estimated to be between six and ten million [3], making it an impossible task to characterize the negative class completely. Therefore, during deployment, the quantification model will have to face examples from categories (species) that were disregarded. Even if we are not interested in quantifying these classes, their instances can significantly degrade the quantifier performance.

Quantification literature has mostly ignored the open-set scenario, indicating that although a significant body of research is available, there is a considerable long path to make this task a mature technology. Several techniques have shown accurate results in typical open-set applications domains. However, we show that such good results are mostly due to the contrived closed-set evaluation setups, instead of the actual merits of the proposals. In our opinion, considering its importance and diversity of applications, quantification is the most under-researched task in Machine Learning.

In this paper, for the *first* time in quantification literature, we define the task of **one-class quantification** and propose methods for quantifying a class of interest among a possibly open-set of negative classes. Although one-class quantification crosses the limits of open-set applications, we restrict ourselves to open-set scenarios to concretely demonstrate the applicability of our findings. We show that our methods can learn exclusively from positive class examples, and are more accurate when facing unseen classes than the state-of-the-art.

This paper is organized as follows. In Sect. 2, we formally define the tackled problem; in Sect. 3, we review the related work in quantification and open-set recognition; in Sect. 4 we introduce our proposals of algorithms for one-class quantification; in Sect. 5 we present and discuss our experimental evaluation; and in Sect. 6, we make our conclusions and prospects for future work.

2   Background and Definitions

In supervised tasks, we are interested in learning from a dataset $D = \{(\mathbf{x}_1, y_1), \ldots, (\mathbf{x}_n, y_n)\}$, in which $\mathbf{x}_i \in \mathcal{X}$ is a vector of m attributes in the feature space $\mathcal{X}$, and $y_i \in \mathcal{Y} = \{c_1, \ldots, c_l\}$ is the respective class label.

In classification, the objective is to predict the class labels of examples based on the observation of their features as stated in Definition 1.

Definition 1. *A **classifier** is a model h induced from D, such that*

$$h : \mathcal{X} \longrightarrow \{c_1, \ldots, c_l\}$$

which aims at predicting the classes of unlabeled examples correctly.

In classification, we assume the examples are independent and identically distributed (*i.i.d*). This assumption states that the examples are independent of each other and the unlabeled (test or deployment) set comes from the same distribution of the labeled (training) set.

Although quantification and classification share similarities, their objectives differ. A quantifier is not interested in individual predictions, rather in the overall quantity of examples of a specific class or a set of classes. Formally, we can define a quantifier according to Definition 2.

Definition 2. *A **quantifier** is a model that predicts the prevalence of each class in a sample, more formally,*

$$q : \mathbb{S}^{\mathcal{X}} \longrightarrow [0, 1]^l$$

$\mathbb{S}^{\mathcal{X}}$ *denotes a sample from* $\mathcal{X}$. *The quantifier output is a vector,* $\hat{p}$, *that estimates the probability for each class, such that* $\sum_{j=1}^{l} \hat{p}(j) = 1$. *The objective is* $[\hat{p}(1), \ldots, \hat{p}(l)]$ *to be as close as possible to the true class probabilities* $[p(1), \ldots, p(l)]$.

In quantification, we still assume the examples from $\mathbb{S}^{\mathcal{X}}$ are independent. However, training and deployment sets are not identically distributed, since we expect the class distributions to differ significantly.

This paper proposes, for the first time, one-class quantifiers, which are quantification models that learn from single-class datasets, according to Definition 3.

Definition 3. *A **one-class quantifier** is a quantification model induced from a single-class dataset, in which all available labeled examples belong to the same class, say the positive one,* $D^{\oplus} = \{(\boldsymbol{x}_1, \oplus), \ldots, (\boldsymbol{x}_n, \oplus)\}$, *and*

$$q^{\oplus} : \mathbb{S}^{\mathcal{X}} \longrightarrow [0, 1]$$

The one-class quantifier outputs a single probability estimate $\hat{p}(\oplus) \in [0, 1]$ *of the positive class prevalence. Notice, however, that* $q^{\oplus}$ *operates over* $\mathbb{S}^{\mathcal{X}}$, *i.e., a sample with all occurring classes.*

Since we evaluate our proposals in open-set scenarios, we also define the open-set quantification task in Definition 4.

Definition 4. ***Open-set quantification** is the task of quantifying one or a small set of classes for problems with a large number of classes. Therefore, we can assume that, at training time, we know only a subset of k classes, k ≪ l. After deployment, other classes (not used for training) may appear.*

Although we are not interested in quantifying those unseen classes, they can significantly hinder the performance of the quantifier.

We can address open-set quantification tasks with one-class quantifiers. In this case, we are interested in quantifying a single positive class and collapse all other categories as a negative class. We highlight that only positive instances are used to induce the model. The negative class depicts the universe composed of several subclasses. We can trivially extend this approach for more than one category of interest with a set of one-class quantifiers, one for each positive class.

3   Related Work

In this section, we review some concepts and methods related to quantification and open-set recognition. In particular, we briefly discuss some of the techniques employed in our empirical comparison. Since we are proposing a new task, there are no existing methods that can directly compare to ours. However, there is a considerable body of research that we can adapt to the open-set quantification scenario and can, therefore, use in our experimental setup.

3.1   Closed-Set Quantification

Although quantification and classification seek different objectives, they are related tasks and can benefit from each other. One example is the most straight-forward method of achieving quantification: the Classify and Count (CC) algorithm [7]. As its name suggests, CC classifies all observations and counts how many belong to each class.

Figure 1 illustrates the score distribution of a classification process using a scorer and a decision threshold. A scorer is a model that, once induced, produces a numeric value (score) for each unlabeled observation. A higher score means more confidence that the example belongs to the positive class. A scorer can produce multiple classifiers by setting a threshold and assigning the observations according to which side of the threshold they are: examples are labeled as negative if scored lower than the threshold, and as positive otherwise.

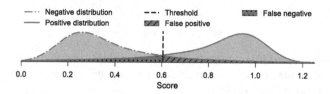

Fig. 1. Threshold and score distribution for a binary classification process.

When the number of false positives and the number false negatives are the same, the classification errors cancel out each other and CC quantifies flawlessly, independently of the classification accuracy. In other words, it is not necessary to have a perfect scorer to reach a flawless quantification, being that an ideal scorer ranks all negative observations lower than the lowest scored positive example. Furthermore, the quantification error committed by CC is the absolute difference between the number of false positives and the number of false negatives.

Quantifying a class always produces a relative scalar: the proportion of a class in comparison with the whole unlabeled data. Therefore, quantification, in its nature, is only useful if we expect this proportion to vary from sample to sample. Hence, we expect the negative and positive distributions to shrink and expand concerning each other. Thus, the ratio between false positive and false negative areas of the distributions may vary as well and, consequently, the quantification error of CC is expected to change.

The most considered quantifier, Adjusted Classify and Count (ACC) [7], avoids this change in CC errors by accounting for the expected numbers of true and false positives. To this end, ACC requires precise estimation of true positive and false positive rates (TPR and FPR, respectively), which can be measured independently of a particular configuration for the class proportions. Considering $\hat{p}(\oplus)$ the proportion of the positive class estimated by CC, ACC's adjustment is as follows:

$$\mathrm{ACC}(\hat{p}(\oplus), \mathrm{TPR}, \mathrm{FPR}) = \min\left\{1, \frac{\hat{p}(\oplus) - \mathrm{FPR}}{\mathrm{TPR} - \mathrm{FPR}}\right\}$$

ACC is proven to provide a perfect quantification when the true TPR and FPR are known. However, computing the true TPR and FPR would require a labeled test set. In practice, researchers estimate such quantities using the training set and some error estimation sampling, such as cross-validation. Nevertheless, if the scores for one of the classes follow a nonstationary distribution, it is very likely that TPR or FPR or both will also change. In this paper, we assume the negative distribution to be composed of a mixture of partially known subclasses that have different degrees of similarity with the positive class. As the proportions of these subclasses vary, FPR also changes, limiting our chances of accurately estimating it using the training set.

Another family of quantification algorithms called Mixture Models (MM) [7] does not rely on a base classifier, but usually requires a scorer. In general, MM's consider the scores obtained on an unlabeled set to follow a parametric mixture between two known distributions (one for each class). The computation of the parameters of this mixture leads to the quantification estimative.

One example of MM is the HDy algorithm [11]. HDy represents each distribution as a histogram. A weighted vectorial sum of their histograms gives the mixture between the known positive and negative distributions (H^+ and H^-, respectively), where the weights sum up to 1. The weights that minimize the Hellinger Distance (HD) between the mixture and the unlabeled distribution (H^T) are considered to be the proportion of the corresponding classes in the unlabeled sample. Figure 2 illustrates the application of HD inside HDy.

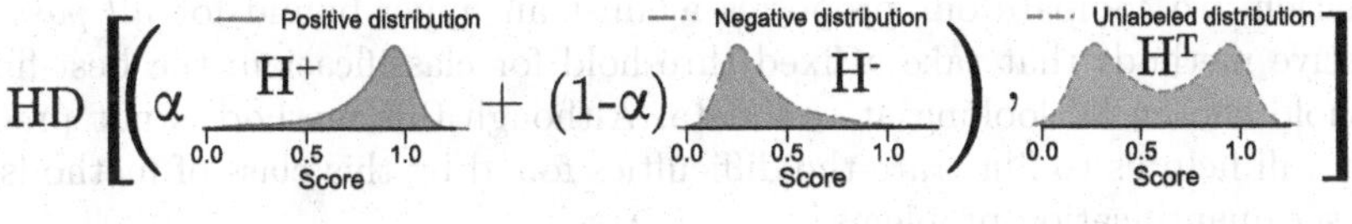

Fig. 2. HDy uses the Hellinger Distance (HD) to measure the similarity between a mixture of positive and negative score distributions and an unlabelled score distribution, where α is the proportion of the positive class. HDy searches for an α that minimizes the Hellinger Distance.

Although there are many more approaches in literature [10], we reviewed some of the principal paradigms in quantification in this section. However, all approaches studied so far expect the acquisition of enough negative examples to obtain a full estimation of its distribution. This expectation is unrealistic for open-set scenarios.

3.2 Open-Set Quantification

To the best of our knowledge, no method the literature has explicitly addressed the task of open-set quantification. Nevertheless, we can adapt existing methods in quantification and open-set recognition literature to this task. This section summarizes some possibilities that we explore in the experimental evaluation.

A first approach is to reduce the open-set to a binary quantification problem and to tackle it with a binary quantifier. Any of the methods explained in Sect. 3.1 fit this role, where the training data for the negative class is a mixture of all known negative subclasses. However, we argue that:

1. We cannot reliably measure FPR in open-set scenarios. The appearance or disappearance of new or existing negative sub-classes, with different levels of similarity with the positive class, may make the classification problem more difficult or easier. Therefore, the number of negative cases misclassified as positive will change, altering FPR and harming all methods that rely on that information, such as ACC;
2. The binary quantifiers assume the knowledge of the negative distribution. We show in the experimental section that their performance depends on which and how many subclasses are in the negative class, and how many of those are present in the training set. Such performance loss is expected to happen to all quantifiers that assume they know all negative subclasses ($k = l$).

A second approach is to adapt existing methods in the open-set classification literature to quantification. The research in open-set recognition is extensive, and we reserve a more comprehensive study about this class of approaches to future publications. In this paper, we compare our proposals to a family of methods that choose an ideal threshold value for a one-class scorer based on the available subclasses [14]. We can trivially adapt those methods to quantification by applying a Classify and Count (CC) with the chosen threshold. In our experimental evaluation, we compare our proposals against an upper-bound for *all possible* inductive methods that take a fixed threshold for classification: the best-fixed threshold chosen by looking at test data. Although this method is not practical, it will help us to illustrate the difficulties found by this class of methods in open-set quantification problems.

Finally, we observe that all previously explained methods require negative observations for the induction of the quantification model. Therefore, they cannot be categorized as one-class quantifiers. This research paper is the first to address one-class quantification and open-set quantification explicitly. We propose two efficient methods that use only positive observations.

4   Proposals

In this section, we introduce two simple one-class quantification algorithms. Both algorithms require a base one-class scorer, such as one-class SVM, and a training set of positive examples to be induced upon. Although they share the same objective and requirements, their rationale diverge.

4.1   Passive Aggressive Threshold ACC

Our first and more straightforward proposal, Passive Aggressive Threshold ACC (PAT-ACC or PAT, for short), draws inspiration from Adjusted Classify and Count and Conservative Average Quantifier [8]. ACC depends on accurate estimates for TPR and FPR. In many applications, however, we cannot reliably measure TPR and FPR because the score distribution for *negative* observations varies from sample to sample. The influence of the negative distribution on ACC stems from the fact that the most suitable thresholds for *classification* usually cut through the negative distribution.

In PAT, we deliberately select a very conservative classification threshold that tries to minimize FPR. We set this threshold according to a quantile q for the scores of positive observations in a training set. Finally, we estimate $\widehat{\text{TPR}} = 1 - q$ and assume $\widehat{\text{FPR}} \approx 0$. Figure 3 illustrates this process.

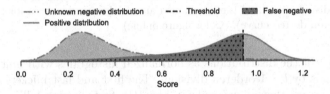

Fig. 3. Expected behavior for a conservative threshold.

After the threshold is set, we perform ACC as usual: we classify all observations in the test sample of size n according to this conservative threshold, count the number of positive instances n_+, estimate the positive proportion $\hat{p}(\oplus) = \frac{n_+}{n}$, and readjust it as follows:

$$\hat{p}'(\oplus) = \text{PAT}(\hat{p}(\oplus), q) = \text{ACC}(\hat{p}(\oplus), 1 - q, 0) = \min\left\{1, \frac{\hat{p}(\oplus)}{1 - q}\right\}$$

In PAT, q is an important parameter. Ideally, it should be set as high as possible so that we can be more confident about the assumption of FPR ≈ 0, even for non-stationary negative distributions. How high it can be set depends on the sample size, since higher values imply greater extrapolation with fewer observations. We also predict the performance to be similar to CC when q approaches 0, as the extrapolation is reduced. We validate the latter statement experimentally and also shows that, although important, q is not a sensitive parameter:

a broad range of possible values lead to similar quantification errors. Therefore, we used this fact and reported all quantification errors in the experimental evaluation with a fixed $q = 0.25$. Alternatively, we note that a strategy similar to Median Sweep [8] can be applied. In this case, one could perform PAT with many different thresholds and consider the median of the estimates.

4.2   One Distribution Inside

Our second proposal, One Distribution Inside (ODIn), is a Mixture Model (MM). Usually, MMs search for parametrization to mix two known distributions, whereas our proposal only knows one distribution that represents the positive class. ODIn looks for the maximum possible scale factor $\mathfrak{s}$, $0 \leq \mathfrak{s} \leq 1$, for the known score distribution of **positive** training instances, so that it fits inside the distribution of scores of test instances with an overflow no greater than a specified limit. The overflow is the area of the scaled positive distribution that transposes the test distribution, as Fig. 4 illustrates.

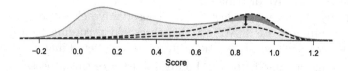

Fig. 4. Rationale behind ODIn. The dotted curves represent candidate scale factors for the positive distribution, and the red-shaded area is the overflow area for the greater scale factor (top dotted curve). (Color figure online)

We represent the distributions as normalized histograms with unit area and b buckets, split by $b - 1$ ordered divisions. The first and last buckets are open-ended. This means that all scores lower than the first division fall into the first bucket, and all scores higher than the last division fall into the last bucket. In our experiments, we set the bucket divisions, $i.e.$, the values of the scores that split the buckets, as score percentiles obtained from the **positive training observations**. The first and last divisions are set as the estimates for, respectively, the 0th and 100th percentiles of the scores. The remaining divisions are set at every ih percentile, $0 < ih < 100$, $i \in \mathbb{N}$, where h is a parameter. For instance, if $h = 10$, the divisions are at the percentiles $0, 10, 20, \ldots, 90, 100$. Although score wise the buckets do not share the same width, they are expected to be equally filled by observations from the positive distribution, thus sharing the same weight. Exceptions are the first and last buckets, which are expected to have weights close to zero. Figure 5 illustrates this process.

The overflow committed by a histogram H^I, at a scale factor α, inside a histogram H^O, where both histogram are normalized so that $\sum_{1 \leq i \leq b} H_i^I = \sum_{1 \leq i \leq b} H_i^O = 1$, is formally defined as follows:

$$\text{OF}(\alpha, H^I, H^O) = \sum_{i=1}^{b} \max\left\{0, \alpha H_i^I - H_i^O\right\}$$

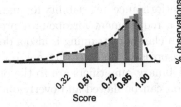

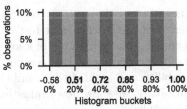

(a) Training examples distribution over scores.

(b) Training examples distribution
over histogram bins.

Fig. 5. The thresholds for the histogram bins are not uniformly distributed over the scores (5a), and yet each bin is filled with the same proportion of examples (5b).

Given an overflow limit $\mathcal{L}$, the histogram H^+ with scores for positive training observations, and a histogram H^T with scores for the unlabeled test sample T, ODIn estimates the proportion of positive observations $\hat{p}(\oplus)$ in T as:

$$\hat{p}(\oplus) = \mathfrak{s} - \text{OF}\left(\mathfrak{s}, H^+, H^T\right) \text{ where } \mathfrak{s} = \sup_{0 \leq \alpha \leq 1} \left\{\alpha | \text{OF}\left(\alpha, H^+, H^T\right) \leq \alpha \mathcal{L}\right\}$$

The parameter $\mathfrak{s}$ can be estimated through Binary Search, within a time complexity of $O\left(\frac{100}{h} \log_2 \epsilon^{-1}\right)$, where ϵ is the expected precision. Also, although $\mathcal{L}$ is a parameter, it can be automatically defined using only positive observations. To this end, we estimate the mean $\hat{\mu}$ and standard deviation $\hat{\sigma}$ of OF for pairs of histograms derived from samples with only positive observations, at scale factor 1, and set $\mathcal{L} = \hat{\mu} + d\hat{\sigma}$, where d is a parameter. Although we are actively replacing one parameter with another one, d has a clearer semantic and its value is domain independent: it is the number of standard deviations of the expected average overflow. In all experiments we used $d = 3$.

Finally, choosing h, although a non-trivial task, is not devoid of useful insights. Histograms with too many bins are negatively affected by two aspects. First, if the sample size is not large enough, large histograms can become too sparse, each bin can have too low weight, and ultimately, the OF can face the curse of dimensionality. Second, a large number of bins has the implicit assumption of high precision for the scores. On the other hand, if there are too few bins, we may be unable to differentiate distributions. In all of our experiments, we used $h = 10$, which means that the thresholds cut the positive distribution at the 0th, 10th, ..., 90th, and 100th percentiles.

5   Experimental Evaluation

In this section, we make a comprehensible experimental evaluation of our proposals. We divided the evaluation into three parts.

First, we compare our proposals with binary quantification methods as we increase the number of known negative subclasses at training and the number of unknown negative subclasses at the test. Performance invariability for an increasing number of unknown classes at test is a requirement for open-set problems. Lower dependence on the number of known classes at training is also a desirable trait. Thus, the objective of this comparison is to demonstrate our proposal's high invariability and highlight how unfit binary quantifiers are to the open-set task. Additionally, we show how evaluation can be flawed and overly optimistic in open-set problems.

Second, as we entirely disregard binary quantifiers due to the obtained results in the previous part, we compare PAT and ODIn against an upper bound for the open-set recognition methods that choose a fixed threshold with training data. Although such methods depend on negative subclasses, this upper bound can serve as a non-trivial baseline for our approaches.

Finally, we compare the quantification performances for different positive class ratios in the test data. We start this section describing the datasets and general experimental setup decisions.

5.1 Experimental Setup

In our experimental evaluation, we used nine real datasets[1]. For most of them, we fixed the size of the test samples as 500 observations for all experiments. Exceptions were made for datasets with a smaller number of entries since we need enough observations to not only assemble the test samples with varying positive proportion, but also to allow for variability among samples with a same positive proportion. The number of observations for training depended on which experiment was in place, and is explained later. Each dataset is described next.

Insects contains information about the flight of 14 species of insects. As some are discriminated further by sex, the dataset has 18 classes. The positive class is female *Aedes aegypti*. The data has $166,880$ records represented by 27 features;

Arabic Digit contains $8,800$ entries described by 26 features for the human speech of Arabic digits. There are 10 classes, and the target class is the digit 0. Test sample size is 400. This versions sets a fixed number of features for every record [12,15];

BNG (Japanese Vowels) benchmark dataset with speech data regarding Japanese Vowels. There are $1,000,000$ entries, represented by 12 features, for 9 speakers. The speaker #1 is the class of interest [19];

Anuran Calls (MFCCs) contains 22 features to represent the sound produced by different species of Anurans (frogs). As the data size is restricted, we only considered the two biggest families of frogs as the classes of the data, ending up with $6,585$ entries. The positive class is the *Hylidae* family, and the negative class is the *Leptodactylidae* family [6,15];

---

[1] Data and code for the proposed algorithms are available at this paper's supplementary material website at https://github.com/denismr/One-class-Quantification.

Handwritten contains 63 features that represent the handwritten lowercase letters q, p and g. The data has $6,014$ entries and the chosen positive class is the letter q. Test sample size is 400 [18];

Letter describes the appearance of the 26 uppercase letters of the alphabet on a black and withe display with 16 features. It contains $20,000$ entries and the class of interest is the letter W. Test sample size is 200 [9,19];

Pen-Based Recognition of Handwritten Digits handwritten digits represented by 16 features. The digit 5 is the target class. There are $10,992$ entries [1,15];

H RU2 Pulsar candidates collected during the HTRU survey, where pulsars are a type of star. It contains two classes, Pulsar (positive) and not-Pulsar (negative), across $17,898$ entries described by 63 features [15,16];

Wine Quality contains 11 features that describe two types of wine (white and red). The quality information was disregarded, and the target class is red wine. The dataset contains $6,497$ entries [5,15].

While the positive class was predefined, all remaining classes in each dataset were considered to be negative subclasses. Each dataset was uniformly and randomly split into two halves. The first half contains observations used for training the models according to the procedures described in each experimental phase, and the second contains the observations from which we sampled the test sets with the number of observations detailed above.

As our objective using a quantifier is to estimate the proportion of positive observations in a sample, we fabricate samples with varying positive class proportion using the examples from the test half. Due to our limited data, different samples can share examples. However, one example is not observed more than once in the same sample. The positive class proportion of each test sample is chosen uniformly at random from 0% to 100%. The remaining of the set is allocated for the negative subclasses, each of those also having a random proportion.

We measure the performance according to the Mean Absolute Error (MAE) for the estimated positive proportion. In other words, we average the absolute difference between the actual positive proportion and the estimated one, across all test samples.

5.2   Proposals Versus Binary Quantifiers

In the first part of our experimental evaluation, we evaluate the performance of binary quantifiers when submitted to an open-set scenario by varying the number of negative subclasses that are known at training and the number of unknown subclasses at test. This evaluation highlights the problems carried by such classifiers in open-set applications: high dependence on negative subclasses at training and unstable and unreliable evaluation at test.

Due to lack of space, we present results[2] only for Insects dataset (which is the motivating application for this work). Except for the Letter dataset, experiments

---

[2] All results are available in our supplemental material repository.

on all other datasets carry similar findings. The assessed algorithms were CC and ACC, both with SVM scores, and HDy with scores obtained with one-class SVM, using the same algorithm to define the thresholds for bins as in ODIn. We performed the latter experiment with this type of histogram because HDy requires the scores to be bounded both sides, and the SVM scores are not strictly bounded. This makes it difficult to set an uniformly distributed histogram. Furthermore, scores from sets with more than one class usually form a multimodal distribution with varying space between modes, which would harm histograms based on percentiles of class scores that are not one-class.

We used 10-fold cross-validation with the training data to estimate TPR and FPR (required by ACC), and to generate one-class SVM training scores. The training set has in total 500 observations for the positive class and 500 for each known negative subclass. Preliminarily experiments showed that results for larger training sets did not differ statistically, as well as results when replacing SVM with Random Forest or ADA Boost. In the experiments where we varied the number of known negative subclasses, test samples always had all 17 negative subclasses included. When varying the number of unknown negative subclasses, the training set always included one negative random subclass. Which negative subclass was available for training varied from test sample to test sample. For each number of known/unknown subclasses at training/test, we repeated the experiment 5,000 times and report the average MAE over these samples.

Figure 6-*left* presents the performance of the binary quantifiers for varying number of known negative subclasses present in the training set. We observe the performance to increase as more classes are included, except for CC. Particularly, CC peaked its performance with seven known subclasses and started to degrade steadily. We suspect that the increase in the number of negative observations made the model lose generality and raise the false negatives.

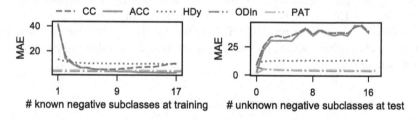

Fig. 6. Quantification mean absolute error (MAE) of binary quantifiers, ODIn and PAT for varying number of negative subclasses in the training set (*left*), and number of unknown negative subclasses at test (*right*). Insects dataset. Notice on the right that binary quantifiers curves stop at 16 negative subclasses since one is used for training.

The findings presented by Fig. 6-*left* must be interpreted cautiously. The increase in performance is not only due to the greater number of known subclasses in the training set but also due to the respective smaller number of unknown subclasses in the test. We cannot tell, only from this figure, which factor is more relevant for the performance: more informed training set or easier test set.

In Fig. 6-*right* we isolate the latter factor, as it sets a fixed number of subclasses in training and increases the number of unknown subclasses in the test. We notice a steep decrease in performance as the test set includes more subclasses. However, the most problematic aspect in using the binary quantifiers is that their performances did not stabilize before we ran out of negative subclasses in our data, making it challenging to create expectations regarding performance after the model is deployed in the real application. In contrast, the two proposed methods have a remarkable stable performance across a wide range of unknown subclasses, which shows their suitability for open-set quantification.

Our results show that comparing those methods against one-class methods is a parametric evaluation, in which the relevant parameters are the number of known classes at training and unknown classes at test. Of the utmost importance, though, is the fact that we are unable to reliably predict the performance of the tested binary quantifiers after deployment in open-set quantification, as we do not know how many unknown classes will appear. However, *if* Fig. 6 approaches deployment behavior, it suggests that our methods outperform all tested binary quantifiers. For these reasons, in the following sections of our evaluation, we disregard further comparisons against binary quantifiers, as we showed they are not suitable for open-set quantification problems.

5.3 Proposals Versus One-Class and Open-Set Approaches

In the second part of our experimental evaluation, we compare our proposals against one-class classification and open-set approaches adapted to quantification. To this end, we compare them against:

OCSVM. CC using the classifications issued by one-class SVM [17], to illustrate a straightforward approach using a state-of-the-art technique;

BFT. Best Fixed Threshold, CC using, as base classifier, an upper bound for *all* algorithms that inductively set a fixed classification threshold on the scores. To achieve this upper bound for each dataset, we report the best experimental results obtained by a threshold T that results in the lowest quantification error (MAE). We search T uniformly from the 0th percentile to the 100th percentile of the positive distribution, where $T = \min\left\{401, 1 + \left\lfloor \frac{3N}{100} \right\rfloor 100\right\}$ and N is the number of available training observations.

We evaluate the scores used by BFT, PAT, and ODIn from two different sources. The first one is one-class SVM trained with positive examples. The second is the Mahalanobis distance from the positive class examples. We include the Mahalanobis distance as an example of a simple and non-inductive scorer. As higher scores mean more confidence in the positive class, we negated the Mahalanobis distance in our experiments. As the tested methods depend on scores for training, we obtained these scores by performing 10-fold cross-validation with the training data. The generated test samples included all subclasses in the assessed dataset (with random proportions). Each reported MAE is an average of $5,000|Y_-^T|$ samples, where $|Y_-^T|$ is the number of negative subclasses.

Table 1 presents the quantification performance obtained in the second part of our experimental evaluation. Most of the lowest errors were obtained by our proposals. A relevant choice to be made by a practitioner is which base scorer should be used: the best performance rotated between one-class SVM and Mahalanobis, including considerable differences, depending on the dataset.

Table 1. MAE as percentages, for each dataset. Standard deviations are in parentheses.

	OCSVM CC	PAT		ODIn		BFT	
		OCSVM	Mahalanobis	OCSVM	Mahalanobis	OCSVM	Mahalanobis
I	23.57(15.06)	4.56(3.04)	10.97(6.86)	**3.84**(2.58)	8.83(5.09)	6.27(3.80)	9.82(5.87)
A	23.72(16.04)	12.13(8.50)	**2.22**(1.52)	12.46(6.47)	2.97(2.19)	11.65(6.94)	5.02(3.13)
B	22.53(15.48)	11.79(8.10)	10.33(6.50)	8.96(5.60)	**7.36**(4.16)	11.46(6.77)	9.96(5.91)
C	24.02(14.05)	6.15(3.52)	3.69(1.80)	**2.98**(2.65)	3.53(2.44)	9.58(5.96)	7.99(4.94)
H	23.51(13.97)	**2.28**(1.49)	3.15(1.97)	2.76(2.35)	2.42(1.86)	5.36(3.41)	4.13(2.91)
L	27.46(16.01)	**1.82**(1.50)	2.92(2.22)	3.46(2.73)	3.78(3.37)	2.07(1.37)	2.34(1.53)
P	49.75(28.92)	40.38(23.49)	2.46(1.59)	49.46(28.60)	2.37(1.48)	37.25(21.91)	**1.77**(1.16)
R	50.35(29.22)	15.91(10.21)	7.60(5.16)	41.24(26.89)	**3.38**(2.33)	20.34(14.29)	10.24(6.10)
W	28.74(16.93)	2.13(1.44)	**1.06**(0.88)	2.82(2.43)	1.61(1.38)	6.13(3.62)	4.15(2.48)

From now on, we consider the best scorer between using one-class SVM and Mahalanobis, for each approach. OCSVM CC was consistently worse than all other approaches and is disregarded from now on. Individually, both PAT and ODIn statistically diverged from BFT, according to the paired t-test, with p-values of 0.015 and 0.023, respectively. PAT and ODIn did not statistically diverge (p-value = 0.344). However, we notice that which one is the best depends on which dataset was evaluated, presenting significant individual differences. BFT presented inferior performance than the best between PAT and ODIn for all but one dataset.

Although BFT presented the best quantification performance for the dataset Pen Digits, we note that BFT is an upper bound for a category of algorithms that chooses a fixed threshold and subsequently perform CC. Figure 7 illustrates that it only takes a slightly badly chosen threshold to heavily impact the performance of the quantifier. Meanwhile, PAT keeps similar performance for a wide range of thresholds. This behavior is recurrent across all datasets[3], including Pen Digits. At last, ODIn does not have an equivalent parameter.

5.4   Evaluation for a Range of Positive Class Distributions

In this part, we analyze the variation of the positive class proportion from 0% to 100% with 1% increments. For each positive ratio, we measured the respective MAE with 5,000 test samples, including all negative subclasses (with random

---

[3] All results are available in our supplemental material repository.

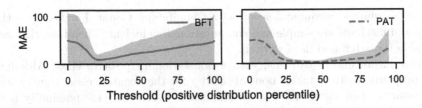

Fig. 7. Impact of different parametrization for BFT and PAT on quantification mean absolute error, dataset Anuran Calls, OCSVM as scorer. The shaded area corresponds to a window of four standard deviations (two towards each side of the curve).

proportions). Figure 8 illustrates our proposals stability for different positive class proportions. The figure also suggests that a better stability in the extreme distributions could be achieved by ensembling both approaches since they behave symmetrically in these regions. Additionally, we can observe the instability of OCSVM and even BFT. This instability was expected since their underlying approach is CC and the error is expected to vary for different class proportions.

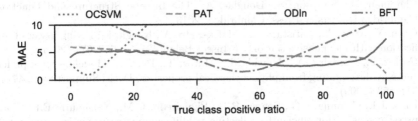

Fig. 8. Quantification mean absolute error (MAE) for different true positive ratios, Insects dataset, one-class SVM scorer.

Although this result is for the Insects dataset, other datasets showed similar results. Due to lack of space, we included additional results in the paper website.

6 Conclusion

This article is the first to explicitly approach the one-class quantification and make a comprehensible evaluation in open-set scenarios. We illustrate the issues of using traditional quantification methods in such situations and propose two novel techniques that can learn solely with positive observations. The available negative examples are still useful for two main reasons: estimating the expected performance after deployment and searching for the best parameterization.

In our experiments, the proposed methods outperformed binary quantifiers as we increased the number of unknown classes in the test set. Our proposals also performed better than the Best Fixed Threshold (BFT) and One-class SVM classifiers adapted to quantification with Classify and Count. Furthermore, the proposed methods are simple and computationally efficient. Therefore, they are suitable for batch and data streams.

Our future efforts will deal with a nonstationary positive class. Although our proposals actively tackle nonstationarity for the negative class, they expect the positive class distribution to be immutable over time. One possibility is to require the positive class to be among a limited set of known distributions [18].

Acknowledgement. The authors thank CAPES (PROEX-6909543/D), CNPq (306 631/2016-4) and FAPESP (2016/04986-6). This material is based upon work supported by the United States Agency for International Development under Grant No AID-OAA-F-16-00072.

References

1. Alimoglu, F., Alpaydin, E., Denizhan, Y.: Combining multiple classifiers for pen-based handwritten digit recognition (1996)
2. Chan, Y.S., Ng, H.T.: Estimating class priors in domain adaptation for word sense disambiguation. In: COLING ACL, pp. 89–96 (2006)
3. Chapman, R., Simpson, S., Douglas, A.: The Insects: Structure and Function. Cambridge University Press, Cambridge (2013)
4. Chen, Y., Why, A., Batista, G.E., Mafra-Neto, A., Keogh, E.: Flying insect classification with inexpensive sensors. J. Insect Behav. **27**(5), 657–677 (2014)
5. Cortez, P., Cerdeira, A., Almeida, F., Matos, T., Reis, J.: Modeling wine preferences by data mining from physicochemical properties. Decis. Support Syst. **47**(4), 547–553 (2009)
6. Diaz, J.J., Colonna, J.G., Soares, R.B., Figueiredo, C.M., Nakamura, E.F.: Compressive sensing for efficiently collecting wildlife sounds with wireless sensor networks. In: CCCN, Munich, pp. 1–7 (2012)
7. Forman, G.: Counting positives accurately despite inaccurate classification. In: Gama, J., Camacho, R., Brazdil, P.B., Jorge, A.M., Torgo, L. (eds.) ECML 2005. LNCS (LNAI), vol. 3720, pp. 564–575. Springer, Heidelberg (2005). https://doi.org/10.1007/11564096_55
8. Forman, G.: Quantifying trends accurately despite classifier error and class imbalance. In: SIGKDD, Philadelphia, pp. 157–166 (2006)
9. Frey, P.W., Slate, D.J.: Letter recognition using holland-style adaptive classifiers. Mach. Learn. **6**(2), 161–182 (1991)
10. González, P., Castaño, A., Chawla, N.V., Coz, J.J.D.: A review on quantification learning. ACM Comput. Surv. **50**(5), 74 (2017)
11. González-Castro, V., Alaiz-Rodríguez, R., Alegre, E.: Class distribution estimation based on the hellinger distance. Inf. Sci. **218**, 146–164 (2013)
12. Hammami, N., Bedda, M.: Improved tree model for arabic speech recognition. ICCSIT **5**, 521–526 (2010)
13. Hopkins, D.J., King, G.: A method of automated nonparametric content analysis for social science. Am. J. Polit. Sci. **54**(1), 229–247 (2010)
14. Jain, L.P., Scheirer, W.J., Boult, T.E.: Multi-class open set recognition using probability of inclusion. In: Fleet, D., Pajdla, T., Schiele, B., Tuytelaars, T. (eds.) ECCV 2014. LNCS, vol. 8691, pp. 393–409. Springer, Cham (2014). https://doi.org/10.1007/978-3-319-10578-9_26

15. Lichman, M.: UCI m.l. repository (2013). http://archive.ics.uci.edu/ml
16. Lyon, R., Stappers, B., Cooper, S., Brooke, J., Knowles, J.: Fifty years of pulsar candidate selection: from simple filters to a new principled real-time classification approach. MNRAS **459**(1), 1104–1123 (2016)
17. Pedregosa, F., et al.: Scikit-learn: machine learning in Python. JMLR **12**, 2825–2830 (2011)
18. dos Reis, D., Maletzke, A., Batista, G.: Unsupervised context switch for classification tasks on data streams with recurrent concepts. In: SAC (2018)
19. Vanschoren, J., van Rijn, J.N., Bischl, B., Torgo, L.: OpenML: networked science in machine learning. SIGKDD Explor. **15**(2), 49–60 (2013)

Deep F-Measure Maximization in Multi-label Classification: A Comparative Study

Stijn Decubber[1,2], Thomas Mortier[2($\boxtimes$)], Krzysztof Dembczyński[3], and Willem Waegeman[2]

[1] ML6, Esplanade Oscar Van De Voorde 1, 9000 Ghent, Belgium
stijn.decubber@ml6.eu
[2] Department of Data Analysis and Mathematical Modelling,
Ghent University, Coupure links 653, 9000 Ghent, Belgium
{thomasf.mortier,willem.waegeman}@ugent.be
[3] Institute of Computing Science, Poznań University of Technology,
Piotrowo 2, 60-965 Poznań, Poland
krzysztof.dembczynski@cs.put.poznan.pl

Abstract. In recent years several novel algorithms have been developed for maximizing the instance-wise F_β-measure in multi-label classification problems. However, so far, such algorithms have only been tested in tandem with shallow base learners. In the deep learning landscape, usually simple thresholding approaches are implemented, even though it is expected that such approaches are suboptimal. In this article we introduce extensions of utility maximization and decision-theoretic methods that can optimize the F_β-measure with (convolutional) neural networks. We discuss pros and cons of the different methods and we present experimental results on several image classification datasets. The results illustrate that decision-theoretic inference algorithms are worth the investment. While being more difficult to implement compared to thresholding strategies, they lead to a better predictive performance. Overall, a decision-theoretic inference algorithm based on proportional odds models outperforms the other methods. Code related to this paper is available at: https://github.com/sdcubber/f-measure.

Keywords: F_β-measure · Bayes optimal classification
Multi-label image classification · Convolutional neural networks

1 Introduction

Amongst other utility measures, the F_β-measure is commonly used as a performance metric for multi-label classification (MLC) problems, especially in the case of imbalanced label occurrences. Given a prediction $\boldsymbol{h}(\boldsymbol{x}) = (h_1(\boldsymbol{x}), \ldots, h_m(\boldsymbol{x}))^T$ of an instance $\boldsymbol{x}$ with m-dimensional binary label vector $\boldsymbol{y} = (y_1, \ldots, y_m)^T$,

M. Berlingerio et al. (Eds.): ECML PKDD 2018, LNAI 11051, pp. 290–305, 2019.
https://doi.org/10.1007/978-3-030-10925-7_18

where both $h(x)$ and y belong to $\{0,1\}^m$, the F_β-measure is usually computed in an instance-wise manner:

$$F_\beta(y, h(x)) = \frac{(1+\beta^2)\sum_{i=1}^m y_i h_i(x)}{\beta^2 \sum_{i=1}^m y_i + \sum_{i=1}^m h_i(x)} \in [0,1] \ , \qquad (1)$$

where $0/0 = 1$ by definition. Alternative ways of computing the F_β-measure are macro-averaging, in which the F_β-measure is not computed per instance, but per label, and micro-averaging, in which the computation is done over the whole instance-label matrix for a predefined dataset. The instance-wise F_β-measure will be the focus of this work. It is a very relevant measure for many practical MLC problems.

In recent years, specialized algorithms have been developed for optimizing the instance-wise F_β-measure. Roughly speaking, existing methods can be subdivided into two categories: utility maximization methods and decision-theoretic approaches. Algorithms in the first category intend to minimize a specific loss during the training phase. Many of those algorithms seek for thresholds on scoring functions [1–4], but also a few more complicated approaches have been proposed [5,6]. For the related problem of binary classification, F_β-measure maximization at training time can be achieved via extensions of logistic regression [7], boosting [8] or support vector machines [9,10]. However, F_β-measure maximization is simpler in binary classification than in multi-label classification, because predictions for subsequent instances are independent, while predictions for subsequent labels are not.

Decision-theoretic methods depart from a different perspective. These methods usually fit a probabilistic model $P(y \mid x)$ to the data during training, followed by an inference procedure at prediction time. This inference procedure consists of optimizing the following optimization problem:

$$h_F(x) = \operatorname*{argmax}_{h \in \{0,1\}^m} \mathbb{E}_{Y \mid x} [F_\beta(Y, h)] = \operatorname*{argmax}_{h \in \{0,1\}^m} \sum_{y \in \{0,1\}^m} P(y \mid x) F_\beta(y, h), \quad (2)$$

in which the ground-truth is a vector of random variables $Y = (Y_1, Y_2, \ldots, Y_m)$, $\mathbb{E}_{Y \mid x}$ denotes the expectation for an underlying probability distribution P over $\{0,1\}^m$, and h denotes a potential prediction. This is a non-trivial optimization problem without closed-form solution. Moreover, a brute-force search requires checking all 2^m combinations of h and summing over an exponential number of terms in each combination and is hence infeasible for moderate values of m [11].

For solving (2), one can distinguish approximate inference algorithms, such as those of [12–17], and Bayes optimal methods [18–20]. Approximate algorithms depart from the assumption of independence of the Y_i, i.e.,

$$P(y \mid x) = \prod_{i=1}^m (p_i(x))^{y_i} (1 - p_i(x))^{1-y_i}, \qquad (3)$$

with $p_i(x) = P(y_i = 1 \mid x)$. In contrast, exact algorithms do not require the independence assumption, which is not realistic for many MLC problems. Optimization problem (2) seems to require information about the entire joint distribution

$P(\boldsymbol{y} \mid \boldsymbol{x})$. However, exact algorithms have been proposed that solve the problem in an efficient way, by estimating only a quadratic instead of an exponential (with respect to m) number of parameters of the joint distribution.

The main goal of this article is to provide additional insights on how the instance F_β-measure can be optimized in the context of (convolutional) neural networks. Multi-label classification methods are commonly used in image analysis, for classical tasks such as tagging, segmentation or edge detection. In such studies the F_β-measure is often reported as a performance measure that reflects the practical performance of a classifier in a realistic way. However, the F_β-measure maximization methods that are discussed above have only been tested on simple MLC problems with shallow base learners that do not involve feature learning. Likewise, deep convolutional neural networks, which dominate the image classification landscape, usually only consider crude solutions when optimizing the F_β-measure. Researchers often stick to simple approaches that are easy to implement, while ignoring the shortcomings of those approximations. In a recent Kaggle competition which involved the multi-label classification of satellite images[1], one could observe that almost all top-scoring submissions applied simple thresholding strategies, which are known to be suboptimal. Only one author in the top ten reported improvement gains by testing something different than thresholding strategies. It is therefore interesting to investigate in a more systematic way how the instance-wise F_β-measure can be maximized in the context of deep neural networks.

This article is organized as follows. In Sect. 2, we will introduce neural network extensions of different algorithms, including several thresholding strategies, and approximate and exact inference methods. Moreover, we introduce a new model based on proportional odds to estimate the set of parameters of the joint label distribution, required to perform exact inference with existing methods. All those methods have pros and cons, which will be discussed without imposing sympathy for one particular method from the beginning. In Sect. 3, we present the results of a comparative experimental study on four image classification datasets, illustrating the behavior of the methods that we introduce. Our proportional odds model outperforms the alternatives in almost all scenarios. We end with a few clear conclusions.

2   Algorithms for Deep F_β-Measure Maximization

In this section we present six different algorithms that can be applied in tandem with (convolutional) neural networks to optimize the F_β-measure. To this end, we make a major distinction between three utility maximization methods and three decision-theoretic methods.

2.1   Utility Maximization Methods

When optimizing F_β-measure during training with (deep) neural networks, engineers usually consider thresholding strategies on marginal probabilities via a

---

[1] https://www.kaggle.com/c/planet-understanding-the-amazon-from-space.

simple line search. Other existing utility maximization methods usually lead to constrained optimization problems, making them not immediately applicable to neural network training. We present three algorithms that seek to optimize the F_β-measure by means of applying specific thresholds to the predicted marginal probabilities $p_1(x), \ldots, p_m(x)$. To this end, we assume that those marginals are modelled with a (convolutional) neural network with one output neuron per label, obtained via a logistic output layer:

$$p_i(x) = \frac{\exp(w_i^T \phi(x; \psi))}{1 + \exp(w_i^T \phi(x; \psi))}, \tag{4}$$

in which w_i represents parameter vectors, and ϕ denotes the map from the input layer to the one-but-last layer, parameterized by a parameter set ψ.

This approach is in multi-label classification often referred to as binary relevance (BR). In the results section, the three BR-inspired algorithms will be referred to as threshold averaging (BR_t^{avg}), global thresholding ($\text{BR}_t^{\text{glob}}$) and threshold stacking ($\text{BR}_t^{\text{stack}}$), respectively.

Threshold Averaging (BR_t^{avg}). The first thresholding approach consists of computing a specific optimal threshold $\theta_*^{(i)}$ for each instance $x^{(i)}$ during training time. The algorithm passes over the data exactly once and considers the marginal probabilities $p_1(x^{(i)}), \ldots, p_m(x^{(i)})$ in decreasing order as candidate thresholds. At test time, the average optimal threshold over the training dataset is applied as a common threshold. Algorithm 1 provides pseudocode for a single instance; the algorithm can be applied on an entire training dataset with $\mathcal{O}(mn)$ time complexity, by vectorizing the counter variables.

Algorithm 1. Threshold Averaging

1: **Input**: a training instance (x, y), predictions $p(x) = (p_1(x), \ldots, p_m(x))$
2: **compute** $s^y = \sum_{i=1}^m y_i$
3: $p \leftarrow \textbf{sort}(p(x))$ s.t. $p_1 \geq p_2 \geq \ldots \geq p_m$, store the sorting indices in a list L
4: **set** $s^h = 0$, $s^{yh} = 0$, $F_{max} = 0$, $\theta_* = p_1$
5: **for** $k = 1$ to $m - 1$ **do**
6:     $s^{yh} \leftarrow s^{yh} + y_{L[k]}$
7:     **compute** $F = (1 + \beta^2) \frac{s^{yh}}{\beta^2 s^y + k}$
8:     **if** $F > F_{max}$ **then**
9:         $F_{max} \leftarrow F$, $\theta_* \leftarrow p_{k+1}$
10:    **end if**
11: **end for**
12: **return** optimal threshold θ_*

Global Thresholding ($\text{BR}_t^{\text{glob}}$). Algorithm 2 directly finds a single global optimal threshold θ_* at training time. The method acts on the entire training data set by concatenating all marginal probabilities $p_1(x), \ldots, p_m(x)$ for different x, and considering each value as candidate threshold. This second thresholding method seeks to improve over the previous method by considering much more candidate thresholds. However, this comes at the expense of an increasing time complexity. Sorting the vector of all marginals takes $\mathcal{O}(mn \log(mn))$ time, and the computation of the optimal threshold takes $\mathcal{O}(m^2 n)$. Each of those two factors might be dominating, depending on m and n. Let us remark that Algorithm 2 could be substantially simplified if the macro or micro F_β-measure would be optimized instead of the instance-wise F_β-measure. For the instance-wise F_β-measure one needs to keep track of the score for every instance individually for different thresholds, resulting in a higher time complexity compared to the micro and macro F_β-measures. Algorithmically, too, threshold-based optimization of the latter two measures is easier.

Algorithm 2. Global Thresholding

1: **Input:** tr. data $\{(\boldsymbol{x}^{(1)}, \boldsymbol{y}^{(1)}), \dots, (\boldsymbol{x}^{(n)}, \boldsymbol{y}^{(n)})\}$, predictions $\{p(\boldsymbol{x}^{(1)}), \dots, p(\boldsymbol{x}^{(n)}))\}$
2: $\boldsymbol{q} \leftarrow \textbf{concatenate}(\{p(\boldsymbol{x}^{(1)}), \dots, p(\boldsymbol{x}^{(n)})\})$
3: $\boldsymbol{r} \leftarrow \textbf{concatenate}(\{\boldsymbol{y}^{(1)}, \dots, \boldsymbol{y}^{(n)}\})$
4: $\boldsymbol{q} \leftarrow \textbf{sort}(\boldsymbol{q})$ s.t. $q_1 \geq q_2 \geq \dots \geq q_{n \times m}$
5: **compute** $\boldsymbol{s}^y : s_j^y = \sum_{i=1}^m y_i^{(j)}$ $\forall j \in \{1, \dots, n\}$
6: **set** $\boldsymbol{s}^h = \boldsymbol{0}_n$, $\boldsymbol{s}^{yh} = \boldsymbol{0}_n$, $F_{max} = 0$, $\theta_* = q_1$
7: **for** $k = 1$ to $(n \times m) - 1$ **do**
8:     $j \leftarrow$ index of the training instance that corresponds to q_k
9:     $s_j^h \leftarrow s_j^h + 1$, $s_j^{yh} \leftarrow s_j^{yh} + r_j$
10:     **compute** $F = \frac{1}{n} \sum_{i=1}^n (1 + \beta^2) \frac{s_i^{yh}}{\beta^2 s_i^y + s_i^h}$
11:     **if** $F > F_{max}$ **then**
12:         $F_{max} \leftarrow F$, $\theta_* \leftarrow q_{k+1}$
13:     **end if**
14: **end for**
15: **return** global optimal threshold θ_*

Threshold Stacking ($\text{BR}_t^{\text{stack}}$). The final thresholding method presented here tries to predict the instance-wise optimal thresholds for each test instance, in an approach similar to stacking, see e.g. [21,22]. A set of marginal probabilities and optimal thresholds $\{(p(\boldsymbol{x}^{(1)}), \theta_*^{(1)}), \dots, (p(\boldsymbol{x}^{(n)}), \theta_*^{(n)})\}$ is obtained via Algorithm 1 and serves as training data to learn a mapping from probability vectors to thresholds. As such, one ends up with a stacked model structure:

$$\boldsymbol{x} \mapsto p_1(\boldsymbol{x}), \dots, p_m(\boldsymbol{x}) \mapsto \theta_*(\boldsymbol{x}).$$

The first mapping consists of a (convolutional) neural network that predicts marginal probabilities, and the second mapping will be a ridge regression model that transforms the distribution over marginals to a distribution-specific threshold. The distribution of marginal probabilities depends on $\boldsymbol{x}$, so one can argue that the predicted threshold is instance-specific.

2.2   Decision-Theoretic Methods

We present in total three algorithms that optimize the F_β-measure in a decision-theoretic perspective using so-called plug-in classifiers, i.e. classifiers that fit a probabilistic model at training time, followed by an inference phase at test time. We first mention an approach that departs from marginal probabilities and optimizes (2) in an approximate way by assuming label independence. This approach will be referred to as the label independence F_β plug-in classifier (LFP). Subsequently, we introduce two methods that do not have this restriction and provide exact solutions for (2). These methods do not require the plugin of m estimated marginal probabilities but rather a set of $m^2 + 1$ parameters of the joint distribution. To this end, we propose a neural network architecture with an output layer that is modified compared to the models that are typically used for BR estimation of marginal probabilities. The two exact methods differ in the hypothesis class that is considered.

All the methods in this section rely on solving (2) via outer and inner maximization. Let H_k denote the space of all possible predictions that contain exactly k positive labels: $H_k = \{\boldsymbol{h} \in \{0,1\}^m \mid \sum_{i=1}^m h_i = k\}$. The inner maximization then solves

$$\boldsymbol{h}_k(\boldsymbol{x}) = \underset{\boldsymbol{h} \in H_k}{\operatorname{argmax}} \ \mathbb{E}_{\boldsymbol{Y} \mid \boldsymbol{x}} \left[F_\beta(\boldsymbol{Y}, \boldsymbol{h}) \right], \tag{5}$$

for each k. Subsequently, the outer maximization seeks to find the F_β-maximizer $\boldsymbol{h}_F$:

$$\boldsymbol{h}_F(\boldsymbol{x}) = \underset{\boldsymbol{h} \in \{\boldsymbol{h}_0(\boldsymbol{x}), \dots, \boldsymbol{h}_m(\boldsymbol{x})\}}{\operatorname{argmax}} \mathbb{E}_{\boldsymbol{Y} \mid \boldsymbol{x}} \left[F_\beta(\boldsymbol{Y}, \boldsymbol{h}) \right]. \tag{6}$$

The solution to (6) is found by checking all $m + 1$ possibilities. The algorithms discussed below differ in the way they solve the inner maximization (5).

Label Independence F_β Plug-In Classifier (LFP). By assuming independence of the random variables $Y_1, \dots, Y_m$, optimization problem (5) can be substantially simplified. It has been shown independently in [12] and [14] that the optimal solution then always contains the labels with the highest marginal probabilities, or no labels at all.

Theorem 1 [12]. *Let* $Y_1, Y_2, \dots, Y_m$ *be independent Bernoulli variables with parameters* $p_1, p_2, \dots, p_m$ *respectively. Then, for all* $j, k \in \{1, \dots, m\}$, $h_{F,j} = 1$ *and* $h_{F,k} = 0$ *implies* $p_j \geq p_k$.

As a consequence, only a few hypotheses $\boldsymbol{h}$ ($m+1$ instead of 2^m) need to be examined, and the computation of the expected F_β-measure can be performed in an efficient way. [13–16] have proposed exact procedures for computing the F_β-maximizer under the assumption of label independence. All those methods take as input predicted marginal probabilities $(p_1, p_2, \dots, p_m)$ with shorthand notation $p_i = p_i(\boldsymbol{x})$, and they all obtain the same solution. In what follows we only discuss the method of [16], which is the most efficient among the four implementations. This method only works for rational β^2; in other cases a less efficient algorithm can be used.

As a starting point, let us assume that the labels are sorted according to the marginal probabilities and let $h_k(x)$ be the prediction that returns a one for the labels with the k highest marginal probabilities and zero for the other labels. Furthermore, let $s_{i:j}^y = \sum_{l=i}^j y_l$, then one can observe that

$$
\mathbb{E}\left[F_\beta(Y, h_k(x))\right] = \sum_{y \in \{0,1\}^m} F_\beta(y, h_k(x)) P(y \mid x) \tag{7}
$$

$$
= \sum_{\substack{0 \le k_1 \le k \\ 0 \le k_2 \le m-k}} \frac{P(s_{1:k}^y = k_1) P(s_{k+1:m}^y = k_2)(1+\beta^2)k_1}{k + \beta^2(k_1 + k_2)}
$$

$$
= \sum_{k_1=0}^{k} (1+\beta^{-2}) k_1 P(s_{1:k}^y = k_1) s(k, k\beta^{-2} + k_1),
$$

where $s(k,\alpha) = \sum_{k_2=0}^{m-k} P(s_{k+1:m}^y = k_2)/(\alpha + k_2)$. Now observe that

$$
P(s_{k:m}^y = i) = p_k P(s_{k+1:m}^y = i-1) + (1 - p_k) P(s_{k+1:m}^y = i).
$$

As a result, the s-values for different values of k in (7) can be computed recursively:

$$
s(k-1,\alpha) = \sum_{k_2=0}^{m-k+1} \frac{P(s_{k:m}^y = k_2)}{\alpha + k_2}
$$

$$
= p_k \sum_{k_2=0}^{m-k+1} \frac{P(s_{k+1:m}^y = k_2 - 1)}{\alpha + k_2} + (1 - p_k) \sum_{k_2=0}^{m-k+1} \frac{P(s_{k+1:m}^y = k_2)}{\alpha + k_2}
$$

$$
= p_k \sum_{k_2=0}^{m-k} \frac{P(s_{k+1:m}^y = k_2)}{\alpha + k_2 + 1} + (1 - p_k) \sum_{k_2=0}^{m-k} \frac{P(s_{k+1:m}^y = k_2)}{\alpha + k_2}
$$

$$
= p_k s(k, \alpha + 1) + (1 - p_k) s(k, \alpha),
$$

with $s(m,\alpha) = 1/\alpha$ and $s(k,\alpha) = 0$ when $k < 0$ or $k > m$. Remark that the transition from the second to the third line follows from an index change.

The recursive formula suggests a dynamic programming implementation with k ranging from $k = m$ to $k = 1$, as given in Algorithm 3. Here we first introduce a list of lists, using double indexing, such that $L[k][j] = P(s_{1:k}^y = j)$ with $j \in \{-1, 0, \dots, k+1\}$. This data structure can also be initialized via dynamic programming:

$$
L[k][j] = p_k P(s_{1:k-1}^y = j - 1) + (1 - p_k) P(s_{1:k-1}^y = j)
$$

$$
= p_k L[k-1][j-1] + (1 - p_k) L[k-1][j]
$$

using $L[1] = [0, (1-p_1), p_1, 0]$ and $L[k][-1] = L[k][k+1] = 0$. After initializing those lists, one can proceed with computing $s(k,\alpha)$ for rational β^2. To this end, we introduce $S[i] = s(k, i/q)$ with $\beta^2 = q/r$, which leads to the implementation given in Algorithm 3. Further speed-ups can be obtained via Taylor series approximations, which might be useful when m becomes very large.

Algorithm 3. LFP – Ye et al. (2012)

1: **Input**: predictions $p(x) = (p_1(x), \ldots, p_m(x))$, $\beta^2 = q/r$
2: $p \leftarrow \mathbf{sort}(p)$ s.t. $p_1 \geq \ldots \geq p_m$
3: **initialize** $L \leftarrow$ a list of m empty lists
4: **set** $L[1] = [0, (1 - p_1), p_1, 0]$
5: **for** $k = 2$ to m **do**
6:    $L[k] \leftarrow$ a list of $k + 3$ zeros with index starting at -1
7:    **for** $j = 0$ to k **do**
8:       $L[k][j] \leftarrow p_k \times L[k-1][j-1] + (1 - p_k) \times L[k-1][j]$
9:    **end for**
10: **end for**
11: For $1 \leq i \leq (q + r)m$: $S[i] \leftarrow q/i$
12: **for** $k = m$ to 1 **do**
13:    $\mathbb{E}\left[F_\beta(Y, h_k(x))\right] \leftarrow \sum_{k_1=0}^{k}(1 + r/q)k_1 L[k][k_1]S[rk + qk_1]$
14:    **for** $i = 1$ to $(q + r)(k - 1)$ **do**
15:       $S[i] \leftarrow (1 - p_k)S[i] + p_k S[i + q]$
16:    **end for**
17: **end for**
18: $k \leftarrow \operatorname{argmax}_k \mathbb{E}\left[F_\beta(Y, h_k^*(x))\right]$
19: **return** $h_F(x)$ by setting $h_i = 1$ for the k labels with the highest p_i

---

General F_β Maximizer (GFM). The algorithm that was explained in the previous section assumed that the labels are independent, so that only marginals need to be modelled in order to solve inner problem (5). In what follows we discuss two different extensions of an alternative algorithm that does not assume label independence [19]. The algorithm is Bayes optimal for any probability distribution, but the price one has to pay for this is that more parameters of $P(y \mid x)$ must be estimated. As a starting point, we introduce the following shorthand notations:

$$s^y = s_{1:m}^y \quad \text{and} \quad \Delta_{ik} = \sum_{y:y_i=1} \frac{P(y \mid x)}{\beta^2 s^y + k}.$$

By plugging (1) into (5), one can write

$$h_k = \operatorname*{argmax}_{h \in H_k} \sum_{y \in \{0,1\}^m} \frac{(1 + \beta^2) \sum_{i=1}^{m} y_i h_i P(y \mid x)}{\beta^2 s^y + k}. \tag{8}$$

Swapping the sums in (8) leads to

$$
\begin{aligned}
h_k &= \underset{h \in H_k}{\operatorname{argmax}}(1 + \beta^2) \sum_{i=1}^{m} h_i \sum_{y \in \{0,1\}^m} \frac{y_i P(y \mid x)}{\beta^2 s^y + k} \\
&= \underset{h \in H_k}{\operatorname{argmax}}(1 + \beta^2) \sum_{i=1}^{m} h_i \Delta_{ik} \,.
\end{aligned}
\tag{9}
$$

The inner maximization is solved by setting $h_i = 1$ for the top k values of Δ_{ik}. For each h_k, $\mathbb{E}\left[F_\beta(Y, h_k)\right]$ is stored and used to solve the outer maximization. For the specific case of h_0, $\mathbb{E}\left[F_\beta(Y, h_0)\right]$ equals $P(y = 0 \mid x)$, which needs to be estimated separately. Algorithm 4 provides pseudocode for the complete procedure. This algorithm requires Δ_{ik} for $1 \le i, k \le m$ and $P(y = 0 \mid x)$, that is, $m^2 + 1$ parameters to obtain h_F. With these parameters, the solution can be obtained in $\mathcal{O}(m^2)$ time, i.e., the dominating part of the procedure is the inner maximization: for each k, a selection of the top k elements must be done, which can be accomplished in linear time. Thus, compared to the approach that assumed label independence, more parameters need to be estimated. The advantage of not imposing any distributional assumptions brings a more difficult estimation problem as disadvantage. Depending on the distributional properties of a specific dataset, it can therefore be the case that one algorithm outperforms the other, or the other way around.

Algorithm 4. General F_β Maximizer (GFM)

1: **Input:** matrix Δ with elements Δ_{ik} and $P(y = 0)$
2: set $\mathbb{E}\left[F_\beta(Y, h_0)\right] = P(y = 0)$
3: **for** $k = 1$ to m **do**
4:     **solve inner maximization:** $h_k(x) = \operatorname{argmax}_{h \in H_k} \mathbb{E}\left[F_\beta(Y, h)\right]$
      by setting $h_i = 1$ for the k labels with the highest Δ_{ik}
5:     set $\mathbb{E}\left[F_\beta(Y, h_k)\right] \leftarrow (1 + \beta^2) \sum_{i=1}^{m} h_i \Delta_{ik}$
6: **end for**
7: **for** $k = 0$ to m **do**
8:     **solve outer maximization:** $h_F = \operatorname{argmax}_{h \in \{h_0(x), \dots, h_m(x)\}} \mathbb{E}\left[F_\beta(Y, h)\right]$
9: **end for**
10: **return** $h_F(x)$

Estimating Δ with Multinomial Regression (GFM$_{\mathbf{MR}}$). [19] proposed the following scheme to estimate the probabilities Δ_{ik}. Let P and W denote two $m \times m$ matrices with elements

$$
p_{is} = P(y_i = 1, s^y = s \mid x), \quad w_{rk} = (\beta^2 r + k)^{-1},
$$

respectively. Then, the $m \times m$ matrix Δ with elements Δ_{ik} can be obtained by

$$
\Delta = PW.
$$

When using simple base learners, one can proceed to estimate $\boldsymbol{P}$ by reducing the problem to m independent problems, each with up to $m + 1$ classes. Each subproblem i involves the estimation of

$$P(y = [\![y_i = 1]\!] \cdot s^y \mid \boldsymbol{x}), \quad \forall y \in \{0, \ldots, m\}, \tag{10}$$

which sum to one. The subproblems can hence be solved with multinomial regression. For $y = \{1, \ldots, m\}$, these probabilities make up the elements of the rows of $\boldsymbol{P}$. Similarly as for the deep neural network that estimated marginal probabilities, we model the i-th row of $\boldsymbol{P}$ via a softmax layer:

$$p_{is}(\boldsymbol{x}) = \frac{\exp(\boldsymbol{w}_{is}^T \phi(\boldsymbol{x}; \boldsymbol{\psi}))}{\sum_{s=0}^{m} \exp(\boldsymbol{w}_{is}^T \phi(\boldsymbol{x}; \boldsymbol{\psi}))},$$

with $i = 1, \ldots, m$, $\boldsymbol{w}_{is}$ parameter vectors, and ϕ the map that originates from the feature learning phase, again parameterized by parameter set $\boldsymbol{\psi}$.

It should be noted that s^y equals m only in the worst case where an instance is attributed with all possible labels. This is rarely encountered in practice. Let $s_m = \max_{1 \leq j \leq n} \sum_{i=1}^{m} y_i^{(j)}$, then the total number of output classes for each subproblem (10) can be reduced to $s_m + 1$. Nevertheless, fitting each of multinomial regression problems independently is undesirable when the cost of training the base learners becomes higher, as with deep (convolutional) neural networks, especially for large m. We propose the natural solution of estimating $\boldsymbol{P}$ in its entirety as the output of a single neural network. The two-dimensional final layer of the network should contain m rows of $(s_m + 1)$ output neurons, where a row-wise soft-max transformation is applied. Then the loss to be minimized during training is composed of m cross-entropy losses, which can be minimized using stochastic gradient descent. The m^2 entries required for $\boldsymbol{P}$ can be obtained from the output of the network by discarding the first column and by adding $m - s_m$ columns with zeros.

Estimating Δ with Ordinal Regression (GFM$_{\text{OR}}$). Additionally, we propose to reformulate the problem of estimating the elements of Δ as an ordinal regression problem. The key insight is to factorize the probabilities p_{is} as follows:

$$p_{is} = P(y_i = 1, s^y = s \mid \boldsymbol{x}) = P(s^y = s \mid y_i = 1, \boldsymbol{x}) \, P(y_i = 1 \mid \boldsymbol{x}).$$

As before, $P(y_i = 1 \mid \boldsymbol{x})$ can be estimated by means of BR. In the conditional probability $P(s^y = s \mid y_i = 1, \boldsymbol{x})$, s^y can take on values from 1 to s_m. By exploiting the ordinal nature of the variable s^y, one can estimate the conditional probability with proportional odds models, while reducing the number of parameters, compared to GFM$_{\text{MR}}$ [23]. After estimating these conditional probabilities, they can be multiplied with the marginals to obtain the probabilities p_{is} required for GFM.

Taking into account the conditioning on $y_i = 1$, one can choose to estimate m independent proportional odds models. However, we will consider a *global* proportional odds model, consisting of m proportional odds submodels which are optimized jointly in a multi-task learning way. As such, the i-th submodel is characterized by s_m classes, a parameter vector $\boldsymbol{w}_i$ and a vector of bias terms $\boldsymbol{b}^{(i)} = (b_0^{(i)}, b_2^{(i)}, \ldots, b_{s_m}^{(i)})$, subject to

$$b_0^{(i)} < b_1^{(i)} < \cdots < b_{s_m}^{(i)}, \tag{11}$$

with $b_0^{(i)} = -\infty$ and $b_{s_m}^{(i)} = \infty$.

Formally speaking, the i-th proportional odds model will estimate the cumulative probabilities

$$P(s^y \le s \mid y_i = 1, \boldsymbol{x}) = \frac{\exp(\boldsymbol{w}_i^T \phi(\boldsymbol{x}; \boldsymbol{\psi}) - b_s^{(i)})}{1 + \exp(\boldsymbol{w}_i^T \phi(\boldsymbol{x}; \boldsymbol{\psi}) - b_s^{(i)})}, \text{ for } i \in \{1, \ldots, m\}, \tag{12}$$

where we depart from some learnable feature representation $\phi(\boldsymbol{x}; \boldsymbol{\psi})$, as in the other methods. Consequently, the conditional distribution of s^y can then be retrieved as follows:

$$P(s^y = s \mid y_i = 1, \boldsymbol{x}) = P(s^y \le s \mid y_i = 1, \boldsymbol{x}) - P(s^y \le s - 1 \mid y_i = 1, \boldsymbol{x}).$$

Furthermore, we estimate the model parameters in (12) jointly for $i \in \{1, \ldots, m\}$, by minimizing the following log-likelihood function:

$$\underset{\boldsymbol{W}, \boldsymbol{B}, \boldsymbol{\psi}}{\operatorname{argmin}} \left(-\sum_n \sum_{i=1}^{m} \sum_{s=1}^{s_m} I_{nis} \, T\Big(P(s^y = s \mid y_i = 1, \boldsymbol{x})\Big) \right), \tag{13}$$

with $\boldsymbol{W} = (\boldsymbol{w}_1, \ldots, \boldsymbol{w}_m)$, $\boldsymbol{B} = (\boldsymbol{b}^{(1)}, \ldots, \boldsymbol{b}^{(m)})$ and I_{nis} a binary indicator, which is one when the n-th training instance $(\boldsymbol{x}, \boldsymbol{y})$ has $y_i = 1$ and $s^y = s$. T is a transformation function

$$T(z; \epsilon) = \begin{cases} \log \epsilon & \text{if } z \le 0 \\ \log z & \text{if } z > 0 \end{cases},$$

for $\epsilon > 0$, that defines a truncated log-likelihood.

This transformation can be seen as the modified negative log-likelihood of the proportional odds model. It is needed to guarantee numerical stability of the optimization algorithm, in case $P(s^y = s \mid y_i = 1, \boldsymbol{x})$ becomes negative. This might happen in the early optimization steps, as (11) is not necessarily obeyed. Moreover, when the ordering constraint on the thresholds is not fulfilled, this will be directly penalized by the truncated log-likelihood, provided that ϵ is chosen sufficiently small, e.g. $\epsilon = 1e^{-10}$. The truncated log-likelihood will hence yield a similar effect as logarithmic barrier penalty terms, which are sometimes used to enforce monotonicity as in (11).

Although GFM$_{\text{OR}}$ needs less parameters to estimate p_{is} than GFM$_{\text{MR}}$, it requires m values for the marginals as additional input. In case a separate model is used to estimate the marginals (starting from the same feature representation of size d), the parameter requirements for BR + GFM$_{\text{OR}}$ are $dm + m + dm + m(s_m - 1)$, which boils down to $m \times (2d + s_m)$. This number will still be lower than the number of parameters required for GFM$_{\text{MR}}$, which can be rewritten as $m \times ((s_m + 1)d + s_m + 1)$.

3 Empirical Analysis

3.1 Experimental Setup

We compare the discussed methods by means of empirical evaluation on real-world datasets. Estimates of the marginal probability vectors are made by means of BR in the form of a convolutional neural network with m output nodes subject to a sigmoid non-linearity in the output layer, as given in Eq. 4. Our results include the F_β-measure scores obtained with BR, without any form of F_β-measure maximization. The marginal probabilities for the training data obtained by BR serve as input for the thresholding methods BR_t^{avg}, BR_t^{glob} and BR_t^{stack}. GFM_{MR} and GFM_{OR} estimate the $m^2 + 1$ parameters of $P(\boldsymbol{y} \mid \boldsymbol{x})$, required for GFM, with multinomial regression and proportional odds, respectively. The GFM algorithm is then used in tandem with these methods to obtain optimal predictions. Finally, the LFP method starts from the marginal probabilities obtained with BR for the test data.

We report both the F_1 and F_2-measure scores obtained on four publicly available multi-label classification image datasets: PASCAL VOC 2007 [24], PASCAL VOC 2012 [25], Microsoft COCO [26] and the Kaggle Planet dataset [27]. We use the recommended *train-val-test* split for VOC 2007 and perform custom training/validation splits for the other datasets. Table 1 provides some summarizing statistics. All experiments were carried out on a single NVIDIA GTX 1080Ti GPU. All algorithms were implemented in Python using TensorFlow [28], Keras [29] and Pytorch [30].

When it comes to the experiments, for each dataset, the features are vectors of size 512 obtained by resizing the images to 224×224 pixels and passing them through the convolutional part of an entire VGG16 architecture, including a max-pooling operation [31]. The final fully connected classification layers from the original architecture are replaced by a single fully connected layer with 128 neurons (ReLu activation), followed by either a single-layer BR, GFM_{MR} or GFM_{OR} classifier, as described in Sect. 2. The weights and biases of this architecture were set to those obtained by training the network on ImageNet; these are publicly available and accessible through the Keras API. First, the convolutional layers are fixed and the fully connected classification layers are trained until convergence. Then, the entire network is fine-tuned with a lower learning rate. In both stages, early stopping is applied, similarly as before. Moreover, the BR estimator consists of a single-layer neural network with m output nodes. Likewise, the GFM_{BR} and GFM_{OR} models consist of single-layer neural networks parameterized as described in the previous section. A small amount of dropout regularization was applied (dropout probability 0.2) at the input level. All models were trained with the Adam optimization algorithm (learning rate $1e^{-3}$), where early stopping was applied with a five epochs patience counter.

3.2 Experimental Results

The results for the conducted experiments are presented in Table 2. As expected, BR without any attempt at maximizing the F_β-measure leads to the worst

Table 1. Summary statistics for the four datasets. m is the number of labels, s_m the maximum number of labels attributed to a single instance in the training data.

	n_{train}	n_{val}	n_{test}	m	s_m
PLANET	32383	8096	61191	17	9
VOC 2007	2501	2510	4952	20	7
VOC 2012	4859	858	5823	20	6
COCO	65665	16416	40137	80	18

performance in almost all cases. Rather surprising is the fact that BR_t^{avg} performs worse than BR for the F_1-measure in several cases, meaning that the average optimal threshold for the training instances is not better than just 0.5 as a threshold. This is especially true for the Planet dataset, which has the smallest m and an imbalanced label distribution. Conversely, this does not occur for the COCO dataset, where, due to a larger number of labels, more candidate thresholds are considered for each instance by Algorithm 1. BR_t^{glob} consistently outperforms both BR and BR_t^{avg}. This is as expected, since BR_t^{glob} considers all $m \times n$ predicted marginal probabilities as candidates. However, this comes at the cost of higher time complexity, as discussed in Sect. 2.

The performance of BR_t^{stack} varies across datasets and seems to depend on whether F_1 or F_2 is the measure of interest. For F_1 it performs substantially worse than BR_t^{glob}, whereas it even becomes competitive with the decision-theoretic approaches for F_2. Figure 1 gives further insights w.r.t. the behavior of BR_t^{stack}. It shows for training data the empirical distribution of instance-wise thresholds obtained by Algorithm 1, as well as the thresholds predicted by BR_t^{stack}. One can observe that for all four datasets the two distributions differ substantially, indicating that the threshold stacking method is not always capable of predicting a good threshold. The empirical distribution of instance-wise thresholds obtained by Algorithm 1 is here considered as the ground truth. The dotted line indicates the threshold that will be returned by Algorithm 1 after training.

More generally, the decision-theoretic approaches seem to outperform the thresholding methods on all datasets. The GFM algorithm, which is the only algorithm that does not require the assumption of label independence, is the best algorithm in all but one setting. In almost all cases the proportional odds model outperforms the multinomial regression model, which might indicate that the assumption of ordinality for s^y is a valid assumption. However, the differences between both methods are small, so the benefit of a more parsimonious model structure is limited. In addition, the LFP method also yields rather good results. Therefore, we hypothesize that for the analyzed datasets the dependence among the labels is not very strong. Moreover, even though LFP assumes independence, it requires less parameters than the GFM methods.

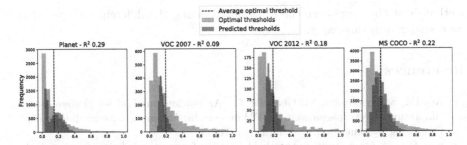

Fig. 1. Empirical distributions of instance-wise thresholds obtained by Algorithm 1 (blue, optimal thresholds), as well as the thresholds predicted by BR_t^{stack} (red, predicted thresholds). Here, the thresholds for training data are shown, and F_2 is the performance measure. The thresholds are obtained by using the convolutional part of a high-quality pre-trained VGG16 architecture (R^2 indicates the quality of the predictions). The dotted line indicates the mean optimal instance-wise threshold, which is returned by BR_t^{avg} after training. See main text for more details. (Color figure online)

Table 2. Comparison of the different methods, with training strategy described in Sect. 3.1.

	Planet		VOC 2007		VOC 2012		COCO	
	F_1	F_2	F_1	F_2	F_1	F_2	F_1	F_2
BR	0.8997	0.8918	0.7398	0.7282	0.7539	0.7405	0.6534	0.6179
BR_t^{avg}	0.8787	0.9135	0.7241	0.8055	0.7286	0.8064	0.6852	0.7170
BR_t^{glob}	0.9017	0.9149	0.7701	0.7973	0.7769	0.8020	0.6846	0.7163
BR_t^{stack}	0.8772	0.9091	0.7121	0.8011	0.7142	0.8007	0.6752	0.7174
GFM_{MR}	**0.9044**	0.9164	0.7918	0.8108	0.7988	0.8154	0.6955	0.7230
GFM_{OR}	0.9026	**0.9172**	0.8005	**0.8211**	**0.8035**	**0.8205**	**0.7040**	**0.7316**
LFP	0.9023	0.9107	**0.8007**	0.8177	0.8032	0.8184	0.7011	0.7177

4   Conclusion

In this article we introduced extensions of utility maximization and decision-theoretic methods that can optimize the F_β-measure with (convolutional) neural networks. We discussed pros and cons of the different methods and we presented experimental results on several image classification datasets. The results illustrate that decision-theoretic inference algorithms are worth the investment. While being more difficult to implement compared to thresholding strategies, they lead to a superior predictive performance. This is a surprising result, given the popularity of thresholding in deep neural networks. For most of the datasets, the inferior performance of thresholding strategies was remarkable, while also big differences could be observed among the different ways of defining a threshold. Overall, the best performance was obtained with an exact decision-theoretic method based on proportional odds models. This is interesting, because this method is at the same time the most novel among the different methods that were analyzed in this paper.

References

1. Keerthi, S., Sindhwani, V., Chapelle, O.: An efficient method for gradient-based adaptation of hyperparameters in SVM models. In: Advances in Neural Information Processing Systems, vol. 19. MIT Press (2007)
2. Fan, R., Lin, C.: A study on threshold selection for multi-label classification. Technical report, Department of Computer Science, National Taiwan University (2007)
3. Zhang, X., Graepel, T., Herbrich, R.: Bayesian online learning for multi-label and multi-variate performance measures. In: Proceedings of the Conference on Artificial Intelligence and Statistics (AISTATS), pp. 956–963 (2010)
4. Lipton, Z.C., Elkan, C., Naryanaswamy, B.: Optimal thresholding of classifiers to maximize F1 measure. In: Calders, T., Esposito, F., Hüllermeier, E., Meo, R. (eds.) ECML PKDD 2014, Part II. LNCS (LNAI), vol. 8725, pp. 225–239. Springer, Heidelberg (2014). https://doi.org/10.1007/978-3-662-44851-9_15
5. Petterson, J., Caetano, T.: Reverse multi-label learning. In: Advances in Neural Information Processing Systems, vol. 24 (2010)
6. Petterson, J., Caetano, T.: Submodular multi-label learning. In: Advances in Neural Information Processing Systems, vol. 25 (2011)
7. Jansche, M.: Maximum expected F-measure training of logistic regression models. In: Proceedings of the Human Language Technology Conference and the Conference on Empirical Methods in Natural Language Processing (HLT/EMNLP), pp. 736–743 (2005)
8. Kokkinos, I.: Boundary detection using F-Measure-, filter- and feature- (f^3) boost. In: Daniilidis, K., Maragos, P., Paragios, N. (eds.) ECCV 2010, Part II. LNCS, vol. 6312, pp. 650–663. Springer, Heidelberg (2010). https://doi.org/10.1007/978-3-642-15552-9_47
9. Musicant, D., Kumar, V., Ozgur, A.: Optimizing F-measure with support vector machines. In: Proceedings of the International FLAIRS Conference, Haller, pp. 356–360. AAAI Press (2003)
10. Joachims, T.: A support vector method for multivariate performance measures. In: Proceedings of the International Conference on Machine Learning (ICML), pp. 377–384 (2005)
11. Waegeman, W., Dembczyński, K., Jachnik, A., Cheng, W., Hüllermeier, E.: On the Bayes-optimality of F-measure maximizers. J. Mach. Learn. Res. 15(1), 3333–3388 (2014)
12. Lewis, D.: Evaluating and optimizing autonomous text classification systems. In: Proceedings of the International ACM Conference on Research and Development in Information Retrieval (SIGIR), pp. 246–254 (1995)
13. Chai, K.: Expectation of F-measures: tractable exact computation and some empirical observations of its properties. In: Proceedings of the International ACM Conference on Research and Development in Information Retrieval (SIGIR) (2005)
14. Jansche, M.: A maximum expected utility framework for binary sequence labeling. In: Proceedings of the Annual Meetings of the Association for Computational Linguistics (ACL), pp. 736–743 (2007)
15. Quevedo, J., Luaces, O., Bahamonde, A.: Multilabel classifiers with a probabilistic thresholding strategy. Pattern Recogn. 45, 876–883 (2012)

16. Ye, N., Chai, K., Lee, W., Chieu, H.: Optimizing F-measures: a tale of two approaches. In: Proceedings of the International Conference on Machine Learning (2012)
17. Dembczyński, K., Kotłowski, W., Koyejo, O., Natarajan, N.: Consistency analysis for binary classification revisited. In: Proceedings of the International Conference on Machine Learning (ICML), vol. 70. PMLR (2017)
18. Dembczyński, K., Waegeman, W., Cheng, W., Hüllermeier, E.: An exact algorithm for F-measure maximization. In: Advances in Neural Information Processing Systems, vol. 25 (2011)
19. Dembczyński, K., Jachnik, A., Kotlowski, W., Waegeman, W., Hüllermeier, E.: Optimizing the F-measure in multi-label classification: plug-in rule approach versus structured loss minimization. In: Proceedings of the International Conference on Machine Learning (ICML) (2013)
20. Gasse, M., Aussem, A.: F-Measure maximization in multi-label classification with conditionally independent label subsets. In: Frasconi, P., Landwehr, N., Manco, G., Vreeken, J. (eds.) ECML PKDD 2016, Part I. LNCS (LNAI), vol. 9851, pp. 619–631. Springer, Cham (2016). https://doi.org/10.1007/978-3-319-46128-1_39
21. Wolpert, D.H.: Original contribution: stacked generalization. Neural Netw. **5**(2), 241–259 (1992)
22. Cheng, W., Hüllermeier, E.: Combining instance-based learning and logistic regression for multilabel classification. Mach. Learn. **76**(2–3), 211–225 (2009)
23. Agresti, A.: Categorical Data Analysis, 3rd edn. Wiley, Hoboken (2013)
24. Everingham, M., Van Gool, L., Williams, C.K.I., Winn, J., Zisserman, A.: The PASCAL Visual Object Classes Challenge 2007 (VOC2007) Results (2007). http://www.pascal-network.org/challenges/VOC/voc2007/workshop/index.html
25. Everingham, M., Van Gool, L., Williams, C.K.I., Winn, J., Zisserman, A.: The PASCAL Visual Object Classes Challenge 2012 (VOC2012) Results (2012). http://www.pascal-network.org/challenges/VOC/voc2012/workshop/index.html
26. Lin, T.-Y., et al.: Microsoft COCO: common objects in context. In: Fleet, D., Pajdla, T., Schiele, B., Tuytelaars, T. (eds.) ECCV 2014, Part V. LNCS, vol. 8693, pp. 740–755. Springer, Cham (2014). https://doi.org/10.1007/978-3-319-10602-1_48
27. Kaggle: Planet: Understanding the amazon from space (2017). https://www.kaggle.com/c/planet-understanding-the-amazon-from-space
28. Abadi, M., Agarwal, A., Barham, P., et al.: TensorFlow: Large-scale machine learning on heterogeneous systems (2015). tensorflow.org
29. Chollet, F., et al.: Keras (2015). https://github.com/keras-team/keras
30. Paszke, A., Gross, S., Chintala, S., et al.: Automatic differentiation in pytorch. In: NIPS-W (2017)
31. Simonyan, K., Zisserman, A.: Very deep convolutional networks for large-scale image recognition. arXiv preprint arXiv:1409.1556 (2014)

Ordinal Label Proportions

Rafael Poyiadzi[1]([✉]), Raúl Santos-Rodríguez[1], and Tijl De Bie[2]

[1] Department of Engineering Mathematics, University of Bristol,
Bristol, UK
{rp13102,enrsr}@bristol.ac.uk
[2] Department of Electronics and Information Systems, IDLab,
Ghent University, Ghent, Belgium
Tijl.DeBie@ugent.be

Abstract. In Machine Learning, it is common to distinguish different degrees of supervision, ranging from fully supervised to completely unsupervised scenarios. However, lying in between those, the Learning from Label Proportions (LLP) setting [19] assumes the training data is provided in the form of bags, and the only supervision comes through the proportion of each class in each bag. In this paper, we present a novel version of the LLP paradigm where the relationship among the classes is ordinal. While this is a highly relevant scenario (e.g. customer surveys where the results can be divided into various degrees of satisfaction), it is as yet unexplored in the literature. We refer to this setting as Ordinal Label Proportions (OLP). We formally define the scenario and introduce an efficient algorithm to tackle it. We test our algorithm on synthetic and benchmark datasets. Additionally, we present a case study examining a dataset gathered from the Research Excellence Framework that assesses the quality of research in the United Kingdom.

Keywords: Label Proportions · Ordinal classification
Discriminant learning

1 Introduction

According to the nature of their output, the two dominating tasks in Machine Learning are those of regression and classification. Attracting an increasing interest, Ordinal Classification (also termed Ordinal Regression) [2,5,6,12] falls somewhere in between the two. Similarly to multiclass classification tasks, the practitioner is provided with a set of data points with their corresponding labels coming from a discrete set $\mathcal{C} = \{r_1, \cdots, r_k\}$, but opposed to its nominal sibling, in ordinal classification, the labels exhibit a natural ordering: $r_1 \prec r_2 \prec \cdots \prec r_k$. There is an abundance of examples, ranging from categorizing responses to questions such as "how much do you like Greek food?" to movie ratings or grade prediction. The difference between the labels in these tasks, e.g. {*very bad, bad, good, great, excellent*} and the standard categorical labels, e.g. {*car, pedestrian, bicycle*}, is clear.

© Springer Nature Switzerland AG 2019
M. Berlingerio et al. (Eds.): ECML PKDD 2018, LNAI 11051, pp. 306–321, 2019.
https://doi.org/10.1007/978-3-030-10925-7_19

Let us consider opinion polls where acceptable outcomes range from strongly agree to strongly disagree on a given topic. There is a clear ordinal relationship among the outcomes. However, for privacy reasons, it is often not possible to publish each individual's opinion. On the other hand, it may be possible to aggregate the results over several demographic subsets (e.g., by region). Therefore, this is data that naturally comes in the forms of bags and where, although the ground truth labels might not be available, we might have access to the proportion of each class in each bag. We can also argue that, in many cases, individuals' opinions are not as relevant as an accurate prediction of the bag proportions. A specific example of such tasks is that of rating research papers according to quality, mapping each of them to a set of predefined categories. In the United Kingdom, publicly funded research is regularly assessed under the Research Excellence Framework (REF)[1]. In order to preserve anonymity, based on the papers submitted by the different research units, REF provides a histogram stating how many of the papers submitted were placed in each category, *without revealing which specific paper was in each of these classes*. As before, individual paper ratings are sensitive, but aggregates per submission are fine to publish. Importantly, funding levels are then based on these histograms. A difficulty for universities is that REF does not rely on a public and formal procedure to classify papers, but on the judgment of a panel. Therefore, although this can be cast as an ordinal classification task, unfortunately, the ground truth labels are not available.

As an aggregate supervision is accessible through the histograms, this problem sits in between fully supervised and unsupervised learning. In the non-ordinal case, this has been studied under the name of *learning from label proportions* [9,14,15,20], where the data is assumed to be given in the form of bags (e.g., research units) and only the proportion of each class in each bag is given (histograms). Up to the authors' knowledge, learning a classifier with this level of supervision, in the ordinal setting, has not yet been explored. We call this setting Ordinal Label Proportions (OLP). The OLP methodology developed in this work is able to efficiently make use of the ordinal constrains in order to learn from the past REF data and unveil how the expert panel operated by inferring a scoring function for papers as a function of various properties, such as journal and conference rankings, citation half life, impact factor, number of authors, or Google scholar citation count at time of submission.

The contributions of this paper are threefold. Firstly, we introduce and rigorously define the new learning paradigm of learning from Ordinal Label Proportions. Secondly, we present an efficient algorithm for learning a classifier in this setting based on discriminant learning. Thirdly, we produce a dataset for REF and present our analysis.

The paper is structured as follows. In Sect. 2 we review the related work. In Sect. 3 we introduce the basic formulation of Linear Discriminant Analysis (LDA) in the ordinal setting and show how it can be adopted to be trained

---

[1] https://www.ref.ac.uk/.

with label proportions. In Sect. 4 we respectively present the empirical analysis in both real and synthetic datasets. Section 5 is devoted to the conclusions.

1.1 Problem Formulation

We assume that we have access to a set of observations $\mathbf{X} = \{\mathbf{x}_1, \ldots, \mathbf{x}_n\}$ where $\mathbf{x}_i \in \mathbb{R}^d$. The true labels, $\mathbf{y} = \{y_1, \ldots, y_n\}$ with y_i being the label of observation $\mathbf{x}_i$, also exist but are hidden. Also, the class labels $\{r_1, \cdots, r_k\}$ have a natural order, i.e. $r_1 \prec r_2 \prec \ldots \prec r_k$. The set $\mathbf{X}$ is separated into distinct bags $\mathbf{X} = \bigcup_{k=1}^{K} \mathbf{B}_k$, where each $\mathbf{B}_k$ corresponds to the subset of points assigned to the k-th bag, and $\mathbf{B}_k \cap \mathbf{B}_j = \emptyset, \forall k, j \in [K]$. Moreover, for each bag $\mathbf{B}_k$ we have access to its class proportions, $\boldsymbol{\pi}_k = \{\pi_{k,1}, \ldots, \pi_{k,c}\}$, where $\sum_{h=1}^{c} \pi_{k,h} = 1$, $\pi_{k,h} \geq 0$, with $\pi_{k,h}$ corresponding to the proportion of class h in bag $\mathbf{B}_k$ and c being the number of classes ($c = 2$ being the binary classification setting).

The OLP task is then cast as minimizing the following objective:

$$
\begin{aligned}
d(\boldsymbol{\pi}, \hat{\boldsymbol{\pi}}[\boldsymbol{s}]) + \lambda R[\boldsymbol{s}] \\
s.t.\ \boldsymbol{s}(\boldsymbol{x}_i) \geq \boldsymbol{s}(\boldsymbol{x}_j), \forall i, j \in C
\end{aligned}
\tag{1}
$$

where $\hat{\boldsymbol{\pi}}$ and $\boldsymbol{R}$ are functionals of a scoring function $\boldsymbol{s}(.)$. The former is the estimate of the bag proportions, while the later acts as a regularizer, and λ controls the strength of the penalty term. C is the set of all pairwise ordinal relationships that should hold, i.e. $\boldsymbol{s}(\boldsymbol{x}_i) \geq \boldsymbol{s}(\boldsymbol{x}_j)$ for $y_i = r_c$ and $y_j = r_h$, with $r_c \succ r_h$. The functional $\boldsymbol{d}(.,.)$ provides a measure of distance between the true and estimated proportions.

2 Related Work

In this section we review related work for both ordinal classification and learning from label proportions.

2.1 Ordinal Classification

For a paper length discussion of the approaches to ordinal classification and their taxonomy we refer the reader to [6] and the references therein. Here, we briefly outline the main approaches.

The assumption of a natural ordering of the labels, which underlies ordinal classification is a strong one, as it states that the ordinal structure of the labels is also present in the feature space, or as stated in [8] "the ordinal class structure induces an ordinal instance structure". One could of course reduce the ordinal classification problem to a nominal one and make use of plenty of existing algorithms, but this would amount to ignoring available information about the structure of the data, that could otherwise be used to improve performance, reduce computation and in general help in building a more consistent classifier. On the other hand, the task could also be transformed to a regression problem by mapping the class labels onto the real line (while respecting the ordering) and then proceed by applying standard regression techniques. A technique in this category is that of training a regression tree [2]. However, one disadvantage of this method is that there is no principled way of choosing the map [6]. In the taxonomy of ordinal classification approaches, presented in [6], this falls under the *naive approaches* category, as they are basically the result of other standard techniques in machine learning. An alternative approach, still naive but more advanced, is that of cost-sensitive classification, where the order of the classes is taken into account in the sense that not all mistakes carry equal weight [16].

The second group of techniques is referred to as *Ordinal Binary Decompositions* (OBD). In most cases multiclass classification is tackled through the use of one-vs-one or one-vs-all voting schemes. In the OBD group, some of the schemes used are one-vs-next, one-vs-followers and one-vs-previous, which clearly make explicit the ordering of the classes (consult [6] for a longer discussion of these schemes and for their properties). One such models is presented in [5], where the original problem is decomposed into a series of binary classification tasks.

The third and final group includes the *threshold models*, which are based on the assumption of a latent continuous variable underlying the ordered discrete labels [6]. These methods have two main ingredients; a function trained to estimate the latent variable and a set of thresholds that distinguish between the classes (in the ordered setting). The reader would be right in noting the similarity of these models with the naive regression approaches. The difference between the two categories is that in the threshold models, there is no mapping from discrete (ordered) labels onto the real line (which, as previously discussed would require prior knowledge about the distances of the classes), but rather thresholds are being used, which are learned during training.

One of the first attempts was the proportional odds model [12], which extends logistic regression to the ordinal setting. The Support Vector Machine is one of the most eminent machine learning techniques due to its generalization performance, and has therefore inevitably seen many adaptations to the ordinal classification setting [2,7]. Finally, and also belonging to the category of threshold models, discriminative learning [17] will be discussed in Sect. 3.

2.2 Learning from Label Proportions

The level of supervision of bag proportions is very similar to the one of Multiple-Instance Learning [4,11], where the practitioner is provided with logical statements indicating the presence of a class in a bag. For example, in binary classification, a bag would have a positive label if it had at least one positive point in it, while it would be labeled as negative if all of the points belonging to it were negative.

Existing algorithms designed for learning from label proportions fall in three main categories. Bayesian approaches such as [9] approach the problem by generating labels consistent with bag proportions. In [14] the authors propose an algorithm that relies on the properties of exponential families and the convergence of the class mean operator, computed from the means and label proportions of each bag. Lastly, maximum-margin approaches [15,20] pose the problem as either an extension of maximum-margin clustering [18] or Support Vector Regression.

Conditional Exponential Families. The notation, as well as the overall treatment, in this section follow from [14]. For further clarification please consult the original paper. Let $\mathcal{X}$ and $\mathcal{Y}$ denote the space of the observations and the (discrete) label space respectively, and let $\phi(\boldsymbol{x}, y) : \mathcal{X} \times \mathcal{Y} \rightarrow \mathcal{H}$ be a feature map into a Reproducing Kernel Hilbert Space $\mathcal{H}$ with kernel $k((\mathbf{x}, y), (\mathbf{x}', y'))$. A conditional exponential family is stated as follows:

$$p(\mathbf{y}|\mathbf{x}, \boldsymbol{\theta}) = exp\big(\phi(\boldsymbol{x}, y)^T \boldsymbol{\theta} - g(\boldsymbol{\theta}|\boldsymbol{x})\big) \quad \text{with}$$
$$g(\boldsymbol{\theta}|\boldsymbol{x}) = log \sum_{y \in \mathcal{Y}} exp(\phi(\boldsymbol{x}, y)^T, \boldsymbol{\theta})$$

where $g(\boldsymbol{\theta}|\boldsymbol{x})$ is a log-partition function and $\boldsymbol{\theta}$ is the parameter of the distribution. Under the assumption that $\{(\boldsymbol{x}_i, y_i)_{i=1}^n\}$ are drawn independently and identically distributed by the distribution $p(x, y)$, one usually optimizes for $\boldsymbol{\theta}$ by minimizing the regularized negative conditional log-likelihood:

$$\boldsymbol{\theta}^* = \arg\min_{\boldsymbol{\theta}} \left\{ \sum_{i=1}^n [g(\boldsymbol{\theta}|\boldsymbol{x}_i)] - n\boldsymbol{\mu}_{XY}^T \boldsymbol{\theta} + \lambda||\boldsymbol{\theta}||^2 \right\},$$

where $\boldsymbol{\mu}_{XY} := \frac{1}{n}\sum_{i=1}^n \phi(\boldsymbol{x}_i, y_i)$. Unfortunately, in the LLP setting we cannot compute this quantity directly, as the labels are unknown.

MeanMap. In [14] the authors build upon conditional exponential families and present MeanMap, which exploits the theoretical guarantees of uniform convergence of the expectation operator to its expected value, $\mu_{xy} := \boldsymbol{E}_{(x,y) \sim p(x,y)}[\phi(\boldsymbol{x}, y)]$. Expanding we get:

$$\mu_{xy} = \sum_{y \in \mathcal{Y}} p(y) \boldsymbol{E}_{x \sim p(x|y)}[\phi(\boldsymbol{x}, y)] \tag{2}$$

A critical assumption is that conditioned on its label, a point is independent of its bag assignment, that is, $p(x|y, i) = p(x|y)$. Based on this we get $p(x|i) = \sum_y p(x|y)\pi_{iy}$ and subsequently

$$\mu_x^{set}[i, y'] = \boldsymbol{E}_{x \sim p(x|i)}[\phi(\boldsymbol{x}, y')] = \sum_y \pi_{iy} \boldsymbol{E}_{x \sim p(x|y)}[\phi(\boldsymbol{x}, y)]$$
$$= \sum_y \pi_{iy} \mu_x^{class}[y']$$

Putting these in matrix notation we get to $\boldsymbol{M}_x^{set} = \pi \boldsymbol{M}_x^{class}$. Assuming π has full column-rank, we can obtain $\mu_x^{class} = (\pi^T \pi)^{-1} \pi^T \mu_x^{set}$, to be used as an approximation of $\boldsymbol{E}_{x \sim p(x|y)}[\phi(\boldsymbol{x}, y)]$ in Eq. 2.

Maximum Margin Approaches. The maximum margin principle has been widely used in both supervised and semi-supervised learning [3]. In [18] it was also introduced to the unsupervised setting under the name Maximum Margin Clustering (MMC).

Informally, the labels are arranged in a way such that, had an SVM been trained on the (labeled) data, it would achieve a maximum margin solution. A treatment of MMC can be found in [10]. In [20] the authors present αSVM, based on MMC with an extra term in the loss function, depending on the provided and estimated bag proportions.

In [15] the authors follow the maximum margin principle by developing a model based on the Support Vector Regression. They present *Inverse Calibration* (InvCal) that replaces the actual dataset with *super-instances* [19], one for each bag, with soft-labels corresponding to their bag-proportions.

3 Discriminant Learning with Ordinal Label Proportions

In this section we first present some necessary background on Linear Discriminant Analysis (LDA), then proceed with the adaptation to the ordinal setting and finally introduce our algorithm.

3.1 Preliminaries

LDA is one of the main approaches in supervised dimensionality reduction, but is also widely used as a classification technique. LDA aims at finding a projection of the data that both minimizes the within-class variance and maximizes the between-class variance.

Following [17], let us define the within-class and between-class scatter matrices (denoted by the w and b subscripts, respectively):

$$S_w = \frac{1}{N} \sum_{k=1}^{K} \sum_{x \in C_k} (x - m_k)(x - m_k)^T \tag{3}$$

where the first sum runs over the K classes, and the second over the elements in each class (where C_k is used to denote the set of data-points in each class) and where $m_k = \frac{1}{N_k} \sum_{x \in C_k} x$ denotes the mean of each class.

The between-class scatter matrix is defined as:

$$S_b = \frac{1}{N} \sum_{k=1}^{K} N_k (m_k - m)(m_k - m)^T \tag{4}$$

where $m = \frac{1}{N} \sum_{i=1}^{N} x_i$ is used to denote the global mean.

The projection is found by minimizing the following generalized Rayleigh quotient:

$$w^* = \arg\min_{w} J(w), \quad \text{where} \quad J(w) = \frac{w^T S_w w}{w^T S_b w} \tag{5}$$

This can be solved as a generalized eigenvalue problem. As with many popular techniques, such as Support Vector Machines and Principal Component Analysis, LDA can be kernelized as well and give rise to Kernel Discriminant Analysis (see for example, [1,13]).

3.2   Kernel Discriminant Learning for Ordinal Regression (KDLOR)

As mentioned earlier, in the Ordinal Classification setting the classes exhibit a natural ordering. This statement can be easily formulated as a constraint to an optimization problem. Similarly to LDA, the projection should be such that the between-class scatter is high and within-class scatter is small. This gives rise to the following problem [17]:

$$\min \; J(\boldsymbol{w},\rho) = \boldsymbol{w}^T \boldsymbol{S}_w \boldsymbol{w} - C\rho$$
$$s.t. \; \boldsymbol{w}^T(\boldsymbol{m}_{k+1} - \boldsymbol{m}_k) \geq \rho, \;\; \text{for} \;\; k = 1, \cdots, K-1 \tag{6}$$

where C can be understood as the parameter controlling the penalty on the margin between the means and where $\rho > 0$ defines the margin between the class means. Also, without loss of generality, we have assumed the class numbering (the subscript) is in accordance with the natural ordering of the classes. It can be easily seen that this problem gives rise to a projection that abides to the desired properties. We want our projection to have: (1) small within-class variance, (2) large distances between the means of the classes, and (3) a projection that respects the inherent ordering.

To solve the above problem we proceed by forming the Lagrangian as follows:

$$\mathcal{L}(\boldsymbol{w},\rho,\alpha) = \boldsymbol{w}^T \boldsymbol{S}_w \boldsymbol{w} - C\rho - \sum_{k=1}^{K-1} \alpha_k \big(\boldsymbol{w}^T(\boldsymbol{m}_{k+1} - \boldsymbol{m}_k) - \rho \big) \tag{7}$$

where $\alpha_k \geq 0$ are the Lagrange multipliers. Differentiating with respect to $\boldsymbol{w}$ and ρ we get:

$$\frac{\partial \mathcal{L}}{\partial \boldsymbol{w}} = 0 \rightarrow \boldsymbol{w} = \frac{1}{2} \boldsymbol{S}_w^{-1} \sum_{k=1}^{K-1} \alpha_k (\boldsymbol{m}_{k+1} - \boldsymbol{m}_k)$$

$$\frac{\partial \mathcal{L}}{\partial \rho} = 0 \rightarrow \sum_{k=1}^{K-1} \alpha_k = C$$

The so-called dual problem is formulated as follows:

$$\min \; f(\alpha) = \sum_{k=1}^{K} \alpha_k (\boldsymbol{m}_{k+1} - \boldsymbol{m}_k)^T \boldsymbol{S}_w^{-1} \sum_{k=1}^{K} \alpha_k (\boldsymbol{m}_{k+1} - \boldsymbol{m}_k)$$
$$s.t. \; \alpha_k \geq 0, \; k = 1, \cdots, K-1$$
$$\sum_{k=1}^{K} \alpha_k = C \tag{8}$$

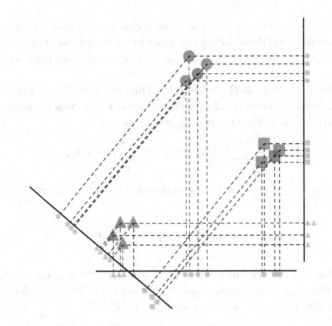

Fig. 1. Ordinary label proportions toy setup. Shape is as indication of class assignment, while colour is an indication of bag assignment. On the figure, we also see three possible projections, that all allow for perfect separation of the data. (Color figure online)

This is an example of (convex) Quadratic Programming with linear constraints and can be solved via a variety of methods. After solving this optimization program for $\boldsymbol{\alpha}^*$, the projection can be obtained using

$$\boldsymbol{w}^* = \frac{1}{2}\,\boldsymbol{S}_w^{-1}\sum_{k=1}^{K-1}\alpha_k^*(\boldsymbol{m}_{k+1}-\boldsymbol{m}_k) \qquad (9)$$

and the derived decision rule is as follows,

$$f(x) = \min_{k\in\{1,\cdots,K\}}\{k : \boldsymbol{w}^{*T}\boldsymbol{x} - b_k < 0\} \qquad (10)$$

where $b_k = \boldsymbol{w}^T\frac{N_{k+1}\boldsymbol{m}_{k+1}+N_k\boldsymbol{m}_k}{N_{k+1}+N_k}$.

3.3 Discriminant Learning for Ordinal Label Proportions (DL-OLP)

The algorithm presented in the previous subsection (KDLOR) is suitable for the fully supervised learning setting. Though, when the level of supervision is that of learning with label proportions, KDLOR cannot be employed as the class means are required for both the main problem Eq. 8 and for the computation of the within-class scatter matrix in Eq. 3. To that end, we choose to estimate the class means building upon [14]. Figure 1 demonstrates the setting through a toy example, where we have three (well) separated clusters (as shown by shape) with the only supervision available being the label proportions (as shown by colour). In the Figure we also see three possible projections, which all fully separate the clusters.

Key to this derivation is the underlying (but often realistic) assumption of conditional independence of a data point and its bag assignment, given its class assignment. Formally, $p(x|y, i) = p(x|y)$, which gives us:

$$p(x|i) = \sum_y p(x|y, i)p(y|i) = \sum_y p(x|y)\pi_{iy} \qquad (11)$$

where $\pi_{iy} = p(y|i)$. Let $\boldsymbol{\mu}_k$ denote the mean of x in bag k and $\boldsymbol{m}_c$ the mean of class c. Following Eq. 11,

$$\boldsymbol{\mu}_k := \boldsymbol{E}_{x \sim p(x|i)}[\boldsymbol{\phi}(\boldsymbol{x})] = \sum_y \pi_{iy}\boldsymbol{m}_y \qquad (12)$$

Putting these in matrix form, with $\boldsymbol{M}^{bag}$ and $\boldsymbol{M}^{class}$ denoting the matrices of means for the bags and classes, respectively, we have $\boldsymbol{M}^{bag} = \boldsymbol{\pi}\boldsymbol{M}^{class}$, from which we can obtain a least squares estimate of $\boldsymbol{M}^{class}$:

$$\hat{\boldsymbol{M}}^{class} = \boldsymbol{\pi}^+ \boldsymbol{M}^{bag} \qquad (13)$$

where the $+$ superscript denotes the Moore–Penrose pseudo-inverse. (For the sake of clarity, it should be noted that $\boldsymbol{\pi}$ denotes a *matrix*, and *not* a vector).

Having shown how to estimate the class means, let us make explicit how to compute the within-class scatter matrix, as it requires a sum over data points in each class.

$$\boldsymbol{S}_w = \frac{1}{N} \sum_{k=1}^{K} \sum_{x \in C_k} (\boldsymbol{x} - \boldsymbol{m}_k)(\boldsymbol{x} - \boldsymbol{m}_k)^T \qquad (14)$$

$$= \frac{1}{N} \sum_{k=1}^{K} \sum_{x \in C_k} \boldsymbol{x}\boldsymbol{x}^T - 2\boldsymbol{x}\boldsymbol{m}_k^T + \boldsymbol{m}_k\boldsymbol{m}_k^T$$

$$= \frac{1}{N} \sum_{x \in \mathcal{X}} \boldsymbol{x}\boldsymbol{x}^T + \frac{1}{N} \sum_{k=1}^{K} N_k\boldsymbol{m}_k\boldsymbol{m}_k^T - \frac{2}{N} \sum_{k=1}^{K} \boldsymbol{m}_k \sum_{x \in C_k} \boldsymbol{x}$$

$$= \frac{1}{N} \sum_{x \in \mathcal{X}} \boldsymbol{x}\boldsymbol{x}^T - \frac{1}{N} \sum_{k=1}^{K} N_k\boldsymbol{m}_k\boldsymbol{m}_k^T$$

The procedure is finally summarized in Algorithm 1. When new instances are observed, one can plug in $\boldsymbol{w}^*$ into Eq. 10 to obtain the corresponding prediction.

4   Experiments

In our experiments we consider both synthetic and real-world datasets. We first describe the datasets and then present the empirical results. In our experiments

---

Algorithm 1. LDA for OLP

Input: π, X, *bag assignments*
Output: w^*
1 Compute the means of each bag, M^{bag}.
2 Compute the means of each class using Eq. 13.
3 Compute the within-class scatter matrix, S_w using Eq. 14.
4 Solve the problem defined by Eq. 8 for α^*.
5 Obtain the projection w^* using Eq. 9

---

we want to test the relevance of the ordinal constraints as well as the trade-off in accuracy when using aggregated proportions instead of the actual labels. To that end we compare against: KDLOR - trained with the actual labels, MeanMap - trained with bag proportions. Additionally, we use Clustering as a baseline. For clustering, we first run k-means, with the number of components corresponding to the number of classes. Then, in order to classify each cluster, we consider a voting scheme, where each data point's vote is its corresponding bag's proportions.

Regarding the first two types of experiments (Synthetic and Benchmark datasets) we make the following notes.

Evaluation. In our experiments (except REF) we consider the test data to be provided *without* bag proportions. In the case of the test data set being provided in the form of a bag, one could do the training in the exact same manner as presented in the paper and during testing, after the predictions have been generated, sort the data points of each bag according to their scores and re-arrange predictions to account for the provided bag proportions. For the synthetic and benchmark datasets we have access to the ground truth (the true labels) and our evaluation is based on those.

Results. These should be read as follows. The first column is the name of the dataset. The rest of the values should be read as *mean(one standard deviation)*.

4.1 Synthetic Dataset

Data. In our experiments we consider one synthetic dataset configuration as shown in Fig. 2. The data samples $(100, 1000)$ were generated as coming from three Gaussian distributions with means lying on a line on equal intervals and identity covariance matrix.

Model Setup. In our experiments we focus on problems involving three classes of equal size, and three bags, again, of equal size – but different proportions. The proportions used are: $\{(0.25, 0.25, 0.50, 0.50, 0.25, 0.25, 0.25, 0.50, 0.25)\}$. The data is first generated according to the desired total size and then separated into the bags, respecting the desired bag proportions.

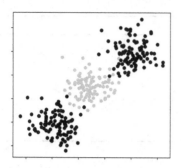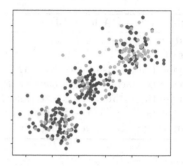

Fig. 2. Configuration used for the synthetic dataset. On the left, colour indicates class, while on the right, colour indicates bag assignment. (Color figure online)

Results. Results for the synthetic dataset are shown in the first two rows of Tables 1 and 2. Table 1 shows results on Zero-One accuracy while Table 2 shows results on Mean Absolute Error (MAE) loss. This simple example is only intended to show the difference between the two evaluation metrics. In terms of MAE, as expected the best performing method is the inherently ordinal KLDOR, while DL-OLP outperforms MeanMap. The Clustering baseline works particularly well due to the underlying distribution of the data.

4.2   Benchmark Datasets

Data. In our experiments we also consider various benchmark datasets for ordinal classification. The datasets can be found in the repository provided[2].

Model Setup. For these datasets, three classes were chosen and then three bags were created by randomly picking points from the classes to fulfill the desired bag proportions. The bag proportions used are: {(0.25, 0.25, 0.50, 0.50, 0.25, 0.25, 0.25, 0.50, 0.25)}.

Results. Results for the benchmark datasets are shown in Tables 1 and 2. Again, in most cases, in terms of MAE, as expected the best performing method is the inherently ordinal KLDOR, while DL-OLP makes better use of the label proportions than MeanMap by respecting the ordinal constraints.

4.3   REF Dataset

The final experiment is a real-world case study illustrating the effectiveness of DL-OLP in a problem of actual importance, both for accurate prediction and for interpretation of the prediction model (i.e., the weight vector).

---

[2] The benchmark datasets are also available at http://www.gagolewski.com/resources/data/ordinal-regr.

Table 1. Zero-one accuracy

	KDLOR	MeanMap	Clustering	DL-OLP
Synthetic100	0.98(0.014)	0.49(0.056)	0.97(0.018)	0.97(0.026)
Synthetic1000	0.98(0.003)	0.53(0.052)	0.98(0.006)	0.97(0.002)
Cali/Housing	0.67(0.015)	0.35(0.032)	0.29(0.011)	0.53(0.018)
Cement-Strength	0.93(0.004)	0.72(0.012)	0.66(0.013)	0.87(0.026)
Fireman	0.83(0.011)	0.70(0.008)	0.64(0.019)	0.81(0.023)
Kinematics	0.66(0.016)	0.62(0.012)	0.68(0.015)	0.62(0.022)
Skill	0.66(0.028)	0.55(0.042)	0.59(0.04)	0.59(0.012)
Stockord	0.68(0.042)	0.40(0.049)	0.59(0.061)	0.80(0.078)

Table 2. Mean absolute error

	KDLOR	MeanMap	Clustering	DL-OLP
Synthetic100	0.023(0.016)	0.66(0.090)	0.26(0.018)	0.031(0.026)
Synthetic1000	0.024(0.003)	0.61(0.031)	0.024(0.006)	0.028(0.009)
Cali/Housing	0.35(0.016)	0.69(0.036)	0.77(0.025)	0.51(0.022)
Cement-Strength	0.073(0.004)	0.28(0.012)	0.35(0.013)	0.14(0.034)
Fireman	0.17(0.011)	0.32(0.010)	0.38(0.029)	0.20(0.023)
Kinematics	0.40(0.016)	0.44(0.015)	0.36(0.017)	0.42(0.013)
Skill	0.35(0.034)	0.52(0.057)	0.44(0.068)	0.42(0.011)
Stockord	0.36(0.086)	0.61(0.056)	0.42(0.086)	0.16(0.057)

Data. In the United Kingdom, the Research Excellence Framework (REF) assesses the research of all university departments. This is done by asking all departments to send in a list of the 4 best publications for each academic staff member. Then a panel decides how many of these submissions are awarded 4* (world-leading), 3* (internationally recognized), 2* (nationally leading), and 1* (nationally recognized) or unclassified. At the end of the process, the outcome is a histogram for each department stating how many research outputs were in each category, but not revealing which paper was in each of these classes. Funding levels are then based on this histogram. As the panel do not reveal how they classify papers, in deciding which papers to submit, the departments have to guess which papers the panel would rank most highly. This also affects the individuals' and departmental publication strategy. In this experiments we aim to reverse engineer the panel's thinking, and work out a classifier that mimics the way the panel operates.

The data is online for all university departments: all the papers each department submitted, and the histogram saying how many were in each class for each department. We use the 2008 submission for all Computer Science departments in the country to compile our REF dataset[3]. For each book, book chapter, conference or journal paper, we collect the following features whenever available:

1. `Number of citations`: total number of citations for a given paper at the time of submission from Google Scholar (integer).
2. `Number of authors`: total number of authors for a given paper (integer).
3. `IsMultidisciplinary`: whether the outcome is categorized by the author as a multidisciplinary piece of research (boolean).
4. `Ranking ERA`: ranking provided by the Excellence in Research for Australia (ERA)[4] for journals and conferences (ordinal categorical).
5. `JCR total citations`: total number of citations for every article published in the journal from the Journal Citation Reports (JCR)[5] (integer).
6. `JCR impact factor`: frequency with which an average article from a journal is cited in a particular year (positive real).
7. `JCR Immediacy index`: frequency with which the average article from a journal is cited within the same year as publication (positive real).
8. `JCR total number of articles of the journal`: total number of articles published in the publication for a specific year (integer).
9. `JCR Cited half-life`: median age of the articles in the journal that were cited in the year (positive real).
10. Additionally, we compute a feature based on the product of the `JCR impact factor` and the `JCR cited half-life`, as this was traditionally thought to be a good proxy for the behaviour of the panel.

This leads to a total of 10 features for the 4966 research outputs over the 81 Computer Science departments.

For non-journal papers, the JCR measures (features 5–10) are not available, and feature 4 is not available for contributions other than conference and journal papers. There are many possible approaches for filling out these missing values. For the sake of simplicity, we set all missing feature values to zero. While this is clearly not very sophisticated or well-founded (and alternative approaches are subject of ongoing investigations), we will show later in the discussion that it nonetheless leads to informative and interpretable results.

Model Setup. For the REF dataset, we consider each department as being one bag and each paper's (hidden) rating to range between 4* and no-stars. We therefore have a total of 81 bags and 5 classes. Even though we do not include experiments using the kernel extension of the algorithm for interpretability purposes, we do provide some guidelines as to how one could use it. The standard approach to learning parameters of a model is through k-fold cross-validation (where one splits the dataset in k folds, trains the model on $k - 1$ of them, and tests on the k-th one). In the LLP setting one does not have access to the true labels, so the standard CV procedure cannot be employed. One can though

---

[3] https://www.rae.ac.uk/pubs/2008/01/.

[4] http://www.arc.gov.au/excellence-research-australia.

[5] http://jcr.incites.thomsonreuters.com/.

adapt it to this setting by using the bags as folds. In order to evaluate the performance on the 'test-bag', the practitioner can consider how well the estimated bag proportions match the provided ones.

Discussion. For the REF data, the ground truth (the actual rating for each paper) is not available and therefore our evaluation is limited. Therefore, we focus on discussing our model parameters. With regards to MAE on the bag proportions, our approach achieves a value of 12.14, outperforming clustering which achieves a value of 19.76.

Based on the empirical long-tail distribution of Number of Citations, it was believed sensible to apply a log-transform on the feature (the difference between 0 and 10 citations is a lot more important than the difference between 1000 and 1010 citations). The rest of the features are standardized. The weight vector obtained by DL-OLP is as follows (numbers are given in the same order as the numbered list of features shown before):

$$[0.179, -0.025, 0.158, 0.026, -0.020, -0.095, -0.074, 0.032, -0.110, -0.081].$$

Not only does DL-OLP allow one to predict the histogram for a given REF submission well – the weight vector also provides insight into what the panel values. The (positive) dominating features are the Number of citations, together with multidisciplinary nature of the submission. The importance of the number of citations comes at no surprise. The latter is in accordance with the widely held belief that multidisciplinary contributions are valued more highly by the REF panels.

It is worth noting that the number of authors has a negative weight. Thus, a large number of authors is perceived as lowering the quality of a submission (perhaps as one's contribution to it is considered inversely proportional to the number of authors). However, many authors are justified from a REF optimization perspective if necessitated by the paper's multidisciplinary nature. To further illustrate this point, we proceeded by combining these two features into one, by multiplying Number of authors with +1 in the case of isMultidisciplinary being true, and −1 otherwise. The new weight vector is shown below (N/A is inserted for the third feature, as it is now combined with the second):

$$[0.102, 0.153, N/A, -0.019, -0.019, -0.051, -0.033, 0.002, -0.104, -0.036].$$

Interestingly, the new feature now dominates the weight vector, with a positive effect, with the number of citations coming second.

A final observation is that the JCR features (with the exception of the 8'th feature, total number of articles in the journal), are negative. This may seem counter intuitive at first, but there are two logical explanations. The first is that journal-level metrics are only proxies for a paper's actual impact, which is better quantified by the actual number of citations (which does have a large positive weight). To test this explanation, we also ran our method after removing the first feature, yielding the following weights (again with N/A for the first as this one is removed):

$$[\text{N/A}, -0.004, 0.078, 0.019, -0.000, -0.068, -0.072, 0.090, 0.003, -0.055].$$

We do indeed see that the JCR measures (features 5–10) all increase, yet, most of them remain negative. This can be explained by our approach of filling in missing values for non-journal papers. Actually, these have JCR measures that are set to zero (i.e. the minimum possible). The result thus implies that the REF panels do value strong non-journal contributions, contrary to popular belief.

5   Conclusion

In this paper we have introduced a new learning task which we have called Ordinal Label Proportions. We have also presented a method to tackle the problem based on discriminant learning and a sound estimation of class means. The method aims to find a projection that minimizes the within-class scatter while also respecting the natural ordering of the classes. Our approach compares favourably with MeanMap, that does not exploit the ordinal nature of the data. Moreover, even though DL-OLP has the benefit of training with the true labels, instead of label proportions, only a minor setback is observed empirically. In the future we wish to examine more real world data sets that exhibit the characteristics of Ordinal Label Proportions.

Acknowledgements. The research leading to these results has received funding from the European Research Council under the European Union's Seventh Framework Programme (FP7/2007-2013)/ERC Grant Agreement no. 615517, from the FWO (project no. G091017N, G0F9816N), and from the European Union's Horizon 2020 research and innovation programme and the FWO under the Marie Sklodowska-Curie Grant Agreement no. 665501. Additionally, this study was supported by EPSRC and MRC through the SPHERE IRC (EP/K031910/1) and CUBOID (MC/PC/16029) respectively.

References

1. Bishop, C.: Pattern Recognition and Machine Learning. Springer, Boston (2006). https://doi.org/10.1007/978-1-4615-7566-5
2. Chu, W., Keerthi, S.S.: Support vector ordinal regression. Neural Comput. **19**(3), 792–815 (2007)
3. Cristianini, N., Shawe-Taylor, J.: An Introduction to Support Vector Machines and Other Kernel-Based Learning Methods. Cambridge University Press, Cambridge (2000)
4. Dietterich, T.G., Lathrop, R.H., Lozano-Pérez, T.: Solving the multiple instance problem with axis-parallel rectangles. Artif. Intell. **89**(1–2), 31–71 (1997)
5. Frank, E., Hall, M.: A simple approach to ordinal classification. In: De Raedt, L., Flach, P. (eds.) ECML 2001. LNCS (LNAI), vol. 2167, pp. 145–156. Springer, Heidelberg (2001). https://doi.org/10.1007/3-540-44795-4_13

6. Gutiérrez, P.A., Perez-Ortiz, M., Sanchez-Monedero, J., Fernandez-Navarro, F., Hervas-Martinez, C.: Ordinal regression methods: survey and experimental study. IEEE Trans. Knowl. Data Eng. **28**(1), 127–146 (2016)
7. Herbrich, R., Graepel, T., Obermayer, K.: Support vector learning for ordinal regression. In: International Conference on Artificial Neural Networks. IET (1999)
8. Huhn, J.C., Hullermeier, E.: Is an ordinal class structure useful in classifier learning? Int. J. Data Min. Model. Manag. **1**(1), 45–67 (2008)
9. Kuck, H., de Freitas, N.: Learning about individuals from group statistics (2012). arXiv preprint: arXiv:1207.1393
10. Li, Y.F., Tsang, I.W., Kwok, J., Zhou, Z.H.: Tighter and convex maximum margin clustering. In: Artificial Intelligence and Statistics, pp. 344–351 (2009)
11. Maron, O., Ratan, A.L.: Multiple-instance learning for natural scene classification. In: Proceedings of the Fifteenth International Conference on Machine Learning, pp. 341–349. Morgan Kaufmann Publishers Inc., San Francisco (1998)
12. McCullagh, P.: Regression models for ordinal data. J. R. Stat. Soc. Ser. B (Methodol.) **42**, 109–142 (1980)
13. Mika, S., Rätsch, G., Weston, J., Schölkopf, B., Müller, K.R.: Fisher discriminant analysis with kernels. In: Proceedings of the 1999 IEEE Signal Processing Society Workshop, Max-Planck-Gesellschaft, vol. 9, pp. 41–48. IEEE (1999)
14. Quadrianto, N., Smola, A.J., Caetano, T.S., Le, Q.V.: Estimating labels from label proportions. J. Mach. Learn. Res. **10**, 2349–2374 (2009)
15. Rueping, S.: SVM classifier estimation from group probabilities. In: Proceedings of the 27th International Conference on Machine Learning, pp. 911–918 (2010)
16. Santos-Rodríguez, R., Guerrero-Curieses, A., Aláiz-Rodríguez, R., Cid-Sueiro, J.: Cost-sensitive learning based on Bregman divergences. Mach. Learn. **76**, 14 (2009)
17. Sun, B.Y., Li, J., Wu, D.D., Zhang, X.M., Li, W.B.: Kernel discriminant learning for ordinal regression. IEEE Trans. Knowl. Data Eng. **22**(6), 906–910 (2010)
18. Xu, L., Neufeld, J., Larson, B., Schuurmans, D.: Maximum margin clustering. In: Advances in Neural Information Processing Systems, pp. 1537–1544 (2005)
19. Yu, F.X., Choromanski, K., Kumar, S., Jebara, T., Chang, S.F.: On learning from label proportions (2014). arXiv preprint: arXiv:1402.5902
20. Yu, F.X., Liu, D., Kumar, S., Jebara, T., Chang, S.F.: $\propto$ SVM for learning with label proportions (2013). arXiv preprint: arXiv:1306.0886

AWX: An Integrated Approach to Hierarchical-Multilabel Classification

Luca Masera$^{(\boxtimes)}$ and Enrico Blanzieri

University of Trento, 38123 Trento, Italy
{luca.masera,enrico.blanzieri}@unitn.it

Abstract. The recent outbreak of works on artificial neural networks (ANNs) has reshaped the machine learning scenario. Despite the vast literature, there is still a lack of methods able to tackle the hierarchical multilabel classification (HMC) task exploiting entirely ANNs. Here we propose AWX, a novel approach that aims to fill this gap. AWX is a versatile component that can be used as output layer of any ANN, whenever a fixed structured output is required, as in the case of HMC. AWX exploits the prior knowledge on the output domain embedding the hierarchical structure directly in the network topology. The information flows from the leaf terms to the inner ones allowing a jointly optimization of the predictions. Different options to combine the signals received from the leaves are proposed and discussed. Moreover, we propose a generalization of the true path rule to the continuous domain and we demonstrate that AWX's predictions are guaranteed to be consistent with respect to it. Finally, the proposed method is evaluated on 10 benchmark datasets and shows a significant increase in the performance over plain ANN, HMC-LMLP, and the state-of-the-art method CLUS-HMC. Code related to this paper is available at: https://github.com/lucamasera/AWX.

Keywords: Hierarchical multilabel classification
Structured prediction · Artificial neural networks

1 Introduction

The task of multilabel classification is an extension of binary classification, where more then one label may be assigned to each example [17]. However, if the labels are independent, the task can be reduced without loss of generality to multiple binary tasks. Of greater interest is the case where there is an underlying structure that forces relations through the labels. These relations define a notion of consistency in the annotations, that can be exploited in the learning process to improve the prediction quality. This task goes under the name of hierarchical multilabel classification (HMC) and can be informally defined as the task of assigning a subset of consistent labels to each example in a dataset [21].

Knowledge is organized in hierarchies in a wide spectrum of applications, ranging from content-categorization [16,19] to medicine [14] and biology [4,10,13].

© Springer Nature Switzerland AG 2019
M. Berlingerio et al. (Eds.): ECML PKDD 2018, LNAI 11051, pp. 322–336, 2019.
https://doi.org/10.1007/978-3-030-10925-7_20

Hierarchies can be described by trees or direct acyclic graphs (DAG), where the nodes are the labels (we will refer to them as terms in the rest of the paper) and the edges represent *is_a* relations that occurs between a child node and its parents. These relations can be seen as a logical implication, because if a term is true then also its parents must be true. In other words, *"the pathway from a child term to its top-level parent(s) must always be true"* [4]. This concept was introduced by "The Gene Ontology Consortium" (GO) under the name of "true path rule" (TPR) to guarantee the consistency of the ontology with respect to the annotations, such that, whenever a gene product is found to break the rule, the hierarchy is remodelled consequently. Besides guaranteeing the consistency of the annotation space, the TPR can be forced also on the predictions. Inconsistencies in the predictions have been shown to be confusing for the final user, who will likely not trust and reject them [11]. Even though there are circumstances where inconsistencies are accepted, we will focus on the strict case, where the TPR should hold for predictions as well.

HMC has a natural application in bioinformatics, where ontologies are widely used as annotation space in predictive tasks. The critical assessment of functional annotation (CAFA) [7,12], for example, is the reference challenge for the protein function prediction community and uses the GO terms as annotations. The ontology comprises thousands of terms organized in three DAGs and the concepts expressed by some of those terms are so specific that just few proteins have been experimentally found belonging to them. Therefore, even though a perfect multilabel classification on the leaf nodes would solve the problem, the lack of examples forces the participants to exploit the hierarchical structure, by learning a single model [15] or by correcting the output of multiple models *a posteriori* [5].

HMC methods can be characterized in terms of local (top-down) or global (one-shot)approaches. The former [1,2,5] rely on traditional classifiers, training multiple models for each or subset of the labels, and applying strategies for selecting training examples or correcting the final predictions. Global methods [15,21], on the other hand, are specific classification algorithms that learn a single global model for the whole hierarchy. Vens *et al.* [21] compare the two approaches, and propose a global method called CLUS-HMC, which trains one decision-tree to cope with the entire classification problem. The proposed method is then compared with its naïve version CLUS-SC, which trains a decision-tree for each class of the hierarchy, ignoring the relationships between classes, and with CLUS-HSC, which explores the hierarchical relationships between the classes to induce a decision-tree for each class. The authors performed the experiments using biological datasets, and showed that the global method was superior both in the predictive performance and size of the induced decision-tree. CLUS-HMC has been shown to have state-of-the-art performance, as reported in the study by Triguero *et al.* [20].

More recently, the introduction of powerful GPU architectures brought artificial neural networks (ANNs) back to the limelight [6,9,18]. The possibility to scale the learning process with highly-parallel computing frameworks allowed the community to tackle completely new problems or old problems with completely new and complex ANNs' topologies. However, ANN-based methods that account for HMC have not yet evolved consequently. Attempts to integrate ANNs and HMC have been conducted by Cerri *et al.* [1,2]. They propose HMC-LMLP, a local model where for each term in the hierarchy is trained an ANN, that is fed with both the original input and with the output of models built for the parent terms. The performance are comparable with CLUS-HMC, however, because of the many models trained, the proposed approach is not scalable with deep learning architectures that requires a considerable amount of time for training. To the best of our knowledge there are no better model that exploits ANNs in the training process.

In this work we present AWX (Adjacency Wrapping matriX), a novel ANN output component. We aim at filling the gap between HMC and ANNs left open in the last years, enabling HMC tasks to be tackled with the power of deep learning approaches. The proposed method incorporates the knowledge on the output-domain directly in the learning process, in form of a matrix that propagates the signals coming from the previous layers. The information flows from the leaves, up to the root, allowing a jointly optimization of the predictions. We propose and discuss two approaches to combine the incoming signals, the first is based on the `max` function, while the second on ℓ-norms. AWX can be incorporated on top of any ANN, guaranteeing the consistency of the results with respect to the TPR. Moreover, we provide formal description of the HMC task, and propose a generalization of the TPR to the continuous case. Finally AWX is evaluated on ten benchmark datasets and compared against CLUS-HMC, that is the state-of-the-art, HMC-LMLP and the simple multi-layer perceptron MLP.

2   The HMC Task

This section formalizes the task of hierarchical multilabel classification (HMC) and introduces the notation used in the paper. Consider the hierarchy involved in the classification task described by a DAG $H = (\mathcal{T}, E)$, where $\mathcal{T} = \{t_1, \dots, t_m\}$ is a set of m terms and $E = \{\mathcal{T} \times \mathcal{T}\}$ is a set of directed edges. In particular the edges in E represent "*is_a*" relations, i.e. given a sample x and $\langle t_u, t_v \rangle \in E$, t_u *is_a* t_v means that t_u implies t_v, $t_u(x) \implies t_v(x)$ for all x.

Here follows a set of relevant definitions.

child t_u is a child of t_v iff $\langle t_u, t_v \rangle \in E$, $children(t_v)$ returns the children of t_v;
parent t_v is a parent of t_u iff $\langle t_u, t_v \rangle \in E$, $parents(t_u)$ returns the parents of t_u;
root a term t_v such that $parents(t_v) = \emptyset$;
leaf a term t_u such that $children(t_u) = \emptyset$, $\mathcal{F} = \{t_u | child(t_u) = \emptyset\}$ is the set of leaves;
ancestors the set of terms belonging to all the paths starting from a term to the root, $ancestors(t_v)$ returns the set of ancestors of t_v;
descendants the set of terms belonging to the paths in the transposed graph H^T starting from a term to the leaves.

Let $\mathbf{X}$ be a set of i.i.d. samples in $\mathbb{R}^d$ drawn from an unknown distribution, and $\mathbf{Y}$ the set of the assignments $\{\mathbf{y}_1, \dots, \mathbf{y}_n\}$ of an unknown labelling function

$y : \mathbf{X} \rightarrow \mathcal{P}(\mathcal{T})^1$, namely $\mathbf{y}_i = y(\mathbf{x}_i)$. The function y is assumed to be consistent with the TPR (formalized in the next paragraph). Let $\mathcal{D}$ be the dataset $\mathcal{D} = \{\langle \mathbf{x}_1, \mathbf{y}_1 \rangle, \ldots, \langle \mathbf{x}_n, \mathbf{y}_n \rangle\}$ where $\mathbf{x}_i \in \mathbf{X}$, and $\mathbf{y}_i \in \mathbf{Y}$. For convenience the labels $\mathbf{y}_i$ assigned to the sample $\mathbf{x}_i$ are expressed as a vector in $\{0, 1\}^m$ such that the j-th element of $\mathbf{y}$ is 1 iff $t_j \in y(\mathbf{x}_i)$. The hierarchical multilabel classification can be defined as the task of finding an estimator $\hat{y} : \mathbf{X} \rightarrow \{0, 1\}^m$ of the unknown labelling function. The quality of the estimator can be assessed with a loss function $L : \mathcal{P}(\mathcal{T}) \times \mathcal{P}(\mathcal{T}) \rightarrow \mathbb{R}$, whose minimization is often the objective in the learning process.

2.1 True Path Rule

The TPR plays a crucial role in the hierarchical classification task, imposing a consistency over the predictions. The definition introduced in Sect. 1 can now be formalized within our framework. The *ancestors* function, that returns the terms belonging to all the paths starting from a node up to the root, can be computed by

$$ancestors(t_u) = \begin{cases} \left(\displaystyle\bigcup_{t_k \in par(t_u)} anc(t_k) \right) \cup par(t_u) & \text{if } par(t_u) \neq \emptyset \\ \emptyset & \text{otherwise} \end{cases} \tag{1}$$

where *anc* and *par* are shorthand abbreviations for the *parents* and *ancestors* functions.

Definition 1. *The labelling function y observes the TPR iff*

$$\forall t_u \in \mathcal{T}, t_u \in y(\mathbf{x}_i) \implies ancestors(t_u) \subset y(\mathbf{x}_i).$$

Generalized TPR. The above definition holds for binary annotations, but, as we will see in Sect. 2.2, hierarchical classifiers often do not set thresholds and predictions are evaluated based only on the output scores order. We introduce here a generalized notion of TPR, namely the generalized TPR (gTPR), that expands the TPR to this setting. Intuitively it can be defined by imposing a partial order over the DAG of predictions' scores. In sthis way the TPR is respected for each global threshold.

Definition 2. *The gTPR is respected iff $\forall \langle t_u, t_v \rangle \in E$ is true that $\hat{y}_v \geq \hat{y}_u$.*

This means that for each couple of terms in a *is_a* relation, the prediction scores for the parent term must be grater or equal to the one of the child. In the extreme case of predictions that have only binary values, the gTPR clearly coincide with the TPR.

---

[1] $\mathcal{P}(\cdot)$ is the power set of a given set.

2.2   Evaluation Metrics

Multilabel classification requires a dedicated class of metrics for performance evaluation. Zang *et al.* [22] reports an exhaustive set of those metrics highlighting properties and use cases. We report here the definitions of the metrics required to evaluate AWX and compare it with the state-of-the-art.

Selecting optimal thresholds in the setting of HMC is not trivial, due to the natural unbalance of the classes. Indeed, by the TPR, classes that lay in the upper part of the hierarchy will have more annotated examples with respect to one on the leaves. Metrics that do not set thresholds, such as the area under the curve (AUC), are therefore very often used. In particular we will use the area under the precision recall curve, with three different averaging variants, each one highlighting different aspects of the methods.

The micro-averaged area under the precision recall curve $(AUC(\overline{PR}))$ computes the area under a single curve, obtained computing the micro-averaged precision and recall of the m classes

$$
\begin{aligned}
\overline{Prec} &= \frac{\sum_i^m TP_i}{\sum_i^m TP_i + \sum_i^m FP_i} \\
\overline{Rec} &= \frac{\sum_i^m TP_i}{\sum_i^m TP_i + \sum_i^m FN_i}
\end{aligned}
\tag{2}
$$

where TP_i, FP_i and FN_i are respectively the number of true positives, the false positives and the false negatives of the i-th term. It gives a global snapshot of the prediction quality but is not sensitive to the size of the classes.

To take more into account the classes with fewer examples, we use also macro-averaged $(\overline{AUCPR})$ and weighted $(\overline{AUCPR}_w)$ area under the precision recall curve. Both compute $AUCPR_i$ for each class $i \in \{1, \ldots, m\}$, which are then averaged uniformly by the former and proportionally by the latter.

$$
\begin{aligned}
\overline{AUCPR} &= \frac{1}{m} \sum_i^m AUCPR_i \\
\overline{AUCPR}_w &= \sum_i^m w_i \cdot AUCPR_i
\end{aligned}
\tag{3}
$$

where $w_i = v_i / \sum_j^m v_j$ with v_i the frequency of the i-th class in the dataset.

3   Model Description

This section describes the AWX hierarchical output layer, that we propose in this paper. Consider an artificial neural network with L hidden layers and the DAG representing the hierarchy $H = (\mathcal{T}, E)$. Let

$$
E' = \{\langle t_u, t_v \rangle | t_u \in \mathcal{F}, t_u = t_v \vee t_v \in ancestors(t_u)\}.
$$

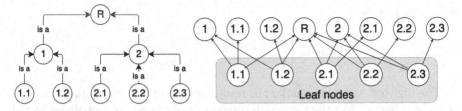

Fig. 1. The figure on the left shows a hierarchical tree structure, i.e. a sub-tree of the FunCat [13] annotation tree. On the right, the adjacency scheme described by E' of the nearby tree.

Note that for each $\langle t_u, t_v \rangle \in E'$ holds that $t_u \in \mathcal{F}$. Let $\mathbf{R}$ be a $|\mathcal{F}| \times m$ matrix that represents the information in E', where $r_{i,j} = 1$ iff $\langle t_i, t_j \rangle \in E'$ and 0 otherwise. Figure 1 shows an example of the topology described by E'.

Now, let $\mathbf{y}^L$, $\mathbf{W}^L$ and $\mathbf{b}$ denote respectively the output, the weight matrix and the bias vector of the last hidden layer in the network and $\mathbf{r}_i$ the i-th column vector of $\mathbf{R}$. The AWX hierarchical layer is then described by the following equation

$$\mathbf{z} = \mathbf{W}^L \cdot \mathbf{y}^L + \mathbf{b}^L ,$$
$$\hat{y}_i = \max(\mathbf{r}_i \circ (f(z_1), \dots, f(z_{|\mathcal{F}|}))^T) \tag{4}$$

where $\circ$ is the symbol of the Hadamard product, $\mathtt{max}$ is the function returning the maximum component of a vector, and f is an activation function $f : \mathbb{R} \to [0, 1]$ (e.g. the sigmoid function). This constraint on the activation function is required to guarantee the consistency of the predicted hierarchy as we will show in Sect. 3.1.

Being $\mathbf{R}$ binary by definition, the Hadamard product in Eq. 4, acts as a mask, selecting the entries of $\mathbf{z}$ corresponding to the non-zero elements of $\mathbf{r}_i$.

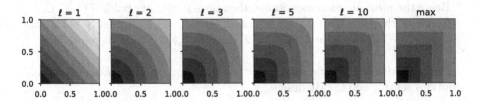

Fig. 2. Shape of the function $z = ||(x, y)^T||_\ell$ at different values of l and the comparison with the $\mathtt{max}$ function. Darker colors corresponds to lower values of z, while brighter ones to higher.

The $\mathtt{max}$ represents a straightforward way of propagating the predictions through the hierarchy, but it complicates the learning process. Indeed, the error can be back-propagated only through the maximum component of $\mathbf{r}_i$, leading to local minima. The $\mathtt{max}$ function can be approximated by the ℓ-norm of the incoming signals as follows

$$\hat{y}_i = \begin{cases} ||\mathbf{r}_i \circ f(\mathbf{z})||_\ell, & \text{if } < 1 \\ 1, & \text{otherwise.} \end{cases} \tag{5}$$

The higher ℓ, the more similar the results will be to the ones obtained by the max, the closer is ℓ to 1 the more each component of the vector contributes to the result. Figure 2 shows a two-dimensional example, and we can see that with $\ell = 5$ the norm is already a good approximation of the max. On the other hand, we can notice that, even if the input is in $[0,1]$, the output exceeds the range and must therefore be clipped to 1. Especially with ℓ close to 1, the ℓ-norm diverges from the max, giving output values that can be much higher than the single components of the input vector.

3.1   The gTPR Holds for AWX

In this section we prove the consistency of the AWX output layer with respect to the gTPR, introduced in Sect. 2.1.

We want to show that $\forall < t_u, t_v > \in E, \hat{y}_v \geq \hat{y}_u$ holds for $\hat{y}_v, \hat{y}_u$ in Eq. 4.

Proof. Note that Eq. 4 can be rewritten as $\hat{y}_v = max(\mathcal{C}_v)$, where

$$\mathcal{C}_v = \{f(z_u)| < t_u, t_v > \in E'\}$$

is the set of the contributions to the predictions coming from the leaf terms. In the special case of leaf terms, $t_u \in \mathcal{F}$, by construction, $\mathcal{C}_u = \{f(z_u)\}$ therefore $\hat{y}_u = f(z_u)$. We can express the statement of the thesis as $\forall < t_u, t_v > \in E$,

$$\hat{y}_v = max(\mathcal{C}_v) \geq max(\mathcal{C}_u) = \hat{y}_u \tag{6}$$

Being the max function monotonic, the above inequality holds if $\mathcal{C}_u \subseteq \mathcal{C}_v$.

Consider the base case $\langle t_u, t_v \rangle \in E$ such that $t_u \in \mathcal{F}$. It clearly holds that $\mathcal{C}_u = \{f(z_u)\} \subseteq \mathcal{C}_v$, because if $\langle t_u, t_v \rangle \in E$ then $\langle t_u, t_v \rangle \in E'$ and therefore $f(z_u) \in \mathcal{C}_v$.

Now consider two generic terms in a "*is_a*" relation $\langle t_u, t_v \rangle \in E$ and their contributions sets $\mathcal{C}_u$ and $\mathcal{C}_v$. By design

$$\forall t_k \in \mathcal{F}, \langle t_k, t_u \rangle \in E' \implies \langle t_k, t_v \rangle \in E'$$

and therefore

$$\{f(z_k)|\langle t_k, t_u \rangle \in E'\} \subseteq \{f(z_k)|\langle t_k, t_v \rangle \in E'\}$$
$$\mathcal{C}_u \subseteq \mathcal{C}_v,$$

and Eq. 6 holds. □

The reasoning proceeds similarly for the estimator $\hat{y}_i$ in Eq. 5, but in order to guarantee the consistency the input must be in $[0, +\infty)$ since the ℓ-norm is monotonic only in that interval.

3.2 Implementation

The model has been implemented within the Keras [3] framework and a public version of the code is available at https://github.com/lucamasera/AWX. The choice of Keras was driven by the will of integrating AWX into deep-learning architectures, and at the time of writing Keras represents a widely-used framework in the area.

An important aspect to consider is that AWX is independent from the underlying network, and can therefore been applied to any ANN that requires a consistent hierarchical output.

4   Experimental Setting

In order to assess the performance of the proposed model, an extensive comparison was performed on the standard benchmark datasets[2]. The datasets cover different experiments [21] conducted on *S. cerevisiae* annotated with two ontologies, i.e. FunCat and GO. For each dataset are provided the train, the validation and the test splits of size respectively of *circa* 1600, 800, and 1200. The only exception is the Eisen dataset where, the examples per split are *circa* 1000, 500, and 800. FunCat is structured as a tree with almost 500 term, while GO is composed by three DAGs, comprising more then 4000 terms. The details of the datasets are reported in Table 1 while Fig. 2 reports the distribution of terms and leaves per level. Despite having many more terms and being deeper, most of the GO terms lay above the sixth level, depicting an highly unbalance structure with just few branches reaching the deepest levels (Fig. 3).

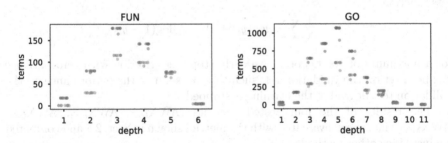

Fig. 3. The figures show the distribution of terms and leaves by level. Each dataset has a blue marker for the number of terms and an orange one for the number of leaves.

AWX has been tested both with the formulation in Eq. 4 (AWX_{MAX}) and with the one of Eq. 5 ($AWX_{\ell=k}$ with $k \in \{1, 2, 3\}$). The overall scheme of the network employed in the experiments is reported in Table 2 and consists in just an hidden layer with 1000 units and AWX as output layer. The model has been trained with

---

[2] https://dtai.cs.kuleuven.be/clus/hmcdatasets/.

Table 1. The table reports the details of the benchmark datasets used in the experiments. d, $|\mathcal{T}|$ and $|\mathcal{F}|$ reports respectively the dimensionality of the dataset, the number of terms and how many of them are leaves.

Dataset	d	FunCat		GO	
		$|\mathcal{T}|$	$|\mathcal{F}|$	$|\mathcal{T}|$	$|\mathcal{F}|$
Cellcycle	77	499	324	4122	2041
Church	27	499	324	4122	2041
Derisi	63	499	324	4116	2037
Eisen	79	461	296	3570	1707
Expr	551	499	324	4128	2043
Gasch1	173	499	324	4122	2041
Gasch2	52	499	324	4128	2043
Hom	47034	499	324	4128	2043
Seq	478	499	324	4130	2044
Spo	80	499	296	4116	2037

Table 2. ANN architecture used in the experiments.

fully_connected(size=1000, activation $=$ relu, l2 $= 10^{-3}$)
awx(activation $=$ sigmoid, l2 $= 10^{-3}$)

the ADAM optimizer algorithm [8] ($lr = 10^{-5}$, $\beta_1 = 0.9$, and $\beta_2 = 0.999$) and loss function

$$L = \frac{1}{N} \sum_i^N y_i \log(\hat{y}_i) + (1 - y_i)\log(1 - \hat{y}_i)$$

for a maximum of 1000 epochs. An early stopping criterion with zero patience has been set on the validation set, such that, as soon as the performance on the validation test degrades, the learning is stopped.

The results of the four tested variants ($AWX_{\ell=1}$, $AWX_{\ell=2}$, $AWX_{\ell=3}$, AWX_{MAX}) have been evaluated with the metrics shown in Sect. 2.2 and compared against three other methods.

1. **HMC-clus** [21]: state-of-the-art model based on decision trees. The results reported are taken from the original work.
2. **HMC-LMLP** [1,2]: The model is trained level by level, enriching the input space with the output of the previous level. As for HMC-clus, the reported results are taken from the original paper.
3. **MLP**$_{leaves}$: ANN trained only on the leaf terms with the same parameters of AWX. The prediction for the non-leaf terms are obtained by taking the max of the predictions for underlying terms in the hierarchy.

The comparison with MLP$_{leaves}$ is crucial, because it highlights the impact of jointly learning the whole hierarchy with respect to inferring the prediction after the learning phase.

Both AWX and MLP$_{leaves}$ are based on ANNs, so, in order to mitigate the effect of the random initialization of the weight matrix, the learning process has been repeated 10 times. We report the average results of the 10 iterations and the standard deviation ranges are reported in the caption of the tables. We performed a t-test with $\alpha = 0.05$ to assess the significativity of the difference with respect to the state-of-the-art and marked with * the results that passed the test.

No parameter tuning has been performed for the trained methods and the validation split has been used only for the early stopping criterion.

5   Results

In this section are reported and discussed the results obtained by the proposed method on ten benchmark datasets. Besides the comparison with the state-of-the-art, we will show the impact of AWX highlighting the differences with respect to MLP$_{leaves}$.

Table 3 reports the micro-averaged area under the precision recall curve (AUC($\overline{PR}$)). AWX$_{\ell=1}$ has a clear edge over the competitors, in both the ontologies. With the FunCat annotation, it is significantly better then CLUS-HMC six out of ten times and worse just in the Hom dataset, while with GO it wins nine out of ten times. AWX$_{\ell=1}$ clearly outperforms also the other AWX versions and MLP$_{leaves}$ in all the datasets. We can notice that the performance tends to decrease for higher values of ℓ, and reaches the minimum with AWX$_{MAX}$. This can be explained by the distribution of the example-annotation: due to the TPR, the terms placed close to the root are more likely to be associated with more examples then the lower terms. With $\ell = 1$ each leaf, among the descendants, contributes equally to the prediction, boosting the prediction values of the upper terms.

Of great interest is the comparison between MLP$_{leaves}$ and AWX$_{MAX}$. We can notice that AWX$_{MAX}$, despite being the worst performing version of AWX, always outperforms MLP$_{leaves}$. Remember that the main difference between the two architectures is that with AWX$_{MAX}$ prediction are propagated at learning time and consequently optimized, while in MLP$_{leaves}$ the predictions of the non-leaf terms are inferred offline.

Table 4 reports the macro-averaged area under the precision recall curves ($\overline{AUCPR}$). Unfortunately HMC-LMLP does not provide the score for this metric, but AWX clearly outperforms CLUS-HMC, both with the FunCat and with the GO annotations. We can notice that the differences are significant in all the datasets, with the exception of Hom, where the variance of the computed results is above 0.04 with FunCat and 0.007 with GO. Within AWX is instead more difficult to identify a version that performs clearly better then the others, their performance are almost indistinguishable. Focusing on the differences between MLP$_{leaves}$ and AWX$_{MAX}$, we can see that the former is outperformed fourteen out of twenty times by the latter. The advantage of the AWX layer in this case

Table 3. AUC($\overline{PR}$). Bold values show the best preforming method on the dataset. The standard deviation of the computed methods is in the range $[0.001, 0.005]$ and $[0.002, 0.008]$ respectively for FunCat and GO, with the only exception of Hom, where is an order of magnitude bigger.

		CLUS-HMC	HMC-LMLP	MLP$_{leaves}$	AWX$_{\ell=1}$	AWX$_{\ell=2}$	AWX$_{\ell=3}$	AWX$_{MAX}$
FunCat	Cellcycle	0.172	0.185	0.148*	**0.205***	0.189*	0.181*	0.174
	Church	0.170	0.164	0.102*	**0.173**	0.150*	0.136*	0.127*
	Derisi	**0.175**	0.170	0.112*	**0.175**	0.152*	0.142*	0.136*
	Eisen	0.204	0.208	0.196*	**0.252***	0.243*	0.234*	0.225*
	Expr	0.210	0.196	0.201	**0.262***	0.236*	0.229*	0.223*
	Gasch1	0.205	0.196	0.182	**0.238***	0.227*	0.217*	0.209
	Gasch2	0.195	0.184	0.150*	**0.211***	0.195	0.186	0.178*
	Hom	**0.254**	0.192	0.100*	0.107*	0.109*	0.106*	0.127*
	Seq	0.211	0.195	0.188*	**0.253***	0.234*	0.227*	0.218
	Spo	**0.186**	0.172	0.117*	0.179	0.159*	0.150*	0.143*
GO	Cellcycle	0.357	0.365	0.315*	**0.441***	0.406*	0.385*	0.362
	Church	0.348	0.347	0.272*	**0.440***	0.378*	0.355*	0.329
	Derisi	0.355	0.349	0.274*	**0.424***	0.376*	0.352	0.335*
	Eisen	0.380	0.403	0.347*	**0.481***	0.449*	0.426*	0.410*
	Expr	0.368	0.384	0.357*	**0.480***	0.437*	0.418*	0.407*
	Gasch1	0.371	0.384	0.346*	**0.468***	0.437*	0.416*	0.401*
	Gasch2	0.365	0.369	0.328*	**0.454***	0.417*	0.394*	0.379
	Hom	**0.401**		0.203*	0.264*	0.256*	0.242*	0.238*
	Seq	0.386	0.384	0.347*	**0.472***	0.429*	0.412*	0.397*
	Spo	0.352	0.345	0.278*	**0.420***	0.378*	0.355	0.336

is not as visible as in terms of AUC($\overline{PR}$), because $\overline{AUCPR}$ gives equal weight to the curve for each class, ignoring the number of annotated examples. Within our setting, where most of the classes have few examples, this evaluation metric tends to flatten the results, and may not be a good indicator of the performance.

Table 5 reports the weighted-average of the area under the precision recall curves ($\overline{AUCPR_w}$). AWX has solid performance also considering this evaluation metric, outperforming significantly CLUS-HMC in most of the datasets. HMC-LMLP provides results only for the FunCat-annotated datasets, but appears to be not competitive with our method. Within the proposed variants of AWX, $\ell = 2$ has an edge over $\ell = 1$ and MAX, while is almost indistinguishable from $\ell = 3$. Moreover, comparing AWX$_{MAX}$ and MLP$_{leaves}$, we can see that the former is systematically better then the latter.

The results reported for AWX (in all its variants) and MLP were not obtained tuning the parameters on the validation sets, but rather setting them *a priori*. The lack of parameter-tuning is clear on the Hom dataset. With all the considered evaluation metrics, this dataset significantly diverges with respect to the others. This behaviour is common also to CLUS-HMC, but unlikely the proposed methods, it has the best performance on this dataset. An explanation of this anomaly

Table 4. $\overline{\text{AUCPR}}$. Bold values show the best preforming method on the dataset. HMC-LMLP provides no results for this metric, so it has been removed from the table. The standard deviation of the computed methods is in the range $[0.001, 0.005]$ for both Fun-Cat and GO, with the only exception of Hom, where it is an order of magnitude bigger.

		CLUS-HMC	MLP$_{leaves}$	AWX$_{\ell=1}$	AWX$_{\ell=2}$	AWX$_{\ell=3}$	AWX$_{\text{MAX}}$
FunCat	Cellcycle	0.034	0.068*	0.075*	0.076*	0.077*	0.076*
	Church	0.029	0.040*	0.040*	**0.041***	0.040*	0.041*
	Derisi	0.033	0.047*	0.047*	0.048*	**0.049***	0.048*
	Eisen	0.052	0.095*	0.103*	0.104*	**0.106***	**0.106***
	Expr	0.052	0.114*	0.120*	**0.121***	**0.121***	0.120*
	Gasch1	0.049	0.101*	0.108*	0.110*	**0.111***	0.109*
	Gasch2	0.039	0.069*	0.080*	**0.078***	0.077*	0.078*
	Hom	0.089	0.116	0.095	0.086	0.112	**0.164**
	Seq	0.053	0.126*	0.121*	**0.126***	**0.126***	0.124*
	Spo	0.035	0.043*	0.045*	0.045*	0.045*	0.045*
GO	Cellcycle	0.021	0.057*	0.059*	0.057*	0.057*	0.057*
	Church	0.018	0.034*	0.030*	**0.032***	0.031*	0.031*
	Derisi	0.019	0.038*	0.041*	**0.040***	**0.040***	0.039*
	Eisen	0.036	0.082*	**0.091***	0.089*	0.088*	0.088*
	Expr	0.029	0.092*	**0.104***	0.102*	**0.104***	0.102*
	Gasch1	0.030	0.076*	0.086*	0.086*	0.086*	0.085*
	Gasch2	0.024	0.065*	0.067*	0.066*	0.067*	0.066*
	Hom	0.051	**0.071**	0.042	0.046	0.042	0.053
	Seq	0.036	**0.130***	**0.130***	**0.130***	0.128*	0.128*
	Spo	0.026	0.037*	0.038*	0.039*	**0.040***	0.038*

can be found in Table 1, where we can see the difference in the dimensionality. Hom dataset features are indeed two order of magnitude more then the second biggest dataset, i.e. Expr. The sub-optimal choice of the model parameters and the dimensionality could therefore explain the deviation of this dataset from the performance trend, and these aspects should be explored in the future.

AWX has solid overall performance. AUC($\overline{\text{PR}}$) is the most challenging evaluation metric, where the improvement over the state-of-the-art is smaller, while with the last two metrics, AWX appears to have a clear advantage. The choice of the value for ℓ depends on the metric we want to optimize, indeed AWX performs the best with $\ell = 1$ considering AUC($\overline{\text{PR}}$), while if we consider $\overline{\text{AUCPR}}$) or $\overline{\text{AUCPR}}_w$ values of $\ell = 2$ or $\ell = 3$ have an edge over the others. The direct comparison between MLP$_{leaves}$ and AWX$_{\text{MAX}}$ is clearly in favour of the latter, which wins almost on all the datasets. This highlights the clear impact of the proposed approach, that allows a jointly optimization of all the classes in the datasets.

Table 5. $\overline{\text{AUCPR}}_w$. Bold values show the best preforming method on the dataset. HMC-LMLP provides no data for the GO datasets. The standard deviation of the computed methods is in the range $[0.001, 0.005]$ for both FunCat and GO, with the only exception of Hom, where it is an order of magnitude bigger.

		CLUS-HMC	HMC-LMLP	MLP$_{leaves}$	AWX$_{\ell=1}$	AWX$_{\ell=2}$	AWX$_{\ell=3}$	AWX$_{MAX}$
FunCat	Cellcycle	0.142	0.145	0.186*	0.200*	0.204*	**0.205***	0.203*
	Church	0.129	0.118	0.132	**0.139***	**0.139***	0.139*	0.138
	Derisi	0.137	0.127	0.144*	0.150*	**0.152***	**0.152***	0.151*
	Eisen	0.183	0.163	0.229*	0.246*	**0.254***	**0.254***	0.252*
	Expr	0.179	0.167	0.239*	**0.260***	**0.260***	0.258*	0.255*
	Gasch1	0.176	0.157	0.217*	0.234*	**0.241***	0.240*	0.237*
	Gasch2	0.156	0.142	0.183*	0.200*	**0.202***	**0.202***	0.200*
	Hom	**0.240**	0.159	0.185	0.185	0.188	0.193	0.222
	Seq	0.183	0.112	0.232*	0.260*	**0.263***	0.262*	0.258*
	Spo	0.153	0.129	0.152	0.159	**0.161***	**0.161***	0.160*
GO	Cellcycle	0.335		0.372*	0.379*	0.384*	**0.385***	0.380*
	Church	0.316		0.325*	0.328*	**0.331***	0.329*	0.327*
	Derisi	0.321		0.331*	0.334*	**0.338***	0.337*	0.336*
	Eisen	0.362		0.402*	0.418*	**0.426***	0.423*	0.418*
	Expr	0.353		0.407*	0.424*	**0.430***	**0.430***	0.427*
	Gasch1	0.353		0.397*	0.410*	**0.421***	0.419*	0.416*
	Gasch2	0.347		0.379*	0.386*	**0.394***	0.393*	0.388*
	Hom	**0.389**		0.354*	0.342*	0.349	0.345*	0.356
	Seq	0.373		0.408*	0.431*	**0.436***	0.434*	0.430*
	Spo	0.324		0.333*	0.332	**0.339***	**0.339***	0.338*

6   Conclusion

In this work we have proposed a generalization to the continuous domain of the true path rule and presented a novel ANN layer, named AWX. The aim of this component is to allow the user to compute consistent hierarchical predictions on top of any deep learning architecture. Despite the simplicity of the proposed approach, it appears clear that AWX has an edge over the state-of-the-art method CLUS-HMC. Significant improvements can be seen almost on all the datasets for the macro and weighted averaged area under the precision recall curves evaluation metric, while the advantage in terms of $\text{AUC}(\overline{\text{PR}})$ is significant in six out of ten datasets.

Part of improvement could be attributed to the power and flexibility of ANN over decision trees, but we have shown that AWX systematically outperforms also HMC-LMLP, based on ANN, and MLP$_{leaves}$, that has exactly the same architecture as AWX, but with sigmoid output layer just for the leaf terms.

Further work will be focused to test the proposed methods with real-world challenging datasets, integrating AWX in deep learning architectures. Another interesting aspect will be to investigate the performance of AWX with semantic-based metrics, both globally or considering only the leaf terms.

References

1. Cerri, R., Barros, R.C., de Carvalho, A.C.P.L.F.: Hierarchical classification of Gene Ontology-based protein functions with neural networks. In: 2015 International Joint Conference on Neural Networks, pp. 1–8 (2015)
2. Cerri, R., Barros, R.C., de Carvalho, A.C.P.L.F.: Hierarchical multi-label classification using local neural networks. J. Comput. Syst. Sci. **80**(1), 39–56 (2014)
3. Chollet, F., et al.: Keras (2015)
4. Gene Ontology Consortium: Creating the gene ontology resource: design and implementation. Genome Res. **11**(8), 1425–33 (2001)
5. Gong, Q., Ning, W., Tian, W.: GoFDR: a sequence alignment based method for predicting protein functions. Methods **93**, 3–14 (2015)
6. Hinton, G., et al.: Deep neural networks for acoustic modeling in speech recognition: the shared views of four research groups. IEEE Signal Process. Mag. **29**(6), 82–97 (2012)
7. Jiang, Y., et al.: An expanded evaluation of protein function prediction methods shows an improvement in accuracy. Genome Biol. **17**(1), 184 (2016)
8. Kingma, D., Ba, J.: Adam: a method for stochastic optimization. arXiv preprint arXiv:1412.6980 (2014)
9. Krizhevsky, A., Sutskever, I., Hinton, G.E.: Imagenet classification with deep convolutional neural networks. In: Advances in Neural Information Processing Systems, pp. 1097–1105 (2012)
10. Murzin, A.G., Brenner, S.E., Hubbard, T., Chothia, C.: SCOP: a structural classification of proteins database for the investigation of sequences and structures. J. Mol. Biol. **247**(4), 536–540 (1995)
11. Obozinski, G., Lanckriet, G., Grant, C., Jordan, M.I., Noble, W.S.: Consisten probabilistic outputs for protein function prediction. Genome Biol. **9**(65), 1–19 (2008)
12. Radivojac, P., et al.: A large-scale evaluation of computational protein function prediction. Nat. Methods **10**(3), 221–227 (2013)
13. Ruepp, A., et al.: The FunCat, a functional annotation scheme for systematic classification of proteins from whole genomes. Nucl. Acids Res. **32**(18), 5539–5545 (2004)
14. Schriml, L.M., et al.: Disease ontology: a backbone for disease semantic integration. Nucl. Acids Res. **40**(D1), D940–D946 (2011)
15. Sokolov, A., Ben-Hur, A.: Hierarchical classification of gene ontology terms using the gostruct method. J. Bioinform. Comput. Biol. **8**(02), 357–376 (2010)
16. Soricut, R., Marcu, D.: Sentence level discourse parsing using syntactic and lexical information. In: Proceedings of the 2003 Conference of the North American Chapter of the Association for Computational Linguistics on Human Language Technology-Volume 1, pp. 149–156. Association for Computational Linguistics (2003)
17. Sorower, M.S.: A literature survey on algorithms for multi-label learning, vol. 18. Oregon State University, Corvallis (2010)
18. Srivastava, N., Hinton, G., Krizhevsky, A., Sutskever, I., Salakhutdinov, R.: Dropout: a simple way to prevent neural networks from overfitting. J. Mach. Learn. Res. **15**(1), 1929–1958 (2014)
19. Sun, A., Lim, E.-P.: Hierarchical text classification and evaluation. In: Proceedings IEEE International Conference on Data Mining, ICDM 2001, pp. 521–528. IEEE (2001)
20. Triguero, I., Vens, C.: Labelling strategies for hierarchical multi-label classification techniques. Pattern Recognit. **56**, 1–14 (2015)

21. Vens, C., Struyf, J., Schietgat, L., Džeroski, S., Blockeel, H.: Decision trees for hierarchical multi-label classification. Mach. Learn. **73**(2), 185–214 (2008)
22. Zhang, M.L., Zhou, Z.H.: A review on multi-label learning algorithms. IEEE Trans. Knowl. Data Eng. **26**(8), 1819–1837 (2014)

Clustering and Unsupervised Learning

Clustering in the Presence
of Concept Drift

Richard Hugh Moulton[1]([✉]), Herna L. Viktor[1], Nathalie Japkowicz[2],
and João Gama[3]

[1] School of Electrical Engineering and Computer Science,
University of Ottawa, Ottawa, ON, Canada
{rmoul026,hviktor}@uottawa.ca
[2] Department of Computer Science, American University, Washington DC, USA
nathalie.japkowicz@american.edu
[3] Faculty of Economics, University of Porto, Porto, Portugal
jgama@fep.up.pt

Abstract. Clustering naturally addresses many of the challenges of data streams and many data stream clustering algorithms (DSCAs) have been proposed. The literature does not, however, provide quantitative descriptions of how these algorithms behave in different circumstances. In this paper we study how the clusterings produced by different DSCAs change, relative to the ground truth, as quantitatively different types of concept drift are encountered. This paper makes two contributions to the literature. First, we propose a method for generating real-valued data streams with precise quantitative concept drift. Second, we conduct an experimental study to provide quantitative analyses of DSCA performance with synthetic real-valued data streams and show how to apply this knowledge to real world data streams. We find that large magnitude and short duration concept drifts are most challenging and that DSCAs with partitioning-based offline clustering methods are generally more robust than those with density-based offline clustering methods. Our results further indicate that increasing the number of classes present in a stream is a more challenging environment than decreasing the number of classes. Code related to this paper is available at: https://doi.org/10.5281/zenodo.1168699, https://doi.org/10.5281/zenodo.1216189, https://doi.org/10.5281/zenodo.1213802, https://doi.org/10.5281/zenodo.1304380.

Keywords: Data streams · Clustering · Concept drift

The authors acknowledge that research at the University of Ottawa is conducted on traditional unceded Algonquin territory. This research was supported by the Natural Sciences and Engineering Research Council of Canada and the Province of Ontario. J. Gama is partially funded by the ERDF through the COMPETE 2020 Programme within project POCI-01-0145-FEDER-006961, and by National Funds through the FCT as part of project UID/EEA/50014/2013.

Electronic supplementary material The online version of this chapter (https://doi.org/10.1007/978-3-030-10925-7_21) contains supplementary material, which is available to authorized users.

© Springer Nature Switzerland AG 2019
M. Berlingerio et al. (Eds.): ECML PKDD 2018, LNAI 11051, pp. 339–355, 2019.
https://doi.org/10.1007/978-3-030-10925-7_21

1   Introduction

Data streams are challenging learning environments: their size is unbounded [3], the probabilities underlying the data stream can change [10,12,21] and labelled data is not readily available [20,22]. Clustering addresses these challenges as a means of summarization [17] and as an unsupervised learning technique. Lacking in the literature, however, is a discussion of how data stream clustering algorithms (DSCAs) are expected to perform. Since clustering is useful in dealing with data streams, understanding DSCA behaviour will help develop effective machine learning techniques for data streams. We use Webb et al.'s framework for quantitatively describing concept drift [23] to analyse DSCA performance.

We make two contributions in this paper. First, we propose a method for generating real-valued data streams with precise quantitative concept drift. This method uses mixture models and the Hellinger distance between concepts to mathematically model data streams. Second, we conduct quantitative analyses of DSCAs in experimental settings to determine the effect that different concept drifts have on the clusterings produced by different DSCAs. We also demonstrate how to use these findings to guide the selection of a DSCA in real-world applications. The research question we address is "how do the clusterings produced by different DSCAs change, relative to the ground truth, as quantitatively different types of concept drift are encountered?" Of particular interest is whether different DSCAs react differently in the presence of concept drift.

In the remainder of this paper we review the literature concerning concept drift and clustering in data streams (Sect. 2), describe the Mixture Model Drift Generator (Sect. 3), lay out our experimental framework (Sect. 4), present our results (Sect. 5) and identify potential future research in our conclusion.

2   Literature Review

In this section we describe the quantitative models used to formalize the data stream environment. We also discuss the types of concept drift identified in the literature and how they can be described using mathematical formalisms as well. Finally, we review the data stream clustering task and survey algorithms proposed for this purpose.

2.1   Concept Drift in Data Streams

Webb et al. describe data streams as data sets with a temporal aspect and generated by some underlying process. This process can be modelled as a random variable, χ, and the data stream's instances as objects drawn from this random variable. An object, o, is a pair $\langle x, y \rangle$ where x is the object's feature vector and y is the object's class label. Each is drawn from a different random variable, X and Y: $x \in \text{dom}(X)$, $y \in \text{dom}(Y)$ and $o \in \text{dom}(X, Y) = \text{dom}(\chi)$ [23].

Many authors have conceptualized concept drift qualitatively [12,25]. Abrupt concept drift is when one concept is immediately replaced by another, e.g. a computer network expands. Gradual concept drift is an alternation between concepts, e.g. a network's computers are slowly upgraded to a new OS. Incremental concept drift, instead sees a series of intermediate concepts, e.g. an office's computer usage evolves over a project's lifetime (Fig. 1).

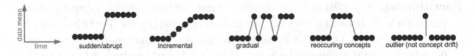

Fig. 1. Qualitatively different types of concept drift (from Gama et al. [12])

Described quantitatively, a data stream's concept at time t is the probability associated with its underlying generative process; Definition 1 [12,26]. Concept drift occurs between points of time t and u in the data stream when $P_t(X, Y) \neq P_u(X, Y)$. This could occur because of changes in $P(X)$ or in $P(Y|X)$. As class labels are not available for the clustering task, we are only concerned with the former. Concept evolution, when a novel pattern emerges, is a special case of concept drift [10,20]. Although a changing number of classes is usually discussed in the context of supervised learning [11,20], $P(X)$ is also likely to affected and so we consider concept evolution here. This might occur when certain activities only occur at certain times, e.g. system updates download overnight.

Definition 1 (Concept). $Concept = P_t(X, Y) = P_t(\chi)$

In the quantitative paradigm, a concept drift's magnitude is the distance between the concepts at times t and u as measured by a distribution distance function, D, (Definition 2) [23]. This is not captured by the types shown in Fig. 1, which do not have the language to describe how far apart two concepts are. Using drift magnitude allows us to distinguish cases with overlapping concepts from cases with divergent concepts.

Definition 2 (Drift magnitude). $Magnitude_{t,u} = D(P_t(\chi), P_u(\chi))$

The duration of a concept drift is the time during which the drift occurs (Definition 3). This distinguishes drifts of different lengths and illustrates that the boundary between abrupt and extended concept drift is a threshold value [23].

Definition 3 (Drift duration). $Duration_{t,u} = u - t$

2.2 Data Stream Clustering Algorithms

DSCAs cluster the instances within data streams as they occur. Barbará argued that they must compactly represent instances; process instances incrementally; and clearly identify outliers [4]. Most DSCAs produce clusterings with an online component to summarize the data stream's instances and an offline component similar to traditional clustering algorithms [14,21]. Although DSCAs use different approaches, we note four general methods of clustering data streams that are discussed in the literature: partitioning, density-based, hierarchical and grid-based. These match Han et al.'s methods of clustering static data sets [16, p. 450].

Partitioning. A partition is produced so that similar objects are in the same partition and dissimilar objects are in different partitions. Partitions can be defined by mean points, representative points, or map nodes. A common drawback is that these methods are only capable of producing hypersphere-shaped clusters and have difficulty representing arbitrarily shaped clusters.

Density-based. Clusters are produced as dense regions separated by less dense regions, allowing them to represent arbitrarily shaped clusters. For example, DenStream [6] models objects with microclusters. Depending on their density, these microclusters are labelled as either core, potentially core or outlier. A key to density-based methods is defining what constitutes a dense region; this is usually done by user-defined parameters.

Hierarchical. Objects are organized into a hierarchical structure, like a tree. In this way, objects are closer in the tree structure to objects that are similar and further away in the tree from those that are dissimilar. The hierarchical nature of the tree structure allows different clusterings to be produced by inspecting the tree at different levels, but may also require the computationally expensive rebuilding of the tree.

Grid-based. The feature space is divided into grid cells, which summarize the data stream objects by acting as bins. D-Stream takes advantage of this to put an upper bound on computations: no matter how many objects arrive, they are represented with a constant number of grids [7]. The grid's fineness (or coarseness) represents a trade off between precision of results and cost of computations; this is generally defined by user supplied parameters.

Silva et al. [21] characterize DSCAs using seven aspects. The first is whether the clustering task is object-based or attribute-based. We focus on the former as its applications are more commonplace [21]. The second is the number of user-defined parameters, e.g. window sizes, decay rates and thresholds. Most notably: some DSCAs require the number of clusters to find - k - which handicaps an algorithm's ability to deal with a data stream's dynamic behaviour [21].

Online Component. A DSCA's online component allows it to process new instances quickly and incrementally; it incorporates three of Silva et al.'s aspects [21]. The third aspect is the data structure used to represent the unbounded instances in a compact manner. One possibility is a feature vector which summarizes N instances, $x_1, x_2, ..., x_N$, as $\langle N, LS, SS \rangle$, where LS is the linear sum of those instances ($\sum_{i=1}^{N} x_i$) and SS is the square sum of those instances ($\sum_{i=1}^{N} x_i^2$) [24]. Prototype arrays summarize partitions using medoids or centroids; these prototypes can later be summarized themselves [21]. Nodes in a self-organizing map or neurons in a growing neural gas are also potential summaries [13]. Alternatively, coreset trees organize $2m$ instances into a tree from which a coreset of

m instances is extracted. Two coresets are reduced by building another coreset tree from their union [1]. Finally, dividing the feature space into grid cells allows each cell to summarize its respective objects [7].

Reasoning that recent objects are more relevant, the fourth aspect is the window model used to passively forget old concepts. Sliding windows store instances in a queue and remove the oldest instance every time a new one is added. Damped windows weight instances by age, often by using an exponential function, until they are forgotten. Meaningful time-based or stream-based landmarks can be used to break the stream into non-overlapping chunks - landmark windows [21].

The fifth aspect is the outlier detection mechanism. The algorithm must decide if a point that doesn't fit the existing clusters represents a new cluster or an outlier to be discarded [4]. Mechanisms to do so include buffering outliers until it is possible to include them in the summarization [6,24] and deleting microclusters or grids with below-threshold density or relevance [2,6,7].

Offline Component. A DSCA's offline component is called when a clustering is required. The summarization is often treated as a static data set and traditional clustering algorithms are used. Silva et al.'s last two aspects are seen here: the offline clustering algorithm used and the resulting clusters' shape [21].

One popular approach for the offline clustering algorithm is the k-means family of clustering algorithms. This includes k-means applied to the statistical summary or to a weighted statistical summary, selecting medoids with k-medoids, and using an initial seeding with k-means++. As expected, DSCAs making use of these algorithms result in hypersphere-shaped clusters. The other popular approach is to use density-based clustering algorithms, such as DBSCAN, applied to feature vectors, grid cells or frequent states. DSCAs that use density-based clustering have the ability to find arbitrarily-shaped clusters [21].

3   The Mixture Model Drift Generator

In order to conduct experiments using real-valued data streams with precise controlled concept drift, we propose a Mixture Model Drift Generator[1] based on Webb et al.'s categorical data generator[2] [23]. The Mixture Model Drift Generator models the periods before and after concept drift - stable concepts - as mixture models with one distribution for each class and a probability vector for choosing between the classes. We use multivariate normal distributions (MVNDs), which are defined by a mean point and a covariance matrix.

3.1   Generating the Underlying Probabilities

The generator requires the number of classes present before, n_0, and after, n_1, the concept drift, the stream's dimensionality, a, the drift magnitude, m, the tolerance for the drift magnitude, ϵ, and the drift duration, d.

---

[1] Available https://doi.org/10.5281/zenodo.1168699.
[2] Available https://doi.org/10.5281/zenodo.35005.

Data: n_0, n_1, a, m, ϵ
Result: Two mixture models M_0 and M_1
Generate M_0: a mixture model of n_0 a-dimensional MVNDs;
do
| Generate M_1: a mixture model of n_1 a-dimensional MVNDs;
while $H(M_0, M_1) \neq m \pm \epsilon$;

Algorithm 1. Mixture Model Drift Generator

Although any distribution distance functions could be used to measure drift magnitude, we use the Hellinger Distance because it is symmetrical and takes values between 0 and 1, inclusively [23]. The Hellinger distance between real valued probability density functions, $f(x)$ and $g(x)$, is shown in (1). From the last form of (1), the Hellinger distance is equal to 0 when the two functions are identical, $f(x) \equiv g(x)$, and equal to 1 when there is no overlap between them, i.e. $(f(x) \neq 0 \implies g(x) = 0) \wedge (g(x) \neq 0 \implies f(x) = 0)$.

$$H^2(f(x), g(x)) = \frac{1}{2} \int \left(\sqrt{f(x)} - \sqrt{g(x)} \right)^2 dx = 1 - \int \sqrt{f(x)g(x)}dx \quad (1)$$

We are unaware of a method to solve the second mixture model's parameters given the first mixture model and m. Instead, mixture models are generated and their Hellinger distance from the first mixture model is calculated until an appropriate second mixture model is found.

3.2 Drawing Instances from the Generator

During a stable concept, sampling the mixture model provides the attributes, x, while the specific MVND that was selected provides the class label, y. Together, these form the data stream object $o \in Dom(\chi)$ as introduced in Sect. 2.

Figure 2 illustrates data streams produced by the Mixture Model Drift Generator. Figures 2a and c depict the initial stable concept for two different data streams; different classes are identified by different colours. These mixture models have identical parameters but the instances drawn from each are different. Figures 2b and d depict the final stable concepts for the data streams. The mixture models are separated by a Hellinger distance of 0.4 in the first case and are separated by a Hellinger distance of 0.8 in the second.

During concept drift, generating instances is based on the qualitative type of the drift. For gradual concept drift, the generator draws the instance from one of the stable concepts with the probability of selecting the original concept decreasing over time. For incremental concept drift, the generator draws instances of the same class from both concepts. These instances are weighted, with the original concept's weight decreasing over time, and returned as the object o.

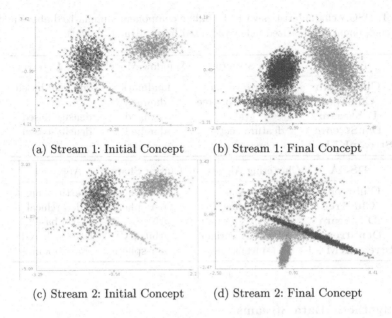

(a) Stream 1: Initial Concept   (b) Stream 1: Final Concept

(c) Stream 2: Initial Concept   (d) Stream 2: Final Concept

Fig. 2. Initial and final concepts for two data streams. $M_1 = 0.4$ and $M_2 = 0.8$

4   Experimental Evaluation

Our research question is "how do the clusterings produced by different DSCAs change, relative to the ground truth, as quantitatively different types of concept drift are encountered?" To answer this, we apply different DSCAs to synthetic real-valued data streams with different concept drifts. We then use these results to select a DSCA for a real word data stream.

4.1   Data Stream Clustering Algorithms

The five algorithms selected (Table 1) were listed by Silva et al. as among the 13 most relevant DSCAs [21]. Importantly, they also cover the four methods identified in Sect. 2.2.

The MOA 17.06 [5] implementation was used for each DSCA. A modified version of ClusTree[3] was used to properly implement the offline k-means clustering algorithm. A modified version of D-Stream[4] was used to permit the specification of grid widths for numerical attributes. Parameters chosen for each algorithm are included in the supplemental materials.

---

[3] Available https://doi.org/10.5281/zenodo.1216189.
[4] Available https://doi.org/10.5281/zenodo.1213802.

Table 1. DSCA characteristics with the online component summarized above and the offline component summarized below (adapted from Silva et al. [21])

DSCA	Data Structure	Window Model	Outlier Detection
CluStream	feature vector	landmark	statistical-based
ClusTree	feature vector tree	damped	-
D-Stream	grid	damped	density-based
DenStream	feature vector	damped	density-based
StreamKM++	coreset tree	landmark	-

DSCA	Clustering Algorithm	Cluster Shape	Approach
CluStream	k-means	hyper-sphere	Partitioning
ClusTree	k-means	hyper-sphere	Hierarchical
D-Stream	DBSCAN variant	arbitrary	Grid-based
DenStream	DBSCAN variant	arbitrary	Density-based
StreamKM++	k-means++	hyper-sphere	Partitioning

4.2 Synthetic Data Streams

Three experiments were conducted using synthetic data streams. One hundred two-dimensional data streams were produced for each experimental setting using the Mixture Model Drift Generator. For all data streams, the initial stable concept occurs for 2,000 instances to allow the DSCAs to achieve a stable clustering. The final stable concept occurs from the end of concept drift until the end of the data stream, again allowing a stable clustering.

The DSCA's clustering quality is the dependent variable in each experiment. Experiment A maintains four classes and has a drift duration of 1 – this is abrupt concept drift. Drift magnitude is the independent variable with values of 0.4, 0.5, 0.6, 0.7, 0.8 or 0.9. Experiment B has four classes before and after concept drift and a drift magnitude of 0.6. Drift duration is the independent variable with values of 1,000, 5,000 or 9,000 instances for both incremental and gradual concept drift. Experiment C involves a drift duration of 1, a drift magnitude of 0.6 and four class prior to concept drift. The number of post-concept drift classes represents concept evolution and varies between 2, 3, 5 or 6 classes.

4.3 Real World Data Streams

Four data streams were built using the ADFA-LD Anomaly Detection dataset, introduced by Creech et al. to replace the classic KDD 1999 Network Intrusion Detection dataset [8]. Haider et al. described four features they extracted from this dataset and showed a nearest neighbour approach using them for anomaly detection [15], suggesting that sensible clusters exist in this low-dimensional feature space. Our features, based on Haider et al.'s [15], are shown in Table 2.

We used the dataset's validation instances as the stream's normal behaviour. Attacks begin sporadically, dominate the data stream starting at instance 2000

Table 2. ADFA-LD Anomaly Detection features (based on Haider et al. [15])

Description
1 System call that appears with the highest frequency in the trace
2 Lowest valued system call that appears in the trace
3 Highest valued system call that appears in the trace
4 Number of distinct valued system calls that appear in the trace
5 Ratio of the number of appearances of the most repeated system call in the trace to the total number of system calls in the trace
6 Ratio between the range from lowest frequency to highest frequency of appearance in the trace to the total number of system calls in the trace

and continue for 250–500 instances before returning to their original frequency. This is abrupt concept drift as the two underlying probabilities are swapped immediately. Concept evolution may also occur as the clusters that make up the normal and attack behaviours may not be present throughout.

4.4 Performance Measure

As we have access to each data stream's ground truth, we use an external measure of cluster quality. Many of these exist in the literature, including purity, Rand statistic and Cluster Mapping Measure (CMM). We choose CMM because it is constrained to $[0, 1]$ (0 is the worst clustering, 1 is the best), accounts for different kinds of faults and performed well in Kremer et al.'s experiments [19].

CMM builds a clustering's fault set, its missed points, misassigned points and noise points assigned to clusters, and evaluates each fault point's connectivity to its true and assigned clusters [19]. An object's connectivity to a cluster is the ratio of its average k-neighbourhood distance ($knhDist$, Definition 4) to the average k-neighbourhood distance of the cluster. Each object's penalty is exponentially weighted by its age [19]. Definitions 4–7 are adapted from Kremer et al. [19].

Definition 4 (average k-neighbourhood distance). *The average distance of point p to its k neighbours in C is: $knhDist(p, C) = \frac{1}{k} \sum_{o \in knh(p,C)} dist(p, o)$. The average distance for a cluster C is: $knhDist(C) = \frac{1}{|C|} \sum_{p \in C} knhDist(p, C)$.*

Definition 5 (Connectivity). *Connectivity between object o and cluster C is:*

$$con(o, C) = \begin{cases} 1 & \textit{if } knhDist(o, C) < knhDist(C) \\ 0 & \textit{if } C = \emptyset \\ \frac{knhDist(C)}{knhDist(o,C)} & \textit{else} \end{cases}$$

Definition 6 (Penalty). $Cl(\cdot)$ *returns the ground truth class of the argument object and* $map(\cdot)$ *returns the ground truth class to which the argument cluster is mapped. The penalty for an object* $o \in \mathcal{F}$ *assigned to cluster* C_i *is:*

$$pen(o, C_i) = con(o, Cl(o)) \cdot (1 - con(o, map(C_i)))$$

Definition 7 (Cluster Mapping Measure). *Given an object set* $\mathcal{O}^+ = \mathcal{O} \cup Cl_{noise}$, *a ground truth* $\mathcal{CL}^+ = \mathcal{CL} \cup \{Cl_{noise}\}$, *a clustering* $\mathcal{C} = \{C_1, ..., C_k, C_\emptyset\}$, *and the fault set* $\mathcal{F} \subseteq \mathcal{O}^+$, *the Cluster Mapping Measure between* $\mathcal{C}$ *and* $\mathcal{CL}^+$ *is defined using the point weight* $w(o)$, *overall penalty* $pen(o, C)$ *and connectivity* $con(o, Cl(o))$ *as:*

$$CMM(\mathcal{C}, \mathcal{CL}) = 1 - \frac{\sum_{o \in \mathcal{F}} w(o) \cdot pen(o, C)}{\sum_{o \in \mathcal{F}} w(o) \cdot con(o, Cl(o))}$$

and if $\mathcal{F} = \emptyset$, *then* $CMM(\mathcal{C}, \mathcal{CL}) = 1$.

We used the implementation of CMM in MOA 17.06 [5] to evaluate the clustering produced by a DSCA every 100 instances.

5   Results and Discussion

The results show the average CMM for each setting's 100 data streams.[5] Algorithms that require the number of clusters were given $k = 4$. We use the Friedman test to compare multiple algorithms across multiple domains because it is non-parametric, doesn't assume the samples are drawn from a normal distribution and doesn't assume the sample variances are equal [18, pp. 247–248].

If the Friedman test leads us to conclude that the difference in algorithm performance is statistically significant, we conduct post-hoc Nemenyi tests. This regime is recommended by Japkowicz and Shah [18, p. 256] as well as Demšar [9]; statistical testing used the *scmamp* package in R[6]. Post-hoc Nemenyi test results are shown graphically; algorithms that are not significantly different (at $p = 0.05$) are linked.

5.1   Experiment A - Abrupt Concept Drift

These data streams exhibited abrupt concept drift at instance 2000. Each algorithm's maximum change in cluster quality for a given setting was calculated using the 1500 instances after concept drift. This change was compared to the algorithm's change in quality for the baseline data streams, controlling for unrelated changes in cluster quality, e.g. StreamKM++'s characteristic decrease in quality. To ease interpretation, only magnitudes 0.0, 0.4, 0.6 and 0.8 are shown.

The algorithms' results (Fig. 3) divide into two qualitative groups. CluStream and ClusTree are largely invariant to abrupt concept drift for all magnitudes.

---

[5] Results for additional cases available: https://doi.org/10.5281/zenodo.1304380.
[6] https://cran.r-project.org/package=scmamp.

The other three algorithms' cluster quality changes due to concept drift, with DenStream and StreamKM++ sensitive when the magnitude of the concept drift is larger. All three algorithms' results behave the same, however: abrupt concept drift is met with a decrease in cluster quality, an extreme cluster quality is reached and then a new stable cluster quality is established for the remainder of the data stream. Larger concept drift magnitudes results in larger decreases in cluster quality for all three algorithms. We also note that the magnitude of the concept drift does not affect the stable cluster quality for the second concept.

Using Friedman's test we conclude that among the five algorithms there is a significant difference in the change of cluster quality due to concept drift ($p < 0.01$). Post-hoc Nemenyi test results are shown in Fig. 4 where the highest ranked algorithm had the highest (most positive) change in cluster quality and the lowest ranked algorithm had the lowest (most negative) change.

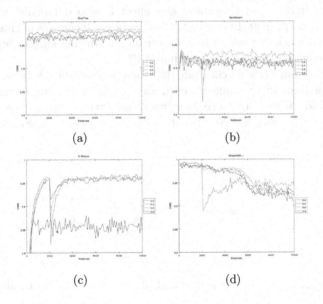

Fig. 3. Experiment A - Data streams with abrupt concept drift

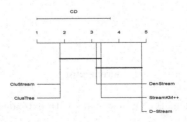

Fig. 4. Nemenyi test results for Experiment A

5.2   Experiment B - Extended Concept Drift

These data streams exhibited either gradual concept drift or incremental concept drift starting at instance 2000 and continuing for the specified duration. Each algorithm's time to reach and time to recover from its extreme CMM value was determined using average performance. Results are shown in Fig. 5.

Qualitatively, the algorithms' results are divided into two groups. CluStream and StreamKM++ are invariant for all durations and for both types. ClusTree, DenStream and D-Stream show changes in cluster quality that are affected by the concept drift's duration and type. DenStream and D-Stream exhibit the same general behaviour from Experiment A: concept drift results in a decrease in cluster quality, an extreme cluster quality is reached and then a new stable cluster quality is established. Longer concept drift durations soften this effect, as seen when comparing the 1,000 duration drift with the 9,000 duration drift for both DenStream and D-Stream; this effect is also dampened when facing incremental concept drift. In contrast, ClusTree's small changes in cluster quality take longer to reach the new stable cluster quality during longer concept drifts and for incremental compared to gradual drift.

Using Friedman's test we conclude that there is a significant difference among the five algorithms in the time to reach an extreme value due to concept drift ($p < 0.01$) and in the time to recover from that extreme value to a new stable quality ($p < 0.01$); post-hoc Nemenyi test results are shown in Fig. 6.

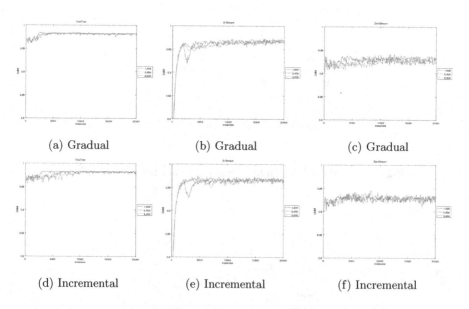

(a) Gradual   (b) Gradual   (c) Gradual

(d) Incremental   (e) Incremental   (f) Incremental

Fig. 5. Experiment B - Data streams with extended concept drift

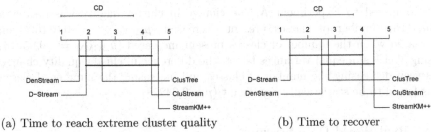

(a) Time to reach extreme cluster quality   (b) Time to recover

Fig. 6. Experiment B - Nemenyi test results.

5.3 Experiment C - Concept Evolution

These data streams exhibited concept evolution. That is, concept drift caused the number of classes present in the data stream to either increase or decrease. The change in cluster quality was measured the same way as for Experiment A; results are shown in Fig. 7.

CluStream and StreamKM++, the algorithms with a k parameter, suffered from increasing the number of classes, though neither suffered from decreasing the number of classes, i.e. representing two classes by splitting them across four clusters still resulted in better clusters than attempting to merge six classes into four clusters. DenStream and D-Stream, both algorithms without a k parameter, were also affected by concept evolution.

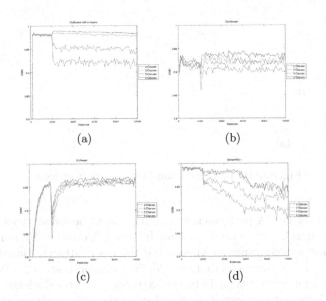

Fig. 7. Experiment C - Data streams exhibiting concept evolution

Compared to Experiment A, the change in cluster quality was decreased when the number of classes present decreased ($\mu_{\Delta CMM} = -0.01169$) and increased when the number of classes present increased ($\mu_{\Delta CMM} = -0.05016$). Using Welch's unequal variances t-test, the difference in cluster quality changes between decreasing the number of classes and increasing the number of classes was found to be statistically significant ($p = 0.04278$).

5.4   Real World Data Streams

Algorithms that required the number of clusters had the parameter k set to 2 by inspection. Cluster quality was generally similar to the quality obtained for the synthetic data streams, with the exception of DenStream. Though more volatile, each algorithm's overall performance is similar to its performance on synthetic data streams with abrupt drift, in line with how we described these data streams in Sect. 4.3. ClusTree, CluStream and StreamKM++ each show stable quality in the face of the first concept drift, although each becomes more volatile after the return to normal behaviour. D-Stream exhibits some volatility throughout and produces, on average, slightly lower quality clusters. DenStream produces consistently lower quality clusters than the other algorithms, though its volatility makes further interpretation difficult (Fig. 8).

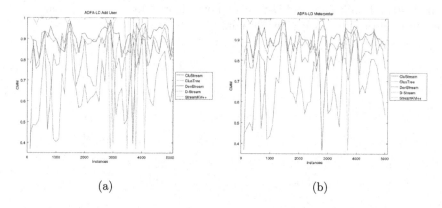

(a)   (b)

Fig. 8. ADFA-LD Intrusion Detection data streams

We note that DSCA performance is largely the same for all types of attack. This is a promising characteristic as an Intrusion Detection System must be robust against a wide range of known and unknown attacks. Based on these results, a sample recommendation would be to incorporate ClusTree as the DSCA chosen for an Intrusion Detection System. ClusTree does not require k, is generally invariant to concept drift and it consistently produces high quality clusters.

6   Conclusion

The length, speed and dynamic behaviour of data streams each present challenges for learning. Although clustering can be used to address these challenges, this paper highlighted the literature's gap in understanding the behaviour of DSCAs for concept drifting data streams. We proposed a method for generating real-valued data streams with precise quantitative concept drift and used these to conduct an experimental study.

Concept drifts with larger magnitudes and shorter durations were the most difficult for DSCAs, with each resulting in a larger decrease in cluster quality. We observe that the key factor in determining how a DSCA will respond to concept drift is the algorithm it uses in the offline component. DSCAs that incorporate partitioning-based clustering methods are more robust to concept drift while those that incorporate density-based clustering methods are more sensitive. We also observe that an increase in the number of classes present in a data stream is more challenging for DSCAs than a decrease and that this was true even for DSCAs that did not require an explicit k parameter.

Applying these DSCAs to a real world data stream in the domain of intrusion detection, we observed similar behaviour as in our experiments with synthetic real-valued data streams. This provides evidence that these findings can be taken from laboratory settings and used to predict behaviour in real world domains. As discussed, however, the dimensionality and data types of the data stream will also play a role in the DSCAs' abilities to produce quality results.

Future work using these findings could include: studying the performance of DSCAs when faced with complex concept drift; using these insights to develop an anomaly detection framework that incorporates clustering; and using an internal measure of cluster quality as a concept drift detection method.

References

1. Ackermann, M.R., Märtens, M., Raupach, C., Swierkot, K., Lammersen, C., Sohler, C.: StreamKM++: a clustering algorithm for data streams. ACM J. Exp. Algorithmics **17**(2), 2–4 (2012). https://doi.org/10.1145/2133803.2184450
2. Aggarwal, C.C., Han, J., Wang, J., Yu, P.S.: A framework for clustering evolving data streams. In: 29th Very Large Database Conference, p. 12, Berlin (2003)
3. Ahmed, M., Naser Mahmood, A., Hu, J., Mahmood, A.N., Hu, J.: A survey of network anomaly detection techniques. J. Netw. Comput. Appl. **60**, 19–31 (2016). https://doi.org/10.1016/j.jnca.2015.11.016
4. Barbará, D.: Requirements for clustering data streams. ACM SIGKDD Explor. Newsl. **3**(2), 23–27 (2002). https://doi.org/10.1145/507515.507519
5. Bifet, A., Holmes, G., Kirkby, R., Pfahringer, B.: Moa: massive online analysis. J. Mach. Learn. Res. **11**, 1601–1604 (2010)
6. Cao, F., Ester, M., Qian, W., Zhou, A.: Density-based clustering over an evolving data stream with noise. In: Ghosh, J., Lambert, D., Skillicorn, D., Srivastava, J. (eds.) Proceedings of the 2006 SIAM International Conference on Data Mining, pp. 328–339, Bethesda (2006). https://doi.org/10.1137/1.9781611972764.29

7. Chen, Y., Tu, L.: Density-based clustering for real-time stream data. In: Proceedings of the 13th ACM SIGKDD International Conference on Knowledge Discovery and Data Mining, pp. 133–142, San Jose, USA (2007)
8. Creech, G., Hu, J.: Generation of a new IDS test dataset: time to retire the KDD collection. In: 2013 IEEE Wireless Communications and Networking Conference (WCNC), pp. 4487–4492. IEEE (2013)
9. Demšar, J.: Statistical comparisons of classifiers over multiple data sets. J. Mach. Learn. Res. **7**, 1–30 (2006). https://doi.org/10.1016/j.jecp.2010.03.005
10. Faria, E.R., Gonçalves, I.J., de Carvalho, A.C., Gama, J.: Novelty detection in data streams. Artif. Intell. Rev. **45**(2), 235–269 (2016). https://doi.org/10.1007/s10462-015-9444-8
11. de Faria, E.R., de Leon, P., Ferreira Carvalho, A.C., Gama, J.: MINAS: multiclass learning algorithm for novelty detection in data streams. Data Min. Knowl. Disc. **30**(3), 640–680 (2015). https://doi.org/10.1007/s10618-015-0433-y
12. Gama, J., Žliobaitė, I., Bifet, A., Pechenizkiy, M., Bouchachia, A.: A survey on concept drift adaptation. ACM Comput. Surv. **46**(4), 1–37 (2014). https://doi.org/10.1145/2523813
13. Ghesmoune, M., Azzag, H., Lebbah, M.: G-Stream: growing neural gas over data stream. In: Loo, C.K., Yap, K.S., Wong, K.W., Teoh, A., Huang, K. (eds.) ICONIP 2014. LNCS, vol. 8834, pp. 207–214. Springer, Cham (2014). https://doi.org/10.1007/978-3-319-12637-1_26
14. Ghesmoune, M., Lebbah, M., Azzag, H.: State-of-the-art on clustering data streams. Big Data Anal. **1**(1), 13 (2016). https://doi.org/10.1186/s41044-016-0011-3
15. Haider, W., Hu, J., Xie, M.: Towards reliable data feature retrieval and decision engine in host-based anomaly detection systems. Proceedings of the 2015 10th IEEE Conference on Industrial Electronics and Applications, ICIEA 2015, pp. 513–517 (2015). https://doi.org/10.1109/ICIEA.2015.7334166
16. Han, J., Pei, J., Kamber, M.: Data Mining: Concepts and Techniques, 3rd edn. Morgan Kaufmann Publishers, San Francisco (2011)
17. Hoplaros, D., Tari, Z., Khalil, I.: Data summarization for network traffic monitoring. J. Netw. Comput. Appl. **37**(1), 194–205 (2014). https://doi.org/10.1016/j.jnca.2013.02.021
18. Japkowicz, N., Shah, M.: Evaluating Learning Algorithms: A Classification Perspective. Cambridge University Press, New York (2011). https://doi.org/10.1017/CBO9780511921803
19. Kremer, H., Kranen, P., Jansen, T., Seidl, T., Bifet, A., Holmes, G.: An effective evaluation measure for clustering on evolving data streams. In: Proceedings of the 17th ACM SIGKDD International Conference on Knowledge Discovery and Data Mining (KDD 2011), pp. 868–876. ACM Press, New York (2011)
20. Masud, M., Gao, J., Khan, L., Han, J., Thuraisingham, B.M.: Classification and novel class detection in concept-drifting data streams under time constraints. IEEE Trans. Knowl. Data Eng. **23**(6), 859–874 (2011). https://doi.org/10.1109/TKDE.2010.61
21. Silva, J.A., Faria, E.R., Barros, R.C., Hruschka, E.R., De Carvalho, A.C., Gama, J.: Data stream clustering: a survey. ACM Comput. Surv. **46**(1), 1–31 (2013). https://doi.org/10.1145/2522968.2522981

22. Souza, V.M.A., Silva, D.F., Gama, J., Batista, G.E.A.P.A.: Data stream classification guided by clustering on nonstationary environments and extreme verification latency. In: Proceedings of the 2015 SIAM International Conference on Data Mining, pp. 873–881. Society for Industrial and Applied Mathematics, Philadelphia, PA, June 2015. https://doi.org/10.1137/1.9781611974010.98
23. Webb, G.I., Hyde, R., Cao, H., Nguyen, H.L., Petitjean, F.: Characterizing concept drift. Data Min. Knowl. Disc. **30**(4), 964–994 (2016). https://doi.org/10.1007/s10618-015-0448-4
24. Zhang, T., Ramakrishnan, R., Livny, M.: BIRCH: an efficient data clustering databases method for very large. ACM SIGMOD Rec. **25**(2), 103–114 (1996). https://doi.org/10.1145/233269.233324
25. Žliobaitė, I.: Learning under concept drift: an overview. Vilnius University, Technical report (2010)
26. Žliobaitė, I., Bifet, A., Pfahringer, B., Holmes, G.: Active learning with drifting streaming data. IEEE Trans. Neural Netw. Learn. Syst. **25**(1), 27–39 (2014). https://doi.org/10.1109/TNNLS.2012.2236570

Time Warp Invariant Dictionary Learning for Time Series Clustering: Application to Music Data Stream Analysis

Saeed Varasteh Yazdi[1]([✉]), Ahlame Douzal-Chouakria[1], Patrick Gallinari[2], and Manuel Moussallam[3]

[1] Univ. Grenoble Alpes, CNRS, Grenoble INP, LIG, Grenoble, France
{Saeed.Varasteh,Ahlame.Douzal}@imag.fr
[2] Université Pierre et Marie Curie, Paris, France
Patrick.Gallinari@lip6.fr
[3] Deezer, Paris, France
Manuel.Moussallam@deezer.com

Abstract. This work proposes a time warp invariant sparse coding and dictionary learning framework for time series clustering, where both input samples and atoms define time series of different lengths that involve variable delays. For that, first an l_0 sparse coding problem is formalised and a time warp invariant orthogonal matching pursuit based on a new cosine maximisation time warp operator is proposed. A dictionary learning under time warp is then formalised and a gradient descent solution is developed. Lastly, a time series clustering based on the time warp sparse coding and dictionary learning is presented. The proposed approach is evaluated and compared to major alternative methods on several public datasets, with an application to DEEZER music data stream clustering. Data related to this paper are available at: The link to the data and the evaluating algorithms are provided in the paper. Code related to this paper is available at: The link will be provided at the first author personal website (http://ama.liglab.fr/~varasteh/).

Keywords: Time series clustering · Dictionary learning
Sparse coding

1 Introduction

Sparse coding and dictionary learning become popular methods in machine learning and pattern recognition for a variety of tasks as feature extraction, reconstruction and classification. The aim of sparse coding methods is to represent input samples as a linear combination of few basis functions called *atoms* composing a given dictionary. Sparse coding problem is formalised basically as an optimisation problem that minimises the error of the reconstruction under l_0 or l_1 sparsity constraints. The l_0 constraint leads to a non convex and NP-hard problem, that can be solved efficiently by pursuit methods such as orthogonal

© Springer Nature Switzerland AG 2019
M. Berlingerio et al. (Eds.): ECML PKDD 2018, LNAI 11051, pp. 356–372, 2019.
https://doi.org/10.1007/978-3-030-10925-7_22

matching pursuit OMP [15]. Relaxing the sparsity constraint from l_0 to l_1 norm yields a convex sparse coding problem, also known as LASSO problem [14]. The dictionary for the sparse representation can be selected among pre-specified family of basis functions (e.g. gabor) or be learned from the training data to sparse represent the input data. A two-step strategy is commonly used: (1) keep the dictionary fixed and find a sparse representation by using a sparse approximation, (2) keep the representation fixed and update the dictionary, either all the atoms at once [6] or one atom at a time [2]. In the context of classification, the dictionary is generally learned from the labeled training dataset and the sparse codes of the testing samples are used for their classification [9,10,17]. In clustering setting, we distinguish two principal sparse coding and dictionary learning approaches. The first category of approaches assumes the data structured into a union of subspaces [3,5,16,18] where each sample may be represented as a linear combination of some input samples that ideally belong to the same subspace. Several of these approaches are related to sparse subspace clustering where samples are first sparse coded based on the input set as a dictionary [5,18], then a spectral clustering [11] of the sparse codes is used to cluster the data. The number of subspaces as well as their dimension may be fixed beforehand or induced from the affinity graph. In the second category of approaches, a sparse coding and dictionary learning framework is proposed to simultaneously learn a set of dictionaries, one for each cluster, to sparse represent and cluster the data [12].

For temporal data analysis, sparse coding and dictionary learning are especially effective to extract class specific latent temporal features, reveal salient primitives and sparsely represent complex temporal features. However, what makes temporal data particularly challenging is that salient events may arise with varying delays, be related to a part of observations that may appear at different time stamps. This work addresses the problem of time series clustering under sparse coding and dictionary learning framework, where both input samples and atoms define time series that may involve varying delays and be of different lengths. For that, in the first part, an l_0 sparse coding problem is formalised and a time warp invariant orthogonal matching pursuit based on a new cosine maximisation time warp operator is proposed. Subsequently, the dictionary learning under time warp is formalised and a gradient descent solution is developed. In the second part, a time series clustering approach based on the time warp invariant sparse coding and dictionary learning is proposed. The main contributions of the paper are:

1. We propose a time series clustering approach under sparse coding and dictionary learning setting.
2. We propose a tractable solution for time warp invariant orthogonal matching pursuit based on a new cosine maximisation time warp operator.
3. We provide a sparse representation of the clustered time series and learn, for each cluster, a sub-dictionary composed of the most discriminative primitives.
4. We conduct experiments on several public and real datasets to compare the proposed approach to the major alternative approaches, with an application to DEEZER music data stream clustering.

The reminder of the paper is organised as follows. Section 2 formalises the time series clustering problem under sparse coding and dictionary learning setting. Section 3 proposes a solution for sparse coding and dictionary learning under time warp, then presents the time series clustering method. Finally, Sect. 4 presents the experiments and discusses the results obtained.

2   Problem Statement

This Section formalises the time warp invariant sparse coding and dictionary learning for time series clustering. In the following, bold lower case letters are used for vectors and upper case letters for matrices. Let $X = \{x_i\}_{i=1}^{N}$ be a set of N input time series $x_i = (x_{i1}, ..., x_{iq_i})^t \in \mathbb{R}^{q_i}$ of length q_i. We formalise the problem of time series clustering under the sparse coding and dictionary learning setting as the estimation of: (a) the partition $C = \{C_l\}_{l=1}^{K}$ of X into K clusters and (b) the K sub-dictionaries $\{D_l\}_{l=1}^{K}$, to minimise the inertia goodness criterion (i.e., the error of reconstruction) as:

$$\min_{C,D} \sum_{l=1}^{K} \sum_{x_i \in C_l} E(x_i, D_l) \tag{1}$$

where $D_l = \{d_j^l\}_{j=1}^{K_l}$ the sub-dictionary of C_l is composed of K_l time series atoms $d_j^l \in \mathbb{R}^{p_j}$. Note that, both input samples x_i and atoms d_j^l define time series of different lengths that may involve varying delays. $E(x_i, D_l)$ the error of reconstruction, under time warp, of x_i based on the sub dictionary $D_l = \{d_j^l\}_{j=1}^{K_l}$ is formalised as:

$$E(x_i, D_l) = \min_{\alpha_i} \|x_i - \mathcal{F}_i(D_l)\alpha_i^l\|_2^2 \quad \text{s.t.} \|\alpha_i^l\|_0 \leq \tau. \tag{2}$$

where $\mathcal{F}_i(D_l) = [f_i(d_1^l), ..., f_i(d_{K_l}^l)] \in \mathbb{R}^{q_i \times K_l}$ is the transformation of D_l to a new dictionary composed of warped atoms $f_i(d_j^l) \in \mathbb{R}^{q_i}$ aligned to x_i to resorb the involved delays w.r.t x_i. $\alpha_i^l = (\alpha_{1i}^l, ..., \alpha_{K_l i}^l)^t$ is the sparse codes of x_i under D_l and τ the sparsity factor under the l_0 norm.

3   Proposed Solution

To resolve the clustering problem defined in Eq. 1, we use a two steps iterative refinement process, as in standard $k$means clustering. In the *cluster assignment* step, D_l's are assumed fixed and the problem remains to resolve the sparse coding based on the warped dictionary $\mathcal{F}_i(D_l)$ defined in Eq. 2. The cluster assignment is then obtained by assigning each x_i to the cluster C_l whose sub dictionary D_l minimises the reconstruction error. In the *dictionary update* step, the learned sparse codes and the clusters C_l are that time fixed and the problem in Eq. 1 defines a dictionary learning problem to minimise the clustering inertia criterion and represent sparsely samples within clusters. For the *cluster assignment*, we propose in Sect. 3.1, a time warp invariant orthogonal matching pursuit based on a new cosine maximisation time warp operator. In Sect. 3.2, a gradient descent solution for dictionary learning under time warp is developed, then the clustering algorithm for time series under sparse coding and dictionary learning setting is given in Sect. 3.3.

3.1 Time Warp Invariant Sparse Coding

For the sparse coding under time warp problem given in Eq. 2, we define $\mathcal{F}_i(D_l)$ as a linear transformation of D_l based on the warping function $f_i(\boldsymbol{d}_j^l) = \Delta_{ij}^l \boldsymbol{d}_j^l$, where the projector $\Delta_{ij}^l \in \{0,1\}^{q_i \times p_j}$ specifies the temporal alignment that resorbs the delays between $\boldsymbol{x}_i$ and $\boldsymbol{d}_j^l$. The problem given in Eq. 2 is then formalised as:

$$\min_{\boldsymbol{\alpha}_i, \Delta_i^l} \|\boldsymbol{x}_i - \sum_{j=1}^{K_l} \Delta_{ij}^l \, \boldsymbol{d}_j^l \, \alpha_{ji}^l\|_2^2 \tag{3}$$

$$\text{s.t. } \|\boldsymbol{\alpha}_i^l\|_0 \leq \tau, \ \Delta_{ij}^l \in \{0,1\}^{q_i \times p_j}, \ \Delta_{ij}^l \mathbf{1}_{p_j} = \mathbf{1}_{q_i}.$$

with $\Delta_i^l = \{\Delta_{ij}^l\}_{j=1}^{K_l}$. The last constraint is a row normalisation of the estimated Δ_{ij}^l to ensure for $\boldsymbol{x}_i$ equally weighted time stamps. To resolve this problem, we propose an extended variant of OMP that can be mainly summarised in the following steps:

1. For each $\boldsymbol{d}_j^l$, estimate Δ_{ij}^l by dynamic programming to maximise the cosine between $\boldsymbol{x}_i$ and $\boldsymbol{d}_j^l$.
2. Use the projector Δ_{ij}^l to align $\boldsymbol{d}_j^l$ to $\boldsymbol{x}_i$.
3. Estimate the sparse codes $\boldsymbol{\alpha}_i^l$ based on the aligned atoms.

For that and to estimate Δ_i^l, we propose a new operator COSTW to estimate the cosine between two time series under time warp. To the best of our knowledge, it is the first time that the cosine operator is generalised to time series under time warp. Then, we present a time warp invariant OMP (TWI-OMP), that extends the standard OMP approach to sparse code time series under non linear time warping transformations.

Cosine Maximisation Time Warp (COSTW): The problem of estimating the cosine between two time series comes to find an alignment between two time series that maximises their cosine. Let $\boldsymbol{x} = (x_1, ..., x_{q_x})$, $\boldsymbol{y} = (y_1, ..., y_{q_y})$ be two time series of length q_x and q_y. An alignment $\boldsymbol{\pi}$ of length $|\boldsymbol{\pi}| = m$ between $\boldsymbol{x}$ and $\boldsymbol{y}$ is defined as the set of m increasing couples:

$$\boldsymbol{\pi} = ((\pi_1(1), \pi_2(1)), (\pi_1(2), \pi_2(2)), ..., (\pi_1(m), \pi_2(m)))$$

where the applications π_1 and π_2 defined from $\{1, ..., m\}$ to $\{1, .., q_x\}$ and $\{1, .., q_y\}$ respectively obey the following boundary and monotonicity conditions:

$$1 = \pi_1(1) \leq \pi_1(2) \leq ... \leq \pi_1(m) = q_x$$
$$1 = \pi_2(1) \leq \pi_2(2) \leq ... \leq \pi_2(m) = q_y$$

and $\forall l \in \{1, ..., m\}$,

$$\pi_1(l + 1) \leq \pi_1(l) + 1 , \ \pi_2(l + 1) \leq \pi_2(l) + 1,$$
$$(\pi_1(l + 1) - \pi_1(l)) + (\pi_2(l + 1) - \pi_2(l)) \geq 1$$

Intuitively, an alignment π between x and y describes a way to associate each element of x to one or more elements of y and vice-versa. Such an alignment can be conveniently represented by a path in the $q_x \times q_y$ grid, where the above monotonicity conditions ensure that the path is neither going back nor jumping. We will denote $\mathcal{A}$ as the set of all alignments between two time series. The cosine maximisation time warp can be formalised as:

$$\text{COSTW}(x, y) = s(\pi^*) \tag{4}$$
$$\pi^* = \underset{\pi \in \mathcal{A}}{\arg \max} \, s(\pi)$$
$$s(\pi) = \frac{\sum_{i=1}^{|\pi|} x_{\pi_1(i)} \, y_{\pi_2(i)}}{\sqrt{\sum_{i=1}^{|\pi|} x_{\pi_1(i)}^2} \sqrt{\sum_{i=1}^{|\pi|} y_{\pi_2(i)}^2}}$$

where $s(\pi)$ is the cost function of the alignment π. The solution for COSTW is obtained by dynamic programming thanks to the recurrence relation detailed here after.

Let $x_{q_x-1} = (x_1, ..., x_{q_x-1})$, $y_{q_y-1} = (y_1, ..., y_{q_y-1})$ be two sub-time series composed of the $q_x - 1$ and $q_y - 1$ first elements of x and y, respectively. In the case of aligned time series, that do not include delays and with the same length (*i.e.*, $q_x = q_y$) the following incremental property of the standard cosine can be established:

$$\cos(x_{q_x-1}, y_{q_y-1}) = f(< x_{q_x-1}, y_{q_y-1} >, \|x_{q_x-1}\|^2, \|y_{q_y-1}\|^2)$$
$$\cos(x, y) = f((< x_{q_x-1}, y_{q_y-1} >, \|x_{q_x-1}\|^2, \|y_{q_y-1}\|^2) \oplus (x_{q_x}, y_{q_y})) \tag{5}$$

where f is a real function defined as $f(a, b, c) = \frac{a}{\sqrt{b}\sqrt{c}}$ with $(b, c \in \mathbb{R}_+^*)$ and $\oplus$ is an operator that associates to a triplet (a, b, c) and a couple (u, v) a new triplet as:

$$(a, b, c) \oplus (u, v) = (a + uv, b + u^2, c + v^2)$$

For time series including delays and based on the incremental property given in Eq. 5, let us introduce the computation and recurrence relation that allows to estimate the alignment π^* that maximises COSTW(x, y) in Eq. 4.

Computation and Recurrence Relation: Let us define $M \in \mathbb{R}^{q_x \times q_y}$ the matrix mapping x and y of general term $M_{i,j} = (a_{i,j}, b_{i,j}, c_{i,j})$. Based on the incremental property established in Eq. 5, computing recursively for $(i, j) \in \{1, ..., q_x\} \times \{1, ..., q_y\}$ the terms $M_{i,j}$ as:

$$\forall \, i \geq 2, j = 1 \ M_{i,1} = (a_{i-1,1}, b_{i-1,1}, c_{i-1,1}) \oplus (x_i, y_1)$$
$$\forall \, j \geq 2, i = 1 \ M_{1,j} = (a_{1,j-1}, b_{1,j-1}, c_{1,j-1}) \oplus (x_1, y_j)$$

Algorithm 1. $MaxTriplet(u, v, z)$

Input: u, v and z.
1: **if** $f(u) \geq f(v)$ and $f(u) \geq f(z)$ **then**
2:     return u;
3: **else if** $f(v) \geq f(u)$ and $f(v) \geq f(z)$ **then**
4:     return v;
5: **else**
6:     return z;
7: **end if**

Algorithm 2. TWI-OMP$(\boldsymbol{x}, D, \tau)$

Input: $\boldsymbol{x}$, $D = \{\boldsymbol{d}_j\}_{j=1}^{K}$, τ
Output: α, Δ
1: $\boldsymbol{r} = \boldsymbol{x}, \Omega = \{\phi\}$
2: **while** $|\Omega| \leq \tau$ **do**
3:     Select the atom $\boldsymbol{d}_j$ $(j \notin \Omega)$ that maximizes $|\text{COSTW}(\boldsymbol{r}, \boldsymbol{d}_j)|$
4:     Update the set of selected atoms $\Omega = \Omega \cup \{j\}$ and $S_\Omega = [\Delta_j \boldsymbol{d}_j]_{j \in \Omega}$
5:     Update the coefficients: $\alpha_\Omega = (S_\Omega^T S_\Omega)^{-1}(S_\Omega^T \boldsymbol{x})$
6:     Estimate the residual: $\boldsymbol{r} = \boldsymbol{x} - S_\Omega \alpha_\Omega$
7: **end while**

---

and $\forall\, i \geq 2, j \geq 2$

$$M_{i,j} = MaxTriplet \begin{cases} (a_{i,j-1}, b_{i,j-1}, c_{i,j-1}) \oplus (x_i, y_j) \\ (a_{i-1,j}, b_{i-1,j}, c_{i-1,j}) \oplus (x_i, y_j) \\ (a_{i-1,j-1}, b_{i-1,j-1}, c_{i-1,j-1}) \oplus (x_i, y_j) \end{cases}$$

and $M_{1,1} = (x_1 y_1, x_1^2, y_1^2)$, we obtain $\text{COSTW}(\boldsymbol{x}, \boldsymbol{y}) = f(M_{q_x, q_y})$ with a quadratic complexity of $O(q_x q_y)$. The two first equations give the first row and column updates, the third equation gives the recurrence formula that ensures the cosine maximisation at each $M_{i,j}$ cell and $MaxTriplet$ function (Algorithm 1) retains the triplet that maximises the cosine at $M_{i,j}$.

Time Warp Invariant OMP (TWI-OMP): Based on the defined COSTW, let us present the time warp invariant OMP (TWI-OMP) to sparse code a given time series $\boldsymbol{x}$ based on a dictionary $D = \{\boldsymbol{d}_j\}_{j=1}^{K}$ under time warp conditions (Algorithm 2). The proposed TWI-OMP follows the three steps given in the previous section. First, perform a COSTW between $\boldsymbol{x}$ and each $\boldsymbol{d}_j$ to estimate $\Delta = \{\Delta_j\}_{j=1}^{K}$ and find the atom $\boldsymbol{d}_j$ that maximises $\text{COSTW}(\boldsymbol{x}, \boldsymbol{d}_j)$ (line 3 in Algorithm 2). Then, update the set Ω of the yet selected projected atoms and the dictionary $S_\Omega = [\Delta_j \boldsymbol{d}_j]_{j \in \Omega}$ of the yet selected warped atoms (line 4). The updated S_Ω is then used to estimate the sparse coefficients of $\boldsymbol{x}$ (lines 5-6). The process is reiterated on the residuals of $\boldsymbol{x}$ until the sparsity factor τ is reached.

3.2   Time Warp Invariant Dictionary Learning

For the dictionary learning step, the problem in Eq. 1 becomes to learn the dictionary D under time warp where, that time, the sparse codes α_i^l and Δ_i^l are assumed fixed as:

$$\min_D \sum_{l=1}^{K} \sum_{x_i \in C_l} \|x_i - \sum_{j=1}^{K_l} \Delta_{ij}^l \, d_j^l \, \alpha_{ji}^l\|_2^2 \quad \text{s.t.} \|d_j^l\|_2 = 1. \qquad (6)$$

This problem is then resolved as K single dictionary learning problems to learn each sub-dictionary D_l that minimises the inertia of the cluster C_l:

$$\mathcal{J}_l = \min_{D_l} \sum_{x_i \in C_l} \|x_i - \sum_{j=1}^{K_l} \Delta_{ij}^l \, d_j^l \, \alpha_{ji}^l\|_2^2 \qquad (7)$$

which is equivalent to

$$\mathcal{J}_l = \min_{D_l} \sum_{x_i \in C_l} \sum_{t=1}^{q_i} (x_{it} - \sum_{j=1}^{K_l} \alpha_{ji}^l \sum_{(t,t') \in \pi_{ij}^*} d_{jt'}^l)^2 \qquad (8)$$

$$\text{s.t.} \ \|d_j^l\|_2 = 1$$

where x_{it} is the tth time instant of x_i and π_{ij}^* denotes the optimal alignment path between x_i and d_j^l. To resolve the Eq. 8, we propose a gradient descend method based on the following update rule at iteration m for the atom d_j^l:

$$d_{jt'}^{l(m+1)} = d_{jt'}^{l(m)} - \eta^m \frac{\partial \mathcal{J}_l}{\partial d_{jt'}^{l(m)}} \qquad (9)$$

$$d_j^{l(m+1)} = \frac{d_j^{l(m+1)}}{\|d_j^{l(m+1)}\|_2}$$

with,

$$\frac{\partial \mathcal{J}_l}{\partial d_{jt'}^l} = \sum_{x_i \in C_l} \sum_{t=1}^{q_i} -2\alpha_{ji}^l (x_{it} - \alpha_{ji}^l d_{jt'}^l - \alpha_{ji}^l \sum_{\substack{(t,t'') \in \pi_{ij}^* \\ (t'' \neq t')}} d_{jt''}^l \qquad (10)$$

$$- \sum_{j' \neq j} \alpha_{j'i}^l \sum_{(t,t'') \in \pi_{ij'}^*} d_{j't''}^l)$$

where η is the learning rate. In the following section, we show how the time warp invariant OMP and dictionary learning are involved for time series clustering.

Algorithm 3. TWI-DLCLUST(X, K, τ)

   Input: $X = \{x_i\}_{i=1}^{N}$, K, τ.
   Output: $\{C_1, ..., C_K\}$, $\{D_1, ..., D_K\}$
1: *{Clustering Initialisation:}*
2:    Define the affinity matrix $S \in \mathbb{R}^{N \times N}$ of general term:
3:                $s_{ii'} = \text{COSTW}(x_i, x_{i'})$
4:    Apply the affinity propagation (or spectral clustering) to cluster S into
5:    K clusters: $C_1, ..., C_K$
6: *{Sub-dictionary initialisation:}*
7: **for** $l = 1, ..., K$ **do**
8:      Initialise D_l randomly from C_l
9:      **repeat**
10:         Sparse code each $x_i \in C_l$: $[\alpha_i^l, \Delta_i^l] = \text{TWI-OMP}(x_i, D_l, \tau)$
11:         Update each $d_j^l \in D_l$ by using Eq. 9 and 10.
12:      **until** Convergence (stopping rule)
13: **end for**
14: **repeat**
15:     *{Cluster assignment:}*
16:     Sparse code each $x_i \in X$ based on each D_l $(l = 1, ..., K)$:
17:        $[\alpha_i^l, \Delta_i^l] = \text{TWI-OMP}(x_i, D_l, \tau)$
18:     Assign x_i to the cluster C_l whose D_l minimises $E(x_i, D_l)$:
19:        $C_l = \{x_i \ / \ l = \min_{l'} \|x_i - \sum_{l'=1}^{K_{l'}} \Delta_{ij}^{l'} d_j^{l'} \alpha_{ji}^{l'}\|_2^2\}$
20:     *{Dictionaries update:}*
21:     **for** $l = 1, ..., K$ **do**
22:        Update each $d_j^l \in D_l$ by using Eq. 9 and 10.
23:     **end for**
24: **until** Convergence (no changes in cluster assignments)

3.3   Time Warp Invariant Dictionary Learning for Time Series Clustering

For time series clustering, the clustering criterion given in Eq. 1 is minimised by an iterative process involving, respectively, time warp invariant sparse coding (TWI-OMP) and dictionary learning for cluster assignments and dictionary update steps (Algorithm 3). In the initialisation step, a clustering (e.g., spectral clustering, affinity propagation) is performed on the COSTW matrix S to determine an initial partition $\{C_l\}_{l=1}^{K}$ of X (lines 1–5). A sparse coding and a dictionary learning are then performed on the samples of each cluster to initialise the sub-dictionaries $\{D_l\}_{l=1}^{K}$ (lines 6–13). Based on the initial partition $\{C_l\}_{l=1}^{K}$ and sub-dictionaries $\{D_l\}_{l=1}^{K}$, the cluster assignment step consists to perform a sparse coding of each input sample based on each dictionary D_l, then to assign it the cluster whose dictionary minimises its reconstruction error (lines 15–19). Subsequently, in the dictionary update step, the atoms d_j^l of each dictionary are updated by using the formula given in Eqs. 9 and 10 (lines 20–23).

4   Experimental Study

In this section, we evaluate the proposed time series clustering under dictionary learning setting (TWI-DLCLUST) on several synthetic and real datasets, including multivariate and univariate time series, that may involve varying delays and be of different or equal lengths. The proposed TWI-DLCLUST clustering method is compared to two major alternative approaches, the subspace sparse clustering (SSC) [5] and the Dictionary Learning with Structured Incoherence (DLSI) [12]. For SSC, two variants SSC-BP [5] and SSC-OMP [18] are studied for a sparse coding under l_0 and l_1 norms, where an orthogonal matching pursuit and a basis pursuit methods are used respectively. For DLSI, both sample-based and atom-based affinity matrix initialisations proposed in [12] are studied. The Matlab codes of these methods are available online[1].

4.1   Data Description

We have considered in Table 1 two groups of datasets. The first group is composed of the top 12 datasets for which the ground truth clustering is given. The four first datasets are composed of public multivariate time series that have different lengths and involve varying delays. In particular, DIGITS, LOWER, and UPPER datasets give the description of 2-D air-handwritten motion gesture of digits, upper and lower case letters performed on a Nintendo (R) Wii device by several writers [4]. The CHAR-TRAJ dataset gives the 2-dimensional handwritten character trajectory performed on a Wacom tablet by the same user [1]. The ECG-MIT dataset was obtained from the MIT-BIH Arrhythmia [7] database where the heartbeats represented by QRS complexes. The 7 remaining datasets are composed of univariate time series of the same lengths that involve significant delays [8]. The last two datasets are provided by DEEZER[2], the online music streaming service that offers access to the music content of nearly 40 million licensed tracks. DEEZER data, for which we have no ground truth, give the description of streaming data of music albums, randomly selected among 10^5 French user streams and recorded from October 2016 to September 2017. They are composed of univariate time series that give the daily total number of streams per album from its release date to September 2017; this study consider only the streams of a duration ≥ 30 s. In particular, DEEZER15 and DEEZER30 are provided for the streams analysis over the crucial early period after the album release date. They give the description of the prefix time series on the early period covering a cumulative number of 10^3 streams (in red in Fig. 1). In addition, for the pertinence of the analysis, the prefix time series of length <7 days are extended to 15 days in DEEZER15 and to 30 days in DEEZER30. Table 1 gives some characteristics of the studied datasets: the size of the clusters when the ground truth is known, the size of the validation and evaluation sets and the length of the time series that may be variable or fixed.

---

[1] SSC-OMP: https://goo.gl/E6khsq, SSC-BP: https://goo.gl/719pvx and DLSI: https://goo.gl/X5nZgE.

[2] https://www.deezer.com/fr/.

Table 1. Data description

Dataset	Nb. class	Valid. set size	Eval. set size	Length range
DIGITS	10	100	100	29–218
LOWER	26	130	260	27–163
UPPER	26	130	260	27–412
CHAR-TRAJ	20	100	200	109–205
ECG-MIT	4	40	160	541
CBF	3	30	900	128
FACEFOUR	4	24	88	350
LIGHTNING2	2	60	61	637
LIGHTNING7	7	70	73	319
CC	6	300	300	60
TRACE	4	100	100	275
ECG200	2	100	100	96
DEEZER15	-	-	281	15–301
DEEZER30	-	-	278	30–301

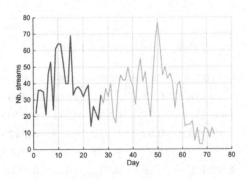

Fig. 1. An album streaming time series, in red the prefix time series covering a cumulative number of 10^3 streams. (Color figure online)

4.2 Validation Protocol

For the top 12 datasets in Table 1, for which the ground truth partition is known, the proposed method TWI-DLCLUST as well as the alternative clustering approaches are applied to cluster the data. For alternative methods, time series of different lengths are zero padded beforehand. The adjusted Rand index [13] is then used to evaluate the goodness of the obtained clusterings. The Rand index lies between 0 and 1, it measures the agreement between the obtained clusters and the ground truth ones. The higher the index, the better the agreement is. In particular, the maximum value "1" of the Rand index is reached when the obtained partition and the ground truth one are identical. For DEEZER datasets, the ground truth being unknown, a DTW-based within-class W_r ratio[3] is used. The lower the within-class ratio W_r, the better the clustering is. W_r is as well used to select the optimal number of clusters. Finally, the parameters related to each studied method, indicated in Table 2, are learned by line/grid search on the validation set, the best parameters are then used to perform the clustering on the evaluation set. The process is iterated over 10 runs and the averaged performances are reported in Tables 3 and 4.

Table 2. Parameter line/Grid values

Method	Line/Grid values	Desc.
SSC-OMP	$\tau \in \{1, 2, 3, 4, 5\}$	l_0 sparsity threshold
SSC-BP	$\lambda \in \{0.001, 0.01\}$, lag of 0.01	l_1 sparsity regularisation
DLSI	$\lambda \in \{0.001, 0.01\}$, lag of 0.01	l_1 sparsity regularisation
	$\eta \in \{0, 0.1, 0.01\}$	dictionary incoherence regularisation
	$K_l = 5, \forall l \in \{1, ..., K\}$	Sub-dictionary D_l size
TWI-DLCLUST	$sc \in [0, 100]$, lag of 10	Sakoe-Chiba band width
	$\tau \in \{1, 2, 3, 4, 5\}$	l_0 sparsity threshold
	$K_l = 5, \forall l \in \{1, ..., K\}$	Sub-dictionary D_l size

Table 3. Adjusted rand index

Dataset	SSC-OMP (τ)	SSC-BP	DLSI-S	DLSI-A	TWI-DLCLUST (τ)
DIGITS	0.839 (2)	0.856	0.854	0.841	**0.940** (1)
LOWER	0.935 (3)	0.943	0.937	0.934	**0.970** (1)
UPPER	*0.940* (2)	**0.942**	*0.940*	*0.938*	**0.942** (1)
CHAR-TRAJ	0.947 (5)	*0.977*	**0.978**	*0.971*	*0.965* (3)
ECG-MIT	0.327 (2)	*0.789*	*0.772*	*0.773*	**0.792** (2)
CBF	0.558 (2)	0.668	0.599	0.601	**0.770** (2)
FACEFOUR	**0.810** (5)	0.722	0.767	0.769	0.776 (3)
LIGHTNING2	**0.559** (2)	**0.559**	**0.559**	0.519	**0.559** (2)
LIGHTNING7	*0.793* (2)	*0.808*	0.724	0.747	**0.814** (3)
CC	0.736 (5)	0.630	0.813	0.791	**0.910** (1)
TRACE	0.680 (5)	0.752	0.755	0.753	**0.805** (1)
ECG200	0.547 (4)	0.631	**0.689**	*0.664*	*0.653* (3)
Nb. Best	2	2	3	0	**9**
Avg. Rank	4.00	2.83	2.92	3.58	**1.67**

---

[3] $W_r = \dfrac{\sum_{l=1}^{K} \sum_{x,y \in C_l} \text{DTW}(x,y)}{\sum_{x,y \in X} \text{DTW}(x,y)}$.

Table 4. Within-class Ratio W_r per Number of Clusters K

Dataset	K	SSC-OMP ($\tau = 5$)	SSC-BP	DLSI-S	TWI-DLCLUST ($\tau = 2$)
DEEZER15	2	*0.266*	0.310	**0.245**	*0.262*
	3	0.201	0.253	0.177	**0.145**
	4	0.212	0.146	*0.114*	**0.112**
	5	0.188	*0.118*	*0.112*	**0.096**
	6	0.133	0.106	*0.074*	**0.069**
DEEZER30	2	0.339	0.348	0.322	**0.303**
	3	0.226	0.292	0.273	**0.173**
	4	0.175	0.241	*0.133*	**0.127**
	5	0.154	*0.119*	*0.110*	**0.096**
	6	*0.100*	**0.080**	*0.085*	*0.085*
Nb. Best		0	1	1	**8**
Avg. Rank		3.40	3.30	2.05	**1.25**

4.3   Results and Discussion

Table 3 gives for the top 12 datasets the obtained adjusted Rand index values. The best values are indicated in bold, the non significantly different ones from the best (t-test at 5% risk) are in italic and the remaining results are significantly different from the bold values. For the two l_0 sparse coding methods SSC-OMP and TWI-DLCLUST, the learned sparsity coefficient τ is given between brackets. The two last rows give, over all the datasets, the number of times a method reaches the best value as well as its average ranking. From Table 3, we can see that the proposed TWI-DLCLUST reaches the best clustering results with 9 times (9 out of 12) as the best values, 2 times as significantly non different from the best and obtained the lowest average ranking. The second best results are obtained by SSC-BP and DLSI-S, followed by SSC-OMP. Although the l_1 sparse coding models (here SSC-BP and DLSI-S) are known to be more efficient than the l_0 models, TWI-DLCLUST even involving an l_0 sparse coding leads to the best results. While TWI-DLCLUST and DLSI-S involve smaller size sub-dictionaries ($K_l = 5$), SSC-OMP and SSC-BP are based on larger dictionary of the size of the evaluation set (Table 1). Finally, by comparing the two l_0 sparse coding methods SSC-OMP and TWI-DLCLUST, we can see that TWI-DLCLUST leads for all datasets to sparser solutions with a lower or equal sparsity coefficient τ than SSC-OMP. For DEEZER data we have performed each clustering method for several number of clusters and the within-class ratio of the obtained partitions reported in Table 4. For simplicity, the DLSI approach is conducted only with DLSI-S variant, DLSI-A being highly equivalent in Table 3. We can see easily that, for both datasets and almost all the number of clusters, the best values are reached by TWI-DLCLUST, followed by the l_1 sparse code approaches SSC-BP and DLSI-S, then by SSC-OMP. Finally, note that from both Tables 3 and 4, SSC-OMP and SSC-BP lead to the lowest performances with a slightly better results for SSC-BP as using an l_1 norm sparse coding. These results may be partly explained by the fact that both SSC-OMP and SSC-BP are purely sparse coding methods based on one global dictionary fixed beforehand, unlike DLSI and TWI-DLCLUST that learn one sub-dictionary per cluster.

In the second study, we analyse more closely the obtained clusterings. For instance, based on Fig. 2 that displays the progression of the within-class ratio w.r.t the number of clusters, a partitioning into four clusters is performed on DEEZER30. Accordingly, Fig. 3, shows for each of the four clusters (each row), the profile of the medoids (in the first column), the closest albums to the medoid in the second column and at the third column, the atom that most contributes to sparse represent the cluster's samples.

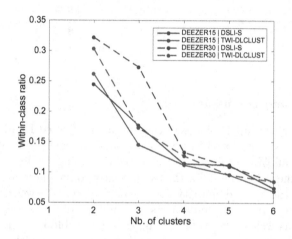

Fig. 2. Number of clusters K vs. Within-class ratio W_r.

DEEZER data provide additional album descriptive features as "Full" (composed of several tracks) or "Single" (composed of one track), if it is a "Deluxe" edition, namely a re-edition of the album featuring extra contents related to the album, as well as the artist popularity before and after the release. The analysis of some album characteristics brings meaningful interpretation of the extracted clusters (Fig. 3).

The first cluster is composed of 71% of "Full" albums and 15% of "Deluxe" editions. It corresponds to album releases with flat stream profiles. Such behaviour usually occurs when the content has already been published ("Deluxe" versions) or for lesser-known artists, as assessed by the cluster medoid "Empereur du Sale" album of the rapper "Lorenzo" that released several singles a few weeks before the album release date and although not highly popular has still a steady fan base.

In the second cluster, 75% of the albums are "Single". The fast decrease stream profile just after the release date is not surprising for short albums

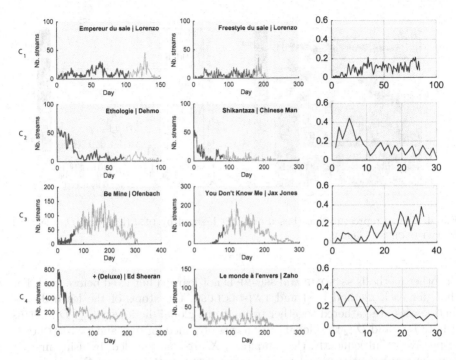

Fig. 3. Four clusters partitioning of DEEZER30: Medoid profile (left column), Nearest album to the medoid (middle column), the most contributing atom to the cluster (right column).

(composed of 1–4 tracks). Indeed, a "Full" album is generally released shortly after the "Single" release, inducing a decrease of streams for the "Single" few weeks after its release. The cluster medoid "Ethologie" is produced by the rapper "Dehmo" that has not released albums since a long period, that may explain the burst of streams for the new content just after its release.

The cluster 3 is composed of 69% of "Full" albums mainly produced by artists that became popular after their album release. This is reflected by the stream profiles that initially evolve at low level then increase significantly several days/weeks after the album release. This is confirmed by the medoid album "Be Mine" a single produced by "Ofenbach" that was in fact revealed to the public with that album.

Finally, the cluster 4 comprises a majority of "Single" albums (84%) produced by very popular artists with a huge fan base and immediate success. The medoid album "Divide" produced by "Ed Sheeran" was one of the biggest hits of 2017. Although the stream profiles of the clusters 4 and 2 seem similar, albums of cluster 4 concern more established artists in their second/third albums while cluster 2 is more related to emerging works and first successes.

The aim of the last study is to analyse the pertinence of the learned sub-dictionaries $\{D_1, ..., D_K\}$ for both DLSI and TWI-DLCLUST; the dictionary for

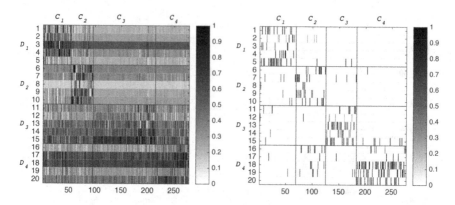

Fig. 4. Sparse representations based on: D_{G1} learned by DLSI (left) and D_{G2} learned by TWI-DLCLUST (right).

the other methods SSC-OMP and SSC-BP is not learned but fixed beforehand. For that, for each method DLSI and TWI-DLCLUST, the atoms of the learned sub-dictionaries are gathered together to built one global dictionary $\cup_{l=1}^{K} D_l$. Let us denote D_{G1} and D_{G2} the global dictionaries obtained for DLSI and TWI-DLCLUST, respectively. Subsequently, the samples in X are sparse coded, by using first an l_1 norm regularisation based on D_{G1}, then a TWI-OMP based on D_{G2}.

For instance, for DEEZER30, Fig. 4 shows for the 278 albums the learned sparse codes based on D_{G1} (on left) and on D_{G2} (on right), organised for interpretation purpose per cluster $\{C_1, ..., C_4\}$ and per sub-dictionary $\{D_1, ..., D_4\}$. It emerges from Fig. 4, that sparse codes based on D_{G2} highlight clearly a block structure that reflects the discriminative performance of the sub-dictionaries composing D_{G2} (learned by TWI-DLCLUST). Indeed, sparse codes show that each sub-dictionary D_l is mainly involved to reconstruct samples of C_l. On the other hand, the structure of the sparse codes based on D_{G1} (learned by DLSI) seems much less sparser and less discriminative. We can note, in particular, that the atoms $d_3^1, d_{15}^3, d_{17}^4$ and d_{18}^4 define common primitives involved to reconstruct the samples of all the clusters.

5   Conclusion

This work proposes a time warp invariant sparse coding and dictionary learning for time series clustering where both input samples and atoms define time series that may have different lengths and involve varying delays. For that, first a time warp invariant orthogonal matching pursuit based on a new cosine maximisation time warp operator is proposed. Then, a dictionary learning approach under time warp is formalised and a gradient descent solution is developed. The proposed time series clustering allows to sparse represent the clustered time series and learn, for each cluster, a sub-dictionary composed of the most discriminative primitives. The conducted experiments show that although TWI-DLCLUST involves an l_0 sparse coding approach based on a very small size sub-dictionaries, it leads to the sparser and the best clustering results, while revealing atoms with a good discriminative capacity to represent the time series of each cluster.

Acknowledgments. This work is supported by the French National Research Agency (ANR-Locust project) and Bpifrance funds in the frame of the French National PIA Program.

References

1. Frank, A., Asuncion, A.: UCI machine learning repository (2010). http://archive. ics.uci.edu/ml/
2. Aharon, M., Elad, M., Bruckstein, A.: k-SVD: an algorithm for designing overcomplete dictionaries for sparse representation. IEEE Trans. Signal Process. **54**(11), 4311–4322 (2006)
3. Bradley, P.S., Mangasarian, O.L.: K-plane clustering. J. Glob. Optim. **16**(1), 23–32 (2000)
4. Chen, M., AlRegib, G., Juang, B.H.: 6DMG: A new 6D motion gesture database. In: Proceedings of the 3rd Multimedia Systems Conference, pp. 83–88. ACM (2012)
5. Elhamifar, E., Vidal, R.: Sparse subspace clustering: algorithm, theory, and applications. IEEE Trans. Pattern Anal. Mach. Intell. **35**(11), 2765–2781 (2013)
6. Engan, K., Aase, S.O., Husoy, J.H.: Method of optimal directions for frame design. In: Proceedings of the 1999 IEEE International Conference on Acoustics, Speech, and Signal Processing, vol. 5, pp. 2443–2446. IEEE (1999)
7. Goldberger, A.L., et al.: Physiobank, physiotoolkit, and physionet. Circulation **101**(23), e215–e220 (2000)
8. Keogh, E.: The UCR time series data mining archive (2006). http://www.cs.ucr. edu/~eamonn/
9. Lee, H., Battle, A., Raina, R., Ng, A.Y.: Efficient sparse coding algorithms. In: Advances in Neural Information Processing Systems, pp. 801–808 (2007)
10. Mairal, J., Bach, F., Ponce, J., Sapiro, G., Zisserman, A.: Discriminative learned dictionaries for local image analysis. In: IEEE Conference on Computer Vision and Pattern Recognition, CVPR 2008, pp. 1–8. IEEE (2008)
11. Ng, A.Y., Jordan, M.I., Weiss, Y.: On spectral clustering: analysis and an algorithm. In: Advances in Neural Information Processing Systems, pp. 849–856 (2002)
12. Ramirez, I., Sprechmann, P., Sapiro, G.: Classification and clustering via dictionary learning with structured incoherence and shared features. In: 2010 IEEE Conference on Computer Vision and Pattern Recognition (CVPR), pp. 3501–3508. IEEE (2010)
13. Rand, W.M.: Objective criteria for the evaluation of clustering methods. J. Am. Stat. Assoc. **66**(336), 846–850 (1971)
14. Tibshirani, R.J., et al.: The LASSO problem and uniqueness. Electron. J. Stat. **7**, 1456–1490 (2013)
15. Tropp, J.A.: Greed is good: algorithmic results for sparse approximation. IEEE Trans. Inf. Theory **50**(10), 2231–2242 (2004)
16. Tseng, P.: Nearest q-flat to m points. J. Optim. Theory Appl. **105**(1), 249–252 (2000)

17. Yazdi, S.V., Douzal-Chouakria, A.: Time warp invariant kSVD: sparse coding and dictionary learning for time series under time warp. Pattern Recognit. Lett. **112**, 1–8 (2018). https://doi.org/10.1016/j.patrec.2018.05.017
18. You, C., Robinson, D., Vidal, R.: Scalable sparse subspace clustering by orthogonal matching pursuit. In: Proceedings of the IEEE Conference on Computer Vision and Pattern Recognition, pp. 3918–3927 (2016)

How Your Supporters and Opponents Define Your Interestingness

Bruno Crémilleux[1], Arnaud Giacometti[2], and Arnaud Soulet[2(✉)]

[1] Normandie Univ, UNICAEN, ENSICAEN, CNRS – UMR GREYC, Caen, France
bruno.cremilleux@unicaen.fr
[2] Université de Tours – LIFAT EA 6300, Blois, France
{arnaud.giacometti,arnaud.soulet}@univ-tours.fr

Abstract. How can one determine whether a data mining method extracts interesting patterns? The paper deals with this core question in the context of unsupervised problems with binary data. We formalize the quality of a data mining method by identifying patterns – the *supporters* and *opponents* – which are related to a pattern extracted by a method. We define a typology offering a global picture of the methods based on two complementary criteria to evaluate and interpret their interests. The quality of a data mining method is quantified via an evaluation complexity analysis based on the number of supporters and opponents of a pattern extracted by the method. We provide an experimental study on the evaluation of the quality of the methods.

1 Introduction

In contrast to a lot of data analysis methods where the goal is to describe all the data with one model, pattern mining focuses on information describing only parts of the data. However, in practice, the number of discovered patterns is huge and patterns have to be filtered or ranked according to additional quality criteria in order to be used by a data analyst. As surveyed by Vreeken and Tatti [26], there exist numerous methods for evaluating the interestingness of extracted patterns, e.g. based on simple measures or the use of statistical testing. However, it remains difficult to clearly identify the advantages and limitations of each approach.

How can one determine whether a data mining method extracts interesting patterns? How can one know and evaluate if a data mining method is better than another for a given task? Our work addresses these core questions. Completely answering those questions is clearly out of the scope of this (or any single) paper but we propose major improvements in these directions in the context of unsupervised problems with binary data.

Our goal is to propose an interestingness theory independent of any assumption about the data such as a model of the data or an expectation using statistical tests. Our key principle to assess the quality of a data mining method extracting a pattern X is to study the relationships between X and the other patterns when X is selected. Roughly speaking, the higher the number of necessary comparisons between X and the other patterns to select X, the higher the quality

© Springer Nature Switzerland AG 2019
M. Berlingerio et al. (Eds.): ECML PKDD 2018, LNAI 11051, pp. 373–389, 2019.
https://doi.org/10.1007/978-3-030-10925-7_23

of the method. As an example, let us consider correlation measures such as the lift [26] and the productive itemset [27]. While calculating lift involves only the individual items contained in X to select X, the productive itemset involves all subsets of X. A pattern selected by the productive itemset must satisfy more tests and the productive itemset is a more effective selector for mining correlated itemsets. Our framework addresses methods to select patterns such as interestingness measures [25], the constraint-based pattern mining imposing constraints on a single pattern [20] or several patterns such as condensed representations of patterns [5] or *top-k* patterns [9]. We call *selector* a data mining method providing patterns. The goal of our framework is to evaluate the quality of a selector and therefore the interestingness of the patterns extracted by the selector. For that purpose, we introduce the notions of *supporter* and *opponent*. A supporter Y of X is a pattern which increases the interestingness of X when only the support of Y increases while all other patterns' support remains unchanged. In other words, when the support of Y increases, it raises the likelihood of X to be selected and therefore Y supports X to be selected. Analogously, an opponent Y of X is a pattern which decreases the interestingness of X when only the support of Y decreases. We show that the number of supporters and opponents and their relations with X (Y is a generalization or a specialization of X, Y and X are incomparable) provide meaningful information about the quality of the selector at hand. Notably, this approach evaluates the quality of a selector only based on the relationships between the patterns from the data without assuming a model or any hypothesis on the data.

This paper formalizes the relationships between patterns to evaluate the quality of a selector through the new notions of supporters and opponents. We present a typology of selectors defined by formal properties and based on two complementary criteria to evaluate and interpret the quality of a selector. This typology offers a global picture of selectors and clarifies their interests and limitations. Highlighting the kinds of patterns' relationships required by a selector helps to compare selectors to each other. We quantify the quality of a selector via an evaluation complexity analysis based on its number of supporters and opponents. This analysis enables us to contrast the quality of a selector with its computing cost. Finally, we conduct an experimental study in the context of correlation measures to evaluate the quality of selectors according to their complexity.

This paper is structured as follows. Section 2 discusses related work. Section 3 introduces preliminaries and defines what a selector is. We present the key notions of supporters and opponents in Sect. 4 and the typology of selectors in Sect. 5. We continue with the analysis of the complexity of the selectors in Sect. 6. Section 7 provides an experimental study on the evaluation of the quality of a few selectors. We round up with discussion and conclusion in Sect. 8.

2   Related Work

As we focus on formal approaches on interestingness, experimental protocols to evaluate the quality of a method like rediscovery [29] or randomization [14] are out of the scope of this related work. The proposal of a general theory of interestingness was already indicated as a challenge for the past decade [8,16].

Several approaches have been proposed in the literature to analyze pattern discovery methods. Regarding condensed representations of patterns, the size of a condensed representation is often used as an objective measure to assess its quality [5]. As a condensed representation based on closed patterns is always more compact than a condensed representation based on free (or key) patterns, closed patterns are deemed most interesting. However one of the most compact condensed representations – non-derivable itemsets (NDI) [4] – is little used. The semantics of NDI, which appears complex, may explain this unpopularity. In this paper, we propose a measure to formally identify the complexity of a selector (cf. Sect. 6). The survey of Vreeken and Tatti [26] presents interestingness measures on patterns by dividing them into two categories, absolute measures and advanced ones. An absolute measure is informally defined as follows: "score patterns using only the data at hand, without contrasting their calculations over the data to any expectation using statistical tests". Advanced measures were introduced to limit redundancy in the results. They are based on statistical models (independence model, partition models, MaxEnt models) having different complexities. Our formalization of complexity classes based on relationships between patterns clarifies the distinction between absolute measures and advanced ones.

A lot of works [12,18,23,24] propose axioms that should be satisfied by an interestingness measure for association rule in order that the measure is considered relevant. These methods state what should be the expected variations of a well-behaved measure under certain conditions (e.g., when the frequency of the body or the head of the association rule increases). More recently, these works were extended to itemsets [15,27] but only by considering their subsets. Our proposal systematizes this approach by taking into account all patterns of the lattice. Besides, the axioms previously introduced in the literature are mainly focused on correlation measures and there are not such axioms for constraints. In this paper, we generalize these principles to constraints (cf. Sect. 5).

There are very few attempts to define interactions between patterns when evaluating an interestingness measure. The concept of global constraints has been informally defined in [6]. This notion has been formalized in [13] by defining a relational algebra extended to pattern discovery. Our framework provides a broader and more precise formal definition, especially to better analyze the interrelationships between the patterns.

3 Preliminaries

Let $\mathcal{I}$ be a set of distinct literals called *items*, an itemset (or pattern) is a subset of $\mathcal{I}$. The language of itemsets corresponds to $\mathcal{L} = 2^{\mathcal{I}}$. A transactional dataset is a multi-set of itemsets of $\mathcal{L}$. Each itemset, usually called *transaction*, is a dataset entry. For instance, Table 2 gives three transactional datasets with 5 transactions $t_1, \ldots, t_5$ each described by 3 items A, B and C. Note that the transaction t_5 in $\mathcal{D}_1$ is empty. $\mathcal{D}$ denotes a dataset and Δ all datasets. The *frequency* of an itemset X, denoted by $freq(X, \mathcal{D})$, is the number of transactions of $\mathcal{D}$ containing X. For simplicity, we write $freq(X)$ when there is no ambiguity.

Constraint-based pattern mining [19] aims at enumerating all patterns occurring at least once in a dataset $\mathcal{D}$ and satisfying a user-defined selection predicate q. A well-known example is the minimal support constraint, based on the frequency measure, which provides the patterns having a support greater than a given minimal threshold. Despite the filtering performed by a constraint, the collection of mined patterns is often too large to be managed and interestingness measures are additionally used to rank patterns and focus on the most relevant ones. There are numerous measures [24], several of which (support, bond, lift, all-confidence) are given in Table 1. In this paper, we consider constraint-based pattern mining imposing constraints on a single pattern or several patterns such as condensed representations of patterns or top-k patterns. Table 1 depicts several examples of constraints. The productive itemset is here defined as a constraint.

Table 1. Itemset mining approaches based on frequency

Name	Definition
Correlation	
Support [1]	$freq(X)/freq(\emptyset)$
All-confidence [21]	$freq(X)/\max_{i \in X} freq(\{i\})$
Bond [21]	$freq(X)/\lvert\{t \in \mathcal{D} : X \cap t \neq \emptyset\}\rvert$
Lift [26]	$supp(X)/\Pi_{i \in X} supp(i)$
Productive itemset [27]	$(\forall Y \subset X)(prod(X) \Rightarrow supp(X) > supp(Y) \times supp(X \setminus Y))$
Condensed Representation (CR)	
Maximal itemset [19]	$(\forall Y \supset X)(max(X) \Rightarrow freq(X) \geq \gamma \wedge freq(Y) < \gamma)$
Free itemset [2]	$(\forall Y \subset X)(free(X) \Rightarrow freq(X) < freq(Y))$
Closed itemset [22]	$(\forall Y \supset X)(closed(X) \Rightarrow freq(X) > freq(Y))$
Non-derivable itemset [4]	$(\forall X \in \mathcal{L})(ndi(X) \Leftrightarrow LB(X, \mathcal{D}) \neq UB(X, \mathcal{D}))$ where $LB(X, \mathcal{D})$ and $UB(X, \mathcal{D})$ are respectively lower and upper bounds derived from subsets of X in $\mathcal{D}$
Other	
Top-k freq. itemset [9]	$(\forall X \in \mathcal{L})(top_k(X) \Leftrightarrow \lvert\{Y \in \mathcal{L} : freq(Y) > freq(X)\}\rvert < k)$
FPOF [17]	$-\sum_{Y \subseteq X} freq(Y)/\sum_{Z \in \mathcal{L}} freq(Z)$

Let $\mathbb{S}$ be a poset. We formally define the notion of *selector* as follows.

Definition 1 (Interestingness Selector). *An interestingness selector s is a function defined from $\mathcal{L} \times \Delta$ to $\mathbb{S}$ that increases when X is more interesting.*

$\mathbb{S}$ is the set of reals $\mathbb{R}$ if the selector is an interestingness measure[1] and booleans $\mathbb{B}$ (i.e. *true* or *false*) with the order *false* < *true* if the selector is a constraint. Clearly, selectors define very different views on what should be a relevant pattern. Relevance may highlight correlations between items (regularity), correlations with a class of the dataset (contrast), removing redundancy (condensed representation), complementarity between several patterns (top-k), outlier detection such as the FPOF measure (cf. Table 1).

4   Framework of Supporters and Opponents

4.1   Fundamental Definitions and Notations

Deciding if a pattern is interesting (and why) generally depends on its frequency, but also on the frequencies of some other patterns. In our framework, we show how the knowledge of those patterns for a given selector makes it possible to qualify this selector and evaluate its quality.

More precisely, in order to isolate the impact of the change in frequency of a pattern Y on the evaluation of the interestingness of a pattern X, we propose to compare the interestingness of the assessed pattern X with respect to two very similar datasets $\mathcal{D}$ and $\mathcal{D}'$, where only the frequency of itemset Y varies. Therefore, we introduce the following definition.

Definition 2 (Increasing at a Point). *Compared to $\mathcal{D}$, a dataset $\mathcal{D}'$ is increasing at a point Y, denoted by $\mathcal{D} <_Y \mathcal{D}'$, iff $freq(Y,\mathcal{D}) < freq(Y,\mathcal{D}')$ and $freq(X,\mathcal{D}) = freq(X,\mathcal{D}')$ for all patterns $X \neq Y$.*

For instance, the first two datasets provided by Table 2 satisfy $\mathcal{D}_1 <_{ABC} \mathcal{D}_2$. It means that patterns $\emptyset$, A, B, C, AB, AC, BC have the same frequency in both datasets, while the frequency of ABC is greater in $\mathcal{D}_2$. Indeed, we have $freq(ABC,\mathcal{D}_1) = 1$, whereas $freq(ABC,\mathcal{D}_2) = 2$. In the same way, we can easily see that $\mathcal{D}_1 <_A \mathcal{D}_3$ due to the addition in $\mathcal{D}_3$ (compared to $\mathcal{D}_1$) of an item A in the fifth transaction t_5. Thus, we have $freq(A,\mathcal{D}_1) = 3$, whereas $freq(A,\mathcal{D}_3) = 4$, and $freq(X,\mathcal{D}_1) = freq(X,\mathcal{D}_3)$ for all other patterns $X \neq A$.

Intuitively, given a selector s, a *supporter* Y of an assessed pattern X is a pattern that increases the interestingness of X when only the support of Y increases (while all other patterns' support remains unchanged). In other words, when the support of Y increases, it raises the likelihood of X to be selected using s. Conversely, if Y is an *opponent* of X, when the support of Y increases, it reduces the likelihood of X to be selected. Using Definition 2, the following definition formalizes these notions of supporter and opponent.

Definition 3 (Supporters and Opponents). *Given a selector s, let X be a pattern in $\mathcal{L}$. Y is a supporter of X for s, denoted by $Y \in s^+(X)$, iff there exist two datasets $\mathcal{D}$ and $\mathcal{D}'$ such that $\mathcal{D}' >_Y \mathcal{D}$ and $s(X,\mathcal{D}') > s(X,\mathcal{D})$.*

---

[1] The choice of the order $<_\mathbb{S}$ has an impact. For instance, in the case of support, $<_\mathbb{S}=<_\mathbb{R}$ (resp. $<_\mathbb{S}=>_\mathbb{R}$) enables us to select the positive (resp. negative) correlations.

Table 2. Three toy datasets with slight variations

$\mathcal{D}_1$		$\mathcal{D}_2$		$\mathcal{D}_3$	
Trans.	Items	Trans.	Items	Trans.	Items
t_1	$A\ B\ C$	t_1	$A\ B\ C$	t_1	$A\ B\ C$
t_2	$A\ B$	t_2	$A\ B\ \mathbf{C}$	t_2	$A\ B$
t_3	$A\quad C$	t_3	A	t_3	$A\quad C$
t_4	$B\ C$	t_4	B	t_4	$B\ C$
t_5		t_5	$\mathbf{C}$	t_5	$\mathbf{A}$

Conversely, Y is an opponent of X for s, denoted by $Y \in s^-(X)$, iff there exist two datasets $\mathcal{D}$ and $\mathcal{D}'$ such that $\mathcal{D}' >_Y \mathcal{D}$ and $s(X, \mathcal{D}') < s(X, \mathcal{D})$.

Given a selector, the strength of the notions of *supporter* and *opponent* is to clearly identify the patterns actually involved in the evaluation of an assessed pattern. Moreover, it is important to note that the set of supporters and opponents of a pattern (given a selector) is not dependent on a specific dataset. They are a property of a given selector.

Considering the datasets given in Table 2, let us illustrate Definition 3 with the all-confidence selector. We already noted that $\mathcal{D}_1 <_{ABC} \mathcal{D}_2$. Additionally, we have $all\text{-}conf(ABC, \mathcal{D}_1) = \frac{freq(ABC, \mathcal{D}_1)}{max_{i \in ABC} freq(i, \mathcal{D}_1)} = \frac{1}{3}$, whereas $all\text{-}conf(ABC, \mathcal{D}_2) = \frac{freq(ABC, \mathcal{D}_2)}{max_{i \in ABC} freq(i, \mathcal{D}_2)} = \frac{2}{3}$. Therefore, we have $\mathcal{D}_1 <_{ABC} \mathcal{D}_2$ and $all\text{-}conf(ABC, \mathcal{D}_1) < all\text{-}conf(ABC, \mathcal{D}_2)$, which means that ABC is a supporter of itself for the all-confidence measure. On the other hand, we have $\mathcal{D}_1 <_A \mathcal{D}_3$, $all\text{-}conf(ABC, \mathcal{D}_1) = \frac{1}{3} > all\text{-}conf(ABC, \mathcal{D}_3) = \frac{1}{4}$. Thus, by Definition 3, A is an opponent of ABC for the all-confidence measure.

More generally, it is possible to show that for all patterns X, $all\text{-}conf^+(X)$ is equal to $\{X\}$, i.e. X has no supporters other than itself, and that $all\text{-}conf^-(X) = \{i : i \in X\}$. In the following Sect. 4.2, we give the set of supporters and opponents for a representative set of usual selectors.

4.2 Supporters and Opponents of Usual Selectors

In this section, we give the sets of supporters s^+ and opponents s^- for a representative set of selectors s. These sets are presented in Table 3 both for interestingness measures (support, all-confidence, bond, lift, etc.) and boolean constraints (productive, free, closed itemset, etc.). Due to lack of space, we do not present a proof for every selector considered in Table 3. Nevertheless, we provide a proof for two examples: the lift measure (see Property 1) and the free constraint (see Property 2). Note that the schema of these proofs could be easily adapted to identify the supporters and opponents of other correlation measures (support, all-confidence, bond, etc.) and other condensed representation constraints (maximal, closed, etc.).

Before detailing the proofs of Properties 1 and 2, given an itemset X, Lemma 1 stresses that it is always possible to build two transactional datasets $\mathcal{D}$

Table 3. Analysis of methods based on supporters and opponents

Selector s	$s^+(X)$	$s^-(X)$	$|s^\pm(X)|$
support	$\{X\}$	$\{\emptyset\}$	$O(1)$
all-confidence	$\{X\}$	sing.	$O(k)$
bond	$\{X\}$	sing.	$O(k)$
lift	$\{X\}$	sing.	$O(k)$
prod. itemset	$\{X\}$	$X^{\downarrow}$	$O(2^k)$
max. itemset	$\{X\}$	$X^{\bar{\uparrow}}$	$O(n-k)$
free itemset	$X^{\underline{\downarrow}}$	$\{X\}$	$O(k)$
closed itemset	$\{X\}$	$X^{\bar{\uparrow}}$	$O(n-k)$
NDI	$X^{\downarrow}$	$X^{\downarrow}$	$O(2^k)$
top-k frequent	$\{X\}$	$X^{\leftrightarrow}$	$O(2^n-2^k$ $-2^{n-k})$
FPOF	$X^{\uparrow}\cup X^{\leftrightarrow}$	$X^{\downarrow}\cup\{X\}$	$O(2^n)$

Notation	Formula
k	$|X|$
n	$|\mathcal{I}|$
$singletons$	$\{\{i\}:i\in X\}$
direct sub. $X^{\underline{\downarrow}}$	$\{X\setminus\{i\}:i\in X\}$
direct sup. $X^{\bar{\uparrow}}$	$\{X\cup\{i\}:i\in\mathcal{I}\setminus X\}$
subsets $X^{\downarrow}$	$2^X\setminus\{X\}$
supersets $X^{\uparrow}$	$\{Y\in\mathcal{L}:X\subset Y\}$
incomp. $X^{\leftrightarrow}$	$\{Y\in\mathcal{L}:Y\not\subseteq X$ $\wedge X\not\subseteq Y\}$
lattice	$\mathcal{L}\setminus\{X\}$

and $\mathcal{D}'$ such that $\mathcal{D}'$ is increasing at X in comparison to $\mathcal{D}$, i.e. $\mathcal{D}' >_X \mathcal{D}$. Note that the sets of transactions $\mathcal{D}_X^-$ and $\mathcal{D}_X^+$ introduced in this lemma are crucial to identify the sets of supporters and opponents of a selector.

Lemma 1. *Given an itemset $X\subseteq\mathcal{I}$, let $\mathcal{D}_X^-$ and $\mathcal{D}_X^+$ be the datasets defined by $\mathcal{D}_X^+ = \{Y\subseteq X : |X\setminus Y|$ is even$\}$ and $\mathcal{D}_X^- = 2^X\setminus\mathcal{D}_X^+$. We have $\mathcal{D}_X^- <_X \mathcal{D}_X^+$.*

For instance, using datasets shown in Table 2, it is easy to see that $\mathcal{D}_2 = \{ABC\}\cup\mathcal{D}_{ABC}^+$ with $\mathcal{D}_{ABC}^+ = \{ABC, A, B, C\}$, and $\mathcal{D}_1 = \{ABC\}\cup\mathcal{D}_{ABC}^-$ with $\mathcal{D}_{ABC}^- = \{AB, AC, BC, \emptyset\}$. Thus, Lemma 1 implies that $\mathcal{D}_1 <_{ABC} \mathcal{D}_2$. Given $\mathcal{D}_0 = \{ABC, AB, AC, BC\}$, we can also check that $\mathcal{D}_1 = \mathcal{D}_0\cup\mathcal{D}_A^-$ with $\mathcal{D}_A^- = \emptyset$, and $\mathcal{D}_3 = \mathcal{D}_0\cup\mathcal{D}_A^+$ with $\mathcal{D}_A^+ = \{A\}$. Thus, Lemma 1 implies that $\mathcal{D}_1 <_A \mathcal{D}_3$.

Proof. First, it is easy to see hat $freq(X,\mathcal{D}_X^-) = 0$ and $freq(X,\mathcal{D}_X^+) = 1$, which shows that $freq(X,\mathcal{D}_X^+) > freq(X,\mathcal{D}_X^-)$. Then, for all itemsets $Y\neq X$, if $Y\not\subseteq X$, we have $freq(Y,\mathcal{D}_X^-) = freq(Y,\mathcal{D}_X^+) = 0$. Otherwise, if $Y\subseteq X$, we can see that $freq(Y,\mathcal{D}_X^-) = freq(Y,\mathcal{D}_X^+) = freq(Y,2^X)/2$ where $freq(Y,2^X) = |\{t\in 2^X : Y\subseteq t\}| = |\{Y\cup t : t\in 2^{X\setminus Y}\}| = 2^{|X|-|Y|}$. Thus, for all $Y\neq X$, $freq(Y,\mathcal{D}_X^-) = freq(Y,\mathcal{D}_X^+)$, which completes the proof that $\mathcal{D}_X^- <_X \mathcal{D}_X^+$. □

Using Lemma 1, we now prove Property 1, which defines the supporters and opponents of the lift measure.

Property 1. For all itemsets X such that $|X| > 1$, $lift^+(X) = \{X\}$ and $lift^-(X) = \{\{i\} : i\in X\}$.

Proof. Given an itemset X such that $|X| > 1$, we distinguish three cases:
1. Let $\mathbf{Y} = \mathbf{X}$ and two datasets $\mathcal{D}'$ and $\mathcal{D}$ such that $\mathcal{D}' >_Y \mathcal{D}$. By definition we have: $freq(X,\mathcal{D}') > freq(X,\mathcal{D})$ and $freq(Z,\mathcal{D}') = freq(Z,\mathcal{D})$ for all $Z\neq (Y = X)$. Because $|X| > 1$, we also have $\{i\}\neq X$ for all $i\in X$. Therefore, $freq(\{i\},\mathcal{D}') = freq(\{i\},\mathcal{D})$ for all $i\in X$, which implies that the denominators of $lift(X,\mathcal{D}')$ and $lift(X,\mathcal{D}')$ are

equal. Finally, we have $lift(X, \mathcal{D}') = \frac{supp(X, \mathcal{D}')}{\prod_{i \in X} supp(\{i\}, \mathcal{D}')} > lift(X, \mathcal{D}) = \frac{supp(X, \mathcal{D})}{\prod_{i \in X} supp(\{i\}, \mathcal{D})}$,
which shows that $X \in lift^+(X)$, whereas $X \notin lift^-(X)$.

2. Let Y be an itemset such that $\mathbf{Y \neq X}$ and $|\mathbf{Y}| > 1$, and two datasets $\mathcal{D}'$ and $\mathcal{D}$ such that $\mathcal{D}' >_Y \mathcal{D}$. By definition, we have: $freq(X, \mathcal{D}') = freq(X, \mathcal{D})$ since $Y \neq X$, and $freq(\{i\}, \mathcal{D}') = freq(\{i\}, \mathcal{D})$ for all $i \in X$ since $Y \neq \{i\}$ (indeed, we assume that $|Y| > 1$). Thus, we necessarily have $lift(X, \mathcal{D}') = lift(X, \mathcal{D})$ for all datasets $\mathcal{D}'$ and $\mathcal{D}$ such that $\mathcal{D}' >_Y \mathcal{D}$. It implies that for all itemsets Y such that $Y \neq X$ and $|Y| > 1$, $Y \notin lift^+(X)$ and $Y \notin lift^-(X)$.

3. Let Y be an itemset such that $\mathbf{Y \neq X}$ and $|\mathbf{Y}| = 1$, and two datasets $\mathcal{D}'$ and $\mathcal{D}$ such that $\mathcal{D}' >_Y \mathcal{D}$. Using the same reasoning as before, it is easy to see that if $Y \not\subset X$, we necessarily have $lift(X, \mathcal{D}') = lift(X, \mathcal{D})$, which implies that $Y \notin lift^+(X)$ and $Y \notin lift^-(X)$. Dually, if $Y \subset X$, because $|Y| = 1$, there exists $j \in X$ such that $Y = \{j\}$. Since $\mathcal{D}' >_Y \mathcal{D}$ and $X \neq Y$, we have $freq(X, \mathcal{D}') = freq(X, \mathcal{D})$ and $\prod_{i \in X} supp(\{i\}, \mathcal{D}') > \prod_{i \in X} supp(\{i\}, \mathcal{D})$ because $freq(\{j\}, \mathcal{D}') > freq(\{j\}, \mathcal{D}')$ and $j \in X$. Thus, we have $lift(X, \mathcal{D}') = \frac{supp(X, \mathcal{D}')}{\prod_{i \in X} supp(\{i\}, \mathcal{D}')} < lift(X, \mathcal{D}) = \frac{supp(X, \mathcal{D})}{\prod_{i \in X} supp(\{i\}, \mathcal{D})}$, which shows that $Y = \{j\} \subset X$ is an opponent of X for the lift measure (and not a supporter). □

We now consider the case of a condensed representation selector, and prove Property 2, which defines the supporters and opponents of the free constraint.

Property 2. For all itemsets X such that $|X| > 1$, $free^+(X) = \{X \setminus \{i\} : i \in X\}$ and $free^-(X) = \{X\}$.

Proof. Let X be an itemset such that $|X| > 1$. We first show that $X^{\perp} \subseteq free^+(X)$, i.e. that for all $k \in X$, $Y = X \setminus \{k\} \in free^+(X)$. By definition, we have to find two datasets $\mathcal{D}$ and $\mathcal{D}'$ such that $\mathcal{D}' >_Y \mathcal{D}$, $free(X, \mathcal{D}) = false$, whereas $free(X, \mathcal{D}') = true$. Let $\mathcal{D} = \{X\} \cup \{X \setminus \{i\} : i \in Y\} \cup \mathcal{D}_Y^-$ and $\mathcal{D}' = \{X\} \cup \{X \setminus \{i\} : i \in Y\} \cup \mathcal{D}_Y^+$. First, it is easy to that $\mathcal{D}' >_Y \mathcal{D}$. Moreover, we have $freq(X, \mathcal{D}) = 1$ and $freq(Y, \mathcal{D}) = 1$ since $Y \subseteq X$ and $Y \notin \mathcal{D}_Y^-$. Thus, X is not a free itemset in $\mathcal{D}$, i.e. $free(X, \mathcal{D}) = false$. Then, we can see that $freq(X, \mathcal{D}') = 1$ and $freq(X \setminus \{i\}, \mathcal{D}') = 2$ for all $i \in X$ (in particular, note that $Y = (X \setminus \{k\} \in \mathcal{D}_Y^+)$. Thus, X is a free itemset in $\mathcal{D}'$, i.e. $free(X, \mathcal{D}) = true$, which completes the proof that $Y = (X \setminus \{k\}) \in free^+(X)$.

We now show that $X \in free^-(X)$. We have to find two datasets $\mathcal{D}$ and $\mathcal{D}'$ such that $\mathcal{D}' >_X \mathcal{D}$, $free(X, \mathcal{D}) = true$, whereas $free(X, \mathcal{D}') = false$. Let $\mathcal{D} = \{X\} \cup \mathcal{D}_X^-$ and $\mathcal{D}' = \{X\} \cup \mathcal{D}_X^+$. By construction (see the definitions of $\mathcal{D}_X^-$ and $\mathcal{D}_X^+$ in the proof of Lemma 1), it is clear that $\mathcal{D}' >_X \mathcal{D}$. Moreover, we have $freq(X, \mathcal{D}) = 1$ and $freq(X \setminus \{k\}, \mathcal{D}) = 2$ for all $k \in X$ (because $X \setminus \{k\} \subseteq X \in \mathcal{D}$, and $X \setminus \{k\} \in \mathcal{D}_X^-$). Therefore, X is a free itemset in $\mathcal{D}$, i.e. $free(X, \mathcal{D}) = true$. Then, we can also check that $freq(X, \mathcal{D}') = 2$ and $freq(X \setminus \{k\}, \mathcal{D}) = 2$ for all $k \in X$. Thus, X is not a free itemset in $\mathcal{D}'$, which completes the proof that $X \in free^-(X)$.

To complete the proof, we have to show that any other pattern $Y \notin \{X\} \cup X^{\perp}$ cannot be a supporter or an opponent of X. In particular, we have to show that for all $Y \subset X$, if $Y \notin X^{\perp}$, then $Y \notin free^+(X)$, i.e. that for all databases $\mathcal{D}$ and $\mathcal{D}'$ such that $\mathcal{D}' >_Y \mathcal{D}$, it is impossible to have $free(X, \mathcal{D}) = false$ and $free(X, \mathcal{D}') = true$. If $free(X, \mathcal{D}) = false$, it means that there exists $k \in X$ such that $freq(X, \mathcal{D}) = freq(X \setminus \{k\}, \mathcal{D})$. Moreover, because $\mathcal{D}' >_Y \mathcal{D}$ and $Y \notin X^{\perp}$, we have $freq(X, \mathcal{D}') = freq(X \setminus \{k\}, \mathcal{D}')$, which shows that X cannot be free in $\mathcal{D}'$, i.e. $free(X, \mathcal{D}') = false$, and contradicts the hypothesis. Thus, we have shown that only the direct subsets of X are supporters of X for the free constraint. The rest of the proof is omitted for lack of space.

To conclude this section, we stress that the strength of the concept of supporters and opponents is to clearly identify the patterns actually involved in the evaluation of a selector. For instance, whereas the definition of free itemsets given in Table 1 involves *all strict subsets* of X (with $\forall Y \subset X$), we can see that *only direct subsets* of X are supporters. In the following sections, we show how supporters and opponents can be used to compare selectors (see Sect. 5), and how the number of supporters and opponents of a selector is related to its effectiveness to select interesting patterns (see Sect. 6).

5 Typology of Interestingness Selectors

5.1 Polarity of Interestingness Selectors

We distinguish two broad categories of selectors according to whether they aim at discovering over-represented phenomena in the data (e.g., positive correlation) or under-represented phenomena in the data (e.g., outlier detection). Naturally, the characterization of these categories is related to the evaluation of the frequency on the pattern to assess. For instance, it is well-known that the interestingness of a pattern X increases with its frequency for finding correlations between items. In order that an interestingness selector s will be sensitive to this variation, it is essential that s increases with the frequency of X. This principle has first been proposed for association rules [23] (Property P2) and after, extended to correlated itemsets [15,27]. Conversely, a selector for outlier detection will favor patterns whose frequency decreases. Indeed, a pattern is more likely to be abnormal as it is not representative of the dataset i.e., its frequency is low.

We formalize these two types of patterns thanks to reflexivity property:

Definition 4 (Positive and Negative Reflexive). *An interestingness selector s is positive (resp. negative) reflexive iff any pattern is its own supporter i.e., $(\forall X \in \mathcal{L})(X \in s^+(X))$ (resp. opponent i.e., $(\forall X \in \mathcal{L})(X \in s^-(X))$).*

As $all\text{-}conf^+(X) = \{X\}$, the all-confidence selector is positive reflexive. Conversely, the free selector is negative reflexive because $free^-(X) = \{X\}$ (when frequency of X increases, X is less likely to be free because its frequency becomes closer to that of its subsets).

This clear separation based on reflexive property constitutes the first analysis axis of our selector typology. Table 4 schematizes this typology where the polarity is the vertical axis of analysis. The horizontal axis (semantics) will be described in the next section. Note that the correlation measures and the closed itemsets are in the same column. Several works in the literature have shown that closed itemsets maximize classification measures [11] and correlation measures [10]. For instance, the lift of a closed pattern has the highest value of its equivalence class because the frequency of X remains the same (numerator) while the denominator decreases.

Table 4. Typology of interestingness selectors

POLARITY

		Positive $X \in s^+(X)$	Negative $X \in s^-(X)$	
SEMANTICS	**Subsets** $X^{\downarrow}$	C1: $X^{\downarrow} \cap s^-(X) \neq \emptyset$ (all-confidence, bond, productive itemsets, NDI, FPOF, lift)	C2: $X^{\downarrow} \cap s^+(X) \neq \emptyset$ (free itemset, NDI, negative border)	QC 2: Completeness $s^+(X) \cup s^-(X) = \mathcal{L}$
	Supersets $X^{\uparrow}$	C3: $X^{\uparrow} \cap s^-(X) \neq \emptyset$ (closed itemsets, maximal itemsets)	$X^{\uparrow} \cap s^+(X) \neq \emptyset$ (FPOF)	
	Incomparable sets $X^{\leftrightarrow}$	$X^{\leftrightarrow} \cap s^-(X) \neq \emptyset$ (top-k frequent itemsets)	$X^{\leftrightarrow} \cap s^+(X) \neq \emptyset$ (FPOF)	

QC 1: Soundness

$$s^+(X) \cap s^-(X) = \emptyset$$

Of course, it should not be possible for an interestingness selector to both isolate over-represented phenomena (i.e., positive) and under-represented phenomena (i.e., negative). For this reason, a selector should never be both positive and negative. Besides, the behavior of an interestingness selector is easier to understand for the end user if the change in frequency of a pattern Y still impacts $s(X)$ in the same way. In other words, the increase of $freq(X)$ should not decrease $s(X)$ in some cases and increase $s(X)$ in others.

Quality Criterion 1 (Soundness). *An interestingness selector s is sound iff no pattern is at the same time a supporter and an opponent of another pattern:* $\forall X \in \mathcal{L}, s^+(X) \cap s^-(X) = \emptyset$.

When Quality Criterion 1 is violated, it makes difficult to interpret a mined pattern. For instance, frequent free itemset mining is not sound. There are two opposite reasons for explaining that a pattern is not extracted: its frequency is too low (non-frequent rejection), or its frequency is too high (non-free rejection). Conversely, for frequent closed patterns, a pattern is not extracted if and only if its frequency is too low (whatever the underlying cause: the pattern is not frequent or non-closed). It means that frequent closed pattern mining is sound. We therefore think that the violation of Quality Criterion 1 (where $s^+(X) = s^-(X) = X^{\downarrow}$) could partly explain the failure of NDI (non-derivable itemsets) even if they form an extremely compact condensed representation.

Recommendation. A well-behaving pattern mining method should not mix interestingness selectors with opposite polarities or make possible the existence of patterns that are supporters and opponents of the same pattern.

Before describing the semantics axis of our typology, Table 4 classifies all the selectors presented in Table 1. As expected, all selectors seeking to isolate over-represented phenomena are in the Positive column.

5.2 Semantics of Interestingness Selectors

This section presents three complementary criteria to identify the nature of an interestingness selector. The key idea is to focus on the relationships between patterns to qualify the semantics of the selector. More precisely, the meaning of a positive selector (whose primary objective is to find over-represented patterns) depends strongly on the set of opponents that can lead to the rejection of the assessed pattern. Conversely, a negative reflexive selector relies often on supporters to better isolate under-represented phenomena. For this reason, the positive (resp. negative) column of Table 4 involves opponents $s^-(X)$ (resp. supporters $s^+(X)$).

Furthermore, for two selectors of the same polarity, it is possible to distinguish their goals (e.g., correlation or condensed representation) according to the opponents/supporters that they involve. Thus, we break down the semantics axis into three parts: subsets $X^\downarrow = \{Y \subset X\}$, supersets $X^\uparrow = \{Y \supset X\}$ and incomparable sets $X^\leftrightarrow = \{Y \in \mathcal{L} : Y \not\subseteq X \wedge Y \not\supseteq X\}$. This decomposition of the lattice of the opponents and the lattice of the supporters is useful to redefine coherent classes of usual selectors (these classes are indicated in Table 4):

Definition 5 (Selector Classes). *An interestingness selector s belongs to:*

- **C1 (Positive correlation)** *iff* $(\forall X \in \mathcal{L})(X^\downarrow \cap s^-(X) \neq \emptyset)$
- **C2 (Minimal condensed representation)** *iff* $(\forall X \in \mathcal{L})(X^\downarrow \cap s^+(X) \neq \emptyset)$
- **C3 (Maximal condensed representation)** *iff* $(\forall X \in \mathcal{L})(X^\uparrow \cap s^-(X) \neq \emptyset)$

Intuitively a pattern is a set of correlated items (or correlated in brief) when its frequency is higher than what was expected by considering the frequency of some of its subsets (this set of opponents varies depending on the statistical model). This means that the increase of the frequency of one of these subsets may lead to the rejection of the assessed pattern. In other words, a correlation measure is based on subsets as opponents. This observation has already been made in the literature for association rules [23] (with Property P3) and itemsets [15, 27]. Table 4 shows that most of correlation measures in the literature satisfy $(\forall X \in \mathcal{L})(X^\downarrow \cap s^-(X) \neq \emptyset)$. The extraction of NDI, classified as a condensed representation, also meets this criterion. It is intriguingly since the NDI selector is not usually used as a correlation measure.

A condensed representation is a reduced collection of patterns that can regenerate some properties of the full collection of patterns. Typically, frequent closed patterns enable to retrieve the exact frequency of any frequent pattern. Most approaches are based on the notion of equivalence class where two patterns are equivalent if they have the same value for a function f and if they are comparable. The equality for f and the comparability result in an interrelation between the assessed pattern and its subsets/supersets. Class C3 (i.e., maximal condensed representations) includes the measures that remove the assessed pattern when a more specific pattern provides more information. Closed patterns and maximal patterns satisfy this criterion: $(\forall X \in \mathcal{L})(X^{\uparrow} \cap s^{-}(X) \neq \emptyset)$. Minimal condensed representations are in the dual class (i.e., Class C2).

Unlike the polarity that opposes two types of irreconcilable patterns, the three parts of the semantics axis (i.e., subsets, supersets and incomparable sets) are simultaneously satisfiable. We think that an ideal pattern extraction method should always belong to these three parts:

Quality Criterion 2 (Completeness). *A selector s is complete iff all patterns are either supporter or opponent:* $\forall X \in \mathcal{L}, s^{+}(X) \cup s^{-}(X) = \mathcal{L}$.

Let us illustrate the principle behind this quality criterion by considering an ideal pattern mining method that isolates correlations. Of course, this method relies on a selector s that belongs to the class of correlations (for example, the lift). At equal frequency, the longer pattern will be preferred because it will maximize lift. This property corresponds to the criterion $X^{\uparrow} \cap s^{-}(X) \neq \emptyset$. At this stage, two incomparable patterns can cover the same set of transactions. To retain only one, we must add a new selection criterion that verifies the criterion $X^{\leftrightarrow} \cap s^{-}(X) \neq \emptyset$. This approach is at the heart of many proposals in the literature [3,10,28]: (i) use of a correlation measure, (ii) elimination of non-closed patterns, (iii) elimination of incomparable redundant patterns.

Recommendation. All patterns should be either supporters or opponents in a well-behaving pattern mining method. It is often necessary to combine a measure with local and global redundancy reduction techniques.

6   Evaluation Complexity of Interestingness Selectors

As Quality Criterion 2 is often violated, we propose to measure its degree of satisfaction to evaluate and compare interestingness selectors. More precisely, we measure the quality of an interestingness selector considering its degree of satisfaction of the semantics criterion. Let us consider the correlation family, it is clear that to detect correlations, support is a poorer measure than lift which is itself less effective than productivity. Whatever the part of the lattice, the more numerous the opponents/supporters of a selector, the better its quality. In other words, a selector is more effective to assess the interestingness of a pattern X when the number of supporters and opponents of X is very high.

Definition 6 (Evaluation Complexity). *The evaluation complexity of an interestingness selector s is the asymptotic behavior of the cardinality of its supporters/opponents.*

The evaluation complexity of a selector usually depends on the cardinality of the assessed pattern (denoted by $k = |X|$) and the cardinality of the set of items $\mathcal{I}$ (denoted by $n = |\mathcal{I}|$). For instance, $|all\text{-}conf^{\pm}(X)| = |all\text{-}conf^{+}(X)| \cup |all\text{-}conf^{-}(X)| = 1 + k$. Therefore, the behavior of the number of evaluations of all-confidence is linear with respect to itemset size. Similarly, the evaluation complexity of productive itemsets is exponential with respect to the size of the assessed pattern since all subsets are involved in the evaluation of this constraint. According to the evaluation complexity, we say that the quality of the constraint of productive itemsets is better than that of the all-confidence, because the opponents are more numerous. More generally, this complexity allows to compare several interestingness selectors to each other. The column $|s^{\pm}(X)|$ (where $s^{\pm}(X)$ is the total number of supporters and opponents of X) in Table 3 indicates the evaluation complexity of each measure or constraint defined in Table 1. Three main complexity classes emerge: constant, linear and exponential. Although Table 1 is an extremely small sample of measures, we observe that the evaluation complexity of pattern mining methods has increased over the past decades. Interestingly, we also note that the evaluation complexity of global constraints [6,13] (or advanced measures [26]) is greater than those of local constraints (or absolute measures).

For Classes C2 and C3, the most condensed representations (among those that enable to regenerate the frequency of each pattern) are also those with the greatest evaluation complexity. Indeed, the free itemsets are more numerous than the closed ones, themselves more numerous than the NDIs. For Class C1, it is clear that measures based on more sophisticated statistical models require more relationships [26]. They have therefore an higher evaluation complexity. We will also experimentally verify this hypothesis in the next section.

7   Experimental Study

Our goal is to verify whether the quality of the correlated pattern selectors follow the evaluation complexity. In other words, if a correlation measure has a greater evaluation complexity than another measure, it is expected to be more effective.

To verify this hypothesis, we rely on the experimental protocol inspired by [14]. The idea is to compare the extracted patterns in an original dataset $\mathcal{D}$ with the same randomized dataset $\mathcal{D}^*$. Specifically, in the randomized dataset $\mathcal{D}^*$, a large number of items are randomly swapped two by two in order to clear any correlation. Nevertheless, this dataset $\mathcal{D}^*$ retains the same characteristics (transaction length and frequency of each item). So, if a pattern X extracted in the original dataset $\mathcal{D}$ is also extracted in the randomized dataset $\mathcal{D}^*$, X is said to be false positive (FP). Its presence in $\mathcal{D}^*$ is not due to the correlation between items but due to the distribution of items in data. Then we evaluate for each selector how many false positive patterns are extracted on average by repeating the protocol on 10 randomized datasets. Experiments were conducted on datasets coming from the UCI ML Repository [7]. Given a minimum support threshold, we compare 4 selectors: **Support** (all frequent patterns); **All-confidence** (all frequent patterns having at least 5 as all-confidence); **Lift** (all frequent patterns having at least 1.5 as lift); and **Productivity** (all frequent patterns having at least 1.5 as productivity).

Even if arbitrary thresholds are used for the last three selectors, the results are approximately the same with other thresholds because we use the FP rate as

evaluation measure. This normalized measure is a ratio, it returns the proportion
of FP patterns among all the mined patterns.

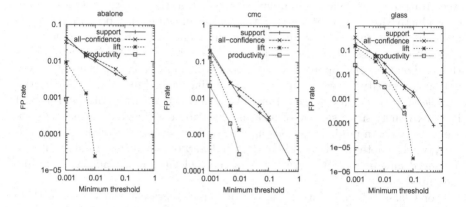

Fig. 1. FP rate with minimum support threshold

Figure 1 plots the FP rate of each selector on `abalone`, `cmc` and `glass`. Since
lift and productivity measures sometimes do not return FP patterns, there are
missing points because the scale is logarithmic. For each dataset, we observe that
the FP rate increases when the minimum support threshold decreases regardless
of the measure. The evolution of the FP rate for the all-confidence is very similar
to that of the support even if the all-confidence has a greater complexity in
evaluation. For the other selectors, there is a clear ranking from worst to best:
support, lift and productivity. This ranking also corresponds to the classes of
complexity from worst to best: constant, linear, exponential. Our framework
could be refined to consider a set of patterns as an opponent (or a supporter).
Then, the relation between a pattern and its supporters (or opponents) would
become a relation between a pattern and a *set* of supporters (or opponents).
That would make possible to capture refinements between selectors as follows:
the all-confidence depends only on one item at once (due to the maximum of
its denominator) whereas the lift can vary according to a set of items (due to
the multiplication). Nevertheless, our experiments show that on the whole the
correlation measures with the highest evaluation complexity are also the best
ones according to the FP rate.

8   Conclusion and Discussion

In this paper, we have addressed the question of the quality of a data mining
method in the context of unsupervised problems with binary data. A key concept
is to study the relationships between a pattern X and the other patterns when X
is selected by a method. These relationships are formalized through the notions of
supporters and opponents. We have presented a typology of methods defined by

formal properties and based on two complementary criteria. This typology offers a global picture and a methodology helping to compare methods to each other. Besides, if a new method is proposed, its quality can be immediately compared to the quality of the other methods according to our framework. Finally, the quality of a method is quantified via an evaluation complexity analysis based on the number of supporters and opponents of a pattern extracted by the method.

Two recommendations can be drawn from this work. We think that the result of a data mining operation should be understandable by the user. So, our first recommendation is a data mining method should not simultaneously extract over-represented phenomena and under-represented phenomena because mixing these two kinds of phenomena obstructs the understandability of the extracted patterns. This recommendation is formally defined by our soundness criterion. Most of methods satisfy this property, but there are a few exceptions such as the constraints extracting NDI and frequent free patterns. The violation of this recommendation might explain why these patterns are of little use.

Another recommendation is a data mining method should extract patterns for which all patterns contribute to the quality of an extracted pattern. This recommendation is formalized by our completeness criterion stating that all patterns must be either supporters or opponents of a pattern extracted by a method. In practice, this recommendation is not satisfied by a lot of methods. However, a few methods are endowed with this behavior, such as [3,10,28] in the context of the correlations. We think that a goal of pattern mining should be to design methods following this recommendation which is more often reached by pattern sets [3] as illustrated by the previous examples [3,10,28].

A perspective of this work is to study an interestingness theory for methods producing pattern sets. Pattern sets are a promising avenue since the interestingness of a pattern also depends on the interestingness of the other patterns of the pattern set, thus providing a global quality of the method. Finally, it is important to note that our framework can be generalized to other pattern languages (sequence, graph, etc.) and other basic functions, e.g., observing the variation of support in a target class would extend our approach to the supervised context.

Acknowledgements. The authors thank Albrecht Zimmermann for highly valuable discussions. This work has been partly supported by the QCM-BioChem project (CNRS Mastodons).

References

1. Agrawal, R., Srikant, R.: Fast algorithms for mining association rules in large databases. In: VLDB, pp. 487–499. Morgan Kaufmann, Burlington (1994)
2. Boulicaut, J.F., Bykowski, A., Rigotti, C.: Free-sets: a condensed representation of boolean data for the approximation of frequency queries. Data Min. Knowl. Discov. **7**(1), 5–22 (2003)
3. Bringmann, B., Zimmermann, A.: The chosen few: on identifying valuable patterns. In: ICDM, pp. 63–72. Omaha, NE (2007)
4. Calders, T., Goethals, B.: Non-derivable itemset mining. Data Min. Knowl. Discov. **14**(1), 171–206 (2007)

5. Calders, T., Rigotti, C., Boulicaut, J.-F.: A survey on condensed representations for frequent sets. In: Boulicaut, J.-F., De Raedt, L., Mannila, H. (eds.) Constraint-Based Mining and Inductive Databases. LNCS (LNAI), vol. 3848, pp. 64–80. Springer, Heidelberg (2006). https://doi.org/10.1007/11615576_4

6. Crémilleux, B., Soulet, A.: Discovering knowledge from local patterns with global constraints. In: Gervasi, O., Murgante, B., Laganà, A., Taniar, D., Mun, Y., Gavrilova, M.L. (eds.) ICCSA 2008, Part II. LNCS, vol. 5073, pp. 1242–1257. Springer, Heidelberg (2008). https://doi.org/10.1007/978-3-540-69848-7_99

7. Dheeru, D., Taniskidou, E.K.: UCI machine learning repository (2017). http://archive.ics.uci.edu/ml

8. Fayyad, U.M., Piatetsky-Shapiro, G., Uthurusamy, R.: Summary from the kdd-03 panel: data mining: the next 10 years. ACM SIGKDD Explor. 5(2), 191–196 (2003)

9. Wai-chee Fu, A., Wang-wai Kwong, R., Tang, J.: Mining N-most interesting itemsets. In: Raś, Z.W., Ohsuga, S. (eds.) ISMIS 2000. LNCS (LNAI), vol. 1932, pp. 59–67. Springer, Heidelberg (2000). https://doi.org/10.1007/3-540-39963-1_7

10. Gallo, A., De Bie, T., Cristianini, N.: MINI: mining informative non-redundant itemsets. In: Kok, J.N., Koronacki, J., Lopez de Mantaras, R., Matwin, S., Mladenič, D., Skowron, A. (eds.) PKDD 2007. LNCS (LNAI), vol. 4702, pp. 438–445. Springer, Heidelberg (2007). https://doi.org/10.1007/978-3-540-74976-9_44

11. Garriga, G.C., Kralj, P., Lavrač, N.: Closed sets for labeled data. J. Mach. Learn. Res. 9, 559–580 (2008)

12. Geng, L., Hamilton, H.J.: Interestingness measures for data mining: a survey. ACM Comput. Surv. 38(3), 9 (2006). https://doi.org/10.1145/1132960.1132963

13. Giacometti, A., Marcel, P., Soulet, A.: A relational view of pattern discovery. In: Yu, J.X., Kim, M.H., Unland, R. (eds.) DASFAA 2011, Part I. LNCS, vol. 6587, pp. 153–167. Springer, Heidelberg (2011). https://doi.org/10.1007/978-3-642-20149-3_13

14. Gionis, A., Mannila, H., Mielikäinen, T., Tsaparas, P.: Assessing data mining results via swap randomization. TKDD 1(3), 14 (2007)

15. Hämäläinen, W.: Efficient search for statistically significant dependency rules in binary data. Ph.D. thesis, University of Helsinki (2010)

16. Han, J., Cheng, H., Xin, D., Yan, X.: Frequent pattern mining: current status and future directions. Data Min. Knowl. Discov. 15(1), 55–86 (2007)

17. He, Z., Xu, X., Huang, Z.J., Deng, S.: FP-outlier: frequent pattern based outlier detection. Comput. Sci. Inf. Syst. 2(1), 103–118 (2005)

18. Lenca, P., Meyer, P., Vaillant, B., Lallich, S.: On selecting interestingness measures for association rules: user oriented description and multiple criteria decision aid. EJOR 184(2), 610–626 (2008)

19. Mannila, H., Toivonen, H.: Levelwise search and borders of theories in knowledge discovery. Data Min. Knowl. Discov. 1(3), 241–258 (1997)

20. Morik, K., Boulicaut, J.-F., Siebes, A. (eds.): Local Pattern Detection. LNCS (LNAI), vol. 3539. Springer, Heidelberg (2005). https://doi.org/10.1007/b137601

21. Omiecinski, E.: Alternative interest measures for mining associations in databases. IEEE Trans. Knowl. Data Eng. 15(1), 57–69 (2003)

22. Pasquier, N., Bastide, Y., Taouil, R., Lakhal, L.: Efficient mining of association rules using closed itemset lattices. Inf. Syst. 24(1), 25–46 (1999)

23. Piatetsky-Shapiro, G.: Discovery, analysis, and presentation of strong rules. In: Knowledge Discovery in Databases, pp. 229–248. AAAI/MIT Press, Cambridge (1991)

24. Tan, P.N., Kumar, V., Srivastava, J.: Selecting the right objective measure for association analysis. Inf. Syst. 29(4), 293–313 (2004)

25. Tew, C.V., Giraud-Carrier, C.G., Tanner, K.W., Burton, S.H.: Behavior-based clustering and analysis of interestingness measures for association rule mining. Data Min. Knowl. Discov. **28**(4), 1004–1045 (2014)
26. Vreeken, J., Tatti, N.: Interesting patterns. In: Aggarwal, C.C., Han, J. (eds.) Frequent Pattern Mining, pp. 105–134. Springer, Cham (2014). https://doi.org/10.1007/978-3-319-07821-2_5
27. Webb, G.I., Vreeken, J.: Efficient discovery of the most interesting associations. ACM Trans. Knowl. Discov. Data **8**(3), 15:1–15:31 (2013)
28. Xin, D., Cheng, H., Yan, X., Han, J.: Extracting redundancy-aware top-k patterns. In: KDD, pp. 444–453. ACM (2006)
29. Zimmermann, A.: Objectively evaluating condensed representations and interestingness measures for frequent itemset mining. J. Intell. Inf. Syst. **45**(3), 299–317 (2015)

Deep Learning

Efficient Decentralized Deep Learning by Dynamic Model Averaging

Michael Kamp[1,2,3]([envelope]), Linara Adilova[1,2], Joachim Sicking[1,2], Fabian Hüger[4], Peter Schlicht[4], Tim Wirtz[1,2], and Stefan Wrobel[1,2,3]

[1] Fraunhofer IAIS, Sankt Augustin, Germany
{michael.kamp,linara.adilova,joachim.sicking,
tim.wirtz,stefan.wrobel}@iais.fraunhofer.de
[2] Fraunhofer Center for Machine Learning, Sankt Augustin, Germany
[3] University of Bonn, Bonn, Germany
{kamp,wrobel}@cs.uni-bonn.de
[4] Volkswagen Group Research, Wolfsburg, Germany
{fabian.huger,peter.schlicht}@volkswagen.de

Abstract. We propose an efficient protocol for decentralized training of deep neural networks from distributed data sources. The proposed protocol allows to handle different phases of model training equally well and to quickly adapt to concept drifts. This leads to a reduction of communication by an order of magnitude compared to periodically communicating state-of-the-art approaches. Moreover, we derive a communication bound that scales well with the hardness of the serialized learning problem. The reduction in communication comes at almost no cost, as the predictive performance remains virtually unchanged. Indeed, the proposed protocol retains loss bounds of periodically averaging schemes. An extensive empirical evaluation validates major improvement of the trade-off between model performance and communication which could be beneficial for numerous decentralized learning applications, such as autonomous driving, or voice recognition and image classification on mobile phones. Code related to this paper is available at: https://bitbucket.org/Michael_Kamp/decentralized-machine-learning.

1 Introduction

Traditionally, deep learning models are trained on a single system or cluster by centralizing data from distributed sources. In many applications, this requires a prohibitive amount of communication. For gradient-based training methods, communication can be reduced by calculating gradients locally and communicating the sum of gradients periodically [7], instead of raw data. This mini-batch

---

M. Kamp, L. Adilova and J. Sicking—These authors contributed equally.

Electronic supplementary material The online version of this chapter (https://doi.org/10.1007/978-3-030-10925-7_24) contains supplementary material, which is available to authorized users.

© Springer Nature Switzerland AG 2019
M. Berlingerio et al. (Eds.): ECML PKDD 2018, LNAI 11051, pp. 393–409, 2019.
https://doi.org/10.1007/978-3-030-10925-7_24

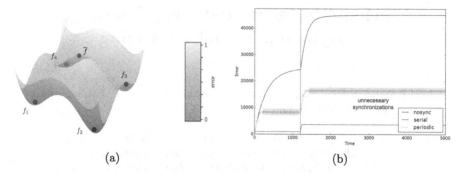

Fig. 1. (a) Illustration of the problem of averaging models in non-convex problems: each of the models $f_1, \ldots, f_4$ has reached a local minimum, but their average $\bar{f}$ has a larger error than each of them. (b) Cumulative error over time for a serial learning algorithm and two decentralized learning algorithms with 10 learners, one that does not communicate (nosync) and one that communicates every 50 time steps (periodic). The vertical line indicates a concept drift, i.e., a rapid change in the target distribution.

approach performs well on tightly connected distributed systems [5,6,33] (e.g., data centers and clusters). For many applications, however, centralization or even periodic sharing of gradients between local devices becomes infeasible due to the large amount of necessary communication.

For decentralized systems with limited communication infrastructure it was suggested to compute local updates [35] and average models periodically, instead of sharing gradients. Averaging models has three major advantages: (i) sending only the model parameters instead of a set of data samples reduces communication[1]; (ii) it allows to train a joint model without exchanging or centralizing privacy-sensitive data; and (iii) it can be applied to a wide range of learning algorithms, since it treats the underlying algorithm as a black-box.

This approach is used in convex optimization [21,27,34]. For non-convex objectives, a particular problem is that the average of a set of models can have a worse performance than any model in the set—see Fig. 1(a). For the particular case of deep learning, McMahan et al. [22] empirically evaluated model averaging in decentralized systems and termed it **Federated Learning**.

However, averaging periodically still invests communication independent of its utility, e.g., when all models already converged to an optimum. This disadvantage is even more apparent in case of concept drifts: periodic approaches cannot react adequately to drifts, since they either communicate so rarely that the models adapt too slowly to the change, or so frequently that they generate an immense amount of unnecessary communication in-between drifts.

---

[1] Note that averaging models requires the same amount of communication as sharing gradients, since the vector of model parameters is of the same dimension as the gradient vector of the loss function.

In Kamp et al. [12] the authors proposed to average models dynamically, depending on the utility of the communication. The main idea is to reduce communication without losing predictive performance by investing the communication efficiently: When local learners do not suffer loss, communication is unnecessary and should be avoided (see Fig. 1(b)); similarly, when they suffer large losses, an increased amount of communication should be invested to improve their performances. The problem setting and a criterion for efficient approaches is defined in Sect. 2. This approach, denoted **dynamic averaging**, was proposed for online learning convex objectives [12,15]. We adapt dynamic averaging to the non-convex objectives of deep learning in Sect. 3.

Our contribution is the description and evaluation of a general method for decentralized training of deep neural networks that (i) substantially reduces communication while retaining high predictive performance and (ii) is in addition well-suited to concept drifts in the data. To that end, Sect. 4 shows that, for common learning algorithms, dynamic averaging is an efficient approach for non-convex problems, i.e., it retains the predictive performance of a centralized learner but is also adaptive to the current hardness of the learning problem.

A natural application for dynamic decentralized machine learning is **in-fleet learning** of autonomous driving functionalities: concept drifts occur naturally, since properties central for the modeling task may change—changing traffic behavior both over time and different countries or regions introduce constant and unforeseeable concept drifts. Moreover, large high-frequency data streams generated by multiple sensors per vehicle renders data centralization prohibitive in large fleets. Section 5 provides an extensive empirical evaluation of the dynamic averaging approach on classical deep learning tasks, as well as synthetic and real-world tasks with concept drift, including in-fleet learning of autonomous driving functionalities. The approach is compared to periodically communicating schemes, including **Federated Averaging** [22], a state-of-the-art approach for decentralized deep learning—more recent approaches are interesting from a theoretical perspective but show no practical improvement [11], or tackle other aspects of federated learning, such as non-iid data [31] or privacy aspects [23].

Section 6 discusses properties and limitations of dynamic averaging and puts it into context of related work, followed by a conclusion in Sect. 7.

2   Preliminaries

We consider a decentralized learning setting with $m \in \mathbb{N}$ **local learners**, where each learner $i \in [m]$ runs the same **learning algorithm** $\varphi \colon \mathcal{F} \times 2^X \times 2^Y \to \mathcal{F}$ that trains a **local model** f^i from a **model space** $\mathcal{F}$ using local samples from an **input space** X and **output space** Y. We assume a streaming setting, where in each round $t \in \mathbb{N}$ each learner $i \in [m]$ observes a sample $E_t^i \subset X \times Y$ of size $|E_t^i| = B$, drawn iid from the same time variant distribution $P_t \colon X \times Y \to \mathbb{R}_+$. The local learner uses its local model to make a prediction whose quality is measured by a **loss function** $\ell \colon \mathcal{F} \times X \times Y \to \mathbb{R}_+$. We abbreviate the loss of the local model of learner i in round t by $\ell_t^i \left(f_t^i \right) = \sum_{(x,y) \in E_t^i} \ell \left(f_t^i, x, y \right)^2$. The goal of decentralized learning is to minimize the **cumulative loss** up to a time horizon $T \in \mathbb{N}$, i.e.,

$$L(T, m) = \sum_{t=1}^{T} \sum_{i=1}^{m} \ell_t^i \left(f_t^i \right). \tag{1}$$

Guarantees on the predictive performance, measured by the cumulative loss, are typically given by a **loss bound** $\mathbf{L}(T, m)$. That is, for all possible sequences of losses it holds that $L(T, m) \leq \mathbf{L}(T, m)$.

In each round $t \in \mathbb{N}$, local learners use a **synchronization operator** $\sigma \colon \mathcal{F}^m \to \mathcal{F}^m$ that transfers the current set of local models, called the current **model configuration** $\mathbf{f}_t = \{f_t^1, \ldots, f_t^m\}$, into a single stronger **global model** $\sigma(\mathbf{f}_t)$ which replaces the local models. We measure the performance of the operator in terms of communication by the **cumulative communication**, i.e.,

$$C(T, m) = \sum_{t=1}^{T} c(\mathbf{f}_t),$$

where $c \colon \mathcal{F}^m \to \mathbb{N}$ measures the number of bytes required by the protocol to synchronize the models $\mathbf{f}_t$ at time t. We investigate synchronization operators that aggregate models by computing their average [21,22,27,34,35], i.e., $\bar{f} = 1/m \sum_{i=1}^{m} f^i$. In the case of neural networks, we assume that all local models have the same architecture, thus their average is the average of their respective weights. We discuss the potential use of other aggregation operations in Sect. 6. We denote the choice of learning algorithm together with the synchronization operator as a **decentralized learning protocol** $\Pi = (\varphi, \sigma)$. The protocol is evaluated in terms of the predictive performance and cumulative communication. In order to assess the efficiency of decentralized learning protocols in terms of the trade-off between loss and communication, Kamp et al. [14] introduced two criteria: consistency and adaptiveness.

Definition 1 (Kamp et al. [14]). *A distributed online learning protocol $\Pi = (\varphi, \sigma)$ processing mT inputs is **consistent** if it retains the loss of the serial online learning algorithm φ, i.e.,*

$$L_\Pi(T, m) \in \mathcal{O}\left(L_\varphi(mT)\right).$$

*The protocol is **adaptive** if its communication bound is linear in the number of local learners m and the loss $L_\varphi(mT)$ of the serial online learning algorithm, i.e.,*

$$C_\Pi(T, m) \in \mathcal{O}\left(m L_\varphi(mT)\right).$$

---

[2] This setup includes online learning ($B = 1$) and mini-batch training $B > 1$. The gradient of ℓ_t^i is the sum of individual gradients. Our approach and analysis also apply to heterogeneous sampling rates B^i for each learner i.

A decentralized learning protocol is **efficient** if it is both consistent and adaptive. Each one of the criteria can be trivially achieved: A non-synchronizing protocol is adaptive but not consistent, a protocol that centralizes all data is consistent but not adaptive. Protocols that communicate periodically are consistent [7,35], i.e., they achieve a predictive performance comparable to a model that is learned centrally on all the data. However, they require an amount of communication linear in the number of learners m and the number of rounds T, independent of the loss. Thus they are not adaptive.

In the following section, we recapitulate dynamic averaging and apply it to the non-convex problem of training deep neural networks. In Sect. 4 we discuss in which settings it is efficient as in Definition 1.

3   Dynamic Averaging

In this section, we recapitulate the dynamic averaging protocol [15] for synchronizations based on quantifying their effect (Algorithm 1). Intuitively, communication is not well-invested in situations where all models are already approximately equal—either because they were updated to similar models or have merely changed at all since the last averaging step—and it is more effective if models are diverse. A simple measure to quantify the effect of synchronizations is given by the **divergence** of the current model configuration, i.e.,

$$\delta(\mathbf{f}) = \frac{1}{m} \sum_{i=1}^{m} \left\| f^i - \overline{f} \right\|^2. \tag{2}$$

Using this, we define the dynamic averaging operator that allows to omit synchronization in cases where the divergence of a model configuration is low.

Definition 2 (Kamp et al. [12]). *A **dynamic averaging operator** with positive divergence threshold $\Delta \in \mathbb{R}_+$ and batch size $b \in \mathbb{N}$ is a synchronization operator $\sigma_{\Delta,b}$ such that $\sigma_{\Delta,b}(\mathbf{f}_t) = \mathbf{f}_t$ if $t \mod b \neq 0$ and otherwise: (i) $\overline{f}_t = \overline{\sigma_{\Delta,b}(\mathbf{f}_t)}$, i.e., it leaves the mean model invariant, and (ii) $\delta\left(\sigma_{\Delta,b}(\mathbf{f}_t)\right) \leq \Delta$, i.e., after its application the model divergence is bounded by Δ.*

An operator adhering to this definition does not generally put all nodes into sync (albeit we still refer to it as *synchronization* operator). In particular it allows to leave all models untouched as long as the divergence remains below Δ or to only average a subset of models in order to satisfy the divergence constraint.

The **dynamic averaging protocol** $\mathcal{D} = (\varphi, \sigma_{\Delta,b})$ synchronizes the local learners using the dynamic averaging operator $\sigma_{\Delta,b}$. This operator only communicates when the model divergence exceeds a **divergence threshold** Δ. In order to decide when to communicate locally, at round $t \in \mathbb{N}$, each local learner $i \in [m]$ monitors the **local condition** $\|f_t^i - r\|^2 \leq \Delta$ for a **reference model** $r \in \mathcal{F}$ [30] that is common among all learners (see [10,16,17,19,29] for a more general description of this method). The local conditions guarantee that if none of them is violated, i.e., for all $i \in [m]$ it holds that $\|f_t^i - r\|^2 \leq \Delta$, then the

---

Algorithm 1. Dynamic Averaging Protocol

Input: divergence threshold Δ, batch size b

Initialization:
  local models $f_1^1, \ldots, f_1^m \leftarrow$ one random f
  reference vector $r \leftarrow f$
  violation counter $v \leftarrow 0$

Round t at node i:
  observe $E_t^i \subset X \times Y$
  update f_{t-1}^i using the learning algorithm φ
  if $t \mod b = 0$ **and** $\|f_t^i - r\|^2 > \Delta$ **then**
    send f_t^i to coordinator (violation)

At coordinator on violation:
  let $\mathcal{B}$ be the set of nodes with violation
  $v \leftarrow v + |\mathcal{B}|$
  if $v = m$ **then** $\mathcal{B} \leftarrow [m]$, $v \leftarrow 0$
  while $\mathcal{B} \neq [m]$ **and** $\left\| \frac{1}{\mathcal{B}} \sum_{i \in \mathcal{B}} f_t^i - r \right\|^2 > \Delta$ **do**
    augment $\mathcal{B}$ by augmentation strategy
    receive models from nodes added to $\mathcal{B}$
  send model $\overline{f} = \frac{1}{\mathcal{B}} \sum_{i \in \mathcal{B}} f_t^i$ to nodes in $\mathcal{B}$
  if $\mathcal{B} = [m]$ also set new reference vector $r \leftarrow \overline{f}$

---

divergence does not exceed the threshold, i.e., $\delta(\mathbf{f}_t) \leq \Delta$ [12, Theorem 6]. The closer the reference model is to the true average of local models, the tighter are the local conditions. Thus, the first choice for the reference model is the average model from the last synchronization step. The local condition is checked every $b \in \mathbb{N}$ rounds. This allows using the common mini-batch approach [3] for training deep neural networks.

If one or more local conditions are violated, all local models can be averaged—an operation referred to as **full synchronization**. However, on a local violation the divergence threshold is not necessarily crossed. In that case, the violations may be locally balanced: the coordinator incrementally queries other local learners for their models; if the average of all received models lies within the safe zone, it is transferred back as new model to all participating nodes. If all nodes have been queried, the result is equivalent to a full synchronization and the reference vector is updated. In both cases, the divergence of the model configuration is bounded by Δ at the end of the balancing process, because all local conditions hold. Also, it is easy to check that this protocol leaves the global mean model unchanged. Hence, it is complying to Definition 2. In the following Section, we theoretically analyze the loss and communication of dynamic averaging.

4   Efficiency of Dynamic Averaging

In order to assess the predictive performance and communication cost of the dynamic averaging protocol for deep learning, we compare it to a periodically averaging approach: Given a learning algorithm φ, the **periodic averaging protocol** $\mathcal{P} = (\varphi, \sigma_b)$ synchronizes the current model configuration $\mathbf{f}$ every $b \in \mathbb{N}$ time steps by replacing all local models by their joint average $\overline{f} = 1/m \sum_{i=1}^{m} f^i$. That is, the synchronization operator is given by

$$\sigma_b(\mathbf{f}_t) = \begin{cases} (\overline{f}_t, \dots, \overline{f}_t), & \text{if } b \equiv O(t) \\ \mathbf{f}_t = (f_t^1, \dots, f_t^m), & \text{otherwise} \end{cases}.$$

A special case of this is the **continuous averaging protocol** $\mathcal{C} = (\varphi, \sigma_1)$, synchronizing every round, i.e., for all $t \in \mathbb{N}$, the synchronization operator is given by $\sigma_1(\mathbf{f}_t) = (\overline{f}_t, \dots, \overline{f}_t)$. As base learning algorithm we use mini-batch SGD algorithm $\varphi_{B,\eta}^{\mathrm{mSGD}}$ [7] with mini-batch size $B \in \mathbb{N}$ and learning rate $\eta \in \mathbb{R}_+$ commonly used in deep learning [3]. One step of this learning algorithm given the model $f \in \mathcal{F}$ can be expressed as

$$\varphi_{B,\eta}^{\mathrm{mSGD}}(f) = f - \eta \sum_{j=1}^{B} \nabla \ell^j(f).$$

Let $\mathcal{C}^{\mathrm{mSGD}} = (\varphi_{B,\eta}^{\mathrm{mSGD}}, \sigma_1)$ denote the continuous averaging protocol using mini-batch SGD. For $m \in \mathbb{N}$ learners with the same model $f \in \mathcal{F}$, mB training samples $(x_1, y_1), \dots, (x_{mB}, y_{mB})$, and corresponding loss functions $\ell^i(\cdot) = \ell(\cdot, x_i, y_i)$, one step of $\mathcal{C}^{\mathrm{mSGD}}$ is

$$\sigma_1\left((\varphi_{B,\eta}^{\mathrm{mSGD}}(f), \dots, \varphi_{B,\eta}^{\mathrm{mSGD}}(f))\right) = \frac{1}{m} \sum_{i=1}^{m} \left(f - \eta \sum_{j=1}^{B} \nabla \ell^{(i-1)B+j}(f)\right).$$

We compare $\mathcal{C}^{\mathrm{mSGD}}$ to the serial application of mini-batch SGD. It can be observed that continuous averaging with mini-batch SGD on $m \in \mathbb{N}$ learners with mini-batch size B is equivalent to serial mini-batch SGD with a mini-batch size of mB and a learning rate that is m times smaller.

Proposition 3. *For $m \in \mathbb{N}$ learners, a mini-batch size $B \in \mathbb{N}$, mB training samples $(x_1, y_1), \dots, (x_{mB}, y_{mB})$, corresponding loss functions $\ell^i(\cdot) = \ell(\cdot, x_i, y_i)$, a learning rate $\eta \in \mathbb{R}_+$, and a model $f \in \mathcal{F}$, it holds that*

$$\sigma_1\left((\varphi_{B,\eta}^{mSGD}(f), \dots, \varphi_{B,\eta}^{mSGD}(f))\right) = \varphi_{mB,\eta/m}^{mSGD}(f).$$

Proof.

$$\sigma_1\left((\varphi_{B,\eta}^{\mathrm{mSGD}}(f), \dots, \varphi_{B,\eta}^{\mathrm{mSGD}}(f))\right) = \frac{1}{m} \sum_{i=1}^{m} \left(f - \eta \sum_{j=1}^{B} \nabla \ell^{(i-1)B+j}(f)\right)$$

$$= \frac{1}{m} mf - \frac{1}{m}\eta \sum_{i=1}^{m} \sum_{j=1}^{B} \nabla \ell^{(i-1)B+j}(f) = f - \frac{1}{m}\eta \sum_{j=1}^{mB} \nabla \ell^j(f) = \varphi_{mB,\eta/m}^{\mathrm{mSGD}}(f)$$

$$\square$$

In particular, Proposition 3 holds for continuous averaging with a mini-batch size of $B = 1$, i.e., classic stochastic gradient descent. From Proposition 3 it follows that continuous averaging is consistent as in Definition 1, since it retains the loss bound of serial mini-batch SGD and classic SGD. If the loss function is locally convex in an $\mathcal{O}(\Delta)$-radius around the current average—a non-trivial but realistic assumption [18,25]—Theorem 2 in Boley et al. [2] guarantees that for SGD, dynamic averaging has a predictive performance similar to any periodically communicating protocol, in particular to σ_1 (see Appendix B in the supplementary material for details). For this case it follows that dynamic averaging using SGD for training deep neural networks is consistent. Theorem 2 in Kamp et al. [14] shows that the cumulative communication of the dynamic averaging protocol using SGD and a divergence threshold Δ is bounded by

$$ C(T,m) \in \mathcal{O}\left(\frac{c(\mathbf{f})}{\sqrt{\Delta}} L(T,m) \right), $$

where $c(\mathbf{f})$ is the number of bytes required to be communicated to average a set of deep neural networks. Since each neural network has a fixed number of weights, $c(\mathbf{f})$ is in $\mathcal{O}(m)$. It follows that dynamic averaging is adaptive. Thus, using dynamic averaging with stochastic gradient descent for the decentralized training of deep neural networks is efficient as in Definition 1.

Note that the synchronization operator can be implemented using different assumptions on the system's topology and communication protocol, i.e., in a peer-to-peer fashion, or in a hierarchical communication scheme. For simplicity, in our analysis of the communication of different synchronization operators we assume that the synchronization operation is performed by a dedicated coordinator node. This coordinator is able to poll local models, aggregate them and send the global model to the local learners.

5   Empirical Evaluation

This section empirically evaluates dynamic averaging for training deep neural networks. To emphasize the theoretical result from Sect. 4, we show that dynamic averaging indeed retains the performance of periodic averaging with substantially less communication. This is followed by a comparison of our approach with a state-of-the-art communication approach. The performance is then evaluated in the presence of concept drifts. Combining the aforementioned aspects, we apply our protocol to a non-convex objective with possible concept drifts from the field of autonomous driving.

Throughout this section, if not specified separately, we consider mini-batch SGD $\varphi_{B,\eta}^{\mathrm{mSGD}}$ as learning algorithm, since recent studies indicate that it is particularly suited for training deep neural networks [32]. That is, we consider communication protocols $\Pi = (\varphi_{B,\eta}^{\mathrm{mSGD}}, \sigma)$ with various synchronization operators σ. The hyper-parameters of the protocols and the mini-batch SGD have been optimized on an independent dataset. Details on the experiments, including network architectures, can be found in the Appendix A in the supplementary material.

Dynamic Averaging for Training Deep Neural Networks: To evaluate the performance of dynamic averaging in deep learning, we first compare it to periodic averaging for training a convolutional neural network (CNN) on the MNIST classification dataset [20]. We furthermore compare both protocols to a non-synchronizing protocol, denoted **nosync**, and a serial application of the learning algorithm on all data, denoted **serial**.

Figure 2 shows the cumulative error of several setups of dynamic and periodic averaging, as well as the nosync and serial baselines. The experiment confirms that for each setup of the periodic averaging protocol a setup of dynamic averaging can be found that reaches a similar predictive performance with substantially less communication (e.g., a dynamic protocol with $\sigma_{\Delta=0.7}$ reaches a performance comparable to a periodic protocol with $\sigma_{b=1}$ using only half of the communication). The more learners communicate, the lower their cumulative loss, with the serial baseline performing the best.

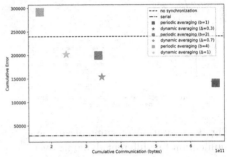

Fig. 2. Cumulative loss and communication of distributed learning protocols with $m = 100$ (similar to McMahan et al. [22]) learners with mini-batch size $B = 10$, each observing $T = 14000$ samples (corresponding to 20 epochs for the serial baseline).

The advantage of the dynamic protocols over the periodic ones in terms of communication is in accordance with the convex case. For large synchronization periods, however, synchronizing protocols ($\sigma_{b=4}$) have even larger cumulative loss than the nosync baseline. This behavior cannot happen in the convex case, where averaging is always superior to not synchronizing [12]. In contrast, in the non-convex case local models can converge to different local minima. Then their average might have a higher loss value than each one of the local models (as illustrated in Fig. 1(a)).

Comparison of the Dynamic Averaging Protocol with FedAvg: Having shown that dynamic averaging outperforms standard periodic averaging, we proceed by comparing it to a highly communication-efficient variant of periodic averaging, denoted **FedAvg** [22], which poses a state-of-the-art for decentralized deep learning under communication-cost constraints.

Using our terminology, FedAvg is a periodic averaging protocol that uses only a randomly sampled subset of nodes in each communication round. This subsampling leads to a reduction of total communication by a constant factor compared to standard periodic averaging. In order to compare dynamic averaging to FedAvg, we repeat the MNIST classification using CNNs and multiple configurations of dynamic averaging and FedAvg.

Figure 3 shows the evolution of cumulative communication during model training comparing dynamic averaging to the optimal configuration of FedAvg

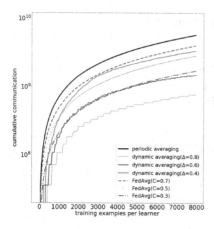
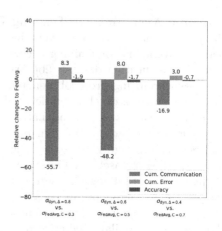

Fig. 3. Evolution of cumulative communication for different dynamic averaging and FedAvg protocols on $m = 30$ learners using a mini-batch size $B = 10$.

Fig. 4. Comparison of the best performing settings of the dynamic averaging protocol with their FedAvg counterparts.

with $b = 5$ and $C = 0.3$ for MNIST (see Sect. 3 in McMahan et al. [22]) and variants of this configuration. We find noteworthy spreads between the communication curves, while all approaches have comparable losses. The communication amounts of all FedAvg variants increase linearly during training. The smaller the fraction of learners, $C \in (0, 1]$, involved in synchronization, the smaller the amount of communication. In contrast, we observe step-wise increasing curves for all dynamic averaging protocols which reflect their inherent irregularity of communication. Dynamic averaging with $\Delta = 0.6$ and $\Delta = 0.8$ beat the strongest FedAvg configuration in terms of cumulative communication, the one with $\Delta = 0.8$ even with a remarkable margin. We find these improvements of communication efficiency to come at almost no cost: Fig. 4 compares the three strongest configurations of dynamic averaging to the best performing FedAvg ones, showing a reduction of over 50% in communication with an increase in cumulative loss by only 8.3%. The difference in terms of classification accuracy is even smaller, dynamic averaging is only worse by 1.9%. Allowing for more communication improves the loss of dynamic averaging to the point where dynamic averaging has virtually the same accuracy as FedAvg with 16.9% less communication.

Adaptivity to Concept Drift: The advantage of dynamic averaging over any periodically communicating protocol lies in the adaptivity to the current hardness of the learning problem, measured by the in-place loss. For fixed target distributions, this loss decreases over time so that the dynamic protocol reduces the amount of communication continuously until it reaches quiescence, if no loss is suffered anymore. In the presence of concept drifts, such quiescence can never be reached; after each drift, the learners have to adapt to the new target. In order to investigate the behavior of dynamic and periodic averaging in this setting, we perform an experiment on a synthetic dataset generated by a random graphical model [4]. Concept drifts are simulated by generating a new random graphical model. Drifts are triggered at random with a probability of 0.001 per round.

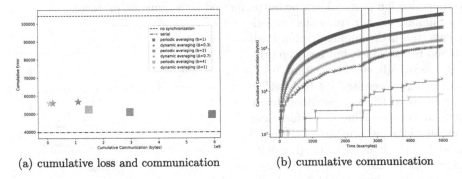

(a) cumulative loss and communication   (b) cumulative communication

Fig. 5. Experiment with periodic and dynamic averaging protocols on $m = 100$ learner after training on 5000 samples per learner from a synthetic dataset with concept drifts (indicated by vertical lines in (b)).

Figure 5(a) shows that in terms of predictive performance, dynamic and periodic averaging perform similarly. At the same time, dynamic averaging requires up to an order of magnitude less communication to achieve it. Examining the cumulative communication over time in Fig. 5(b), one can see that dynamic averaging communicates more after each concept drift and decreases communication until the next drift. This indicates that dynamic averaging invests communication when it is most impactful and can thereby save a substantial amount of communication in between drifts.

Case Study on Deep Driving: After having studied dynamic averaging in contrast to periodic approaches and FedAvg on MNIST and a synthetic dataset with concept drifts, we analyze how the suggested protocol performs in the realistic application scenario of in-fleet training for autonomous driving introduced in Sect. 1. One of the approaches in autonomous driving is direct steering control of a constantly moving car via a neural network that predicts a steering angle given an input from the front view camera. Since one network fully controls the car this approach is termed **deep driving**. Deep driving neural networks can be trained on a dataset generated by recording human driver control and corresponding frontal view [1,9,26].

For our experiments we use a neural network architecture suggested for deep driving by Bojarski et al. [1]. The learners are evaluated by their driving ability following the qualitative evaluation made by Bojarski et al. [1] or Pomerleau [26] as well as techniques used in the automotive industry. For that, we developed a custom loss together with experts for autonomous driving that takes into account the time the vehicle drives on track and the frequency of crossing road sidelines.

Figure 6 shows the measurements of the custom loss against the cumulative communication. The principal difference from the previous experiments is the evaluation of the resulting models without taking into account cumulative training loss. All the resulting models as well as baseline models were loaded to the simulator and driven with a constant speed. The plot shows that each periodic communication protocol can be outperformed by a dynamic protocol.

Similar to our previous experiments, too little communication leads to bad performance, but for deep driving, very high communication ($\sigma_{b=1}$ and $\sigma_{\Delta=0.01}$) results in a bad performance as well. On the other hand, proper setups achieve performance similar to the performance of the serial model (e.g. dynamic averaging with $\Delta = 0.1$ or $\Delta = 0.3$). This raises the question, how much diversity is beneficial in-between averaging steps and how diverse models should be initialized. We discuss this question and other properties of dynamic averaging in the following section.

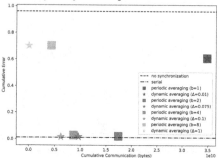

Fig. 6. Performance in the terms of the custom loss for the models trained according to a set of communication protocols and baseline models.

6   Discussion

A popular class of parallel learning algorithms is based on stochastic gradient descent, both in convex and non-convex learning tasks. As for all gradient-based algorithms, the gradient computation can be parallelized 'embarrassingly' [24] easily. For convex problems, the best so far known algorithm, in terms of predictive performance, in this class [28] is the distributed mini-batch algorithm [7]. For the non-convex problem of training (deep) neural networks, McMahan et al. [22] have shown that periodic averaging performs similar to the mini-batch algorithm. Section 4 substantiates these results from a theoretical perspective. Sub-sampling learners in each synchronization allows to further reduce communication at the cost of a moderate loss in predictive performance.

Note that averaging models, similar to distributed mini-batch training, requires a common architecture for all local models since the goal is to jointly train a single global model distributedly using observations from local data streams—which also sets it apart from ensemble methods.

For the convex case, Kamp et al. [15] have shown that dynamic averaging retains the performance of periodic averaging and certain serial learning algorithms (including SGD) with substantially less communication. Section 4 proves that these results are applicable to the non-convex case as well. Section 5 indicates that these results also hold in practice and that dynamic averaging indeed outperforms periodic averaging, both with and without sub-sampling of learners.

This advantage is even amplified in the presence of concept drifts. Additionally, dynamic averaging is a black-box approach, i.e., it can be applied with arbitrary learning algorithms (see Appendix A.5 in the supplementary material for a comparison of using dynamic averaging with SGD, ADAM, and RMSprop).

However, averaging models instead of gradients has the disadvantage of being susceptible to outliers. That is, without a bound on the quality of local models, their average can be arbitrarily bad [13,28]. More robust approaches are computationally expensive, though, e.g., the geometric median [8]. Others are not directly applicable to non-convex problems, e.g., the Radon point [13]. Thus, it remains an open question whether robust methods can be applied to decentralized deep learning.

Another open question is the choice of the divergence threshold Δ for dynamic averaging. The model divergence depends on the expected update steps (e.g., in the case of SGD on the expected norm of gradients and the learning rate), but the threshold is not intuitive to set. A good practice is to optimize the parameter for the desired trade-off between predictive performance and communication on a small subset of the data. It is an interesting question whether the parameter can also be adapted during the learning process in a theoretically sound way.

In dynamic averaging, the amount

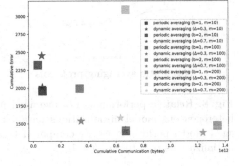

Fig. 7. Cumulative loss and cumulative communication of learning protocols for a different amount of learners. Training is performed on MNIST for 2, 20 and 40 epochs for $m = 10$, $m = 100$, $m = 200$ setups correspondingly.

of communication not only depends on the actual divergence of models, but also on the probability of local violations. Since the local conditions can be violated without the actual divergence crossing the threshold, these false alarms lead to unnecessary communication. The more learners in the system, the higher the probability of such false alarms. In the worst case, though, dynamic averaging communicates as much as periodic averaging. Thus, it scales at least as well as current decentralized learning approaches [11,22]. Moreover, using a resolution strategy that tries to balance violations by communicating with just a small number of learners partially compensates for this problem. Indeed, experiments on the scalability of the approach show that dynamic averaging scales well with the number of learners (see Fig. 7 and Appendix A.6 in the supplementary material for details).

A general question when using averaging is how local models should be initialized. McMahan et al. [22] suggest using the same initialization for all local models and report that different initializations deteriorate the learning process when models are averaged only once at the end. Studying the transition from homogeneously initialized and converging model configurations to heterogeneously initialized and failing ones reveals that, surprisingly, for multiple rounds of aver-

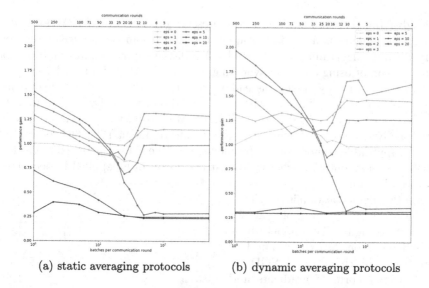

(a) static averaging protocols   (b) dynamic averaging protocols

Fig. 8. Relative performances of averaged models on MNIST obtained from various heterogeneous model initializations parameterized by ϵ and various $b \in \mathbb{N}$. All averaged model performances are compared to an experiment with homogeneous model initializations ($\epsilon = 0$) and $b = 1$.

aging different initializations can indeed be beneficial. Figure 8 shows the performances of dynamic and periodic averaging for different numbers of rounds of averaging and different levels of inhomogeneity in the initializations. The results confirm that for one round of averaging, strongly inhomogeneous initializations deteriorate the learning process, but for more frequent rounds of averaging mild inhomogeneity actually improves training. For large heterogeneities, however, model averaging fails as expected. This raises an interesting question about the regularizing effects of averaging and its potential advantages over serial learning in case of non-convex objectives.

7 Conclusion

In decentralized deep learning there is a natural trade-off between learning performance and communication. Averaging models periodically allows to achieve a high predictive performance with less communication compared to sharing data. The proposed dynamic averaging protocol achieves similarly high predictive performance yet requires substantially less communication. At the same time, it is adaptive to concept drifts. The method is theoretically sound, i.e., it retains the loss bounds of the underlying learning algorithm using an amount of communication that is bound by the hardness of the learning problem.

Acknowledgements. This research has been supported by the Center of Competence Machine Learning Rhein-Ruhr (ML2R).

References

1. Bojarski, M., et al.: End to end learning for self-driving cars. CoRR abs/1604.07316 (2016)
2. Boley, M., Kamp, M., Keren, D., Schuster, A., Sharfman, I.: Communication-efficient distributed online prediction using dynamic model synchronizations. In: BD3@ VLDB, pp. 13–18 (2013)
3. Bottou, L.: Stochastic gradient learning in neural networks. Proc. Neuro-Nımes **91**(8), 12 (1991)
4. Bshouty, N.H., Long, P.M.: Linear classifiers are nearly optimal when hidden variables have diverse effects. Mach. Learn. **86**(2), 209–231 (2012)
5. Chen, J., Monga, R., Bengio, S., Jozefowicz, R.: Revisiting distributed synchronous SGD. In: International Conference on Learning Representations Workshop Track (2016)
6. Dean, J., et al.: Large scale distributed deep networks. In: Advances in Neural Information Processing Systems, pp. 1223–1231 (2012)
7. Dekel, O., Gilad-Bachrach, R., Shamir, O., Xiao, L.: Optimal distributed online prediction using mini-batches. J. Mach. Learn. Res. **13**, 165–202 (2012)
8. Feng, J., Xu, H., Mannor, S.: Outlier robust online learning. CoRR abs/1701.00251 (2017)
9. Fernando, T., Denman, S., Sridharan, S., Fookes, C.: Going deeper: autonomous steering with neural memory networks. In: Proceedings of the IEEE Conference on Computer Vision and Pattern Recognition, pp. 214–221 (2017)
10. Gabel, M., Keren, D., Schuster, A.: Communication-efficient distributed variance monitoring and outlier detection for multivariate time series. In: Proceedings of the 28th International Parallel and Distributed Processing Symposium, pp. 37–47. IEEE (2014)
11. Jiang, Z., Balu, A., Hegde, C., Sarkar, S.: Collaborative deep learning in fixed topology networks. In: Advances in Neural Information Processing Systems, pp. 5904–5914 (2017)
12. Kamp, M., Boley, M., Keren, D., Schuster, A., Sharfman, I.: Communication-efficient distributed online prediction by dynamic model synchronization. In: Calders, T., Esposito, F., Hüllermeier, E., Meo, R. (eds.) ECML PKDD 2014. LNCS (LNAI), vol. 8724, pp. 623–639. Springer, Heidelberg (2014). https://doi.org/10.1007/978-3-662-44848-9_40
13. Kamp, M., Boley, M., Missura, O., Gärtner, T.: Effective parallelisation for machine learning. In: Advances in Neural Information Processing Systems, pp. 6480–6491 (2017)
14. Kamp, M., Boley, M., Mock, M., Keren, D., Schuster, A., Sharfman, I.: Adaptive communication bounds for distributed online learning. In: 7th NIPS Workshop on Optimization for Machine Learning (2014)
15. Kamp, M., Bothe, S., Boley, M., Mock, M.: Communication-efficient distributed online learning with kernels. In: Frasconi, P., Landwehr, N., Manco, G., Vreeken, J. (eds.) ECML PKDD 2016. LNCS (LNAI), vol. 9852, pp. 805–819. Springer, Cham (2016). https://doi.org/10.1007/978-3-319-46227-1_50

16. Keren, D., Sagy, G., Abboud, A., Ben-David, D., Schuster, A., Sharfman, I., Deligiannakis, A.: Geometric monitoring of heterogeneous streams. IEEE Trans. Knowl. Data Eng. **26**(8), 1890–1903 (2014)
17. Keren, D., Sharfman, I., Schuster, A., Livne, A.: Shape sensitive geometric monitoring. IEEE Trans. Knowl. Data Eng. **24**(8), 1520–1535 (2012)
18. Keskar, N.S., Mudigere, D., Nocedal, J., Smelyanskiy, M., Tang, P.T.P.: On large-batch training for deep learning: generalization gap and sharp minima. In: International Conference on Learning Representations (2017)
19. Lazerson, A., Sharfman, I., Keren, D., Schuster, A., Garofalakis, M., Samoladas, V.: Monitoring distributed streams using convex decompositions. Proc. VLDB Endow. **8**(5), 545–556 (2015)
20. LeCun, Y.: The mnist database of handwritten digits (1998). http://yann.lecun.com/exdb/mnist/
21. Mcdonald, R., Mohri, M., Silberman, N., Walker, D., Mann, G.S.: Efficient large-scale distributed training of conditional maximum entropy models. In: Advances in Neural Information Processing Systems, pp. 1231–1239 (2009)
22. McMahan, B., Moore, E., Ramage, D., Hampson, S., y Arcas, B.A.: Communication-efficient learning of deep networks from decentralized data. In: Artificial Intelligence and Statistics, pp. 1273–1282 (2017)
23. McMahan, B., Ramage, D., Talwar, K., Zhang, L.: Learning differentially private recurrent language models. In: International Conference on Learning Representations (2018)
24. Moler, C.: Matrix computation on distributed memory multiprocessors. Hypercube Multiprocessors **86**(181–195), 31 (1986)
25. Nguyen, Q., Hein, M.: The loss surface of deep and wide neural networks. In: International Conference on Machine Learning, pp. 2603–2612 (2017)
26. Pomerleau, D.A.: Alvinn: an autonomous land vehicle in a neural network. In: Advances in Neural Information Processing Systems, pp. 305–313 (1989)
27. Shamir, O.: Without-replacement sampling for stochastic gradient methods. In: Advances in Neural Information Processing Systems, pp. 46–54 (2016)
28. Shamir, O., Srebro, N., Zhang, T.: Communication-efficient distributed optimization using an approximate newton-type method. In: International Conference on Machine Learning, pp. 1000–1008 (2014)
29. Sharfman, I., Schuster, A., Keren, D.: A geometric approach to monitoring threshold functions over distributed data streams. Trans. Database Syst. **32**(4), 301–312 (2007)
30. Sharfman, I., Schuster, A., Keren, D.: Shape sensitive geometric monitoring. In: Proceedings of the ACM SIGMOD-SIGACT-SIGART Symposium on Principles of Database Systems, pp. 301–310. ACM (2008)
31. Smith, V., Chiang, C.K., Sanjabi, M., Talwalkar, A.S.: Federated multi-task learning. In: Advances in Neural Information Processing Systems, pp. 4424–4434 (2017)
32. Zhang, C., Bengio, S., Hardt, M., Recht, B., Vinyals, O.: Understanding deep learning requires rethinking generalization. In: Proceedings of the International Conference on Learning Representations (2017)
33. Zhang, S., Choromanska, A.E., LeCun, Y.: Deep learning with elastic averaging sgd. In: Advances in Neural Information Processing Systems, pp. 685–693 (2015)

34. Zhang, Y., Wainwright, M.J., Duchi, J.C.: Communication-efficient algorithms for statistical optimization. In: Advances in Neural Information Processing Systems, pp. 1502–1510 (2012)
35. Zinkevich, M., Weimer, M., Smola, A.J., Li, L.: Parallelized stochastic gradient descent. In: Advances in Neural Information Processing Systems, pp. 2595–2603 (2010)

Using Supervised Pretraining to Improve Generalization of Neural Networks on Binary Classification Problems

Alex Yuxuan Peng[1(✉)], Yun Sing Koh[1], Patricia Riddle[1],
and Bernhard Pfahringer[2]

[1] University of Auckland, Auckland, New Zealand
ypen260@aucklanduni.ac.nz, {ykoh,pat}@cs.auckland.ac.nz
[2] University of Waikato, Hamilton, New Zealand
bernhard@cs.waikato.ac.nz

Abstract. Neural networks are known to be very sensitive to the initial weights. There has been a lot of research on initialization that aims to stabilize the training process. However, very little research has studied the relationship between initialization and generalization. We demonstrate that poorly initialized model will lead to lower test accuracy. We propose a supervised pretraining technique that helps improve generalization on binary classification problems. The experimental results on four UCI datasets show that the proposed pretraining leads to higher test accuracy compared to the he_normal initialization when the training set is small. In further experiments on synthetic data, the improvement on test accuracy using the proposed pretraining reaches more than 30% when the data has high dimensionality and noisy features. Code related to this paper is available at: https://github.com/superRookie007/supervised_pretraining.

Keywords: Neural network · Pretraining · Initialization
Generalization

1 Introduction

Neural networks have attracted a lot of attention from both academia and industry due to their success in different machine learning tasks. Recurrent neural networks (RNNs) have been successfully applied to speech recognition and natural language translation [13,17,18]. Convolutional neural networks (CNNs) have been the winning models for many image understanding challenges [10–12]. Fully-connected neural networks (FNNs) were shown to be very effective at identifying exotic particles without features hand-engineered by physicists [9].

Apart from novel network architectures, the success of neural networks in so many applications is a result of many advancements in the basic components such as activation functions and initialization. During the training process deep neural networks are sensitive to the initial weights. A poor choice of initial weights

© Springer Nature Switzerland AG 2019
M. Berlingerio et al. (Eds.): ECML PKDD 2018, LNAI 11051, pp. 410–425, 2019.
https://doi.org/10.1007/978-3-030-10925-7_25

can result in very slow training, "vanishing gradient", "dead neurons" or even numerical problems. A few initialization methods have been proposed to stabilize the training process [1,5,6]. However, they do not aim to improve the generalization of neural networks. It is possible that even though these methods can successfully improve the stability and speed of training, they are not necessarily the best for improving the generalization of neural networks. Initialization of neural networks is a very difficult problem to study due to the complex nature of the cost surface and the optimization dynamics. However, more research needs to be devoted to exploring the connection between the initialization and the generalization of neural networks. We argue that the initialization affects the generalization of neural networks. Intuitively speaking, the initial weights of a neural network can be thought of as the prior of a model. If this intuition is correct, the choice of initial weights will have a big effect on which model we get after training, when there is a lack of labelled training data. Hence, studying the effect of initialization on generalization is very valuable to domains where labelled data is not easy and cheap to collect.

The **main contribution** of this paper is a supervised pretraining that improves the generalization of fully-connected neural networks (FNNs) on binary classification problems. We firstly demonstrate that poorly initialized models lead to poorer generalization. We then propose a supervised pretraining method that improves the generalization when labelled data is limited, by taking advantage of unlabelled data. The proposed method trains a neural network to distinguish real data from shuffled data. The learned weights are reused as initial weights when training on the labelled training data. The intuition is that during the pretraining, the model has to learn joint distributions of the real data in order to identify real data points from shuffled ones. And the learned weights might be a better prior than standard initialization methods that sample values from normal or uniform distributions. The improvement in generalization by using the proposed method is shown to be more obvious when there is a lack of labelled data and the class separation is small. Experimental results also suggest that the proposed pretraining works best when the data has high dimensionality and contains noisy features. We only consider binary classification problems in this work. We measure generalization using test accuracy in all our experiments.

The rest of the paper is organized as follows. In Sect. 2, we review related works. In Sect. 3, we demonstrate that initialization of neural networks affects generalization. In Sect. 4, we propose a supervised pretraining method to improve the generalization of FNNs on binary classification problems. We conducted experiments on both UCI datasets and synthetic datasets to evaluate the proposed method in Sect. 5. In Sect. 6, we discuss some issues with the proposed pretraining method. Finally, we conclude our work and propose future works in Sect. 7.

2   Related Works

Glorot and Bengio [6] derived a way to initialize weights depending on the number of input units and output units of a layer. They derived this method by assuming only linear activation functions are used, and attempting to initialize the weights in such a way that the variances of activation values and gradient across all layers are the same at the beginning of training. Despite the unrealistic assumption of a linear activation function, this initialization works well in practice. Using the same idea, He et al. [1] derived an initialization method specifically for the ReLU activation function depending on the number of input units of a layer. It draws weight values from a normal distribution with mean value of 0 and the standard deviation $sqrt(2/fan\_in)$, where fan_in is the number of input units of a layer. We call this initialization method he_normal. They showed that he_normal performed well even for really deep networks using ReLU while the Glorot and Bengio method [6] failed.

Saxe et al. [4] recommended initializing weights to random orthogonal matrices. However one needs to carefully choose the gain factor when a nonlinear activation function is used. Mishkin and Matas [5] proposed a method called layer-sequential unit-variance (LSUV) initialization. LSUV firstly initializes the weights with orthonormal matrices, and then normalizes the activation values of each layer to have variance of 1 through a few forward passes.

Unsupervised methods such as restricted Boltzmann machines and autoencoders were used as layer-wise pretraining to initialize deep neural networks [14–16]. This was done to solve the "vanishing gradient" problem in training deep neural networks, before piece-wise linear activation functions were widely adopted. However, it was later discovered that unsupervised pretraining actually improves generalization when labelled training data is limited [8]. When labelled data for a task is limited, supervised pretraining on a different but related task was found to improve the generalization on the original task [3].

3   Demonstration that Initialization Affects Generalization

Most machine learning problems can be reduced to search problems, especially when an iterative learning algorithm is used to find the final model. A search problem usually consists of a search space, an initial state, a target state and a search algorithm. Figure 1 describes training neural networks as a search process. The area inside the ellipse represents the representation capacity of a neural network. The larger the neural network, the bigger the representation capacity and the larger the search space. The square represents the initial state of the model and the triangle is the final state of the model. The circular dots are the model states the learning algorithm has visited. The model search space can be discrete or continuous. But in neural networks, the search space is usually continuous. And the search space or representation capacity of a neural network is not completely independent of the search process. The bigger and more complex a neural network is, the more difficult it is to train. So the representation capacity can potentially affect where the the model will end up while holding the initial state and learning algorithm constant.

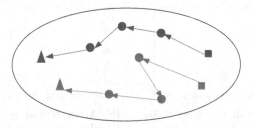

Fig. 1. Machine learning represented as a search process. The area inside the elliptical boundary is the representation capacity of the model chosen. The circular dots are different model (parameter) states a learning algorithm has visited. The square is the initial state of the model while the triangle is the final state of the model.

Unlike many search problems where the target state is known, we usually do not know the target state in machine learning. And in the case of neural networks, we cannot analytically find the optimal model state due to the non-convex loss function. So iterative optimization algorithms are used in training neural networks. The most common learning algorithms in training neural networks are stochastic gradient descent and its variants. The first order derivatives of the model parameters are used as a heuristic for determining the next step. Obviously, different learning algorithms can result in different final models even if both the search space and initial state are kept the same.

Another factor that affects the final state of the search space is the initial state of the model. Different initial weights can lead a neural network to converging to different states, as illustrated in Fig. 1. We argue that initialization affects the generalization of neural networks, especially when the training data is limited. The initial weights of a neural network can be seen as the prior of the model. The choice of the initial weights does not affect the generalization much if there is plenty of training data, as long as the initial weights do not lead to training problems. However, when there is a lack of training data, the prior will have a larger effect on the final state of the model. Figure 2 shows the plots of four fully connected networks trained to fit the same two data points but with different initial weights and activation functions. The two data points were (0.1, 0.1) and (0.5, 0.5). All four of the models had 128 units in each of the first two layers and one output unit. The initial weights were drawn from truncated normal distribution with mean of 0 and different standard deviation values (0.004 and 0.4). Any value that was more than two standard deviations away from the mean was discarded. All biases were set to 0. We used both rectified linear unit (ReLU) and hyperbolic tangent (tanh) as activation functions. We trained the model until the training loss became 0 and plotted the model. It can be seen that when the standard deviation of the initial weights is small, the learned function is relatively simple. When the standard deviation of the initial weights is big, the final model is much more "complex" and less likely to generalize well on unseen data.

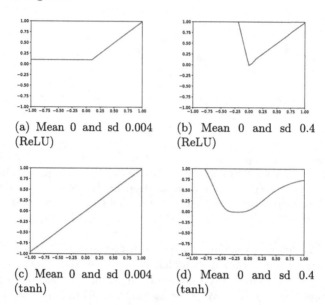

(a) Mean 0 and sd 0.004 (ReLU)

(b) Mean 0 and sd 0.4 (ReLU)

(c) Mean 0 and sd 0.004 (tanh)

(d) Mean 0 and sd 0.4 (tanh)

Fig. 2. Plots of a three-layer neural network trained on two data points (0.1, 0.1) and (0.5, 0.5). The activation functions used are ReLU and tanh. The initial weights were drawn from normal distributions with different mean and standard deviation values. When the initial weights have a small standard deviation, the learned model is more linear ("simpler"). The learned model becomes more complex and less likely to fit unseen data well, when the initial weights have a higher standard deviation.

Similar experiments cannot be applied to real datasets and larger networks, because initial weights drawn from a normal distribution with a large standard deviation leads to various training problems such as "dead neurons" and exploded activation values. Inspired by transfer learning [3], we decided to learn inappropriate initial weights for the original classification task by pretraining a model on randomly shuffled data. More specifically, we shuffled the attribute values of the original data and randomly attached labels to the shuffled data according to the uniform distribution. We obtained a set of weights by training a model on this shuffled data. We expected this set of weights to be a very poor prior for the original classification problem and thus would lead to lower test accuracy. When training a model on the shuffled data, we used he_normal as the initialization method. These learned weights were then used as initial weights when training on the original data. We then compared the model initialized with these transferred weights against the model initialized using he_normal [1].

We carried out the experiments on four different datasets: BanknoteID, MAGIC, Waveform and Waveform_noise. Details of these datasets can be found in Sect. 5.1. We used a five-layer fully connected network for all the experiments. Each layer has 1024 units except the last layer. The last layer is a softmax layer. The activation function used was ReLU. We used the full-batch gradient descent with a constant learning rate of 0.01. We chose the number of epochs to make sure both models had converged. The number of epochs we used for each dataset can be found in the second column of Table 3. The training on the shuffled data was run 10000 epochs for all the cases. We ran each experiment 10 times. Figure 3 shows the distribution of test accuracies obtained using the transferred initial weights and he_normal on four datasets. As can be seen, in all four cases, the model initialized with the transferred weights performed worse.

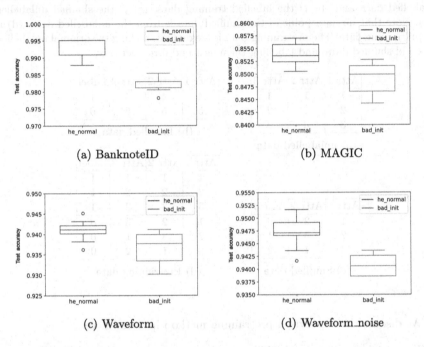

(a) BanknoteID   (b) MAGIC

(c) Waveform   (d) Waveform_noise

Fig. 3. Comparison of the model initialized with transferred weights (red) against the model initialized with he_normal (blue) in terms of test accuracy. We ran each case 10 times. (Color figure online)

In this section, we demonstrated that a poorly initialized model does indeed lead to lower test accuracy. In the next section, we propose a supervised pretraining method that improves generalization, especially when labelled training data is limited.

4   Supervised Pretraining

We propose a supervised pretraining method that takes advantage of unlabelled data to improve the generalization of the original classification problem. The underlying assumption of this method is that the weights learned during the supervised pretraining are a good prior for the original classification problem. This can happen if the pretrained model contains information about the joint distributions of the attributes in the original data.

The pretraining phase is basically a binary classification that attempts to identify real data from shuffled data. Shuffled data does not provide us with real patterns but instead are considered as noise. The learned weights are then reused in the original classification problem.

Table 1. An example of how the pretraining data is generated. (a) is the original unlabelled data and (b) is the labelled training data. (c) is the shuffled unlabelled data. Note that no data point in the shuffled data is from the unlabelled data. (d) is the pretraining data. The pretraining data is created by stacking the original unlabelled data and shuffled data, and labelling them as 1 and 0 respectively.

Attr 1	Attr 2	Attr 3
1	1	1
2	2	2
3	3	3

(a) unlabelled data

Attr 1	Attr 2	Attr 3	Label
4	4	4	1
5	5	5	0

(b) Labelled data

Attr 1	Attr 2	Attr 3
1	2	3
2	3	1
3	1	2

(c) Shuffled data

Attr 1	Attr 2	Attr 3	Label
1	1	1	1
2	2	2	1
3	3	3	1
1	2	3	0
2	3	1	0
3	1	2	0

(d) Pretraining data

We discuss the supervised pretraining method in detail here.

1. We start by randomly shuffling the attribute values of the unlabelled data across rows. Each attribute is shuffled independently from other attributes. This is to break the joint distributions of the original data, but to keep the distributions of individual attributes unchanged. Table 1(a) shows an example of unlabelled data and Table 1(c) shows an example of shuffled data. Note that the shuffling is random, so there can be many different examples of shuffled data generated from the same unlabelled data. In our implementation, we ensured that there was no original real data points in the shuffled data due to chance collision. Any data points in the shuffled data that could be found in the original data were replaced by new shuffled data points. This was done to avoid confusing the model during the pretraining process. Note that labelled training data can be added to the unlabelled data after taking out the labels, in order to create a larger set of unlabelled data. A shuffled dataset can then be created from this enlarged unlabelled dataset.
2. The unlabelled data and shuffled data are stacked together. Then all the real data points are assigned one label and all the shuffled data points are assigned a different label. This is the pretraining data used in our method. The specific values being used as labels do not matter, as long as all the real data points

are assigned the same label and all the shuffled data points are assigned a different label. This is because one hot encoding is used to encode the labels in our models. In our implementation, the real data is labelled as 1 and the shuffled data is labelled as 0. The pretraining data always has a balanced class distribution. Table 1(d) shows an example of pretraining data.

3. The pretraining data is then split into a training set and a validation set using stratified sampling. In our implementation, 70% of the pretraining data was used as training data and the rest was used as a validation set.

4. During the pretraining, a neural network model is trained on the pretraining data. The validation set is used to stop the pretraining early. For the pretraining, the model was initialized using he_normal in our implementation.

5. Finally, the weights learned in the pretraining are reused as initial weights when training a model on the labelled training data shown in Table 1(b). This is sometimes called transfer learning, where weights learned in one problem are transferred to another problem [3]. In our implementation, all the layers except the last one were transferred from the pretrained model. The last layer was replaced with weights initialized using he_normal. We did not freeze any layers during training.

Note that the negative instances in the pretraining data are generated by shuffling the attribute values of the original data. So the the distributions of the attributes in the shuffled data are exactly the same as the original data, but the joint distributions of the attributes are different. By training the model to distinguish real data from the shuffled data, we hypothesize that the model has to learn the joint distributions between attributes. We also postulate that if the weights learned during pretraining are a better prior than commonly used initialization methods such as he_normal, then they should lead to better generalization.

5 Experiments

In this section, we present the proposed supervised pretraining experiments for binary classification. We evaluate the proposed method against he_normal initialization on UCI datasets in Sect. 5.1, and investigate when this method provides an increase in generalization using synthetic data in Sect. 5.2. In our experiments, we simulate a learning environment where labelled training data is limited but there is plenty of unlabelled data. The source code and data can be found on GitHub: github.com/superRookie007/supervised_pretraining.

5.1 Evaluation on UCI Datasets

We conducted experiments on four UCI datasets to evaluate the proposed supervised pretraining against the he_normal [1] initialization method. The goal of the experiments is to test if the proposed supervised pretraining leads to better test accuracy compared to the he_normal initialization. All the datasets were obtained from the UCI repository [2].

Table 2. Data information

Dataset	# Training examples	Class 1	Class 2	# Attributes
BanknoteID	288	55.5%	44.5%	4
BanknoteID_small	10	55.5%	44.5%	4
MAGIC	3995	64.8%	35.2%	10
MAGIC_small	10	64.8%	35.2%	10
Waveform	702	49.3%	50.7%	21
Waveform_small	10	49.3%	50.7%	21
Waveform_noise	695	50.0%	50.0%	40
Waveform_noise_small	10	50.0%	50.0%	40

Datasets. BanknoteID is a dataset for identifying genuine banknotes from the forged banknotes. It has four continuous variables and 1372 data points in total. MAGIC is a simulated dataset for discovering primary gammas from the hadronic showers initiated by cosmic rays. It has 10 continuous variables and 19020 data points in total. Both Waveform and Waveform_noise are datasets for identifying different types of waves. Waveform has 20 attributes while Waveform_noise has 19 additional noisy variables drawn from a standard normal distribution. Both of them have 5000 examples in total. The original datasets of Waveform and Waveform_noise have three classes. We extracted two classes (class 1 and class 2) from the original datasets and treated them as binary classification problems. So the Waveform and Waveform_noise datasets in our experiments have 3343 and 3308 instances in total respectively.

Preprocessing. Thirty percent (30%) of the data was used as the test set. The remaining data was split into labelled training data (21%) and unlabelled data (49%). Additionally, we created a smaller training set with only 10 instances for each of the four datasets by sampling the labelled training data. When sampling the datasets, we used stratified sampling. Table 2 shows some data characteristics of the training sets. When creating the pretraining data for our experiments, we combined the training data (ignoring the labels) and unlabelled data, and then applied the method described earlier in Sect. 4 to this combined data. This gave us an even bigger pretraining dataset. The pretraining data was scaled to a range [0, 1], then the same scaling parameters were applied to the test set.

Setup. We used a five-layer fully connected network for all the experiments. Each layer has 1024 units except the last layer. The last layer is a softmax layer. The activation function used was ReLU. We used the standard full-batch gradient descent with a constant learning rate. Because the learning rate can potentially affect the generalization of neural networks, we used the same constant learning rate of 0.01 for all the experiments. We chose the number of epochs to make sure both models had converged. We consider the model has

Table 3. Comparison of the model initialized using the supervised pretraining method (Pretrained) against the model initialized with he_normal (Base). We ran each case 10 times. The training accuracy reached 100% for all the runs.

Data	Epochs	Method	Train loss	Test loss	Test acc.
BanknoteID	10000	**Base**	$9.78E-04 \pm 2.57E-05$	$5.79E-02 \pm 1.38E-03$	**99.32%** $\pm$ 0.28%
		Pretrained	$1.04E-02 \pm 2.99E-04$	$4.30E-02 \pm 5.53E-03$	98.74% $\pm$ 0.39%
BanknoteID_small	5000	Base	$1.93E-04 \pm 5.40E-06$	$9.02E-01 \pm 3.16E-02$	82.18% $\pm$ 0.80%
		Pretrained	$1.96E-02 \pm 7.88E-04$	$6.13E-01 \pm 2.67E-02$	**82.45%** $\pm$ 0.81%
MAGIC	100000	Base	$4.15E-03 \pm 4.88E-04$	$9.15E-01 \pm 1.71E-02$	85.42% $\pm$ 0.23%
		Pretrained	$8.59E-03 \pm 3.21E-03$	$7.12E-01 \pm 2.54E-02$	**85.69%** $\pm$ 0.27%
MAGIC_small	5000	Base	$2.37E-04 \pm 3.93E-06$	$1.22E+00 \pm 5.87E-02$	78.12% $\pm$ 0.20%
		Pretrained	$2.04E-02 \pm 7.59E-04$	$7.20E-01 \pm 3.54E-02$	**78.62%** $\pm$ 0.51%
Waveform	20000	Base	$5.71E-04 \pm 3.82E-05$	$3.02E-01 \pm 6.99E-03$	94.10% $\pm$ 0.25%
		Pretrained	$9.89E-03 \pm 6.39E-03$	$2.36E-01 \pm 1.24E-02$	**94.38%** $\pm$ 0.25%
Waveform_small	10000	Base	$8.97E-02 \pm 2.83E-01$	$4.06E-01 \pm 3.02E-02$	88.48% $\pm$ 0.58%
		Pretrained	$1.02E-02 \pm 2.20E-04$	$4.13E-01 \pm 2.96E-02$	**89.10%** $\pm$ 0.67%
Waveform_noise	20000	**Base**	$3.13E-04 \pm 1.70E-05$	$3.15E-01 \pm 9.78E-03$	**94.72%** $\pm$ 0.30%
		Pretrained	$1.38E-02 \pm 6.81E-03$	$2.23E-01 \pm 9.31E-03$	94.54% $\pm$ 0.34%
Waveform_noise_small	10000	Base	$4.85E-05 \pm 1.35E-06$	$6.33E-01 \pm 4.43E-02$	82.38% $\pm$ 0.57%
		Pretrained	$1.01E-02 \pm 4.17E-04$	$5.03E-01 \pm 1.58E-01$	**86.97%** $\pm$ 3.33%

converged if the training loss stops decreasing and the training accuracy reaches 100%. In practice, a validation set is usually used to terminate the training early to avoid overfitting. However, we trained the models until convergence because we wanted a fair stopping criterion for both cases, and we wanted to eliminate the generalization effect and uncertainty of early stopping in the results. The number of epochs we used for each training set can be found in the second column of Table 3. Furthermore, the pretraining data was split into a training set (70%) and a validation set (30%). The pretraining stopped if the validation accuracy stopped increasing for the subsequent 100 epochs, and the model with the best validation accuracy was saved. We did not fine tune which layers to reuse or freeze the reused weights for a few epochs before updating them. We always transferred all the weights except the last layer and did not freeze any layer during training. The same experiment was run 10 times on each dataset. The experiments were all implemented using Keras [19] with the Tensorflow [20] backend.

Results. The metric we used to measure generalization was accuracy on the test set. We have evaluated all the models with F1-score, AUC-ROC and Kappa coefficient, but they always moved in line with the test accuracy in this experiment. They did not add interesting information to the results, so they are not reported. The experimental results are shown in Table 3. The higher mean test accuracy is shown in bold, but the difference is not necessarily statistically significant. As expected, the test accuracy is generally higher when the training set is larger. And the standard deviation of the test accuracy tends to be larger when limited training data is available. The base models tend to have much lower training loss while the pretrained models tend to have lower test loss. When there is not a lack of labelled training data, there is no clear improvement on generalization using the supervised pretraining. On BanknoteID and Waveform_noise, the pretraining actually hurt the generalization. However, when the training set is very small, only 10 instances in this case, the mean test accuracy for BanknoteID and Waveform_noise of the pretrained model became higher than that of the base model. It is interesting that the improvement in generalization by using the pretraining was big on Waveform_noise_small, but this was not the case on Waveform_small. Recall the difference between Waveform_noise and Waveform is that Waveform_noise has 19 additional noisy features drawn from the standard normal distribution. It is not very clear if the difference in the results was due to the increased dimensionality or the characteristics of the additional noisy features. We conducted further experiments on synthetic data in order to investigate when the supervised pretraining helps improve generalization. These experiments are presented in the next section.

5.2 When to Use the Supervised Pretraining

The experiments discussed here investigate the circumstances, where the proposed pretraining holds an advantage over the he_normal initialization in terms of test accuracy. More specifically, we test the effect of data dimensionality, different types of noisy features, size of labelled training data and distance between class clusters on the performance of the proposed pretraining. All the experiments were conducted on synthetic datasets generated using the *make_classification* API in *sklearn* library. This is an adapted implementation of the algorithm used to generate the "Madelon" dataset [7].

Firstly, we test how data dimensionality, noisy features and redundant features affect the effectiveness of the supervised pretraining compared to he_normal. We generated 6 different datasets. All of the datasets have only two classes and have balanced class distribution. Each of the raw datasets had 10000 instances. The *class_sep* parameter was set to 0.01 for all datasets (the smaller the *class_sep*, the more difficult the classification). We applied the same preprocessing as the previous experiment. But in this experiment, we sampled a training set with 50 instances for each dataset. All the other experiment setups were exactly the same as described in Sect. 5.1. The characteristics of the 6 datasets are described below.

- Informative_10 has 10 attributes and all of the attributes are informative for the classification task.
- Informative_20 has 20 attributes and all of the them are informative.
- Redundant_20 has 20 attributes, but 10 of them are the exact copy of the other 10 informative attributes.
- In Std_normal_20, 10 of the 20 attributes are random values drawn from the standard normal distribution (mean 0, standard deviation 1), while the other 10 are informative.

- In Normal_20, 10 of the attributes are drawn from normal distributions whose means and standard deviations are kept the same as the other 10 informative attributes.
- In Shuffled_20, instead of drawn randomly from normal distributions, the 10 noisy attributes are simply shuffled informative attributes (the distribution of each of the noisy attributes is the same as the corresponding informative attribute).

Table 4. Investigating the effect of dimensionality, noisy and redundant features on the performance of the supervised pretraining. We ran each case 10 times and all of the models were trained for 10000 epochs.

Data	Method	Train loss	Train acc.	Test loss	Test acc.
Informative_10	Base	6.50E−04 ± 2.46E−05	100.00%	5.79E−01 ± 6.89E−02	84.48% ± 1.37%
	Pretrained	1.02E−02 ± 1.41E−04	100.00%	1.43E−01 ± 2.69E−02	**96.24%** ± 0.77%
Informative_20	Base	3.56E−04 ± 1.76E−05	100.00%	1.82E+00 ± 7.84E−02	63.61% ± 0.72%
	Pretrained	1.03E−02 ± 2.87E−04	100.00%	4.71E−01 ± 1.02E−01	**87.19%** ± 2.64%
Redundant_20	Base	6.07E−04 ± 2.50E−05	100.00%	5.86E−01 ± 4.82E−02	84.41% ± 0.93%
	Pretrained[a]	1.10E−02 ± 7.33E−04	100.00%	4.64E−01 ± 4.32E−02	**85.48%** ± 1.35%
Std_normal_20	Base	3.87E−04 ± 2.01E−05	100.00%	2.40E+00 ± 7.50E−02	59.94% ± 0.64%
	Pretrained	1.02E−02 ± 2.83E−04	100.00%	2.13E−01 ± 3.53E−02	**94.18%** ± 1.09%
Normal_20	Base	4.12E−04 ± 1.43E−05	100.00%	1.24E+00 ± 5.77E−02	65.40% ± 0.92%
	Pretrained	1.02E−02 ± 1.60E−04	100.00%	1.84E−01 ± 2.92E−02	**94.93%** ± 0.87%
Shuffled_20	Base	3.42E−04 ± 1.35E−05	100.00%	2.42E+00 ± 8.56E−02	57.95% ± 0.39%
	Pretrained	1.02E−02 ± 1.15E−04	100.00%	1.74E−01 ± 2.34E−02	**95.19%** ± 0.55%

[a]When running the pretrained model of Redundant_20, 4 out of 10 runs failed to converge, the results shown here were collected from the 6 converged runs.

The results are shown in Table 4. The base models for Informative_10 and Redundant_20 performed very similarly. This is not too surprising considering Redundant_20 basically concatenated two copies of the features in Informative_10 together. Compared to Informative_10, Redundant_20 neither has any additional information nor lacks any useful information. The improvement on generalization using the supervised pretraining was around 11% on Informative_10. But for Redundant_20, the pretrained model failed to converge in 4 out of 10 runs. And there is no obvious improvement on test accuracy when pretraining is used. The generalization gain on Informative_20 was more than 20%, while the gain on the Std_normal_20 was around 34%. The biggest gain (around 37%) occurred in Shuffled_20 dataset. The results suggest that both the increased dimensionality and the noisiness of the additional features both contribute to the generalization improvement we observed on Waveform_noise_small in the previous experiment.

Next, we investigated how the size of labelled training data and the class separation would affect the generalization advantage of the proposed pretraining over the he_normal initialization. The experimental setup was exactly the same as the previous experiments, except the datasets used. When testing the effect of the size of labelled training data, we held all other factors constant including the class separation ($class\_sep = 0.01$) and unlabelled data. The data had 10 attributes

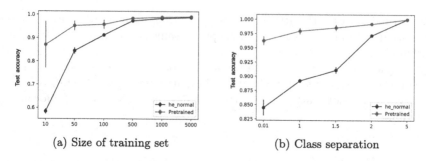

(a) Size of training set   (b) Class separation

Fig. 4. Comparison of the model trained with supervised pretraining (red) against the model initialized with he_normal (blue) in terms of test accuracy. The plots show the distributions of test accuracy. Each experiment was run 10 times. (Color figure online)

and all of them were informative. We tested six labelled training sets with 10, 50, 100, 500, 1000 and 5000 instances respectively. We ran each experiment 10 times. The results can be found in Fig. 4(a). The results show that as the size of labelled training data increases, the advantage of the pretraining gets smaller. And as expected, as the training set gets larger, the standard deviation of test accuracy shrinks. Finally, we tested the effect of class separation. Class separation in this case means the distance between classes. The smaller the class separation, the harder the classification. We created three datasets with three different levels of class separation by setting the *class_sep* parameter to 0.01, 1, 1.5, 2 and 5 respectively. All of the datasets had only 10 informative attributes. The labelled training sets had 50 instances for these experiments. Again, all the other experiment setups were the same as described earlier. Figure 4(b) shows the results of these experiments. The pretraining had the biggest gain in test accuracy over he_normal when the class separation is 0.01. However, as the class separation gets larger the advantage becomes smaller. Both of these experiments support the hypothesis that what we are learning during pretraining is a good prior for the model. The advantage of a good prior is larger when the data size is small and when the classification problem is difficult.

We proposed a new supervised pretraining method to help improve generalization for binary classification problems. The experiments on both UCI and synthetic datasets suggest that the proposed pretraining method provides more generalization gain when there is a lack of labelled training data. Further experiments on synthetic data suggest that this method works best when the input has high dimensionality and contains noisy attributes. Finally, the proposed method has bigger advantage over he_normal in terms of test accuracy when the class separation is small.

6   Discussion

By inspecting the results closely, we noticed that the standard deviation of the test accuracy for the pretrained model was bigger than the base model without pretraining for a few cases. This is contradictory to the findings in unsupervised pretraining. Unsupervised pretraining was shown to help lower the variance of test accuracy [8]. And the results in Table 3 are not statistically significant. One possible explanation for this might be the weights learned during the pretraining across different runs were very different. Recall that we used a pretraining validation set to stop the pretraining. Another way to do this is to run the pretraining using different epochs and choose the epoch that gives the best validation accuracy on the original classification problem. Then we can run the experiments multiple times using this fixed number of epochs. This may be a more stable method. Another possible explanation for the large standard deviation is the small size of the training data. The standard deviation of the test accuracy tends to be larger when training data is small. The difficulty of the classification problem may also affect the statistical significance of the results. As the experiments on synthetic data indicate, the advantage of the proposed pretraining method disappears when the clusters are far apart from each other. The complexity of the target decision boundary that separates different classes may also affect the effectiveness of the pretraining method. Lastly, we also expect label noise in the data to reduce the advantage provided by the pretraining. The exact reasons why the proposed method works better on certain datasets need to be explored further.

The pretraining phase means more training time. However, the proposed supervised pretraining only adds a small amount of additional training time. For instance, the pretraining phase for the MAGIC dataset took on average 241 s, about 4.67% of the average total training time of 5061 s. Note that the exact training time in practice depends on the dataset, tuning of training parameters and the number of epochs used. We just want to show that despite of the additional pretraining phase the proposed method is still practical.

Note that in all of our experiments, we did not fine tune the pretraining or the weight transferring process. One can possibly improve the results further by carefully and tediously choosing when to stop the pretraining, which layers to reuse and whether to freeze some layers for a certain number of epochs. Our goal here is mainly to show that the proposed method can help improve generalization when only limited training data is available, instead of achieving state-of-the-art result on a particular dataset. We used the same experimental setup across all the experiments (except number of epochs) to not bias any particular setting.

7   Conclusion and Future Works

Current default initialization methods such as he_normal are designed to stabilize the training, especially for deep neural networks. However, they are not necessarily optimal for generalization. When labelled data is limited, the choice of initial weights becomes very important in deciding what final model we will end up with. We demonstrated that inappropriate initial weights of neural networks do indeed lead to lower test accuracy. We then proposed a supervised pretraining method to improve the generalization in binary classification problems. During the pretraining, a model is learned to identify real data from shuffled data. Then the learned weights are reused in the original problem. Based on the experimental results on four UCI datasets and synthetic datasets, the supervised pretraining leads to better test accuracy than he_normal initialization, when there is a lack of labelled training data and the class separation is small. The experiments using synthetic datasets showed that this supervised pretraining works best on datasets with higher dimensionality and noisy features.

A very important future work is to understand why this supervised pretraining works in some of the cases and why it fails in other situations. We only focused on binary classification problems in this work, we would like to evaluate how well the proposed method works on multi-class problems. Lastly, it would be interesting to explore how to apply the same idea to image and text data.

References

1. He, K., Zhang, X., Ren, S., Sun, J.: Delving deep into rectifiers: surpassing human-level performance on imagenet classification. In: Proceedings of the IEEE International Conference on Computer Vision, pp. 1026–1034 (2015)
2. Dheeru, D., Karra Taniskidou, E.: UCI Machine Learning Repository. University of California, School of Information and Computer Science, Irvine, CA (2017). http://archive.ics.uci.edu/ml
3. Yosinski, J., Clune, J., Bengio, Y., Lipson, H.: How transferable are features in deep neural networks? In: Advances in Neural Information Processing Systems, pp. 3320–3328 (2014)
4. Saxe, A.M., McClelland, J.L., Ganguli, S.: Exact solutions to the nonlinear dynamics of learning in deep linear neural networks (2013). arXiv preprint: arXiv:1312.6120
5. Mishkin, D., Matas, J.: All you need is a good init. In: Proceedings of the International Conference on Learning Representations (2016)
6. Glorot, X., Bengio, Y.: Understanding the difficulty of training deep feedforward neural networks. In: Proceedings of the Thirteenth International Conference on Artificial Intelligence and Statistics, pp. 249–256 (2010)
7. Guyon, I.: Design of experiments of the NIPS 2003 variable selection benchmark (2003)
8. Erhan, D., Bengio, Y., Courville, A., Manzagol, P.A., Vincent, P., Bengio, S.: Why does unsupervised pre-training help deep learning? J. Mach. Learn. Res. 11, 625–660 (2010)
9. Baldi, P., Sadowski, P., Whiteson, D.: Searching for exotic particles in high-energy physics with deep learning. Nat. Commun. 5, 4308 (2014)
10. Krizhevsky, A., Sutskever, I., Hinton, G.E.: Imagenet classification with deep convolutional neural networks. In: Advances in Neural Information Processing Systems, pp. 1097–1105 (2012)
11. Gulshan, V., et al.: Development and validation of a deep learning algorithm for detection of diabetic retinopathy in retinal fundus photographs. JAMA 316(22), 2402–2410 (2016)
12. He, K., Zhang, X., Ren, S., Sun, J.: Deep residual learning for image recognition. In: Proceedings of the IEEE Conference on Computer Vision and Pattern Recognition, pp. 770–778 (2016)

13. Sutskever, I., Vinyals, O., Le, Q.V.: Sequence to sequence learning with neural networks. In: Advances in Neural Information Processing Systems, pp. 3104–3112 (2014)
14. Ranzato, M., Poultney, C., Chopra, S., Cun, Y.L.: Efficient learning of sparse representations with an energy-based model. In: Advances in Neural Information Processing Systems, pp. 1137–1144 (2007)
15. Bengio, Y., Lamblin, P., Popovici, D., Larochelle, H.: Greedy layer-wise training of deep networks. In: Advances in Neural Information Processing Systems, pp. 153–160 (2007)
16. Hinton, G.E., Osindero, S., Teh, Y.W.: A fast learning algorithm for deep belief nets. Neural Comput. **18**(7), 1527–1554 (2006)
17. Graves, A., Mohamed, A.r., Hinton, G.: Speech recognition with deep recurrent neural networks. In: 2013 IEEE International Conference on Acoustics, Speech and Signal Processing, pp. 6645–6649. IEEE (2013)
18. Sak, H., Senior, A., Rao, K., Beaufays, F.: Fast and accurate recurrent neural network acoustic models for speech recognition. In: Sixteenth Annual Conference of the International Speech Communication Association (2015)
19. Chollet, F., et al.: Keras (2015). https://github.com/fchollet/keras
20. Abadi, M., et al.: TensorFlow: Large-scale machine learning on heterogeneous systems (2015). https://www.tensorflow.org/

Towards Efficient Forward Propagation on Resource-Constrained Systems

Günther Schindler[1]([✉]), Matthias Zöhrer[2], Franz Pernkopf[2],
and Holger Fröning[1]

[1] Institute of Computer Engineering, Ruprecht Karls University,
Heidelberg, Germany
{guenther.schindler,holger.froening}@ziti.uni-heidelberg.de
[2] Signal Processing and Speech Communication Laboratory,
Graz University of Technology, Graz, Austria
{matthias.zoehrer,pernkopf}@tugraz.at

Abstract. In this work we present key elements of DeepChip, a framework that bridges recent trends in machine learning with applicable forward propagation on resource-constrained devices. Main objective of this work is to reduce compute and memory requirements by removing redundancy from neural networks. DeepChip features a flexible quantizer to reduce the bit width of activations to 8-bit fixed-point and weights to an asymmetric ternary representation. In combination with novel algorithms and data compression we leverage reduced precision and sparsity for efficient forward propagation on a wide range of processor architectures. We validate our approach on a set of different convolutional neural networks and datasets: ConvNet on SVHN, ResNet-44 on CIFAR10 and AlexNet on ImageNet. Compared to single-precision floating point, memory requirements can be compressed by a factor of 43, 22 and 10 and computations accelerated by a factor of 5.2, 2.8 and 2.0 on a mobile processor without a loss in classification accuracy. DeepChip allows trading accuracy for efficiency, and for instance tolerating about 2% loss in classification accuracy further reduces memory requirements by a factor of 88, 29 and 13, and speeds up computations by a factor of 6.0, 4.3 and 5.0. Code related to this paper is available at: https://github.com/UniHD-CEG/ECML2018.

1 Introduction

Deep Neural Networks (DNNs) are widely used for many applications including object recognition [16], speech recognition [12] and robotics [17]. While DNNs deliver excellent prediction performance on many classification tasks, they come at the cost of extremely high computational complexity and memory requirements. Both are scarce for resource-constrained systems like mobile processors, which due to their ubiquity otherwise would be prime target for classification tasks. However, DNNs are typically over-parameterized, which motivates various recent efforts in academia and industry to reduce this redundancy and deploy them on embedded systems.

© Springer Nature Switzerland AG 2019
M. Berlingerio et al. (Eds.): ECML PKDD 2018, LNAI 11051, pp. 426–442, 2019.
https://doi.org/10.1007/978-3-030-10925-7_26

It seems natural to improve resource efficiency by removing this redundancy of over-parameterized neural networks. There are mainly three approaches to remove redundant information within network topologies: the first main approach is to design novel network topologies (e.g. SqueezeNet [14]), which leverage less parameter to reduce computational complexity and memory requirements. This approach, however, requires expert knowledge in multiple disciplines and is considered extremely time consuming. Second, Hinton et al. [11] propose distilling the knowledge in a neural network which transfers knowledge from an ensemble or from a large highly regularized model into a smaller, distilled model. The third main approach is to use quantization (reduce bit-width of weights and activations) and pruning (reduce the number of connections) on state-of-the-art network topologies (e.g. ResNet [10]).

In this work, we focus on quantization and the implications on forward propagation. Particularly, we present a flexible quantizer that is used in our DeepChip framework (Fig. 1). DeepChip enables resource-efficient DNNs and easy prototyping for resource-constrained systems.

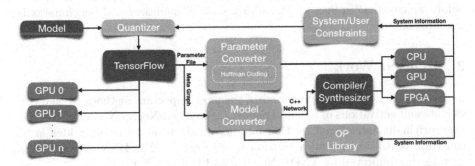

Fig. 1. Overview of the proposed DeepChip framework including design flow and key elements (green: DeepChip, dark grey: third party). (Color figure online)

The proposed quantizer uses system and user constraints to optimize a given input model. The parameter and model converters transform the trained model into a compilable source file. The Operator (OP) library includes highly optimized back-ends for a set of different processors. This design implicates several demands on the used quantization strategy:

- **Preserving accuracy:** the quantization needs to achieve full-precision accuracy while still being more efficient in execution. In other terms, only redundancy should be removed, but not non-redundant parameters.
- **Accuracy trade-offs:** real-time requirements and hardware constraints (e.g. on-chip memory capacity) need to be satisfied (possibly at the cost of accuracy loss).
- **Architectural feasibility:** the implementation of forward propagation under the given constraints (real-time, resources) needs to be feasible on a wide range of processor architectures.

G. Schindler et al.

In this work, we cover the quantizer, the parameter converter and the core of the OP library. The main contributions of this work are as follows:

- A novel combination of existing quantization strategies that suits modern computer architectures and aims to achieve full-precision accuracy.
- A novel sparse matrix-matrix multiplication algorithm that leverages reduced precision and sparsity in the proposed quantization strategy for efficient forward propagation.
- A novel data structure in combination with compression based on Huffman coding.
- A set of prototypes in the form of convolutional neural networks, which show the potential of the proposed approach on modern computer architectures.

The remainder of this work is structured as follows: related work is reviewed in Sect. 2, before we reason about our quantization decision in Sect. 3. Section 4 describes how forward propagation can be implemented on different processor architectures. We report experimental results in Sect. 5, in combination with a detailed comparison to other approaches. We discuss our approach in Sect. 6, before we conclude in Sect. 7. The source code of quantizer and benchmarks is available online[1].

2 Related Work

In this section we briefly review some of the most important methods to quantize weights and activations of a neural network. Binarized Neural Networks (BNNs) were originally introduced by Courbariaux et al. [6] and the basic idea is to constrain both weights and activation to +1 and −1. The BNN approach was then further improved by XNOR-Net [19] and DoReFa-Net [24].

As can be seen in Table 1, XNOR-Net outperforms BNN by a large margin and DoReFa-Net slightly improves upon XNOR-Net. From an architectural perspective, the inference of these methods can be implemented very efficiently as 1-bit scalar products of two vectors $\mathbf{x}$ and $\mathbf{y}$ of length N can be computed by bit-wise *xnor()* operations, followed by counting the number of set bits:

$$\mathbf{x} \cdot \mathbf{y} = N - 2 \cdot bitcount(xnor(\mathbf{x}, \mathbf{y})), x_i, y_i \in [-1, +1] \forall i \qquad (1)$$

Besides lower computational complexity, the memory footprint is reduced as weights require only 1 bit. Using this approach for inference has been proven to be very successful on FPGAs [21], GPUs and ARM CPUs [20]. While the binarization approach only has a negligible accuracy loss to full-precision networks on simple datasets like MNIST or SVHN, binarization on more complex datasets like CIFAR-10 or ImageNet shows severe accuracy degradation.

In order to overcome accuracy degradation, Cai et al. [4] (HWGQ) propose to increase the bit width of activations to 2 bit while still binarizing the weights. This reduces the gap to full-precision networks, for instance there is a 5.8%

---

[1] https://github.com/UniHD-CEG/ECML2018.

Table 1. Accuracy (Top-1, Top-5) comparison of state-of-the-art quantization approaches of AlexNet on ImageNet for different bit-width combinations of Activations (A) and Weights (W).

A-W		DC	BNN	XNOR	DoReFa	TWN	TTQ	HWGQ
32-32	Top-1	57.2	56.6	56.6	57.2	57.2	57.2	58.5
	Top-5	80.3	80.2	80.2	80.3	80.3	80.3	81.5
32-8/5	Top-1	57.2	–	–	–	–	–	–
	Top-5	80.3	–	–	–	–	–	–
32-2	Top-1	–	–	–	–	54.5	57.5	–
	Top-5	–	–	–	–	76.8	79.7	–
32-1	Top-1	–	–	–	53.9	–	–	–
	Top-5	–	–	–	76.3	–	–	–
2-1	Top-1	–	–	–	–	–	–	52.7
	Top-5	–	–	–	–	–	–	76.3
1-1	Top-1	–	27.9	44.2	45.4	–	–	–
	Top-5	–	50.4	69.2	69.3	–	–	–

Top-1 and 5.2% Top-5 accuracy loss for ImageNet. Similar to a 1-bit scalar product, few-bit scalar products can be computed by a bit-serial (or bit-plane) implementation, which uses a bit-wise *and()* operation, followed by population count.

Whereas activations require a relatively high amount of bits to perform as well as full-precision activations, weights seem to require only few bits. For full-precision activations and binary weights, DoReFa has only 3.3% Top-1 and 4.0% Top-5 accuracy loss on ImageNet. For identical activation quantization but ternary weights, Ternary Weight Networks (TWN) [18] achieve 2.7% Top-1 and 3.5% Top-5 accuracy loss. Finally, Trained Ternary Quantization (TTQ) [25] even improves Top-1 accuracy by 0.3% and looses only 0.5% Top-5 accuracy.

Less extreme quantization approaches use 8-bit integer or fixed point representations for activations and weights in order to avoid accuracy loss. The work by Benoit et al. [15] lowers the inference to integer-only computations based on 8-bit matrix-matrix multiplications and is employed as part of TensorFlow [2].

Deep Compression [9] is a compression approach that uses a combination of pruning, trained quantization and Huffman coding. It reduces storage requirements of neural networks significantly without a loss of accuracy. Han et al. [8] showed that this approach can be implemented efficiently on a specialized processor (ASIC). However, the combination of 5-/8-bit weights and sparsity poorly matches general-purpose processors.

3   Quantization

Our quantizer is based on 8-bit activations and 2-bit weights. We quantize weights based on the inspiring concept of TTQ [25] and use fixed-point quantization for the activations. Using this concept, we second the opinion of [4]: while it is possible to quantize weights successfully with only a few bits, activations favor more bits for large-scale classification tasks. Additionally, we show in Sect. 5 that weight and activation quantization do not interfere each other.

3.1   Quantization of Weights

TTQ uses two scaling coefficients W_l^p and W_l^n for positive and negative weights in each layer l. Both coefficients are independent, asymmetric parameters and are trained together with other parameters (further explained in [25]). The quantized weights w_l^i are calculated by thresholding the full-precision weights on the basis of $\pm\Delta_l$ and are set to either W_l^p, 0 or W_l^n:

$$w_l^i = \begin{cases} W_l^p : w_l > \Delta_l \\ 0 : |w_l| \leq \Delta_l \\ -W_l^n : w_l < -\Delta_l \end{cases} \tag{2}$$

The value of $\pm\Delta_l$ is calculated by using the maximum absolute value of the weights and the hyper-parameter t:

$$\Delta_l = t \cdot max(|w|) \tag{3}$$

We control the sparsity within the weight matrix in our experiments by tuning the threshold hyper-parameter $t \in [0, 1]$.

3.2   Quantization of Activations

Fixed-point representation is superior to integer representation when quantizing activations, because activations are probabilities in the interval of $[0, 1]$ and, thus, would need scaling when implemented with integer representation. Contrary, a fixed-point representation can easily express the probability interval by a variable length fractional part. Fixed-point arithmetic, however, is usually not supported by general-purpose processors and can potentially cause a significant overhead when computed with integer hardware.

We use fixed-point representation to quantize activations and reason in Sect. 4 why integer hardware is sufficient for our approach. First, we pass the pre-activation $\tilde{a}_i$ through a bounded activation (bounded ReLU) function to ensure that the range of an activation is $a_i \in [0, 1]$:

$$a_i = \begin{cases} 0 : \tilde{a}_i \leq 0 \\ \tilde{a}_i : 0 < \tilde{a}_i < 1 \\ 1 : \tilde{a}_i \leq 1 \end{cases} \tag{4}$$

Next, we adopt the approach of Zhou et al. [24] and use Eq. 5 to quantize floating-point activations a_i into k-bit fixed-point activations a_i^q:

$$a_i^q = \frac{1}{2^k - 1} round((2^k - 1)a_i)),\qquad(5)$$

where we use 8-bit fixed-point ($k = 8$) for all quantized layers because of two reasons: first, our experiments show that fewer bits cause degradation of classification accuracy on complex datasets. Second, an 8-bit representation is well suited for most computer architectures (further discussions follow in Sect. 4). However, the bit-width of activations is configurable.

4   Architectural Feasibility

The core operation in the forward propagation of neural networks is the scalar product, which takes two vectors **a** and **b** of the same length and returns a scalar value c. It consists of multiplying connected elements of those two vectors and accumulating the product to the output element:

$$c = \sum_{i=1}^{N} a_i \cdot b_i, \quad a_i, b_i \in \mathbb{R} \quad \forall i \qquad(6)$$

In particular, Multiply and Accumulate (MAC) operations are the essence of such a scalar product.

According to Eq. 2, the weight matrix of a layer consists of elements $\{W_l^p, 0, W_l^n\}$. As a consequence, positive and negative weighted values can be treated independently and zero-weighted values can be skipped. As shown in Eq. 7, this results in only two multiplications per scalar product and reduces the major part of the computation to additions.

$$c = W_l^p \cdot \sum_{i \in \mathbf{i}_l^p} a_i + W_l^n \cdot \sum_{i \in \mathbf{i}_l^n} a_i, \quad where \qquad(7)$$

$$\mathbf{i}_l^p = \{i | b_i = W_l^p\} \quad and \quad \mathbf{i}_l^n = \{i | b_i = W_l^n\} \qquad(8)$$

This has several advantages since accumulations are extremely efficient (Sect. 4.1) and the underlying scalar-product algorithm can easily exploit sparsity (Sect. 4.2). Furthermore, the constrained range of the weight matrices should allow for a highly effective compression (Sect. 4.3). We leverage these advantages in the following sections for the example of two-dimensional matrices and matrix-matrix multiplication (which build the core of fully-connected layers). Obviously the same approach applies to convolutional layers if they are lowered into matrix multiplications [5].

4.1   Fixed-Point Arithmetic Using Integer Hardware

Representing real numbers with fractional part on modern computer architectures is a trade-off between precision and efficiency. The generalized fixed-point representation is $[Q_I.Q_F]$, where Q_I corresponds to the integer part and Q_F corresponds to the fractional part of the number. A common way to leverage integer hardware for inference is based on the fact that activations can be represented by only a fractional part (e.g. $Q0.8$) and weights can be represented by only an integer part (e.g. $Q8.0$). Vanhoucke et al. [22] utilize this approach to accelerate inference on CPUs by using 16-bit integer Fused-Multiply-Accumulate (FMA) instructions:

$$(Q8.8)Accu = (Q8.8)Accu + (Q8.8)((Q0.8)I \cdot (Q8.0)W) \tag{9}$$

$$(int16)Accu = (int16)Accu + (int16)((int8)I \cdot (int8)W) \tag{10}$$

Our approach exploits integer hardware in a similar way, with the difference that we can avoid the multiplication, but instead simply add the 8-bit fractional inputs ($Q0.8$) to a 16-bit ($Q8.8$) accumulator:

$$(Q8.8)Accu = (Q8.8)Accu + (Q0.8)I \tag{11}$$

$$(int16)Accu = (int16)Accu + (int8)I \tag{12}$$

As a result, neither floating-point units nor FMA operations are required, improving the viability of such computations on simple processors like FPGAs and DSPs. Furthermore, integer arithmetic (especially the addition) is faster and more energy efficient than floating-point arithmetic (Table 2).

Table 2. Cycles and energy per MAC and addition (ADD) operation.

Instruction	Cycles (normalized) [3]	Energy (pJ) [13]
float32 FMA	8	4.60
int16 FMA	3	1.60
int16 ADD	1.5	0.05

4.2   Algorithm

The heavy lifting of computing the forward propagation is performed by General-Matrix-Vector-Multiplication (GEMV) or General-Matrix-Matrix-Multiplication (GEMM). Here, GEMM is used for batched inference and GEMV for single-input inference. Whereas sparsity can be exploited rather easily for GEMV, leveraging sparsity for GEMM efficiently is difficult because many code optimizations are not applicable.

In this section, we describe our approach of implementing matrix-vector and matrix-matrix multiplication, which build the core of the Operator (OP) library (see Fig. 1). The basic idea is to split positive and negative scaling coefficients into two separate arrays and to remove zeros (as shown in Eq. 7). We transpose the original weight matrix $\mathbf{W}_l$ into $\mathbf{W}_l^T$ in order to increase data locality during runtime. Now we extract all indices i for which the respective element in the weight matrix $\mathbf{W}_l^T$ is either W_l^p or W_l^n and store the indices in $\mathbf{I}_l^p$ or $\mathbf{I}_l^n$ respectively ($\mathbf{I}_l^p = \{i | w_l^i = W_l^p\}$ and $\mathbf{I}_l^n = \{i | w_l^i = W_l^n\}$). Processing a weight matrix $\mathbf{W}_l$ into $\mathbf{I}_l^p$ and $\mathbf{I}_l^n$ is exemplary illustrated in Eqs. 13 and 14.

$$\mathbf{W}_l^T = \begin{pmatrix} 0 & W_l^p & W_l^p & 0 & W_l^n \\ W_l^n & 0 & 0 & W_l^p & 0 \\ W_l^p & W_l^n & 0 & W_l^n & W_l^n \end{pmatrix} \tag{13}$$

$$\mathbf{I}_l^p = \begin{pmatrix} 1\ 2 \\ 3 \\ 0 \end{pmatrix} \quad \text{and} \quad \mathbf{I}_l^n = \begin{pmatrix} 4 \\ 0 \\ 1\ 3\ 5 \end{pmatrix} \tag{14}$$

This transformation is performed after training, by the parameter converter shown in Fig. 1, and does not cause additional processing at inference time. Algorithm 1 implements the sparse matrix-matrix multiplication based on Eq. 7.

Algorithm 1. Sparse matrix-matrix multiplication.

Input: $(int8)Input, (int16)\mathbf{I}_l^p, (float32)\mathbf{W}_l^p, (int16)\mathbf{I}_l^n, (float32)\mathbf{W}_l^n$
Output: $(float32)Output$
1: **for** $row := 0$ to $Input.rows()$ **do**
2: **for** $el := 0$ to $\mathbf{I}_l^p[row].elements()$ **do**
3:    $index \leftarrow \mathbf{I}_l^p[row][el]$
4:    **for** $col := 0$ to $Input.cols()$ **do**
5:      $(int16)\text{Accu}_p[col] \leftarrow (int16)\text{Accu}_p[col] + (int8)Input[row][index]$
6:    **end for**
7: **end for**
8: **for** $el := 0$ to $\mathbf{I}_l^n[row].elements()$ **do**
9:    $index \leftarrow \mathbf{I}_l^n[row][el]$
10:   **for** $col := 0$ to $Input.cols()$ **do**
11:     $(int16)\text{Accu}_n[col] \leftarrow (int16)\text{Accu}_n[col] + (int8)Input[row][index]$
12:   **end for**
13: **end for**
14: **for** $col := 0$ to $Input.cols()$ **do**
15:    $(float32)\text{Res}_p \leftarrow castFixedPoint16ToFloat32((int16)\text{Accu}_p[col])$
16:    $(float32)\text{Res}_n \leftarrow castFixedPoint16ToFloat32((int16)\text{Accu}_n[col])$
17:    $Output[row][col] \leftarrow (float32)\mathbf{W}_l^p \cdot \text{Res}_p + (float32)\mathbf{W}_l^n \cdot \text{Res}_n$
18: **end for**
19: **end for**

As can be seen, the two accumulator arrays $Accu_p$ and $Accu_n$ are used to accumulate the 8-bit fixed-point inputs on the basis of the previously obtained index arrays $\mathbf{I}_l^p$ and $\mathbf{I}_l^n$. Here, $Input$ can be either a vector (batch size = 1) or a matrix (batch size > 1). The accumulations are performed using 16-bit integer addition (as discussed in Sect. 4.1). We saturate, and therefore approximate, the results of both accumulators to 16 bit since 8 bit might be not sufficient for the integer bits. After the accumulations are computed, the results are cast to single-precision floating point, multiplied with the scaling coefficients W_l^p and W_l^n, and summed up to the final result. The input array is stored in column-major order for batched inputs to increase spatial locality.

We optimized this algorithm for ARM architectures, however, the same optimizations are also applicable to other processors. The algorithm is multithreaded using OpenMP (in Operation 1) without the need of synchronization. Furthermore, the accumulations (in Operation 5 and 11) are vectorized using the ARM NEON processor extension.

4.3   Compression

Memory requirements are usually the limiting factor when deploying neural networks on specialized processors like FPGAs. For instance, models like AlexNet and ResNet-18 require 244 MB and 50 MB, respectively, while small to mid-sized FPGAs feature only 180 kB–4.3 MB of on-chip memory (Xilinx Spartan7 series). Being able to stash all parameters on on-chip memory allows using a streaming architecture that can fully utilize computational resources. Furthermore, accessing data from on-chip memory is about 100 times more energy efficient than off-chip memory [13].

In order to reduce memory requirements, we flatten the transposed weight matrices $\mathbf{W}_l^T$, store the signs together with their indices, and apply Huffman coding. Both indices and signs are organized as vectors ($\mathbf{i}_l$ respectively $\mathbf{s}_l$). In $\mathbf{s}_l$, only a single bit per element is required to distinguish between positive and negative scaling coefficient. For the compression, we first determine indices and signs of the scaling coefficients by evaluating all non-zero elements in the weight matrix ($\mathbf{i}_l = \{i | w_l^i \neq 0\}$). Then, we calculate the distance vector $\mathbf{d}_l$ based on the distances between consecutive elements of the previously obtained index vector ($\mathbf{d}_l^j = \mathbf{i}_l^{j-1} - \mathbf{i}_l^j$, with $\mathbf{i}_l^{-1} = 0$). This step decreases the amount of possible values and increases the frequency of appearance of those values. Calculating $\mathbf{s}_l$ and $\mathbf{d}_l$ is exemplary illustrated in Eqs. 15, 16, 17 and 18.

$$\mathbf{W}_l^T = \begin{pmatrix} 0 & W_l^p & W_l^p & 0 & W_l^n \\ W_l^n & 0 & 0 & W_l^p & 0 \\ W_l^p & W_l^n & 0 & W_l^n & W_l^n \end{pmatrix} \tag{15}$$

$$\mathbf{s}_l = \begin{pmatrix} 0 \ 0 \ 1 \ 1 \ 0 \ 0 \ 1 \ 1 \ 1 \end{pmatrix} \tag{16}$$

$$\mathbf{i}_l = \begin{pmatrix} 1 \ 2 \ 4 \ 5 \ 8 \ 10 \ 11 \ 13 \ 14 \end{pmatrix} \tag{17}$$

$$\mathbf{d}_l = \begin{pmatrix} 1 \ 1 \ 2 \ 1 \ 3 \ 2 \ 1 \ 2 \ 1 \end{pmatrix} \tag{18}$$

We finally compress $\mathbf{d}_l$ using Huffman coding [1], which leverages the frequency of appearance of values within a given dataset. The principle is to use fewer bits to encode values with a high frequency of appearance. In order to reduce the search space, we use a single codebook that contains the codes for all layers. Then, decompression is simply a matter of looking up the values in the codebook.

5   Experiments

The evaluation of our approach is done on three different convolutional neural networks and datasets: ConvNet [23] on SVHN, ResNet-44 [10] on CIFAR10 and AlexNet [16] on ImageNet. The reported metrics are test accuracy, memory footprint and inference rate in Frames Per Second (FPS).

5.1   Methodology

The inference of convolutional neural networks is usually lowered to matrix-matrix multiplications (fully-connected layer and convolutional layer using the image2col approach). Additionally, for the used models, computations for batch normalization, activation and pooling are insignificant to overall inference rate. Hence, we evaluate throughput by executing all matrix multiplications - required for inference of the respective model - sequentially.

Comparison to Related Work: we compare these results to a full-precision implementation (baseline), a binarized implementation (BNN) and an 8-bit integer implementation (Int8):

1. The **baseline** implementation uses single-precision floating point for weights and activations. In detail, we rely on the GEMM operator of the Eigen library, as it offers (to the best of our knowledge) the fastest floating-point matrix multiplication on CPUs.
2. DoReFa-Net [24] is selected as representative of **BNNs**. This comparison shows upper bounds regarding inference rate and memory footprint, as it is currently the most effective way (without considering an accuracy drop) to perform inference. We employ an extended version of the Eigen library (from previous work [20]) for such binarized computations.
3. The **Int8** implementation is based on the integer-arithmetic-only approach of Benoit et al. [15] (8-bit activations and weights), omitting reproducing accuracy results as we do not assume a loss in accuracy. We compare against this approach as it is a reliable quantization that is the foundation of mobile inference in TensorFlow Light. Calculations are done using the 8-bit integer GEMM operator of Google's gemmlowp library[2].

In general, the first and last layer of the neural networks are not quantized for DeepChip and BNN, but all hidden layers. Furthermore, all GEMM operations are vectorized and parallelized in order to guarantee a fair comparison. Last, all reported results are based on equal models, including identical hyper-parameters and epochs for training.

---

[2] https://github.com/google/gemmlowp.

5.2   Experimental Setup

Training and quantization is performed with Tensorpack [23] on NVIDIA K80 GPUs. As we see an increasing interest in mobile inference using ARM CPUs (TensorFlow Mobile, TensorFlow Light[3] and Baidu's mobile-deep-learning[4]), we use an ARM Cortex-A57 processor to evaluate Algorithm 1 in terms of inference rate. However, note that our approach is not limited to ARM CPUs.

5.3   ConvNet on SVHN

Training on the SVHN dataset is performed with a single GPU, a batch size of 128 over 200 epochs, and the threshold hyper-parameter set to $t = 0.30$. The achieved results are summarized in Table 3.

Table 3. Results for ConvNet on the SVHN dataset.

	Baseline	BNN	Int8	DeepChip
Accuracy	97.5%	97.0%	–[a]	97.5%
Sparsity	0%	0%	0%	89%
Inference rate	258 FPS	1,409 FPS	368 FPS	1,337 FPS
Memory footprint	8,321 kB	299 kB	2,080 kB	192 kB

[a]We assume, as claimed by the authors [15], no loss in classification accuracy.

Compared to the baseline implementation, we achieve full-precision accuracy while reducing the memory footprint by a factor of 43 and increase the inference rate by a factor of 5.2. These results are similar to those of BNNs, however with better accuracy and less memory requirements, but a slightly worse inference rate. Also, we significantly outperform the Int8 approach.

5.4   ResNet-44 on CIFAR-10

Training on the CIFAR-10 dataset is performed with two GPUs and a batch size of 128 over 400 epochs. For this dataset, we use parameters of a pre-trained ResNet-44 (full-precision weights) before re-training the neural network again with quantization. The threshold hyper-parameter is set to $t = 0.25$. The results are shown in Table 4.

Compared to the baseline implementation, at the cost of a slight accuracy loss (-0.2%), DeepChip reduces memory requirements by a factor of 22 and increase the inference rate by a factor of 2.8. By contrast, BNN reduces memory requirements by a factor of 32 and increases the inference rate by 7.7, but results in a significant accuracy loss (-5.0%). Again, DeepChip outperforms Int8 by a large margin. Less sparsity and the accuracy drop of BNN indicate that ResNet on CIFAR-10 is less over-parameterized than ConvNet on SVHN.

---

[3] https://www.tensorflow.org/mobile/.

[4] https://github.com/baidu/mobile-deep-learning.

Table 4. Results for ResNet-44 on CIFAR-10 dataset.

	Baseline	BNN	Int8	DeepChip
Accuracy	92.6%	87.6%	–[a]	92.4%
Sparsity	0%	0%	0%	58%
Inference rate	69 FPS	532 FPS	100 FPS	191 FPS
Memory footprint	2,622 kB	82 kB	655 kB	117 kB

[a]We assume, as claimed by the authors [15], no loss in classification accuracy.

5.5 AlexNet on ImageNet

Training on the ImageNet dataset is performed with two GPUs and a batch size of 64 over 100 epochs. The threshold hyper-parameter is set to $t = 0.05$. The achieved results are summarized in Table 5.

Table 5. Results for AlexNet on ImageNet dataset.

	Baseline	BNN	Int8	DeepChip
Top-1 accuracy	56.2%	45.4%	–[a]	56.4%
Top-5 accuracy	78.3%	56.4%	–[a]	79.0%
Sparsity	0%	0%	0%	63%
Inference rate	4 FPS	22 FPS	7 FPS	8 FPS
Memory footprint	244 MB	24 MB	61 MB	25 MB

[a]We assume, as claimed by the authors [15], no loss in classification accuracy.

Compared to the baseline implementation, DeepChip improves memory requirements by a factor of 10, increases the inference rate by a factor of 2.0 and achieves 0.2% better top-1 accuracy. BNN, however, achieves roughly the same improvement regarding memory requirements but increases the inference rate by a factor of 5.5 at the cost of 10.8% top-1 accuracy loss. Even though DeepChip achieves a sparsity of 63% on AlexNet, the improvements over baseline and Int8 approach are less compared to ConvNet or ResNet. This is due to the fact that we currently quantize only hidden layers, but the output layers of AlexNet contributes about 16 MB. Still, we believe that this is not a major issue since input and output layer of more recent networks (e.g. ResNet or Inception) only have a small impact to overall network size.

5.6 Accuracy vs. Efficiency Considerations

In this section we analyze the correlation among quantization strategy, accuracy and efficiency.

Trading Accuracy with Efficiency. Using neural networks in resource-constrained systems or under real-time requirements enforces certain demands on memory footprint, latency and inference rate. In this experiment, we increase the threshold hyper-parameter t and evaluate the impact on these metrics.

Increasing the threshold from $t = 0.30$ to $t = 0.35$ on the SVHN dataset results in only 0.7% accuracy loss while the achieved sparsity increases from 89% to 96%. Due to this sparsity increase, the memory footprint is reduced to 95 kB and inference rate is increased to 1,538 FPS. A similar behavior can be observed for ResNet-44 and AlexNet, as shown in Table 6 respectively Table 7.

Table 6. Results for ResNet-44 on CIFAR-10 for a varying threshold-parameter t.

	$t = 0.25$	$t = 0.30$	$t = 0.35$
Accuracy	92.4%	91.1%	90.1
Sparsity	58%	63%	71%
Inference rate	191 FPS	255 FPS	293 FPS
Memory footprint	117 kB	108 kB	90 kB

Table 7. Results for AlexNet on ImageNet for a varying threshold-parameter t.

	$t = 0.05$	$t = 0.10$	$t = 0.15$
Top-1 accuracy	56.4%	54.7%	53.7%
Top-5 accuracy	79.0%	77.5%	77.0%
Sparsity	63%	78%	88%
Inference rate	4 FPS	17 FPS	20 FPS
Memory footprint	25 MB	22 MB	19 MB

Equalizing Accuracy. Quantization using the BNN approach is highly effective in terms of inference rate and memory requirements but can cause accuracy degradation. This degradation can be compensated by either making the neural network wider (increasing the amount of neurons per layer) or deeper (increasing the amount of layers). In this section, we deepen the ResNet model on CIFAR-10 with BNN quantization until the respective quantization roughly reaches full-precision accuracy (ResNet-44 baseline). Our results show, that BNN requires an increase from 44 to 272 layers to achieve 91.2% test accuracy. As a result, DeepChip achieves a memory footprint of 108 kB with 1.5% test accuracy loss (see Table 6), while BNN after scaling results in a footprint of 507 kB with 1.4% test accuracy loss. Furthermore, the additional layers increase training time roughly by a factor of 6.5, compared to baseline ResNet-44.

5.7   Comparing to Deep Compression

In this section, we compare our results with Deep Compression on the example of AlexNet (Table 8). As mentioned earlier, Deep Compression is a highly efficient approach of quantizing and compressing neural networks without any accuracy degradation for specialized processors. As can be seen, Deep Compression out-performs our approach in terms of memory footprint (7.4% better compression ratio) and sparsity (26% more sparsity) on AlexNet. This is mainly because we currently quantize only hidden layers and not input and output layers (the output layer of AlexNet alone has a size of 16 MB). This advantage diminishes when comparing only the quantized layers, e.g. only 1.2% better compression ratio. Still, Deep Compression achieves about 22% more sparsity, which directly translates into fewer operations but relies on 5-/8-bit multiplications. An FPGA synthesis report (Vivado design suite) indicates that for one Processing Element (PE), consisting of 8 channels, and using 8 × 8 inputs, a multiply accumulate unit requires roughly 10 times more Configurable Logic Blocks (CLBs) than an accumulator as it is required for DeepChip.

Table 8. Comparison of deep compression and DeepChip (memory footprint and sparsity refer to AlexNet on ImageNet).

	Deep compression [9]	DeepChip
Top-1 accuracy	57.2%	56.4%
Top-5 accuracy	80.3%	79.0%
Memory footprint (all layers)	6.9 MB	25.2 MB
Memory footprint (hidden layers only)	5.9 MB	8.7 MB
Compression ratio (all layers)	97.1%	89.7%
Compression ratio (hidden layers only)	97.4%	96.2%
Sparsity (all layers)	89%	63%
Sparsity (hidden layers only)	90%	68%
Energy costs per PE (using data from [13])	153.6 pJ	15.4 pJ
CLB LUTs per PE	765	64
Maximum frequency of PEs	384 MHz	400 MHz

Furthermore, Deep Compression is targeting real-time inference in which batching is usually not used (batch size of 1). Han et al. [9] report excellent speedups for Deep Compression when comparing to dense and sparse GEMV. However, when using batched inputs, sparse GEMV performs roughly equal or even worse than dense GEMM. On the contrary, we target batched inference as well as real-time inference, as efficient processing of frame sequences is of fundamental importance for many applications.

6   Discussion

The concept of ternary weights and 8-bit fixed-point activations, in combination with DeepChip's inference architecture achieve promising performance and is highly flexible. Of particular importance is the fact that redundancy is removed without a loss in test accuracy, and that the concept and architecture are based on a generic processor. As a result, we expect an efficient application to a variety of processor architectures, in particular because our architecture results in a very low amount of rather simple computations (see Sect. 5.7).

Currently, we do not quantize input and output layers of the neural networks, because quantizing the weights of these layers in the same way as we quantize the hidden layers would result in accuracy drops. Instead, we are considering using either integer or more than just two scaling coefficients for the input and output layer. The sparse matrix-multiplication algorithm and compression approach is also applicable to multiple scaling coefficients. This might be relevant for Recurrent Neural Network (RNNs) too, as related work [7] reports that RNNs require more bits for weight representations than CNNs.

We performed all experiments using 8 bit fixed-point activation representations, because 8 bit suits general-purpose processors well and fewer bits cause accuracy degradation on complex datasets. The negligible accuracy loss of BNNs on SVHN, however, indicates that efficient quantization is dependent on data set complexity. Even though we did not evaluate fewer bits for activations, we anticipate that this will further improve computations on specialized processors like FPGAs.

7   Conclusion

We have introduced key elements of DeepChip, a framework that targets efficient inference on resource-constrained computing systems and is flexible/configurable for various quantizations, different target architectures, and allows for trade-offs of complexity, accuracy and efficiency. The quantizer generates variable-length fixed-point activations and asymmetric ternary weights. A combination of a simple but highly efficient matrix-multiplication algorithm and compression technique exploits this representation by leveraging reduced precision as well as sparsity to improve memory footprint, latency and inference rate. DeepChip is able to achieve single-precision floating point accuracy while removing redundancy of neural networks. Furthermore, trading accuracy with efficiency allows to satisfy real-time, system and user constraints. Finally, a detailed comparison to other approaches highlights its efficiency and potential regarding processor-agnostic implementations.

Acknowledgments. We gratefully acknowledge the valuable contributions of Andreas Kugel and Andreas Melzer. We also acknowledge funding by the German Research Foundation (DFG) under the project number FR3273/1-1 and the Austrian Science Fund (FWF) under the project number I2706-N31.

References

1. Huffman, D.A.: A method for the construction of minimum-redundancy codes. Resonance **11**, 91–99 (2006)
2. Abadi, M.: Tensorflow: Large-scale machine learning on heterogeneous distributed systems. CoRR abs/1603.04467 (2016)
3. ARM: Cortex-a9 neon media - technical reference manual. Technical report (2008)
4. Cai, Z., He, X., Sun, J., Vasconcelos, N.: Deep learning with low precision by half-wave gaussian quantization. CoRR abs/1702.00953 (2017)
5. Chetlur, S., et al.: cuDNN: efficient primitives for deep learning. CoRR abs/1410.0759 (2014)
6. Courbariaux, M., Bengio, Y.: Binarynet: training deep neural networks with weights and activations constrained to +1 or −1. CoRR (2016)
7. Han, S., et al.: ESE: efficient speech recognition engine with compressed LSTM, on FPGA. CoRR abs/1612.00694 (2016)
8. Han, S., et al.: EIE: efficient inference engine on compressed deep neural network. CoRR abs/1602.01528 (2016)
9. Han, S., Mao, H., Dally, W.J.: Deep compression: compressing deep neural network with pruning, trained quantization and Huffman coding. CoRR abs/1510.00149 (2015)
10. He, K., Zhang, X., Ren, S., Sun, J.: Deep residual learning for image recognition. CoRR abs/1512.03385 (2015)
11. Hinton, G., Dean, J., Vinyals, O.: Distilling the knowledge in a neural network. In: NIPS 2014 Deep Learning Workshop, pp. 1–9 (2014)
12. Hinton, G., et al.: Deep neural networks for acoustic modeling in speech recognition: the shared views of four research groups. IEEE Signal Process. Mag. **29**, 82–97 (2012)
13. Horowitz, M.: 1.1 computing's energy problem (and what we can do about it), vol. 57, pp. 10–14 (2014)
14. Iandola, F.N., Moskewicz, M.W., Ashraf, K., Han, S., Dally, W.J., Keutzer, K.: Squeezenet: Alexnet-level accuracy with 50x fewer parameters and <1mb model size. CoRR abs/1602.07360 (2016)
15. Jacob, B., et al.: Quantization and training of neural networks for efficient integer-arithmetic-only inference. eprint arXiv:1712.05877 (2017)
16. Krizhevsky, A., Sutskever, I., Hinton, G.E.: Imagenet classification with deep convolutional neural networks. In: Proceedings of the 25th NIPS, NIPS 2012, pp. 1097–1105. Curran Associates Inc., USA (2012)
17. Lenz, I.: Deep Learning for Robotics (2016)
18. Li, F., Liu, B.: Ternary weight networks. CoRR abs/1605.04711 (2016)
19. Rastegari, M., Ordonez, V., Redmon, J., Farhadi, A.: Xnor-net: imagenet classification using binary convolutional neural networks. CoRR abs/1603.05279 (2016)
20. Schindler, G., Mücke, M., Fröning, H.: Linking application description with efficient SIMD Code generation for low-precision signed-integer GEMM. In: Heras, D.B., Bougé, L. (eds.) Euro-Par 2017. LNCS, vol. 10659, pp. 688–699. Springer, Cham (2018). https://doi.org/10.1007/978-3-319-75178-8_55
21. Umuroglu, Y., et al.: FINN: a framework for fast, scalable binarized neural network inference. CoRR abs/1612.07119 (2016)
22. Vanhoucke, V., Senior, A., Mao, M.Z.: Improving the speed of neural networks on CPUs. In: Deep Learning and Unsupervised Feature Learning Workshop, NIPS 2011 (2011)

23. Wu, Y., et al.: Tensorpack (2016)
24. Zhou, S., Ni, Z., Zhou, X., Wen, H., Wu, Y., Zou, Y.: Dorefa-net: training low bitwidth convolutional neural networks with low bitwidth gradients. CoRR abs/1606.06160 (2016)
25. Zhu, C., Han, S., Mao, H., Dally, W.J.: Trained ternary quantization. CoRR (2016)

Auxiliary Guided Autoregressive Variational Autoencoders

Thomas Lucas[(✉)] and Jakob Verbeek

Université. Grenoble Alpes, Inria, CNRS, Grenoble INP, LJK, 38000 Grenoble, France
{thomas.lucas,jakob.verbeek}@inria.fr

Abstract. Generative modeling of high-dimensional data is a key problem in machine learning. Successful approaches include latent variable models and autoregressive models. The complementary strengths of these approaches, to model global and local image statistics respectively, suggest hybrid models that encode global image structure into latent variables while autoregressively modeling low level detail. Previous approaches to such hybrid models restrict the capacity of the autoregressive decoder to prevent degenerate models that ignore the latent variables and only rely on autoregressive modeling. Our contribution is a training procedure relying on an auxiliary loss function that controls which information is captured by the latent variables and what is left to the autoregressive decoder. Our approach can leverage arbitrarily powerful autoregressive decoders, achieves state-of-the art quantitative performance among models with latent variables, and generates qualitatively convincing samples.

1 Introduction

Unsupervised modeling of complex distributions with unknown structure is a landmark challenge in machine learning. The problem is often studied in the context of learning generative models of the complex high-dimensional distributions of natural image collections. Latent variable approaches can learn disentangled and concise representations of the data [3], which are useful for compression [11] and semi-supervised learning [14,22]. When conditioned on prior information, generative models can be used for a variety of tasks, such as attribute or class-conditional image generation, text and pose-based image generation, image colorization, *etc.* [6,20,23,26]. Recently significant advances in generative (image) modeling have been made along several lines, including adversarial networks [1,10], variational autoencoders [16,24], autoregressive models [21,23], and non-volume preserving variable transformations [8].

In our work we seek to combine the merits of two of these lines of work. Variational autoencoders (VAEs) [16,24] can learn latent variable representations that abstract away from low-level details, but model pixels as conditionally independent given the latent variables. This renders the generative model computationally efficient, but the lack of low-level structure modeling leads to overly smooth and blurry samples. Autoregressive models, such as pixelCNNs [21], on

M. Berlingerio et al. (Eds.): ECML PKDD 2018, LNAI 11051, pp. 443–458, 2019.
https://doi.org/10.1007/978-3-030-10925-7_27

the other hand, estimate complex translation invariant conditional distributions among pixels. They are effective to model low-level image statistics, and yield state-of-the-art likelihoods on test data [25]. This is in line with the observations of [17] that low-level image details account for a large part of the likelihood. These autoregressive models, however, do not learn a latent variable representations to support, *e.g.*, semi-supervised learning.

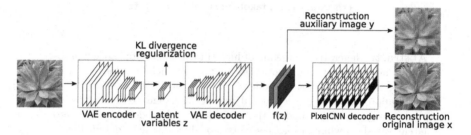

Fig. 1. Schematic illustration of our auxiliary guided autoregressive variational autoencoder (AGAVE). The objective function has three components: KL divergence regularization, per-pixel reconstruction with the VAE decoder, and autoregressive reconstruction with the pixelCNN decoder.

The complementary strengths of VAEs and pixelCNNs, modeling global and local image statistics respectively, suggest hybrid approaches combining the strengths of both. Prior work on such hybrid models needed to limit the capacity of the autoregressive decoder to prevent degenerate models that completely ignore the latent variables and rely on autoregressive modeling only [5,12]. In this paper we describe Auxiliary Guided Autoregressive Variational autoEncoders (AGAVE), an approach to train such hybrid models using an auxiliary loss function that controls which information is captured by the latent variables and what is left to the AR decoder. That removes the need to limit the capacity of the latter. See Fig. 1 for a schematic illustration of our approach.

Using high-capacity VAE and autoregressive components allows our models to obtain quantitative results on held-out data that are on par with the state of the art in general, and set a new state of the art among models with latent variables. Our models generate samples with both global coherence and low-level details. See Fig. 2 for representative samples of VAE and pixelCNN models.

2   Related Work

Generative image modeling has recently taken significant strides forward, leveraging deep neural networks to learn complex density models using a variety of approaches. These include the variational autoencoders and autoregressive models that form the basis of our work, but also generative adversarial networks (GANs) [1,10] and variable transformation with invertible functions [8].

While GANs produce visually appealing samples, they suffer from mode dropping and their likelihood-free nature prevents measuring how well they model held-out test data. In particular, GANs can only generate samples on a non-linear manifold in the data space with dimension equal to the number of latent variables. In contrast, probabilistic models such as VAEs and autoregressive models generalize to the entire data space, and likelihoods of held-out data can be used for compression, and to quantitatively compare different models. The non-volume preserving (NVP) transformation approach of [8] chains together invertible transformations to map a basic (*e.g.* unit Gaussian) prior on the latent space to a complex distribution on the data space. This method offers tractable likelihood evaluation and exact inference, but obtains likelihoods on held-out data below the values reported using state-of-the-art VAE and autoregressive models. Moreover, it is restricted to use latent representations with the same dimensionality as the input data, and is thus difficult to scale to model high-resolution images.

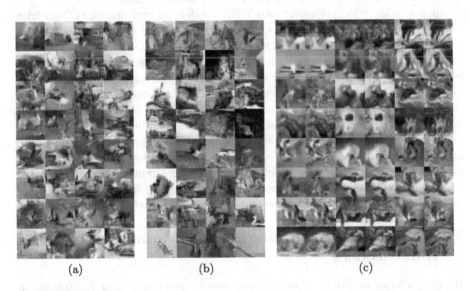

(a)   (b)   (c)

Fig. 2. Randomly selected samples from unsupervised models trained on 32×32 CIFAR10 images: (a) IAF-VAE [15], (b) pixelCNN++ [25], and (c) our hybrid AGAVE model. For our model, we show the intermediate high-level representation based on latent variables (left), that conditions the final sample based on the pixelCNN decoder (right).

Autoregressive density estimation models, such as pixelCNNs [21], admit tractable likelihood evaluation, while for variational autoencoders [16,24] accurate approximations can be obtained using importance sampling [4]. Naively combining powerful pixelCNN decoders in a VAE framework results in a degenerate model which ignores the VAE latent variable structure, as explained through the lens of bits-back coding by [5]. To address this issue, the capacity of the the autoregressive component can be restricted. This can, for example, be achieved by reducing its depth and/or field of view, or by giving the pixelCNN only access to grayscale values, *i.e.* modeling $p(x_i|\mathbf{x}_{<i}, \mathbf{z}) = p(x_i|\text{gray}(\mathbf{x}_{<i}), \mathbf{z})$ [5,12]. This forces the model to leverage the latent variables $\mathbf{z}$ to model part of the dependencies among the pixels. This approach, however, has two drawbacks. (i) Curbing the capacity of the model is undesirable in unsupervised settings where training data is abundant and overfitting unlikely, and is only a partial solution to the problem. (ii) Balancing what is modeled by the VAE and the pixelCNN by means of architectural design choices requires careful hand-design and tuning of the architectures. This is a tedious process, and a more reliable principle is desirable. To overcome these drawbacks, we propose to instead control what is modeled by the VAE and pixelCNN with an auxiliary loss on the VAE decoder output before it is used to condition the autoregressive decoder. This allows us to "plug in" powerful high-capacity VAE and pixelCNN architectures, and balance what is modeled by each component by means of the auxiliary loss.

In a similar vein, [17] force pixelCNN models to capture more high-level image aspects using an auxiliary representation $\mathbf{y}$ of the original image $\mathbf{x}$, *e.g.* a low-resolution version of the original. They learn a pixelCNN for $\mathbf{y}$, and a conditional pixelCNN to predict $\mathbf{x}$ from $\mathbf{y}$, possibly using several intermediate representations. This approach forces modeling of more high-level aspects in the intermediate representations, and yields visually more compelling samples. [23] similarly learn a series of conditional autoregressive models to upsample coarser intermediate latent images. By introducing partial conditional independencies in the model they scale the model to efficiently sample high-resolution images of up to 512×512 pixels. [11] use a recurrent VAE model to produces a sequence of RGB images with increasing detail derived from latent variables associated with each iteration. Like our work, all these models work with intermediate representations in RGB space to learn accurate generative image models.

3   Auxiliary Guided Autoregressive Variational Autoencoders

We give a brief overview of variational autoencoders and their limitations in Sect. 3.1, before we present our approach to learning variational autoencoders with autoregressive decoders in Sect. 3.2.

3.1   Variational Autoencoders

Variational autoencoders [16,24] learn deep generative latent variable models using two neural networks. The "decoder" network implements a conditional distribution $p_\theta(\mathbf{x}|\mathbf{z})$ over observations $\mathbf{x}$ given a latent variable $\mathbf{z}$, with parameters θ. Together with a basic prior on the latent variable $\mathbf{z}$, *e.g.* a unit Gaussian, the generative model on $\mathbf{x}$ is obtained by marginalizing out the latent variable:

$$p_\theta(\mathbf{x}) = \int p(\mathbf{z})p_\theta(\mathbf{x}|\mathbf{z})\,d\mathbf{z}. \tag{1}$$

The marginal likelihood can, however, not be optimized directly since the non-linear dependencies in $p_\theta(\mathbf{x}|\mathbf{z})$ render the integral intractable. To overcome this problem, an "encoder" network is used to compute an approximate posterior distribution $q_\phi(\mathbf{z}|\mathbf{x})$, with parameters ϕ. The approximate posterior is used to define a variational bound on the data log-likelihood, by subtracting the Kullback-Leibler divergence between the true and approximate posterior:

$$\ln p_\theta(\mathbf{x}) \geq \mathcal{L}(\theta, \phi; \mathbf{x}) = \ln(p_\theta(\mathbf{x})) - D_{\mathrm{KL}}(q_\phi(\mathbf{z}|\mathbf{x}) \| p_\theta(\mathbf{z}|\mathbf{x})) \qquad (2)$$

$$= \underbrace{\mathbb{E}_{q_\phi}[\ln(p_\theta(\mathbf{x}|\mathbf{z})]}_{\text{Reconstruction}} - \underbrace{D_{\mathrm{KL}}(q_\phi(\mathbf{z}|\mathbf{x}) \| p(\mathbf{z}))}_{\text{Regularization}}. \qquad (3)$$

The decomposition in (3) interprets the bound as the sum of a reconstruction term and a regularization term. The first aims to maximize the expected data log-likelihood $p_\theta(\mathbf{x}|\mathbf{z})$ given the posterior estimate $q_\phi(\mathbf{z}|\mathbf{x})$. The second term prevents $q_\phi(\mathbf{z}|\mathbf{x})$ from collapsing to a single point, which would be optimal for the first term.

Variational autoencoders typically model the dimensions of $\mathbf{x}$ as conditionally independent,

$$p_\theta(\mathbf{x}|\mathbf{z}) = \prod_{i=1}^{D} p_\theta(x_i|\mathbf{z}), \qquad (4)$$

for instance using a factored Gaussian or Bernoulli model, see *e.g.* [15,16,26]. The conditional independence assumption makes sampling from the VAE efficient: since the decoder network is evaluated only once for a sample $\mathbf{z} \sim p(\mathbf{z})$ to compute all the conditional distributions $p_\theta(x_i|\mathbf{z})$, the x_i can then be sampled in parallel.

A result of relying on the latent variables to account for all pixel dependencies, however, is that all low-level variability must also be modeled by the latent variables. Consider, for instance, a picture of a dog, and variants of that image shifted by one or a few pixels, or in a slightly different pose, with a slightly lighter background, or with less saturated colors, *etc.* If these factors of variability are modeled using latent variables, then these low-level aspects are confounded with latent variables relating to the high-level image content. If the corresponding image variability is not modeled using latent variables, it will be modeled as independent pixel noise. In the latter case, using the mean of $p_\theta(\mathbf{x}|\mathbf{z})$ as the synthetic image for a given $\mathbf{z}$ results in blurry samples, since the mean is averaged over the low-level variants of the image. Sampling from $p_\theta(\mathbf{x}|\mathbf{z})$ to obtain synthetic images, on the other hand, results in images with unrealistic independent pixel noise.

3.2 Autoregressive Decoders in Variational Autoencoders

Autoregressive density models, see *e.g.* [9,19], rely on the basic factorization of multi-variate distributions,

$$p_\theta(\mathbf{x}) = \prod_{i=1}^{D} p_\theta(x_i | \mathbf{x}_{<i}) \tag{5}$$

with $\mathbf{x}_{<i} = x_1, \ldots, x_{i-1}$, and model the conditional distributions using a (deep) neural network. For image data, PixelCNNs [20,21] use a scanline pixel ordering, and model the conditional distributions using a convolution neural network. The convolutional filters are masked so as to ensure that the receptive fields only extend to pixels $\mathbf{x}_{<i}$ when computing the conditional distribution of x_i.

PixelCNNs can be used as a decoder in a VAE by conditioning on the latent variable $\mathbf{z}$ in addition to the preceding pixels, leading to a variational bound with a modified reconstruction term:

$$\mathcal{L}(\theta, \phi; \mathbf{x}) = \mathbb{E}_{q_\phi} \left[\sum_{i=1}^{D} \ln p_\theta(x_i | \mathbf{x}_{<i}, \mathbf{z}) \right] - D_{\mathrm{KL}}(q_\phi(\mathbf{z}|\mathbf{x}) \| p(\mathbf{z})). \tag{6}$$

The regularization term can be interpreted as a "cost" of using the latent variables. To effectively use the latent variables, the approximate posterior $q_\phi(\mathbf{z}|\mathbf{x})$ must differ from the prior $p(\mathbf{z})$, which increases the KL divergence.

[5] showed that for loss (6) and a decoder with enough capacity, it is optimal to encode no information about x in z by setting $q(z|x) = p(z)$. To ensure meaningful latent representation learning [5,12] restrict the capacity of the pixelCNN decoder. In our approach, in contrast, it is always optimal for the autoregressive decoder, regardless of its capacity, to exploit the information on $\mathbf{x}$ carried by z. We rely on two decoders in parallel: the first one reconstructs an auxiliary image $\mathbf{y}$ from an intermediate representation $f_\theta(\mathbf{z})$ in a non-autoregressive manner. The auxiliary image can be either simply taken to be the original image ($\mathbf{y} = \mathbf{x}$), or a compressed version of it, *e.g.* with lower resolution or with a coarser color quantization. The second decoder is a conditional autoregressive model that predicts $\mathbf{x}$ conditioned on $f_\theta(\mathbf{z})$. Modeling $\mathbf{y}$ in a non-autoregressive manner ensures a meaningful representation $\mathbf{z}$ and renders $\mathbf{x}$ and $\mathbf{z}$ dependent, inducing a certain non-zero KL "cost" in (6). The uncertainty on $\mathbf{x}$ is thus reduced when conditioning on $\mathbf{z}$, and there is no longer an advantage in ignoring the latent variable for the autoregressive decoder. We provide a more detailed explanation of why our auxiliary loss ensures a meaningful use of latent variables in powerful decoders in Sect. 3.3. To train the model we combine both decoders in a single objective function with a shared encoder network:

$$\mathcal{L}(\theta, \phi; \mathbf{x}, \mathbf{y}) = \underbrace{\mathbb{E}_{q_\phi} \left[\sum_{i=1}^{D} \ln p_\theta(x_i | \mathbf{x}_{<i}, \mathbf{z}) \right]}_{\text{Primary Reconstruction}} + \underbrace{\mathbb{E}_{q_\phi} \left[\sum_{j=1}^{E} \ln p_\theta(y_j | \mathbf{z}) \right]}_{\text{Auxiliary Reconstruction}}$$

$$- \underbrace{\lambda \, D_{\mathrm{KL}}(q_\phi(\mathbf{z}|\mathbf{x}) \| p(\mathbf{z}))}_{\text{Regularization}}. \tag{7}$$

Treating $\mathbf{x}$ and $\mathbf{y}$ as two variables that are conditionally independent given a shared underlying latent variable $\mathbf{z}$ leads to $\lambda = 1$. Summing the lower bounds in Eqs. (3) and (6) of the marginal log-likelihoods of $\mathbf{y}$ and $\mathbf{x}$, and sharing the encoder network, leads to $\lambda = 2$. Larger values of λ result in valid but less tight lower bounds of the log-likelihoods. Encouraging the variational posterior to be closer to the prior, this leads to less informative latent variable representations.

Sharing the encoder across the two decoders is the key of our approach. The factored auxiliary VAE decoder can only model pixel dependencies by means of the latent variables, which ensures that a meaningful representation is learned. Now, given that the VAE encoder output is informative on the image content, there is no incentive for the autoregressive decoder to ignore the intermediate representation $f(\mathbf{z})$ on which it is conditioned. The choice of the regularization parameter λ and auxiliary image $\mathbf{y}$ provide two levers to control *how much* and *what type* of information should be encoded in the latent variables.

3.3   It Is Optimal for the Autoregressive Decoder to Use z

Combining a VAE with a flexible decoder (for instance an autoregressive one) leads to the latent code being ignored. This problem could be attributed to optimization challenges: at the start of training $q(\mathbf{z}|\mathbf{x})$ carries little information about $\mathbf{x}$, the KL term pushes the model to set it to the prior to avoid any penalty, and training never recovers from falling into that local minimum. [5] have proposed extensive explanations showing that the problem goes deeper: if a sufficiently expressive decoder is used, ignoring the latents actually is the optimal behavior. The gist of the argument is based on bits-back coding as follows: given an encoder $q(\mathbf{z}|\mathbf{x})$, a decoder $p(\mathbf{x}|\mathbf{z})$ and a prior $p(\mathbf{z})$, $\mathbf{z} \sim q(\mathbf{z}|\mathbf{x})$ can be encoded in a lossless manner using $p(\mathbf{z})$, and $\mathbf{x}$ can be encoded, also losslessly, using $p(\mathbf{x}|\mathbf{z})$. Once the receiver has decoded $\mathbf{x}$, $q(\mathbf{z}|\mathbf{x})$ becomes available and a secondary message can be decoded from it. This yields and average code length of:

$$C_{BitsBack} = \mathbb{E}_{\mathbf{x}\sim D, \mathbf{z}\sim q(.|\mathbf{x})}[\log(q(\mathbf{z}|\mathbf{x})) - \log(p(\mathbf{z})) - \log(p(\mathbf{x}|\mathbf{z}))].$$

$C_{BitsBack}$ corresponds to the standard VAE objective. A lower-bound on the expected code length for the data being encoded is given by the Shannon entropy: $\mathcal{H}(D) = \mathbb{E}_{\mathbf{x}\sim D}[-\log p_D(\mathbf{x})]$, which yields:

$$C_{BitsBack} = \mathbb{E}_{\mathbf{x}\sim D}[-\log(p(\mathbf{x})) + D_{KL}(q(\mathbf{z}|\mathbf{x})\|p(\mathbf{z}|\mathbf{x}))]$$
$$\geq \mathcal{H}(D) + \mathbb{E}_{\mathbf{x}\sim D}[D_{KL}(q(\mathbf{z}|\mathbf{x})\|p(\mathbf{z}|\mathbf{x}))].$$

If $p(.|\mathbf{x}_{j<i})$ is expressive enough, or if $q(.|\mathbf{x})$ is poor enough, the following inequality can be verified:

$$\mathcal{H}(D) \leq \mathbb{E}_{\mathbf{x}\sim D}[-\log p(\mathbf{x}|\mathbf{x}_{j<i})] < \mathcal{H}(D) + \mathbb{E}_{\mathbf{x}\sim D}[D_{KL}(q(\mathbf{z}|\mathbf{x})\|p(\mathbf{z}|\mathbf{x}))]$$

This is always true in the limit of infinitely expressive autoregressive decoders. In that case, any use of the latents that p might decrease performance. The optimal behavior is to set $q(\mathbf{z}|\mathbf{x}) = p(\mathbf{z})$ to avoid the extra KL cost. Then $\mathbf{z}$ becomes independent from $\mathbf{x}$ and no information about $\mathbf{x}$ is encoded in $\mathbf{z}$. Therefore, given an encoder, the latent variables will only be used if the capacity of the autoregressive decoder is sufficiently restricted. This is the approach taken by [5,12]. This approach works: it has obtained competitive quantitative and qualitative performance. However, it is not satisfactory in the sense that autoregressive models cannot be used to the full extent of their potential, while learning a meaningful latent variable representation.

In our setting, both (Y, X) have to be sent to and decoded by the receiver. Let us denote C_{VAE} the expected code length required to send the auxiliary message, $\mathbf{y}$. Once $\mathbf{y}$ has been sent, sending $\mathbf{x}$ costs: $\mathbb{E}_{\mathbf{z} \sim q(\mathbf{z}|\mathbf{x})}[-\sum_i \log(p(x_i|\mathbf{z}, \mathbf{x}_{j<i}))]$, and we have:

$$C_{VAE} = \mathbb{E}_{\mathbf{x} \sim D, \mathbf{z} \sim q(.|\mathbf{x})}[\log(q(\mathbf{z}|\mathbf{x})) - \log(p(\mathbf{z})) - \log(p(\mathbf{y}|\mathbf{z}))] \tag{8}$$

$$C_{AGAVE} = C_{VAE} + \mathbb{E}_{\mathbf{z} \sim q(\mathbf{z}|\mathbf{x})}[-\sum_i \log(p(x_i|\mathbf{z}, \mathbf{x}_{j<i}))]. \tag{9}$$

Using the fact that the Shannon entropy is the optimal expected code length for transmitting $X|Z$, we obtain $C_{AGAVE} \geq C_{VAE} + \mathcal{H}(X|Z)$.

The entropy of a random variable decreases when it is conditioned on another, i.e. $\mathcal{H}(X|Z) \leq \mathcal{H}(X)$. Therefore, the theoretical lower-bound on the expected code length in our setup is always better when the autoregressive component takes Z into account, no matter its expressivity. In the limit case of an infinitely expressive autoregressive decoder, denoted by $*$, the lower bound is attained and $C^*_{AGAVE} = C_{VAE} + \mathcal{H}(X|Z) \leq C_{VAE} + \mathcal{H}(X)$. In non-degenerate cases, the VAE is optimized to encode information about X into a meaningful Z, with potentially near perfect reconstructions, and there exists $\epsilon > 0$ such that $\mathcal{H}(X|Z) < \mathcal{H}(X) - \epsilon$, making the lower bound strictly better by a possibly big margin.

This analysis shows that in our setup it is theoretically always better for the autoregressive model to make use of the latent and auxiliary representation it is conditioned on. That is true no matter how expressive the model is. It also shows that in theory our model should learn meaningful latent structure.

4 Experimental Evaluation

In this section we describe our experimental setup, and present results on CIFAR10.

4.1 Dataset and Implementation

The CIFAR10 dataset [18] contains 6,000 images of 32×32 pixels for each of the 10 object categories *airplane, automobile, bird, cat, deer, dog, frog, horse, ship, truck*. The images are split into 50,000 training images and 10,000 test images. We train all our models in a completely unsupervised manner, ignoring the class information.

We implemented our model based on existing architectures. In particular we use the VAE architecture of [15], and use logistic distributions over the RGB color values. We let the intermediate representation $f(\mathbf{z})$ output by the VAE decoder be the per-pixel and per-channel mean values of the logistics, and learn per-channel scale parameters that are used across all pixels. The cumulative density function (CDF), given by the sigmoid function, is used to compute probabilities across the 256 discrete color levels, or fewer if a lower quantization level is chosen in $\mathbf{y}$. Using RGB values $y_i \in [0, 255]$, we let b denote the number of discrete color levels and define $c = 256/b$. The probabilities over the b discrete color levels are computed from the logistic mean and variance μ_i and s_i as

$$p(y_i|\mu_i, s_i) = \sigma\left(c + c\lfloor y_i/c\rfloor |\mu_i, s_i\right) - \sigma\left(c\lfloor y_i/c\rfloor |\mu_i, s_i\right). \tag{10}$$

Table 1. Bits per dimension (lower is better) of models on the CIFAR10 test data.

Model	BPD	$.|z$	$.|x_{j<i}$
NICE [7]	4.48	✓	
Conv. DRAW [11]	≤3.58	✓	
Real NVP [8]	3.49	✓	
MatNet [2]	≤3.24	✓	
PixelCNN [21]	3.14		✓
VAE-IAF [15]	≤3.11	✓	
Gated pixelCNN [20]	3.03		✓
Pixel-RNN [21]	3.00		✓
Aux. pixelCNN [17]	2.98		✓
Lossy VAE [5]	≤2.95	✓	✓
AGAVE, $\lambda = 12$ (this paper)	≤2.92	✓	✓
pixCNN++ [25]	2.92		✓

For the pixelCNN we use the architecture of [25], and modify it to be conditioned on the VAE decoder output $f(\mathbf{z})$, or possibly an upsampled version if $\mathbf{y}$ has a lower resolution than $\mathbf{x}$. In particular, we apply standard non-masked convolutional layers to the VAE output, as many as there are pixelCNN layers. We allow each layer of the pixel-CNN to take additional input using non-masked convolutions from the feature stream based on the VAE output. This ensures that the conditional pixelCNN remains autoregressive.

To speed up training, we independently pretrain the VAE and pixelCNN in parallel, and then continue training the full model with both decoders. We use the Adamax optimizer [13] with a learning rate of 0.002 without learning rate decay. We will release our TensorFlow-based code to replicate our experiments upon publication.

4.2   Quantitative Performance Evaluation

Following previous work, we evaluate models on the test images using the bits-per-dimension (BPD) metric: the negative log-likelihood divided by the number of pixels values ($3 \times 32 \times 32$). It can be interpreted as the average number of bits per RGB value in a lossless compression scheme derived from the model.

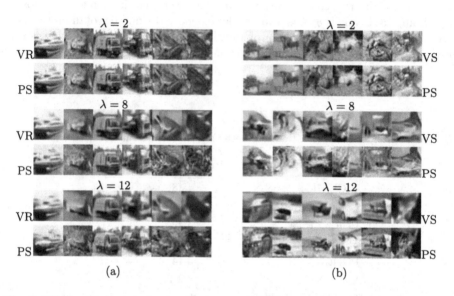

Fig. 3. Effect of the regularization parameter λ. Reconstructions (a) and samples (b) of the VAE decoder (VR and VS, respectively) and corresponding conditional samples from the pixelCNN (PS).

The comparison in Table 1 shows that our model performs on par with the state-of-the-art results of the pixelCNN++ model [25]. Here we used the importance sampling-based bound of [4] with 150 samples to compute the BPD metric for our model.[1] We refer to Fig. 2 for qualitative comparison of samples from our model and pixelCNN++, the latter generated using the publicly available code.

4.3   Effect of KL Regularization Strength

In Fig. 3 we show reconstructions of test images and samples generated by the VAE decoder, together with their corresponding conditional pixelCNN samples for different values of λ. As expected, the VAE reconstructions become less accurate for larger values of λ, mainly by lacking details while preserving the global shape of the input. At the same time, the samples become more appealing for larger λ, suppressing the unrealistic high-frequency detail in the VAE samples

---

[1] The graphs in Figs. 4 and 8 are based on the bound in Eq. (7) to reduce the computational effort.

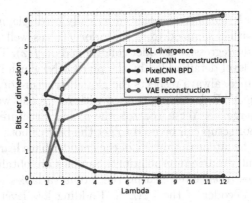

Fig. 4. Bits per dimension of the VAE decoder and pixelCNN decoder, as well as decomposition in KL regularization and reconstruction terms.

obtained at lower values of λ. Note that the VAE samples and reconstructions become more similar as λ increases, which makes the input to the pixelCNN during training and sampling more consistent.

For both reconstructions and samples, the pixelCNN clearly takes into account the output of the VAE decoder, demonstrating the effectiveness of our auxiliary loss to condition high-capacity pixelCNN decoders on latent variable representations. Samples from the pixelCNN faithfully reproduce the global structure of the VAE output, leading to more realistic samples, in particular for higher values of λ.

For $\lambda = 2$ the VAE reconstructions are near perfect during training, and the pixelCNN decoder does not significantly modify the appearance of the VAE output. For larger values of λ, the pixelCNN clearly adds significant detail to the VAE outputs.

Figure 4 traces the BPD metrics of both the VAE and pixelCNN decoder as a function of λ. We also show the decomposition in regularization and reconstruction terms. By increasing λ, the KL divergence can be pushed closer to zero. As the KL divergence term drops, the reconstruction term for the VAE rapidly increases and the VAE model obtains worse BPD values, stemming from the inability of the VAE to model pixel dependencies other than via the latent variables. The reconstruction term of the pixelCNN decoder also increases with λ, as the amount of information it receives drops. However, in terms of BPD which sums KL divergence and pixelCNN reconstruction, a substantial gain of 0.2 is observed increasing λ from 1 to 2, after which smaller but consistent gains are observed.

4.4   Role of the Auxilliary Representation

The Auxilliary Variables are Taken into Account: Sect. 3.3 shows that in theory it is always optimal for the autoregressive decoder to take the latent variables into account. Figure 5 demonstrates this empirically by displaying auxiliary representations $f(\mathbf{z})$ with z sampled from the prior $f(z)$ as well as nine different samples from the autoregressive decoder conditioned on $f(\mathbf{z})$. This qualitatively shows that the low level detail added by the pixelCNN, which is crucial for log-likelihood performance, always respects the global structure of the image being conditioned on. The VAE decoder is trained with $\lambda = 8$ and weights very little in terms of KL divergence. Yet it controls the global structure of the samples, which shows that our setup can be used to get the best of both worlds. Figure 6 demonstrates that the latent variables z of the encoder have learned meaningfull structure with latent variable interpolations. Samples are obtained by encoding ground truth images, then interpolating the latent variables obtained, decoding them with the decoder of the VAE and adding low level detail with the pixelCNN.

$f(\mathbf{z})$ Conditional PixelCNN samples

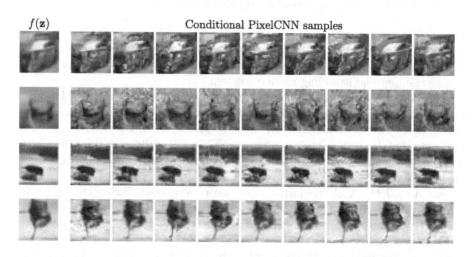

Fig. 5. The column labeled $f(\mathbf{z})$ displays auxiliary representations, with $\mathbf{z}$ sampled from the unit Gaussian prior $p(\mathbf{z})$, accompanied by ten samples of the conditional pixelCNN.

The Auxilliary Loss is Necessary: The fact that the autoregressive decoder ignores the latent variables could be attributed to optimization challenges, as explained in Sect. 3.3. In that case, the auxilliary loss could be used as an initialization scheme only, to guide the model towards a good use of the latent variables. To evaluate this we perform a control experiment where during training we first optimize our objective function in Eq. (7), *i.e.* including the auxiliary reconstruction term, and then switch to optimize the standard objective function of Eq. (6) without the auxiliary term. We proceed by training the full model to convergence then removing the auxiliary loss and fine-tuning from there. Figure 7 displays ground-truth images, with corresponding auxiliary reconstructions and

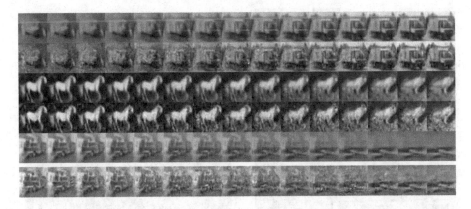

Fig. 6. The first and last columns contain auxilliary reconstructions, images in between are obtained from interpolation of the corresponding latent variables. Odd rows contain auxilliary reconstructions, and even rows contain outputs of the full model.

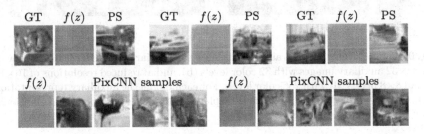

Fig. 7. Auxiliary reconstructions obtained after dropping the auxilliary loss. (GT) denotes ground truth images unseen during training, $f(z)$ is the corresponding intermediate reconstruction, (PS) denotes pixelCNN samples, conditioned on $f(z)$.

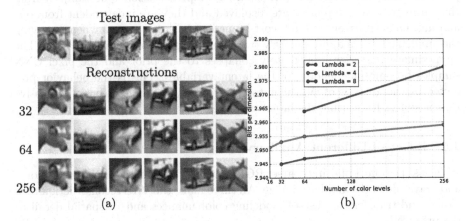

Fig. 8. Impact of the color quantization in the auxiliary image. (a) Reconstructions of the VAE decoder for different quantization levels ($\lambda = 8$). (b) BPD as a function of the quantization level. (Color figure online)

Fig. 9. Samples from models trained with grayscale auxiliary images with 16 color levels (a), 32×32 auxiliary images with 32 color levels (b), and at reduced resolutions of 16×16 (c) and 8×8 pixels (d) with 256 color levels. For each model the auxilliary representation $f(\mathbf{z})$, with $\mathbf{z}$ sampled from the prior, is displayed above the corresponding conditional pixelCNN sample. (Color figure online)

conditional samples, as well as pure samples. The reconstructions have become meaningless and independent from the ground truth images. The samples display the same behavior: for each auxiliary representation four samples from the autoregressive component are displayed and they are independent from one another. Quantitatively, the KL cost immediately drops to zero when removing the auxiliary loss, in approximately two thousand steps of gradient descent. The approximate posterior immediately collapses to the prior and the pixel CNN samples become independent of the latent variables. This is the behavior predicted by the analysis of [5]: the autoregressive decoder is sufficiently expressive that it suffers from using the latent variables.

4.5   Effect of Different Auxiliary Images

We assess the effect of using coarser RGB quantizations, lower spatial resolutions, and grayscale in the auxiliary image. All three make the VAE reconstruction task easier, and transfer the task of modeling color nuances and/or spatial detail to the pixelCNN.

The VAE reconstructions in Fig. 8(a) obtained using coarser color quantization carry less detail than reconstructions based on the original images using 256 color values, as expected. To understand the relatively small impact of the quantization level on the reconstruction, recall that the VAE decoder outputs the continuous means of the logistic distributions regardless of the quantization level. Only the reconstruction loss is impacted by the quantization level via the computation of the probabilities over the discrete color levels in Eq. (10). In Fig. 8(b) we observe small but consistent gains in the BPD metric as the number of color bins is reduced, showing that it is more effective to model color nuances using the pixelCNN, rather than the latent variables. We trained models with auxiliary images down-sampled to 16×16 and 8×8 pixels, which yield 2.94 and 2.93 BPD, respectively. This is comparable to the 2.92 BPD obtained using our best model at scale 32×32. We also trained models with 4-bit per pixel grayscale auxiliary images, as in [17]. While the grayscale auxilliary images are subjectively the ones that have the best global structure, the results are still qualitatively inferior to those obtained by [17] with a pixelCNN modelling grayscale images. Our model does, however, achieve better quantitative performance at 2.93 BPD. In Fig. 9(a) we show samples obtained using models trained with 4-bit per pixel grayscale auxiliary images, in Fig. 9(b) with 32 color levels in the auxiliary image, and in Fig. 9(c) and (d) with auxiliary images of size 16×16 and 8×8. The samples are qualitatively comparable, showing that in all cases the pixelCNN is able to compensate the less detailed outputs of the VAE decoder and that our framework can be used with a variety of intermediate reconstruction losses.

5   Conclusion

We presented a new approach to training generative image models that combine a latent variable structure with an autoregressive model component. Unlike prior approaches, it does not require careful architecture design to trade-off how much is modeled by latent variables and the autoregressive decoder. Instead, this trade-off can be controlled using a regularization parameter and choice of auxiliary target images. We obtain quantitative performance on par with the state of the art on CIFAR10, and samples from our model exhibit globally coherent structure as well as fine details.

Acknowledgments. This work has been partially supported by the grant ANR-16-CE23-0006 "Deep in France" and LabEx PERSYVAL-Lab (ANR-11-LABX-0025-01).

References

1. Arjovsky, M., Chintala, S., Bottou, L.: Wasserstein generative adversarial networks. In: ICML (2017)
2. Bachman, P.: An architecture for deep, hierarchical generative models. In: NIPS (2016)
3. Bengio, Y., Courville, A., Vincent, P.: Representation learning: a review and new perspectives. PAMI **35**(8), 1798–1828 (2013)
4. Burda, Y., Salakhutdinov, R., Grosse, R.: Importance weighted autoencoders. In: ICLR (2016)

5. Chen, X., et al.: Variational lossy autoencoder. In: ICLR (2017)
6. Deshpande, A., Lu, J., Yeh, M.C., Chong, M., Forsyth, D.: Learning diverse image colorization. In: CVPR (2017)
7. Dinh, L., Krueger, D., Bengio, Y.: NICE: non-linear independent components estimation. In: ICLR (2015)
8. Dinh, L., Sohl-Dickstein, J., Bengio, S.: Density estimation using real NVP. In: ICLR (2017)
9. Germain, M., Gregor, K., Murray, I., Larochelle, H.: MADE: masked autoencoder for distribution estimation. In: ICML (2015)
10. Goodfellow, I., et al.: Generative adversarial nets. In: NIPS (2014)
11. Gregor, K., Besse, F., Rezende, D., Danihelka, I., Wierstra, D.: Towards conceptual compression. In: NIPS (2016)
12. Gulrajani, I., et al.: PixelVAE: a latent variable model for natural images. In: ICLR (2017)
13. Kingma, D., Ba, J.: Adam: a method for stochastic optimization. In: ICLR (2015)
14. Kingma, D., Rezende, D., Mohamed, S., Welling, M.: Semi-supervised learning with deep generative models. In: NIPS (2014)
15. Kingma, D., Salimans, T., Jozefowicz, R., Chen, X., Sutskever, I., Welling, M.: Improved variational inference with inverse autoregressive flow. In: NIPS (2016)
16. Kingma, D., Welling, M.: Auto-encoding variational Bayes. In: ICLR (2014)
17. Kolesnikov, A., Lampert, C.: PixelCNN models with auxiliary variables for natural image modeling. In: ICML (2017)
18. Krizhevsky, A.: Learning multiple layers of features from tiny images. Master's thesis, University of Toronto (2009)
19. Larochelle, H., Murray, I.: The neural autoregressive distribution estimator (2011)
20. van den Oord, A., Kalchbrenner, N., Vinyals, O., Espeholt, L., Graves, A., Kavukcuoglu, K.: Conditional image generation with PixelCNN decoders. In: NIPS (2016)
21. Oord, A.v.d., Kalchbrenner, N., Kavukcuoglu, K.: Pixel recurrent neural networks. In: ICML (2016)
22. Rasmus, A., Berglund, M., Honkala, M., Valpola, H., Raiko, T.: Semi-supervised learning with ladder networks. In: NIPS (2015)
23. Reed, S., et al.: Parallel multiscale autoregressive density estimation. In: ICML (2017)
24. Rezende, D., Mohamed, S., Wierstra, D.: Stochastic back propagation and approximate inference in deep generative models. In: ICML (2014)
25. Salimans, T., Karpathy, A., Chen, X., Kingma, D.: Pixelcnn++: improving the pixel CNN with discretized logistic mixture likelihood and other modifications. In: ICLR (2017)
26. Yan, X., Yang, J., Sohn, K., Lee, H.: Attribute2image: conditional image generation from visual attributes. In: ECCV (2016)

Cooperative Multi-agent Policy Gradient

Guillaume Bono[1]([✉])(iD), Jilles Steeve Dibangoye[1], Laëtitia Matignon[1,2],
Florian Pereyron[3], and Olivier Simonin[1]

[1] Univ Lyon, INSA Lyon, INRIA, CITI, 69621 Villeurbanne, France
{guillaume.bono, jillessteeve.dibangoye, laetitia.matignon,
olivier.simonin}@inria.fr
[2] Univ Lyon, Université Lyon 1, LIRIS, CNRS, UMR5205,
69622 Villeurbanne, France
[3] Volvo Group, Advanced Technology and Research, 69800 Saint-Priest, France
florian.pereyron@volvo.com

Abstract. Reinforcement Learning (RL) for decentralized partially observable Markov decision processes (Dec-POMDPs) is lagging behind the spectacular breakthroughs of single-agent RL. That is because assumptions that hold in single-agent settings are often obsolete in decentralized multi-agent systems. To tackle this issue, we investigate the foundations of policy gradient methods within the centralized training for decentralized control (CTDC) paradigm. In this paradigm, learning can be accomplished in a centralized manner while execution can still be independent. Using this insight, we establish policy gradient theorem and compatible function approximations for decentralized multi-agent systems. Resulting actor-critic methods preserve the decentralized control at the execution phase, but can also estimate the policy gradient from collective experiences guided by a centralized critic at the training phase. Experiments demonstrate our policy gradient methods compare favorably against standard RL techniques in benchmarks from the literature. Code related to this paper is available at: https://gitlab.inria.fr/gbono/coop-ma-pg.

Keywords: Decentralized control
Partial observable Markov decision processes
Multi-agent systems · Actor critic

1 Introduction

The past years have seen significant breakthroughs in agents that can gain abilities through interactions with the environment [23,24], thus promising spectacular advances in the society and the industry. These advances are partly due to single-agent (deep) RL algorithms. That is a learning scheme in which the agent

Electronic supplementary material The online version of this chapter (https://doi.org/10.1007/978-3-030-10925-7_28) contains supplementary material, which is available to authorized users.

© Springer Nature Switzerland AG 2019
M. Berlingerio et al. (Eds.): ECML PKDD 2018, LNAI 11051, pp. 459–476, 2019.
https://doi.org/10.1007/978-3-030-10925-7_28

describes its world as a Markov decision process (MDP), other agents being part of that world, and assumptions at both learning and execution phases being identical [31]. In this setting, policy gradient and (natural) actor-critic variants demonstrated impressive results with strong convergence guarantees [1,8,17,32]. These methods directly search in the space of parameterized policies of interest, adjusting the parameters in the direction of the policy gradient. Unfortunately, extensions to cooperative multi-agent systems have restricted attention to either independent learners [28,35] or multi-agent systems with common knowledge about the world [38], which are essentially single-agent systems.

In this paper, we instead consider cooperative multi-agent settings where we accomplished learning in a centralized manner, but execution must be independent. This paradigm allows us to break the independence assumption in decentralized multi-agent systems but only during the training phase, while still preserving the ability to meet it during the execution phase. In many real-world cooperative multi-agent systems, conditions at the training phase do not need to be as strict as those at the execution phase. During rehearsal, for example, actors can read the script, take breaks, or receive feedback from the director, but none of these will be possible during the show [19]. To win matches, a soccer coach develops (before the game) tactics players will apply during the game. So, it is natural to wonder whether the policy gradient approach in such a paradigm could be as successful as for the single-agent learning paradigm.

The CTDC paradigm has been successfully applied in planning methods for Dec-POMDPs, *i.e.*, a framework of choice for sequential decision making by a team of cooperative agents [5,9,16,26,33]. In the literature of game theory, Dec-POMDPs are partially observable stochastic games with identical payoffs. They subsume many other collaborative multi-agent models, including multi-agent MDPs [7]; stochastic games with identical payoffs [30]; to cite a few. The critical assumption that makes Dec-POMDPs significantly different from MDPs holds only at the execution phase: agents can neither see the real state of the world nor explicitly communicate with one another their noisy observations. Nonetheless, agents can share their local information at the training phase, as long as they act at the execution phase based solely on their individual experience. Perhaps surprisingly, this insight has been neglected so far, explaining the formal treatment of CTDC received little attention from the RL community [19]. When this centralized training takes place in a simulator or a laboratory, one can exploit information that may not be available at the execution time, *e.g.*, hidden states, local information of the other agents, etc. Recent work in the (deep) multi-agent RL community builds upon this paradigm to design domain-specific methods [14,15,22], but the theoretical foundations of decentralized multi-agent RL are still in their infancy.

This paper investigates the theoretical foundations of policy gradient methods within the CTDC paradigm. In this paradigm, among policy gradient algorithms, actor-critic methods can train multiple independent actors (or policies) guided by a centralized critic (Q-value function) [14]. Methods of this family differ only through how they represent and maintain the centralized critic. The primary result of this article generalizes the policy gradient theorem and compatible function approximations from (PO)MDPs to Dec-POMDPs. In particular, these results show the compatible centralized critic is the sum of individual critics, each of which is linear in the "features" of its corresponding individual policy. Even more interestingly, we derive update rules adjusting individual critics in the direction of the gradient of the centralized critic. Experiments demonstrate our policy gradient methods compare favorably against techniques from standard RL paradigms in benchmarks from the literature. Proofs of our results are provided in the companion research report [6].

We organized the rest of this paper as follows. Section 2 gives formal definitions of POMDPs and Dec-POMDPs along with useful properties. In Sect. 3, we review the policy gradient methods for POMDPs, then pursue the review for cooperative multi-agent settings in Sect. 4. Section 5 develops the theoretical foundations of policy gradient methods for Dec-POMDPs and derives the algorithms. Finally, we present empirical results in Sect. 6.

2 Backgrounds

2.1 Partially Observable Markov Decision Processes

Consider a (centralized coordinator) agent facing the problem of influencing the behavior of a POMDP as it evolves through time. This setting often serves to formalize cooperative multi-agent systems, where all agents can explicitly and instantaneously communicate with one another their noisy observations.

Definition 1. *Let $M_1 \doteq (\mathcal{X}, \mathcal{U}, \mathcal{Z}, p, r, T, s_0, \gamma)$ be a POMDP, where X_t, U_t, Z_t and R_t are random variables taking values in $\mathcal{X}$, $\mathcal{U}$, $\mathcal{Z}$ and $\mathbb{R}$, and representing states of the environment, controls the agent took, observations and reward signals it received at time step $t = 0, 1, \ldots, T$, respectively. State transition and observation probabilities $p(x', z'|x, u) \doteq \mathbb{P}(X_{t+1} = x', Z_{t+1} = z'|X_t = x, U_t = u)$ characterize the world dynamics. $r(x, u) \doteq \mathbb{E}[R_{t+1}|X_t = x, U_t = u]$ is the expected immediate reward. Quantities s_0 and $\gamma \in [0, 1]$ define the initial state distribution and the discount factor.*

We call t^{th} history, $o_t \doteq (o_{t-1}, u_{t-1}, z_t)$ where $o_0 \doteq \emptyset$, a sequence of controls and observations the agent experienced up to time step $t = 0, 1, \ldots, T$. We denote $\mathcal{O}_t$ the set of histories of the agent might experience up to time step t.

Definition 2. *The agent selects control u_t through time using a parametrized policy $\pi \doteq (a_0, a_1, \ldots, a_T)$, where $a_t(u_t|o_t) \doteq \mathbb{P}_{\theta_t}(u_t|o_t)$ denotes the decision rule at time step $t = 0, 1, \ldots, T$, with parameter vector $\theta_t \in \mathbb{R}^{\ell_t}$ where $\ell_t \ll |\mathcal{O}_t|$.*

In practice, we represent policies using a deep neural network; a finite-state controller; or a linear approximation architecture, *e.g.*, Gibbs. Such policy representations rely on different (possibly lossy) descriptions of histories, called internal states. It is worth noticing that when available, one can use p to calculate a unique form of internal-states, called *beliefs*, which are sufficient statistics of histories [3]. If we let $b^o \doteq \mathbb{P}(X_t | O_t = o)$ be the current belief induced by history o, with initial belief $b^\emptyset \doteq s_0$; then, the next belief after taking control $u \in \mathcal{U}$ and receiving observation $z' \in \mathcal{Z}$ is:

$$b^{o,u,z'}(x') \doteq \mathbb{P}\big(X_{t+1} = x' | O_{t+1} = (o, u, z')\big) \propto \sum_{x \in \mathcal{X}} p(x', z' | x, u) b^o(x), \quad \forall x' \in \mathcal{X}.$$

Hence, using beliefs instead of histories in the description of policies preserves the ability to act optimally, while significantly reducing the memory requirement. Doing so makes it possible to restrict attention to stationary policies, which are particularly useful for infinite-horizon settings, *i.e.*, $T = \infty$. Policy π is said to be stationary if $a_0 = a_1 = \ldots = a$ and $\theta_0 = \theta_1 = \ldots = \theta$; otherwise, it is non-stationary.

Through interactions with the environment under policy π, the agent generates a trajectory of rewards, observations, controls and states $\omega_{t:T} \doteq (x_{t:T}, z_{t:T}, u_{t:T})$. Each trajectory produces return $R(\omega_{t:T}) \doteq \gamma^0 r(s_t, u_t) + \cdots + \gamma^{T-t} r(s_T, u_T)$. Policies of interest are those that achieve the highest expected return starting at s_0

$$J(s_0; \theta_{0:T}) \doteq \mathbb{E}_{\pi, M_1}[R(\Omega_{0:T})] = \int \mathbb{P}_{\pi, M_1}(\omega_{0:T}) R(\omega_{0:T}) \mathrm{d}\omega_{0:T} \qquad (1)$$

where $\mathbb{P}_{\pi, M_1}(\omega_{0:T})$ denotes the probability of generating trajectory $\omega_{0:T}$ under π. Finding the best way for the agent to influence M_1 consists in finding parameter vector $\theta^*_{0:T}$ that satisfies: $\theta^*_{0:T} \in \arg\max_{\theta_{0:T}} J(s_0; \theta_{0:T})$.

It will prove useful to break the performance under policy π into pieces to exploit the underlying structure—*i.e.*, the performance of π from time step t onward depend on earlier controls only through the current states and histories. To this end, the following defines value, Q-value and advantage functions under π. The Q-value functions under π is given by:

$$Q^\pi_t : (x, o, u) \mapsto \mathbb{E}_{\pi, M_1}[R(\Omega_{t:T}) | X_t = x, O_t = o, U_t = u], \quad \forall t = 0, 1, \ldots \quad (2)$$

where $Q^\pi_t(x, o, u)$ denotes the expected return of executing u starting in x and o at time step t and then following policy π from time step $t+1$ onward. The value functions under π is given by:

$$V^\pi_t : (x, o) \mapsto \mathbb{E}_{a_t}[Q^\pi_t(x, o, U_t)], \qquad \forall t = 0, 1, \ldots \quad (3)$$

where $V^\pi_t(x, o)$ denotes the expected return of following policy π from time step t onward, starting in x and o. Finally, the advantage functions under π is given by:

$$A^\pi_t : (x, o, u) \mapsto Q^\pi_t(x, o, u) - V^\pi_t(x, o), \qquad \forall t = 0, 1, \ldots \quad (4)$$

where $A^\pi_t(x, o, u)$ denotes the relative advantage of executing u starting in x and o at time step t and then the following policy π from time step $t+1$ onward. The nice property of these functions is that they satisfy certain recursions.

Lemma 3 (Bellman equations [4]). *Q-value functions under π satisfy the following recursion:* $\forall t = 0, 1, \ldots, T, \forall x \in \mathcal{X}, o \in \mathcal{O}_t, u \in \mathcal{U}$,

$$Q_t^\pi(x, o, u) = R(x, u) + \gamma \mathbb{E}_{a_{t+1}, p}[Q_{t+1}^\pi(X_{t+1}, O_{t+1}, U_{t+1}) | X_t = x, O_t = o, U_t = u]$$

Lemma 3 binds altogether $V_{0:T}^\pi$, $Q_{0:T}^\pi$ and $A_{0:T}^\pi$, including overall performance $J(s_0; \theta_{0:T}) = \mathbb{E}_{s_0}[V_0^\pi(X_0, \emptyset)]$.

So far we restricted our attention to systems under the control of a single agent. Next, we shall generalize to settings where multiple agents cooperate to control the same system in a decentralized manner.

2.2 Decentralized Partially Observable Markov Decision Processes

Consider a slightly different framework in which n agents cooperate when facing the problem of influencing the behavior of a POMDP, but can neither see the state of the world and nor communicate with one another their noisy observations.

Definition 4. *A Dec-POMDP $M_n \doteq (\mathcal{I}_n, \mathcal{X}, \mathcal{U}, \mathcal{Z}, p, R, T, \gamma, s_0)$ is such that $i \in \mathcal{I}_n$ indexes the ith agent involved in the process; $\mathcal{X}, \mathcal{U}, \mathcal{Z}, p, R, T, \gamma$ and s_0 are as in M_1; $\mathcal{U}^i$ is an individual control set of agent i, such that $\mathcal{U} = \mathcal{U}^1 \times \cdots \times \mathcal{U}^n$ specifies the set of controls $u = (u^1, \ldots, u^n)$; $\mathcal{Z}^i$ is an individual observation set of agent i, where $\mathcal{Z} = \mathcal{Z}^1 \times \cdots \times \mathcal{Z}^n$ defines the set of observations $z = (z^1, \ldots, z^n)$.*

We call the individual history of agent $i \in \mathcal{I}_n$, $o_t^i = (o_{t-1}^i, u_{t-1}^i, z_t^i)$ where $o_0^i = \emptyset$, the sequence of controls and observations up to time step $t = 0, 1, \ldots, T$. We denote $\mathcal{O}_t^i$, the set of individual histories of agent i at time step t.

Definition 5. *Agent $i \in \mathcal{I}_n$ selects control u_t^i at the tth time step using a parametrized policy $\pi^i \doteq (a_0^i, a_1^i, \ldots, a_T^i)$, where $a_t^i(u_t^i | o_t^i) \doteq \mathbb{P}_{\theta_t^i}(u_t^i | o_t^i)$ is a parametrized decision rule, with parameter vector $\theta_t^i \in \mathbb{R}^{\ell_t^i}$, assuming $\ell_t^i \ll |\mathcal{O}_t^i|$.*

Similarly to M_1, individual histories grow every time step, which quickly becomes untractable. The only sufficient statistic for individual histories known so far [9,11] relies on the *occupancy state* given by: $s_t(x, o) \doteq \mathbb{P}_{\theta_{0:T}^{1:n}, M_n}(x, o)$, for all $x \in \mathcal{X}$ and $o \in \mathcal{O}_t$. The individual occupancy state induced by individual history $o^i \in \mathcal{O}_t^i$ is a conditional distribution probability: $s_t^i(x, o^{-i}) \doteq \mathbb{P}(x, o^{-i} | o^i, s_t)$, where o^{-i} is the history of all agents except i. Learning to map individual histories to internal states close to individual occupancy states is hard, which limits the ability to find optimal policies in M_n. One can instead restrict attention to stationary individual policies, by mapping the history space into a finite set of possibly lossy representations of individual occupancy states, called internal states $\varsigma \doteq (\varsigma^1, \ldots, \varsigma^n)$, e.g., nodes in finite-state controllers or hidden state of a Recurrent Neural Network (RNN). We define transition rules prescribing the next internal state given the current internal state, control and next observation as follows: $\psi: (\varsigma, u, z') \mapsto (\psi^1(\varsigma^1, u^1, z'^1), \ldots, \psi^n(\varsigma^n, u^n, z'^n))$ where

$\psi^i: (\varsigma^i, u^i, z'^i) \mapsto \varsigma'^i$ is an individual transition rule. In general, ψ and $\psi^{1:n}$ are stochastic transition rules. In the following, we will consider these rules fixed a-priori.

The goal of solving M_n is to find a joint policy $\pi \doteq (\pi^1, \ldots, \pi^n)$, i.e., a tuple of individual policies, one for each agent—that achieves the highest expected return, $\theta_{0:T}^{*,1:n} \in \arg\max_{\theta_{0:T}^{1:n}} J(s_0; \theta_{0:T}^{1:n})$, starting at initial belief s_0: $J(s_0; \theta_{0:T}^{1:n}) \doteq \mathbb{E}_{\pi, M_n}[R(\Omega_{0:T})]$. M_n inherits all definitions introduced for M_1, including functions $V_{0:T}^\pi$, $Q_{0:T}^\pi$ and $A_{0:T}^\pi$ for a given joint policy π.

3 Policy Gradient for POMDPs

In this section, we will review the literature of policy gradient methods for centralized single-agent systems. In this setting, the policy gradient approach consists of a centralized algorithm which searches the best $\theta_{0:T}$ in the parameter space. Though, we restrict attention to non-stationary policies, methods discussed here easily extend to stationary policies when $a_t = a$, i.e. $\theta_t = \theta$, for all $t = 0, 1, \ldots, T$. Assuming π is differentiable w.r.t. its parameter vector, $\theta_{0:T}$, the centralized algorithm updates $\theta_{0:T}$ in the direction of the gradient:

$$\Delta\theta_{0:T} = \alpha \frac{\partial J(s_0; \theta_{0:T})}{\partial \theta_{0:T}}, \tag{5}$$

where α is the step-size. Applying iteratively such a centralized update rule, assuming a correct estimation of the gradient, $\theta_{0:T}$ can usually converge towards a local optimum. Unfortunately, correct estimation of the gradient may not be possible. To overcome this limitation, one can rely on an unbiased estimation of the gradient, actually restricting (5) to stochastic gradient: $\Delta\theta_{0:T} = \alpha R(\omega_{0:T}) \frac{\partial}{\partial \theta_{0:T}} \log \mathbb{P}_{\pi, M_n}(\omega_{0:T})$. We compute $\frac{\partial}{\partial \theta_{0:T}} \log \mathbb{P}_{\pi, M_n}(\omega_{0:T})$ with no knowledge of the trajectory distribution $\mathbb{P}_{\pi, M_n}(\omega_{0:T})$. Indeed $\mathbb{P}_{\pi, M_n}(\omega_{0:T}) \doteq s_0(x_0) \prod_{t=0}^{T} p(x_{t+1}, z_{t+1}|x_t, u_t) a_t(u_t|o_t)$ implies:

$$\frac{\partial \log \mathbb{P}_{\pi, M_n}(\omega_{0:T})}{\partial \theta_{0:T}} = \frac{\partial \log a_0(u_0|o_0)}{\partial \theta_0} + \ldots + \frac{\partial \log a_T(u_T|o_T)}{\partial \theta_T}.$$

3.1 Likelihood Ratio Methods

Likelihood ratio methods, e.g., Reinforce [36], exploit the separability of parameter vectors $\theta_{0:T}$, which leads to the following update rule:

$$\Delta\theta_t = \alpha \mathbb{E}_{\mathcal{D}}\left[R(\omega_{0:T}) \frac{\partial \log a_t(u_t|o_t)}{\partial \theta_t} \right] \qquad \forall t = 0, 1, \ldots, T \tag{6}$$

where $\mathbb{E}_{\mathcal{D}}[\cdot]$ is the average over trajectory samples $\mathcal{D}$ generated under policy π. The primary issue with this centralized update-rule is the high-variance of $R(\Omega_{0:T})$, which can significantly slow down the convergence. To somewhat mitigate this high-variance, one can exploit two observations. First, it is easy to see that future actions do not depend on past rewards, *i.e.*, $\mathbb{E}_{\mathcal{D}}[R(\omega_{0:t-1})\frac{\partial}{\partial \theta_t} \log a_t(u_t|o_t)] = 0$. This insight allows us to use $R(\omega_{t:T})$ instead of $R(\omega_{0:T})$ in (6), thereby resulting in a significant reduction in the variance of the policy gradient estimate. Second, it turns out that the absolute value of $R(\omega_{t:T})$ is not necessary to obtain an unbiased policy gradient estimate. Instead, we only need a relative value $R(\omega_{t:T}) - \beta_t(x_t, o_t)$, where $\beta_{0:T}$ can be any arbitrary value function, often referred to as *a baseline*.

3.2 Actor-Critic Methods

To moderate even more the variance for the gradient estimate in (6), the policy gradient theorem [32] suggests replacing $R(\omega_{t:T})$ by $Q_t^{\mathrm{w}}(x_t, o_t, u_t)$, *i.e.*, an approximate value of taking control u_t starting in state x_t and history o_t and then following policy π from time step $t + 1$ onward: $Q_t^{\mathrm{w}}(x_t, o_t, u_t) \approx Q_t^{\pi}(x_t, o_t, u_t)$, where $\mathrm{w}_t \in \mathbb{R}^{l_t}$ is a parameter vector with $l_t \ll |\mathcal{X}||\mathcal{O}_t||\mathcal{U}|$. Doing so leads us to the actor-critic algorithmic scheme, in which a centralized algorithm maintains both parameter vectors $\theta_{0:T}$ and parameter vectors $\mathrm{w}_{0:T}$: $\forall t = 0, 1, \ldots, T$,

$$\Delta \mathrm{w}_t = \alpha \mathbb{E}_{\mathcal{D}} \left[\delta_t \frac{\partial \log a_t(u_t|o_t)}{\partial \theta_t} \right] \tag{7}$$

$$\Delta \theta_t = \alpha \mathbb{E}_{\mathcal{D}} \left[Q_t^{\mathrm{w}}(x_t, o_t, u_t) \frac{\partial \log a_t(u_t|o_t)}{\partial \theta_t} \right] \tag{8}$$

where $\delta_t \doteq \widehat{Q}_t^{\pi}(x_t, o_t, u_t) - Q_t^{\mathrm{w}}(x_t, o_t, u_t; w_t)$ and $\widehat{Q}_t^{\pi}(x_t, o_t, u_t)$ is an unbiased estimate of true Q-value $Q_t^{\pi}(x_t, o_t, u_t)$.

The choice of parameter vector $\mathrm{w}_{0:T}$ is critical to ensure the gradient estimation remains unbiased [32]. There is no bias whenever Q-value functions $Q_{0:T}^{\mathrm{w}}$ are *compatible* with parametrized policy π. Informally, a compatible function approximation $Q_{0:T}^{\mathrm{w}}$ of $Q_{0:T}^{\pi}$ should be linear in "features" of policy π, and its parameters $\mathrm{w}_{0:T}$ are the solution of a linear regression problem that estimates $Q_{0:T}^{\pi}$ from these features. In practice, we often relax the second condition and update parameter vector $\mathrm{w}_{0:T}$ using Monte-Carlo or temporal-difference learning methods.

3.3 Natural Actor-Critic Methods

Following the direction of the gradient might not always be the right option to take. In contrast, the natural gradient suggests updating the parameter vector $\theta_{0:T}$ in the steepest ascent direction w.r.t. the Fisher information metric

$$\boldsymbol{\Phi}(\theta_t) \doteq \mathbb{E}_{\mathcal{D}} \left[\frac{\partial \log a_t(u_t|o_t)}{\partial \theta_t} \left(\frac{\partial \log a_t(u_t|o_t)}{\partial \theta_t} \right)^{\top} \right]. \tag{9}$$

This metric is invariant to re-parameterizations of the policy. Combining the policy gradient theorem with the compatible function approximations and then taking the steepest ascent direction, $\mathbb{E}_{\mathcal{D}}[\boldsymbol{\Phi}(\theta_t)^{-1}\boldsymbol{\Phi}(\theta_t)w_t]$, results in natural actor-critic algorithmic scheme, which replaces the update rule (8) by: $\Delta \theta_t = \alpha \mathbb{E}_{\mathcal{D}}[w_t]$.

4   Policy Gradient for Multi-Agent Systems

In this section, we review extensions of single-agent policy gradient methods to cooperative multi-agent settings. We shall distinguish between three paradigms: centralized training for centralized control (CTCC) *vs* distributed training for decentralized control (DTDC) *vs* centralized training for decentralized control (CTDC), illustrated in Fig. 1.

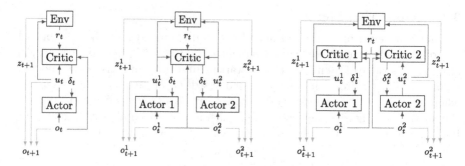

Fig. 1. Best viewed in color. For each paradigms—(*left*) CTCC; (*center*) CTDC; and (*right*) DTDC—we describe actor-critic algorithmic schemes. We represent in blue, green and red arrows: forward control flow; the aggregation of information for the next time step; and the feedback signals back-propagated to update all parameters, respectively.

4.1   Centralized Training for Centralized Control (CTCC)

Some cooperative multi-agent applications have cost-free instantaneous communications. Such applications can be modeled as POMDPs, making it possible to use single-agent policy gradient methods (Sect. 3). In such a CTCC paradigm, *see* Fig. 1 *(left)*, centralized single-agent policy gradient methods use a single critic and a single actor. The major limitation of this paradigm is also its strength: the requirement for instantaneous, free and noiseless communications among all agents till the end of the process both at the training and execution phases.

4.2   Distributed Training for Decentralized Control (DTDC)

Perhaps surprisingly, the earliest multi-agent policy gradient method aims at learning in a distributed manner policies that are to be executed in a decentralized way, *e.g.*, distributed `Reinforce` [28]. In this DTDC paradigm, *see* Fig. 1 *(right)*, agents simultaneously but independently learn via `Reinforce` their individual policies using multiple critics and multiple actors. The independence of parameter vectors $\theta_{0:T}^1, \ldots, \theta_{0:T}^n$, leads us to the following distributed update-rule:

$$\Delta\theta_t^i = \alpha \mathbb{E}_{\mathcal{D}}\left[R(\omega_{0:T}) \frac{\partial \log a_t^i(u_t^i|o_t^i)}{\partial \theta_t^i} \right], \qquad \forall t = 0, 1, \ldots, T, \forall i \in I_n \qquad (10)$$

Interestingly, the sum of individual policy gradient estimates is an unbiased estimate of the joint policy gradient. However, how to exploit insights from actor-critic methods (Sect. 3) to combat high-variance in the joint policy gradient estimate remains an open question. Distributed `Reinforce` restricts to on-policy setting, off-policy methods instead can significantly improve the exploration, *i.e.*, learns target joint policy π while following and obtaining trajectories from behavioral joint policy $\bar{\pi}$ [8].

4.3 Centralized Training for Decentralized Control (CTDC)

The CTDC paradigm has been successfully applied in planning [2,5,9–11,13, 16,26,27,33,34] and learning [12,19–21] for M_n. In such a paradigm, a centralized coordinator agent learns on behalf of all agents at the training phase and then assigns policies to corresponding agents before the execution phase takes place. Actor-critic algorithms in this paradigm, *see* Fig. 1 *(center)*, maintain a centralized critic but learn multiple actors, one for each agent.

Recent work in the (deep) multi-agent RL builds upon this paradigm [14, 15,22], but lacks theoretical foundations, resulting in different specific forms of centralized critics, including: individual critics with shared parameters [15]; or counterfactual-regret based centralized critics [14]. Theoretical results similar to ours were previously developed for *collective* multi-agent planning domains [25], *i.e.*, a setting where all agents have the same policy, but their applicability to general Dec-POMDPs remain questionable.

5 Policy Gradient for Dec-POMDPs

In this section, we address the limitation of both CTCC and DTDC paradigms and extend both 'vanilla' and natural actor-critic algorithmic schemes from M_1 to M_n.

5.1 The Policy Gradient Theorem

Our primary result is an extension of the policy gradient theorem [32] from M_1 to M_n. First, we state the partial derivatives of value functions $V_{0:T}^\pi$ w.r.t. the parameter vectors $\theta_{0:T}^{1:n}$ for finite-horizon settings.

Lemma 6. *For any arbitrary M_n, target joint policy $\pi \doteq (a_0, \ldots, a_T)$ and behavior joint policy $\bar{\pi} \doteq (\bar{a}_0, \ldots, \bar{a}_T)$, the following holds, for any arbitrary $t = 0, 1, \ldots, T$, and agent $i \in \mathcal{I}_n$, hidden state $x_t \in \mathcal{X}$, and joint history $o_t \in \mathcal{O}_t$:*

$$\frac{\partial V_t^\pi(x_t, o_t)}{\partial \theta_t^i} = \mathbb{E}_{\bar{a}_t}\left[\frac{a_t(U_t|o_t)}{\bar{a}_t(U_t|o_t)} Q_t^\pi(x_t, o_t, U_t) \frac{\partial \log a_t^i(U_t^i|o_t^i)}{\partial \theta_t^i} \right]. \tag{11}$$

We are now ready to state the main result of this section.

Theorem 7. *For any arbitrary M_n, target joint policy $\pi \doteq (a_0, \ldots, a_T)$ and behavior joint policy $\bar{\pi} \doteq (\bar{a}_0, \ldots, \bar{a}_T)$, the following holds:*

1. *for finite-horizon settings $T < \infty$, any arbitrary $t = 0, 1, \ldots, T$ and $i \in \mathcal{I}_n$,*

$$\frac{\partial J(s_0; \theta_{0:T}^{1:n})}{\partial \theta_t^i} = \gamma^t \mathbb{E}_{\bar{a}_t, M_n} \left[\frac{a_t(U_t|O_t)}{\bar{a}_t(U_t|O_t)} Q_t^\pi(X_t, O_t, U_t) \frac{\partial \log a_t^i(U_t^i|O_t^i)}{\partial \theta_t^i} \right].$$

2. *for finite-horizon settings $T = \infty$, and any arbitrary agent $i \in \mathcal{I}_n$,*

$$\frac{\partial J(s_0; \theta^{1:n})}{\partial \theta^i} = \mathbb{E}_{\bar{s}, \bar{a}} \left[\frac{a(U|\Sigma)}{\bar{a}(U|\Sigma)} Q^\pi(X, \Sigma, U) \frac{\partial \log a^i(U^i|\Sigma^i)}{\partial \theta^i} \right],$$

where $\bar{s}(x, \varsigma) \doteq \sum_{t=0}^{\infty} \gamma^t \mathbb{P}_{\bar{a}, \psi, M_n}(X_t = x, \Sigma_t = \varsigma)$.

While the policy gradient theorem for M_1 [32] assumes a single agent learning to act in a (PO)MDP, Theorem 7 applies to multiple agents learning to control a POMDP in a decentralized manner. Agents act independently, but their policy gradient estimates are guided by a centralized Q-value function $Q_{0:T}^\pi$. To use this property in practice, one needs to replace $Q_{0:T}^\pi$ with a function approximation of $Q_{0:T}^\pi$. To ensure this function approximation is compatible—*i.e.*, the corresponding gradient still points roughly in the direction of the real gradient, we carefully select its features. The following addresses this issue for M_n.

5.2   Compatible Function Approximations

The main result of this section characterizes compatible function approximations $V_{0:T}^\sigma$ and $A_{0:T}^\nu$ for both the value function $V_{0:T}^\pi$ and the advantage function $A_{0:T}^\pi$ of any arbitrary M_n, respectively. These functions together shall provide a function approximation for $Q_{0:T}^\pi$ assuming $Q_t^\pi(x_t, o_t, u_t) \doteq V_t^\pi(x_t, o_t) + A_t^\pi(x_t, o_t, u_t)$, for any time step $t = 0, 1, \ldots, T$, state x_t, joint history o_t and joint control u_t.

Theorem 8. *For any arbitrary M_n, function approximations $V_{0:T}^\sigma$ and $A_{0:T}^\nu$, with parameter vectors $\sigma_{0:T}^{1:n}$ and $\nu_{0:T}^{1:n}$ respectively, are compatible with parametric joint policy $\pi \doteq (a_0, \ldots, a_T)$, with parameter vector $\theta_{0:T}^{1:n}$, if one of the following holds: $\forall t = 0, 1, \ldots, T$*

1. *for any state $x_t \in \mathcal{X}$, joint history $o_t \in \mathcal{O}_t$, and agent $i \in \mathcal{I}_n$,*

$$\frac{\partial V_t^\sigma(x_t, o_t)}{\partial \sigma_t^i} = \mathbb{E}_{a_t^i} \left[\frac{\partial \log a_t^i(U_t^i|o_t^i)}{\partial \theta_t^i} \right]. \tag{12}$$

   and σ minimizes the MSE $\mathbb{E}_{\pi, M_n}[\epsilon_t(X_t, O_t, U_t)^2]$

2. *for any state $x_t \in \mathcal{X}$, joint history $o_t \in \mathcal{O}_t$, joint control $u_t \in \mathcal{U}$, and agent $i \in \mathcal{I}_n$,*

$$\frac{\partial A_t^\nu(x_t, o_t, u_t)}{\partial \nu_t^i} = \frac{\partial \log a_t^i(u_t^i|o_t^i)}{\partial \theta_t^i} \tag{13}$$

   and ν minimizes the MSE $\mathbb{E}_{\pi, M_n}[\epsilon_t(X_t, O_t, U_t)^2]$

where $\epsilon_t(x,o,u) \doteq Q_t^\pi(x,o,u) - V_t^\sigma(x,o) - A_t^\nu(x,o,u)$. Then, $\frac{\partial}{\partial\theta_t^i}V_t^\pi(x_t,o_t)$ follows

$$\mathbb{E}_{\bar{a}_t}\left[\frac{a_t(U_t|o_t)}{\bar{a}_t(U_t|o_t)}\left(V_t^\sigma(x_t,o_t) + A_t^\nu(x_t,o_t,U_t)\right)\frac{\partial\log a_t^i(U_t^i|o_t^i)}{\partial\theta_t^i}\right], \quad (14)$$

for any behavior joint policy $\bar{\pi} \doteq (\bar{a}_0,\ldots,\bar{a}_T)$.

We state Theorem 8 for non-stationary policies and $T < \infty$, but the result naturally extends to infinite-horizon and stationary policies. The theorem essentially demonstrates how compatibility conditions generalize from M_1 to M_n. Notable properties of a compatible centralized critic include the *separability* w.r.t. individual approximators:

$$V_t^\sigma : (x_t,o_t) \mapsto \sum_{i\in I_n}\mathbb{E}_{a_t^i}\left[\frac{\partial\log a_t^i(U_t^i|o_t^i)}{\partial\theta_t^i}\right]^\top \sigma_t^i + \beta_t(x_t,o_t), \quad (15)$$

$$A_t^\nu : (x_t,o_t,u_t) \mapsto \sum_{i\in I_n}\left(\frac{\partial\log a_t^i(u_t^i|o_t^i)}{\partial\theta_t^i}\right)^\top \nu_t^i + \tilde{\beta}_t(x_t,o_t,u_t), \quad (16)$$

where $\beta_{0:T}$ and $\tilde{\beta}_{0:T}$ are baselines independent of $\theta_{0:T}^{1:n}$, $\nu_{0:T}^{1:n}$ and $\sigma_{0:T}^{1:n}$. Only one of (12) or (13) needs to be verified to preserve the direction of the policy gradient. Similarly to the compatibility theorem for M_1, the freedom granted by the potentially unconstrained approximation and the baselines can be exploited to reduce the variance of the gradient estimation, but also take advantage of extra joint or hidden information unavailable to the agents at the execution phase. We can also benefit from the separability of both approximators at once to decrease the number of learned parameters and speed up the training phase for large-scale applications. Finally, the separability of function approximators does not allow us to independently maintain individual critics, the gradient estimation is still guided by a centralized critic.

5.3 Actor-Critic for Decentralized Control Algorithms

In this section, we derive actor-critic algorithms for M_n that exploit insights from Theorem 8, as illustrated in Algorithm 1, namely *Actor-Critic for Decentralized Control* (ACDC). This algorithm is model-free, centralized[1], off-policy and iterative. Each iteration consists of policy evaluation and policy improvement. The policy evaluation composes a mini-batch based on trajectories sampled from $\mathbb{P}_{\bar{\pi},M_n}(\Omega_{0:T})$ and the corresponding temporal-difference errors, *see* lines (6–11). The policy improvement updates θ, ν, and σ by taking the average over mini-batch samples and exploiting compatible function approximations, *see* lines (12–16), where $\phi_t^i(o_t,u_t) \doteq \frac{\partial}{\partial\theta_{t,h}^i}\log a_t^i(u_t^i|o_t^i)$.

---

[1] One can easily extend this algorithm to allow agents to collaborate during the training phase by exchanging their local information, and hence makes it a distributed algorithm.

Algorithm 1: Actor-Critic for Decentralized Control (ACDC).

```
1  ACDC()
2  │  Initialize θ₀, ν₀, σ₀ arbitrarily and h ← 0.
3  │  while θₕ has not converged do
4  │  │  evaluation() and improvement()
5  │  └  h ← h + 1

6  evaluation()
7  │  Initialize 𝒟ₕ₀:ₜ ← ∅
8  │  for j = 1 ... m and t = 0 ... T do
9  │  │  Sample trajectory step (xₜ:ₜ₊₁, oₜ:ₜ₊₁, uₜ) ~ āₜ, p
10 │  │  Evaluate δₜ ← rₜ + γVσₜ₊₁(xₜ₊₁, oₜ₊₁) − Vσₜ(xₜ, oₜ)
11 │  │  Compute weighting factor ρₜ(oₜ, uₜ) ← aₜ(uₜ|oₜ)/āₜ(uₜ|oₜ)
12 │  └  Compose batch 𝒟ₜ,ₕ ← {(oₜ, uₜ, δₜ, ρₜ(uₜ, oₜ))} ∪ 𝒟ₜ,ₕ

13 improvement()
14 │  for i = 1 ... n and t = 0 ... T do
15 │  │  Baseline σⁱₜ,ₕ₊₁ ← σⁱₜ,ₕ + αᵍₕ𝔼_{𝒟ₜ,ₕ}{δₜρₜ(oₜ, uₜ)φⁱₜ,ₕ(oⁱₜ, uⁱₜ)}
16 │  │  Critic νⁱₜ,ₕ₊₁ ← νⁱₜ,ₕ + ανₕ𝔼_{𝒟ₜ,ₕ}{δₜρₜ(oₜ, uₜ)φⁱₜ,ₕ(oⁱₜ, uⁱₜ)}
17 │  └  Actor θⁱₜ,ₕ₊₁ ← θⁱₜ,ₕ + αθₕ𝔼_{𝒟ₜ,ₕ}{ρₜ(oₜ, uₜ)φⁱₜ,ₕ(oⁱₜ, uⁱₜ)(Aνₜ(oₜ, uₜ)+Vσₜ(oₜ))}
```

The step-sizes α_h^θ, α_h^ν and α_h^σ should satisfy the standard Robbins and Monro's conditions for stochastic approximation algorithms [29], *i.e.*, $\sum_{h=0}^\infty \alpha_h = \infty$, $\sum_{h=0}^\infty \alpha_h^2 < \infty$. Moreover, according to [18], they should be scheduled such that we update θ at a slower time-scale than ν and σ to ensure convergence. To ease the maximum improvement of a joint policy for a constant fixed change of its parameters, the method of choice is the natural policy gradient [1,17]. The natural ACDC (NACDC) differs from ACDC only in the update of the actors: $\theta_{t,h+1}^i \leftarrow \theta_{t,h}^i + \alpha_h^\theta \mathbb{E}_{\mathcal{D}_{t,h}}[\frac{a_t(u_t|o_t)}{\bar{a}_t(u_t|o_t)}\nu_t^i]$. We elaborate on this analysis of natural Policy Gradient in our companion research report [6].

We conclude this section with remarks on theoretical properties of ACDC algorithms. First, they are guaranteed to converge with probability one under mild conditions to local optima as they are true gradient descent algorithms [8]. The basic argument is that they minimize the mean square projected error by stochastic gradient descent, *see* [8] for further details. They further terminate with a local optimum that is also a Nash equilibrium, *i.e.*, the partial derivatives of the centralized critic w.r.t. any parameter is zero only at an equilibrium point.

6   Experiments

In this section, we empirically demonstrate and validate the advantage of CTDC over CTCC and DTDC paradigms. We show that ACDC methods compare favorably w.r.t. existing algorithms on many decentralized multi-agent domains from the literature. We also highlight limitations that preclude the current implementation of our methods to achieve better performances.

6.1 Experimental Setup

As discussed throughout the paper, there are many key components in actor-critic methods that can affect their performances. These key components include: training paradigms (CTCC *vs* DTDC *vs* CTDC); policy representations (stationary *vs* non-stationary policies); approximation architectures (linear approximations *vs* deep recurrent neural networks); history representations (truncated histories *vs* hidden states of deep neural networks). We implemented three variants of actor-critic methods that combine these components. Unless otherwise mentioned, we will refer to actor-critic methods from: the acronym of the paradigm in which they have been implemented, *e.g.*, CTDC for ACDC; plus the key components, "*CTDC_TRUNC(K)*" for ACDC where we use K last observations instead of histories (non-stationary policy); or "*DTDC_RNN*" for distributed Reinforce where we use RNNs (stationary policy), see Fig. 2.

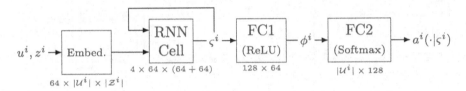

Fig. 2. Best viewed in color. Recurrent neural network architecture used to represent actors of agent $i \in I_n$. The blue boxes are standard neural network layers, red text denotes intermediate tensors computed during forward pass, and green text indicates the number of parameters in each layer. An LSTM cell maintains an internal state updated using an embedding of the action-observation pair. A fully connected layer followed by an ReLU generates a feature vector ϕ^i, which are combined by a second FC layer then normalized by Softmax to get conditional decision rule $a^i(\cdot|\varsigma^i)$.

We conducted experiments on a Dell Precision Tower 7910 equipped with a 16-core, 3 GHz Intel Xeon CPU, 16 GB of RAM and a 2 GB nVIDIA Quadro K620 GPU. We run simulations on standard benchmarks from Dec-POMDP literature, including *Dec. Tiger, Broadcast Channel, Mars, Box Pushing, Meeting in a Grid*, and *Recycling Robots*, see http://masplan.org. For the sake of conciseness, we report details on hyper-parameters in the companion research report [6].

6.2 History Representation Matters

In this section, we conducted experiments with the goal of gaining insights on how the representation of histories affects the performance of ACDC methods. Figure 3 depicts the comparison of truncated histories *vs* hidden states of deep neural networks. Results obtained using an ϵ-optimal planning algorithm called FB-HSVI [9] are included as reference. For short planning horizons, *e.g.*, $T = 10$, *CTDC_RNN* quickly converges to good solutions in comparison to *CTDC_TRUNC(1)* and *CTDC_TRUNC(3)*. This suggests CTDC rnn learns more useful and concise representations of histories than the truncated representation. However, for some of the more complex tasks such as *Dec. Tiger*, *Box Pushing* or *Mars*, no internal representation was able to perform optimally.

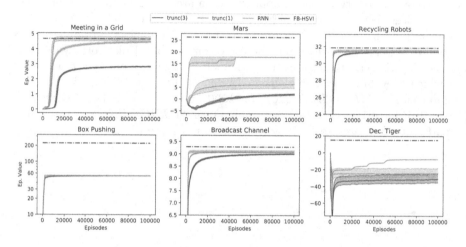

Fig. 3. Comparison of different structures used to represent histories.

Overall, our experiments on history representations show promising results for RNNs, which have the advantage over truncated histories to automatically learn equivalence classes and compact internal representations based on the gradient back-propagated from the reward signal. Care should be taken though, as some domain planning horizons and other specific properties might cause early convergence to poor local optima. We are not entirely sure which specific features of the problems deteriorate performances, and we leave for future works to explore better methods to train these architectures.

6.3   Comparing Paradigms Altogether

In this section, we compare paradigms, CTCC, DTDC, and CTDC. We complement our experiments with results from other Dec-POMDP algorithms: an ϵ-optimal planning algorithm called FB-HSVI [9]; and a sampling-based planning algorithm called Monte-Carlo Expectation-Maximization (MCEM) algorithm [37], which shares many similarities with actor-critic methods. It is worth noticing that we are not competing against FB-HSVI as it is model-based. As for MCEM, we reported performances[2] recorded in [37].

In almost all tested benchmarks, CTDC seems to take the better out of the two other paradigms, for either $T = 10$ (Fig. 4) or $T = \infty$ (Fig. 5). CTCC might

---

[2] Two results in MCEM [37] were above optimal values, so we reported optimal values instead.

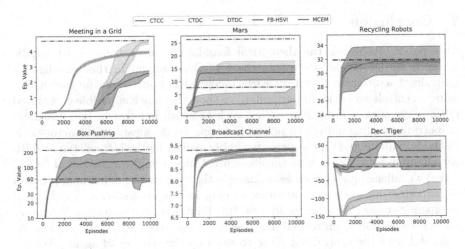

Fig. 4. Comparison of the three paradigms for $T = 10$.

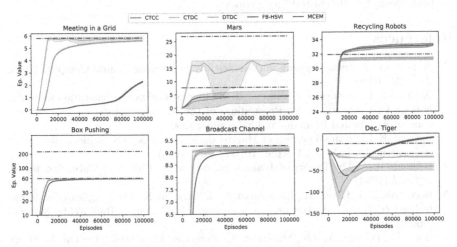

Fig. 5. Comparison of the three paradigms for $T = \infty$.

suffer from the high dimensionality of the joint history space, and fail to explore it efficiently before the learning step-sizes become negligible, or we reached the predefined number of training episodes. Our on-policy sampling evaluation certainly amplified this effect. Having a much smaller history space to explore, CTDC outperforms CTCC in these experiments. Compared to DTDC which also explores smaller history space, there is a net gain to consider a compatible centralized critic in the CTDC paradigm, resulting in better performances. Even if CTDC achieves performances better or equal to the state of the art MCEM algorithm, there is still some margins of improvements to reach the global optima given by FB-HSVI in every benchmark. As previously mentioned, this is partly due to inefficient representations of histories.

7 Conclusion

This paper establishes the theoretical foundations of centralized actor-critic methods for Dec-POMDPs within the CTDC paradigm. In this paradigm, a centralized actor-critic algorithm learns independent policies, one for each agent, using a centralized critic. In particular, we show that the compatible centralized critic is the sum of individual critics, each of which is linear in the "features" of its corresponding individual policy. Experiments demonstrate our actor-critic methods, namely ACDC, compares favorably against methods from standard RL paradigms in benchmarks from the literature. Current implementations of ACDC reveal a challenging and open issue, namely the representation learning problem of individual histories, *e.g.*, learning to map individual histories to individual occupancy states. We plan to address this limitation in the future. Whenever the representation of individual histories is not an issue, ACDC can exploit the separability of the centralized critic to scale up the number of agents. We are currently investigating a large-scale decentralized multi-agent application, where we plan to exploit this scalability property.

References

1. Amari, S.I.: Natural gradient works efficiently in learning. Neural Comput. **10**(2), 251–276 (1998)
2. Amato, C., Dibangoye, J.S., Zilberstein, S.: Incremental policy generation for finite-horizon DEC-POMDPs. In: Proceedings of the Nineteenth International Conference on Automated Planning and Scheduling (2009)
3. Aström, K.J.: Optimal control of Markov decision processes with incomplete state estimation. J. Math. Anal. Appl. **10**, 174–205 (1965)
4. Bellman, R.E.: The Theory of dynamic programming. Bull. Am. Math. Soc. **60**(6), 503–515 (1954)
5. Bernstein, D.S., Givan, R., Immerman, N., Zilberstein, S.: The complexity of decentralized control of Markov decision processes. Math. Oper. Res. **27**(4), 819–840 (2002)
6. Bono, G., Dibangoye, J.S., Matignon, L., Pereyron, F., Simonin, O.: On the Study of Cooperative Multi-Agent Policy Gradient. Research Report RR-9188, INSA Lyon, INRIA (2018)
7. Boutilier, C.: Planning, learning and coordination in multiagent decision processes. In: Proceedings of the Sixth Conference on Theoretical Aspects of Rationality and Knowledge (1996)
8. Degris, T., White, M., Sutton, R.S.: Linear off-policy actor-critic. In: Proceedings of the 29th International Conference on ML, ICML 2012, Edinburgh, Scotland, UK, 26 June–1 July 2012 (2012)
9. Dibangoye, J.S., Amato, C., Buffet, O., Charpillet, F.: Optimally solving Dec-POMDPs as continuous-state MDPs. J. AI Res. **55**, 443–497 (2016)
10. Dibangoye, J.S., Amato, C., Buffet, O., Charpillet, F.: Optimally solving Dec-POMDPs as continuous-state MDPs. In: Proceedings of the Twenty-Fourth International Joint Conference on Artificial Intelligence (2013)
11. Dibangoye, J.S., Amato, C., Buffet, O., Charpillet, F.: Optimally solving Dec-POMDPs as Continuous-State MDPs: Theory and Algorithms. Research Report RR-8517 (2014)

12. Dibangoye, J.S., Buffet, O.: Learning to Act in Decentralized Partially Observable MDPs. Research report, INRIA, Jun 2018. https://hal.inria.fr/hal-01809897
13. Dibangoye, J.S., Buffet, O., Charpillet, F.: Error-bounded approximations for infinite-horizon discounted decentralized POMDPs. In: Calders, T., Esposito, F., Hüllermeier, E., Meo, R. (eds.) ECML PKDD 2014. LNCS (LNAI), vol. 8724, pp. 338–353. Springer, Heidelberg (2014). https://doi.org/10.1007/978-3-662-44848-9_22
14. Foerster, J., Farquhar, G., Afouras, T., Nardelli, N., Whiteson, S.: Counterfactual multi-agent policy gradients (2018)
15. Gupta, J.K., Egorov, M., Kochenderfer, M.: Cooperative multi-agent control using deep reinforcement learning. In: Sukthankar, G., Rodriguez-Aguilar, J.A. (eds.) AAMAS 2017. LNCS (LNAI), vol. 10642, pp. 66–83. Springer, Cham (2017). https://doi.org/10.1007/978-3-319-71682-4_5
16. Hansen, E.A., Bernstein, D.S., Zilberstein, S.: Dynamic programming for partially observable stochastic games. In: Proceedings of the Nineteenth National Conference on Artifical intelligence (2004)
17. Kakade, S.: A natural policy gradient. In: Advances in Neural Information Processing Systems 14 (NIPS 2001) (2001)
18. Konda, V.R., Tsitsiklis, J.N.: Actor-critic algorithms. In: Advances in Neural Information Processing Systems 12 (2000)
19. Kraemer, L., Banerjee, B.: Multi-agent reinforcement learning as a rehearsal for decentralized planning. Neurocomputing 190, 82–94 (2016)
20. Liu, M., Amato, C., Anesta, E.P., Griffith, J.D., How, J.P.: Learning for decentralized control of multiagent systems in large, partially-observable stochastic environments. In: AAAI (2016)
21. Liu, M., Amato, C., Liao, X., Carin, L., How, J.P.: Stick-breaking policy learning in Dec-POMDPs. In: International Joint Conference on Artificial Intelligence (IJCAI 2015). AAAI (2015)
22. Lowe, R., WU, Y., Tamar, A., Harb, J., Pieter Abbeel, O., Mordatch, I.: Multi-agent actor-critic for mixed cooperative-competitive environments. In: Advances in Neural Information Processing Systems 30 (2017)
23. Mnih, V., et al.: Human-level control through deep reinforcement learning. Nature 518(7540), 529 (2015)
24. Moravčík, M., et al.: DeepStack: expert-level artificial intelligence in heads-up no-limit poker. Science 356(6337), 508–513 (2017)
25. Nguyen, D.T., Kumar, A., Lau, H.C.: Policy gradient with value function approximation for collective multiagent planning. In: Advances in Neural Information Processing Systems 30 (2017)
26. Oliehoek, F.A., Spaan, M.T.J., Amato, C., Whiteson, S.: Incremental clustering and expansion for faster optimal planning in Dec-POMDPs. J. AI Res. 46, 449–509 (2013)
27. Oliehoek, F.A., Spaan, M.T.J., Dibangoye, J.S., Amato, C.: Heuristic search for identical payoff Bayesian games. In: Proceedings of the Ninth International Conference on Autonomous Agents and Multiagent Systems (2010)
28. Peshkin, L., Kim, K.E., Meuleau, N., Kaelbling, L.P.: Learning to cooperate via policy search. In: Sixteenth Conference on Uncertainty in Artificial Intelligence (UAI-2000) (2000)
29. Robbins, H., Monro, S.: A stochastic approximation method. Ann. Math. Stat. 22(3), 400–407 (1951)
30. Shoham, Y., Leyton-Brown, K.: Multiagent Systems: Algorithmic, Game-Theoretic, and Logical Foundations. Cambridge University Press, New York (2008)

31. Sutton, R.S., Barto, A.G.: Introduction to Reinforcement Learning, 2nd edn. MIT Press, Cambridge (2016)
32. Sutton, R.S., McAllester, D., Singh, S., Mansour, Y.: Policy Gradient Methods for Reinforcement Learning with Function Approximation. In: Proceedings of the 12th International Conference on Neural Information Processing Systems, Cambridge, MA, USA (1999)
33. Szer, D., Charpillet, F.: An optimal best-first search algorithm for solving infinite horizon DEC-POMDPs. In: Gama, J., Camacho, R., Brazdil, P.B., Jorge, A.M., Torgo, L. (eds.) ECML 2005. LNCS (LNAI), vol. 3720, pp. 389–399. Springer, Heidelberg (2005). https://doi.org/10.1007/11564096_38
34. Szer, D., Charpillet, F., Zilberstein, S.: MAA*: a heuristic search algorithm for solving decentralized POMDPs. In: Proceedings of the Twenty-First Conference on Uncertainty in Artificial Intelligence (2005)
35. Tan, M.: Multi-agent reinforcement learning: independent vs. cooperative agents. In: Readings in Agents, San Francisco, CA, USA (1998)
36. Williams, R.J.: Simple statistical gradient-following algorithms for connectionist reinforcement learning. Mach. Learn. **8**(3), 229–256 (1992)
37. Wu, F., Zilberstein, S., Jennings, N.R.: Monte-Carlo expectation maximization for decentralized POMDPs. In: Proceedings of the Twenty-Fourth International Joint Conference on Artificial Intelligence (2013)
38. Zhang, X., Aberdeen, D., Vishwanathan, S.V.N.: Conditional random fields for multi-agent reinforcement learning. In: Proceedings of the 24th International Conference on Machine Learning (2007)

Parametric t-Distributed Stochastic Exemplar-Centered Embedding

Martin Renqiang Min[1]([⊠]), Hongyu Guo[2], and Dinghan Shen[3]

[1] NEC Labs America Princeton, Princeton, NJ 08540, USA
renqiang@nec-labs.com
[2] National Research Council Canada Ottawa, Ottawa, ON K1A 0R6, Canada
hongyu.guo@nrc-cnrc.gc.ca
[3] Duke University Durham, Durham, NC 27708, USA
dinghan.shen@duke.edu

Abstract. Parametric embedding methods such as parametric t-distributed Stochastic Neighbor Embedding (pt-SNE) enables out-of-sample data visualization without further computationally expensive optimization or approximation. However, pt-SNE favors small mini-batches to train a deep neural network but large mini-batches to approximate its cost function involving all pairwise data point comparisons, and thus has difficulty in finding a balance. To resolve the conflicts, we present parametric t-distributed stochastic exemplar-centered embedding. Our strategy learns embedding parameters by comparing training data only with precomputed exemplars to indirectly preserve local neighborhoods, resulting in a cost function with significantly reduced computational and memory complexity. Moreover, we propose a shallow embedding network with high-order feature interactions for data visualization, which is much easier to tune but produces comparable performance in contrast to a deep feedforward neural network employed by pt-SNE. We empirically demonstrate, using several benchmark datasets, that our proposed method significantly outperforms pt-SNE in terms of robustness, visual effects, and quantitative evaluations.

1 Introduction

Unsupervised nonlinear dimensionality reduction methods, which embed high-dimensional data to a low-dimensional space, have been extensively deployed in many real-world applications for data visualization. Data visualization is an important component of data exploration and data analytics, as it helps data analysts to develop intuitions and gain deeper understanding about the mechanisms underlying data generation. Comprehensive surveys about dimensionality reduction and data visualization methods can be found in van der Maaten et al. [13] and Burges [3]. Among these approaches, nonparametric neighbor embedding methods such as t-SNE [12] and Elastic Embedding [4] are widely adopted. They generate low-dimensional latent representations by preserving neighboring probabilities of high-dimensional data in a low-dimensional space, which involves

© Springer Nature Switzerland AG 2019
M. Berlingerio et al. (Eds.): ECML PKDD 2018, LNAI 11051, pp. 477–493, 2019.
https://doi.org/10.1007/978-3-030-10925-7_29

pairwise data point comparisons and thus has quadratic computational complexity with respect to the size of a given data set. This prevents them from scaling to any dataset with a size beyond several thousand. Moreover, these methods are not designed for readily generating the embedding of out-of-sample data that are prevalent in modern big data analytics. To generate out-of-sample data embedding given an existing sample embedding, computationally expensive numerical optimization or Nyström approximation is often performed, which is undesirable in practice [2,5,26].

Parametric embedding methods, such as parametric t-SNE (pt-SNE) [11] employing a deep neural network (DNN), learn an explicit parametric mapping function from a high-dimensional data space to a low-dimensional embedding space, which can readily generate the embedding of out-of-sample data. The objective function of pt-SNE is the same as that of t-SNE with quadratic computational complexity. Fortunately, owing to the explicit mapping function defined by the DNN, optimization methods such as stochastic gradient descent or conjugate gradient descent based on mini-batches can be deployed when pt-SNE is applied to large-scale datasets.

However, on one hand, the objective function of pt-SNE is a sum of a quadratic number of terms over pairwise data points, which requires mini-batches with fairly large batch sizes to achieve a reasonably good approximation to the original objective; On the other hand, optimizing the parameters of the DNN in pt-SNE also requires careful choices of batch sizes, which is often best served with small batch sizes to avoid being stuck in a bad local minimum. These conflicting choices of batch sizes make the optimization of pt-SNE hard and render its performance sensitive to the chosen batch size. In addition, to approximate the loss function defined over all pairwise data points, pt-SNE independently computes pairwise neighboring probabilities of high-dimensional data for each mini-batch, so it often produces dramatically different embeddings with different choices of user-defined perplexities that are coupled with batch sizes. Finally, although the mapping function of pt-SNE parameterized by a DNN is powerful, it is very hard to learn and requires complicated procedures such as tuning network architectures and tuning many hyper-parameters. For data embedding and visualization purposes, most users are reluctant to go through these complicated procedures.

To address the aforementioned problems, in this paper, we present unsupervised parametric t-distributed stochastic exemplar-centered embedding. Instead of modeling pairwise neighboring probabilities, our strategy learns embedding parameters by comparing high-dimensional data only with precomputed representative high-dimensional exemplars, resulting in an objective function with linear computational and memory complexity with respect to the number of exemplars. The exemplars are identified by a small number of iterations of k-means updates, taking into account both local data density distributions and global clustering patterns of high-dimensional data. These nice properties make the parametric exemplar-centered embedding insensitive to batch size and scalable to large-scale datasets. All the exemplars are repeatedly included into each mini-batch, and the choice of the perplexity hyper-parameter only concerns the expected number of neighboring exemplars calculated globally, independent of batch sizes. Therefore, the perplexity is much easier to choose by the user and much more robust to produce good embedding performance. We further use noise contrastive samples to avoid comparing data points with all exemplars, which further reduces computational/memory complexity and increases scalability. Although comparing training data points only with representative exemplars indirectly preserves similarities between pairwise data points in each local neighborhood, it is much better than randomly sampling small mini-batches in pt-SNE whose coverages are too small to capture all pairwise similarities on a large dataset.

Moreover, we propose a shallow embedding network with high-order feature interactions for data visualization, which is much easier to tune but produces comparable performance in contrast to a deep neural network employed by pt-SNE. Experimental results on several benchmark datasets show that, our proposed parametric exemplar-centered embedding methods for data visualization significantly outperform pt-SNE in terms of robustness, visual effects, and quantitative evaluations. We call our proposed deep t-distributed stochastic exemplar-centered embedding method dt-SEE and high-order t-distributed exemplar-centered embedding method hot-SEE.

Our contributions in this paper are summarized as follows: (1) We propose a scalable unsupervised parametric data embedding strategy with an objective function of significantly reduced computational complexity, avoiding pairwise training data comparisons in existing methods; (2) With the help of exemplars, our methods eliminate the instability and sensitivity issues caused by batch sizes and perplexities haunting other unsupervised embedding approaches including pt-SNE; (3) Our proposed approach hot-SEE learns a simple shallow high-order parametric embedding function, beating state-of-the-art unsupervised deep parametric embedding method pt-SNE on several benchmark datasets in terms of both qualitative and quantitative evaluations.

2   Related Work

Dimensionality reduction and data visualization have been extensively studied in the last twenty years [3,13]. SNE [9], its variant t-SNE [12], and Elastic Embedding [4] are among the most successful approaches. To efficiently generate the embedding of out-of-sample data, SNE and t-SNE were, respectively, extended to take a parametric embedding form of a shallow neural network [15] and a deep neural network [11]. As is discussed in the introduction, the objective functions of neighbor embedding methods have $O(n^2)$ computational complexity for n data points, which limits their applicability only to small datasets. Recently, with the growing importance of big data analytics, several research efforts have been devoted to enhancing the scalability of nonparametric neighbor embedding methods [23,24,26,27]. These methods mainly borrowed ideas from efficient approximations developed for N-body force calculations based on Barnes-Hut trees [23] or fast multipole methods [7]. Iterative methods with auxiliary variables and second-order methods have been developed to optimize the objective functions of neighbor embedding approaches [5,25,26]. Particularly, the alternating optimization method with auxiliary variables was shown to achieve faster convergence than mini-batch based conjugate gradient method for optimizing the objective function of pt-SNE. All these scalability handling and optimization research efforts are orthogonal to our development in this paper, because all these methods are designed for the embedding approaches modeling the neighboring relationship between pairwise data points. Therefore, they still have the sensitivity and instability issues, and we can readily borrow these speedup methods to further accelerate our approaches modeling the relationship between data points and exemplars.

Our proposed method hot-SEE learns a shallow parametric embedding function by considering high-order feature interactions. High-order feature interactions have been studied for learning Boltzmann Machines, autoencoders, structured outputs, feature selection, and biological sequence classification [8,10,14, 16,17,19–22]. To the best of our knowledge, our work here is the first successful one to model input high-order feature interactions for unsupervised data embedding and visualization.

Our work in this paper is also related to a recent supervised data embedding method called en-HOPE [17]. Unlike en-HOPE, our proposed methods here are unsupervised and have a completely different objective function with different motivations.

3   Methods

In this section, we introduce the objective of pt-SNE at first. Then we describe the parametric embedding functions of our methods based on a deep neural network as in pt-SNE and a shallow neural network with high-order feature interactions. Finally, we present our proposed parametric stochastic exemplar-centered embedding methods dt-SEE and hot-SEE with low computational cost.

3.1   Parametric t-SNE Using a Deep Neural Network and a Shallow High-Order Neural Network

Given a set of data points $\mathcal{D} = \{\mathbf{x}^{(i)} : i = 1,\ldots,n\}$, where $\mathbf{x}^{(i)} \in \mathbb{R}^H$ is the input feature vector. pt-SNE learns a deep neural network as a nonlinear feature transformation from the high-dimensional input feature space to a low-dimensional latent embedding space $\{f(\mathbf{x}^{(i)}) : i = 1,\ldots,n\}$, where $f(\mathbf{x}^{(i)}) \in \mathbb{R}^h$, and $h < H$. For data visualization, we set $h = 2$.

pt-SNE assumes, $p_{j|i}$, the probability of each data point i chooses every other data point j as its nearest neighbor in the high-dimensional space follows a Gaussian distribution. The joint probabilities measuring the pairwise similarities between data points $\mathbf{x}^{(i)}$ and $\mathbf{x}^{(j)}$ are defined by symmetrizing two conditional probabilities, $p_{j|i}$ and $p_{i|j}$, as follows,

$$p_{j|i} = \frac{\exp(-||\mathbf{x}^{(i)} - \mathbf{x}^{(j)}||^2/2\sigma_i^2)}{\sum_{k \neq i} \exp(-||\mathbf{x}^{(i)} - \mathbf{x}^{(k)}||^2/2\sigma_i^2)}, \tag{1}$$

$$p_{i|i} = 0, \tag{2}$$

$$p_{ij} = \frac{p_{j|i} + p_{i|j}}{2n}, \tag{3}$$

where the variance of the Gaussian distribution, σ_i, is set such that the perplexity of the conditional distribution P_i equals a user-specified perplexity u that can be interpreted as the expected number of nearest neighbors of data point i. With the same u set for all data points, σ_i's tend to be smaller in regions of higher data densities than the ones in regions of lower data densities. The optimal value of σ_i for each data point i can be easily found by a simple binary search [9]. Although the user-specified perplexity u makes the variance σ_i for each data point i adaptive, the embedding performance is still very sensitive to this hyperparameter, which will be discussed later. In the low-dimensional space, pt-SNE assumes, the neighboring probability between pairwise data points i and j, q_{ij}, follows a heavy-tailed student t-distribution. The student t-distribution is able to, on one hand, measure the similarities between pairwise low-dimensional points, on the other hand, allow dissimilar objects to be modeled far apart in the embedding space, avoiding crowding problems.

$$q_{ij} = \frac{(1 + ||f(\mathbf{x}^{(i)}) - f(\mathbf{x}^{(j)})||^2)^{-1}}{\sum_{kl:k \neq l}(1 + ||f(\mathbf{x}^{(k)}) - f(\mathbf{x}^{(l)})||^2)^{-1}}, \tag{4}$$

$$q_{ii} = 0. \tag{5}$$

To learn the parameters of the deep embedding function $\mathbf{f}(.)$, pt-SNE minimizes the following Kullback-Leibler divergence between the joint distributions P and Q using Conjugate Gradient descent,

$$\ell = KL(P||Q) = \sum_{ij:i \neq j} p_{ij} \log \frac{p_{ij}}{q_{ij}}. \tag{6}$$

The above objective function has $O(n^2)$ terms defined over pairwise data points, which is computationally prohibitive and prevents pt-SNE from scaling to a fairly big dataset. To overcome such scalability issue, heuristic mini-batch approximation is often used in practice. However, as will be shown in our experiments, pt-SNE is unstable and highly sensitive to the chosen batch size to achieve reasonable performance. This is due to the dilemma of the quadratic cost function approximation and DNN optimization through mini-batches: approaching the true objective requires large batch sizes but finding a good local minimum benefits from small batch sizes.

Although pt-SNE based on a deep neural network has a powerful nonlinear feature transformation, parameter learning is hard and requires complicated procedures such as tuning network architectures and tuning many hyperparameters. Most users who are only interested in data embedding and visualization are reluctant to go through these complicated procedures. Here we propose to use high-order feature interactions, which often capture structural knowledge of input data, to learn a shallow parametric embedding model instead of a deep model. The shallow model is much easier to train and does not have many hyperparameters. In the following, the shallow high-order parametric embedding function will be presented. We expand each input feature vector $\mathbf{x}$ to have an additional component of 1 for absorbing bias terms, that is, $\mathbf{x}' = [\mathbf{x}; 1]$, where $\mathbf{x}' \in \mathbb{R}^{H+1}$. The O-order feature interaction is the product of all possible O features $\{x_{i_1} \times \ldots \times x_{i_t} \times \ldots \times x_{i_O}\}$ where, $t \in \{1, \ldots, O\}$, and $i_t \in \{1, \ldots, H\}$. Ideally, we want to use each O-order feature interaction as a coordinate and then learn a linear transformation to map all these high-order feature interactions to a low-dimensional embedding space. However, it's very expensive to enumerate all possible O-order feature interactions. For example, if $H = 1000, O = 3$, we must deal with a 10^9-dimensional vector of high-order features. We approximate a Sigmoid-transformed high-order feature mapping $\mathbf{y} = f(\mathbf{x})$ by constrained tensor factorization as follows,

$$y_s = \sum_{k=1}^{m} V_{sk}\sigma(\sum_{f=1}^{F} W_{fk}(\mathbf{C}_f^T\mathbf{x}')^O + b_k), \qquad (7)$$

where b_k is a bias term, $\mathbf{C} \in \mathbb{R}^{(H+1)\times F}$ is a factorization matrix, $\mathbf{C}_f$ is the f-th column of $\mathbf{C}$, $\mathbf{W} \in \mathbb{R}^{F\times m}$ and $\mathbf{V} \in \mathbb{R}^{h\times m}$ are projection matrices, y_s is the s-th component of $\mathbf{y}$, F is the number of factors, m is the number of high-order hidden units, and $\sigma(x) = \frac{1}{1+e^{-x}}$. Because the last component of $\mathbf{x}'$ is 1 for absorbing bias terms, the full polynomial expansion of $(\mathbf{C}_f^T\mathbf{x}')^O$ essentially captures all orders of input feature interactions up to order O. Empirically, we find that $O = 2$ works best for all datasets we have and set $O = 2$ for all our experiments. The hyperparameters F and m are set by users. Combining Eqs. 1, 4, 6 and the feature transformation function in Eq. 7 leads to a method called high-order t-SNE (hot-SNE). As pt-SNE, the objective function of hot-SNE involves comparing pairwise data points and thus has quadratic computational complexity with respect to the sample size. The parameters of hot-SNE are learned by Conjugate Gradient descent as in pt-SNE.

3.2   Parametric t-Distributed Stochastic Exemplar-Centered Embedding

To address the instability, sensitivity, and unscalability issues of pt-SNE, we present deep t-distributed stochastic exemplar-centered embedding (dt-SEE) and high-order t-distributed stochastic exemplar-centered embedding (hot-SEE) building upon pt-SNE and hot-SNE for parametric data embedding described earlier. The resulting objective function has significantly reduced computational complexity with respect to the size of training set compared to pt-SNE. The underlying intuition is that, instead of comparing pairwise training data points, we compare training data only with a small number of representative exemplars in the training set for neighborhood probability computations. To this end, we simply precompute the exemplars by running a fixed number of iterations of k-means with scalable k-means++ seeding on the training set, which has at most linear computational complexity with respect to the size of training set [1].

Formally, given the same dataset $\mathcal{D}$ with formal descriptions as introduced in Sect. 3.1, we perform a fixed number of iterations of k-means updates on the training data to identify z exemplars from the whole dataset, where z is a user-specified free parameter and $z \ll n$ (please note that k-means often converges within a dozen iterations and shows linear computational cost in practice). Before performing k-means updates, the exemplars are carefully seeded by scalable k-means++, which will make our methods robust under abnormal conditions, although our experiments show that random seeding works equally well. We denote these exemplars by $\{\mathbf{e}^{(j)} : j = 1, \ldots, z\}$. The high-dimensional neighboring probabilities is calculated through a Gaussian distribution,

$$p_{j|i} = \frac{\exp(-||\mathbf{x}^{(i)} - \mathbf{e}^{(j)}||^2/2\sigma_i^2)}{\sum_k \exp(-||\mathbf{x}^{(i)} - \mathbf{e}^{(k)}||^2/2\sigma_i^2)}, \tag{8}$$

$$p_{j|i} = \frac{p_{j|i}}{n}, \tag{9}$$

where $i = 1, \ldots, n, j = 1, \ldots, z$, and the variance of the Gaussian distribution, σ_i, is set such that the perplexity of the conditional distribution P_i equals a user-specified perplexity u that can be interpreted as the expected number of nearest exemplars, not neighboring data points, of data instance i. Since the high-dimensional exemplars capture both local data density distributions and global clustering patterns, different choices of perplexities over exemplars will not change the embedding too much, resulting in much more robust visualization performance than that of other embedding methods insisting on modeling local pairwise neighboring probabilities.

Similarly, the low-dimensional neighboring probabilities is calculated through a t-distribution,

$$q_{j|i} = \frac{(1 + d_{ij})^{-1}}{\sum_{i=1}^{n} \sum_{k=1}^{z} (1 + d_{ik})^{-1}}, \tag{10}$$

$$d_{ij} = ||f(\mathbf{x}^{(i)}) - f(\mathbf{e}^{(j)})||^2, \tag{11}$$

where $f(\cdot)$ denotes a deep neural network for dt-SEE or the high-order embedding function as described in Eq. 7 for hot-SEE.

Then we minimize the following objective function to learn the embedding parameters Θ of dt-SEE and hot-SEE while keeping the exemplars $\{\mathbf{e}^{(j)}\}$ fixed,

$$\min \ell(\Theta, \{\mathbf{e}^{(j)}\}) = \sum_{i=1}^{n} \sum_{j=1}^{z} p_{j|i} \log \frac{p_{j|i}}{q_{j|i}} \tag{12}$$

where i indexes training data points, j indexes exemplars, Θ denotes the high-order embedding parameters $\{\{b_k\}_{k=1}^{m}, \mathbf{C}, \mathbf{W}, \mathbf{V}\}$ in Eq. 7.

Note that unlike the probability distribution in Eq. 4, $q_{j|i}$ here is computed only using the pairwise distances between training data points and exemplars. This small modification has significant benefits. Because $z \ll n$, compared to the quadratic computational complexity with respect to n of Eq. 6, the objective function in Eq. 12 has a significantly reduced computational cost, considering that the number of representative exemplars is often much much smaller than n for real-world large datasets in practice.

3.3 Further Reduction on Computational Complexity and Memory Complexity by Noise Contrastive Estimation

We can even further reduce the computational complexity and memory complexity of dt-SEE and hot-SEE using noise contrastive estimation (NCE). Instead of computing neighboring probabilities between each data point i and all z exemplars, we can simply only compute the probabilities between data point i and its z_e nearest exemplars for both P and Q. For high-dimensional probability distribution P_i, we simply set the probabilities between i and other exemplars 0; for low-dimensional probability distribution Q_i, we randomly sample z_n non-neighboring exemplars outside of these z_e neighboring exemplars, and use the sum of these z_n non-neighboring probabilities multiplied by a constant K_e and the z_e neighboring probabilities to approximate the normalization terms involving data point i in Eq. 10. Since this strategy based on noise contrastive estimation eliminates the need of computing neighboring probabilities between data points and all exemplars, it further reduces computational and memory complexity.

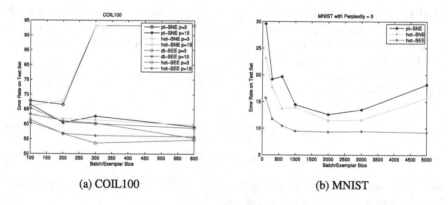

(a) COIL100

(b) MNIST

Fig. 1. The sensitivity test of batch size/exemplar size on COIL100 and MNIST.

4   Experiments

In this section, we evaluate the effectiveness of dt-SEE and hot-SEE by comparing them against state-of-the-art unsupervised parametric embedding method pt-SNE based upon three datasets, *i.e.*, COIL100, MNIST, and Fashion.

The COIL100 data[1] contains 7200 images with 100 classes, where 3600 samples for training and 3600 for test. The MNIST dataset[2] consists of 60,000 training and 10,000 test gray-level 784-dimensional images. The Fashion dataset[3] has the same number of classes, training and test data points as that of MNIST, but is designed to classify 10 fashion products, such as boot, coat, and bag, where each contains a set of pictures taken by professional photographers from different aspects of the product, such as looks from front, back, with model, and in an outfit.

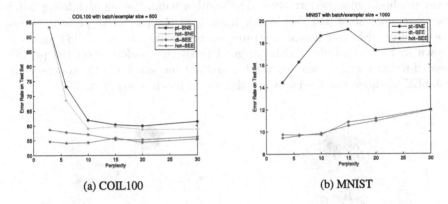

(a) COIL100   (b) MNIST

Fig. 2. The sensitivity test of perplexity on COIL100 and MNIST.

To make computational procedures and tuning procedures for data visualization simpler, none of these models was pre-trained using any unsupervised learning strategy, although hot-SNE, hot-SEE, dt-SEE, and pt-SNE could all be pretrained by autoencoders or variants of Restricted Boltzmann Machines [10,18].

For hot-SNE and hot-SEE, we set $F = 800$ and $m = 400$ for all the datasets used. For pt-SNE and dt-SEE, we set the deep neural network architecture to input dimensionality H-500-500-2000-2 for all datasets, following the architecture design in van der Maaten [11]. For hot-SEE and dt-SEE, when the exemplar size is smaller than 1000, we set batch size to 100; otherwise, we set it 1000. With the above architecture design, the shallow high-order neural network used in hot-SNE and hot-SEE is as fast as 2.5 times of the deep neural network used in pt-SNE and dt-SEE for embedding 10,000 MNIST test data.

For all the experiments, the predictive accuracies were obtained by the 1NN approach on top of the 2-dimensional representations generated by different methods. The error rate was calculated by the number of misclassified test data points divided by the total number of test data points.

---

[1] http://www1.cs.columbia.edu/CAVE/software/softlib/coil-100.php.
[2] http://yann.lecun.com/exdb/mnist/.
[3] https://github.com/zalandoresearch/fashion-mnist.

4.1 Performance Comparisons with Different Batch Sizes and Perplexities on COIL100 and MNIST

Our first experiment aims at examining the robustness of different testing methods with respect to the batch size and the perplexity used. Figures 1 and 2 depict our results on the COIL100 and MNIST datasets when varying the batch size and perplexity, respectively, used by the testing methods.

Figure 1 suggests that, for the COIL100 data, the pt-SNE was very sensitive to the selection of the batch size; efforts were needed to find a right batch size in order to obtain good performance. On the other hand, the use of different batch sizes/exemplar sizes had very minor impact on the predictive performance of both the dt-SEE and hot-SEE strategies. Similarly, for the MNIST data, as shown in Fig. 2, in order to obtain good predictive performance, the pt-SNE needed to have a batch size not too big and not too small. On the contrary, the hot-SEE methods was insensitive to the size of batch larger than 300.

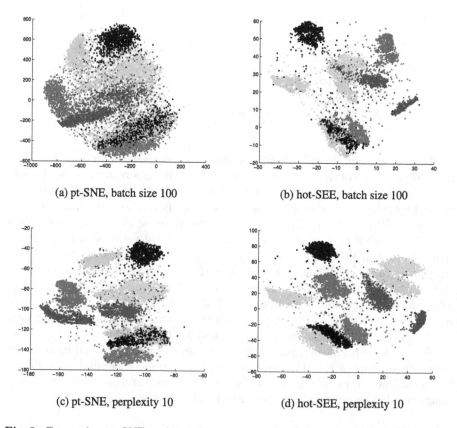

(a) pt-SNE, batch size 100   (b) hot-SEE, batch size 100

(c) pt-SNE, perplexity 10   (d) hot-SEE, perplexity 10

Fig. 3. Comparing pt-SNE to hot-SEE with a small batch size = 100 (perplexity = 3) or a reasonable perplexity = 10 (batch size = 1000) to illustrate pt-SNE's unstable visual performance on MNIST.

Based on the results in Fig. 1, we selected the best batch sizes for both the COIL100 and MNIST data sets, with 600 and 1000, respectively, but we varied the values of the perplexities used. In Fig. 2, one can observe that, the performance of the pt-SNE and hot-SNE could dramatically change due to the use of different perplexities, but that was not the case for both the dt-SEE and hot-SEE. Similarly, for the MNIST data, as depicted in Fig. 2, in order to obtain good predictive performance, one would need to carefully tune for the right perplexity. On the contrary, both the dt-SEE and hot-SEE methods performed quite robust with respect to different selected perplexities.

Because the choices of batch size and perplexity are coupled in a complicated way in pt-SNE as explained in the introduction, we run additonal experiments to show the advantages of dt-see and hot-see. When we set perplexity to 10 and batch size to 100, 300, 600, 1000, 2000, 3000, 5000, 10000, the test error rate of pt-SNE on MNIST is, respectively, 32.97%, 22.1%, 24.00%, 16.30%, 12.41%, 12.28%, 13.09%, 16.43%, which still varies a lot. In contrast, the error rates of dt-SEE or hot-SEE using 1000 exemplars are consistently below 10% with the same

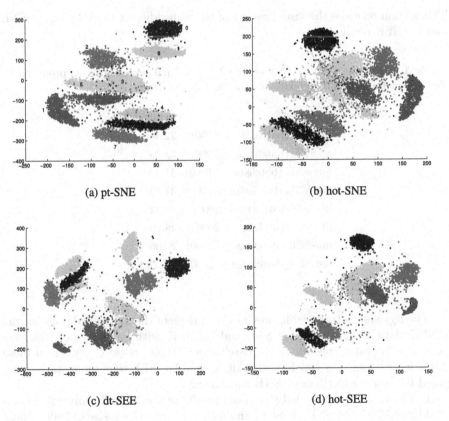

(a) pt-SNE

(b) hot-SNE

(c) dt-SEE

(d) hot-SEE

Fig. 4. The best MNIST embeddings generated by pt-SNE, hot-SNE, dt-SEE, and hot-SEE.

batch size ranging from 100 to 10000 and perplexity 3 and 10, which again shows shat exemplar-centered embedding dt-see and hot-see are much more robust than pt-SNE.

4.2   Experimental Results on the Fashion Dataset

We also further evaluated the predictive performance of the testing methods using the Fashion data set. We used batch sizes of 1000 and 2000, along with perplexity of 3 in all the experiments since both pt-SNE and hot-SNE favored these settings as suggested in Figs. 1 and 2. The achieved accuracies are shown in Table 1.

Results in Table 1 further confirmed the superior performance of our methods. Both the dt-SEE and hot-SEE significantly outperformed the pt-SNE and hot-SNE.

4.3   Two-Dimensional Visualization of Embeddings

This section provides the visual results of the embeddings formed by the pt-SNE and hot-SEE methods.

Table 1. Error rates (%) by 1NN on the 2-dimensional representations produced by different methods with perplexity = 3 on the Fashion dataset.

Methods	Error rates
pt-SNE (batchsize = 1000)	32.48
pt-SNE (batchsize = 2000)	32.04
hot-SNE (batchsize = 1000)	31.29
hot-SNE (batchsize = 2000)	31.82
dt-SEE (batchsize = 1000)	29.42
dt-SEE (batchsize = 2000)	28.30
hot-SEE (batchsize = 1000)	29.06
hot-SEE (batchsize = 2000)	**28.18**

The top and bottom subfigures in Fig. 3 depicts the 2D embeddings on the MNIST data set created by pt-SNE and hot-SEE, with batch size of 100 (perplexity = 3) and perplexity of 10 (batch size = 1000), respectively. From these visual figures, one may conclude that dt-see and hot-SEE were more stable compared to their competitors pt-SNE and hot-SNE.

In Fig. 4, we also provided the visual results of the MNIST embeddings created by pt-SNE, hot-SNE, dt-SEE, and hot-SEE, with batch size of 2000. These results imply that the dt-SEE and hot-SEE produced the best visualization:

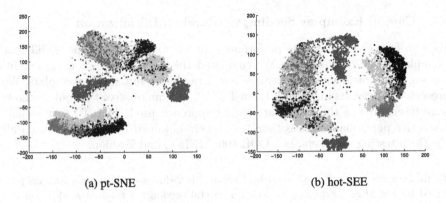

(a) pt-SNE   (b) hot-SEE

Fig. 5. The embeddings generated by pt-SNE and hot-SEE on the Fashion dataset.

the data points in each cluster were close to each other but with large separation between different clusters, compared to that of the pt-SNE and hot-SNE methods.

Also, in Fig. 5, we depicted our visual 2D embedding results on the Fashion data set. These figures further confirmed the better clustering quality generated by the hot-SEE method, compared to that of the pt-SNE strategy.

4.4   Noise Contrastive Estimation

In this section, we evaluated the performance of the noise contrastive estimation (NCE) strategy applied to our method hot-SEE with perplexity 3 and 2000 exemplars. We set $z_e = z_n = 100$ and $K_e = 18$. Table 2 show the error rates (%) obtained by 1NN on the two-dimensional representations produced by hot-SEE with or without NCS, respectively, on the MNIST and Fashion datasets.

Table 2. Error rates (%) obtained by 1NN on the two-dimensional representations produced by hot-SEE (perplexity = 3 and 2000 exemplars) with or without further computational complexity reduction based on Noise Contrastive Estimation (NCE), respectively, on the MNIST and Fashion datasets.

MNIST		Fashion	
Standard	w/NCE	Standard	w/NCE
9.30	9.69	28.18	28.19

Results in Table 2 suggest that the NCE was able to further reduce the computational and memory complexity of our method without sacrificing the predictive performance. As shown in the table, the accuracy difference of the hot-SEE method with and without NCE was less than 0.4% for both the MNIST and Fashion data sets.

4.5 Careful Exemplar Seeding vs. Random Initialization

We also further evaluate the performance of our methods in terms of different exemplar initializations used. We compared the performance of using careful seeding based on scalable K-means++ and randomly initialized exemplars. We presented the results in Table 3. From Table 3, one can observe that our methods were insensitive to the exemplar seeding approach used. That is, very similar predictive performances (less than 0.4%) were obtained by our methods on all the three testing data sets, i.e., COIL100, MNIST, and Fashion.

Table 3. Error rates (%) obtained by 1NN on the 2-dimensional representations produced by hot-SEE (perplexity = 3) with careful seeding or random seeding on the COIL100 (with 600 exemplars), MNIST (with 2000 exemplars), and Fashion (with 2000 exemplars) datasets.

COIL100		MNIST		Fashion	
Careful seeding	Random seeding	Careful seeding	Random seeding	Careful seeding	Random seeding
58.67	58.44	9.30	9.19	28.18	28.53

4.6 Comparing Evaluation Metrics of kNN ($k \geq 1$) and Quality Score

We believe that the evaluation metric based on 1NN test error rate used in the previous experimental sections is more appropriate than kNN test error rate with $k > 1$. The reason is that the 1NN performance exactly shows how accurately our exemplar-based embedding methods catpure very local neighborhood information, which is more challenging for our proposed methods. Because exemplars are computed globally, it is much easier for dt-see and hot-see to achieve better performance based on kNN with $k > 1$. On the MNIST dataset, we show the best training and test error rates of kNN with $k \geq 1$ using the two-dimensional embedding generated by different methods in Table 4, which consistently shows that dt-see and hot-see significantly outperforms pt-SNE and supports our claims above.

Another evaluation metric based on Quality Score was used by a recent method called kernel t-SNE (kt-SNE) [6]. The Quality Score metric computes the k (neighborhood size) nearest neighbors of each data point, respectively, in the high-dimensional space and in the low-dimensional space, and the metric calculates the preserved percentage of the high-dimensional neighborhood in the low-dimensional neighborhood averaged over all test data points as the Quality Score, with respect to different neighborhood size k. In Table 5, we compute the quality scores of different methods on the MNIST test data for preserving their neighborhood on the training data with neighborhood size ranging from 1 to 100. These results also show that hot-see and dt-see consistently outperform pt-SNE.

Table 4. The training error rates (_tr) and test error rates (_te) of kNN with different k's using the two-dimensional embedding generated by different methods on MNIST.

Method	The number of nearest neighbors k in kNN									
	1	2	3	4	5	6	7	8	9	10
pt-sne_tr	12.49	12.49	9.26	8.84	8.45	8.30	8.18	8.12	8.08	8.08
pt-sne_te	12.55	12.55	9.79	9.48	9.12	8.95	8.83	8.72	8.72	8.69
hot-see_tr	8.87	8.87	6.31	6.05	5.83	5.68	5.64	5.63	5.60	5.58
hot-see_te	9.19	9.19	7.21	6.76	6.61	6.42	6.41	6.41	6.42	6.36
dt-see_tr	**7.19**	**7.19**	**5.09**	**4.90**	**4.72**	**4.67**	**4.62**	**4.62**	**4.56**	**4.56**
dt-see_te	**8.80**	**8.80**	**6.69**	**6.45**	**6.25**	**6.17**	**6.02**	**6.02**	**5.94**	**5.96**

Table 5. Quality scores (%, the higher the better) for different embedding methods computed on the test set against the training set on MNIST.

Method	Neighborhood size										
	1	10	20	30	40	50	60	70	80	90	100
pt-sne	0.55	4.01	6.68	8.76	10.56	12.17	13.62	14.93	16.06	17.19	18.23
hot-see	1.12	5.25	8.22	10.53	12.48	14.19	15.69	17.04	18.27	19.41	20.44
dt-see	**1.14**	**6.74**	**10.68**	**13.52**	**15.78**	**17.56**	**19.03**	**20.22**	**21.31**	**22.27**	**23.17**

We find that Kernel t-SNE is also capable of embedding out-of-sample data. To have a similar experiment setting on MNIST as that used in kernel t-SNE, we randomly choose 2000 data points as held-out test set from the original test set (size = 10000) to get 10 different test sets with size 2000, the test error rates of our methods compared to kernel t-SNE are, kernel t-SNE: 14.2%, fisher kernel t-SNE: 13.7%, hot-see: 9.11% ± 0.43%, dt-see: 8.74% ± 0.37%. Our methods hot-see and dt-see significantly outperform (fisher) kernel t-SNE.

5   Conclusion and Future Work

In this paper, we present unsupervised parametric t-distributed stochastic exemplar-centered data embedding and visualization approaches, leveraging a deep neural network or a shallow neural network with high-order feature interactions. Owing to the benefit of a small number of precomputed high-dimensional exemplars, our approaches avoid pairwise training data comparisons and have significantly reduced computational cost. In addition, the high-dimensional exemplars reflect local data density distributions and global clustering patterns. With these nice properties, the resulting embedding approaches solved the important problem of embedding performance being sensitive to hyper-parameters such as batch sizes and perplexities, which have haunted other neighbor embedding methods for a long time. Experimental results on several benchmark datasets demonstrate that our proposed methods significantly outperform state-of-the-art unsupervised deep parametric embedding method pt-SNE in terms of robustness, visual effects, and quantitative evaluations.

In the future, we plan to incorporate recent neighbor-embedding speedup developments based on efficient N-body force approximations into our exemplar-centered embedding framework.

Acknowledgement. We thank Hans Peter Graf, Farley Lai and Yitong Li for helpful discussions. We thank anonymous reviewers for valuable comments.

References

1. Bahmani, B., Moseley, B., Vattani, A., Kumar, R., Vassilvitskii, S.: Scalable k-means++. Proc. VLDB Endow. **5**(7), 622–633 (2012)
2. Bengio, Y., Paiement, J.F., Vincent, P., Delalleau, O., Roux, N.L., Ouimet, M.: Out-of-sample extensions for LLE, isomap, MDS, eigenmaps, and spectral clustering. In: Advances in Neural Information Processing Systems, pp. 177–184 (2004)
3. Burges, C.J.: Dimension reduction: a guided tour, January 2010
4. Carreira-Perpinán, M.A.: The elastic embedding algorithm for dimensionality reduction. In: ICML, vol. 10, pp. 167–174 (2010)
5. Carreira-Perpinán, M.A., Vladymyrov, M.: A fast, universal algorithm to learn parametric nonlinear embeddings. In: Advances in Neural Information Processing Systems, pp. 253–261 (2015)
6. Gisbrecht, A., Schulz, A., Hammer, B.: Parametric nonlinear dimensionality reduction using kernel t-SNE. Neurocomputing **147**, 71–82 (2015)
7. Greengard, L., Rokhlin, V.: A fast algorithm for particle simulations. J. Comput. Phys. **73**(2), 325–348 (1987)
8. Guo, H., Zhu, X., Min, M.R.: A deep learning model for structured outputs with high-order interaction. CoRR abs/1504.08022 (2015)
9. Hinton, G., Roweis, S.: Stochastic neighbor embedding. In: Advances in Neural Information Processing Systems, vol. 15, pp. 833–840 (2003)
10. Kuksa, P.P., Min, M.R., Dugar, R., Gerstein, M.: High-order neural networks and kernel methods for peptide-MHC binding prediction. Bioinformatics **31**(22), 3600–3607 (2015)
11. van der Maaten, L.: Learning a parametric embedding by preserving local structure. In: Proceedings of the 12th International Conference on Artificial Intelligence and Statistics, pp. 384–391 (2009)
12. van der Maaten, L., Hinton, G.: Visualizing data using t-SNE. J. Mach. Learn. Res. **9**, 2579–2605 (2008)
13. van der Maaten, L., Postma, E.O., van den Herik, H.J.: Dimensionality reduction: a comparative review (2008)
14. Memisevic, R.: Gradient-based learning of higher-order image features. In: ICCV, pp. 1591–1598 (2011)
15. Min, M.R.: A non-linear dimensionality reduction method for improving nearest neighbour classification. In: Master Thesis. Department of Computer Science, University of Toronto (2005)
16. Min, M.R., Chowdhury, S., Qi, Y., Stewart, A., Ostroff, R.: An integrated approach to blood-based cancer diagnosis and biomarker discovery. In: Pacific Symposium on Biocomputing (PSB), pp. 87–98 (2014)

17. Min, M.R., Guo, H., Song, D.: Exemplar-centered supervised shallow parametric data embedding. In: Proceedings of the Twenty-Sixth International Joint Conference on Artificial Intelligence, IJCAI 2017, Melbourne, Australia, 19–25 August 2017, pp. 2479–2485 (2017)
18. Min, M.R., van der Maaten, L., Yuan, Z., Bonner, A.J., Zhang, Z.: Deep supervised t-distributed embedding. In: Proceedings of the 27th International Conference on Machine Learning, pp. 791–798 (2010)
19. Min, M.R., Ning, X., Cheng, C., Gerstein, M.: Interpretable sparse high-order Boltzmann machines. In: Proceedings of the Seventeenth International Conference on Artificial Intelligence and Statistics, pp. 614–622 (2014)
20. Purushotham, S., Min, M.R., Kuo, C.C.J., Ostroff, R.: Factorized sparse learning models with interpretable high order feature interactions. In: KDD, New York, USA (2014)
21. Ranzato, M., Hinton, G.E.: Modeling pixel means and covariances using factorized third-order Boltzmann machines. In: CVPR (2010)
22. Ranzato, M., Krizhevsky, A., Hinton, G.E.: Factored 3-way restricted Boltzmann machines for modeling natural images. In: Proceedings of the Thirteenth International Conference on Artificial Intelligence and Statistics, Chia Laguna Resort, Sardinia, Italy, 13–15 May 2010, pp. 621–628 (2010)
23. Van Der Maaten, L.: Barnes-hut-sne. arXiv preprint arXiv:1301.3342 (2013)
24. Van Der Maaten, L.: Accelerating t-SNE using tree-based algorithms. J. Mach. Learn. Res. **15**(1), 3221–3245 (2014)
25. Vladymyrov, M., Carreira-Perpinan, M.: Partial-hessian strategies for fast learning of nonlinear embeddings. arXiv preprint arXiv:1206.4646 (2012)
26. Vladymyrov, M., Carreira-Perpinan, M.: Linear-time training of nonlinear low-dimensional embeddings. In: Artificial Intelligence and Statistics, pp. 968–977 (2014)
27. Yang, Z., Peltonen, J., Kaski, S.: Scalable optimization of neighbor embedding for visualization. In: International Conference on Machine Learning, pp. 127–135 (2013)

Joint Autoencoders: A Flexible Meta-learning Framework

Baruch Epstein$^{(\boxtimes)}$, Ron Meir, and Tomer Michaeli

Viterbi Faculty of Electrical Engineering, Technion - Israel Institute of Technology,
Haifa, Israel
baruch.epstein@gmail.com, {rmeir,tomer.m}@ee.technion.ac.il

Abstract. We develop a framework for learning multiple tasks simultaneously, based on sharing features that are common to all tasks, achieved through the use of a modular deep feedforward neural network consisting of shared branches, dealing with the common features of all tasks, and private branches, learning the specific unique aspects of each task. Once an appropriate weight sharing architecture has been established, learning takes place through standard algorithms for feedforward networks, e.g., stochastic gradient descent and its variations. The method deals with meta-learning (such as domain adaptation, transfer and multi-task learning) in a unified fashion, and can deal with data arising from different modalities. Numerical experiments demonstrate the effectiveness of learning in domain adaptation and transfer learning setups, and provide evidence for the flexible and task-oriented representations arising in the network. In particular, we handle transfer learning between multiple tasks in a straightforward manner, as opposed to many competing state-of-the-art methods, that are unable to handle more than two tasks. We also illustrate the network's ability to distill task-specific and shared features.

Keywords: Autoencoders · Meta-learning
Weakly-supervised learning

1 Introduction

A major goal of inductive learning is the selection of a rule that generalizes well based on a finite set of examples. It is well-known [20], and quantified rigorously in precise terms (e.g., chapter 7 in [10]), that inductive learning is impossible unless some regularity assumptions are made about the world. Such assumptions,

This work was supported by the Technion's Skillman Chair in Biomedical Sciences and by the Ollendorff Center of the Viterbi Faculty of Electrical Engineering, Technion.

Electronic supplementary material The online version of this chapter (https://doi.org/10.1007/978-3-030-10925-7_30) contains supplementary material, which is available to authorized users.

© Springer Nature Switzerland AG 2019
M. Berlingerio et al. (Eds.): ECML PKDD 2018, LNAI 11051, pp. 494–509, 2019.
https://doi.org/10.1007/978-3-030-10925-7_30

by their nature, go beyond the data, and are based on prior knowledge achieved through previous interactions with 'similar' problems. Following its early origins [3,32], the incorporation of prior knowledge into learning has become a major effort recently, and is gaining increasing success by relying on the rich representational flexibility available through current deep learning schemes [5]. Various aspects of prior knowledge are captured in different settings of meta-learning, such as learning-to-learn, domain adaptation, transfer learning, multi-task learning, etc. (e.g., [18]). In this work, we consider the setup of multi-task learning, first formalized in [3], where a set of tasks is available for learning, and the objective is to extract knowledge from a subset of tasks in order to facilitate learning of other, related, tasks. Within the framework of representation learning, the core idea is that of shared representations, allowing a given task to benefit from what has been learned from other tasks, since the shared aspects of the representation are based on more information ([37]).

We consider both unsupervised and semi-supervised learning setups. In the former setting we have several related datasets, arising from possibly different domains, and aim to compress each dataset based on features that are shared between the datasets, and on features that are unique to each problem. Neither the shared nor the individual features are given apriori, but are learned using a deep neural network architecture within an autoencoding scheme. While such a joint representation could, in principle, serve as a basis for supervised learning, it has become increasingly evident that representations should contain some information about the output (label) identity in order to perform well, and that using pre-training based on unlabeled data is not always advantageous (e.g., chap. 15 in [18]). However, since unlabeled data is far more abundant than labeled data, much useful information can be gained from it. We therefore propose a joint encoding-classification scheme where both labeled and unlabeled data are used for the multiple tasks, so that internal representations found reflect both types of data, but are learned simultaneously.

The main contributions of this work are: *(i)* A generic and flexible **modular setup** for combining unsupervised, supervised and transfer learning. *(ii)* Efficient **end-to-end transfer learning** using mostly unsupervised data (i.e., very few labeled examples are required for successful transfer learning), capable of **handling multiple tasks** with ease. *(iii)* Explicit extraction of **task-specific** and **shared representations**.

2   Related Work

Previous related work can be broadly separated into two classes of models: *(i)* Generative models attempting to learn the input representations. *(ii)* Non-generative methods that construct separate or shared representations in a bottom-up fashion driven by the inputs.

We first discuss several works within the non-generative setting. The Deep Domain Confusion (DDC) algorithm in [34] studies the problems of unsupervised domain adaptation based on sets of unlabeled samples from the source and target domains, and supervised domain adaptation where a (usually small) subset of the target domain is labeled. By incorporating an adaptation layer and a domain confusion loss they learn a representation that optimizes both classification accuracy and domain invariance, where the latter is achieved by minimizing an appropriate discrepancy measure. By maintaining a small distance between the source and target representations, the classifier makes good use of the relevant prior knowledge. The algorithm suggested in [14] augments standard deep learning with a domain classifier that is connected to the feature extractor, and acts to modify the gradient during backpropagation. This adaptation promotes the similarity between the feature distributions in a domain adaptation task. The Deep Reconstruction Classification Network (DRCN) in [16] tackles the unsupervised domain adaptation task by jointly learning a shared encoding representation of the source and target domains based on minimizing a loss function that balances between the classification loss of the (labeled) source data and the reconstruction cost of the target data. The shared encoding parameters allow the target representation to benefit from the ample source supervised data. In addition to these mostly algorithmic approaches, a number of theoretical papers have attempted to provide a deeper understanding of the benefits available within this setting [4,29].

Next, we mention some recent work within the generative approach, briefly. Recent work has suggested several extensions of the increasingly popular Generative Adversarial Networks (GAN) framework [19]. GANs aim to learn generative models of the data distribution by mapping specified latent distributions to highly complex data distributions. This basic approach is formulated as a minimax game, based on a generator network that aims to construct the data distribution from arbitrary latent distributions, while a discriminator tries to distinguish between the generated and the true distributions. The Coupled Generative Adversarial Network (CoGAN) framework in [25] aims to generate pairs of corresponding representations from inputs arising from different domains. They propose learning joint distributions over two domains based only on samples from the marginals. This yields good results for small datasets, but is unfortunately challenging to achieve for large adaptation tasks, and is computationally cumbersome. The Adversarial Discriminative Domain Adaptation (ADDA) approach [33] subsumes some previous results within the GAN framework of domain adaptation. The approach learns a discriminative representation using the data in the labeled source domain, and then learns to adapt the model for use in the (unlabeled) target domain through a domain adversarial loss function. The idea is implemented through a minimax formulation similar to the original GAN setup. Other relevant work in this direction includes [11,15] and [12].

The extraction of shared and task-specific representations is the subject of a number of works, such as [13,31,36]. However, works in this direction typically require inputs of the same dimension and for the sizes of their shared and task-specific features to be the same. A more flexible approach can be seen in [9], proposing to augment the feature spaces of the source and target problems. Namely, the augmented source (target) version of the data includes general and source-specific (target-specific) features. In a sense, our work can be considered an extension of this strategy. However, [9] has no unsupervised learning component, does not discuss learning multiple tasks simultaneously (as opposed to transfer learning), and concerns itself with experiments from the language processing domain, solely.

A great deal of work has been devoted to multi-modal learning where the inputs arise from different modalities. Exploiting data from multiple sources (or views) to extract meaningful features, is often done by seeking representations that are sensitive only to the common variability in the views and are indifferent to view-specific variations. Many methods in this category attempt to maximize the correlation between the learned representations, as in the linear canonical correlation analysis (CCA) technique and its various nonlinear extensions [2,30]. Other methods use losses based on both correlation and reconstruction error in an auto-encoding like scheme [35], or employ diffusion processes to reveal the common underlying manifold [24]. However, all multi-view representation learning algorithms rely on *paired examples* from the two views. This setting is thus very different from transfer learning, multi-task learning, or domain adaptation, where one has access only to *unpaired samples* from each of the domains.

While GANs provide a powerful approach to multi-task learning and domain adaptation, they are often hard to train and fine tune [17]. Our approach offers a complementary non-generative perspective, and operates in an end-to-end fashion allowing the parallel training of multiple tasks, incorporating both unsupervised, supervised and transfer settings within a single architecture. This simplicity allows the utilization of standard optimization techniques for regular deep feedforward networks, so that any advances in that domain translate directly into improvements in our results. The approach does not require paired inputs and can operate with inputs arising from entirely different domains, such as speech and audio (although this has not been demonstrated empirically here). Our work is closest to [7] which shares with us the separation into common and private branches as well as unsupervised learning. They base their optimization on several loss functions beyond the reconstruction and classification losses, enforcing constraints on intermediate representations. Specifically, they penalize differences between the common and private branches of the same task, and encourage similarity between the different representations of the source and target in the common branch. This multiplicity of loss functions adds several free parameters to the problem that require further fine-tuning. Our framework uses only losses penalizing reconstruction and classification errors, thereby directly focusing on the task without adding internal constrains. Moreover, since DSN does not use a classification error for the target it cannot use labeled targets, and thus can only perform unsupervised transfer learning. Also, due to the internal loss functions, it is not clear how to extend DSN to multi-task learning, which is immediate in our formalism. Practically, the proposed DSN architecture is costly; it is larger by more than on order of magnitude than either the models we have studied or ADDA. Thus it is computationally challenging as well as relatively struggling to deal with small datasets.

3   Joint Autoencoders

In this section, we introduce *joint autoencoders* (JAE), a general method for multi-task learning by unsupervised extraction of features shared by the tasks as well as features specific to each task. We begin by presenting a simple case, point out the various possible generalizations, and finally describe two transfer and multi-task learning procedures utilizing joint autoencoders.

3.1   Joint Autoencoders for Reconstruction

Consider a multi-task learning scenario with T tasks $t^1, ..., t^T$ defined by domains $\left\{ \left(\mathcal{X}^i \right) \right\}_{i=1}^T$. Each task t^i is equipped with a set of unlabeled samples $\left\{ x_n^i \in \mathcal{X}^i \right\}_{n=1}^{N^{i,u}}$, where $N^{i,u}$ denotes the size of the unlabeled data set, and with a reconstruction loss function $\ell_r^i \left(x_n^i, \tilde{x}_n^i \right)$, where $\tilde{x}_n^i$ is the reconstruction of the sample x_n^i. Throughout the paper, we will interpret ℓ_r^i as the L_2 distance between x_n^i and $\tilde{x}_n^i$, but in principle ℓ_r^i can represent any unsupervised learning goal. The tasks are assumed to be related, and we are interested in exploiting this similarity to improve the reconstruction. To do this, we make the following two observations:

(i) Certain aspects of the unsupervised tasks we are facing may be similar, but other aspects may be quite different (e.g., when two domains contain color and grayscale images, respectively).

(ii) The similarity between the tasks can be rather "deep". For example, cartoon images and natural images may benefit from different low-level features, but may certainly share high-level structures. To accommodate these two observations, we associate with each task t^i a pair of functions: $f_p^i \left(x; \theta_p^i \right)$, the "private branch", and $f_s^i \left(x; \theta_s^i, \tilde{\theta}_s \right)$, the "shared branch". The functions f_p^i are responsible for the task-specific representations of t^i and are parametrized by parameters θ_p^i. The functions f_s^i are responsible for the shared representations, and are parametrized, in addition to parameters θ_s^i, by $\tilde{\theta}_s$ shared by all tasks. The key idea is that the weight sharing forces the common branches to learn to represent the common features of the two sources. Consequently, the private branches are implicitly forced to capture only the features that are not common to the other task. We aim at minimizing the cumulative weighted loss

$$\mathcal{L}_r = \sum_{i=1}^T w_r^i \sum_{n=1}^{N^{i,u}} \ell_r^i \left(x_n^i, f_p^i \left(x_n^i; \theta_p^i \right), f_s^i \left(x_n^i; \theta_s^i, \tilde{\theta}_s \right) \right). \tag{1}$$

In practice, we implement all functions as autoencoders and the shared parameters $\tilde{\theta}_s$ as the bottleneck of the shared branch of each task, with identical weights across the tasks. Our framework, however, supports more flexible sharing as well, such as sharing more than a single layer, or even partially shared layers. The resulting network can be trained with standard backpropagation on all reconstruction losses simultaneously. Figure 1(a) illustrates a typical autoencoder for the MNIST dataset [23], and Fig. 1(b) illustrates the architecture obtained from implementing all branches in the formal description above with such autoencoders (AE). We call this architecture a *joint autoencoder* (JAE).

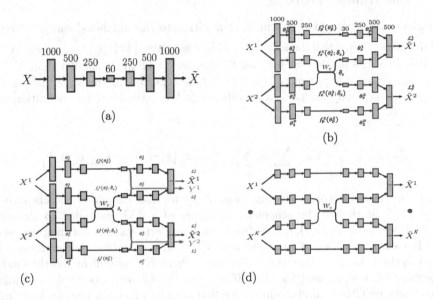

Fig. 1. (a) An example of an MNIST autoencoder (b) The joint autoencoder constructed out of the AE in (a), where $X^1 = \{0,1,2,3,4\}$ and $X^2 = \{5,6,7,8,9\}$. Each layer is a fully connected one, of the specified size, with ReLU activations. The weights shared by the two parts are denoted by W_c. The pairs of the top fully connected layers of dimension 500 are concatenated to create a layer of dimension 1000 which is then used directly to reconstruct the input of size 784. (c) A schematic depiction of a JAE architecture extended for supervised learning. The parameters and functions in figures (b) and (c) are explained in the main text. (d) A schematic depiction of a JAE architecture for K datasets. Each dataset is associated with a 'private' branch, as well as a 'shared' branch that includes the weights W_c shared by all tasks. It is also possible to design JAEs with multiple tied weights W_c^S, each shared by a subset $S \subset K$ of the datasets.

As mentioned before, in this simple example, both inputs are MNIST digits, all branches have the same architecture, and the bottlenecks are single layers of the same dimension. However, this need not be the case. The inputs can be entirely different (e.g., image and text), all branches may have different architectures, the bottleneck sizes can vary, and more than a single layer can be shared. Furthermore, the shared layers need not be the bottlenecks, in general. Finally, the generalization to more than two tasks is straightforward - we simply add a pair of autoencoders for each task, and share some of the layers of the common-feature autoencoders (Fig. 1(d)). Weight sharing can take place between subsets of tasks, and can occur at different levels for the different tasks.

3.2 Joint Autoencoders for Multi-task, Semi-supervised and Transfer Learning

Consider now a situation in which, in addition to the unlabeled samples from all domains $\mathcal{X}^i$, we also have datasets of labeled pairs $\left\{ \left(x_k^i, y_k^i \right) \right\}_{k=1}^{N^{i,l}}$ where $N^{i,l}$ is the size of the labeled set for task t^i, assumed to be much smaller than $N^{i,u}$. The supervised component of each task t^i is reflected in the supervised loss $\ell_c^i \left(y_n^i, \tilde{y}_n^i \right)$, typically multi-class classification. We extend our loss definition in Eq. 1 to be

$$\mathcal{L} = \mathcal{L}_r + \mathcal{L}_c = \mathcal{L}_r + \sum_{i=1}^{T} w_c^i \sum_{n=1}^{N^{i,l}} \ell_c^i \left(y_n^i, f_p^i \left(x_n^i; \theta_p^i \right), f_s^i \left(x_n^i; \theta_s^i, \tilde{\theta}_s \right) \right), \quad (2)$$

where we now interpret the functions f_s^i, f_p^i to also output a classification. Figure 1(c) illustrates the schematic structure of a JAE extended to include supervised losses. Note that this framework supports various learning scenarios. Indeed, if a subset of the tasks has $N^{i,l} = 0$, the problem becomes one of unsupervised domain adaptation. The case where $N^{i,l}$ are all or mostly small describes semi-supervised learning. If some of the labeled sets are large while the others are either small or empty, we find ourselves facing a transfer learning challenge. Finally, when all labeled sets are of comparable sizes, this is multi-task learning, either supervised (when $N^{i,l}$ are all positive) or unsupervised (when $N^{i,l} = 0$).

We describe two strategies to improve supervised learning by exploiting shared features.

Common-Branch Transfer. In this approach, we first train joint autoencoders on both source and target tasks simultaneously, using all available unlabeled data. Then, for the source tasks (the ones with more labeled examples), we fine-tune the branches up to the shared layer using the sets of labeled samples, and freeze the learned shared layers. Finally, for the target tasks, we use the available labeled data to train only its private branches while fixing the shared layers fine-tuned on the source data (see the supplementary material for a quantitative comparison to the alternative).

End-to-End Learning. The second, *end-to-end* approach, combines supervised and unsupervised training. Here we extend the JAE architecture by adding new layers, with supervised loss functions for each task; see Fig. 1(c). We train the new network using all losses from all tasks simultaneously - reconstruction losses using unlabeled data, and supervised losses using labeled data. When the size of the labeled sets is highly non-uniform, the network is naturally suitable for transfer learning. When the labeled sample sizes are roughly of the same order of magnitude, the setup is suitable for semi-supervised learning. We find that this end-to-end strategy is not only simpler but also outperforms the alternative consistently, and use it in all subsequent experiments.

3.3 On the Depth of Sharing

It is common knowledge that similar *low-level features* are often helpful for similar tasks. For example, in many vision applications, CNNs exhibit the same Gabor-type filters in their first layer, regardless of the objects they are trained to classify. This observation makes low-level features immediate candidates for sharing in multi-task learning settings. However, unsurprisingly, sharing low-level features is not as beneficial when working with domains of different nature (e.g., handwritten digits vs. street signs). Our approach allows to share weights in deeper layers of a neural net, while leaving the shallow layers un-linked. The key idea is that by forcing all shared-branch nets to share deep weights, their preceding shallow layers must learn to transform the data from the different domains into a common form. We support this intuition through several experiments. As our preliminary results in Sect. 4.2 show, for similar domains, sharing deep layers provides the same performance boost as sharing shallow layers. Thus, we pay no price for relying only on "deep similarities". But for domains of a different nature, sharing deep layers has a clear advantage.

4 Experiments

All experiments were implemented in Keras [8] over Tensorflow [1]. The code will be made available on publication, and the network architectures and training parameters used are given in detail in the supplementary material.

4.1 Unsupervised Learning

We present experimental results demonstrating the improvement in unsupervised learning of multiple tasks on the MNIST, CIFAR-10 [22] and celebA [26] datasets. For CIFAR-10 we trained the baseline autoencoder on single-class subsets of the database (e.g., all airplane images) and trained the JAE on pairs of such subsets. Table 1 shows a few typical results, demonstrating a consistent advantage for JAEs. Besides the lower reconstruction error, we can see that visually similar image classes, enjoy a greater boost in performance. For instance, the pair deer-horses enjoyed a performance boost of 37%, greater than the typical boost of 33–35%. The autoencoders had the same cumulative bottleneck size as the JAE, to ensure the same hidden representation size. To ensure we did not benefit solely from increased capacity, we also compared the AEs to a JAE with the same total number of parameters as the baseline, obtained by reducing the size of each layer by $\sqrt{2}$. We achieved a 22–24% boost, retaining most of the advantage over the baseline. Thus, the observed improvement is clearly not a result of mere increased network capacity.

For the experiments with the celebA dataset, a collection of faces with diverse attributes, we collected 12,000 faces of men and women each, and trained a JAE on both these subsets. The training sets contained 10,000 samples, while the test sets contained the remaining 2000 samples. When enforcing the same number of parameters in the JAE and the pair of baseline AEs, we observed a performance boost of 24%, consistent with our findings from CIFAR-10.

For the MNIST experiment, we have separated the training images into two subsets: X^1, containing the digits $\{0 - 4\}$ and X^2, containing the digits $\{5 - 9\}$. We compared the L_2 reconstruction error achieved by the JAE to a baseline of a pair of AEs trained on each dataset with architecture identical to a single branch of the JAE. The joint autoencoder (MSE =5.4) out-performed the baseline (MSE = 5.6) by 4%. The autoencoders had the same cumulative bottleneck size as the JAE, to ensure the same hidden representation size. As with CIFAR-10, we also compared the pair of autoencoders to a JAE with the same total number of parameters (obtained by $\sqrt{2}$ size reduction of each layer). This model achieved an MSE of 5.52, a 1.4% improvement over the baseline despite the simplicity of each single task.

Table 1. JAE reconstruction performance for CIFAR-10

	A-D	A-H	A-S	D-H	D-S	H-S
AE error	20.8	18.5	16.2	20.6	18.2	16.0
JAE error	13.9	12.2	10.8	13.2	11.4	10.6
JAE-reduced error	16.2	14.2	12.6	15.6	14.0	12.3
JAE Improvement	33%	34%	33%	37%	35%	34%
JAE-reduced Improvement	22%	23%	22%	24%	23%	23%

Performance of JAEs and JAEs reduced by a $\sqrt{2}$ factor vs standard AEs in terms of reconstruction MSE on pairs of objects in CIFAR-10: airplanes (A), deer (D), horses (H), ships(S). For each pair of objects, we give the mean AE error, JAE and JAE-reduced error and the improvement percentage.

To further understand the features learned by the shared and private branches, we have visualized the activations of the bottlenecks and final reconstructions in the MNIST experiment. First, we considered pairs of typical digits sharing common geometric structure, such as $\{1, 9\}$, $\{3, 8\}$ and $\{4, 7\}$. Figure 2(b) highlights the pixels in each such digit that correspond to peaks in the activity of the common and private bottleneck layers. We see that the private bottleneck focuses on the elements specific to each digit, such as the circle in '9', and the left side curves in '8', absent in '1' and '3', respectively. In contrast, the common features are sensitive to elements shared by both digits, such as vertical line in$\{1, 9\}$ and the vertical-line-with-horizontal-branch pattern in $\{4, 7\}$. Next, we

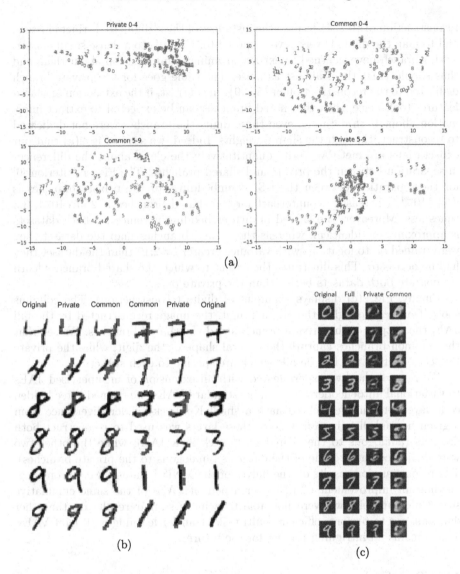

Fig. 2. MNIST visualizations: (a) t-SNE visualizations of the responses of each bottleneck to images from {0 − 4} (red) and {5 − 9} (green) MNIST digits. (b) Pairs of digits sharing geometric features, with the areas corresponding to bottleneck activation peaks highlighted in red. (c) From left to right: original digits, reconstruction by the JAE, reconstruction by the private branch, reconstruction by the shared branch. (Color figure online)

present the $2D$ t-SNE embeddings [28] of the responses of each bottleneck to the digits 0 − 9. The digits in first half, {0 − 4}, are plotted in red, while the digits in {5 − 9}, are plotted in green. We expect the private branch dedicated to the reconstruction of $\{0-4\}$ to map the instances of these digits to distinct clusters, and to map the digits it was never exposed to, $\{5-9\}$, to a dense small cluster, as this branch never learned to extract meaningful information about them and thus should treat them almost as noise. The same goes for the private branch dedicated to the reconstruction of $\{5-9\}$. In contrast, if the extraction of shared features took place, then the shared branches can be expected to extract information about both datasets, even those digits the branches were not dedicated to reconstruct, through the shared weights. Indeed, we see in Fig. 2(a) that our expectations are met. We verify quantitatively the claim about the differences in separation between the private and shared branches. The Fisher criterion [6] for the separation between the t-SNE embeddings of the private branches is $9.14 \cdot 10^{-4}$, whereas its counterpart for the shared branches is $2.77 \cdot 10^{-4}$, 3.3 times less. Moreover, the shared branch embedding variance for both datasets is approximately identical, whereas the private branches map the dataset they were trained on to locations with variance greater by 2.12 than the dataset they had no access to. This illustrates the extent to which the shared branches learn to separate both datasets better than the private ones.

Finally, Fig. 2(c) displays examples of digits reconstructions. The columns show (from left to right) the original digit, the image reconstructed by the full JAE, the output of the private branches and the shared branches. We see that the common branches capture the general shape of the digit, while the private branches capture the fine details which are specific to each subset.

We remark that we experimented with an extension of unsupervised JAEs to variational autoencoders [21]. Unlike standard VAEs, we trained three hidden code layers, requiring each to have a small Kullback-Leibler divergence from a given normal distribution. One of these layers was used to reconstruct both datasets (analogous to the shared bottleneck in a JAE), while the other two were dedicated each to one of the datasets (analogous to the private branches). The reconstruction results on the halves of the MNIST dataset were promising, yielding an improvement of 12% over a pair of VAEs of the same cumulative size. Unfortunately, we were not able to achieve similar results on the other datasets, nor to perform efficient multi-task\ transfer learning with joint VAEs. This remains an intriguing project for the future.

4.2   Transfer learning

Next, we compare the performance on MNIST of the two JAE-based transfer learning methods detailed in Sect. 3.2. For both methods, X^1 contains digits from $\{0-4\}$ and X^2 contains the digits $\{5-9\}$. The source and target datasets comprise 2000 and 500 samples, respectively. All results are measured on the full MNIST test set. The common-branch transfer method yields 92.3% and 96.1% classification precision for the $X^1 \rightarrow X^2$ and $X^2 \rightarrow X^1$ transfer tasks, respectively. The end-to-end approach results in 96.6% and 98.3% scores on the same tasks, which demonstrates the superiority of the end-to-end approach.

Shared Layer Depth. We investigate the influence of shared layer depth on the transfer performance. We see in Table 2 that for highly similar pairs of tasks such as the two halves of the MNIST dataset, the depth has little significance, while for dissimilar pairs such as MNIST-USPS, "deeper is better" - the performance improves with the shared layer depth. We believe that if the inputs are very different, the lower layers transform them into a similar representation using very different features, and then use the shared deep layers to perform the transfer learning (see discussion in Sect. 3.3). Moreover, when the input dimensions differ, early sharing is impossible - the data must first be transformed to have the same dimensions.

Table 2. Shared layer depth and transferability

	1	2	3	4	5
MNIST {0–4} → {5–9}	96.5	95.4	95.8	96.1	96.0
MNIST {5–9} → {0–4}	98.3	97.6	97.8	98.2	98.3
MNIST → USPS			84.8		87.6
USPS → MNIST			83.2		86.9

Influence of the shared layer depth on the transfer learning performance (target test accuracy). For the MNIST-USPS pair, only partial data are available for dimensional reasons.

MNIST, USPS and SVHN Digits Datasets. We have seen that the end-to-end JAE-with-transfer algorithm outperforms the alternative approach. We now compare it to other domain adaptation methods that use little to no target samples for supervised learning, applied to the MNIST, USPS and SVHN digits datasets. The transfer tasks we consider are MNIST→USPS, USPS→MNIST and SVHN→MNIST. Following [27,33], we use 2000 samples for MNIST and 1800 samples from USPS. For SVHN→MNIST, we use the complete training sets. In all three tasks, both the source and the target samples are used for the unsupervised JAE training. In addition, the source samples are used for the source supervised element of the network. We study the weakly-supervised performance of JAE and ADDA allowing access to a **small number of target samples**, ranging from 5 to 50 per digit. For the supervised version of ADDA, we fine-tune the classifiers using the small labeled target sets after the domain adaptation. Figure 3 $(a) - (c)$ provides the results of our experiments. For recent methods such as CoGAN, gradient reversal, domain confusion and DSN, we display results with zero supervision, as these methods do not support weakly-supervised training. For DSN, we provide preliminary results on MNIST↔USPS, without model optimization that is likely to prevent over-fitting.

On all tasks, we achieve results comparable or superior to existing methods using very limited supervision, despite JAE being both conceptually and computationally simpler than competing approaches. In particular, we do not train

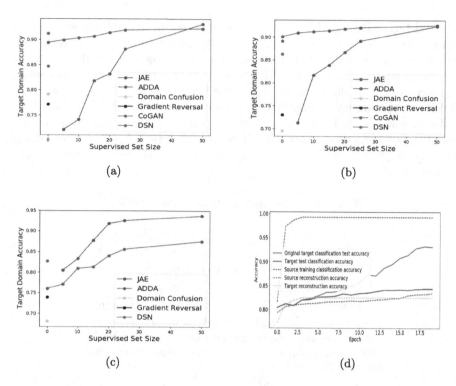

Fig. 3. Transfer learning results on MNIST, USPS and SVHN. (a) MNIST→USPS. 50 samples per digit allow JAE to surpass ADDA and CoGAN, reaching 93.1% accuracy. (b) USPS→MNIST. 25 samples per digit allow JAE to surpass CoGAN. At 50 samples per digit, JAE (92.4%) is comparable to ADDA (92.6%). (c) SVHN→MNIST. JAE achieves state-of-the-art performance, reaching 93.7% accuracy. CoGAN did not converge on this task. (d) Transfer occurring purely due to source unsupervised learning. The green graph is the test classification accuracy achieved without our interference by freezing the source supervised learning, target unsupervised learning and target classifier training. The reconstruction accuracy is measured as the fraction of pixels correctly classified as white\black. Equivalently, it is the complement of the average reconstruction error. (Color figure online)

a GAN as in CoGAN, and require a single end-to-end training period, unlike ADDA that trains three separate networks in three steps. Computationally, the models used for MNIST→USPS and USPS→MNIST have $1.36M$ parameters, whereas ADDA uses over $1.5M$ weights. For SVHN→MNIST, we use a model with $3M$ weights, comparable to the $1.5M$ parameters in ADDA and smaller by an order of magnitude than DSN. The SVHN→MNIST task is considered the hardest (for instance, GAN-based approaches fail to address it) yet the abundance of unsupervised training data allows us to achieve good results, relative to previous methods. We provide further demonstration that knowledge is indeed transferred from the source to the target in the MNIST→USPS transfer task with 50 samples per digit. Source supervised learning, target unsupervised learning and target classifier training are frozen after the source classifier saturates (epoch 4). The subsequent target test improvement by 2% is due solely to the source dataset reconstruction training, passed to the target via the shared bottleneck layer (Fig. 3(d)).

Three-Way Transfer Learning. We demonstrate the ability to extend our approach to multiple tasks with ease by transferring knowledge from SVHN to MNIST and USPS simultaneously. That is, we train a triple-task JAE reconstructing all three datasets, with additional supervised training on SVHN and weakly-supervised training on the target sets. All labeled samples are used for the source, while the targets use 50 **samples per digit**. The results illustrate the benefits of multi-task learning: 94.5% classification accuracy for MNIST, a 0.8% improvement over the SVHN→MNIST task, and 88.9% accuracy in UPS, a 1.2% improvement over SVHN→USPS. This is consistent with unsupervised learning being useful for the classification. USPS is much smaller, thus it has a lower score, but it benefits relatively more from the presence of the other, larger, task. We stress that the extension to multiple tasks was straightforward, and indeed we did not tweak the various models, opting instead for previously used JAEs, with a single shared bottleneck. Most state-of-the-art transfer methods do not allow for an obvious, immediate adaptation for transfer learning between multiple tasks. Indeed, [7] would require a number of loss functions growing quadratically in the task number, and an even more demanding architecture than is already employed, while [25] would require a quadratically growing amount of discriminators, or else a novel idea to perform efficient domain adaptation for multiple tasks. It is even less clear how to extend [33] to such scenarios.

5 Conclusion

We presented a general scheme for incorporating prior knowledge within deep feedforward neural networks for domain adaptation, multi-task and transfer learning problems. The approach is general and flexible, operates in an end-to-end setting, and enables the system to self-organize to solve tasks based on prior or concomitant exposure to similar tasks, requiring standard gradient based optimization for learning. The basic idea of the approach is the sharing of representations for aspects which are common to all domains/tasks while maintaining private branches for task-specific features. The method is, in principle, applicable to data from multiple sources and types, though adapting it for non-image domains is an ongoing research project. It also has the advantage of being able to share weights at arbitrary network levels, enabling abstract levels of sharing.

We demonstrated the efficacy of our approach on several domain adaptation and transfer learning problems, and provided intuition about the meaning of the representations in various branches. In a broader context, it is well known that the imposition of structural constraints on neural networks, usually based on prior domain knowledge, can significantly enhance their performance. The prime example of this is, of course, the convolutional neural network. Our work can be viewed within that general philosophy, showing that improved functionality can be attained by the modular prior structures imposed on the system, while maintaining simple learning rules.

References

1. Abadi, M., et al.: TensorFlow: Large-scale machine learning on heterogeneous systems (2015). http://tensorflow.org/, software available from tensorflow.org
2. Andrew, G., et al.: Deep canonical correlation analysis. In: Proceedings of ICML (2013)
3. Baxter, J.: A model of inductive bias learning. JAIR **12**, 149–198 (2000)
4. Ben-David, S., et al.: A theory of learning from different domains. Mach. Learn. **79**, 151–175 (2009)
5. Bengio, Y., Courville, A., Vincent, P.: Representation learning: a review and new perspectives. IEEE Trans. PAMI **35**(8), 1798–1828 (2013). n2423
6. Bishop, C.M.: Pattern Recognition and Machine Learning. Springer, New York (2006)
7. Bousmalis, K., et al.: Domain separation networks. In: Advances in Neural Information Processing Systems 29 (NIPS 2016) (2016)
8. Chollet, F.: keras (2015). https://github.com/fchollet/keras
9. Daumé III, H.: Frustratingly easy domain adaptation. CoRR, abs/0907.1815 (2009)
10. Devroye, L., Gyoörfi, L., Lugosi, G.: A Probabilistic Theory of Pattern Recognition. Springer, New York (1996). https://doi.org/10.1007/978-1-4612-0711-5
11. Donahue, J., Krähenbühl, P., Darrell, T.: Adversarial feature learning. arXiv preprint arXiv:1605.09782 (2016)
12. Dumoulin, V., et al.: Adversarially learned inference. arXiv preprint arXiv:1606.00704(2017)
13. Evgeniou, T., Pontil, M.: Regularized multi-task learning. In: Proceedings of the tenth ACM SIGKDD International Conference on Knowledge Discovery and Data Mining (2004)
14. Ganin, Y., Lempitsky, V.: Unsupervised domain adaptation by backpropagation. In: International Conference on Machine Learning, pp. 1180–1189 (2015)
15. Ganin, Y., et al.: Domain-adversarial training of neural networks. J. Mach. Learn. Res. **17**(59), 1–35 (2016)
16. Ghifary, M., Kleijn, W.B., Zhang, M., Balduzzi, D., Li, W.: Deep reconstruction-classification networks for unsupervised domain adaptation. In: Leibe, B., Matas, J., Sebe, N., Welling, M. (eds.) ECCV 2016. LNCS, vol. 9908, pp. 597–613. Springer, Cham (2016). https://doi.org/10.1007/978-3-319-46493-0_36
17. Goodfellow, I.: Nips 2016 tutorial: Generative adversarial networks. arXiv preprint arXiv:1701.00160 (2016). n2574 (Tech)
18. Goodfellow, I., Bengio, Y., Courville, A.: Deep Learning. MIT Press, Cambridge (2016)
19. Goodfellow, I., et al.: Generative adversarial nets. In: Advances in Neural Information Processing Systems, pp. 2672–2680 (2014)
20. Hume, D.: An Enquiry Concerning Human Understanding. A. Millar, London (1748)
21. Kingma, D.P., Welling, M.: Auto-encoding variational bayes. CoRR abs/1312.6114 (2014)

22. Krizhevsky, A.: Learning multiple layers of features from tiny images. Technical report, University of Toronto (2009)
23. LeCun, Y., et al.: Gradient-based learning applied to document recognition. Proc. IEEE **86**(11), 2278–2324 (1998)
24. Lederman, R.R., Talmon, R.: Learning the geometry of common latent variables using alternating-diffusion. Appl. Comput. Harmonic Anal. **44**(3), 509–536 (2015)
25. Liu, M.Y., Tuzel, O.: Coupled generative adversarial networks. In: Advances in Neural Information Processing Systems, pp. 469–477 (2016)
26. Liu, Z., Luo, P., Wang, X., Tang, X.: Deep learning face attributes in the wild. In: Proceedings of International Conference on Computer Vision (ICCV) (2015)
27. Long, M., Wang, J., Ding, G., Sun, J., Yu, P.S.: Transfer feature learning with joint distribution adaptation. In: 2013 IEEE International Conference on Computer Vision (2013)
28. van der Maaten, L.J.P., Hinton, G.E.: Visualizing data using t-sne. J. Mach. Learn. Res. **9**, 2579–2605 (2008)
29. Maurer, A., Pontil, M., Romera-Paredes, B.: The benefit of multitask representation learning. J. Mach. Learn. Res. **17**, 2853–2884 (2016)
30. Michaeli, T., et al.: Nonparametric canonical correlation analysis. In: Proceedings of ICML (2016)
31. Parameswaran, S., Weinberger, K.Q.: Large margin multi-task metric learning. In: Advances in Neural Information Processing Systems 23 (NIPS 2010) (2010)
32. Thrun, S., Pratt, L.: Learning to Learn. Kluwer Academic Publishers, Dordrecht (1998)
33. Tzeng, E., Hoffman, J., Saenko, K., Darrell, T.: Adversarial discriminative domain adaptation. CoRR abs/1702.05464 (2017)
34. Tzeng, E., Hoffman, J., Zhang, N., Saenko, K., Darrell, T.: Deep domain confusion: maximizing for domain invariance. CoRR, abs/1412.347 (2014)
35. Wang, W., Arora, R., Livescu, K., Bilmes, J.: On deep multi-view representation learning. In: Proceedings of the 32nd ICML, pp. 1083–1092 (2015)
36. Weston, J., et al.: Deep learning via semi-supervised embedding. In: Proceedings of ICML (2008)
37. Zhang, J., Ghahramani, Z., Yang, Y.: Flexible latent variable models for multi-task learning. Mach. Learn. **73**(3), 221–242 (2008). n2570

Privacy Preserving Synthetic Data Release Using Deep Learning

Nazmiye Ceren Abay[1]([✉]) [iD], Yan Zhou[1] [iD], Murat Kantarcioglu[1,2] [iD],
Bhavani Thuraisingham[1] [iD], and Latanya Sweeney[2] [iD]

[1] University of Texas at Dallas, Richardson, TX, USA
{nca150130,yan.zhou2,muratk,bxt043000}@utdallas.edu
[2] Harvard University, Cambridge, MA, USA
latanya@fas.harvard.edu

Abstract. For many critical applications ranging from health care to
social sciences, releasing personal data while protecting individual privacy is paramount. Over the years, data anonymization and synthetic
data generation techniques have been proposed to address this challenge.
Unfortunately, data anonymization approaches do not provide rigorous
privacy guarantees. Although, there are existing synthetic data generation techniques that use rigorous definitions of differential privacy, to our
knowledge, these techniques have not been compared extensively using
different utility metrics.

In this work, we provide two novel contributions. First, we compare
existing techniques on different datasets using different utility metrics.
Second, we present a novel approach that utilizes deep learning techniques coupled with an efficient analysis of privacy costs to generate differentially private synthetic datasets with higher data utility. We show
that we can learn deep learning models that can capture relationship
among multiple features, and then use these models to generate differentially private synthetic datasets. Our extensive experimental evaluation
conducted on multiple datasets indicates that our proposed approach is
more robust (i.e., one of the top performing technique in almost all type
of data we have experimented) compared to the state-of-the art methods
in terms of various data utility measures. Code related to this paper is
available at: https://github.com/ncabay/synthetic_generation.

Keywords: Differential privacy · Deep learning · Data generation

1 Introduction

Increasingly more data is collected about almost every aspect of human life
ranging from health care delivery to social media. As the amount of collected
data increases, more opportunities have emerged for leveraging this collected
data for important societal purposes.

Since the collected data can be used to offer important services and facilitate
much needed research, many organizations are striving to share the data that

© Springer Nature Switzerland AG 2019
M. Berlingerio et al. (Eds.): ECML PKDD 2018, LNAI 11051, pp. 510–526, 2019.
https://doi.org/10.1007/978-3-030-10925-7_31

they collect. For example, many novel applications, such as smart cities [34] and personalized medicine [27], require the collection and sharing of privacy sensitive micro-data, i.e., information at the level of individual respondents.

Due to the importance of these goals, many organizations advocate for openly sharing data to serve important social, economic and democratic goals. For example, in 2016, the City of Seattle announced an open data policy where the city's data would be "open by preference" except when such data sharing may affect individual privacy. Similarly, National Institute of Health (NIH) requires the sharing of genomic data created as a part of NIH funded research with other researchers.

At the same time, sharing micro data carries inherent risks to individual privacy. For example, a municipal dataset that contains information about bike sharing has been used to identify individuals and their transit patterns [32]. Similarly, a taxi ride data set from New York have been used to identify certain individuals' addresses and their trips to certain night clubs [33]. These examples show that there is an important societal need in sharing micro data while protecting individual privacy.

To address this privacy challenge, solutions have been proposed in two broad categories. In the first category, the data anonymization based approaches (e.g., [30]) try to use various definitions to sanitize data so that it cannot be easily re-identified. Although these approaches have some important use cases, they are not usually based on rigorous privacy definitions that can withstand various types of re-identification attacks. In the second category, synthetic data generation approaches have been proposed to generate realistic synthetic data using rigorous differential privacy definition [12]. Although these approaches have been shown to work in some limited cases, they have not been extensively tested on different types of use cases with different requirements (e.g., high dimensionality, correlation among features). Therefore, it was not clear which technique works well under what conditions for what type of data sets. We answer these questions by conducting extensive experimentation. Furthermore, we provide a new differentially private deep learning based synthetic data generation technique to address the limitations of the existing techniques.

In this paper, we propose an auto-encoder technique (DP-SYN), a generative deep learning technique that generates privacy preserving synthetic data. We test our approach on benchmark datasets and compare the results with other state-of-the-art techniques. We show that our proposed technique outperforms them in terms of three evaluation metrics.

Our contributions can be summarized as follows:

- We test existing techniques using different datasets with different properties using three utility metrics. We show that none of the existing techniques consistently outperforms others on all types of data sharing tasks and datasets.
- We propose a novel differentially private deep learning based synthetic data generation technique that is shown to be robust under different utility metrics with respect to different synthetic data generation tasks.

– We show that our approach does not deteriorate when faced with imbalanced or high dimensional datasets. Due to an inner partitioning of the latent structure, our approach gives more robust results in noise addition and works with both relational and image data.

This work is organized as follows. In Sect. 2, we discuss the related work. In Sect. 3, we provide the preliminaries for our work. In Sect. 4, we discuss our novel differentially private auto-encoder based technique for synthetic data generation. In Sect. 5, we run extensive empirical analyses to understand the relative strengths of the existing techniques and show that our technique works well in almost all of the given datasets. Finally, we conclude in Sect. 6.

2   Related Work

Extensive research has been conducted on publishing private data for preserving privacy. Despite their success in data utility measures, most of the proposed methods in the literature are impractical to be implemented for high dimensional data. In this section, we discuss the related techniques with their strengths and limitations.

In statistical analysis, publishing a marginal table while preserving the privacy has been a fundamental research goal. One of the initial efforts in addressing this problem is proposed by Barak et al. [3]. In this method, a full contingency table constructed on the original data is represented by the Fourier coefficients. The noise is then added to these coefficients in order to construct the desired k-way marginal tables, instead of perturbing the original data. Despite its feasibility and widespread use in low dimensional data, the number of Fourier coefficients, 2^d, grows exponentially with increased dimensionality. This results in intractable computational cost when working with high dimensional data. Another method is designed by Ding et al. [9] to work with high dimensional data such as online analytical processing (OLAP). In this framework, strategic cuboids that are useful to generate other cuboids are chosen first, and a private version of these cuboids is constructed by using differential privacy. The main limitation of this study arises while constructing the strategic cuboids. As all possible cuboids are iteratively traversed and selected, the number of the cuboids grows with the dimensions of the data, resulting in an increased complexity. A more practical and efficient approach, known as PRIVIEW, addresses the high dimensionality problem [24]. PRIVIEW also constructs the private k-way marginal tables for $k \geq 3$. While constructing private marginal tables, PRIVIEW first extracts low-dimensional marginal views from the flat data and adds noise to the views. Next, PRIVIEW applies a reprocessing technique to ensure the consistency of the noisy views. Afterwards, PRIVIEW applies maximum entropy optimization on the views to obtain the k-way marginal tables. PRIVIEW is reported as a more efficient technique in terms of time and space complexity; however, it can be employed on binary data only.

There are other frameworks, designed particularly for differential optimization problems. First, Dwork et al. [11] propose an output perturbation technique that directly add noise to the regularized objective function after optimization. This technique is outperformed by the objective perturbation technique proposed by Chaudhuri et al. [8] which adds noise to the objective function before optimization. We denote this work as PRIVATESVM and compare its results to those of DP-SYN in the experiments section.

Differential privacy has been implemented in a number of data analysis tasks, including regression models [7,37], classification models [17,26,31] and association rule mining [20,35].

Generating artificial data from the original one is another privacy preserving technique for data publication. Here, instead of using the sanitization models discussed previously, Rubin [25] introduces repetitive perturbation of the original data as a substitute to the original data. To execute this technique, Zhang et al. [36] present a synthetic data generation technique, PRIVBAYES. PRIVBAYES is defined as a differential generative model that decomposes high dimensional data into low dimensional marginals by constructing a Bayesian network. Afterwards, noise is injected into these learned low dimensional marginals to ensure differential privacy and the synthetic data is inferred from these noised marginals. Although PRIVBAYES is credited as an effective technique, as we will show in our experiments, our proposed technique has a significant improvement over PRIVBAYES.

Acs et al. [2] model generative neural networks to produce synthetic samples. The authors first cluster the original datasets into k clusters with private kernel k-means. Afterwards, they use generative neural networks for each cluster to create synthetic data. In our experiments, we denote this work with DP-VAE and compare its results to our method.

Bindschaedler et al. [5] present another differential generative framework. The authors introduce an idea of *plausible deniability*, rather than adding noise to the generative model directly. Plausible deniability is ensured by a *privacy threshold* in releasing synthetic data. Here, an adversary cannot tell whether a particular input belongs to the original data by observing synthetic records.

Park et al. [22] propose a private version of the iterative expectation maximization algorithm. They effectively combine differential privacy and expectation maximization algorithm to cluster datasets. Here, we use this approach to discover patterns in latent space. We observed an improvement in the performance of this technique when used with partitioning the original dataset into unique data label groups. We use this modified version in our experiments [22] as DP-EM(SYN) and compare its results in the experiments section.

In some cases, combining differentially private algorithms has been proven to be useful in formulating more complex privacy solutions. However, such combinations may result in degradation of the privacy protection as more information is leaked by multiple usage of the private techniques. To track the total privacy loss while executing such mechanisms, Dwork et al. [12] propose basic and advanced composition theorems. Abadi et al. [1] propose another advanced composition theorem known as the *moments accountant* and verify that it has the best overall privacy bound in the literature. Abadi et al. also utilize moment accountant while constructing a deep learning technique to classify images.

3   Preliminaries

This section briefly recalls the definitions and standards of differential privacy and the principles of deep learning.

3.1   Differential Privacy

Differential privacy is the formal mathematical model that ensures privacy protection, and it is primarily used to analyze and release sensitive data statistics [11]. Differential privacy utilizes randomized algorithms to sanitize sensitive information while bounding the privacy risk of revealing sensitive information.

Theorem 1 ($((\varepsilon, \delta)$-Differential Privacy [11]**).** *For two non-negative numbers ε, δ, a randomized algorithm, $\mathcal{F}$, satisfies (ε, δ)-differential privacy iff for any neighboring pair d, d' and $S \subseteq Range(\mathcal{F})$, the following formula holds:*

$$Pr[\mathcal{F}(d) \in S] \le e^{\varepsilon} Pr[\mathcal{F}(d') \in S] + \delta. \tag{1}$$

Here, the neighboring pair differ from each other with only one entry while the remaining entries are identical. In Theorem 1 [11], δ is a relaxation to ε-*differential privacy* that formulates the probability of privacy leakage. However, to avoid such leakage, Dwork et al. [12] shows that δ must be chosen smaller than $1/n$ for a data of n samples.

Our proposed technique sanitizes sensitive data based on a widely used differentially private technique, the Gaussian mechanism [6]. The deterministic function f takes d as input. $f(d)$ perturbs the input with noise sampled from the normal distribution $\mathcal{N}$, based on ϵ, δ, and s_f which is the sensitivity of f defined as follows:

Definition 1 (Sensitivity [11]**).** *For a given function f, the sensitivity of f is defined as a maximum absolute distance between two neighboring pairs (d, d')*

$$s_f = \max_{(d,d')} \|f(d) - f(d')\|, \tag{2}$$

where $\|.\|$ is L_1 norm.

The (ε, δ)-*differential privacy* of function f over data d is guaranteed by $\mathcal{F}(d)$ with the Gaussian mechanism:

$$\mathcal{F}(d) = f(d) + z, \tag{3}$$

where z is a random variable from distribution $\mathcal{N}(0, \sigma^2 s_f^2)$. Here, when $\varepsilon \in [0, 1]$, the relation among the parameters of Gaussian mechanism [12] is such that

$$\sigma^2 \varepsilon^2 \ge 2 \ln (1.25/\delta) s_f^2.$$

3.2   Deep Learning

Deep learning is a subfield of machine learning that can be either supervised or unsupervised [16]. The power of deep learning comes from discovering essential concepts of data as nested hierarchy concepts where simpler concepts are refined to obtain complex concepts. Deep learning has been applied to many different research areas including computer vision [16], speech recognition [18] and bioinformatics [19]. We focus on the auto-encoder, an unsupervised deep learning technique that outputs a reconstruction of its input.

Similar to an artificial neural network, an auto-encoder is trained by optimizing an objective function. Stochastic gradient descent (SGD) is used as a scalable technique [29] to solve this optimization problem. Rather than iterating over every training instance, SGD iterates over a *mini-batch* of the instances. For a given training set of m samples, $D = \{x_i\}_{i=1}^{m}$ and $x_i \in \mathbb{R}^d$, the objective function is given as:

$$\min_{w} \mathcal{L}(w) = \frac{1}{|B|} \sum_{x_i \in B} \ell(w; x_i), \tag{4}$$

where B is the *mini-batch*, w is auto-encoder model parameter and ℓ is the discrepancy cost of example x_i and its reconstruction $\tilde{x}_i$.

At each step t, model gradient is computed for a given batch B_t and learning parameter η. Then, the model parameter is updated for the next step as follows:

$$w_{t+1} = w_t - \eta \left(\frac{1}{|B_t|} \sum_{x_i \in B_t} \nabla_w \ell(w; x_i) \right). \tag{5}$$

Figure 1 presents two main phases of an auto-encoder: the **encoder** and the **decoder**.

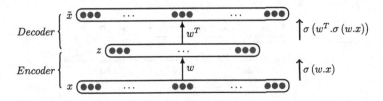

Fig. 1. One hidden layer auto-encoder that encodes the input to the latent space, and decode the latent space to the reconstruction of input.

The **encoder** maps its input to a hidden intermediate layer that usually has less neurons than the input size to get a latent representation of the input. Here, the element-wise activation function σ maps $x \in \mathbb{R}^d$ into $z \in \mathbb{R}^{d'}$ where $d' < d$. On the other hand, the **decoder** takes the latent representation z, and reconstructs $\tilde{x} \in \mathbb{R}^d$.

In Sect. 4, we describe how the SGD model defined in Eq. 5 is used in the inference of model parameter w.

4   Methodology

This section describes the main components of our differentially private synthetic data generation approach. We first introduce our private auto-encoder and explain the private expectation maximization algorithm. Next, we present the privacy analysis of the proposed technique.

4.1   Differentially Private Synthetic Data Generation Algorithm

Our main framework aims to generate synthetic data without sacrificing the utility. A similar approach is proposed in [1] which designs a private convolutional neural network on supervised learning. However, this method can only be used in classification tasks. We combine this method with DP-EM and to create a generative deep learning model.

Assume that we have the sensitive dataset $D = \{(x_1, y_1), \ldots, (x_m, y_m)\}$, where every instance $x \in \mathbb{R}^d$ has a label $y \in \{1, \ldots, k\}$. We partition the sensitive dataset D into k groups denoted as $\hat{D}_1 \ldots \hat{D}_k$ such that every instance x in a group $\hat{D}_i \in D$ has the same label y. The value of k is limited by the number of unique labels in dataset D.

Figure 2 shows the two main steps of our approach. For each data group we build a private generative auto-encoder which are denoted with DP-SYN. The lower pane of the figure shows the inner working of a DP-SYN.

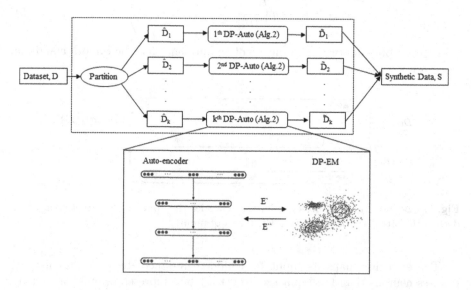

Fig. 2. Differentially private synthetic data generation DP-SYN

The details of our technique are shown in Algorithm 1. After partitioning the dataset D into k groups (Line 1 in Algorithm 1), the noise injected to each group

---

Algorithm 1. DP-SYN: Differentially Private Synthetic Data Generation

Input: D: $\{(x_1, y_1), \ldots, (x_m, y_m)\}$ where $x \in \mathbb{R}^d$, $y \in \{1, \ldots, k\}$; η: learning rate; T: iteration number; ε: privacy budget; δ: gaussian delta; σ: standard deviation; C: clipping constant.

Output: S: Synthetic data.

1: $\{\hat{D}_1 \ldots \hat{D}_k\} \leftarrow$ Partition data records in D based on associated labels
2: $S \leftarrow \{\}$
3: **for** $i = 1$ to k **do**
4:     Partition ε into $\varepsilon_A = \varepsilon/2$, $\varepsilon_H = \varepsilon/2$ and δ into $\delta_A = \delta/2$, $\delta_H = \delta/2$
5:     $W \leftarrow$ DP-Auto $\left(\hat{D}_i, \eta, \mathrm{T}, \varepsilon_A, \delta_A, \sigma, \mathrm{C}\right)$ // see Algorithm 2
6:     $E' \leftarrow encode\left(W, \hat{D}_i\right)$
7:     $E'' \leftarrow$ DP-EM $(E', \varepsilon_H, \delta_H)$ // see [22]
8:     $\tilde{D}_i \leftarrow decode\,(W, E'')$
9:     $S \leftarrow S \cup D'_i$
10: **return** S

---

is also partitioned (Line 4 in Algorithm 1), as specified in the sequential composition theorem [10]. For each previously obtained group we build one private auto-encoder (Line 5 in Algorithm 1), which is detailed in Algorithm 2. Next, we obtain the private latent representation of the group (Line 6 in Algorithm 1), and inject it into a differentially private expectation maximization (DP-EM) function. The DP-EM function is detailed in [22]. The main task of DP-EM is to detect different latent patterns in the encoded data and to generate output data with similar patterns. Here, DP-EM is used to sample encoded data (Line 7 in Algorithm 1) and newly sampled encoded data is decoded with using the model parameter W (Line 8 in Algorithm 1). $\tilde{D}_i$ is the synthetic data associated with an inputted group $\hat{D}_i$ and appended to the S to be output (Line 9 in Algorithm 1).

4.2 Building a Private Auto-Encoder

In this section, we discuss the private auto-encoder given in Algorithm 2.

Our private auto-encoder employs steps to improve the optimization process with gradient computation and clipping. While a gradient is computed for a batch in the standard stochastic training techniques, we compute the gradient for each training instance instead. This approach improves the optimization process since it reduces the sensitivity of the gradient present at each instance [14]. Norms of the gradients define the direction that optimizes the network parameters. However, in some deep networks, the gradients can be unstable and fluctuate in a large range. Such fluctuations can inhibit the learning process due to the increased vulnerability of the networks. To avoid this undesired situation,

---

Algorithm 2. DP-Auto: Differentially Private Auto-encoder

Input: η: Learning rate; T: iteration number; ε_A: privacy budget; δ: gaussian
   delta; σ: standard deviation; C: clipping constant.

Output: w: Model parameter.

1: ℓ is the objective function
2: $\nabla\ell$ is the gradient of objective function
3: initialize w_0 randomly
4: $\varepsilon'_A = 0$
5: **for** $t = 1$ to T **do**
6:    **if** $\varepsilon'_A < \varepsilon_A$ **then**
7:       $B_t \leftarrow$ random batch
8:       $i_t \sim b$ where $x_{i_t} \in B_t$
9:       $z_{i_t} \sim \mathcal{N}(0, \sigma^2 C^2)$
10:       $w_{t+1} \leftarrow w_t - \eta \cdot \left(\dfrac{1}{|B_t|} \sum\limits_{i_t} \left(\nabla\ell(w_t; x_{i_t}) + z_{i_t} \right) \right)$
11:       $\varepsilon'_A \leftarrow$ calculate privacy loss with moments accountant
12: **return** w

---

we bound norms of the previously computed gradients by a clipping constant C [23].

After clipping the gradients, noise is sampled from the Gaussian distribution with zero mean and standard deviation of $\sigma\,C$ and added to the previously clipped gradients (Line 9–10 in Algorithm 2). At the end of each batch, model parameters of the network are updated with the negative direction of the learning rate η multiplied by the averaged noisy gradients. At the end of this step, the private auto-encoder outputs the model parameter w (Line 11 in Algorithm 2).

4.3 Privacy Analysis

The privacy analysis of our proposed technique employs the *moments accountant* approach developed by Abadi et al. [1] to keep track of the privacy cost in multiple iterations. Moments accountant is a combination of both the strong composition theorem [13] and the privacy amplification theorem [4]. Moments accountant has an improvement on estimating of the privacy loss for composing differentially private Gaussian mechanisms, and it is the best for overall estimation of privacy budget in literature [1].

In our proposed work, while training the auto-encoder, we track the privacy loss at the end of each batch iteration. As given in Lines 5–11 of Algorithm 2, we compute the value of current privacy loss ε' that has been spent on private auto-encoder in a given iteration $t \in T$. Training ends when ε' reaches the final privacy budget ε.

According to moments accountant, Algorithm 2 is (ε, δ)-differentially private if the privacy loss for any $\varepsilon' < k_1(|B|/n)^2 T$ is such that for some constants k_1, k_2:

$$\varepsilon' \geq k_2 \frac{|B|/n \sqrt{T \log(1/\delta)}}{\sigma},$$

where T is the number of training steps and $|B|$ is the mini-batch for a data of n samples with a given privacy budget ε, gaussian delta δ and standard deviation σ of the Gaussian distribution.

5 Experiments

In this section, we present the experimental results to demonstrate the efficiency of our proposed approach. We compare our results with other state-of-the-art techniques. To ensure fairness, we also employ the *Gaussian mechanism* in these techniques.

We start the evaluation by explaining the experimental settings. We evaluate the performance with statistical measures, accuracy in machine learning models and agreement rate. For each task, 70% of the data is used as a training set, while the rest is used for testing.

5.1 Experimental Settings

Datasets. We test the proposed approach on nine real datasets. The following is a brief description of each dataset:

(i) The **Adult** [21] dataset contains the information of **45222 individuals**, extracted from the 1994 US census. The dataset shows whether the income of the individuals exceeds fifty thousand US dollars. The dataset contains **15 features**.

(ii) The **Lifesci** [21] dataset contains **26733 records** and **10 principal components** from chemistry and biology experiments.

(iii) The **Optical Digit Recognition** (ODR) [21] dataset contains **5620 handwritten digits of 10**. Each instance is represented by **64 numeric features**.

(iv) The **Spambase** [21] dataset contains **4601 emails**, each of which is labeled as spam or non-spam. Each instance has **58 attributes**.

(v) The **Contraceptive Method Choice** (CMC) [21] dataset contains **9 features** of **1473 married women** to predict their current contraceptive method choice.

(vi) The **German Credit** [21] dataset contains the anonymized information of **1000 customers** with **20 features**. Each customer is classified as good or bad credit risk.

(vii) The **Mammographic Mass** [21] dataset contains the information of **961 patients'** mammographic masses of with **5 attributes**. The class value shows that patient has breast cancer or not based on mammographic mass.

(viii) The **Diabetes** [21] dataset contains the information of female patients who are at least 21 years old. Each patient is classified as diabetic or not diabetic. The dataset has **768 records** with **8 features**.

(ix) The **BreastCancer** [21] dataset contains the information about whether a patient has breast cancer or not. It has **699 patient records** with **10 features**.

In all experiments, we compare our results with four state-of-the-art techniques: PRIVATESVM [8], PRIVBAYES [36], DP-EM(SYN) [22] and DP-VAE [2].

We repeat each experiment 10 times for each task and report the average measurements in our experimental results. In total, our experiments consist of 7840 runs of the mentioned techniques. Here, we only report the best results from each algorithm.

5.2   Accuracy in Machine Learning Models

In this set of experiments, we evaluate the accuracy in a Support Vector Machine (SVM) [15] task. More specifically, we report the percentage of incorrectly classified tuples as the *misclassification rate*. For the PRIVATESVM, out of two proposed approaches we only report results from the objective perturbation approach because it outperforms the output perturbation approach.

For each training set, we generate synthetic data by using each method and construct SVM models on the synthetic data. Performance of these models is evaluated on the test set. PRIVATESVM has a regularization parameter λ for the objective function. We run PRIVATESVM with $\lambda \in \{10^{-3}, 10^{-4}, 10^{-5}, 10^{-6}, 10^{-8}\}$ and pick the model that reports the lowest misclassification rate.

Figure 3 presents the misclassification rate of the techniques for a given (ε, δ) pair. In the figure, the black straight line shows the misclassification rate on the original dataset, i.e., without privacy. Here, it presents the best case to aim for. We now compare the performance of DP-SYN with respect to each state-of-the-art method.

Figure 3 shows that DP-SYN has better performance than PRIVBAYES for eight out of the nine datasets. Only for the Adult dataset, PRIVBAYES performs slightly better than our DP-SYN approach. DP-SYN outperforms DP-VAE for seven datasets. For BreastCancer and Diabetes, DP-VAE has better performance; however, it fails to classify any instance in the GermanCredit dataset. For high dimensional datasets such as Spambase and ODR, the misclassification rate of DP-EM(SYN) is two times bigger than that of DP-SYN. A reason for this high misclassification rate shows that DP-EM(SYN) fails in generating synthetic data task when the dimension of data is more than two dozens. DP-SYN also outperforms PRIVATESVM in five datasets. PRIVATESVM is specifically designed for SVM, and it is expected to have lower misclassification rates in SVM tasks. However, PRIVATESVM cannot be employed in other machine learning tasks easily.

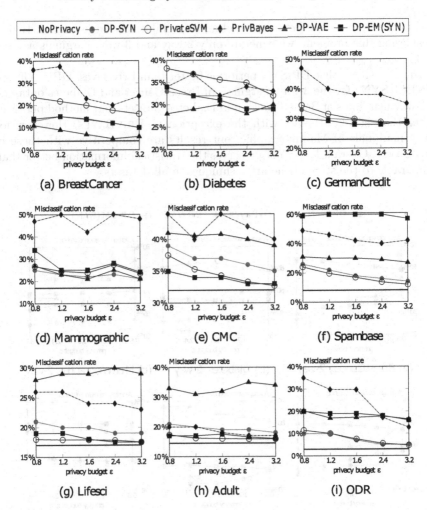

Fig. 3. Misclassification rates for the nine datasets.

Consequently, we point out that on datasets of different types, no single method gives the best misclassification rate consistently. As shown in Fig. 3c with the absence of DP-VAE, some algorithms cannot even classify a dataset if the dataset is highly imbalanced. Considering these issues, Fig. 3 shows that DP-SYN can be employed on all datasets and reports competitive results.

5.3 Statistical Measures

We evaluate the quality of synthetic data in terms of statistical utility. We generate k-way marginals of the dataset and compare the probability distribution of the noisy and original marginals. *Total variation distance* [28] is used to report the statistical difference between the noisy and original marginals. The datasets used in the experiments are large, leading to prohibitively large queries. Hence, considering this problem, we generate only 2-way and 3-way marginals as used in [36].

Figure 4 shows that DP-SYN performs better than PrivBayes, DP-VAE and DP-EM(SYN) for the 3-way marginals of BreastCancer and Diabetes datasets. In 2-way marginals of BreastCancer and Diabetes datasets, our method performs better than state-of-the-art with the exception of PRIVBAYES. However, for 2-way marginals of Mammographic, our results are competitive with those of PRIVBAYES. Overall, DP-SYN preserves the statistical information better than comparable to the state-of-the-art techniques in all datasets.

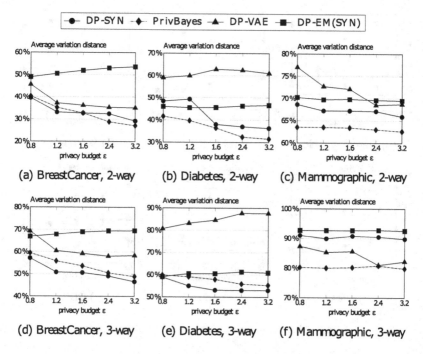

Fig. 4. Statistical difference between the noisy and original k-way marginals.

5.4   Agreement Rate

In this section, we evaluate the quality of the synthetic data in terms of the *agreement rate* in an SVM label prediction task. Specifically the agreement rate is defined as the percentage of records for which the two classifiers make the same prediction [5].

Figure 5 shows the performance of four techniques in terms of SVM agreement rate and its standard deviation which indicates the certainty and consistency in model predictions.

For the BreastCancer dataset, our approach has the highest agreement rate for privacy loss $\varepsilon \in \{0.8, 1.2, 1.6\}$. For the remaining two cases where $\varepsilon \in \{2.4, 3.2\}$, our approach outperforms DP-VAE and PRIVBAYES and it has slightly lower agreement rate than DP-EM(SYN). PRIVBAYES has the lowest agreement rate and the highest standard deviation in most cases. This is expected since PRIVBAYES does not have much improvement on SVM classification of BreastCancer as previously shown in Sect. 5.2.

For the Spambase and Mammographic datasets, our technique achieves significantly higher agreement rate than that of other state-of-art approaches. For Spambase DP-SYN has lowest standard deviation which indicates high consistency with the SVM classifier that runs on original Spambase training set. We expect such a highest agreement rate because the proposed approach outperforms other techniques in terms of SVM accuracy in Fig. 3.

For the Adult dataset, the proposed method outperforms DP-VAE, PRIVBAYES when $\varepsilon \in \{0.8, 1.2\}$. For the remaining cases, the performance of DP-SYN is better than DP-VAE and comparable to DP-EM(SYN) and PRIVBAYES.

In conclusion, our approach exhibits a significant improvement in the majority of the test cases as evident from SVM agreement rate.

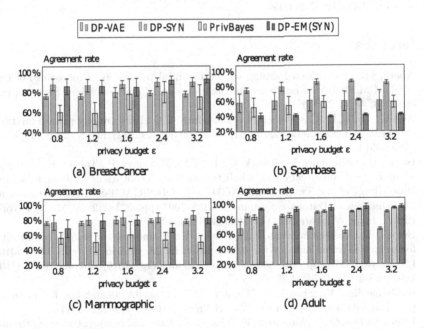

Fig. 5. SVM agreement rate of the four methods reported on the four datasets.

6   Conclusion

We propose a new generative deep learning method, DP-SYN, that produces synthetic data from a dataset while preserving the utility of the original dataset. Our generative auto-encoder method partitions the original data into groups, and then employs the private auto-encoder for each group. Auto-encoder learns the latent structure of each group, and uses expectation maximization algorithm to simulate them. This approach eliminates impurity of groups and results in more accurate representations for each latent group.

We test DP-SYN on nine datasets and compare to four state-of-the art methods in synthetic data generation. Our evaluation process uses statistical, machine learning based and agreement rate based metrics. Although not a single method outperforms others consistently in all tasks, our experiments show that DP-SYN gives robust results across all datasets, and performs better than state-of-the-art in multiple settings for both relational and image based datasets. Furthermore, DP-SYN performance does not deteriorate when the original dataset is imbalanced or high dimensional.

Acknowledgement. The research reported herein was supported in part by NIH award 1R01HG006844, NSF awards CNS-1111529, CICI-1547324, and IIS-1633331 and ARO award W911NF-17-1- 0356.

References

1. Abadi, M., et al.: Deep learning with differential privacy. In: Proceedings of the 2016 ACM SIGSAC Conference on Computer and Communications Security, pp. 308–318. ACM (2016)
2. Ács, G., Melis, L., Castelluccia, C., Cristofaro, E.D.: Differentially private mixture of generative neural networks. CoRR abs/1709.04514 (2017). http://arxiv.org/abs/1709.04514
3. Barak, B., Chaudhuri, K., Dwork, C., Kale, S., McSherry, F., Talwar, K.: Privacy, accuracy, and consistency too: a holistic solution to contingency table release. In: Proceedings of the Twenty-Sixth ACM SIGMOD-SIGACT-SIGART Symposium on Principles of Database Systems, PODS 2007, pp. 273–282. ACM, New York (2007). https://doi.org/10.1145/1265530.1265569
4. Beimel, A., Kasiviswanathan, S.P., Nissim, K.: Bounds on the sample complexity for private learning and private data release. In: Micciancio, D. (ed.) TCC 2010. LNCS, vol. 5978, pp. 437–454. Springer, Heidelberg (2010). https://doi.org/10.1007/978-3-642-11799-2_26
5. Bindschaedler, V., Shokri, R., Gunter, C.A.: Plausible deniability for privacy-preserving data synthesis. Proc. VLDB Endow. **10**(5), 481–492 (2017)
6. Blum, A., Dwork, C., McSherry, F., Nissim, K.: Practical privacy: the SuLQ framework. In: Proceedings of the Twenty-Fourth ACM SIGMOD-SIGACT-SIGART Symposium on Principles of Database Systems, pp. 128–138. ACM (2005)
7. Chaudhuri, K., Monteleoni, C.: Privacy-preserving logistic regression. In: Advances in Neural Information Processing Systems, pp. 289–296 (2009)
8. Chaudhuri, K., Monteleoni, C., Sarwate, A.D.: Differentially private empirical risk minimization. J. Mach. Learn. Res. **12**, 1069–1109 (2011)

9. Ding, B., Winslett, M., Han, J., Li, Z.: Differentially private data cubes: optimizing noise sources and consistency. In: Proceedings of the 2011 ACM SIGMOD International Conference on Management of Data, pp. 217–228. ACM (2011)
10. Dwork, C., Kenthapadi, K., McSherry, F., Mironov, I., Naor, M.: Our data, ourselves: privacy via distributed noise generation. In: Vaudenay, S. (ed.) EUROCRYPT 2006. LNCS, vol. 4004, pp. 486–503. Springer, Heidelberg (2006). https://doi.org/10.1007/11761679_29
11. Dwork, C., McSherry, F., Nissim, K., Smith, A.: Calibrating noise to sensitivity in private data analysis. In: Halevi, S., Rabin, T. (eds.) TCC 2006. LNCS, vol. 3876, pp. 265–284. Springer, Heidelberg (2006). https://doi.org/10.1007/11681878_14
12. Dwork, C., Roth, A., et al.: The algorithmic foundations of differential privacy. Found. Trends Theor. Comput. Sci. **9**(3–4), 211–407 (2014)
13. Dwork, C., Rothblum, G.N., Vadhan, S.: Boosting and differential privacy. In: 2010 51st Annual IEEE Symposium on Foundations of Computer Science (FOCS), pp. 51–60. IEEE (2010)
14. Goodfellow, I.: Efficient per-example gradient computations. arXiv preprint arXiv:1510.01799 (2015)
15. Hearst, M.A., Dumais, S.T., Osuna, E., Platt, J., Scholkopf, B.: Support vector machines. IEEE Intell. Syst. Appl. **13**(4), 18–28 (1998)
16. Hinton, G., et al.: Deep neural networks for acoustic modeling in speech recognition: the shared views of four research groups. IEEE Signal Process. Mag. **29**(6), 82–97 (2012)
17. Jagannathan, G., Pillaipakkamnatt, K., Wright, R.N.: A practical differentially private random decision tree classifier. In: IEEE International Conference on Data Mining Workshops, ICDMW 2009, pp. 114–121. IEEE (2009)
18. Krizhevsky, A., Sutskever, I., Hinton, G.E.: Imagenet classification with deep convolutional neural networks. In: Advances in Neural Information Processing Systems, pp. 1097–1105 (2012)
19. Li, H., et al.: The sequence alignment/map format and samtools. Bioinformatics **25**(16), 2078–2079 (2009)
20. Li, N., Qardaji, W., Su, D., Cao, J.: PrivBasis: frequent itemset mining with differential privacy. Proc. VLDB Endow. **5**(11), 1340–1351 (2012)
21. Lichman, M.: UCI machine learning repository (2013). http://archive.ics.uci.edu/ml
22. Park, M., Foulds, J., Chaudhuri, K., Welling, M.: DP-EM: differentially private expectation maximization. arXiv preprint arXiv:1605.06995 (2016)
23. Pascanu, R., Mikolov, T., Bengio, Y.: On the difficulty of training recurrent neural networks. In: International Conference on Machine Learning, pp. 1310–1318 (2013)
24. Qardaji, W., Yang, W., Li, N.: Priview: practical differentially private release of marginal contingency tables. In: Proceedings of the 2014 ACM SIGMOD International Conference on Management of Data, pp. 1435–1446. ACM (2014)
25. Rubin, D.B.: Discussion statistical disclosure limitation. J. Off. Stat. **9**(2), 461 (1993)
26. Rubinstein, B.I., Bartlett, P.L., Huang, L., Taft, N.: Learning in a large function space: privacy-preserving mechanisms for svm learning. arXiv preprint arXiv:0911.5708 (2009)
27. Schork, N.J.: Personalized medicine: time for one-person trials. Nature **520**(7549), 609–611 (2015)
28. Shah, I.M.: Introduction to nonparametric estimation. Investigación Operacional **30**(3), 284–285 (2009)

29. Song, S., Chaudhuri, K., Sarwate, A.D.: Stochastic gradient descent with differentially private updates. In: Global Conference on Signal and Information Processing (Global-SIP), pp. 245–248. IEEE (2013)
30. Sweeney, L.: k-anonymity: a model for protecting privacy. Int. J. Uncertainty Fuzziness Knowl. Based Syst. 10(05), 557–570 (2002)
31. Vaidya, J., Shafiq, B., Basu, A., Hong, Y.: Differentially private naive Bayes classification. In: Proceedings of the 2013 IEEE/WIC/ACM International Joint Conferences on Web Intelligence (WI) and Intelligent Agent Technologies (IAT), vol. 01, pp. 571–576. IEEE Computer Society (2013)
32. Vogel, P., Greiser, T., Mattfeld, D.C.: Understanding bike-sharing systems using data mining: exploring activity patterns. Procedia Soc. Behav. Sci. 20, 514–523 (2011)
33. Wong, C.: Nyc taxi trips dataset (2017). https://github.com/andresmh/nyctaxitrips
34. Zanella, A., Bui, N., Castellani, A., Vangelista, L., Zorzi, M.: Internet of things for smart cities. IEEE Internet Things J. 1(1), 22–32 (2014)
35. Zeng, C., Naughton, J.F., Cai, J.Y.: On differentially private frequent itemset mining. Proc. VLDB Endow. 6(1), 25–36 (2012)
36. Zhang, J., Cormode, G., Procopiuc, C.M., Srivastava, D., Xiao, X.: PrivBayes: private data release via Bayesian networks. In: Proceedings of the 2014 ACM SIGMOD International Conference on Management of Data, pp. 1423–1434. ACM (2014)
37. Zhang, J., Zhang, Z., Xiao, X., Yang, Y., Winslett, M.: Functional mechanism: regression analysis under differential privacy. Proc. VLDB Endow. 5(11), 1364–1375 (2012)

On Finer Control of Information Flow in LSTMs

Hang Gao$^{(\boxtimes)}$ and Tim Oates$^{(\boxtimes)}$

Computer Science and Electrical Engineering Department, University of Maryland,
Baltimore County, 1000 Hilltop Cir, Baltimore, MD 21250, USA
hanggao1@umbc.edu, oates@cs.umbc.edu

Abstract. Since its inception in 1995, the Long Short-Term Memory (LSTM) architecture for recurrent neural networks has shown promising performance, sometimes state-of-art, for various tasks. Aiming at achieving constant error flow through hidden units, LSTM introduces a complex unit called a memory cell, in which gates are adopted to control the exposure/isolation of information flowing in, out and back to itself. Despite its widely acknowledged success, in this paper, we propose a hypothesis that LSTMs may suffer from an implicit functional binding of information exposure/isolation for the output and candidate computation, i.e., the output gate at time $t - 1$ is not only in charge of the information flowing out of a cell as the response to the external environment, but also controls the information flowing back to the cell for the candidate computation, which is often the only source of nonlinear combination of input at time t and previous cell state at time $t - 1$ for cell memory updates. We propose Untied Long Short Term Memory (ULSTM) as a solution to the above problem. We test our model on various tasks, including semantic relatedness prediction, language modeling and sentiment classification. Experimental results indicate that our proposed model is capable to at least partially solve the problem and outperform LSTM for all these tasks. Code related to this paper is available at: https://github.com/HangGao/ULSTM.git.

Keywords: LSTM · Recurrent neural network · Sequence modeling

1 Introduction

Since its inception in 1995, recurrent neural networks with Long Short Term Memory (LSTM) [1] have shown promising performance on modeling sequential data. Aiming at achieving constant error flow through hidden units, LSTMs are proven to be a scalable method that is both general and effective at capturing long-term temporal dependencies. In fact, LSTMs are widely adopted to advance the state-of-art for many difficult problems in various areas, including handwriting recognition [2,3] and generation [4], language modeling [5–8] and translation [9], image caption generation [10,11], question answering [12], video to text [13] and so on.

© Springer Nature Switzerland AG 2019
M. Berlingerio et al. (Eds.): ECML PKDD 2018, LNAI 11051, pp. 527–540, 2019.
https://doi.org/10.1007/978-3-030-10925-7_32

The key idea behind LSTMs is a complex unit called a memory cell, self-connected and capable of maintaining its state over time, and a set of nonlinear gating units aiming at regulating the information flowing in, out and back to the memory cell. At each time step, as vanilla RNNs, LSTMs are expected to receive a new input, compose it with previous cell state, and then update the cell memory with the guidance of those gates. For standard LSTMs, the fusion of the new input and previous cell state is often mathematically computed as their linear combination followed by a nonlinear transformation (activation). For convenience, we name this fusion as cell input, while referring to the new input as network input. Notice that the cell input is the only source of nonlinear combination involving both network input and cell state, functionally as a candidate for cell update.

However, the above architecture implicitly introduces a bias, that is, the exposure/isolation of information to the external environment and to the generation of the cell input remains the same and can be controlled by the same gate (output gate), i.e., they are functionally tied. This is a strong assumption since the output gate calculated at time $t - 1$ is mathematically independent of the new network input coming at time t, but is expected to guide the information flowing out of and back to the cell to generate the cell input in order to update its memory.

In this paper, we propose Untied Long Short Term Memory (ULSTM) as a solution to the above problem. Our idea is to introduce a new type of gate called a retrieve gate, dependent on the network input at each time step, to replace the output gate in the procedure of cell input generation. We only apply the idea to standard LSTMs in our paper, but it can also be generalized to many LSTM variants, e.g., Convolutional LSTM [14], Dynamic Cortex Memory [15] and Group LSTM [16].

We evaluate the proposed model on various tasks, including semantic relatedness prediction, language modeling and sentiment classification. Our experiment results indicate that the proposed model can outperform standard LSTMs in various conditions and is capable to at least partially solve the problem mentioned above.

2　Long Short Term Memory

Initially Long Short Term Memory (LSTM) proposed by [1] included only memory cells, and input and output gates. Targeting at the goal of constant error flow, LSTMs were carefully designed to protect memory cells from perturbation by irrelevant inputs with input gates, and prevent other units from perturbation by currently irrelevant cell content with output gates. Later, forget gates were introduced by [17] to enable LSTMs to reset their own states instead of growing without bound. In general, the transitions of a standard LSTM are defined as follows:

$$i_t = \sigma(W_i x_t + U_i h_{t-1} + b_i) \tag{1}$$

$$f_t = \sigma(W_f x_t + U_f h_{t-1} + b_f) \tag{2}$$

$$o_t = \sigma(W_o x_t + U_o h_{t-1} + b_o) \tag{3}$$

$$\tilde{c}_t = n(W_n x_t + U_n h_{t-1} + b_n) \tag{4}$$

$$c_t = i_t \odot \tilde{c}_t + f_t \odot c_{t-1} \tag{5}$$

$$h_t = o_t \odot m(c_t) \tag{6}$$

where Ws and Us are weight matrices, bs are bias vectors, x_t is the network input, i_t, f_t and o_t are the input, forget and output gate respectively, $\tilde{c}_t$ is the cell input, c_t/c_{t-1} and h_t/h_{t-1} are corresponding cell state and output state at the current and previous time steps, and m and n are activation functions, often taken as tanh.

2.1 Peephole Connection

Peephole connections were proposed in [18] to allow all gates to inspect current cell state even when output gates are closed. The transitions of a LSTM with peephole connections are:

$$i_t = \sigma(W_i x_t + U_i h_{t-1} + P_i c_{t-1} + b_i) \tag{7}$$

$$f_t = \sigma(W_f x_t + U_f h_{t-1} + P_i c_{t-1} + b_f) \tag{8}$$

$$\tilde{c}_t = n(W_n x_t + U_n h_{t-1} + b_n) \tag{9}$$

$$c_t = i_t \odot \tilde{c}_t + f_t \odot c_{t-1} \tag{10}$$

$$o_t = \sigma(W_o x_t + U_o h_{t-1} + P_o c_t + b_o) \tag{11}$$

$$h_t = o_t \odot m(c_t) \tag{12}$$

where P_i, P_f and P_o are peephole matrices. In practice, we often put constraints on these weight matrices so that they are diagonal, i.e., each gate unit only receives the connection from its own cell. The architecture of a LSTM with peephole connections is presented in Fig. 1.

2.2 Full Gate Recurrence

Mentioned in [19], there is a version of LSTM called Full Gate Recurrence LSTM (FGR-LSTM). The idea is to add connections among all gates. The transitions thus become (with peephole),

$$i_t = \sigma(W_i x_t + U_i h_{t-1} + P_i c_{t-1} + b_i + R_{ii} i_{t-1} + R_{if} f_{t-1} + R_{io} o_{t-1}) \tag{13}$$

$$f_t = \sigma(W_f x_t + U_f h_{t-1} + P_i c_{t-1} + b_f + R_{fi} i_{t-1} + R_{ff} f_{t-1} + R_{fo} o_{t-1}) \tag{14}$$

$$\tilde{c}_t = n(W_n x_t + U_n h_{t-1} + b_n) \tag{15}$$

$$c_t = i_t \odot \tilde{c}_t + f_t \odot c_{t-1} \tag{16}$$

$$o_t = \sigma(W_o x_t + U_o h_{t-1} + P_o c_t + b_o + R_{oi} i_{t-1} + R_{of} f_{t-1} + R_{oo} o_{t-1}) \tag{17}$$

$$h_t = o_t \odot m(c_t) \tag{18}$$

FGR requires nine additional weight matrices, which significantly increases the number of parameters.

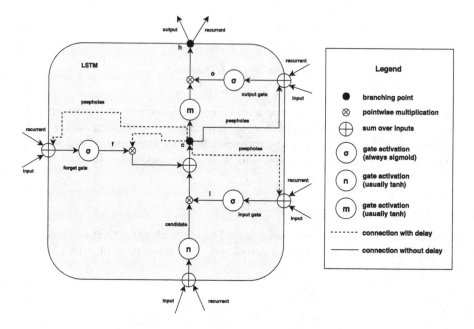

Fig. 1. The architecture of a LSTM with peephole connections

3 Untied Long Short Term Memory

Here we provide the details of our proposed Untied Long Short Term Memory (ULSTM) mentioned above. Although our model can be generalized to many LSTM variants, in this paper, we only focus on the standard LSTM and leave the other variants as future work.

3.1 The Problem

Let us first look at the architecture of a standard LSTM. The information flowing in, back and out of each memory cell is controlled by their corresponding input, forget and output gate. In an alternative view, input gates control the memory write access (W), forget gates manage the memory erase access (E) and output gates determine the memory read access (R). However, if we take a deeper analysis by viewing h_{t-1} as $o_{t-1} \odot \tanh(c_{t-1})$, a summary in Table 1 can be generated.

From that table, one can find that both h_t and c_t have controllers (i_t, f_t or o_t) dependent on current network input x_t, while $\tilde{c}_t$ and all gates are regulated by controllers (o_{t-1}) independent of x_t. The later may lead to a problem we call false exposure/isolation. That is, o_{t-1} incorrectly exposes or isolates information from c_{t-1} in the procedure of computation of $\tilde{c}_t$, i_t, f_t and o_t.

In fact, unlike i_t and f_t, o_t implicitly provides two functions: (1) respond to the external environment at time t; (2) determine the information exposed to

Table 1. Memory cell access at time t

Unit	Direct dependents	Type	Memory accessed	Controller	Controller dependents
$\tilde{c}_t$	x_t, h_{t-1}	R	c_{t-1}	o_{t-1}	x_{t-1}, h_{t-2}
c_t	$\tilde{c}_t, c_{t-1}$	E/W	c_{t-1}	i_t, f_t	x_t, h_{t-1}
h_t	c_t	R	c_t	o_t	x_t, h_{t-1}
i_t	x_t, h_{t-1}	R	c_{t-1}	o_{t-1}	x_{t-1}, h_{t-2}
f_t	x_t, h_{t-1}	R	c_{t-1}	o_{t-1}	x_{t-1}, h_{t-2}
o_t	x_t, h_{t-1}	R	c_{t-1}	o_{t-1}	x_{t-1}, h_{t-2}

the computation of units within the LSTM cell architecture itself at time $t + 1$. o_t in a standard LSTM is more likely to behave incorrectly for the latter since it is independent of x_{t+1}, which may be crucial in the decision on whether the information in the memory cell c_t is valuable or not.

Note that LSTMs with peephole connections already provide a solution for gate computation by introducing fixed connections from cells to gate units. In this paper, we focus on proposing a solution for the cell input $\tilde{c}_t$ computation.

3.2   Is Peephole Connection a Solution?

It is almost natural to consider peephole connections for a solution, as has been done with gates. However, this will probably not work. This is due to the fundamental difference between the functions of the cell input and gates. The latter function as controllers on the access of memory cells while the former, instead, function as the candidate to update those cells. A full inspection of the cell state does not logically prevent gates from closing/opening the path from which the information can flow through a cell, but it logically means the information stored in the cell is fully leaked to the candidate (cell input) computation. To prevent cells from being polluted or perturbed during the update procedure, one can only expect the input gates to be capable to not only determine how much to update the cells, but also separate the leaked information from useful one, which is perhaps an even harder problem.

3.3   Retrieve Gates

Since the problem originates from the implicit binding of the two functions of output gates, it is better to find a solution that focuses on detaching them. Out of many potential alternatives, adding a new set of gates to replace output gates and take over their second function is probably the simplest one. We name these gates as retrieve gates and represent them by z in the following. A LSTM with retrieve gates has exactly the same transitions of Eqs. (1)–(3) and (5), (6) as the standard one, with only the modification to Eq. (4),

$$\tilde{c}_t = n(W_n x_t + U_n(z_t \odot \tanh(c_{t-1})) + b_n) \tag{19}$$

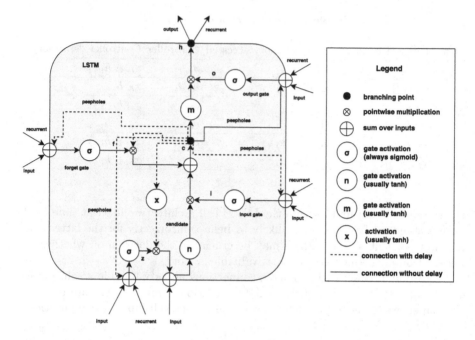

Fig. 2. The architecture of a ULSTM with peephole connections

where,

$$z_t = \sigma(W_z x_t + U_z h_{t-1} + b_z) \tag{20}$$

Because the solution is inspired by untying the two functions of output gates, we call the model Untied Long Short Term Memory (ULSTM). Note that z_t is now dependent on x_t. We present the architecture of ULSTM in Fig. 2.

4   Experiments

Since our paper focuses on the cell input $\tilde{c}_t$ computation, we only perform tests on ULSTM without peephole connections. We leave the peephole version as future work. In addition, we only compare our model to standard LSTMs and LSTMs with only peephole connections in candidate computations (PLSTM in our experiments), which is mentioned as a potential solution in Sect. 3.2, as our purpose is not to introduce a new model aiming at achieving state-of-art performance for specific tasks.

We evaluate our model on three tasks: (1) predicting the semantic relatedness of sentence pairs; (2) word level language modeling and (3) sentiment classification of sentences sampled from movie reviews.

4.1   Semantic Relatedness

For a given pair of sentences, the semantic relatedness task is to predict a human-generated rating of the similarity of the two sentences in meaning. We use the Sentences Involving Compositional Knowledge (SICK) dataset introduced by [20]. Consisting of 9927 sentence pairs, this dataset is pre-split into train/valid/test sets with ratio 4500/500/4927. All sentences are derived from existing image and video description datasets, with each pair annotated with a relatedness score $y \in [1, 5]$. The higher the score, the more related the pair of sentences are. We adopt the same similarity model and objective function as [21], with only a slight difference on the activation function choice for h_s. Instead of,

$$h_s = \sigma(W^{(x)}h_x + W^{(+)}h_+ + b^{(h)}) \tag{21}$$

We use,

$$h_s = ReLU(W^{(x)}h_x + W^{(+)}h_+ + b^{(h)}) \tag{22}$$

Table 2. Evaluation results of LSTM, PLSTM and ULSTM on SICK test data. The best results in each subsection are marked as bold

Model	Hidden size	Parameters	Pearson's γ	Spearman's ρ	MSE
LSTM	150	1009205	0.8545	0.7931	0.2774
PLSTM	150	1009355	0.8542	0.7926	0.2782
ULSTM	150	1077005	**0.8591**	**0.7973**	**0.2686**
LSTM (o=1)	150	<1009205	0.8562	0.7955	0.2747
PLSTM (o=1)	150	<1009355	0.8568	0.7948	0.2737
ULSTM (o=1)	150	<1077005	**0.8604**	**0.7984**	**0.2662**
LSTM	180	1088045	0.8531	0.7896	0.2803
PLSTM	180	1088225	0.8518	0.7900	0.2811
ULSTM	180	1174805	**0.8609**	**0.8020**	**0.2674**
LSTM (o=1)	180	<1088045	0.8550	0.7932	0.2747
PLSTM (o=1)	180	<1088225	0.8566	0.7956	0.2738
ULSTM (o=1)	180	<1174805	**0.8603**	**0.8022**	**0.2674**
LSTM	210	1174085	0.8525	0.7916	0.2794
PLSTM	210	1174295	0.8516	0.7905	0.2825
ULSTM	210	1281605	**0.8630**	**0.8039**	**0.2641**
LSTM (o=1)	210	<1174085	0.8534	0.7906	0.2789
PLSTM (o=1)	210	<1174295	0.8532	0.7936	0.2793
ULSTM (o=1)	210	<1281605	**0.8620**	**0.8028**	**0.2657**
LSTM	245	1283565	0.8476	0.7842	0.2902
PLSTM	245	1283810	0.8490	0.7890	0.2875
ULSTM	245	1417580	**0.8609**	**0.8024**	**0.2664**
LSTM (o=1)	245	<1283565	0.8521	0.7907	0.2816
PLSTM (o=1)	245	<1283810	0.8509	0.7908	0.2845
ULSTM (o=1)	245	<1417580	**0.8537**	**0.7937**	**0.2776**

We adopt the publicly available Glove vectors [22] as the initialization of word embeddings. Following [21], we do not fine tune these embeddings during the training procedure. For the optimization algorithm, we choose Adagrad [23], with a learning rate of 0.03 and a weight decay rate of 0.0001. We set the batch size to be 25 and number of epochs to be 10. For the similarity model, the hidden layer size is set to be 50. We use Pearson's γ, Spearman's ρ and MSE as the evaluation metrics. We adopt early stopping and perform prediction on the test data with the model of the highest Pearson's γ on the validation data. For each set of hyper-parameters, we report the mean of 5 independent runs with random seeds 1234/2341/3451/3651/3851.

For this task, we seek to compare ULSTM, PLSTM and LSTM with (1) varying hidden size; (2) output gates manually fixed to 1 or not; (3) the same number of parameters. We list all evaluation results in Table 2.

Varying Hidden Size. When we look at the results of LSTM, PLSTM and ULSTM with varying hidden size from 150 to 245, we find that ULSTM consistently outperforms LSTM and PLSTM, regardless of whether output gates are fixed to 1 or not. Besides, ULSTM can benefit from increased hidden size until a certain point, while the performance of LSTM keeps dropping as the hidden size increases. PLSTM, on the other hand, shows a similar but weaker trend as LSTM. Since in this task error signals for parameter tuning only come from the end of each sentence, ULSTM seems to be better at exploiting the benefits brought by increased storage capacity, while LSTM and PLSTM, on the contrary, suffer from it due to the lack of error signals to correct memory access management, as the former has lower requirements on the behavior of its output gates or input gates.

Fixed Output Gates or Not. We take another look at the performance of each model with/without fixing output gates. Out of 4 different hidden sizes, when the output gates are not fixed to 1, LSTM noticeably performs worse under 3 cases and PLSTM performs worse under all cases, while for ULSTM, in most scenarios, the performance is roughly the same or even better than the one with fixed output gates. A possible explanation is that for ULSTM, output gates are only expected to affect information exposure to the external environment or gate computation. As discussed above, gates can be correctly opened/closed even with information wrongfully exposed/isolated. So as long as the external environment (the similarity model) is robust, the performance of ULSTM is expected to remain stable or slightly worse. This is not the case for LSTM, in which the output gates play an important role in memory candidate generation or for PLSTM, where information in cells is simply always fully exposed to the generation procedure.

The Same Number of Parameters. By comparing LSTM, PLSTM and ULSTM with different hidden sizes but roughly the same number of parameters, one can conclude that ULSTM, for this task, always performs better than either PLSTM or LSTM, if given similar number of parameters.

4.2   Language Modeling

The goal of a language model is to compute the probability of a sentence or a sequence of words. We use a preprocessed version of the Penn Treebank data set (PTB) introduced in [24], which is also adopted by [5]. PTB has long been a central dataset for evaluation of language models. The data set is preprocessed so that it does not contain capital letters, numbers or punctuation. The vocabulary includes around 10000 unique words, which is small compared to many modern datasets.

Table 3. Valid and Test PPL on PTB of LSTM, PLSTM and ULSTM with varying number of layers. The best results in each subsection are marked as bold

Model	Hidden size	Layers	Parameters	Valid PPL	Test PPL
AWD-ULSTM	400	1	5614000	**84.23**	**81.76**
AWD-PLSTM	400	1	5293600	85.62	83.14
AWD-LSTM	400	1	5293200	86.17	83.73
AWD-ULSTM	400	2	7218000	**70.05**	**67.38**
AWD-PLSTM	400	2	6577200	71.16	68.64
AWD-LSTM	400	2	6576400	72.83	70.18
AWD-ULSTM	400	3	8822000	**68.47**	**65.41**
AWD-PLSTM	400	3	7860800	69.45	67.06
AWD-LSTM	400	3	7859600	69.87	67.03
AWD-ULSTM	400	4	10426000	**69.58**	**66.42**
AWD-PLSTM	400	4	9144400	70.01	67.23
AWD-LSTM	400	4	9142800	70.58	67.51
AWD-ULSTM	400	5	12030000	**70.12**	**67.05**
AWD-PLSTM	400	5	10428000	71.1	68.51
AWD-LSTM	400	5	10426000	71.47	68.78

We adopt an implementation of AWD-LSTM [5], which regulates LSTMs with various techniques that do not make any modification to existing model architectures, such as variable length sequence, DropConnect and so on. For this task, we randomly initialize word embeddings with a uniform distribution of $(-0.1, 0.1)$ with dimension of 400. The maximum sequence length is set to be 70. We apply dropout with rate 0.5 on the decoder, rate 0.4 on embeddings, and rate 0.25 between hidden layers of RNNs. In addition, words from the embedding layer are dropped with rate 0.1 and RNN hidden to hidden weight connections are randomly dropped with rate 0.5. We set the random seed to be 141, L2 regularization rate on RNN activation to be 2, slowness regularization rate on RNN activation to be 1 and weight decay rate to be $1.2e^{-6}$. For the optimization algorithm, we adopt SGD with learning rate of 30 and switch to NT-ASGD [5]

Table 4. Valid and Test PPL on PTB of LSTM, PLSTM and ULSTM with varying hidden size. The best results in each subsection are marked as bold

Model	Hidden size	Layers	Parameters	Valid PPL	Test PPL
AWD-ULSTM	300	2	6467000	**72.01**	**69.01**
AWD-PLSTM	300	2	5976300	73.49	70.52
AWD-LSTM	300	2	5975600	74.55	71.81
AWD-ULSTM	400	2	7218000	**70.50**	**67.38**
AWD-PLSTM	400	2	6577200	71.16	68.64
AWD-LSTM	400	2	6576400	72.83	70.18
AWD-ULSTM	500	2	8069000	69.65	**67.12**
AWD-PLSTM	500	2	7258100	**69.56**	67.44
AWD-LSTM	500	2	7257200	71.06	68.43
AWD-ULSTM	600	2	9020000	68.83	**65.97**
AWD-PLSTM	600	2	8019000	**68.55**	66.30
AWD-LSTM	600	2	8018000	69.42	67.03
AWD-ULSTM	720	2	10293200	67.53	**64.77**
AWD-PLSTM	720	2	9037680	**67.48**	65.23
AWD-LSTM	720	2	9036560	68.63	66.17
AWD-ULSTM	1000	2	13824000	65.49	62.95
AWD-PLSTM	1000	2	11862600	**64.98**	**62.86**
AWD-LSTM	1000	2	11861200	66.09	63.83
AWD-ULSTM	1170	2	16350200	64.55	**62.14**
AWD-PLSTM	1170	2	13883730	**64.30**	62.16
AWD-LSTM	1170	2	13882160	65.14	62.94
AWD-ULSTM	1350	2	19340000	64.04	61.73
AWD-PLSTM	1350	2	16275750	**63.83**	**61.43**
AWD-LSTM	1350	2	16274000	64.48	62.11
AWD-ULSTM	1500	2	22079000	63.97	61.21
AWD-PLSTM	1500	2	18467100	**62.71**	**60.66**
AWD-LSTM	1500	2	18465200	63.68	61.40

if its trigger criterion is satisfied. We set the logging interval to be 1 and the non-monotone interval to be 5 for NT-ASGD. All gradients are clipped with absolute value 0.25. The batch size is set to be 20 and the maximum number of epochs is 500.

For this task, we seek to compare ULSTM, PLSTM and LSTM with (1) different stacking layers; (2) varying hidden size; (3) the same number of parameters. Note that unlike semantic relatedness prediction, in this task, both models may receive error signals at each time step. We list all evaluation results in Tables 3 and 4.

Different Stacking Layers. The comparison among ULSTM, PLSTM and LSTM with fixed hidden size but different number of layers is presented in Table 3. The results indicate that along with the increased number of layers, ULSTM always outperforms LSTM on both validation and test data, while PLSTM is usually better but occasionally worse than LSTM with a small margin.

Varying Hidden Size. We show the comparison among ULSTM, PLSTM and LSTM with fixed number of layers but varying hidden sizes in Table 4. It is clear that both ULSTM and PLSTM outperform LSTM on test data with various hidden sizes, indicating the existence of the proposed problem for standard LSTMs. However, when the hidden size is not very large (<1000), ULSTM outperforms PLSTM while when the hidden size grows large ($\geq$1000), the latter starts to outperform the former. A possible explanation is that the increase of hidden size may reduce the benefits brought by the retrieve gates as cells have more storage capacity for redundant information, thus (1) false isolation is unlikely to happen since it requires all relevant gates to be closed at the same time; (2) false exposure is more likely to occur as it is hard to make sure that not a single relevant gate is open, when the number of cells is large. As a result, as the hidden size increases, the performance gap between ULSTM and LSTM decreases and since it is difficult to prevent false exposure with large hidden size, PLSTM seems to benefit more by simply adding direct peephole connections so that other gates are trained to work with irrelevant noise that is always present. This trend does not occur above because in Sect. 4.1, there is not enough error signal at every time step to correct gate behavior.

The Same Number of Parameters. From Table 4, ULSTM can outperform LSTM with the same number of parameters when the hidden size is still small, but along with the increase of hidden size, LSTM starts to outperform ULSTM, which is probably due to the reasons mentioned above. However, PLSTM usually outperforms LSTM with roughly the same number of parameters, regardless of the number of layers or the hidden sizes.

4.3 Sentiment Classification

In this task, we predict the sentiment of sentences sampled from movie reviews. We use the Stanford Sentiment Treebank [25]. There are two possible subtasks for this dataset, but we only focus on the fine-grained classification task over five classes: very negative, negative, neutral, positive, very positive. We use the train/valid/test split provided by the dataset.

For this task, we do not control variables when comparing ULSTM, PLSTM and LSTM. Instead we perform hyper-parameter search on the number of layers from the set [1, 2, 3] and hidden size from the set [50, 100, 150, 200, 250, 300] to find the best hyper-parameter setting for each model. We initialize word embeddings with Glove vectors [22] and fine-tune them during the training procedure. We use SGD followed by NT-ASGD [5] if the trigger criterion is satisfied. We set the logging interval to be 1 and the non-monotone interval to be 5 for

NT-ASGD. The learning rate is set to be 1. We apply dropout with rate 0.4 on embeddings, rate 0.25 between hidden layers of RNNs. In addition, RNN hidden to hidden weight connections are randomly dropped with rate 0.5. We set the random seed to be 141, L2 regularization rate on RNN activation to be 2, slowness regularization rate on RNN activation to be 1 and weight decay rate to be $1.2e^{-6}$. The number of epochs is set to be 50.

We report the test accuracy of each model with the hyper-parameter set chosen on validation data in Table 5. ULSTM still shows better performance than LSTM for the sentiment classification task, for both validation and test accuracy. Note that both LSTM and ULSTM share the same hyper-parameter set when they achieve the highest validation accuracy, indicating that like other tasks, ULSTM is likely to perform better than LSTM given the same number of layers and hidden size.

However, PLSTM in this experiment performs worse than either LSTM or ULSTM on both validation and test data. Note that similar to Semantic Relatedness task, all models only receive error signals at the end of a sequence from the classifier. It is possible that the lack of error signals makes it difficult for PLSTM to adjust its gate computation to work with noise constantly present, leading to worse results.

Table 5. Results of LSTM, PLSTM and ULSTM on Stanford Sentiment Treebank. The best result is marked as bold

Model	Hidden size	Layers	Valid Acc	Test Acc
LSTM	50	2	48.56	48.28
LSTM	200	2	48.56	47.72
PLSTM	200	1	48.28	46.82
ULSTM	50	2	**49.71**	**48.66**

5   Conclusion and Future Work

In this paper, we address the problem of implicit functional binding of output gates in LSTM, which may lead to false exposure/isolation of information for the cell input computation. We propose a model called Untied Long Short Term Memory (ULSTM) that introduces a new set of gates as a solution. We evaluate our model on three tasks: (1) semantic relatedness prediction; (2) language modeling; (3) sentiment classification. The experimental results indicate: (a) ULSTM consistently outperforms LSTM on all the three tasks, given the same number of layers and hidden size; (b) ULSTM usually outperforms LSTM with the same number of parameters when the hidden size is small, but it is not necessarily true if we increase the hidden size; (c) On the other hand, LSTM may benefit or suffer from large hidden size, depending on the task; (d) Although LSTM with peephole connections in cell input computation (PLSTM) can sometime work better than LSTM, or even ULSTM, it is highly task-dependent and inconsistent.

We seek to generalize our model to large data sets in the future. And it is also possible to apply the idea to other LSTM variants, e.g., Convolutional LSTM. Besides the idea of adding a new set of gates which introduces more parameters, we are also interested in looking for other alternative solutions that keep or reduce the number of parameters, which may be more computationally efficient.

References

1. Hochreiter, S., Schmidhuber, J.: Long short-term memory. Neural Comput. **9**(8), 1735–1780 (1997). https://doi.org/10.1162/neco.1997.9.8.1735
2. Graves, A., et al.: A novel connectionist system for unconstrained handwriting recognition. IEEE Trans. Pattern Anal. Mach. Intell. **31**(5), 855–868 (2009). https://doi.org/10.1109/TPAMI.2008.137
3. Pham, V., et al.: Dropout improves recurrent neural networks for handwriting recognition. In: 2014 14th International Conference on Frontiers in Handwriting Recognition (ICFHR). IEEE (2014). https://doi.org/10.1109/ICFHR.2014.55
4. Graves, A.: Generating sequences with recurrent neural networks. arXiv preprint arXiv:1308.0850 (2013)
5. Merity, S., Keskar, N.S., Socher, R.: Regularizing and optimizing LSTM language models. arXiv preprint arXiv:1708.02182 (2017)
6. Yang, Z., et al.: Breaking the softmax bottleneck: a high-rank RNN language model. arXiv preprint arXiv:1711.03953 (2017)
7. Inan, H., Khosravi, K., Socher, R.: Tying word vectors and word classifiers: a loss framework for language modeling. arXiv preprint arXiv:1611.01462 (2016)
8. Zaremba, W., Sutskever, I., Vinyals, O.: Recurrent neural network regularization. arXiv preprint arXiv:1409.2329 (2014)
9. Luong, M.-T., et al.: Addressing the rare word problem in neural machine translation. arXiv preprint arXiv:1410.8206 (2014)
10. Vinyals, O., et al.: Show and tell: a neural image caption generator. In: 2015 IEEE Conference on Computer Vision and Pattern Recognition (CVPR). IEEE (2015). https://doi.org/10.1109/CVPR.2015.7298935
11. Xu, K., et al.: Show, attend and tell: neural image caption generation with visual attention. In: International Conference on Machine Learning (2015)
12. Wang, D., Nyberg, E.: A long short-term memory model for answer sentence selection in question answering. In: Proceedings of the 53rd Annual Meeting of the Association for Computational Linguistics and the 7th International Joint Conference on Natural Language Processing (Volume 2: Short Papers), vol. 2 (2015)
13. Venugopalan, S., et al.: Sequence to sequence-video to text. In: Proceedings of the IEEE International Conference on Computer Vision (2015). https://doi.org/10.1109/ICCV.2015.515
14. Xingjian, S.H.I., et al.: Convolutional LSTM network: a machine learning approach for precipitation nowcasting. In: Advances in Neural Information Processing Systems (2015)
15. Otte, S., Liwicki, M., Zell, A.: Dynamic cortex memory: enhancing recurrent neural networks for gradient-based sequence learning. In: Wermter, S., et al. (eds.) ICANN 2014. LNCS, vol. 8681, pp. 1–8. Springer, Cham (2014). https://doi.org/10.1007/978-3-319-11179-7_1

16. Kuchaiev, O., Ginsburg, B.: Factorization tricks for LSTM networks. arXiv preprint arXiv:1703.10722 (2017)
17. Gers, F.A., Schmidhuber, J., Cummins, F.: Learning to forget: continual prediction with LSTM, pp. 850–855 (1999)
18. Gers, F.A., Schraudolph, N.N., Schmidhuber, J.: Learning precise timing with LSTM recurrent networks. J. Mach. Learn. Res. **3**, 115–143 (2002)
19. Greff, K., et al.: LSTM: a search space odyssey. IEEE Trans. Neural Netw. Learn. Syst. **28**(10), 2222–2232 (2017). https://doi.org/10.1109/TNNLS.2016.2582924
20. Marelli, M., et al.: Semeval-2014 task 1: evaluation of compositional distributional semantic models on full sentences through semantic relatedness and textual entailment. In: Proceedings of the 8th International Workshop on Semantic Evaluation (SemEval 2014) (2014)
21. Tai, K.S., Socher, R., Manning, C.D.: Improved semantic representations from tree-structured long short-term memory networks. arXiv preprint arXiv:1503.00075 (2015)
22. Pennington, J., Socher, R., Manning, C.: Glove: global vectors for word representation. In: Proceedings of the 2014 Conference on Empirical Methods in Natural Language Processing (EMNLP) (2014). https://doi.org/10.3115/v1/D14-1162
23. Duchi, J., Hazan, E., Singer, Y.: Adaptive subgradient methods for online learning and stochastic optimization. J. Mach. Learn. Res. **12**, 2121–2159 (2011)
24. Mikolov, T., et al.: Recurrent neural network based language model. In: Eleventh Annual Conference of the International Speech Communication Association (2010)
25. Socher, R., et al.: Recursive deep models for semantic compositionality over a sentiment treebank. In: Proceedings of the 2013 Conference on Empirical Methods in Natural Language Processing (2013)

MaxGain: Regularisation of Neural Networks by Constraining Activation Magnitudes

Henry Gouk[1]([✉]), Bernhard Pfahringer[1], Eibe Frank[1], and Michael J. Cree[2]

[1] Department of Computer Science, University of Waikato, Hamilton, New Zealand
hgrg1@students.waikato.ac.nz,
{bernhard.pfahringer,eibe.frank}@waikato.ac.nz
[2] School of Engineering, University of Waikato, Hamilton, New Zealand
michael.cree@waikato.ac.nz

Abstract. Effective regularisation of neural networks is essential to combat overfitting due to the large number of parameters involved. We present an empirical analogue to the Lipschitz constant of a feed-forward neural network, which we refer to as the maximum gain. We hypothesise that constraining the gain of a network will have a regularising effect, similar to how constraining the Lipschitz constant of a network has been shown to improve generalisation. A simple algorithm is provided that involves rescaling the weight matrix of each layer after each parameter update. We conduct a series of studies on common benchmark datasets, and also a novel dataset that we introduce to enable easier significance testing for experiments using convolutional networks. Performance on these datasets compares favourably with other common regularisation techniques. Data related to this paper is available at: https://www.cs.waikato.ac.nz/~ml/sins10/.

1 Introduction

Regularisation is a crucial component in machine learning systems. This is particularly true for neural networks, where the huge number of parameters can lead to extreme overfitting, such as memorising the training set—even in the case where the labels have been randomised [19]. In this work, we investigate a regularisation technique inspired by recent work regarding the Lipschitz continuity of neural networks [7]. Most work in machine learning that deals with the concept of Lipschitz continuity assumes, often implicitly [7,13], that the input domain of the function of interest is $\mathbb{R}^d$—sometimes with the additional assumption that each component in this vector space is bounded in, for example, the range $[-1, 1]$. However, when working with unstructured data—a task at which neural networks excel—a common assumption is that the data lie in a low dimensional manifold embedded in a high dimensional space. This is known as the manifold hypothesis [2]. In this paper, we explore the idea of constraining the Lipschitz continuity of neural network models when they are viewed

© Springer Nature Switzerland AG 2019
M. Berlingerio et al. (Eds.): ECML PKDD 2018, LNAI 11051, pp. 541–556, 2019.
https://doi.org/10.1007/978-3-030-10925-7_33

in this light: as mappings from the subset of $\mathbb{R}^d$ that contains the low dimensional manifold, to some meaningful vector space, such as the distribution over possible classes. The precise structure of the manifold is unknown to us, which makes constraining a function that operates on this manifold difficult. To circumvent this problem, we introduce the concept of *gain*—an empirical analogue to the operator norm technique used by Gouk et al. [7] to compute the Lipschitz constant of a neural network layer.

We present a regularisation scheme that improves the generalisation performance of neural networks by constraining the maximum gain of each layer. This is accomplished using a simple modification to conventional neural network optimisers that applies a stochastic projection function in addition to a stochastic estimate of the gradient. We demonstrate the effectiveness of our regularisation algorithm on several classification datasets. A novel dataset that facilitates significance testing for convolutional network-based classifiers is introduced as part of these experiments. Additionally, we show how our technique performs when used in conjunction with other regularisation methods such as dropout [17] and batch normalisation [9]. We also provide empirical evidence that constraining the gain on the training set results in observing lower gain on the test set compared to when the gain on the training set is not constrained. Details of how the performance of models trained with out regularisation technique as its hyperparameter is varied are also provided.

2   Related Work

Several recent publications have addressed the idea of Lipschitz continuity of neural networks. Most of this work has been on generative adversarial networks (GANs) [6]. Wasserstein GANs [1] are the first GAN variant that require some way of enforcing Lipschitz continuity in order to converge. They accomplish this by clipping each weight whenever its absolute value exceeds some predefined threshold. While this will maintain Lipschitz continuity, the Lipschitz constant will not be known. An alternative to weight clipping is to penalise the norm of the gradient of the critic network [8], which has been shown to improve the stability of training Wasserstein GANs. This technique for constraining Lipschitz continuity is similar to ours, in the sense that it uses an approximate measure of the Lipschitz constant on the training data. It is, however, quite different in the sense that it is not being used for regularisation and that it is applied as a soft constraint using a penalty term. Miyato et al. [13] have also proposed normalising the weights in each layer of the discriminator network of a GAN using the spectral norm of the respective weight matrix, but they provide no evidence showing that their heuristic for applying this to convolutional layers actually constrains the spectral norm. Some recent work has shown how to precisely compute and constrain the Lipschitz constant of a network with respect to the ℓ_1 and ℓ_∞ norms [7] and demonstrated that constraining the Lipschitz constant with respect to these norms has a regularising effect comparable to dropout and batch normalisation.

The idea of constraining the Lipschitz constant of a network is conceptually related to quantifying the flatness of minima. While there is no single formalisation for what constitutes a flat minimum, the unifying intuition is that a minimum is flat when a small perturbation of the model parameters does not have a large impact on the performance of the model. Dinh et al. [4] have shown that Lipschitz continuity is not a reliable tool for quantifying the flatness of minima. However, there is a subtle but very important difference between how they employ Lipschitz continuity, and how it is used by Gouk et al. [7] and in this work. Neural networks are functions parameterised by two distinct sets of variables: the model parameters, and the features. Dinh et al. [4] consider Lipschitz continuity with respect to the model parameters, whereas we consider Lipschitz continuity with respect the features being supplied to the network. The crux of the argument given by Dinh et al. is that the Lipschitz constant of a network with respect to its weights is not invariant to reparameterisation.

Dropout [17] is one of the most widely used methods for regularising neural networks. It is popular because it is efficient and easy to implement, requiring only that each activation is set to zero with some probability, p, during training. An extension proposed by Srivastava et al. [17], known as maxnorm, is to constrain the magnitude of the weight vector associated with each unit in some layer. One can also use multiplicative Gaussian noise, rather than Bernoulli noise. Kingma et al. [11] provide a technique that enables automatic tuning of the amount of noise that should be applied in the case of Gaussian dropout. A similar technique exists for automatically tuning p for Bernoulli dropout—this extension is known as concrete dropout [5].

Batch normalisation [9], which was originally motivated by the desire to improve the convergence rate of neural network optimisers, is often used as a regularisation scheme. It is similar to our technique in the sense that it rescales the activations of a layer, but it does so in a different way: by standardising them and subsequently multiplying them by a learned scale factor. Unlike other regularisation techniques, there is no hyperparameter for batch normalisation that can be tuned to control the capacity of the network. A similar technique, which does not rely on measuring activation statistics over minibatches, is weight normalisation [15]. This approach decouples the length and direction of the weight vector associated with each unit in the network, and enables one to train networks on very small batch sizes, which is a situation where batch normalisation cannot be applied reliably.

3 Lipschitz Continuous Neural Networks

Gouk et al. (2018) recently demonstrated that constraining the Lipschitz continuity of a neural network improves generalisation in the context of classification. We briefly review their technique to aid overall understanding and provide several useful definitions. Recall the definition of Lipschitz continuity:

$$D_B(f(\boldsymbol{x}_1), f(\boldsymbol{x}_2)) \leq k D_A(\boldsymbol{x}_1, \boldsymbol{x}_2) \quad \forall \boldsymbol{x}_1, \boldsymbol{x}_2 \in A, \tag{1}$$

for some real-valued $k \geq 0$, and metrics D_A and D_B. We refer to f as being k-Lipschitz. We are most interested in the smallest possible value of k, which is sometimes referred to as the best Lipschitz constant. A particularly useful property of Lipschitz continuity is that the composition of a k_1-Lipschitz function with a k_2-Lipschitz function is a $k_1 k_2$-Lipschitz function. Given that a feed-forward neural network can be expressed as a series of function compositions,

$$f(\boldsymbol{x}) = (\phi_l \circ \phi_{l-1} \circ \ldots \circ \phi_1)(\boldsymbol{x}), \tag{2}$$

one can compute the Lipschitz constant of the entire network by computing the constant of each layer in isolation and taking the product of these constants:

$$L(f) = \prod_{i=1}^{l} L(\phi_i), \tag{3}$$

where $L(\phi_i)$ indicates the Lipschitz constant of some function, ϕ_i.

Many functions in this product, such as commonly used activation functions and pooling operations, have a Lipschitz constant of one for all vector p-norms on $\mathbb{R}^d$. Other commonly used functions, such as fully connected and convolutional layers, can be expressed as affine transformations,

$$f(\boldsymbol{x}) = W\boldsymbol{x} + \boldsymbol{b}, \tag{4}$$

where W is a weight matrix and $\boldsymbol{b}$ is a bias vector. For fully connected layers, there is no special structure to W. In the case of convolutional layers, W is a block matrix where each block is in turn a doubly block circulant matrix. Batch normalisation layers can also be expressed as affine transformations, where the linear operation is a diagonal matrix. Each element on the diagonal is one of the scaling parameters divided by the standard deviation of the corresponding activation. The Lipschitz constant of an affine function is given by the operator norm of the weight matrix,

$$\|W\|_p = \sup_{\boldsymbol{x} \neq 0} \frac{\|W\boldsymbol{x}\|_p}{\|\boldsymbol{x}\|_p}, \tag{5}$$

for some vector p-norm. For the ℓ_1 and ℓ_∞ vector norms, the matrix operator norms are given by the maximum absolute column sum and maximum absolute row sum norms, respectively. In the case of the ℓ_2 norm, the operator norm of a matrix is given by the spectral norm—the largest singular value. This can be approximated for fully connected layers relatively efficiently using the power iteration method. Once the operator norms have been computed, projected gradient methods can be used to constrain the Lipschitz constant of each layer to be less than a user specified value.

4   Regularisation by Constraining Gain

A common assumption in machine learning is that many types of unstructured data, such as images and audio, lie near a low dimensional manifold embedded in a high dimensional vector space. This is known as the manifold hypothesis. If we assume that the manifold hypothesis holds, then a network will only be supplied with elements of some set $\mathcal{X} \subset \mathbb{R}^d$. As a consequence, the training procedure need only ensure that the network is Lipschitz continuous on $\mathcal{X}$ in order to construct a network with a slowly varying decision boundary. In practice, the exact structure of $\mathcal{X}$ is unknown, but we do have a finite sample of instances, $X \subset \mathcal{X}$, which we can use to empirically estimate various characteristics of $\mathcal{X}$.

4.1   Gain

Lipschitz continuity is not something that can be established empirically. However, one can find a lower bound for k by sampling pairs of points from the training set and determining the smallest value of k that satisfies Eq. 1. This solution, while conceptually simple, has a number of finer details that can greatly impact the result. For example, how should pairs be sampled? If they are chosen randomly, then a very large number of pairs will be required to provide a good estimate of k. On the other hand, if a hard-negative mining approach were employed, fewer pairs would be required, but the amount of computation per pair would be greatly increased.

By restricting our analysis to feed-forward neural networks, we derive a simpler and more computationally efficient approach. Recall that the Lipschitz constant of a feed-forward network is given by the product of the Lipschitz constants associated with each activation function—which are usually less than or equal to one and cannot be changed during training—and the operator norms associated with the linear transformations in the learned layers. We define gain using the fraction from Eq. 5,

$$Gain_p(W, \boldsymbol{x}) = \frac{\|W\boldsymbol{x}\|_p}{\|\boldsymbol{x}\|_p}, \tag{6}$$

for some input instance $\boldsymbol{x}$, and we use the maximum gain observed over some set of input vectors from our manifold of interest as an approximation of the operator norm. This empirical estimate of the operator norm of a matrix has several advantages over computing the true operator norm. Firstly, it fulfills our desire to approximately compute the Lipschitz constant of an affine function on $\mathcal{X}$. It is also well behaved, in the sense that $X = \mathcal{X} \implies \sup_{\boldsymbol{x}} Gain(W, \boldsymbol{x}) = \|W\|_p$. Some more practical advantages include not having to explicitly construct W, but merely requiring a means of computing $W\boldsymbol{x}$—a property that is extremely useful when computing the operator norm of a convolutional layer. Also, because one need not compute a matrix norm directly, it is possible to compute the gain with respect to a p-norm for which it would be NP-hard to compute the induced matrix operator norm.

4.2   MaxGain Regularisation

The crux of our regularisation technique is to limit the gain of each layer in a feed-forward neural network. Each layer is constrained, in isolation, to have a gain less than or equal to a user specified hyperparameter, γ. Put formally, we wish to solve the following optimisation problem:

$$W_{1..l} = \arg\min_{W_{1..l}} \sum_{x_i^1 \in X} L(x_i^1, y_i) \tag{7}$$

$$s.t. \max_{x_i^j} Gain_p(W_j, x_i^j) \leq \gamma \qquad \forall j \in \{1 \dots l\}, \tag{8}$$

where x_i^j indicates the input to the jth layer for instance i, y_i is a label vector associated with instance i, W_j is the weight matrix for layer j, and $L(\cdot)$ is some task-specific loss function. Note that if $\|x_i^j\|_p$ is zero, we set the gain for that particular measurement to zero rather than leaving it undefined.

The conventional approach to solving Eq. 7 without the constraint in Eq. 8 is to use some variant of the stochastic gradient method. For simple constraints, such as requiring W_j to lie in some known convex set, a projection function can be used to enforce the constraint after each parameter update. In our case, applying the projection function after each parameter update would involve propagating the entire training set through the network to measure the maximum gain for each layer. Even for modest sized datasets this is completely infeasible, and it defeats the purpose of using a stochastic optimiser. Instead, we propose the use of a stochastic projection function, where the max in Eq. 8 is taken over the same minibatch used to compute an estimate of the loss function gradient. We reuse the "stale" activations computed before the weight update in order to avoid the extra computation required for propagating all of the instances through the network again. The following projection function is used:

$$\pi(W, \hat{\gamma}, \gamma) = \frac{1}{\max(1, \frac{\hat{\gamma}}{\gamma})} W, \tag{9}$$

where $\hat{\gamma}$ is our estimate of the maximum gain for layer j. If the MaxGain constraint is not violated, then W will be left untouched. If the constraint is violated, W will be rescaled to fix the violation. In the case where the maximum gain is computed exactly, this function will rescale the weight matrix such that the maximum gain is less than or equal to γ. Because we are only approximately computing the maximum gain, this constraint will not be perfectly satisfied on the training set.

During training, batch normalisation applies a transformation to the activations of a minibatch using statistics computed using only the instances contained in that minibatch. Thus, the gain measured for a particular instance is dependent on the other instances in the batch in which it is observed by the network. Specifically, the activations, x, produced by some layer, are standardised:

$$\phi^{bn}(x) = \text{diag}(\frac{\alpha}{\sqrt{\text{Var}[x]}})(x - \text{E}[x]) + \beta, \tag{10}$$

where $\text{diag}(\cdot)$ denotes a diagonal matrix, α and β are learned parameters, and the $\text{Var}[\cdot]$ and $\text{E}[\cdot]$ operations are computed over only the instances in the current minibatch. If the estimated mean and variance values are particularly unstable, then the gain values will also be very unstable and the training procedure will converge very slowly—or possibly not at all. We have found that the high dimensionality of neural network hidden layer activation vectors, and their sparse nature when using the ReLU activation function, coupled with a relatively small batch size, leads to unstable measurements when using MaxGain in conjunction with batch normalisation. We remedy this by recomputing the batch normalisation output in the projection function using the running averages of the standard deviation estimates that are kept for performing test-time predictions. By standardising the minibatch activations using these more stable estimates of the activation statistics, we observed considerably more reliable convergence. Note that the stochastic estimates of the mean and standard deviation of activations are still used for computing the gradient—it is only the projection function that uses the running averages of these values.

Pseudocode for our constrained optimisation algorithm based on stochastic projection is provided in Algorithm 1. The inputs to each layer for each minibatch, $X_{1:l}^{(t)}$, and the results of transforming these by the linear term of the affine transformations, $Z_{1:l}^{(t)}$, are cached during the gradient computation to be reused in the projection function. We use a single hyperparameter, γ, to control the allowed gain of each layer. There is no fundamental reason that a different γ cannot be selected for each layer other than the added difficulty in optimising more hyperparameters. The $update(\cdot, \cdot)$ function can be any stochastic optimisation algorithm commonly used with neural networks. We consider both Adam [10] and SGD with Nesterov momentum.

Algorithm 1. This algorithm makes use of the stochastic gradient method (or some variation thereof) and a stochastic projection function to approximately solve the constrained optimisation problem outlined in Eqs. 7 and 8.

$t \leftarrow 0$
while $W_{1:l}^{(t)}$ not converged **do**
 $t \leftarrow t + 1$
 $(g_{1:l}^{(t)}, X_{1:l}^{(t)}, Z_{1:l}^{(t)}) \leftarrow \nabla_{W_{1:l}} f(W_{1:l}^{(t-1)})$
 $\widehat{W}_{1:l}^{(t)} \leftarrow update(W_{1:l}^{(t-1)}, g_{1:l}^{(t)})$
 for $i = 1$ **to** l **do**
  $\hat{\gamma} \leftarrow 0$
  for $(x_j, W_i^{(t)} x_j)$ **in** $zip(X_i^{(t)}, Z_i^{(t)})$ **do**
   $\hat{\gamma} \leftarrow max(\hat{\gamma}, \frac{\|W_i^{(t)} x_j\|_p}{\|x_j\|_p})$
  end for
  $W_i^{(t)} \leftarrow \pi(\widehat{W}_i^{(t)}, \hat{\gamma}, \gamma)$
 end for
end while

4.3 Compatibility with Dropout

There are two parts to applying dropout regularisation to a network. Firstly, during training, one must stochastically corrupt the activations of some hidden layers, usually by multiplying them with vectors of Bernoulli random variables. Secondly, during test time, the activations are scaled such that the expected magnitude of each activation is the same as what it would have been during training. In the case of standard Bernoulli dropout, this just means multiplying each activation by the probability that it was not corrupted during training. This scaling is known to change the Lipschitz constant of a network over $\mathbb{R}^d$ [7], and the same argument applies to the Lipschitz constant on $\mathcal{X}$. Because many commonly used activation functions are homogeneous, namely ReLU and its many variants, scaling the output activations is equivalent to scaling the output of the affine transformation. This, in turn, has an identical effect to scaling both the weight matrix and bias vector. Due to the homogeneity of norms, this scaling also directly affects the gain. Therefore, one might expect that one needs to increase γ when using our technique in conjunction with dropout.

5 Experiments

The experiments reported in this section aim to demonstrate several aspects of our MaxGain regularisation method. The primary question we wish to answer is whether our technique for constraining the maximum gain of each learned layer in a network is an effective regularisation method. We also demonstrate that constraining the gain on training instances results in observing lower gain on the test, compared to when the gain is not constrained at all. All networks trained with MaxGain regularisation use the same γ parameter for each layer in order to simplify hyperparameter optimisation. While the method we have presented can be used in conjunction with any vector norm, in this work we only investigate how well MaxGain works when using the ℓ_2 vector norm.

Throughout our experiments, we make use of several different datasets. We also introduce a novel dataset larger than some typical benchmark datasets, like CIFAR-10 and MNIST, yet smaller and more manageable than the ImageNet releases used for the Large Scale Visual Recognition challenges. This dataset is designed so that performing significance tests is easy, and a greater degree of confidence can therefore be attributed to conclusions drawn from experiments using this dataset. The pixel intensities of all images have been scaled to lie in the range $[-1, 1]$.

5.1 CIFAR-10

CIFAR-10 [12] is a collection of 60,000 tiny colour images, each labelled with one of 10 classes. In our experiments we follow the standard protocol of using 50,000 images for training and 10,000 images for testing. Additionally, we use a 10,000 image subset of the training set to tune the hyperparameters. We use the VGG-19 network [16] trained using the Adam optimiser [10]. The model is trained for 140 epochs, starting with a learning rate of 10^{-4}, which is decreased to 10^{-5} at epoch 100 and 10^{-6} at epoch 120. We make use of data augmentation in the form of horizontal flips, and padding training images to 40×40 pixels and cropping out a random 32×32 patch.

Results demonstrating how our technique compares with other common regularisation techniques are given in Table 1. Several trends stand out in this table. Firstly, when comparing with each other technique in isolation, our method performs noticeably better than dropout and similarly to batch normalisation. When used in conjunction with batch normalisation the resulting test accuracy improves further. Interestingly, combining the use of dropout with both other regularisation approaches does not seem to have a noticeable cumulative effect.

Table 1. Accuracy of a VGG-19 network trained in CIFAR-10 with different regularisation techniques.

Regulariser	Accuracy
None	88.29%
Dropout	89.71%
Batchnorm	90.80%
Batchnorm + Dropout	90.90%
MaxGain	90.75%
MaxGain + Dropout	90.95%
MaxGain + Batchnorm	91.76%
MaxGain + Batchnorm + Dropout	91.52%

5.2   CIFAR-100

CIFAR-100 [12] is similar to CIFAR-10, on account of containing 60,000 colour images of size 32×32, also split into a predefined set of $50,000$ for training and $10,000$ for testing. It differs in that it contains 100 classes, and exhibits more subtle inter-class variation. We use a Wide Residual Network [18] on this dataset, in order to investigate how well MaxGain works on networks with residual connections. Batch normalisation is applied to all models trained on this dataset. We found convergence to be unreliable when training Wide ResNets without batch normalisation. Stochastic gradient descent with Nesterov momentum is used to train for a total of 200 epochs. We start with a learning rate of 10^{-1} and decrease by a factor of 5 at epochs 60, 120, and 160. We use the same data augmentation as was used for the CIFAR-10 models.

Results for experiments run on CIFAR-100 are given in Table 2. In this case, we can see that our method performs comparably to dropout when both techniques are used in conjunction with batch normalisation. The combination of all three regularisation schemes performs the best.

Table 2. Accuracy of a Wide Residual Network with a depth of 16 and a width factor of four trained on CIFAR-100 with different regularisation techniques.

Regulariser	Accuracy
Batchnorm	75.34%
Batchnorm + Dropout	75.72%
MaxGain + Batchnorm	75.89%
MaxGain + Batchnorm + Dropout	76.44%

5.3   Street View House Numbers (SVHN)

The Street View House Numbers Dataset contains over 600,000 colour images depicting house numbers extracted from Google street view photos. Each image is 32×32 pixels, and the dataset has a predefined train and test split of 604,388 and 26,032 images, respectively. The distributions of the training and test splits are slightly different, in that the majority of the training images are considered less difficult. We train a VGG-style network on this dataset using the Adam [10] optimiser. Likely due to the large size of the dataset, we found that the network only needed to be trained for 17 epochs. We began with a learning rate of 10^{-4} and reduced it by a factor of 10 for the last two epochs.

Table 3 shows how the different models we considered performed on SVHN. An interesting result here is, in isolation, dropout outperforms both MaxGain and batch normalisation in terms of accuracy improvement over the baseline. This is potentially due to the mismatch between the distributions of the training and testing datasets. Despite the lackluster performance of these methods in isolation, they do still provide a benefit when combined with each other and dropout, which is consistent with the results of our other experiments.

Table 3. Accuracy of a VGG-style network on the SVHN dataset when trained with various regularisation techniques.

Regulariser	Accuracy
None	96.99%
Dropout	97.72%
Batchnorm	96.97%
Batchnorm + Dropout	97.86%
MaxGain	97.22%
MaxGain + Dropout	97.89%
MaxGain + Batchnorm	97.31%
MaxGain + Batchnorm + Dropout	97.98%

5.4   Scaled ImageNet Subset (SINS-10)

Many datasets used by the deep learning community consist of a single predefined training and test split. For example, in the previous experiments on CIFAR-10 we stated that a set of 50,000 images was used for training, and another set of 10,000 images was used for testing. In order to perform some sort of significance test, and thus have some degree of confidence in our results and the conclusions we draw from them, we must gather multiple measurements of how well models trained using a particular algorithm configuration perform. To this end, we propose the Scaled ImageNet Subset (SINS-10) dataset, a set of 100,000 colour images retrieved from the ImageNet collection [3]. The images are evenly divided into 10 different classes, and each of these classes is associated with multiple synsets from the ImageNet database. All images were first resized such that their smallest dimension was 96 pixels and their aspect ratio was maintained. Then, the central 96×96 pixel subwindow of the image was extracted to be used as the final instance.

An important difference between the proposed dataset and currently available benchmark datasets is how it has been split into training and testing data. The entire dataset is divided into 10 equal sized predefined folds of 10,000 instances. The first 9,000 images in each fold are intended for training a model, and the remaining 1,000 for testing it. One can then apply a machine learning technique to each fold in the dataset, and repeat the process for techniques one wishes to compare against. This will result in 10 performance measurements for each algorithm. A paired t-test can then be used to determine whether there is a significant difference, with some level of confidence, between the performance of the different techniques.

Note that the protocol for SINS-10 is different to the commonly used cross-validation technique. When performing cross-validation, the training sets overlap significantly, and the measurements for the test fold performance are therefore not independent. To mitigate this, one can use a heuristic for correcting the paired t-test [14]. Rather than use this heuristic, we simply avoid fitting models using overlapping training (or test) sets, and can therefore use the standard paired t-test.

We train a Wide Residual Network with a width factor of four on this dataset. No data augmentation was used and each model was trained for 90 epochs using stochastic gradient descent with Nesterov momentum. The learning rate was started at 10^{-1} and decreased by a factor of five at epochs 60 and 80. For each regularisation scheme, we trained a model on each fold of the dataset. Regularisation hyperparameters, such as γ and the dropout rate, were determined on a per-fold basis using a validation set of 1,000 instances drawn from the training set of the fold under consideration.

Results for the different regularisation schemes trained on this dataset are given in Table 4. We report the mean accuracy across each of the 10 folds, as well as the standard error. Paired t-tests were performed for comparing Batchnorm to MaxGain + Batchnorm, and also for Batchnorm + Dropout versus MaxGain +

Table 4. Performance of the Wide Residual Network on the Scaled ImageNet Subset dataset using various combinations of regularisation techniques. The figures in this table are the mean accuracy ± the standard error, as measured across the 10 different folds.

Regulariser	Accuracy
Batchnorm	70.13% (±0.27)
Batchnorm + Dropout	74.81% (±0.49)
MaxGain + Batchnorm	70.65% (±0.54)
MaxGain + Batchnorm + Dropout	74.80% (±0.51)

Batchnorm + Dropout. Neither of the tests resulted in a statistically significant difference ($p = 0.332$ and $p = 0.976$, respectively).

5.5   Gain on the Test Set

Due to the stochastic nature of the projection function, the technique used to constrain the gain on the training set is only approximate. Therefore, it is important that we verify whether the constraint is fulfilled in practice. Moreover, even if the constraint is satisfied on the training set, that does not necessarily mean it will be satisfied on data not seen during training. To investigate this, we supply plots in Fig. 1 showing the distribution of gains in each layer in the VGG-19 network trained using MaxGain on the CIFAR-10 dataset. We can see that the distributions between the train and test sets are virtually identical, and are never significantly above 2—the value selected for γ when training this network.

In addition to demonstrating that the stochastic projection function does effectively limit the maximum gain on the test set, we find it interesting to visualise gain measurements taken from each layer in a network trained without the MaxGain regulariser. This visualisation is given in Fig. 2. Once again, the distributions of gains measured on the training versus test data are almost identical. Comparing the distributions given in Fig. 2 with those provided in Fig. 1 show that the MaxGain regulariser has a substantial effect on the activation magnitudes produced by each layer.

If there is no constraint on the magnitude of the weights, then once the network can almost perfectly classify the training data, the optimiser can easily decrease the log loss by making the weights bigger. This results in an "exploding activation" effect, similar to the exploding/vanishing gradient phenomenon, which is only curbed when the cost of the small number of instances in the training set that are very confidently classified incorrectly begin to outweigh the increase in confidence on the correct classifications. Because MaxGain constrains the weight sizes of each layer, those that would have had large weights no longer do, and those that would have had small weights will now need larger weights in order to increase the confidence of the model. This results in the far more uniform changes in activation magnitude in Fig. 1 compared to those in Fig. 2.

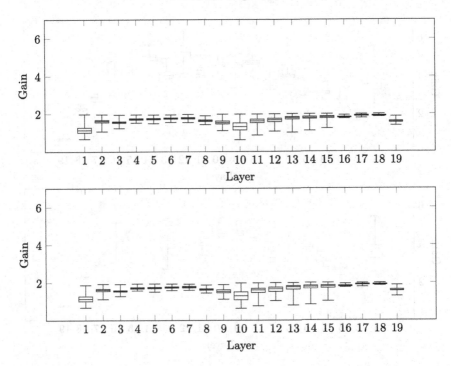

Fig. 1. Boxplots showing the distributions of gains measured on each layer of the MaxGain-regularised VGG-19 network trained on CIFAR-10. The top plot shows the distributions on the training set, and the bottom plot on the test set.

5.6   Sensitivity to γ

The single hyperparameter, γ, that is used to control the capacity of MaxGain-regularised networks should behave similarly to the λ hyperparameter proposed by Gouk et al. [7] which is used to precisely bound the Lipschitz constant. In particular, when γ is set to a small value the model should underfit, and when it is set to a large value one should observe overfitting. We explore this empirically in the context of the VGG-style network trained on SVHN. Figure 3 shows how the performance on the training and test sets of SVHN varies as γ is changed. This plot shows that γ behaves in much the same way as the previously mentioned λ hyperparameter. Specifically, for very low values of γ, the network exhibits low accuracy and high loss for both the train and test splits of the dataset. As the value of γ is increased, the training accuracy goes towards 100% and the loss goes towards zero. The test accuracy peaks and then plateaus, however the loss on the training set continues to increase, indicating that the network is more confidently misclassifying instances rather than misclassifying more instances.

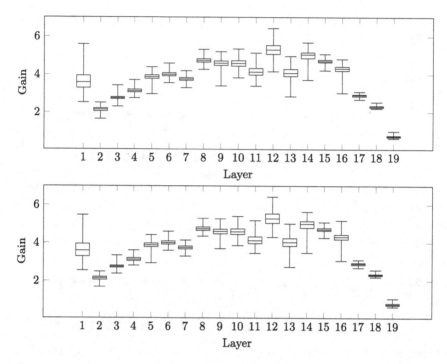

Fig. 2. Boxplots showing the distributions of gains measured on each layer of the unregularised VGG-19 network trained on CIFAR-10. The top plot shows the distributions on the training set, and the bottom plot on the test set.

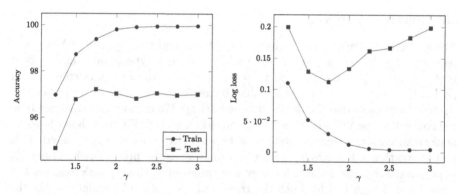

Fig. 3. Accuracy (left) and log loss (right) of the VGG-style model on both the train and test splits of the SVHN dataset as the γ hyperparameter is varied. The legend is shared between both plots.

6   Conclusion

This paper introduced MaxGain, a method for regularising neural networks by constraining how the magnitudes of activation vectors can vary across layers. It was shown how this method can be seen as an approximation to constraining the Lipschitz constant of a network, with the advantage of being usable for any vector norm. The technique is conceptually simple and easy to implement efficiently, thus making it a very practical approach to controlling the capacity of neural networks. We have shown that MaxGain performs competitively with other common regularisation schemes, such as batch normalisation and dropout, when compared in isolation. It was also demonstrated that when these techniques are combined together, further performance gains can be achieved. Some of these results were obtained using a novel dataset with predefined folds that allows for practical significance testing in experiments involving convolutional networks.

References

1. Arjovsky, M., Chintala, S., Bottou, L.: Wasserstein gan. arXiv preprint arXiv:1701.07875 (2017)
2. Cayton, L.: Algorithms for manifold learning. University of California at San Diego, Technical report 12(1–17), 1 (2005)
3. Deng, J., Dong, W., Socher, R., Li, L.-J., Li, K., Fei-Fei, L.: Imagenet: a large-scale hierarchical image database. In: IEEE Conference on Computer Vision and Pattern Recognition, CVPR 2009, pp. 248–255. IEEE (2009)
4. Dinh, L., Pascanu, R., Bengio, S., Bengio, Y.: Sharp minima can generalize for deep nets. In: International Conference on Machine Learning, pp. 1019–1028 (2017)
5. Gal, Y., Hron, J., Kendall, A.: Concrete dropout. In: Advances in Neural Information Processing Systems, pp. 3584–3593 (2017)
6. Goodfellow, I., et al.: Generative adversarial nets. In: Advances in Neural Information Processing Systems, pp. 2672–2680 (2014)
7. Gouk, H., Frank, E., Pfahringer, B., Cree, M.: Regularisation of neural networks by enforcing Lipschitz continuity. Technical report, University of Waikato (2018)
8. Gulrajani, I., Ahmed, F., Arjovsky, M., Dumoulin, V., Courville, A.C.: Improved training of Wasserstein GANs. In: Advances in Neural Information Processing Systems, pp. 5769–5779 (2017)
9. Ioffe, S., Szegedy, C.: Batch normalization: accelerating deep network training by reducing internal covariate shift. In: International Conference on Machine Learning, pp. 448–456 (2015)
10. Kingma, D., Ba, J.: Adam: a method for stochastic optimization. arXiv preprint arXiv:1412.6980 (2014)
11. Kingma, D.P., Salimans, T., Welling, M.: Variational dropout and the local reparameterization trick. In: Advances in Neural Information Processing Systems, pp. 2575–2583 (2015)
12. Krizhevsky, A., Hinton, G.: Learning multiple layers of features from tiny images. Technical report, University of Toronto (2009)
13. Miyato, T., Kataoka, T., Koyama, M., Yoshida, Y.: Spectral normalization for generative adversarial networks. In: 6th International Conference on Learning Representations (2018)

14. Nadeau, C., Bengio, Y.: Inference for the generalization error. In: Advances in Neural Information Processing Systems, pp. 307–313 (2000)
15. Salimans, T., Kingma, D.P.: Weight normalization: a simple reparameterization to accelerate training of deep neural networks. In: Advances in Neural Information Processing Systems, pp. 901–909 (2016)
16. Simonyan, K., Zisserman, A.: Very deep convolutional networks for large-scale image recognition. arXiv preprint arXiv:1409.1556 (2014)
17. Srivastava, N., Hinton, G., Krizhevsky, A., Sutskever, I., Salakhutdinov, R.: Dropout: a simple way to prevent neural networks from overfitting. J. Mach. Learn. Res. **15**(1), 1929–1958 (2014)
18. Zagoruyko, S., Komodakis, N.: Wide residual networks. In: Proceedings of the British Machine Vision Conference (BMVC), September 2016
19. Zhang, C., Bengio, S., Hardt, M., Recht, B., Vinyals, O.: Understanding deep learning requires rethinking generalization. arXiv preprint arXiv:1611.03530 (2016)

Ontology Alignment Based on Word Embedding and Random Forest Classification

Ikechukwu Nkisi-Orji[1]([✉])([iD]), Nirmalie Wiratunga[1], Stewart Massie[1],
Kit-Ying Hui[1], and Rachel Heaven[2]

[1] Robert Gordon University, Aberdeen, UK
{i.o.nkisi-orji,n.wiratunga,s.massie,k.hui}@rgu.ac.uk
[2] British Geological Survey, Nottingham, UK
reh@bgs.ac.uk

Abstract. Ontology alignment is crucial for integrating heterogeneous data sources and forms an important component of the semantic web. Accordingly, several ontology alignment techniques have been proposed and used for discovering correspondences between the concepts (or entities) of different ontologies. Most alignment techniques depend on string-based similarities which are unable to handle the vocabulary mismatch problem. Also, determining which similarity measures to use and how to effectively combine them in alignment systems are challenges that have persisted in this area. In this work, we introduce a random forest classifier approach for ontology alignment which relies on word embedding for determining a variety of semantic similarity features between concepts. Specifically, we combine string-based and semantic similarity measures to form feature vectors that are used by the classifier model to determine when concepts align. By harnessing background knowledge and relying on minimal information from the ontologies, our approach can handle knowledge-light ontological resources. It also eliminates the need for learning the aggregation weights of a composition of similarity measures. Experiments using Ontology Alignment Evaluation Initiative (OAEI) dataset and real-world ontologies highlight the utility of our approach and show that it can outperform state-of-the-art alignment systems. Code related to this paper is available at: https://bitbucket.org/paravariar/rafcom.

Keywords: Ontology alignment · Word embedding
Machine classification · Semantic web

1 Introduction

Ontology alignment or matching deals with the discovery of correspondences between the entities of different ontologies. This has been the subject of various research works over the years with several techniques adopted from methods

© Springer Nature Switzerland AG 2019
M. Berlingerio et al. (Eds.): ECML PKDD 2018, LNAI 11051, pp. 557–572, 2019.
https://doi.org/10.1007/978-3-030-10925-7_34

for integrating heterogeneous databases. The utility of ontologies are enhanced through alignment and the reduced semantic gap enables applications requiring cross-ontology reasoning or data exchange. Interest in ontology alignment is reflected through the Ontology Alignment Evaluation Initiative (OAEI)[1] which provides a platform to assess and compare systems for automated or semi-automated alignment. Also, the Linking Open Data community project[2] which aims to align ontologies on a Web scale currently have hundreds of datasets from different contributors in multiple domains such as DBpedia, WordNet, GeoNames, and MeSH.

The ontology alignment process is challenging, especially when the ontologies are of heterogeneous origins leading to inherent differences between them. Ontologies can vary vastly in levels of formalisation and vocabulary use even when they are of similar domain. The predominant methods for alignment use a composition of multiple string-based similarity metrics on textual features of entities [2]. Semantic matching is essential for discovering correspondences by meaning when the vocabularies of source and target ontologies differ. However, there is a shortage of semantic matching techniques [18,19]. Lexical databases such as WordNet have been leveraged for semantic matching but they lack sufficient coverage and this becomes apparent when dealing with domain-specific terminology. Accordingly, word embedding approaches which are effective at capturing language semantics have been proposed for semantic matching in ontology alignment [21,22]. Semantic matching approaches do not always outperform string-based similarity and effectively combining both strategies in alignment systems remain a challenge [18].

In this work, we introduce a novel matching system that integrates string-based similarity and semantic similarity features using word embedding to build a machine learning model, a random forest classifier. Alignment is completed in two stages by first selecting a set of candidate alignments using basic matching techniques. Afterwards, a machine classifier determines which entity pairs of the candidate alignments are true alignments. The classifier uses feature vectors that are generated from a variety of direct and indirect similarity indicators. Our main contributions are the incorporation of word embedding for semantic match discovery in the alignment process and the introduction of novel features for a machine classifier for alignment. The alignment system relies on minimal information from the ontologies making it suitable for aligning knowledge-light ontological resources. Although it requires training a classifier model, our approach eliminates the need to learn aggregation weights for multiple similarity measures. We evaluate the alignment system on benchmark datasets from OAEI and dataset from EuroVoc (EU's multilingual thesaurus)[3].

The remainder of this paper is organised as follows: Sect. 2 reviews relevant works in literature; Sect. 3 presents our ontology alignment approach; Sect. 4 is an experimental evaluation which compares our approach to alternative approaches; and Sect. 5 concludes with an outline for future work.

---

[1] http://oaei.ontologymatching.org/.

[2] http://linkeddata.org/.

[3] European Union, 2018, http://eurovoc.europa.eu/.

2   Related Work

Ontology alignment establishes semantic links between the entities of different ontologies which is a solution to the semantic heterogeneity problem [6,19]. Alignment reduces the semantic gap between overlapping representations of a domain and trends show increasing interest in this area [18]. Establishing correspondences between the entities of different ontologies generally follows pairwise comparisons (direct or indirect) to identify best matches. Techniques for matching entities can be element-level or structure-level [18]. Element-level matching uses intrinsic features of entities such as natural language labels and definitions [10]. Instead of exact string matching, edit distance approaches such as Levenshtein and Jaro-Winkler distances are commonly used for fuzzy matching to account for spelling variations and word inflection. Structure-level matching considers the ontological neighbourhood of entities in order to determine similarity. Even when entities share little element-level features, correspondences can be discovered by similarity of structures such as having similar ancestors or descendants [17].

String similarity methods differ and an individual approach cannot be always relied on for effective alignment [2]. Accordingly, most alignment systems use a composition of multiple similarity metrics (basic matchers) which are aggregated sequentially or in parallel [3,10] or form features for a machine learner [5,16]. This leads to a categorisation of research in ontology alignment as matching techniques or matching systems [18]. Matching techniques deal with measures of similarity and strategies that determine the extent to which the concepts of different ontologies relate while matching systems use one or more matching techniques to align ontologies. The choice of matching techniques and determining composition weights for multiple similarity metrics have been subject of several research works [7,13]. As ontologies differ widely, it is not unusual to encounter alignment systems which work well for some alignment tasks and perform weakly on others.

String comparison is less effective for alignment when the vocabulary of ontologies differ. As a result, external knowledge resources such as WordNet and Wikipedia have been used to estimate semantic similarities [8,9,12]. Use of external resources requires anchoring entities being compared to the external resources which is then used for inferencing. By matching by meaning, semantic matching can discover alignments which are omitted by string-based similarity approaches. Yet, semantic matching is rarely used because effective integration of string-based similarity and semantic similarity remains a challenge [18,19]. Recent experiments show that matching using word embedding vectors outperforms use of lexical databases such as WordNet for semantic matching [22]. Word2vec models are popular implementations of word embedding using shallow neural network architecture to embed words in a dense continuous vector space based on their linguistic contexts in a corpus [14]. Word embedding preserves several linguistic regularities and similarity between word vectors have been shown to correlate well with human judgements. The use of word embedding is also promising for cross-lingual alignment by jointly embedding ontologies in a vector space [21]. An even more effective use of word embedding for ontology alignment is a hybrid similarity approach that incorporates string similarity using edit distance [22]. To the best of our knowledge, no other system has extended use of word embedding for alignment beyond a hybrid similarity of edit distance and vector similarity. We extend the hybrid similarity approach by introducing other similarity features which are used by a random forest classifier to align ontologies.

3   Classifier-Based Ontology Alignment

Our approach is based on generating a machine classifier model using a hybrid of element-level string-based features, semantic similarity features, and context-based structure-level similarity features. A high-level overview of the alignment process is presented in Fig. 1 and the rest of this section describes the process in detail. The alignment process starts with the selection of candidate alignments using a variety of basic matching techniques. A feature vector is then generated for each candidate alignment which is passed to the machine classifier. The classifier determines whether the concept pair is accepted as a correspondence or discarded.

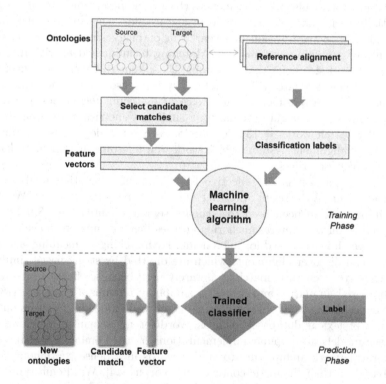

Fig. 1. Overview of ontology alignment process using machine learning.

Notations, Scope and Assumptions. An ontology, θ specifies a set of concepts (or entities), $\theta = \{c_1, ..., c_n\}$. A concept $c \in \theta$ represents the semantic definition of a meaningful entity in a domain. Although some ontologies also specify data properties and object properties, we use this minimal specification to include knowledge-light ontological resources such as thesauri and controlled vocabularies. Let $labels(c)$ return the set of textual labels of a concept including alternative names (or synonyms), $labels(\theta)$ return an ontology's document collection which is all labels of all concepts of θ, and $tok(l)$ return all words from a concept's label, $l \in labels(c)$. To illustrate with Fig. 2, concept #3945 has two labels making $label(\#3945) = \{$"petroleum industry", "oil industry"$\}$, $tok($"petroleum industry"$) = \{$"petroleum", "industry"$\}$, and $label(\theta)$ returns eight labels. We assume that the ontologies being aligned specify some form of subsumption relations between concepts such as "is-a" or "broader-than" relations. This allows for the identification of a concept's semantic context and depth on the ontology structure. The subsumption relation between two concepts c_i and c_j is represented as $c_i \prec c_j$ specifying that c_i is a broader concept of c_j (e.g. #2673 $\prec$ #3945 in Fig. 2).

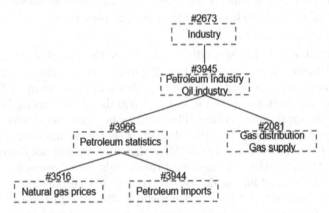

Fig. 2. Example of concepts' hierarchy with textual labels.

The output of the alignment process between the source ontology θ and target ontology θ' is the alignment, A which is a set of correspondences between semantically equivalent concepts of both ontologies. Each correspondence $a \in A$ is a 4-tuple, $a : <c, c', \equiv, s>$ where $c \in \theta$, $c' \in \theta'$, $\equiv$ indicates equivalent relation type between c and c', and s is the confidence of alignment correspondence in $[0.0, 1.0]$ interval. Confidence is either 1 (correspondence) or 0 (no correspondence) for crisp alignment.

3.1 Identification of Alignment Candidates

The objective for selecting candidate alignments is to avoid including concept pairs that have little or no chance of being aligned in subsequent machine classification stage. A pair of concepts being compared become candidate alignments if their similarity exceeds the threshold for any of four similarity measures. Accordingly, similarity thresholds for candidate selection are kept low enough to maximise recall but not very low to select the entire similarity matrix. This avoids having to generate features for concept pairs with very low similarities and also leads to a better class balance for training a classifier. We also use a $Max1$ selection approach for each similarity measure such that if multiple concepts in the target ontology exceed the selection threshold, we only choose the pair(s) with highest similarity value. This is commonly used to enforce a one-to-one correspondence in alignment [19]. A variety of ways in which concepts can be similar were considered in selecting similarity measures for identifying candidate alignments as follows.

1. Hybrid similarity (*hybrid*): combines word embedding and edit distance,
2. Vector space model (*vsm*): cosine similarity of term vectors using term frequency – inverse document frequency (tf-idf) scheme,
3. ISUB similarity (*isub*): string similarity metric designed for ontology alignment, and
4. Similarity of semantic context (*context*): indirect similarity between concepts by comparing their neighbours on the ontology structure.

Hybrid Similarity. Hybrid similarity combines the use of word embedding and edit distance measures [22]. After discarding words which occurred less than 10 times, we embedded a November 2016 database dump of Wikipedia English language articles in vector space of 300 dimensions using Word2vec's continuous skip-gram architecture. The word embedding model was generated using an open-source deep learning library[4]. There is an abundance of literature and software tools on word embedding therefore, we will not discuss details of implementation further. We also used the Google New Corpus model[5] as an alternative word embedding model for comparison. The edit distance component of our hybrid similarity is based on Levenshtein distance. In contrast to [22], a threshold is imposed on the edit distance component. This is because below certain thresholds, similarity by sharing similar characters is no more than a coincidence. Similarity between terms is based on the approach for measuring sentence similarity [11] as shown in Eq. 1.

$$hybrid(c, c') = \tag{1}$$
$$\max_{\{l\in labels(c), l'\in labels(c')\}} \left\{ \frac{1}{maxLen(l, l')} \cdot \sum_{w\in l} \sum_{w'\in l'} max(emb(w, w'), lev(w, w')) \right\}$$

$maxLen(l, l') = max(|tok(l)|, |tok(l')|)$ is length of the longer label, $emb(w, w')$ is the cosine similarity between the embedding vectors of w and w', and $lev(w, w')$ is normalised Levenshtein similarity. First, Levenshtein distance is normalised to

---

[4] https://deeplearning4j.org/word2vec.html.
[5] https://s3.amazonaws.com/dl4j-distribution/GoogleNews-vectors-negative300.bin.gz.

[0.0, 1.0] interval by dividing by the length of the longer string. Similarity is then determined as 1 − normalised distance and is only considered when up to 0.8. In other words, Eq. 1 compares each word from one label with every word in the other label and selects the maximum similarity of either word embedding or edit distance. The sum of best pairwise similarities is then divided by the length of the longer label. For example, in comparing "oil industry" and "petroleum industry", the best similarities are emb(oil, petroleum) = 0.65 using the Google model and lev(industry, industry) = 1.0 giving an overall similarity $\frac{1}{2}$(0.65+1) = 0.825. The most similar labels are used when concepts have multiple labels. A low hybrid similarity threshold of 0.4 was chosen in our experiments to maximise recall.

Vector Space Model. The second similarity measure is based on the vector space model using cosine similarity of tf-idf weights. Each ontology forms a collection, D ($D = labels(\theta)$). The tf-idf weight of each word, w in a document, d (a concept's label) is determined as shown in Eq. 2.

$$\text{tf-idf}(w) = f_{w,d} \cdot \log \frac{|D|}{n_w} \qquad (2)$$

$f_{w,d}$ is the frequency of w in d, and n_w is the number of documents in which w appears. Since multiple documents can belong to a concept, VSM similarity is determined as the maximum similarity of the documents of a concept pair as shown in Eq. 3.

$$\text{vsm}(c, c') = \max_{\{d \in c, d' \in c'\}} \{cosSim(d, d')\} \qquad (3)$$

$cosSim(d, d')$ is the cosine similarity between documents d and d' using their tf-idf weight vectors. By weighing terms such that frequently occurring words in an ontology contribute less to similarity, we discover alignments that will otherwise be missed as observed in [17]. Similarity threshold was set at 0.7 which is low enough for good recall.

ISUB Similarity. The third similarity approach is a string similarity metric which was specifically designed for the purpose of aligning ontologies [20]. The similarity between two strings is determined by the extent of their common substrings which is offset by their differences (Eq. 4).

$$\text{isub}(c, c') = \max_{\{l \in labels(c), l' \in labels(c')\}} \{Comm(l, l') - Diff(l, l') + winkler(l, l')\}$$
$$(4)$$

$Comm(l, l')$ is a function of common substrings, $Diff(l, l')$ is a function of the difference between the strings, and $winkler(l, l')$ is for improving the results. We used an implementation of ISUB similarity in the Alignment API [4].

Context Similarity. When the lexical forms of textual features of a pair of concepts are different, comparing their ontological neighbourhoods can discover correspondences which are missed by direct comparisons. Accordingly, we indirectly measure the similarity of concepts by comparing their semantic contexts. If the parents and children of the concepts being compared are similar, the pair are included in the set of candidate alignments. Let the immediate parent concepts of c be $P(c)$ and its immediate child concepts be $C(c)$, we implemented context similarity as in Eq. 5.

$$\text{context}(c, c') = max \left\{ \frac{1}{2} \cdot (hybrid(c_p, c'_p) + hybrid(c_c, c'_c)) \right\} \qquad (5)$$

max indicates that only the most similar parent and child concepts are used to determine context similarity with $c_p \prec c|c_p \in P(c)$, $c \prec c_c|c_c \in C(c)$, $c'_p \prec c'|c'_p \in P(c')$ and $c' \prec c'_c|c'_c \in C(c')$. We set selection threshold at half of hybrid similarity threshold since Eq. 5 is an average.

3.2   Features for Alignment Classification

In the second stage, feature vectors are generated for candidate alignments which are used by a machine classifier to determine whether they are actual alignments. We introduce various novel features in addition to similarity metrics that are commonly used for basic matching. Features are grouped into three categories (selection, direct similarity, and context features) and summarised in Table 1. Recall that each alignment candidate comprises of a concept from the source ontology ($c \in \theta$) and the most similar concept it in the target ontology ($c' \in \theta'$). We also note the next most similar concept to c in the target ontology ($c'' \in \theta'$) for the purpose of determining features which are related to similarity offsets.

Selection Features. These features are determined during the selection of candidates alignments to reflect the best similarity value (sim), the method of similarity used ($matchType$), and similarity offset to the next most similar concept in target ontology ($simOffset$). $matchType$ is a nominal attribute used to indicate the similarity method that was used to select a candidate alignment. sim is determined as $max(hybrid(c, c'), vsm(c, c'), isub(c, c'), context(c, c'))$. $simOffset$ is determined as $sim(c, c') - sim(c, c'')$ and this captures the distinctiveness of a candidate alignment. High sim and $simOffset$ values are expected to be good indicators of actual alignments. Finally, we also include each of the similarity methods for selecting candidate alignments as a separate feature.

Direct Similarity Features. This category comprises other similarity metrics that directly compare textual labels of concepts. These include five commonly used string-based similarity measures – Levenshtein (lev), Fuzzy Score[6] ($fuzzy$), Longest Common Subsequence (lcs), Sorensen-Dice ($dice$), and Monge-Elkan

---

[6] https://commons.apache.org/proper/commons-text/apidocs/org/apache/commons/text/similarity/FuzzyScore.html.

Table 1. Feature vectors for alignment

Feature category	Feature	Description
Selection	$matchType$	Similarity method to select alignment
	sim	$max(hybrid, vsm, isub, context)$
	$simOffset$	Offset to the next best sim
	$hybrid$	Combines lev and emb
	vsm	Similarity based on vector space model
	$isub$	String similarity for ontology alignment
	$context$	$hybrid$ of semantic contexts
Direct similarity	lev	Similarity based on Levenshtein distance
	$fuzzy$	Fuzzy string score gives bonus points as characters in matched substrings increases
	lcs	Similarity based on Longest Common Subsequence
	$dice$	Similarity based on Sorensen-Dice coefficient
	$mongeElkan$	Monge-Elkan similarity measure
	$prefixOverlap$	Prefix overlap divided by length of shorter string
	$suffixOverlap$	Suffix overlap divided by length of shorter string
	emb	Similarity of word embedding vectors
Context	$parentsOverlap$	Hybrid similarity of parent concepts
	$childrenOverlap$	Hybrid similarity of child concepts
	$contextOverlap$	Hybrid similarity of all context words
	$contextOverlapOffset$	Offset to next best $contextOverlap$
	$hasParents$	Indicates if both, one, or none of the concepts have parent nodes
	$hasChildren$	Indicates if both, one, or none of the concepts have child nodes
	$depthDiff$	Difference in relative depths of concepts

($mongeElkan$) [2,15]. These were chosen to provide a variation of string similarities as each algorithm differs in its approach. Also, we include features for similarity based on word embedding alone (emb) and maximum prefix overlap ($prefixOverlap$) and suffix overlap ($suffixOverlap$) of concept labels. Prefix overlap and suffix overlap are the number of contiguous characters shared at the beginning and ending of strings respectively and are normalised by diving by the length of the shorter string. Most of the string similarity measures were implemented using publicly available API[7].

---

[7] http://github.com/tdebatty/java-string-similarity.

Context Features. Features in this category are determined by the placement of concepts on the ontology structure. These include *parentsOverlap* and *childrenOverlap* which are *hybrid* similarities of parent and child concepts (of candidate nodes) respectively. We also introduce *contextOverlap* which is the *hybrid* similarity between all context words. That is, $contextOverlap(c, c') = hybrid((P(c) \cup C(c)), (P(c') \cup C(c')))$. *contextOverlapOffset* is given as $contextOverlap(c, c') - contextOverlap(c, c'')$. Furthermore, we introduce two features (*hasParents* and *hasChildren*) for additional insight into the neighbourhood of candidate alignments. *hasParents* uses nominal features to indicate whether both concepts in a candidate alignment have parent nodes, only one concept have parent nodes, or none have parent nodes. Similarly with *hasChildren*, we indicate the presence or absence of child nodes. Finally, *depthDiff* is the absolute difference of the relative depths of concepts being compared. The depth of a concept is the number of edges in the shortest path between the root node and that concept. We assume the presence of a top concept (root node) even when an ontology does not specify one. A concept's relative depth is the ratio of its depth to the total number of edges on the concept's path (i.e. from root to leaf passing through the concept). In Fig. 2 for example, the relative depth of concept #3945 is 0.5 since #3945 is halfway down on the shortest path.

3.3 Machine Learning

The final step is the classification of candidate alignments as either true or false correspondences. We use a Random Forest classifier which is an ensemble method using multiple decision trees for improved classification and to avoid overfitting. Each decision tree uses a subset of features and classification is based on majority voting of decision trees' predictions [1]. Decision trees have been previously shown to outperform other machine learning algorithms for aligning ontologies [16]. In the training phase, feature vectors (as in Table 1) are generated for candidate alignments and class labels are determined by the reference alignments. Reference alignments form the gold standard as they specify actual correspondences between source and target ontologies. When a correspondence from the candidate alignments is also present in the reference alignment, it is labelled as a true alignment, otherwise, it is labelled as a false alignment. In the prediction (or classification) phase, the trained model uses generated feature vectors to determine if unseen candidate alignments are true alignments.

4 Evaluation

4.1 Experiment Setup

We perform experiments to evaluate the performance of our approach on two alignment datasets as follows.

Benchmark Dataset. The Conference track of 2016 Ontology Alignment Evaluation Initiative (OAEI)[8] which consists of 7 small to medium-sized ontologies specifying concepts in the domain of conference organisation. The ontologies have heterogeneous origins resulting in differences in structure and vocabulary. The gold standard is 21 reference alignments representing the entire alignment space between ontology pairs.

EuroVoc Dataset. This consists of two large controlled vocabularies – the European Union multilingual thesaurus (EuroVoc)[9] and the GEneral Multilingual Environmental Thesaurus (GEMET)[10] describing 7,234 and 5,220 concepts respectively. The gold standard is 1,126 correspondences between equivalent concepts in both ontologies[11].

Alternative Alignment Approaches

- *StringEquiv*: An OAEI baseline which discovers alignments by exact string matching of concept labels.
- *edna*: Another OAEI baseline which uses edit distance (Levenshtein distance) for approximate string matching of concept labels.
- *WordEmb*: Word embedding approach using Word2Vec's continuous skip-gram model and Wikipedia data dump (version 20161130). Concepts are compared by the cosine similarity their label vectors.
- *Hybrid*: Combines word embedding and edit distance to discover correspondences [22].

Our approach which we refer to as $Rafcom^{12}$ with two variants, $Rafcom_W$ and $Rafcom_G$ for Wikipedia-based and Google News word embedding models respectively. Leave-one-out approach is used for the Conference dataset by leaving a pair of ontologies out in turn while a model is trained using the remaining dataset. The trained model is then used to aligned left out ontologies. Since the EuroVoc dataset have a pair of ontologies only, we use ten-fold cross-validation for evaluation. Alignment performance is based on standard precision, recall and F-measure which are averaged over all the folds for each dataset. Precision is the proportion of set of correspondences returned that are present in the reference alignment. Recall is the proportion of correspondences in the reference alignment that are discovered by an alignment system.

4.2 Results and Discussion

The performances of alignment approaches at best F1-measures are as shown in Tables 2 and 3 for the Conference and EuroVoc datasets respectively.

---

[8] http://oaei.ontologymatching.org/2016/conference/.
[9] http://eurovoc.europa.eu.
[10] http://www.eionet.europa.eu/gemet/en/themes.
[11] http://data.europa.eu/euodp/en/data/dataset/eurovoc/resource/3430afb6-51c7-44d8-b1c7-a1e045ef5696.
[12] https://bitbucket.org/paravariar/rafcom.

Best performances for each evaluation metric are in boldface. Our approach clearly outperformed the others on the Conference dataset for all evaluation metrics with $Rafcom_G$ slightly outperforming $Rafcom_W$. About 84% of true correspondences were discovered in the candidate selection stage and the classifier achieved about 96% accuracy in classifying candidate alignments. Performance differences were more subtle for EuroVoc. In this dataset, $Rafcom_W$ and $Rafcom_G$ had better precisions while $edna$ was best in recall. Similar to the Conference dataset, 84% of true correspondences were included in the candidate alignments selected. However, the classifier achieved about 90% accuracy in telling true alignments and false alignments apart. $edna$ outperformed $StringEquiv$ on both datasets using F1-measures and this is consistent with results at the OAEI challenge and previous works [2]. Also, $hybrid$ outperformed its components as had been expected [22].

Table 2. Performances on OAEI 2016 conference track (classes only)

Approach	Precision	Recall	F1-measure
$StringEquiv$	0.878	0.498	0.635
$edna$	0.880	0.537	0.667
$WordEmb$	0.881	0.544	0.673
$Hybrid$	0.880	0.564	0.687
$Rafcom_W$	0.889	0.680	0.770
$Rafcom_G$	**0.891**	**0.695**	**0.781**

Figure 3 shows results of alignment systems on the Conference dataset at the OAEI challenge ordered by F1-measure. Although the systems may have competed under a different circumstance, our results are promising when compared with the best systems at the challenge.

Influence of Similarity Methods in Discovering Alignment Types. The easiest correspondences to discover are exact string matches. Both $hybrid$ and $isub$ can discover such correspondences. There are observed differences between similarity approaches when concept labels do not match as shown in Table 4. The correspondence "edas#Academic_Event" ≡ "ekaw#Scientific_Event" was found using $hybrid$ because "academic" and "scientific" were embedded in similar vector space for an overall similarity of 0.84. "conference#Track-workshop_chair" ≡ "ekaw#Workshop_Chair" was discovered using $isub$. ISUB similarity puts greater emphasis on common substrings resulting in high similarity of 0.91. The similarity between this pair is 0.6 using the Levenshtein distance approach. The word "conference" appeared multiple times in conference# ontology resulting in a low tf-idf weight. The correspondence "conference#Conference_document" ≡ "ekaw#Document" has a high similarity of 0.94 using vsm highlighting the reduced importance of "conference". Also interesting is the comparison between

Table 3. Performances on EuroVoc dataset (EuroVoc–GEMET alignment)

Approach	Precision	Recall	F1-measure
StringEquiv	0.580	0.746	0.653
edna	0.572	**0.776**	0.659
WordEmb	0.581	0.746	0.653
Hybrid	0.581	0.768	0.662
Rafcom$_W$	**0.714**	0.632	**0.671**
Rafcom$_G$	**0.714**	0.629	0.669

Matcher	Threshold	Precision	F.5-measure	F1-measure	F2-measure	Recall
CroMatch	0	0.78	0.77	0.76	0.75	0.74
AML	0	0.83	0.8	0.76	0.72	0.7
LogMap	0	0.84	0.79	0.73	0.67	0.64
XMap	0	0.86	0.8	0.73	0.67	0.63
LogMapBio	0	0.8	0.76	0.71	0.67	0.64
DKPAOMLite	0	0.82	0.76	0.69	0.63	0.59
DKPAOM	0	0.82	0.76	0.69	0.63	0.59
edna	0	0.88	0.78	0.67	0.59	0.54
NAISC	0.98	0.85	0.77	0.67	0.59	0.55
FCAMap	0	0.75	0.72	0.67	0.63	0.61
LogMapLt	0	0.84	0.76	0.66	0.58	0.54
StringEquiv	0	0.88	0.76	0.64	0.55	0.5
Lily	0	0.59	0.6	0.61	0.62	0.63
LPHOM	0.76	0.89	0.71	0.55	0.45	0.4
Alin	0	0.89	0.65	0.46	0.36	0.31
LYAM	0.97	0.48	0.36	0.26	0.21	0.18

Fig. 3. Performance of alignment systems on OAEI 2016 conference track (classes only) (http://oaei.ontologymatching.org/2016/conference/eval.html).

"edas#Paper" and "iasted#Submission" which returned low similarity scores for all direct comparisons. The concepts have relations "edas#Document" $\prec$ "edas#Paper" and "iasted#Document" $\prec$ "iasted#Submission". Comparing their semantic neighbourhoods using *context* rightly identifies the pair as alignment candidates with 0.76 similarity.

Influence of Feature Categories. We dropped feature categories during classification of candidate alignments to analyse how the features influenced performance. Precision and recall values were observed for each group of feature categories as shown in Fig. 4. Previous experiment configurations were reused and performances were based on 10-fold cross-validation on the Conference dataset.

Classification using all features (group 1) was best but only marginally better than dropping the context features (group 7). Context features contributed least to performance and this is further highlighted by weak performance when context features alone (group 4) are used for classification. We attribute the weak

Table 4. Similarity values for some correspondences discovered

Source concept vs Target concept	Similarity approaches			
	hybrid	isub	vsm	context
conference#Paper vs confOf#Paper	**1.0**	**1.0**	**1.0**	0.28
edas#Academic_Event vs ekaw#Scientific_Event	**0.84**	0.61	0.34	0.72
conference#Track-workshop_chair vs ekaw#Workshop_Chair	0.56	**0.91**	0.42	0.25
conference#Conference_document vs ekaw#Document	0.57	0.81	**0.94**	0.33
edas#Paper vs iasted#Submission	0.18	0.0	0.0	**0.76**

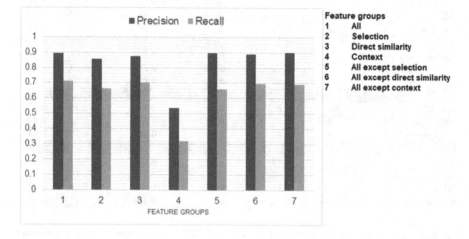

Fig. 4. Impact of excluding features categories.

performance on context features to insufficient data. Analysis showed that only 3% of the candidate alignments identified using context similarity were actual alignments. Accordingly, the classifier model did not learn to effectively use context information due to a significant class imbalance in training data. Also interesting is the slight difference between using direct similarity features alone (group 3) and dropping the direct similarity features (group 6). This suggests that some similarity features are redundant for classifying candidate alignments.

5   Conclusion and Future Work

We introduced a classifier-based approach for ontology alignment which uses a hybrid of string-based similarity features, semantic similarity features, and semantic context features. Word embedding was used to generate semantic features for a random forest classifier in addition to other novel similarity features. Our experiments showed promising results and outperformed previous known approach which incorporates word embedding. Also, comparison with best-performing alignment systems at the OAEI challenge show that it can outperform state-of-the-art systems.

Future work will investigate a systematic determination of similarity thresholds for selecting candidate alignments and how to deal with class imbalances in generating the classifier model. Also, the ability to transfer a trained model to a different domain will be explored. This is particularly useful in the initial stages of alignment where there are no reference alignments with which to generate a classifier model.

Acknowledgement. This work was supported in part by the British Geological Survey (BGS) through the BGS University Funding Initiative (BUFI S291). We are grateful for the valuable comments of our reviewers.

References

1. Breiman, L.: Random forests. Mach. Learn. **45**(1), 5–32 (2001)
2. Cheatham, M., Hitzler, P.: String similarity metrics for ontology alignment. In: Alani, H., et al. (eds.) ISWC 2013, Part II. LNCS, vol. 8219, pp. 294–309. Springer, Heidelberg (2013). https://doi.org/10.1007/978-3-642-41338-4_19
3. Cruz, I.F., Antonelli, F.P., Stroe, C.: AgreementMaker: efficient matching for large real-world schemas and ontologies. Proc. VLDB Endowment **2**(2), 1586–1589 (2009)
4. David, J., Euzenat, J., Scharffe, F., Trojahn dos Santos, C.: The alignment API 4.0. Seman. Web **2**(1), 3–10 (2011)
5. Doan, A., Madhavan, J., Dhamankar, R., Domingos, P., Halevy, A.: Learning to match ontologies on the semantic web. VLDB J. **12**(4), 303–319 (2003)
6. Euzenat, J., Shvaiko, P., et al.: Ontology Matching, vol. 333. Springer, Heidelberg (2007)
7. Gulić, M., Vrdoljak, B., Banek, M.: CroMatcher: an ontology matching system based on automated weighted aggregation and iterative final alignment. Web Seman. Sci. Serv. Agents World Wide Web **41**, 50–71 (2016)
8. Husein, I.G., Akbar, S., Sitohang, B., Azizah, F.N.: Review of ontology matching with background knowledge. In: Data and Software Engineering, pp. 1–6. IEEE (2016)
9. Jain, P., Hitzler, P., Sheth, A.P., Verma, K., Yeh, P.Z.: Ontology alignment for linked open data. In: Patel-Schneider, P.F., et al. (eds.) ISWC 2010, Part I. LNCS, vol. 6496, pp. 402–417. Springer, Heidelberg (2010). https://doi.org/10.1007/978-3-642-17746-0_26
10. Li, J., Tang, J., Li, Y., Luo, Q.: RIMOM: a dynamic multistrategy ontology alignment framework. IEEE Trans. Knowl. Data Eng. **21**(8), 1218–1232 (2009)
11. Li, Y., McLean, D., Bandar, Z.A., O'shea, J.D., Crockett, K.: Sentence similarity based on semantic nets and corpus statistics. IEEE Trans. Knowl. Data Eng. **18**(8), 1138–1150 (2006)
12. Lin, F., Sandkuhl, K.: A survey of exploiting wordNet in ontology matching. In: Bramer, M. (ed.) IFIP AI 2008. ITIFIP, vol. 276, pp. 341–350. Springer, Boston, MA (2008). https://doi.org/10.1007/978-0-387-09695-7_33

13. Martínez-Romero, M., Vázquez-Naya, J.M., Nóvoa, F.J., Vázquez, G., Pereira, J.: A genetic algorithms-based approach for optimizing similarity aggregation in ontology matching. In: Rojas, I., Joya, G., Gabestany, J. (eds.) IWANN 2013, Part I. LNCS, vol. 7902, pp. 435–444. Springer, Heidelberg (2013). https://doi.org/10.1007/978-3-642-38679-4_43

14. Mikolov, T., Chen, K., Corrado, G., Dean, J.: Efficient estimation of word representations in vector space. arXiv preprint arXiv:1301.3781 (2013)

15. Monge, A.E., Elkan, C., et al.: The field matching problem: algorithms and applications. In: KDD, pp. 267–270 (1996)

16. Ngo, D.H., Bellahsene, Z.: YAM++: a multi-strategy based approach for ontology matching task. In: ten Teije, A., et al. (eds.) EKAW 2012. LNCS (LNAI), vol. 7603, pp. 421–425. Springer, Heidelberg (2012). https://doi.org/10.1007/978-3-642-33876-2_38

17. Ngo, D.H., Bellahsene, Z., Todorov, K.: Opening the black box of ontology matching. In: Cimiano, P., Corcho, O., Presutti, V., Hollink, L., Rudolph, S. (eds.) ESWC 2013. LNCS, vol. 7882, pp. 16–30. Springer, Heidelberg (2013). https://doi.org/10.1007/978-3-642-38288-8_2

18. Otero-Cerdeira, L., Rodríguez-Martínez, F.J., Gómez-Rodríguez, A.: Ontology matching: a literature review. Expert Syst. Appl. **42**(2), 949–971 (2015)

19. Shvaiko, P., Euzenat, J.: Ontology matching: state of the art and future challenges. IEEE Trans. Knowl. Data Eng. **25**(1), 158–176 (2013)

20. Stoilos, G., Stamou, G., Kollias, S.: A string metric for ontology alignment. In: Gil, Y., Motta, E., Benjamins, V.R., Musen, M.A. (eds.) ISWC 2005. LNCS, vol. 3729, pp. 624–637. Springer, Heidelberg (2005). https://doi.org/10.1007/11574620_45

21. Sun, Z., Hu, W., Li, C.: Cross-lingual entity alignment via joint attribute-preserving embedding. In: d'Amato, C., et al. (eds.) ISWC 2017, Part I. LNCS, vol. 10587, pp. 628–644. Springer, Cham (2017). https://doi.org/10.1007/978-3-319-68288-4_37

22. Zhang, Y., et al.: Ontology matching with word embeddings. In: Sun, M., Liu, Y., Zhao, J. (eds.) CCL/NLP-NABD-2014. LNCS (LNAI), vol. 8801, pp. 34–45. Springer, Cham (2014). https://doi.org/10.1007/978-3-319-12277-9_4

Domain Adaption in One-Shot Learning

Nanqing Dong[1,2(✉)] and Eric P. Xing[2]

[1] Cornell University, Ithaca, NY 14850, USA
nd367@cornell.edu
[2] Petuum Inc., Pittsburgh, PA 15217, USA

Abstract. Recent advances in deep learning lead to breakthroughs in many machine learning tasks. Due to the data-dri ven nature of deep learning, the training procedure often requires large amounts of manually annotated data, which is often unavailable. One-shot learning aims to categorize the new classes unseen in the training set, given only one example of each new class. Can we transfer knowledge learned by one-shot learning from one domain to another? In this paper, we formulate the problem of domain adaption in one-shot image classification, where the training data and test data come from similar but different distributions. We propose a domain adaption framework based on adversarial networks. This framework is generalized for situations where the source and target domain have different labels. We use a policy network, inspired by human learning behaviors, to effectively select samples from the source domain in the training process. This sampling strategy can further improve the domain adaption performance. We investigate our approach in one-shot image classification tasks on different settings and achieve better results than previous methods. Code related to this paper is available at: https://github.com/NanqingD/DAOSL.

Keywords: One-shot learning · Domain adaption
Adversarial networks · Reinforcement learning
Distance metric learning · Cognitive science

1 Introduction

Convolutional Neural Networks (CNNs) have led significant progress in the domain of computer vision such as image recognition [12], object detection [22] and semantic segmentation [19]. When modern visual recognition systems can benefit from large image datasets like ImageNet [6] and PASCAL VOC [8], deep learning methods still face the obstacle of requiring large amounts of manually annotated data. With the knowledge transfer, humans can tell the difference between up to 30,000 object categories [4]. Especially, children can recognize new objects quickly in their learning phase with proper guidance, even they only see the examples for few times. These motivate the study of one-shot learning, where one annotated example is available for each class to predict. One approach is based on Bayesian statistics. Li et al. [18] proposed a complex framework with

© Springer Nature Switzerland AG 2019
M. Berlingerio et al. (Eds.): ECML PKDD 2018, LNAI 11051, pp. 573–588, 2019.
https://doi.org/10.1007/978-3-030-10925-7_35

strong probabilistic hypothesis using generative object category model and variational Bayesian expectation maximization (VBEM). Another approach is *meta-learning* [27]. Santoro et al. [23] attacked the problem by learning to memorize unseen classes with a Memory Augmented Neural Network (MANN). Ravi and Larochelle [21] utilized a Long Short-Term Memory network (LSTM) [13] as a meta-learner to optimize the learner. There are two challenges in meta-learning approach. The gradient-based optimization usually requires large amounts of labeled data, and the random initiation can have unpredictable effects on the learner. In this work, we focus on a simpler but more efficient approach, the metric-based approach. The metric-based approach projects the raw images into a learned feature space and classifies the image based on a certain distance metric. Due to the simplicity and efficiency, the metric-based approach has been applied in the industry for tasks like face recognition and person re-identification.

The metric-based methods can achieve state-of-the-art performance in one-shot classification tasks, but the accuracy can be easily influenced when the test data comes from a different distribution [24, 29]. Domain adaption means learning a mapping from the source domain to the target domain with the presence of a *shift* between two data distributions, so a predictor trained on the source domain can be applied on the target domain [9, 32]. In our case, a good one-shot learning system can be applied to the target domain with classes unseen in the source domain, just like a student with only basic knowledge in English can differentiate Greek letters with just a glance. In previous domain adaption methods, examples in the source domain are assumed to have equal importance in the training process if there is no prior knowledge. Assume there is a learner wants to learn animals of Canidae family from an incomplete encyclopedia which only includes sections about Felidae and Insecta. Given only a few pictures about Canidae, the learner may find that dogs and cats share more features than bugs. After few trials, the learner should pay more attention to Felidae than Insecta, even though the learner may not have a clear definition of Canidae.

In this paper, we formulate the domain adaption problem in one-shot learning. Fused by recent advances in one-shot learning and domain adaption, we propose an adversarial framework for domain adaption in one-shot learning. We train the one-shot classifier and auxiliary domain discriminator simultaneously. Motivated by the behavior of human learners, we propose to use a policy gradient method [25, 26, 30] to select the samples from the source domain in the training phase, which is different from the traditional random sample selection, By incorporating the reinforced sample selection process in our adversarial framework, we further improve the domain adaption performance in one-shot learning. We also discuss the how the proposed sampling strategy is linked to *distance metric learning* (DML) [31] and *curriculum learning* [3]. The concept is illustrated in Fig. 1. This work focuses on a difficult situation where source domain and target domain do not have any overlap in categories. We investigate our approach in one-shot image classification tasks with different settings. To the best of our knowledge, there is no similar work in either one-shot learning or domain adaption.

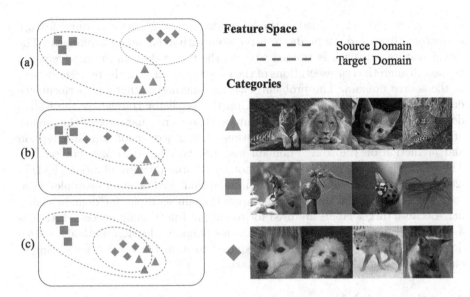

Fig. 1. Illustration of the motivation. Examples are embedded to certain feature spaces under three situations. (a) No domain adaption. (b) Domain adaption with random sample selection. (c) Domain adaption with reinforced sample selection.

2   Related Work

Many works [16,21,23,24,29] have contributed to q-shot learning, here $q > 0$ means the number of labeled examples for the new class unseen in the training set. One-shot learning is an extreme case when there is only one example for each new category. Compared with the Bayesian approach [18] and the meta-learning approach [21,23] in one-shot learning, recently proposed metric-based methods [24,29] achieve state-of-the-art performance with fewer parameters and simpler optimization settings. Given an *episode*, which consists of a query image and a support set of images, a metric-based method computes a certain similarity measure between the embedded query image and each of the embedded support image, and then uses the similarities as weights of a weighted nearest neighbor classifier to predict the label of the query image.

Domain adaption can also be accomplished through adversarial training after Goodfellow et al. first introduced adversarial networks in generative adversarial networks (GANs) [11]. A standard classifier can be decomposed into two parts, a feature extractor, and a label predictor. Domain-adversarial neural network (DANN) proposed a gradient reversal layer to connect an auxiliary domain discriminator with feature extractor for unsupervised domain adaption (UDA). One problem for DANN is that the domain discriminator converges quickly, which can cause the gradient to vanish [28]. Another unsupervised domain adaption method is adversarial discriminative domain adaption (ADDA) [28]. ADDA uses different feature extractors for each domain. The source feature extractor and predictor (classifier) is trained on the source domain first. Then the source feature extractor is fixed and the predictor is replaced with a domain discriminator. The target feature extractor is trained with on the target domain in an adversarial fashion to align the representations of the target domain with the representations of the source domain. The problem with this method is that the performance on target domain is highly dependent on the predictor trained on the source domain [7]. With limited training examples, there is no guarantee of the quality of the predictor. In other words, the optimization objective for domain adaption and prediction on the source domain may not be aligned in one-shot setting. The most related recent work is few-shot adversarial domain adaption (FADA) [20], which focus on supervised domain adaption. FADA pairs examples from source domain with examples from target domain as input for domain classifier. Because target labels are used for pairing in the training process, FADA is a supervised domain adaption. For previous domain adaption methods, source domain and the target domain are required to have the same classes. But in one-shot learning, this constraint is relaxed.

3   Adversarial Domain Adaption with Reinforced Sample Selection

To address the problems listed in Sect. 2, we present our methodology for domain adaption in one-shot learning. Firstly, we formulate the domain adaption problem in metric-based one-shot learning. Secondly, we propose an adversarial domain adaption framework without stage-wise training scheme. Thirdly, we introduce the concept of overgeneralization in domain adaption. Finally, we propose reinforced sample selection as a solution to overgeneralization. The complete pipeline is illustrated in Fig. 2.

3.1   Problem Definition

Given a source domain S as training data and a target domain T as test data, domain adaption learns a mapping between S and T. We denote

$$S = \{(x_1, y_1), ..., (x_{N_S}, y_{N_S})\},$$

where x_i represents an example from S and $y_i \in Y_S$ with $Y_S = \{1, ..., K_S\}$ is the corresponding label. x_i is multi-dimensional, for simplicity, we assume it can be represented as a D-dimension feature vector, $x_i \in \mathbb{R}^D$. We denote

$$T = \{\{(\bar{x}_1, \bar{y}_1), ..., (\bar{x}_t, \bar{y}_t)\}, \{\bar{x}_{t+1}, ..., \bar{x}_{N_T}\}\}f,$$

where $\bar{x}_j$ represents an example from D and $\bar{y}_j \in Y_T$ with $Y_T = \{K_S+1, ..., K_S+K_T\}$. In this paper, we assume $K_S > K_T$ and $N_T \gg t$. We focus on $Y_S \cap Y_T = \emptyset$, which is the most difficult situation for the learner.

A K-way q-shot learning task is defined as: Given q labelled examples for each of K classes that have not been seen before as support set, classifying unlabelled query examples into one of K classes [16,21,23,24,29]. Let f_θ denotes an

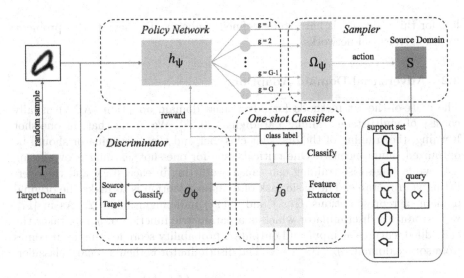

Fig. 2. Illustration of the model architecture. The figure depicts the data flow in the training phase. At the beginning of an episode, a random sample from the target domain goes through the feature extractor and discriminator for the first pass. Then policy network receives the sample and outputs a sampling policy to the sampler. The sampler selects support set and query image from the source domain based on the policy. The one-shot classifier uses the support set and query image to update the feature extractor. The target sample goes through the one-shot classifier with the support set again to calculate the reward. The reward is used to update the policy network. The details are described in Sects. 3.1, 3.2 and 3.4.

embedding function with parameters θ. f_θ embeds the input to a M-dimensional representation, $f_\theta : \mathbb{R}^D \to \mathbb{R}^M$. d denotes a similarity measure function. For $q = 1$ and $k \in \{1, ..., K\}$, with support set $\{(x_k, y_k)\}$ and query example $\mathbf{x}$, the probability of $\mathbf{x}$ belongs to class k is defined as

$$p_\theta(y = k|\mathbf{x}) = \frac{\exp(d(f_\theta(\mathbf{x}), f_\theta(x_k)))}{\sum_{k'=1}^{K} \exp(d(f_\theta(\mathbf{x}), f_\theta(x_{k'})))}. \tag{1}$$

Under this definition, the metric-based one-shot learning problem can be formulated as a standard multiclass classification problem.

In a naive transfer learning setting, the classifier trained on S is finetuned on T to offset the shift between S and T, where t is expected to be much larger than K_T to produce a good result. However, we have $t = K_T$ in one-shot learning. It is not practical to finetune the one-shot classifier with K_T labeled examples. We argue that, in one-shot learning, we can train f_θ on S and use Eq. 1 to predict labels for $\{\bar{x}_{t+1}, ..., \bar{x}_{N_T}\}$ based on $\{(\bar{x}_1, \bar{y}_1), ..., (\bar{x}_t, \bar{y}_t)\}$. f_θ is a projection function and $f_\theta(\bar{x}_j)$ represents the feature vector when $\bar{x}_j$ is projected to some feature space. The objective of domain adaption in one-shot learning can be defined as follows: We want to find the optimal θ, such that $f_\theta(\bar{x}_j)$ has the most discriminative features for a classifier to correctly assign a label to it. The loss for this objective is hard to be defined explicitly. To alleviate this problem, we use adversarial networks.

3.2   Adversarial Domain Adaption

The state-of-the-art methods for adversarial domain adaption (ADA) usually consist of multi-stage training paradigms [20,28]. We argue that, in one-shot learning, the training of the one-shot classifier and the discriminator should be optimized simultaneously. One critical issue for one-shot learning is overfitting [24], while stage-wise training can cause overfitting in each stage and the overfitting is intractable. The basic task of domain adaption is to make the original domain of representations $f_\theta(x_i)$ and $f_\theta(\bar{x}_j)$ indistinguishable [2]. As in [11], we introduce a discriminator which is a parametric function of g_ϕ. g_ϕ takes the embedded features as input and outputs a probability score for the input comes from source domain, $g_\phi : \mathbb{R}^M \to \mathbb{R}$. The discriminator is then a binary classifier,

$$p_\phi(y = 1|f_\theta(x_i)) = \frac{\exp(g_\phi(f_\theta(x_i)))}{1 + \exp(g_\phi(f_\theta(x_i)))}, \tag{2}$$

$$p_\phi(y = 0|f_\theta(\bar{x}_j)) = \frac{1}{1 + \exp(g_\phi(f_\theta(\bar{x}_j)))}. \tag{3}$$

The one-shot classifier and the domain discriminator are optimized alternatively.

Given fixed f_θ, g_ϕ is optimized to maximize the probability of correctly differentiating $f_\theta(\bar{x}_j)$ from $f_\theta(x_i)$. The binary cross entropy loss is defined as

$$J_\phi = -\frac{1}{B_S} \sum_i \log(p_\phi(y = 1|f_\theta(x_i))) - \frac{1}{B_T} \sum_j \log(1 - p_\phi(y = 0|f_\theta(\bar{x}_j))), \tag{4}$$

where B_S is the batch size of samples from S and B_T is the batch size of samples from T for the discriminator updating step.

Given fixed g_ϕ, f_θ is optimized to achieve two goals at the same time. Firstly, we want to train a one-shot classifier on S which can assign the correct label for each query example. The multiclass cross entropy loss is defined as

$$J_{cls} = -\frac{1}{B_S} \sum_i \sum_k y_k \log(p_\theta(y = k|x_i)), \tag{5}$$

where B_S is the batch size of samples from S and y_k is binary, denoting whether class label k is the correct classification for x_i. Secondly, we want to use the embedding function to project examples from both S and T to a feature space that S and T have high similarity. Adversarial networks transform the original problem of how to maximize the similarity between S and T into how to make them indistinguishable. So f_θ is also trained to make discriminator to assign the wrong labels to $f_\theta(\bar{x}_j)$, where the optimization goal is to maximize the loss that $f_\theta(\bar{x}_j)$ is classified from T. Following the practice of [11], the optimization

Algorithm 1. Training an episode of adversarial domain adaption in a K way one-shot learning task, where $K \leq K_S$ and $K \leq K_T$. sample() denotes a function that samples fixed number of elements from a set. sampleSupport() denotes a function that samples the support set of image and label pairs from S and sampleQuery() denotes a function that samples the query image and label from S based on support set, same as [29, 24]. All samples are sampled uniformly without replacement.

Input : S, T, θ, ϕ
Output: θ, ϕ
$Support \leftarrow \{\}$
$Query \leftarrow \{\}$
$Target \leftarrow \text{sample}(\{\bar{x}_j\}, B_T)$
for $b \in \{1, ..., B_S\}$ **do**
   | $\quad support \leftarrow \text{sampleSupport}(S)$; Add $support$ to $Support$
   | $\quad query \leftarrow \text{sampleQuery}(S, support)$; Add $query$ to $Query$
end
Calculate J_ϕ with $Support$, $Query$ and $Target$ by (4)
Update ϕ by minimizing J_ϕ
Calculate J_{cls} with $Support$ and $Query$ by (5)
Calculate J_{adv} with $target$ by (6)
Update θ by minimizing J_θ

---

problem can also been seen as minimization of loss that $f_\theta(\bar{x}_j)$ is classified from S, so this adversarial loss can be defined as

$$J_{adv} = -\frac{1}{B_T} \sum_j \log(p_\phi(y = 1 | f_\theta(\bar{x}_j))), \qquad (6)$$

where B_T is the batch size of samples from T. The total loss for classifier updating step is then

$$J_\theta = J_{cls} + \lambda_{adv} J_{adv}, \qquad (7)$$

where λ_{adv} is a weight for adversarial loss. The training for adversarial domain adaption in one-shot learning is illustrated in Algorithm 1. Note, $\{\bar{y}_j\}$ is not used in the optimization for either θ or ϕ, so adversarial domain adaption in one-shot learning is unsupervised.

3.3 Overgeneralization

Generalization is an important ability for humans and animals to acquire knowledge in one circumstance and apply the knowledge to new situations [10]. In contrast, discrimination is the ability to discriminate different stimuli. Humans can not memorize all the discriminative features with limited memory. Generalization can help humans to save memory in the learning process. Domain adaption in one-shot learning can be seen as a mixture of generalization and discrimination. In this study, we observe a phenomenon that the learner learns too much in S and performs worse on T. We call this phenomenon overgeneralization for domain adaption in one-shot learning.

Overgeneralization can be caused by the misaligned optimization objectives. The learner's goal is to accurately classify the examples from T, while ADA tries to minimize the distance between the distributions of S and T in a projected space [2,11]. There is no supervision for T, thus the extracted features are dependent on S. With limited memory, the learner memorizes more generalized features from S but misses the features that are most discriminative for T, especially when $K_S \gg K_T$. Previous methods [20,28] have shown that ADA performs well when S and T share same categories. One solution is then to find a subset of S, so the distance between the distributions of the subset and T is minimal. Note this subset selection problem is not convex or differentiable. We present our solution, reinforced sample selection.

3.4   Reinforced Sample Selection

Random sample selection has been widely used in many machine learning tasks to reduce variance and avoid overfitting. In supervised learning, more examples usually help the learner to grasp more discriminative features. However, the large sample size of S may not help domain adaption in one-shot learning because S and T can have totally different categories. The unsupervised domain adaption problem is intractable since there are no labels from T. The minimization of J_{cls} can be seen as a regularization of f_θ to learn useful features for one-shot learning task on S. However, there is no guarantee for the performance of T.

We propose to train the learner to learn the sampling strategy through reinforcement learning, which is in contrast to typical random sample selection. In the domain adaption process, the learning system actively selects samples from S when it sees an image from T. To accomplish this, we introduce a policy network to select the categories from S. In each episode, the support set and query image will be sampled from this selected categories. Be more specific, given an image from T, the policy network will output a policy for the sampler, and the sampler will sample examples from a subset of S. The examples sampled from the subset of S are used to train the one-shot classifier and the domain discriminator. Given $\bar{\mathbf{x}}$ from T, assume there are $(x_{sim}, y_{sim}) \in S$ and $(x_{dis}, y_{dis}) \in S$, where $y_{sim} \neq y_{dis}$. Here, sim means x_{sim} and $\bar{\mathbf{x}}$ are similar because they share some attributes in the semantic feature space, e.g. a cat and a dog both have four legs and fur. dis means x_{dis} and $\bar{\mathbf{x}}$ are not similar. Mathematically, f_θ trained in this way should make $f_\theta(\bar{\mathbf{x}})$ close to $f_\theta(x_{sim})$ and distant to $f_\theta(x_{dis})$ in a projected feature space, even without the label information from T. The illustration is presented in Fig. 1(c). We call this sampling mechanism reinforced sample selection (RSS). Since we output one sampling policy at once, RSS is actually a single-step Markov Decision Process [25].

The policy network is parameterized with ψ, denoted as h_ψ. We have $h_\psi : \mathbb{R}^D \to \mathbb{R}^G$, where G is the number of disjoint subsets of S. An intuitive design is to make sampling decision for each category independently, which can be implemented by G independent Bernoulli distributions. The support set is then sampled from the selected categories. However, there are two problems with this design: (1) The number of possible combinations of the selected categories is huge for large G (for $G = 10$, we have $2^{10} > 10^3$); (2) For each combination, there is a large variety which is uncontrollable. Here, constrained by the computational power, we simplify the problem by making the subsets mutually exclusive. Ideally, $G = K_S$, but considering computational complexity when $K_S \gg K_T$, in practice, we can utilize the side information (e.g. superclass), or clustering to G groups through a preprocessing step [31]. For $\bar{\mathbf{x}} \in \{\bar{x}_j\}$, $h_\psi(\bar{\mathbf{x}})$ is a G elements vector. Let $g \in \{1, ..., G\}$, we define

$$p_\psi(y = g|\bar{\mathbf{x}}) = \frac{\exp(h_\psi(\bar{\mathbf{x}})[g])}{\sum_{g'=1}^{G} \exp(h_\psi(\bar{\mathbf{x}})[g'])}, \tag{8}$$

where $[n]$ represents the nth element of a vector. We will decide whether or not to sample from group g based on a multinomial distribution with probabilities $\{p_\psi(y = g|\bar{\mathbf{x}})| \ \forall \ g \in \{1, ..., G\}\}$, the sampling policy is denoted as $\Omega_\psi(\bar{\mathbf{x}})$.

Another key component in reinforcement learning is setting the proper reward. With Euclidean distance defined on f_θ, the optimization objective of f_θ can be formulated as

$$\min \|f_\theta(\bar{\mathbf{x}}) - f_\theta(x_{sim})\|^2, \ \max \|f_\theta(\bar{\mathbf{x}}) - f_\theta(x_{dis})\|^2. \tag{9}$$

This can be further generalized as a deep DML problem with proper constraints [31]. However, the set of *sim* and the set of *dis* are not defined in most situations, and we can not solve the problem directly. Alternatively, we utilize the one-shot classifier. In an episode of a K-way one-shot learning task, we select the subset of S according to $\Omega_\psi(\bar{\mathbf{x}})$ before sampling the support set and query image. After θ and ϕ are updated as in Algorithm 1, if the one-shot classifier correctly predicts the class label for the query image, then we replace the query image with the target image. We perform a one-shot classification with the original support set and updated query image. Note, the label of the query image is still the original label since we do not have the label for the target image. We want to see if the target image can confuse the one-shot classifier. The one-shot classifier is based on nearest neighbor search. If the target query image can be correctly classified, the target image is "close" to the corresponding image in the projected feature space. The reward is defined as

$$R(\Omega_\psi(\bar{\mathbf{x}})) = \begin{cases} 1 & \text{if correct,} \\ -\gamma & \text{otherwise.} \end{cases} \tag{10}$$

where γ is a small positive number. Since $K_S \gg K_T$, the reward will be sparse. In practice, given a support set, we choose to accumulate the reward by repeating the sampling operation for all the possible classes of query images. In other words, after the support set is sampled, we sample the query images for all K classes and for each class, we replace the query image with the target image to perform a one-shot classification. The reward of each query class is added up to calculate the total reward for the sampling action.

Algorithm 2. Training an episode of adversarial domain adaption with reinforced sample selection for a K way one-shot learning task, where $K \leq K_S$ and $K \leq K_D$. The settings and notations are the same as Algorithm 1.

Input : S, T, θ, ϕ, ψ
Output: θ, ϕ, ψ
$Query \leftarrow \{\}$
$Target \leftarrow \text{sample}(\{\bar{x}_j\}, 1)$
Sample $Support$ from S according to $\Omega_\psi(Target)$
for $q \in \{1, ..., K\}$ **do**
| $query \leftarrow \text{sampleQuery}(S, support)$; Add $query$ to $Query$
end
Calculate J_ϕ with $Support$, $Query$ and $Target$ by (4)
Update ϕ by minimizing J_ϕ
Calculate J_{cls} with $Support$ and $Query$ by (5)
Calculate J_{adv} with $Target$ by (6)
Update θ by minimizing J_θ
Calculate J_{pn} with $Support$ and $Target$ by (11)
Update ψ by minimizing J_{pn}

The policy network is trained to maximize the expected reward $\mathbb{E}_{\Omega_\psi}[R]$. We define the loss for policy network as the negative expected reward

$$J_{pn} = -\mathbb{E}_{\Omega_\psi}[R(\Omega_\psi(\bar{\mathbf{x}}))]. \tag{11}$$

The J_{pn} or expected reward can be optimized by policy gradient, based on the REINFORCE rule [30]. The expected gradient is

$$\frac{\partial}{\partial \psi} J_{pn} = -\mathbb{E}_{\Omega_\psi}[R(\Omega_\psi(\bar{\mathbf{x}})) \frac{\partial}{\partial \psi} \log(p(\Omega_\psi(\bar{\mathbf{x}}))] \tag{12}$$

where $\log(p(\Omega_\psi(\bar{\mathbf{x}}))$ means the log probability of sampled policy Ω_ψ when the target image is $\bar{\mathbf{x}}$. Ω_ψ is a multinominal distributions with G possible events, the probability mass function thus can be written as

$$p(\Omega_\psi(\bar{\mathbf{x}})) = \prod_{g=1}^{G} p_\psi(y = g|\bar{\mathbf{x}})^{\mathbf{1_g}}, \tag{13}$$

$\mathbf{1_g}$ is an indicator function indicates whether group g is selected by $\Omega_\psi(\bar{\mathbf{x}})$, $\sum_g \mathbf{1_g} = 1$. RSS can be incorporated into Algorithm 1 with moderate modification. The updated algorithm is illustrated in Algorithm 2.

It is worth noting that RSS can be linked to curriculum learning. Similar to a curriculum, the entry-level courses can give a student general information about the field of study, which is easy to learn. The advanced courses have narrower topics but provide more details, and they are difficult to learn. When h_ψ is randomly initialized, the sampling strategy, similar to random selection, can help f_θ learn more general information. As the learning proceeds, the sampling strategy learned by h_ψ can focus on certain category groups and extract more domain-specific features, thus achieving better domain adaption performance. Similar to trial and error of human learners, RSS updates ψ and adjusts the output policy iteratively. By focusing on more relevant data and neglecting noise, RSS can also be interpreted as a weighted sampling method. A large probability for certain category means there is a higher chance that the category is sampled. The policy network learns the *attention* to category.

4 Experiments

4.1 Basic Settings

Dataset. Hand-written character recognition has been used to evaluate the machine learning algorithms in many works [16,21,23,24,29]. We use Omniglot [17] as the source domain and EMNIST [5] as the target domain (Fig. 3). Omniglot contains 1623 different characters from 50 different languages. Each character is written by 20 different people. Each image has a resolution of 105×105. Because the characters of *Latin* is identical to the characters of *English*, we remove *Latin*. The modified Omniglot has 1597 classes. EMNIST consists of 10 digits, 26 English letter with both uppercase and lowercase. There are 62 classes in total. Each image has a resolution of 28×28. We randomly select 20, 50 and 100 examples for each class to make a balanced subset of EMNIST. All the images are resized to 28×28, same as [21,24,29].

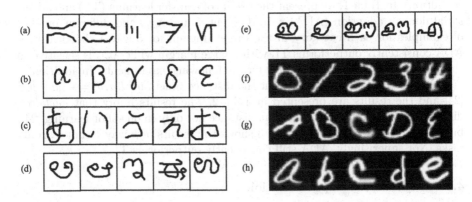

Fig. 3. Examples for hand-written character. Omniglot: (a) Aurebesh (an invented *Star Wars* language); (b) Greek; (c) Japanese (Hiragana); (d) Kannada; (e) Malayalam. EMNIST: (f) digits; (g) English (upper-case); (h) English (lower-case).

Implementation. There is no previous work for domain adaption in one-shot learning. A naive baseline model is training one-shot classifier on the source domain and applying it directly to the target domain, which is the standard transfer learning. We choose Matching Networks (MN) [29] as the backbone architecture. We use Adam optimizer [15] for all experiments. The learning rate is fixed to 10^{-4} for the one-shot classifier, the domain discriminator and the policy network. We choose $B_S = 1$ and $B_T = 1$. We stop the training for all experiments train the one-shot classifier for 100 epochs which each epoch consists of 2000 episodes. The experiments are implemented in TensorFlow framework [1] on a GTX Titan X GPU.

Evaluation. The evaluation follows the protocol of [29]. The evaluation metric is the standard mean accuracy. In the evaluation phase, the support set and the query image are randomly sampled 10000 times. It is worth mentioning that [24,29] only report the accuracy as a single number. With the fixed checkpoints of the model, we repeat the evaluation 100 times to produce the mean and the standard deviation of the accuracy.

4.2   Adversarial Domain Adaption

f_θ is a CNN feature extractor with four identical modules. Same as MN, each module is a sequential operations of two 3×3 convolutions with batch normalization [14] and ReLU, and one 2×2 max-pooling. The number of filters for the four modules are all 64. f_θ is followed by a metric-based non-parametric classifier defined in Eq. 1. Here $d(a, b) = cos(a, b)$. g_ϕ consists of 3 fully-connected layers with number of outputs 64, 64 and 2. For this experiment, the domains differ in the content (language and writing styles) and image quality (color and resolution). In Table 1, we present the results of transfer learning (TL) and adversarial domain adaption (ADA) with different values λ_{adv} and different number of examples for each category in the target domain for 5-way 1-shot learning. ADA consistently outperforms TL with a large margin while λ_{adv} is sensitive to the target data. We fix $\lambda_{adv} = 10^{-3}$ for the following experiments. In addition to $d(a, b) = cos(a, b)$, we try a different the distance functions $d(a, b) = -||a - b||^2$ [24] and the results are presented in Table 2. The results imply that *euclidean* distance is more suitable than *cosine* distance not only in one-shot learning [24], but also in domain adaption. Table 3 shows the results of k-way 1-shot learning tasks. Not superisingly, ADA still consistently outperforms TL.

4.3   Reinforced Sample Selection

Without losing the generality of the proposed method, we simulate the experiments for RSS in the task of hand-written character recognition. The simulated experiments can be easily extended to general object recognition. h_ψ has similar architecture as f_θ, where the four identical modules are followed by one fully-connected layer with number of outputs G. In this experiment, we simulate the *ideal* situation, where the source domain can be split into two disjoint subset, i.e. $G = 2$. There are one *sim* set and one *dis* set, as discussed in Sect. 3.4. Considering the huge computational cost for large K_S, we shrink both

Table 1. Sensitivity of λ_{adv} of ADA in 5-Way 1-Shot Learning. n_T means the number of examples for each category in the target domain.

Model	$n_T = 20$	$n_T = 50$	$n_T = 100$
Random Guess	20	20	20
TL	42.35 ± 0.51	47.48 ± 0.53	47.35 ± 0.47
ADA ($\lambda_{adv} = 1$)	46.23 ± 1.58	49.80 ± 1.68	52.70 ± 1.51
ADA ($\lambda_{adv} = 10^{-1}$)	51.30 ± 1.77	47.55 ± 1.73	51.06 ± 1.56
ADA ($\lambda_{adv} = 10^{-2}$)	52.58 ± 1.55	47.54 ± 1.65	48.58 ± 1.62
ADA ($\lambda_{adv} = 10^{-3}$)	53.73 ± 1.59	48.46 ± 1.54	48.32 ± 1.34
ADA ($\lambda_{adv} = 10^{-4}$)	49.98 ± 1.32	52.96 ± 1.43	48.67 ± 1.54

Table 2. Sensitivity of $d(\cdot, \cdot)$ in ADA in 5-Way 1-Shot Learning.

Model	acc(%) $\pm$ std (%)
TL (*cosine*)	42.35 ± 0.51
ADA (*cosine*)	53.73 ± 1.59
TL (*euclidean*)	50.07 ± 0.45
ADA (*euclidean*)	55.87 ± 1.53

Table 3. ADA in k-Way 1-Shot Learning.

Model	5-way	10-way	20-way
Random Guess	20	10	5
TL	42.35 ± 0.51	31.09 ± 0.45	21.62 ± 0.43
ADA	53.73 ± 1.59	35.22 ± 1.62	23.30 ± 1.29

the source domain and target domain. For the target domain, we only use the capital English characters of EMNIST, i.e. $K_T = 26$. For the source domain, we only select 16 languages with 596 characters in total, i.e. $K_S = 596$. The set of *sim* contains 256 characters from *Anglo-Saxon* (Futhorc), *Armenian*, *Asomtavruli* (Georgian), *Cyrillic*, *Greek*, *Hebrew*, *Latin*, and *Mkhedruli* (Georgian). Because we do not involve the lower-case characters of EMNIST, *Latin* is added back to the source domain. The set of *dis* contains 340 characters from *Balinese, Bengali, Grantha, Gujarati, Gurmukhi, Kannada, Malayalam, Oriya*. See Fig. 3 for an intuitive illustration of the *sim* set, the *dis* set and the target domain. The results are presented in Table 4. Intuitively, training on more training data usually leads to better performance on the test data in the supervised setting, where the training data and the test data have the same categories. However, it may not be true in unsupervised setting, as discussed in Sect. 3.4. In the 5-way 1-shot learning task, we get an opposite result that training on the subset which is similar to the test data produce better results, in both TL and ADA. As discussed in Sect. 3.3, this may caused by overgeneralization. The same phenomenon can also be observed in the ADA of 10-way 1-shot learning task. RSS can even achieve better results than ADA trained on *sim*. RSS tries to utilize the source domain maximally and achieve a balance between the generalization and discrimination.

Table 4. RSS in k-Way 1-Shot Learning.

Model	5-way	10-way
TL	41.17 ± 0.51	36.19 ± 1.42
TL (*dis*)	45.78 ± 1.68	32.23 ± 1.30
TL (*sim*)	46.34 ± 1.34	30.50 ± 1.41
ADA	45.02 ± 1.43	33.99 ± 1.36
ADA (*dis*)	43.11 ± 1.51	30.35 ± 1.49
ADA (*sim*)	51.20 ± 1.64	37.31 ± 1.51
RSS	52.31 ± 1.63	38.41 ± 1.48

5   Conclusions

In this paper, we study the problem of domain adaption in one-shot learning. We review and compare the recent studies in one-shot learning and adversarial domain adaption. We formulate the problem of domain adaption in metric-based one-shot image classification. We propose an adversarial framework and investigate the limitations of adversarial training. Motivated by human learning, we introduce a new sampling strategy called reinforced sample selection to improve the domain adaption performance. We acknowledge that the improvements can be made to the reinforcement learning setting and optimization procedure. Domain adaption in one-shot learning and using reinforcement learning in domain adaption are both underdeveloped. In this work, we have the first trial in this area based on the cognitive science concepts and use experiments to validate the proposed framework. In the future, we will work more on the theoretical analysis of domain adaption in one-shot learning.

References

1. Abadi, M., et al.: Tensorflow: a system for large-scale machine learning. In: Proceedings of the 12th USENIX Conference on Operating Systems Design and Implementation, pp. 265–283. USENIX Association (2016)
2. Ben-David, S., Blitzer, J., Crammer, K., Pereira, F.: Analysis of representations for domain adaptation. In: Advances in Neural Information Processing Systems, pp. 137–144 (2007)

3. Bengio, Y., Louradour, J., Collobert, R., Weston, J.: Curriculum learning. In: In: International Conference on Machine Learning, pp. 41–48. ACM (2009)
4. Biederman, I.: Recognition-by-components: a theory of human image understanding. Psychol. Rev. **94**(2), 115 (1987)
5. Cohen, G., Afshar, S., Tapson, J., van Schaik, A.: Emnist: an extension of mnist to handwritten letters. arXiv preprint arXiv:1702.05373 (2017)
6. Deng, J., Dong, W., Socher, R., Li, L.J., Li, K., Li, F.F.: Imagenet: a large-scale hierarchical image database. In: Proceedings of the IEEE Conference on Computer Vision and Pattern Recognition, pp. 248–255 (2009)
7. Dong, N., Kampffmeyer, M., Liang, X., Wang, Z., Dai, W., Xing, E.: Unsupervised domain adaptation for automatic estimation of cardiothoracic ratio. In: Frangi, A.F., Schnabel, J.A., Davatzikos, C., Alberola-López, C., Fichtinger, G. (eds.) MICCAI 2018. LNCS, vol. 11071, pp. 544–552. Springer, Cham (2018). https://doi.org/10.1007/978-3-030-00934-2_61
8. Everingham, M., Van Gool, L., Williams, C.K., Winn, J., Zisserman, A.: The pascal visual object classes (VOC) challenge. Int. J. Comput. Vis. **88**(2), 303–338 (2010)
9. Ganin, Y., et al.: Domain-adversarial training of neural networks. J. Mach. Learn. Res. **17**(1), 2030–2096 (2016)
10. Gluck, M.A., Mercado, E., Myers, C.E.: Learning and memory: From brain to behavior
11. Goodfellow, I., et al.: Generative adversarial nets. In: Advances in Neural Information Processing Systems, pp. 2672–2680 (2014)
12. He, K., Zhang, X., Ren, S., Sun, J.: Deep residual learning for image recognition. In: Proceedings of the IEEE Conference on Computer Vision and Pattern Recognition, pp. 770–778 (2016)
13. Hochreiter, S., Schmidhuber, J.: Long short-term memory. Neural Comput. **9**(8), 1735–1780 (1997)
14. Ioffe, S., Szegedy, C.: Batch normalization: accelerating deep network training by reducing internal covariate shift. In: In: International Conference on Machine Learning, pp. 448–456 (2015)
15. Kingma, D.P., Ba, J.: Adam: A method for stochastic optimization. In: International Conference on Learning Representations (2015)
16. Koch, G., Zemel, R., Salakhutdinov, R.: Siamese neural networks for one-shot image recognition. In: International Conference on Machine Learning Deep Learning Workshop (2015)
17. Lake, B., Salakhutdinov, R., Gross, J., Tenenbaum, J.: One shot learning of simple visual concepts. In: Proceedings of the Annual Meeting of the Cognitive Science Society, vol. 33 (2011)
18. Li, F.F., Fergus, R., Perona, P.: One-shot learning of object categories. IEEE Trans. Pattern Anal. Mach. Intell. **28**(4), 594–611 (2006)
19. Chen, L.C., Papandreou, G., Kokkinos, I., Murphy, K., Yuille, A.: Semantic image segmentation with deep convolutional nets and fully connected CRFs. In: International Conference on Learning Representations (2015)
20. Motiian, S., Jones, Q., Iranmanesh, S., Doretto, G.: Few-shot adversarial domain adaptation. In: Advances in Neural Information Processing Systems, pp. 6673–6683 (2017)
21. Ravi, S., Larochelle, H.: Optimization as a model for few-shot learning. In: In: International Conference on Learning Representations (2017)
22. Ren, S., He, K., Girshick, R., Sun, J.: Faster R-CNN: Towards real-time object detection with region proposal networks. In: Advances in Neural Information Processing Systems, pp. 91–99 (2015)

23. Santoro, A., Bartunov, S., Botvinick, M., Wierstra, D., Lillicrap, T.: Meta-learning with memory-augmented neural networks. In: International Conference on Machine Learning, pp. 1842–1850 (2016)
24. Snell, J., Swersky, K., Zemel, R.: Prototypical networks for few-shot learning. In: Advances in Neural Information Processing Systems, pp. 4077–4087 (2017)
25. Sutton, R.S., Barto, A.G.: Reinforcement learning: an introduction. IEEE Trans. Neural Netw. **9**(5), 1054–1054 (1998)
26. Sutton, R.S., McAllester, D.A., Singh, S.P., Mansour, Y.: Policy gradient methods for reinforcement learning with function approximation. In: Advances in Neural Information Processing Systems, pp. 1057–1063 (2000)
27. Thrun, S.: Lifelong learning algorithms. In: Thrun, S., Pratt, L. (eds.) Learning to Learn, pp. 181–209. Springer, Boston (1998). https://doi.org/10.1007/978-1-4615-5529-2_8
28. Tzeng, E., Hoffman, J., Saenko, K., Darrell, T.: Adversarial discriminative domain adaptation. In: Proceedings of the IEEE Conference on Computer Vision and Pattern Recognition, pp. 2962–2971. IEEE (2017)
29. Vinyals, O., Blundell, C., Lillicrap, T., Wierstra, D., et al.: Matching networks for one shot learning. In: Advances in Neural Information Processing Systems, pp. 3630–3638 (2016)
30. Williams, R.J.: Simple statistical gradient-following algorithms for connectionist reinforcement learning. Mach. Learn. **8**, 229–256 (1992)
31. Xing, E.P., Jordan, M.I., Russell, S.J., Ng, A.Y.: Distance metric learning with application to clustering with side-information. In: Advances in Neural Information Processing Systems, pp. 521–528 (2003)
32. Zhang, K., Schölkopf, B., Muandet, K., Wang, Z.: Domain adaptation under target and conditional shift. In: International Conference on Machine Learning, pp. 819–827 (2013)

Ensemble Methods

Axiomatic Characterization of AdaBoost and the Multiplicative Weight Update Procedure

Ibrahim Alabdulmohsin[(⊠)]

Business Intelligence Division, Saudi Aramco, Dhahran 31311, Saudi Arabia
`ibrahim.alabdulmohsin@kaust.edu.sa`

Abstract. AdaBoost was introduced for binary classification tasks by Freund and Schapire in 1995. Ever since its publication, numerous results have been produced, which revealed surprising links between AdaBoost and related fields, such as information geometry, game theory, and convex optimization. This remarkably comprehensive set of connections suggests that adaBoost is a *unique* approach that may, in fact, arise out of *axiomatic* principles. In this paper, we prove that this is indeed the case. We show that three natural axioms on adaptive re-weighting and combining algorithms, also called arcing, suffice to construct adaBoost and, more generally, the multiplicative weight update procedure as the unique family of algorithms that meet those axioms. Informally speaking, our three axioms only require that the arcing algorithm satisfies some elementary notions of additivity, objectivity, and utility. We prove that any method that satisfies these axioms must be minimizing the composition of an exponential loss with an additive function, and that the weights must be updated according to the multiplicative weight update procedure. This conclusion holds in the general setting of learning, which encompasses regression, classification, ranking, and clustering.

Keywords: Ensemble methods · Boosting · AdaBoost · Axioms

1 Introduction

In an axiomatic treatment, the goal is to formalize broad intuitive notions into precise mathematical terms, called *axioms*. Often, a collection of axioms would pinpoint unequivocally to a *unique* solution but, in some striking cases, it could result in an *impossibility theorem*, where it is concluded that no solution could possibly satisfy all of the postulated axioms. In both instances, an in-depth insight is provided by axiomatization. Unfortunately, despite the fact that axiomatic methods are quite abundant in related fields, developing an axiomatic basis for artificial intelligence, in general, and machine learning, in particular, is not a common practice today.

Consider, for the sake of comparison, the closely-related field of information theory. In information theory, the Shannon entropy alone has been characterized

© Springer Nature Switzerland AG 2019
M. Berlingerio et al. (Eds.): ECML PKDD 2018, LNAI 11051, pp. 591–604, 2019.
https://doi.org/10.1007/978-3-030-10925-7_36

axiomatically via, at least, seven different approaches [2,11,22,33]. These different axiomatizations are built on intuitive notions, such as continuity, additivity, monotonicity, recursivity and symmetry. Similarly, the maximum-entropy method has been axiomatized in various ways [10,34,35]. Moreover, axiomatic characterizations have been developed for the Kullback-Leibler divergence by Kannappan and Ng, for the Rényi entropies by Daróczy, and for the broad class of f-divergences by Csiszár [11,24].

In machine learning, on the other hand, only a few areas within such a broad discipline have witnessed some axiomatic treatments. These include clustering, rank aggregation, Bayesian inference, and collaborative filtering. Even more, in some of those few axiomatic approaches, the conclusion turned out to be implicitly imposed in the axioms themselves, thus significantly weakening the implications of their results. For instance, [3] proposed an axiomatic approach for defining *relevance* in feature subset selection. Because the two axioms of [3] required that the *mutual information* between the instance and the target be preserved and that the description length be minimized, it is not, perhaps, quite surprising that the proposed definition of relevance was expressed in terms of the mutual information itself. Similarly, the axiomatic basis of [17] for clustering algorithms identified *single-linkage* as the only clustering method that met those axioms. However, as pointed out by [1], this is arguably a consequence of the fact that one of the axioms used in [17] was, implicitly, the optimization objective of that particular clustering algorithm.

Nevertheless, there are celebrated stories of successful axiomatic characterizations in machine learning. One of the most prominent examples is the axiomatization of Bayesian inference using Cox's theorem. In 1946, Cox established that the laws of probability were the *only* method of manipulating one's degrees of belief in a manner that was both consistent and agreed with common sense [9]. Jaynes, in his posthumous book *"Probability Theory: The Logic of Science"*, viewed this theorem as the cornerstone of the Bayesian interpretation of probability theory [18].

A second area in machine learning that has received a fair amount of axiomatic treatment is clustering. In [21], Kleinberg developed an impossibility theorem, showing that no clustering function could achieve scale-invariance and richness, while also simultaneously satisfying a third condition, which he called "consistency". This negative result was interpreted as a formal proof of the ill-defined nature of the clustering task. However, while the first two of Kleinberg's axioms were quite natural, the third axiom was, in fact, quite strong. One, arguably, more natural approach would be to state the third axiom in terms of the "refinement" of the partition under "transformations", rather than requiring the partitions to be identical, *per se*, or, at least, to fix the number of clusters in advance. Indeed, using similar arguments, a different axiomatization was derived in [1] that captured the same principles of Kleinberg while sidestepping his impossibility result.

One particular axiomatic tool that has found many applications in machine learning are the axioms of social choice theory. In [26], an analog to the celebrated Arrow's impossibility theorem was derived for combining the predictions of weak learners in ensemble methods in the multiclass setting, and axiomatic characterizations of weighted averaging and majority voting were provided as well. Similarly, [27] derived analogs of the axioms of social choice theory for rank aggregation, while [25] derived analogs for collaborative filtering.

In this paper, we develop an axiomatic characterization of boosting algorithms that operate via adaptive re-weighting and combining, which are also called arcing methods [5]. By requiring that these procedures satisfy three elementary notions of additivity, objectivity and utility, we prove that adaBoost and its variants are the *unique* family of boosting methods that meet those axioms. More precisely, we prove that any boosting procedure that satisfies these axioms must also be minimizing the composition of an exponential loss with an additive function and that the weights must be updated according to the multiplicative weight update procedure. This conclusion holds even in the general setting of learning [32,36], which encompasses regression, classification, ranking, and clustering, despite the fact that *weak learnability* is hard, if not impossible, to define in such a broad setting. To the best of our knowledge, this is the first axiomatic treatment of boosting in the literature.

Definition 1 (General Setting of Learning). *In the general setting of learning, we have a hypothesis space $\mathcal{H}$ and a stochastic loss $l : \mathcal{H} \to \mathbb{R}$. The learner is provided with m realizations of the stochastic loss $S = \{l_1, \ldots, l_m\}$ drawn i.i.d. from some unknown probability measure $\mathcal{D}$. The goal is to select a hypothesis $h \in \mathcal{H}$ according to the sample S such that the expected risk $\mathbb{E}_{l \sim \mathcal{D}}[l(h)]$ is small.*

There are several families of boosting algorithms that have been proposed in the literature. Two prominent families include gradient boosting [16,23], of which *anyBoost* is a prominent example, and the adaptive re-weighting and combining (arcing) procedure [5], which includes algorithms such as *adaBoost* [14], arc-x4 [5], and the averaging algorithm in [19]. In this paper, we focus on the latter class of algorithms. We will show that while, in principle, many adaptive re-weighting and combining schemes can be devised, as argued by Breiman in [5], there exists a *unique* method that satisfies some natural axioms. This unique approach coincides with adaBoost in the binary classification setting.

AdaBoost was introduced for binary classification by Freund and Schapire in [14]. Ever since its publication, numerous results have been produced, which revealed surprising links between AdaBoost and related fields, such as information geometry, game theory, and convex optimization [29]. This remarkably comprehensive set of connections suggests that adaBoost is a fundamental approach that may, in fact, arise out of *axiomatic* principles. We establish in this paper that this is indeed the case. In the literature, several variants of adaBoost have been proposed that minimize different loss functionals, such as the logistic loss, the hinge loss, the square loss, and so on [23,29,37,39]. Our axiomatic characterization establishes in what sense do those algorithms essentially differ from adaBoost.

AdaBoost is, in turn, a particular instance of a more general learning strategy, called the Multiplicative Weight Update procedure [29]. The Multiplicative Weight Update procedure arises in game theory as a learning algorithm in mixed games with repeated play. Similarly, it arises in the online prediction setting as a generalization of the Weighted Majority Algorithm. As will be clear later, the axioms we present for AdaBoost can also serve as axioms for the Multiplicative Weight Update procedure.

2   The Boosting Framework

Before we present an axiomatic characterization of adaBoost and related algorithms, we need to define what a "boosting (arcing) procedure" is. We begin with an informal description, first.

2.1   Informal Description

Boosting is an instance of a broad category of machine learning algorithms, called *ensemble methods*, which combine weak learners into strong aggregated rules. Ensemble methods differ in how they instantiate weak learning algorithms and how they combine them afterward. For instance, in a *bagging* approach, multiple weak learners are trained on bootstrap subsamples of the training set, which, in turn, are aggregated by averaging or majority voting. The most well-known example of this approach is the *random forests* algorithm introduced by [6]. In a *stacking* approach, on the other hand, the predictions of weak learners form a new representation of the data, and aggregation is carried out by training a weak learner on the newly learned representation [12].

In a boosting approach, by contrast, a weak learner is supplied with *both* a training sample and a set of weights on those training examples. At each stage of the algorithm, the task of a weak learner is to do well with respect to a weighted training set. The predictions of those weak learners are, then, combined using an appropriate aggregation rule. As mentioned by [14,29] and studied empirically in [5], this setting can be relaxed by subsampling from the training set according to the weights. However, since subsampling can be considered as a part of the weak learner's algorithm, we adopt the convention of having weighted training examples here for generality.

In our axiomatic approach, which holds in the general setting of learning (see Definition 1), we will not formally define the notion of "weak learnability"[1]. In fact, we will not even require it! Hence, we will refer to weak learners from now on as *base* learners.

---

[1] Weak learnability, informally speaking, only ensures that the learner performs better than random guessing rather than mandating the learner to achieve an arbitrarily optimal performance.

2.2 Formal Definitions

In this section, we introduce our notation and some preliminary definitions. Let $\mathcal{H}$ be a hypothesis space and let $\mathbb{H}$ be a space that is formed from $\mathcal{H}$ using an appropriate aggregation rule. For example, $\mathbb{H}$ might be the space of all linear combinations of hypotheses in $\mathcal{H}$. We assume throughout this paper that $\mathcal{H} \subseteq \mathbb{H}$ and that $\mathbb{H}$ is a *vector space* on $\mathbb{R}$ (or $\mathbb{C}$). That is, $\mathbb{H}$ is closed under addition and scalar multiplication on $\mathbb{R}$ (or $\mathbb{C}$). In particular, notions such as addition, scalar multiplication and linear maps of hypotheses in $\mathcal{H}$ are meaningful. Examples of hypothesis spaces that satisfy this assumption include the Euclidean plane $\mathbb{R}^d$, such as in linear classification or regression problems, and function spaces, such as Hilbert spaces in kernel methods. Moreover, the probability simplex in $\mathbb{R}^m$ will be denoted $\mathcal{P}^m$.

Definition 2 (Span). *Let $\mathcal{H}$ be a set of hypotheses that reside in some vector space on $\mathbb{R}$ (or $\mathbb{C}$). Then, the span of $\mathcal{H}$ is the set of all possible combinations of finite elements in $\mathcal{H}$. That is:* $\mathbf{Span}(\mathcal{H}) = \{h : \exists h_1, \ldots, h_K \in \mathcal{H} : h = \sum_{k=1}^{K} h_k\}$.

The goal of learning via boosting is to select an aggregated rule $\mathbf{h} \in \mathbb{H}$ that minimalizes some weighted loss $\sum_{i=1}^{m} w_1(i)\, l_i(\cdot)$, for some fixed set of loss functions $l_i : \mathbb{H} \to \mathbb{R}$ and some initial distribution $w_1 \in \mathcal{P}^m$. In the general setting of learning, a loss function $l_i(\cdot)$ is often of the form $l(\cdot, z_i)$, where $z_i \in \mathcal{Z}$ is the i-th training example. In the latter case, $S = \{l_1, \ldots, l_m\}$ is fixed in all rounds of the boosting algorithm because the training sample $(z_1, \ldots, z_m)$ is fixed. This is often referred to as the *boosting-by-sampling* setting [14,31]. Since, for a fixed sample S, only the *weights* on the loss functions determine the outcome of a learning algorithm, one may view a base learner as a mapping from the probability simplex $\mathcal{P}^m$ to the hypothesis space $\mathcal{H}$. This brings us to the following definition:

Definition 3 (Base Learner). *For a fixed set of loss functions $S = \{l_1, \ldots, l_m\}$, a base learner f_S is a (possibly randomized) mapping $f_S : \mathcal{P}^m \to \mathcal{H}$. The set of all base learners will be denoted $\mathcal{F}_S$.*

Informally, we interpret f_S as follows. A base learner f_S is supplied with a distribution $w_t \in \mathcal{P}^m$ on the loss functions in S. Then, the task of f_S is to select a hypothesis $h_t \in \mathcal{H}$ whose weighted loss $\sum_{i=1}^{m} w_t(i)\, l_i(h_t)$ is small. Note that this is merely an informal interpretation since f_S can be any *arbitrary* map. Examples of base learners include the support vector machine (SVM), decision stumps, and the classification and regression tree (CART) algorithm [7,8,29].

Definition 4 (Boosting Procedure). *For a fixed sample $S = \{l_1, \ldots, l_m\}$, a boosting (arcing) procedure is a mapping $g : \mathcal{P}^m \times \mathcal{F}_S \to \mathbb{H}$, which operates as follows. Initially, g is provided with a distribution $w_1 \in \mathcal{P}^m$ and a base learner f_S. Then, g operates sequentially in T rounds. At round t, it assigns $h_t = f_S(w_t)$ and updates the weight $w_t \to w_{t+1}$ according to $(h_1, \ldots, h_t)$. The final output is $\mathbf{h} \in \mathbb{H}$, which is an aggregation of all the base hypotheses $(h_1, \ldots, h_T)$.*

According to Definition 4, a boosting procedure generates several base hypotheses in sequence via adaptive re-weighting. Finally, it combines them into an aggregated rule. Needless to mention, several possible aggregation methods exist, such as by using majority voting, averaging, random sampling, or by simply selecting the single hypothesis obtained in the final round[2].

3   Axiomatic Characterization

As shown in Definition 4, boosting procedures, which operate via adaptive re-weighting and combining, can vary according to how they update the weights w_t and how they combine hypotheses afterwards. In fact, they also vary depending on the choice of the loss functions $S = \{l_1, \ldots, l_m\}$ that they minimalize. Indeed, many boosting algorithms that have been proposed in the past can be categorized along these lines.

For example, Breiman introduced a boosting algorithm, called arc-x4, in which the weight of a training example at a given round is proportional to $1+\epsilon(i)$, where $\epsilon(i)$ is the number of times the i-th training example had been misclassified by hypotheses in previous rounds [5]. This is similar in spirit to adaBoost; it forces the classifier to focus on the training examples that are harder to predict correctly. Similarly, Ji and Ma proposed a different scheme, in which training examples are partitioned into "cares" and "don't-cares", according to whether or not they are classified correctly by the combined classifier [19]. Needless to mention, other possibilities also exist. For instance, [20] studied adaptive re-weighting schemes, where $w_t(i) \propto (1 + \epsilon(i)^n)$ for different choices of $n \in \mathbb{N}$, and so on.

The fact that many adaptive re-weighting schemes can be (and have been) proposed raises the following fundamental question: What natural axioms should a boosting procedure satisfy? And, do such axioms lead to a unique approach? We will answer these questions in the remainder of the paper. We will show that once three natural axioms are imposed, the range of possibilities is greatly reduced. Essentially, the only boosting algorithm that operates via adaptive re-weighting and combining and also satisfies the postulated axioms will turn out to be a slight generalization of adaBoost. Surprisingly, this holds even though (1) we operate in the general setting of learning and (2) we do not impose any constraints on the base learner f_S and the base hypotheses $(h_1, \ldots, h_T)$.

3.1   The Axioms

Our axioms are three: ADDITIVITY, OBJECTIVITY and UTILITY. We describe each axiom in details next.

---

[2] The main results of this paper can be extended to the setting where the base learner f_S is different at each round t. However, we assume in this paper, with no loss of generality, that f_S is fixed to simplify the discussion and notation.

Axiom 1 (ADDITIVITY). *Let $w_T = g_T(h_1, \ldots, h_{T-1})$ be the adaptive weights selected by the boosting procedure g at round T when $h_1, \ldots, h_{T-1}$ are the base hypotheses provided by the base learner $f_S : \mathcal{P}^m \to \mathcal{H}$ at all preceding rounds. Let $h_t = f_S(w_t)$ for all rounds $t \geq 1$. Then, the aggregated rule is $\sum_{t=1}^{T} h_t$.*

Our first axiom states that the boosting procedure is a stagewise *additive* model; at each round, it improves its prediction rule by reweighting the sample and *adding* a new hypothesis to the aggregated rule. These additive models have a long history in statistics and signal processing (see for instance [15] and the references therein). Note that in the case of binary classification problems, where the target set is $\mathcal{Y} = \{-1, +1\}$ and $\mathbf{sign}(\mathbf{h}(\cdot)) : \mathcal{X} \to \mathcal{Y}$ is the prediction rule, the aggregation rule in Axiom 1 reduces to *majority voting*. For regression problems, the aggregation rule in Axiom 1 reduces to *averaging*. Hence, it has a wide applicability.

Axiom 2 (OBJECTIVITY). *For any $T \geq 1$ and any $i \in \{1, \ldots, m\}$, we have $w_T(i) \to 0$ as $w_1(i) \to 0$. More precisely: $\forall (\epsilon \geq 0, i \in \{1, \ldots, m\}, T \geq 1) : \exists \delta_T \geq 0 : \quad w_1(i) \leq \delta_T \Rightarrow w_T(i) \leq \epsilon$.*

Informally, our second axiom requires that the boosting procedure aims at minimalizing its original objective function *only*. More precisely, if a loss function $l_i(\cdot)$ has an initial weight of zero, such as when it is not in the training sample to begin with, then its weight $w_t(i)$ at all rounds $t \geq 1$ will remain zero. Formally, Axiom 2 requires that for any fixed $t \geq 1$, we have $w_t(i) \to 0$ as $w_1(i) \to 0$. In particular, we note that the formal specification of Axiom 2 implies that $w_t(i) = 0$ for all $t \geq 1$ if $w_1(i) = 0$ as described earlier[3].

Axiom 3 (UTILITY). *At any round $T \geq 1$ and for any $h_T \in \mathcal{H}$, we have $\sum_{i=1}^{m} w_T(i) \, l_i(h_T) \propto \sum_{i=1}^{m} w_1(i) \, l_i(\sum_{t=1}^{T} h_t)$, with a proportionality constant that is independent of h_T.*

The last axiom can be intuitively understood in light of the aggregation rule $\mathbf{h} = \sum_{t=1}^{T} h_t$ stipulated by Axiom 1. Prior to round T, the aggregated hypothesis is $\sum_{t=1}^{T-1} h_t$. At round T, the task of the base learner f_S is to select a hypothesis $h_T \in \mathcal{H}$ that performs well according to the objective function $\sum_{i=1}^{m} w_T(i) l_i(\cdot)$. This hypothesis h_T will, then, be added to the aggregated rule, which, ideally, should result in a better aggregated hypothesis with respect to the *original* objective function $\sum_{i=1}^{m} w_1(i) l_i(\mathbf{h})$. Axiom 3 states that the better h_T is for the learning problem at round T, the better it is for the original optimization problem.

To recall, the base hypotheses $(h_1, \ldots, h_T)$ selected by the base learner $f_S : \mathcal{P}^m \to \mathcal{H}$ can be entirely arbitrary. As mentioned earlier, we have not imposed any notion of goodness on f_S, such as weak learnability. However, in

---

[3] Axiom 2 can also be interpreted as a *stability* constraint on the boosting procedure. It can be argued that the main advantage of ensemble methods is their ability to improve stability, which results in a reduced over-fitting risk and an improved generalization [4,5,32].

order for the boosting algorithm to be of any utility, the performance of the base hypothesis h_t at round t should factor into the performance of the overall boosting procedure. In particular, we should impose a condition, which qualitatively states that having a "better" hypothesis at round t would be more "helpful" to the overall boosting algorithm. This is achieved by Axiom 3.

3.2   Proof of Independence

Next, we show that the three axioms are mutually independent.

Proposition 1. *Any two of the three axioms* ADDITIVITY, OBJECTIVITY, *and* UTILITY *can be satisfied without satisfying the third axiom. In other words, the three axioms are independent of each other.*

Proof. First, we show that ADDITIVITY and OBJECTIVITY can be satisfied without satisfying the UTILITY axiom. Let $l_i(h) = \langle z_i, h \rangle$ be a linear cost function. The objective is to minimalize $\sum_{i=1}^{m} w_1(i) \langle z_i, h \rangle$ in some hypothesis space $\mathcal{H}$. Let g be the boosting procedure, which always sets $w_t(i) = w_1(i)$ and combines hypotheses according to Axiom 1. Then, g satisfies the ADDITIVITY and OBJECTIVITY axioms trivially. However:

$$\sum_{i=1}^{m} w_T(i) l_i(h_T) = \sum_{i=1}^{m} w_1(i) \langle z_i, h_T \rangle = \sum_{i=1}^{m} w_1(i) \langle z_i, \sum_{t=1}^{T} h_t \rangle + \beta,$$

where $\beta = -\sum_{i=1}^{m} w_1(i) \langle z_i, \sum_{t=1}^{T-1} h_t \rangle$. Hence, Axiom 3 is not satisfied unless $\beta = 0$, but the value of β is determined by the base learner f_S, not the boosting procedure g.

Second, consider the boosting procedure g that always selects $w_t(i) = w_1(i)$ and uses h_T as the final aggregated hypothesis. In other words, $\mathbb{H} = \mathcal{H}$ and the boosting procedure aggregates the hypotheses $(h_1, \ldots, h_T)$ by selecting h_T only. This is a boosting algorithm that trivially satisfies the OBJECTIVITY and UTILITY axioms, but not ADDITIVITY.

Finally, we show that ADDITIVITY and UTILITY can be satisfied without satisfying the OBJECTIVITY axiom. As will be proved later, AdaBoost satisfies the three axioms. Let g be the boosting procedure that coincides with adaBoost except that if $l_j(h) = l_k(h)$ for some $j, k \in \{1, \ldots, m\}$ and all $h \in \mathcal{H}$, then the boosting procedure always sets $w_t(j) = w_t(k)$ for all $t \geq 2$. In other words, it distributes the weight equally between the two loss functions if they are identical. More precisely, let $\hat{w}_t(j)$ and $\hat{w}_t(k)$ be the weights assigned by the adaBoost procedure at round t and let the corresponding weights assigned by the new boosting procedure g be $w_t(j) = w_t(k) = (\hat{w}_t(j) + \hat{w}_t(k))/2$. Then, g satisfies the ADDITIVITY and UTILITY axioms but without satisfying the OBJECTIVITY axiom because $w_t(k) \to 0$ as $w_1(k) \to 0$ only if $w_1(j) = 0$. Therefore, the three axioms are independent.   □

4   Implications of the Axioms

Before we present our main uniqueness theorem, we elaborate on an important definition first. Suppose that in the fixed set of loss functions $S = \{l_1, \ldots, l_m\}$, two of those functions were, in fact, identical. That is, suppose that there exists $i, j \in \{1, \ldots, m\}$ with $i \neq j$ such that $l_i(\cdot) = l_j(\cdot)$. Then, it is clear that no algorithm can make a *meaningful* distinction between the weights $w_t(i)$ and $w_t(j)$ for all $t \geq 1$. Similar conclusions hold when a loss function $l_i(\cdot)$ can be written as a *linear combination* of the others. Therefore, linearly dependent loss functions pose an *inherent* source of ambiguity. This brings us to the following definition [28].

Definition 5 (Linear Independence). *A set of loss functions* $\{l_1, \ldots, l_m\}$, *with* $l_i : \mathcal{H} \to \mathbb{R}$, *are called linearly independent on* $\mathcal{H}$ *if and only if there exists* $h_1, \ldots, h_m \in \mathcal{H}$ *such that the column vectors* $\boldsymbol{v}_j = (l_1(h_j), l_2(h_j), \ldots, l_m(h_j))^T$ *for all* $j = 1, \ldots, m$ *are linearly independent.*

Informally, the set of loss functions are linearly independent if the training examples are sufficiently different. We analyze the implications of our axioms on linearly independent loss functions, next.

Lemma 1. *Let* $S = \{l_1, \ldots, l_m\}$ *comprises of* m *linearly independent functions on* $\mathcal{H}$. *Let* $w_T = g_T(h_1, \ldots, h_{T-1})$ *be the adaptive weights selected by the boosting procedure* g *at round* T *when* $h_1, \ldots, h_{T-1}$ *are the base hypotheses provided by the base learner* $f_S : \mathcal{P}^m \to \mathcal{H}$ *at all preceding rounds. Let* $h_t = f_S(w_t)$ *for all rounds* $t \geq 1$. *Then, the* ADDITIVITY, OBJECTIVITY, *and* UTILITY *axioms are satisfiable only if* $w_T(i) = \zeta_i(\sum_{t=1}^{T-1} h_t)$ *for some function* $\zeta_i : \mathbf{Span}(\mathcal{H}) \to [0, 1]$.

Proof. First, the statement trivially holds when $T = 1$. Suppose that $T > 1$. Then, by the UTILITY and ADDITIVITY axioms, we have:

$$\sum_{i=1}^{m} w_T(i) \, l_i(h_T) = c \cdot \sum_{i=1}^{m} w_1(i) \, l_i(\sum_{t=1}^{T} h_t), \tag{1}$$

with a proportionality constant c that is independent of h_T. Since $w_T(i)$ is independent of h_T and the loss functions in S are linearly independent, there exists m hypotheses $\tilde{h}_1, \tilde{h}_2, \ldots, \tilde{h}_m$ such that the matrix:

$$L = \begin{bmatrix} l_1(\tilde{h}_1) & l_2(\tilde{h}_1) & \cdots & l_m(\tilde{h}_1) \\ l_1(\tilde{h}_2) & l_2(\tilde{h}_2) & \cdots & l_m(\tilde{h}_2) \\ \cdots & \cdots & \cdots & \cdots \\ l_1(\tilde{h}_m) & l_2(\tilde{h}_m) & \cdots & l_m(\tilde{h}_m) \end{bmatrix}$$

is non-singular. Consequently, Eq. (1) implies that:

$$\begin{bmatrix} w_T(1) \\ w_T(2) \\ \cdots \\ w_T(m) \end{bmatrix} = c \, L^{-1} \cdot \begin{bmatrix} \sum_{i=1}^{m} w_1(i) \, l_i(\sum_{t=1}^{T-1} h_t + \hat{h}_1) \\ \sum_{i=1}^{m} w_1(i) \, l_i(\sum_{t=1}^{T-1} h_t + \hat{h}_2) \\ \cdots \\ \sum_{i=1}^{m} w_1(i) \, l_i(\sum_{t=1}^{T-1} h_t + \hat{h}_m) \end{bmatrix}$$

Therefore, the weights $w_T(i)$ are determined by the sequence of base hypotheses $(h_1, \ldots, h_{T-1})$ only via the aggregated rule $\sum_{t=1}^{T-1} h_t$.   □

Lemma 1 shows that a boosting procedure that satisfies the three axioms is a Markov chain; the future of the boosting procedure is conditionally independent of the sequence of base hypotheses given the aggregated rule. Now, we are ready to state the main uniqueness theorem. To recall, a mapping M is called *additive* if it satisfies $M(x+y) = M(x) + M(y)$. For example, linear mappings are additive.

Theorem 1. *If $S = \{l_1, \ldots, l_m\}$ comprises of m linearly independent functions on $\mathcal{H}$ and $0 \in \mathcal{H}$, then the* ADDITIVITY, OBJECTIVITY, *and* UTILITY *axioms are satisfiable simultaneously for all initial distributions $w_1 \in \mathcal{P}^m$ if and only if the following two conditions hold:*

1. *We have $l_i(h) \propto \exp\{\Lambda_i(h) + \lambda_i\}$ for some additive mapping $\Lambda_i : \mathbf{Span}(\mathcal{H}) \to \mathbb{R}$ and some constant $\lambda_i \in \mathbb{R}$.*
2. *The weights are updated according to the multiplicative weight update procedure:*

$$w_T(i) \propto w_{T-1}(i) \cdot \exp\{\Lambda_i(h_{T-1})\}$$

Proof. [PROOF OF NECESSITY]: Lemma 1 implies that there exists some functions $\zeta_i : \mathbf{Span}(\mathcal{H}) \to [0,1]$ such that:

$$w_T(i) = \zeta_i \Big(\sum_{t=1}^{T-1} h_t \Big), \tag{2}$$

for all $T \geq 1$. In other words, the weights at round T depend on the sequence of hypotheses $(h_1, \ldots, h_{T-1})$ only via their aggregated rule.

In addition, Axiom 3 states that for any $h \in \mathcal{H}$ and any round $T \geq 1$:

$$\sum_{i=1}^{m} w_T(i)\, l_i(h) \propto \sum_{i=1}^{m} w_1(i)\, l_i \Big(h + \sum_{t=1}^{T-1} h_t \Big) \tag{3}$$

However, $\sum_{t=1}^{T-1} h_t$ is arbitrary, so we denote it by $u \in \mathbf{Span}(\mathcal{H})$. Therefore, by (2), we conclude that for any $u \in \mathbf{Span}(\mathcal{H})$ and any $h \in \mathcal{H}$, the following equality must hold for some constant $c > 0$, which is independent of h:

$$\sum_{i=1}^{m} \zeta_i(u)\, l_i(h) = c \cdot \sum_{i=1}^{m} w_1(i) \cdot l_i(h+u) \tag{4}$$

Because c is independent of h and $0 \in \mathcal{H}$, we set $h = 0$ to conclude that:

$$c = \frac{\sum_{i=1}^{m} \zeta_i(u)\, l_i(0)}{\sum_{i=1}^{m} w_1(i) \cdot l_i(u))}$$

Hence, (4) may be rewritten as:

$$\frac{\sum_{i=1}^{m} \zeta_i(u)\, l_i(h)}{\sum_{i=1}^{m} \zeta_i(u)\, l_i(0)} = \frac{\sum_{i=1}^{m} w_1(i) \cdot l_i(h+u)}{\sum_{i=1}^{m} w_1(i) \cdot l_i(u)} \tag{5}$$

Since we require that the axioms hold simultaneously for all initial probability distributions $w_1 \in \mathcal{P}^m$, consider the following initial distribution:

$$w_1^{(i)}(k) = \begin{cases} 1 - \epsilon, & \text{if } k = i \\ \frac{\epsilon}{m-1}, & \text{otherwise} \end{cases}$$

By Axiom 2, we know that $\zeta_i(u) \to 1$ as $\epsilon \to 0^+$. In the latter case, (5) reduces to the functional equation:

$$l_i(h + u) = \frac{l_i(h) \cdot l_i(u)}{l_i(0)} \tag{6}$$

Since we require the axioms to hold simultaneously for all initial weights, the above functional equation must hold as well for all loss functions $l_i(\cdot)$. Now, consider the function $\Lambda_i : \mathbf{Span}(\mathcal{H}) \to \mathbb{R}$ defined by $\Lambda_i(u) = \log l_i(u) - \log l_i(0)$. Then, $\Lambda_i(\cdot)$ satisfies $\Lambda_i(u + h) = \Lambda_i(u) + \Lambda_i(h)$. Hence, $\Lambda_i(\cdot)$ is an additive function on $\mathbf{Span}(\mathcal{H})$, which implies that $l_i(\cdot)$ must be equal to $\exp\{\Lambda_i(w) + \lambda_i\}$ for some additive mapping $\Lambda_i : \mathbf{Span}(\mathcal{H}) \to \mathbb{R}$ and some constant $\lambda_i \in \mathbb{R}$.

Now, (4) can be rewritten as:

$$\forall h \in \mathcal{H} : \sum_{i=1}^{m} \zeta_i(u)\, l_i(h) = c \cdot \sum_{i=1}^{m} \frac{w_1(i) \cdot l_i(u)}{l_i(0)}\, l_i(h), \tag{7}$$

where c does not depend on h. This defines a system of linear equations on $\zeta_i(u)$ for different choices of $h \in \mathcal{H}$. Because the set of loss functions are linearly independent on $\mathcal{H}$, the above condition is satisfiable if and only if $\forall i \in \{1, \ldots, m\} : \zeta_i(u) = c\, w_1(i) \cdot l_i(u)/l_i(0)$. However, $\zeta_i(u)$ is a probability distribution so c is absorbed in the normalization constant, which we can ignore. We have:

$$\zeta_i(u) \propto w_1(i) \cdot \frac{l_i(u)}{l_i(0)} \tag{8}$$

$$= w_1(i)\, e^{\Lambda_i(h_1)} \prod_{t=2}^{T-1} \exp\{\Lambda_i(h_t)\} \propto w_{T-1} \cdot \exp\{\Lambda_i(h_t)\}, \tag{9}$$

where the last line holds by induction and the fact that $\Lambda_i(\cdot)$ is an additive function. This proves that the conditions are necessary.

[PROOF OF SUFFICIENCY]: Next, we prove that the conditions are sufficient. First, from the multiplicative weight update mechanism, it is clear that Axiom 2 is satisfied. Moreover, (9) shows that the weights can be determined at any round T using only the aggregated rule $\sum_{t=1}^{T-1} h_t$. In particular, we have:

$$w_T(i) \propto w_1(i) \cdot e^{\Lambda_i(h_1)} \prod_{t=2}^{T-1} \exp\{\Lambda_i(h_t)\} = w_1(i) \cdot \exp\left\{\Lambda_i\left(\sum_{t=1}^{T-1} h_t\right)\right\}$$

Hence, Axiom 1 is satisfied. Finally, by plugging the functional equation in (6) and the expression in (8) into (5), we deduce that Axiom 3 is satisfied as well. Therefore, the conditions are also sufficient for the three axioms to hold.   □

Theorem 1 reveals an axiomatic characterization of adaBoost and related algorithms, such as the extension of adaBoost to confidence-rated predictions [30], the RankBoost algorithm [13], the Real-AdaBoost for probabilistic classifiers [15], and the Multi-class AdaBoost method [38]. In particular, the ADDITIVITY, OBJECTIVITY and UTILITY axioms are satisfied if and only if the loss functions were of the exponential type and the weights were updated according to the multiplicative weight update mechanism. Therefore, even though many possible adaptive re-weighting methods could be (and have been) proposed, such as the methods studied in [5, 19, 20], the adaptive re-weighting method employed by adaBoost and its variants can be *uniquely* constructed axiomatically. This sheds some insight on the rich set of connections that have been established between adaBoost and related fields, such as information geometry, game theory, and convex optimization [29][4].

As mentioned earlier, the ADDITIVITY, OBJECTIVITY, and UTILITY axioms can also serve as axioms for the more general Multiplicative Weight Update procedure. This follows from the fact that the axioms are satisfied if and only if the multiplicative weight update procedure is used without imposing any additional conditions on the base learners. That is, the base learners are entirely arbitrary.

5   Concluding Remarks

Boosting procedures, which operate via adaptive re-weighting and combining, can vary according to the choice of the loss function they minimalize and how they adaptively update the weights. Not surprisingly, different algorithms for adaptive re-weighting have been proposed in the literature, such as the arc-x4 algorithm [5], its generalization to higher order polynomials [20], and the partitioning scheme in [19]. This raises the fundamental questions: What natural axioms should an adaptive re-weighting and combining procedure satisfy? And, do such axioms lead to a *unique* solution?

In this work, we address these questions. We establish that three natural axioms on boosting algorithms are satisfied *if and only if* the boosting algorithm minimalizes the sum of exponential-additive loss functions and the weights are updated according to the multiplicative weight update procedure. Surprisingly, this conclusion holds even though (1) we operate in the general setting of learning, which encompasses regression, classification, ranking, and clustering, and (2) we do not impose any constraints on the base learner and the base hypotheses.

The fact that the loss functions have to be of the form specified in Theorem 1 might appear to be overly restrictive at first sight. For instance, it is not immediately obvious how one might define a loss function for regression tasks, while also being a composition of the exponential with additive functions. However,

---

[4] The assumption of linear independence in Theorem 1 is required to eliminate an *inherent* source of ambiguity. Without this assumption, no uniqueness theorem can be established. However, this result does not imply that a boosting algorithm must guarantee linear independence. Rather, it states that *up to this inherent source of ambiguity*, adaBoost and its variants arise uniquely out of three natural axioms.

this function class is by no means restrictive. For instance, if the hypothesis space is a subset of $\mathbb{R}^d$, then the Fourier theorem states that *any* desired function in a compact domain can be approximated arbitrarily well using a sum of exponential-additive functions. Hence, the general class of exponential-additive functions is quite rich. Indeed, many extensions of adaBoost have been proposed that also satisfy the EFFICIENCY, STABILITY and LINEAR UTILITY axioms, including the extension of adaBoost to confidence-rated predictions [30], the RankBoost algorithm [13], the Real-AdaBoost for probabilistic classifiers [15], and the Multi-class AdaBoost method [38].

References

1. Ackerman, M., Ben-David, S.: Measures of clustering quality: a working set of axioms for clustering. In: NIPS, pp. 121–128 (2009)
2. Aczél, J., Forte, B., Ng, C.T.: Why the shannon and hartley entropies are natural. Adv. Appl. Probab. **6**(01), 131–146 (1974)
3. Bell, D.A., Wang, H.: A formalism for relevance and its application in feature subset selection. Mach. Learn. **41**(2), 175–195 (2000)
4. Bousquet, O., Elisseeff, A.: Stability and generalization. J. Mach. Learn. Res. **2**, 499–526 (2002)
5. Breiman, L.: Prediction games and arcing algorithms. Neural Comput. **11**(7), 1493–1517 (1999)
6. Breiman, L.: Random forests. Mach. Learn. **45**(1), 5–32 (2001)
7. Breiman, L., Friedman, J., Stone, C.J., Olshen, R.A.: Classification and Regression Trees. CRC Press, Boca Raton (1984)
8. Cortes, C., Vapnik, V.: Support-vector networks. Mach. Learn. **20**(3), 273–297 (1995)
9. Cox, R.T.: Probability, frequency and reasonable expectation. Am. J. Phys. **14**(1), 1–13 (1946)
10. Csiszar, I.: Why least squares and maximum entropy? an axiomatic approach to inference for linear inverse problems. Ann. Statist. **19**, 2032–2066 (1991)
11. Csiszár, I.: Axiomatic characterizations of information measures. Entropy **10**(3), 261–273 (2008)
12. Džeroski, S., Ženko, B.: Is combining classifiers with stacking better than selecting the best one? Mach. Learn. **54**(3), 255–273 (2004)
13. Freund, Y., Iyer, R., Schapire, R.E., Singer, Y.: An efficient boosting algorithm for combining preferences. JMLR **4**, 933–969 (2003)
14. Freund, Y., Schapire, R.E.: A desicion-theoretic generalization of on-line learning and an application to boosting. In: Vitányi, P. (ed.) EuroCOLT 1995. LNCS, vol. 904, pp. 23–37. Springer, Heidelberg (1995). https://doi.org/10.1007/3-540-59119-2_166
15. Friedman, J., Hastie, T., Tibshirani, R., et al.: Additive logistic regression: a statistical view of boosting. Ann. Stat. **28**(2), 337–407 (2000)
16. Friedman, J.H.: Greedy function approximation: a gradient boosting machine. Ann. Stat. **29**, 1189–1232 (2001)
17. Jardine, N., Sibson, R.: The construction of hierarchic and non-hierarchic classifications. Comput. J. **11**(2), 177–184 (1968)
18. Jaynes, E.T.: Probability theory: The Logic of science. Cambridge University Press, Cambridge (2003)

19. Ji, C., Ma, S.: Combined weak classifiers. NIPS **9**, 494–500 (1997)
20. Khanchel, R., Limam, M.: Empirical comparison of arcing algorithms (2005)
21. Kleinberg, J.: An impossibility theorem for clustering. In: NIPS, vol. 15, pp. 463–470 (2002)
22. Lee, P.: On the axioms of information theory. Ann. Math. Stat. **35**(1), 415–418 (1964)
23. Mason, L., Baxter, J., Bartlett, P.L., Frean, M.R.: Boosting algorithms as gradient descent. In: NIPS, pp. 512–518 (1999)
24. Österreicher, F.: Csiszár's f-divergences-basic properties. Technical report (2002)
25. Pennock, D.M., Horvitz, E.: Analysis of the axiomatic foundations of collaborative filtering. Ann Arbor **1001**, 48109–2110 (1999)
26. Pennock, D.M., Maynard-Reid II, P., Giles, C.L., Horvitz, E.: A normative examination of ensemble learning algorithms. In: ICML, pp. 735–742 (2000)
27. Prasad, A., Pareek, H.H., Ravikumar, P.: Distributional rank aggregation, and anaxiomatic analysis. In: ICML, pp. 2104–2112 (2015)
28. Sansone, G.: Orthogonal Functions. Dover Publications, New York (1991)
29. Schapire, R.E., Freund, Y.: Boosting: Foundations and Algorithms. MIT Press, Cambridge (2012)
30. Schapire, R.E., Singer, Y.: Improved boosting algorithms using confidence-rated predictions. Mach. Learn. **37**(3), 297–336 (1999)
31. Servedio, R.A.: Smooth boosting and learning with malicious noise. J. Mach. Learn. Res. (JMLR) **4**, 633–648 (2003)
32. Shalev-Shwartz, S., Shamir, O., Srebro, N., Sridharan, K.: Learnability, stability and uniform convergence. J. Mach. Learn. Res. (JMLR) **11**, 2635–2670 (2010)
33. Shannon, C.: A mathematical theory of communication. Bell Syst. Tech. J. **27**, 379–423 (1948)
34. Shore, J., Johnson, R.: Axiomatic derivation of the principle of maximum entropy and the principle of minimum cross-entropy. IEEE Trans. Inf. Theory **26**(1), 26–37 (1980)
35. Skilling, J.: The axioms of maximum entropy. In: Erickson, G.J., Smith, C.R. (eds.) Maximum-Entropy and Bayesian Methods in Science and Engineering, pp. 173–187. Springer, Dordrecht (1988). https://doi.org/10.1007/978-94-009-3049-0_8
36. Vapnik, V.N.: An overview of statistical learning theory. IEEE Trans. Neural Netw. **10**(5), 988–999 (1999)
37. Wang, Z., et al.: Multi-class hingeboost. Methods Inf. Med. **51**(2), 162–167 (2012)
38. Zhu, J., Zou, H., Rosset, S., Hastie, T.: Multi-class adaboost. Stat. Interface **2**(3), 349–360 (2009)
39. Zou, H., Zhu, J., Hastie, T.: New multicategory boosting algorithms based on multicategory fisher-consistent losses. Ann. Appl. Stat. **2**(4), 1290 (2008)

Modular Dimensionality Reduction

Henry W. J. Reeve[1]([✉]), Tingting Mu[2], and Gavin Brown[2]

[1] University of Birmingham, Edgbaston, Birmingham B15 2TT, UK
henrywjreeve@gmail.com
[2] University of Manchester, Oxford Rd, Manchester M13 9PL, UK

Abstract. We introduce an approach to modular dimensionality reduction, allowing efficient learning of multiple complementary representations of the same object. Modules are trained by optimising an unsupervised cost function which balances two competing goals: Maintaining the inner product structure within the original space, and encouraging structural diversity between complementary representations. We derive an efficient learning algorithm which outperforms gradient based approaches without the need to choose a learning rate. We also demonstrate an intriguing connection with Dropout. Empirical results demonstrate the efficacy of the method for image retrieval and classification.

Keywords: Ensemble learning · Dimensionality reduction · Dropout Kernel principal components analysis

1 Introduction

High dimensional data is a widespread challenge in machine learning applications, from computer vision through to bioinformatics and natural language processing. A natural solution is to find a structure-preserving mapping to a low dimensional space, many techniques for which can be found in the literature, such as kernel PCA, Isomap, LLE and Laplacian Eigenmaps [6,23]. This paper provides a meta-level tool for modular dimensionality reduction, applicable to each of the aforementioned approaches.

We start from the observation that *multiple* abstractions of the same concept can be taken, and may provide complementary views on a task of interest. We therefore propose a *modular* approach to unsupervised dimensionality reduction, in which we learn a diverse collection of low-dimensional representations of the data. Once a modular representation is learned, each module may be used independently – with their respective predictions combined at test time. This procedure is naturally parallelisable in a distributed computing architecture; and, since each representation is low-dimensional, processing for each module is fast and efficient.

Electronic supplementary material The online version of this chapter (https://doi.org/10.1007/978-3-030-10925-7_37) contains supplementary material, which is available to authorized users.

© Springer Nature Switzerland AG 2019
M. Berlingerio et al. (Eds.): ECML PKDD 2018, LNAI 11051, pp. 605–619, 2019.
https://doi.org/10.1007/978-3-030-10925-7_37

In the context of supervised learning, successful ensemble performance emanates from a fruitful trade-off between the accuracy of the individual members of the ensemble and the degree of diversity [4,15]. We carry this insight across to the domain of unsupervised dimensionality reduction, by demonstrating the importance of diversity for a set of representation modules. We introduce an unsupervised loss function for training a *set* of dimensionality reduction modules, which balances two competing objectives. The first objective is for each module to preserve relational structure within the original feature space; the second is for modules to exhibit a *diversity* of relational structures.

The contributions of this paper are as follows:

1. An unsupervised loss function for modular dimensionality reduction.
2. A bespoke optimisation procedure which outperforms gradient based methods such as stochastic gradient descent in our setting.
3. A detailed empirical comparison with competitors.
4. An intriguing connection to the dropout algorithm from deep learning [13].

2 Background

We first review work on dimensionality reduction and ensemble learning.

2.1 Unsupervised Dimensionality Reduction

The canonical approach to unsupervised dimensionality reduction is PCA, and its kernelised generalisation, KPCA [21]. KPCA is a general approach which may be applied to a wide variety of application domains through an appropriate choice of kernel [19,25]. Several manifold learning techniques have also been shown to be special cases of KPCA, with a data-dependent kernel [12].

Classically, KPCA has been viewed as the orthogonal projection which maximises the preserved variance [21]. We shall adopt an alternative perspective in which we view KPCA as a form of *unsupervised similarity learning*, whereby a mapping is chosen so that inner-products in the low dimensional space approximate the kernel. To make this precise we require some notation. We let $\mathcal{X}$ denote our original feature space and let $k : \mathcal{X} \times \mathcal{X} \to \mathbb{R}$ denote a symmetric positive semi-definite kernel function. Take an unlabelled dataset $\mathcal{D} = \{\boldsymbol{x}_1, ..., \boldsymbol{x}_N\} \subset \mathcal{X}$. For simplicity, we assume throughout that k is centred with respect to $\mathcal{D}$ [21]. Let $\mathbb{H}_k$ denote the associated reproducing kernel Hilbert space (RKHS) of real-valued functions. For each $H \in \mathbb{N}$, we let $\mathbb{H}_k^H$ denote the class of H-dimensional mappings $\varphi : \mathcal{X} \to \mathbb{R}^H$ with coordinate functions taken from the RKHS $\mathbb{H}_k$. That is, for each $\varphi \in \mathbb{H}_k^H$ there exists $\varphi^1, \cdots, \varphi^H \in \mathbb{H}_k$ such that for all $x \in \mathcal{X}$, $\varphi(x) = (\varphi^h(x))_{h=1}^H$.

Definition 1 (Inner product loss function). *Given an unsupervised data set $\mathcal{D}$ and a mapping $\varphi \in \mathbb{H}_k^H$, the inner product loss is given by*

$$L_k(\varphi, \mathcal{D}) = \frac{1}{N^2} \sum_{i,j \leq N} \left(\langle \varphi(\boldsymbol{x}_i), \varphi(\boldsymbol{x}_j) \rangle - k(\boldsymbol{x}_i, \boldsymbol{x}_j) \right)^2.$$

We can interpret KPCA as a form of unsupervised similarity learning which minimises the inner product loss. Let $\xi : \mathcal{X} \to \mathbb{H}_k$ denote the canonical embedding given by $\xi(x)(y) = k(x, y)$.

Proposition 1. *The inner product loss $L_k (\varphi, \mathcal{D})$ is minimised by taking φ to be the member of $\mathbb{H}_k^H$ obtained by embedding $\mathcal{D}$ into $\mathbb{H}_k$ via ξ and projecting onto the top H kernel principal components.*

The proofs of all results within the main text are given in the appendices (see supplementary material).

2.2   Ensembles and Diversity

Combining the outputs of multiple predictors often brings both statistical advantages, such as bias or variance reduction, and computational advantages, through parallelism. In order outperform an individual model, ensembles promote a level of diversity or disagreement between the predictions the constituent models [10,15]. Whilst methods such as bagging and boosting encourage diversity through a manipulation of the training data, a more direct approach is the *Negative Correlation Learning* (NCL) algorithm of Liu and Yao [18] in which diversity is targeted explicitly.

Suppose we have a supervised regression ensemble $\mathcal{H} = \{h_m\}_{m=1}^M$ consisting of predictors h_m. In the previous section we used an unsupervised dataset $\mathcal{D} = \{\mathbf{x}_1, ..., \mathbf{x}_N\}$. To distinguish this we use notation $\mathcal{T} = \{(\mathbf{x}_1, y_1), ..., (\mathbf{x}_N, y_N)\}$ for a *supervised* dataset. We let $\mathbb{V}(\cdot)$ denote the empirical variance of a finite sequence. The NCL algorithm can be understood in terms of the following *modular loss function*.

Definition 2 (Modular loss function). *The modular loss E_λ is defined by*

$$E_\lambda(\mathcal{H}, \mathcal{T}) = \frac{1}{NM} \sum_{n=1}^N \sum_{m=1}^M (h_m(\boldsymbol{x}_n) - y_n)^2 - \lambda \cdot \frac{1}{N} \sum_{n=1}^N \mathbb{V}\left(\{h_m(\boldsymbol{x}_n)\}_{m=1}^M\right).$$

The modular loss function consists of two terms: A squared loss term which targets the average individual accuracy of the predictors h_m, combined with a diversity term which encourages disagreement between the predictors. The hyper-parameter λ controls the degree of emphasis placed on the diversity. This has the special property that when $\lambda = 1$, $E_\lambda(\mathcal{H}, \mathcal{T})$ is exactly the squared loss for the ensemble predictor $\frac{1}{M} \sum_m h_m(\mathbf{x})$ from the target y.

The NCL algorithm is equivalent to stochastic gradient descent applied to the modular loss. This perspective differs from original formulation of the NCL algorithm first introduced by Liu and Yao which utilises a multiplicity of interacting cost functions [18]. However, the updates of the two formulations are equal up to a factor of $1/M$ applied to the learning rate.

3 The Modular Inner Product Loss

Our goal is to train a collection of M distinct but complementary representations of the data. With this goal in mind, we introduce the modular inner product loss which combines two contrasting objectives. On the one hand, we seek high quality representations which faithfully preserve the relational structure encoded by the kernel. On the other hand, we would like the relational structure encoded in our different representations to be diverse. Let $\mathcal{F}(H, M)$ denote the class of all M-tuples $\boldsymbol{\Phi} = \{\varphi_m\}_{m=1}^M$ with each $\varphi_m \in \mathbb{H}_k^H$. Recall that $\mathbb{V}(\cdot)$ denotes the empirical variance.

Definition 3 (The modular inner product loss). *Suppose we have an unlabelled data set $\mathcal{D} \subset \mathcal{X}$ and a kernel k. Given $\boldsymbol{\Phi} \in \mathcal{F}(H, M)$, the modular inner product loss is given by*

$$L_k^\lambda (\boldsymbol{\Phi}, \mathcal{D}) = \frac{1}{M} \sum_{m=1}^M L_k (\varphi_m, \mathcal{D}) - \lambda \cdot \frac{1}{N^2} \sum_{1 \leq i.j \leq N} \mathbb{V}\left(\{\varphi_m(\boldsymbol{x}_i) \cdot \varphi_m(\boldsymbol{x}_j)\}_{m=1}^M \right).$$

$$(1)$$

The modular inner product loss is an analogue of the supervised modular loss function (Definition 2), with inner products between a pair of examples in a representation module replacing predictions for a single example, and the target replaced by an unsupervised inner product.

An equivalent reformulation of the modular inner product loss is as a convex combination between the average inner product loss of the individual modules and the inner product loss of a composite representation. Given $\boldsymbol{\Phi} \in \mathcal{F}(H, M)$ we define $\overline{\boldsymbol{\Phi}} \in (\mathbb{H}_k)^{H \cdot M}$ by $\overline{\boldsymbol{\Phi}}(\boldsymbol{x}) = \left(1/\sqrt{M}\right) \cdot \left[\varphi_1(\boldsymbol{x})^T, \cdots, \varphi_M(\boldsymbol{x})^T\right]^T$. Proposition 2 is proved in Appendix 9 (see supplementary material).

Proposition 2. $L_k^\lambda (\boldsymbol{\Phi}, \mathcal{D}) = (1 - \lambda) \cdot \frac{1}{M} \sum_{m=1}^M L_k (\varphi_m, \mathcal{D}) + \lambda \cdot L_k (\overline{\boldsymbol{\Phi}}, \mathcal{D})$.

When $\lambda = 0$ the loss $L_k^\lambda (\boldsymbol{\Phi}, \mathcal{D})$ is minimised by taking each φ_m to be a projection onto the top H kernel principal components, whilst for $\lambda = 1$, $L_k^\lambda (\boldsymbol{\Phi}, \mathcal{D})$ is minimised by taking $\overline{\boldsymbol{\Phi}}$ to be the projection onto the top $M \cdot H$ kernel principal components. Hence, $L_k^\lambda (\boldsymbol{\Phi}, \mathcal{D})$ blends smoothly between training representation modules as individuals and targeting the composite representation.

4 Efficient Optimization

We now introduce the *module-by-module* (MBM) algorithm, which is a form of alternating optimisation designed to minimise the modular inner product loss without the need to choose a learning rate. Our objective is to minimise $L_\lambda (\boldsymbol{\Phi}, \mathcal{D})$ over $\boldsymbol{\Phi} \in \mathcal{F}(H, M)$. We require an empirical kernel map.

Definition 4. *A rank R empirical kernel map is a function $\psi \in \mathbb{H}_k^R$ such that $\psi(\boldsymbol{x}_i)^T \psi(\boldsymbol{x}_j) = k(\boldsymbol{x}_i, \boldsymbol{x}_j)$ for all pairs $(\boldsymbol{x}_i, \boldsymbol{x}_j) \in \mathcal{D}^2$.*

One can always construct an empirical kernel map of rank N by taking $\psi(\boldsymbol{x}) = \boldsymbol{K}(\mathcal{D})^{-\frac{1}{2}}\left[k(\boldsymbol{x}, \boldsymbol{x}_1), \cdots, k(\boldsymbol{x}, \boldsymbol{x}_N)\right]^T$, where $\boldsymbol{K}(\mathcal{D}) = (k(\boldsymbol{x}_i, \boldsymbol{x}_j))_{ij}$ denotes the kernel gram matrix. Moreover, given a kernel k we can often obtain a low rank empirical kernel map ψ for a kernel $\tilde{k}$ which closely approximates k by employing a method such as random Fourier features [11] or the Nyström method [27]. By reasoning analogous to [20] we have the following useful proposition.

Proposition 3. *Given a rank R empirical kernel map ψ, the minimum for $L_k^\lambda(\boldsymbol{\Phi}, \mathcal{D})$ is attained by $\boldsymbol{\Phi} = \{\varphi_m\}_{m=1}^M$ with each φ_m of the form $\varphi_m(\boldsymbol{x}) = \boldsymbol{W}_m \cdot \psi(\boldsymbol{x})$ for some matrix $\boldsymbol{W}_m \in \mathbb{R}^{H \times R}$.*

Hence, our objective reduces to the following matrix optimisation problem: Minimise

$$
\begin{aligned}
C^\lambda(\mathcal{W}, \boldsymbol{\Psi}) :&= \sum_{m=1}^M \left\| \boldsymbol{F}_m^T \boldsymbol{F}_m - \boldsymbol{\Psi}^T \boldsymbol{\Psi} \right\|^2 - \lambda \cdot \sum_{m=1}^M \left\| \boldsymbol{F}_m^T \boldsymbol{F}_m - \frac{1}{M} \sum_{q=1}^M \boldsymbol{F}_q^T \boldsymbol{F}_q \right\|^2 \\
&\propto L_k^\lambda(\boldsymbol{\Phi}_\mathcal{W}, \mathcal{D}),
\end{aligned}
$$

where $\boldsymbol{\Psi} = [\psi(\boldsymbol{x}_1), \cdots, \psi(\boldsymbol{x}_n)] \in \mathbb{R}^{R \times N}$, $\boldsymbol{F}_m = \boldsymbol{W}_m \cdot \boldsymbol{\Psi}$ and $\boldsymbol{\Phi}_\mathcal{W} = \{\boldsymbol{W}_m \cdot \psi\}_{m=1}^M$. We make use the concept of the rank-constrained approximate square root of a symmetric matrix.

Definition 5. *Define $RT_r : \mathbb{R}^{d \times d} \to \mathbb{R}^{r \times d}$ by*

$$
RT_r(\boldsymbol{M}) = argmin_{\boldsymbol{F} \in \mathbb{R}^{r \times d}} \left\{ \left\| \boldsymbol{F}^T \boldsymbol{F} - \boldsymbol{M} \right\|^2 \right\}.
$$

Dax has shown that the rank-constrained approximate square root $RT_r(\boldsymbol{M})$ of any $d \times d$ symmetric matrix $\boldsymbol{M}$ (not necessarily positive semi-definite) may be computed via the singular value decomposition in $O(d^2 \cdot r)$ time and $O(d^2)$ space complexity [7]. The following proposition allows us to optimise the weights of a single module φ_m, whilst leaving the remaining modules fixed.

Proposition 4. *Suppose we take $m \in \{1, \cdots, M\}$, fix $\boldsymbol{W}_q$ for $q \neq m$, and let*

$$
\boldsymbol{T}_m = \frac{M}{M - \lambda \cdot (M-1)} \cdot \boldsymbol{\Psi}^T \left(\boldsymbol{I}_D - \frac{\lambda}{M} \cdot \sum_{q \neq m} \boldsymbol{W}_q^T \boldsymbol{W}_q \right) \boldsymbol{\Psi}.
$$

Take $\boldsymbol{F}_m = RT_H(\boldsymbol{T}_m)$. Setting $\boldsymbol{W}_m = \boldsymbol{F}_m \boldsymbol{\Psi}^\dagger$ minimises $C^\lambda(\mathcal{W}, \boldsymbol{\Psi})$ with respect to $\boldsymbol{W}_m$, under the constraint that $\boldsymbol{W}_q$ remains fixed for $q \neq m$, where $\boldsymbol{\Psi}^\dagger$ denotes the psuedo-inverse of $\boldsymbol{\Psi}$.

Unfortunately, computing $\boldsymbol{F}_m$ via Proposition 4 is $O(N^2 \cdot H)$ which is intractable for large N. The following proposition enables us to reduce the complexity of this optimisation whenever we have access to an empirical kernel map of rank $R \ll N$.

Proposition 5. *Suppose that ψ is an empirical kernel map of rank R. Take $\tilde{\boldsymbol{\Psi}} = (RT_R(\boldsymbol{\Psi}\boldsymbol{\Psi}^T))^T \in \mathbb{R}^{R\times R}$. For all $\mathcal{W} = \{\boldsymbol{W}_m\}_{m=1}^M$ with $\boldsymbol{W}_m \in \mathbb{R}^{H\times R}$ we have $C_\lambda\left(\mathcal{W}, \tilde{\boldsymbol{\Psi}}\right) = C_\lambda\left(\mathcal{W}, \boldsymbol{\Psi}\right)$. Moreover, computing $\tilde{\boldsymbol{\Psi}}$ is $O\left(R^2 \cdot N\right)$ in time complexity and $O(R^2)$ in space complexity.*

Combining Propositions 3, 4 and 5 gives rise to the *module-by-module* algorithm (MBM, Algorithm 1), which is $O(NR^2 + EHR^2)$ in time and $O(NR)$ in space complexity, and has the advantage of reducing the modular inner product loss at every iteration until a critical point is reached.

Inputs: A data set $\mathcal{D} = \{\boldsymbol{x}_1, \cdots, \boldsymbol{x}_N\}$, a rank R empirical kernel map ψ, a number of modules M, a number of dimensions per module H, a diversity parameter λ and $\epsilon > 0$.
Compute $\boldsymbol{\Psi} = [\psi(\boldsymbol{x}_1), \cdots, \psi(\boldsymbol{x}_n)]$;
Update $\boldsymbol{\Psi} = (RT_\rho\left(\boldsymbol{\Psi}\boldsymbol{\Psi}^T\right))^T$;
Randomly initialise $\boldsymbol{F}_m \in \mathbb{R}^{H\times R}$ for $m = 1, \cdots, M$;
Compute $\boldsymbol{Q} = \boldsymbol{\Psi}^T\boldsymbol{\Psi}$ and $\boldsymbol{S} = \sum_{m=1}^M \boldsymbol{F}_m^T\boldsymbol{F}_m$;
Compute $c = ((1-\lambda) + \lambda/M + \epsilon)^{-1}$;
for $e = 1, \cdots, E$ **do**
    for $m = 1, \cdots, M$ **do**
        Compute $\boldsymbol{S}_{-m} = \boldsymbol{S} - \boldsymbol{F}_m^T\boldsymbol{F}_m$;
        Compute $\boldsymbol{T} = c \cdot \left(\boldsymbol{Q} - (\lambda/M)\,\boldsymbol{S}_{-m} + \epsilon \cdot \boldsymbol{F}_m^T\boldsymbol{F}_m\right)$;
        Update $\boldsymbol{F}_m = RT_H\left(\boldsymbol{T}\right)$;
        Update $\boldsymbol{S} = \boldsymbol{S}_{-m} + \boldsymbol{F}_m^T\boldsymbol{F}_m$;
    end
end
Compute $\boldsymbol{W}_m = \boldsymbol{F}_m\boldsymbol{\Psi}^\dagger$ for $m = 1, \cdots, M$;
Output: $\boldsymbol{\Phi} = \{\boldsymbol{W}_m \cdot \psi\}_{m=1}^M$.

Algorithm 1. The module-by-module (MBM) algorithm.

The following theorem justifies the use of the MBM algorithm - it is guaranteed to reduce the modular inner product loss at every epoch until a critical point is reached.

Theorem 1. *Given $E \in \mathbb{N}$, let $\boldsymbol{\Phi}^E \in \mathcal{F}(H, M)$ denote the set obtained by training with Algorithm 1, for E epochs. Then for all $E \in \mathbb{N}$, $L_k^\lambda\left(\boldsymbol{\Phi}^{E+1}, \mathcal{D}\right) < L_k^\lambda\left(\boldsymbol{\Phi}^E, \mathcal{D}\right)$, unless $\boldsymbol{\Phi}^E$ is a critical point of $L_k^\lambda\left(\boldsymbol{\Phi}, \mathcal{D}\right)$, in which case $L_k^\lambda\left(\boldsymbol{\Phi}^{E+1}, \mathcal{D}\right) \leq L_k^\lambda\left(\boldsymbol{\Phi}^E, \mathcal{D}\right)$.*

5   The Dropout Connection

In this section we introduce a surprising connection between the modular inner product loss and the dropout algorithm [13,22]. Dropout is a state of the art approach to regularising deep neural networks in which a random collection of hidden neurons is "dropped out" at each stochastic gradient update. The dropout algorithm can be understood as implicitly minimising the expectation of a stochastic loss function based on predictions from a random sub-network [22, 26]. There is a natural analogue of this, in our setting: to minimise the expectation of a stochastic variant of the inner product loss, based on inner products computed from a random subset of modules. We refer to this analogue as the drop-module (DM) algorithm. To be precise, given an ensemble of feature mappings $\boldsymbol{\Phi} \in \mathcal{F}(H, M)$ each binary vector $\boldsymbol{\eta} = \{\eta_m\}_{m=1}^{M} \in \{0,1\}^M$ corresponds to a 'noisy' representation $\boldsymbol{\Phi}_{\boldsymbol{\eta}}$ given by $\boldsymbol{\Phi}_{\boldsymbol{\eta}}(\boldsymbol{x}) = (1/\sqrt{M}) \cdot (\eta_1 \cdot \boldsymbol{\varphi}_1(\boldsymbol{x}), \cdots, \eta_M \cdot \boldsymbol{\varphi}_M(\boldsymbol{x}))$.

Fix a probability $p \in [0, 1]$ and let $B(p)$ denote the probability measure on $\{0, 1\}$ with $\mathbb{E}_{B(p)}(\eta) = p$. Let $\boldsymbol{\Theta}$ denote the parameters of $\boldsymbol{\Phi}$. The DM algorithm proceeds by randomly sampling $\boldsymbol{x}_i, \boldsymbol{x}_j \in \mathcal{D}$ and $\eta_m \sim B(p)$ and updating

$$\boldsymbol{\Theta} \leftarrow \boldsymbol{\Theta} - \alpha \cdot \frac{\partial}{\partial \boldsymbol{\Theta}} \left(\langle \boldsymbol{\Phi}_{\boldsymbol{\eta}}(\boldsymbol{x}_i), \boldsymbol{\Phi}_{\boldsymbol{\eta}}(\boldsymbol{x}_j) \rangle - k(\boldsymbol{x}_i, \boldsymbol{x}_j) \right)^2 .$$

The DM algorithm implicitly minimises the following stochastic loss function

$$L_{k,p}^{\mathrm{drop}}(\boldsymbol{\Phi}, \mathcal{D}) = \mathbb{E}_{\boldsymbol{\eta} \sim B(p)^M} \left[L_k(\boldsymbol{\Phi}_{\boldsymbol{\eta}}, \mathcal{D}) \right] .$$

Previously, Baldi et al. demonstrated that dropout may be understood as training an exponentially large ensemble with shared weights [1]. In our setting this corresponds to shared weight a ensemble of size 2^M with an ensemble member for each $\boldsymbol{\eta} \in \{0, 1\}^M$. We demonstrate that the DM algorithm can be related to an ensemble of size M, trained via the modular inner product loss (see Definition 3). We emphasise that unlike the shared weight ensembles considered by Baldi et al. [1], here we consider ensembles with separate weights in which the interaction takes place purely via the diversity term in the modular inner product loss.

Theorem 2. *The drop-module inner product loss at p is equivalent to the modular inner product loss at $\lambda = Mp/(1+p(M-1))$. To be precise, for $\boldsymbol{\Phi}_0 \in \mathcal{F}(H, M)$ we have*

$$\left. \frac{\partial L_{k,p}^{drop}(\boldsymbol{\Phi}, \mathcal{D})}{\partial \boldsymbol{\Phi}} \right|_{\boldsymbol{\Phi} = \boldsymbol{\Phi}_0} = p \cdot \left. \frac{\partial L_k^{\lambda}(\boldsymbol{\Phi}, \mathcal{D})}{\partial \boldsymbol{\Phi}} \right|_{\boldsymbol{\Phi} = (\sqrt{p/\lambda}) \cdot \boldsymbol{\Phi}_0} .$$

Theorem 2 implies that if we take $\lambda = Mp/(1 + p(M - 1))$ then the minima of $L_k^{\lambda}(\boldsymbol{\Phi}, \mathcal{D})$ are equal to the minima of $L_{k,p}^{\mathrm{drop}}(\boldsymbol{\Phi}, \mathcal{D})$, up to a constant scaling factor of $\sqrt{p/\lambda}$. In this sense, the minima for the two loss functions are *representationally equivalent*. The relationship between the diversity parameter λ in the MBM algorithm and the probability of keeping module p in the DM algorithm is illustrated in Fig. 1.

6 Experimental Results

In this section we first demonstrate the optimisation performance of the MBM algorithm before comparing our method for other natural approaches for training multiple kernelised representations. The data sets used in all experiments are described in Sect. 12.1 (see supplementary material).

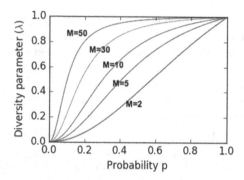

Fig. 1. The diversity parameter λ in MBM vs. the corresponding p in DM.

6.1 Optimisation Performance of the MBM Algorithm

In this section we assess the MBM algorithm (Algorithm 1) in terms of its efficiency at optimising the modular inner product loss. We compare with two gradient based approaches. As a baseline we consider stochastic gradient descent (SGD) applied directly to the modular inner product loss. This is expected to perform poorly in our setting since the modular inner product loss sums over all pairs of examples. We also consider the state of the art Adam optimiser of Kingma et al. [14] (Adam). The Adam optimiser is applied in batch mode, first compressing the data by applying Proposition 5. We set $M = 10$, $H = 10$, $\lambda = 0.9$ and let k to be the Gaussian kernel with γ set using the heuristic of Kwok and Tsang [16, Sect. 4]. In addition, we employ a rank $R = 1000$ Nyström approximation [27]. For SGD and Adam we consider learning rates in the range $\{10^{-6}, \cdots, 10^{2}\}$. We evaluated the algorithms by training for one hour and evaluating the minimum value of the loss function attained during training and the convergence time - the time taken for the loss function to fall within 1% of its minimum. For SGD and Adam we report results corresponding to the learning rate which achieves the lowest minimum loss. The results are shown in Table 1. The SGD method was extremely slow and typically failed to converge within one hour. The compressed Adam method performed much better and typically converged within 30 min. However, the bespoke MBM algorithm achieved the same

Table 1. A comparison of the MBM algorithm with gradient based methods.

Data set	Loss function minimum			Convergence time (seconds)		
	SGD	Adam	MBM	SGD	Adam	MBM
Convex	110.0 ± 21.2	13.4 ± 0.0	13.3 ± 0.0	3369.8 ± 163.8	963.9 ± 368.9	$\mathbf{334.7 \pm 151.2}$
MNIST	427.6 ± 16.3	59.9 ± 0.0	59.9 ± 0.0	3484.7 ± 63.2	1609.1 ± 69.2	$\mathbf{400.6 \pm 69.4}$
NUS Wide	598.6 ± 8.4	73.1 ± 0.0	73.1 ± 0.0	3376.3 ± 185.2	1495.1 ± 149.5	$\mathbf{422.7 \pm 47.7}$
NVDR	110.0 ± 18.5	10.7 ± 0.0	10.6 ± 0.0	1891.3 ± 578.8	1157.2 ± 227.0	$\mathbf{402.4 \pm 35.1}$
Rectangles	29.4 ± 0.4	10.3 ± 0.0	10.3 ± 0.0	1796.1 ± 687.6	746.9 ± 162.0	$\mathbf{256.2 \pm 36.5}$

minimum loss in at most twice the speed on each of the data sets. The MBM algorithm also has the advantage of not requiring the user to set a learning rate.

6.2   Image Retrieval and Classification Performance of MBM Modules

We compare four unsupervised approaches to training multiple kernelised feature mappings:

Partition. We compute the top HM KPCAs and randomly partition these into M sets of H, so that each mapping $\varphi_m \in \mathbb{H}_k^H$ is a projection onto a disjoint subset of the top $H \cdot M$ KPCAs.

Bootstrap. Bagging [3] applied to KPCA. For each $m \in \{1, \cdots, M\}$ take a bootstrap sample $\tilde{\mathcal{D}}_m$, of size N, and let $\varphi_m \in \mathbb{H}_k^H$ be the KPCA projection mapping onto H dimensions for $\tilde{\mathcal{D}}_m$.

Random. A kernelised variant of the widely used technique of random projections [2,8]. For each $m = 1, \cdots, M$ we sample a random matrix $\boldsymbol{R}_m \in \mathbb{R}^{H \times D}$ from an $H \times D$ standard normal distribution, and normalise each row so that it has unit norm. In order to kernelise this technique the feature space for the random matrices is the output of the empirical kernel map ψ (see Sect. 4).

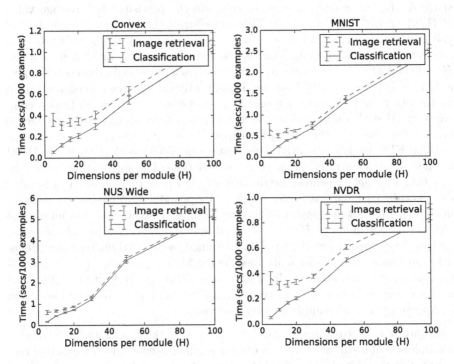

Fig. 2. Test time as vs. H. See Appendix 12.2 for discussion & other data sets (see supplementary material).

MBM. Our proposed approach in which $\boldsymbol{\Phi}$ is trained to minimise the modular inner product loss via MBM algorithm. The diversity parameter λ is set based upon performance on a validation set. We shall consider $H \in \{5, 10, 15, 20, 30, 50, 100\}$ and take M so that $H \cdot M = 300$. We also compare with the following non-modular base line.

Monolithic. A single mapping $\varphi \in \mathbb{H}_k^H$ - the projection onto the top 300 KPCAs.

In each case we take k to be the Gaussian kernel with the γ parameter set via the heuristic of Kwok and Tsang [16, Sect. 4]. For computational efficiency we employ a rank 1000 Nyström approximation [27] in each case. We shall consider two distinct tasks:

Image Retrieval: We shall consider the modular low dimensional representation's capability for efficiently retrieving a set of κ close-by images. Let $\boldsymbol{\Phi} = \{\varphi_m\}_{m=1}^M \in \mathcal{F}(H, M)$ be a modular representation and $\mathcal{D}$ an unlabelled training set. Given a test point $\boldsymbol{x} \in \mathcal{X}$, for each module, we compute the set $\mathcal{I}_{\kappa,n}^{\varphi_m}(\boldsymbol{x}) \subset \mathcal{D}$ of κ-nearest neighbours of $\boldsymbol{x}$ based the distance $\|\varphi_m(\boldsymbol{x}_q) - \varphi_m(\boldsymbol{x})\|_2$. We then extract subset of size κ, $\mathcal{I}_{\kappa,n}^{\boldsymbol{\Phi}}(\boldsymbol{x}) \subset \bigcup_{m=1}^M \mathcal{I}_{\kappa,n}^{\varphi_m}(\boldsymbol{x})$ so that the elements $\boldsymbol{x}_q \in \mathcal{I}_{\kappa,n}^{\boldsymbol{\Phi}}(\boldsymbol{x})$ minimise the average squared distance from the test point $\boldsymbol{x}$ over the low dimensional spaces ie. $(1/M) \cdot \sum_{m=1}^M \|\varphi_m(\boldsymbol{x}_q) - \varphi_m(\boldsymbol{x})\|_2^2$ is minimised. Let $\mathcal{I}_{\kappa,n}(\boldsymbol{x})$ denote the set of κ nearest neighbours as computed in the original space $\mathcal{X}$. To assess performance we compute the precision: the average value of $(1/\kappa) \cdot \# \left(\mathcal{I}_{\kappa,n}^{\boldsymbol{\Phi}}(\boldsymbol{x}) \cap \mathcal{I}_{\kappa,n}(\boldsymbol{x}) \right)$. This procedure is based upon the method of [24] and gives a quantitative assessment of the representation's ability to preserve structural information. The results of the image retrieval task for $\kappa = 10$, $H = 20$ and $M = 15$ are shown in Table 2. On each of the eight data sets the precision attained by the MBM method significantly exceeds the precision attained by the other modular methods: partition, bootstrap and random. Table 3 compares MBM with the monolithic method in which we simply compute the 10 nearest neighbours in the φ-projected space, where φ is the projection onto the top 300 KPCAs. For a relatively modest reduction in performance the MBM method obtains a significant speed up at test time. The speed up is due to the fact that each set of nearest neighbours $\mathcal{I}_{\kappa,n}^{\varphi_m}(\boldsymbol{x})$ may be computed in parallel on a low-dimensional space (see Fig. 2 and Appendix 12.2 (see supplementary material). Figure 3 demonstrates the precision as a function of the number of dimensions per module (H) with $\kappa = 10$ and $H \cdot M = 300$. As H increases, the precision approaches the precision attained by the 300-dimensional Monolithic approach. The precision attained by the MBM approach typically exceeds that attained by the other modular approaches (Bootstrap, Partition, Random) across a range of values of H. Corresponding figures for other data sets are given in Appendix 12.3 (see supplementary material).

Classification. We compare the methods in terms of their capacity for extracting multiple sets of features for use in a classification ensemble. Given a modular representation $\boldsymbol{\Phi} = \{\varphi_m\}_{m=1}^M \in \mathcal{F}(H, M)$, for each m we train a classifier f_m based on the features extracted by φ_m. Given a test point $\boldsymbol{x}$ we combine the outputs

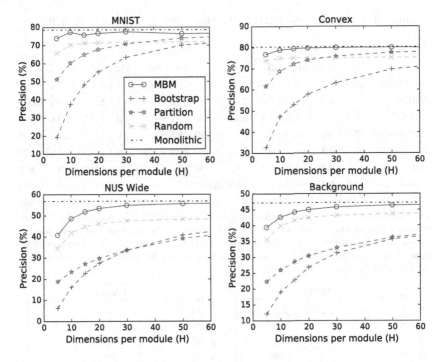

Fig. 3. Precision as a function of H (See Sect. 6.2).

Table 2. A comparison of methods for modular dimensionality reduction. See Sect. 6.2 for details.

Data set	Image retrieval precision (%)				Ensemble classification accuracy (%)			
	Partition	Bootstrap	Random	MBM	Partition	Bootstrap	Random	MBM
NUS Wide	29.8 ± 0.3	27.5 ± 0.3	46.1 ± 0.3	**53.4 ± 0.3**	32.2 ± 0.7	37.7 ± 0.7	42.0 ± 0.7	**43.7 ± 0.7**
NVDR	68.0 ± 0.5	48.1 ± 0.5	73.4 ± 0.5	**77.6 ± 0.4**	**44.9 ± 1.0**	39.6 ± 1.0	43.9 ± 1.0	44.8 ± 1.0
MNIST	67.7 ± 0.1	55.4 ± 0.1	71.5 ± 0.1	**76.6 ± 0.1**	89.2 ± 0.2	95.0 ± 0.2	94.5 ± 0.2	**95.8 ± 0.1**
Background	30.5 ± 0.2	26.8 ± 0.2	42.4 ± 0.2	**45.0 ± 0.2**	41.0 ± 0.4	47.3 ± 0.4	53.8 ± 0.4	**57.3 ± 0.4**
Random	7.0 ± 0.1	10.9 ± 0.1	6.1 ± 0.1	**17.2 ± 0.1**	40.9 ± 0.4	89.5 ± 0.2	53.1 ± 0.4	**90.3 ± 0.2**
Rotations	59.6 ± 0.2	54.3 ± 0.1	68.4 ± 0.1	**75.3 ± 0.1**	72.0 ± 0.3	85.5 ± 0.3	83.8 ± 0.3	**87.4 ± 0.2**
Convex	74.1 ± 0.1	57.8 ± 0.2	75.2 ± 0.1	**79.7 ± 0.1**	55.7 ± 0.4	**65.5 ± 0.3**	60.3 ± 0.4	65.6 ± 0.3
Rectangles	63.3 ± 0.1	44.6 ± 0.1	46.7 ± 0.1	**71.6 ± 0.1**	93.0 ± 0.2	95.5 ± 0.2	93.9 ± 0.2	**98.1 ± 0.1**

of $\{f_m\,(\varphi_m(\boldsymbol{x}))\}_{m=1}^M$ by taking a modal average. Table 2 shows the classification accuracy for ensembles consisting of 15 5-nearest neighbour classifiers trained on 20-dimensional spaces. The MBM approach significantly outperforms the other approaches on five out of eight data sets, and performs comparably or better than the alternatives on every data set. Table 3 compares with the monolithic approach - a single 5-nearest neighbour classifier on 300 KPCAs. The MBM approach is both faster and more accurate than the monolithic method on all but one data set (Fig. 4).

Table 3. Comparing the MBM & the Monolithic approach (See Sect. 6.2).

Data set	Image retrieval				Ensemble classification			
	Monolithic	MBM Δ	Speed	λ	Monolithic	MBM Δ	Speed	λ
NUS Wide	56.8 ± 0.3	-3.4 ± 0.5	10.8×	0.999	40.6 ± 0.7	$+3.1 \pm 1.0$	11.7×	0.990
NVDR	79.0 ± 0.5	-1.4 ± 0.6	7.7×	0.999	40.8 ± 1.0	$+3.9 \pm 1.4$	12.6×	0.999
MNIST	78.6 ± 0.1	-1.9 ± 0.1	9.1×	0.990	95.7 ± 0.1	$+0.1 \pm 0.2$	12.9×	0.900
Background	47.2 ± 0.2	-2.2 ± 0.2	5.8×	0.999	52.5 ± 0.4	$+4.9 \pm 0.5$	8.1×	0.999
Random	23.1 ± 0.1	-5.9 ± 0.1	5.7×	0.950	83.9 ± 0.3	$+6.4 \pm 0.3$	7.1×	0.500
Rotations	76.5 ± 0.1	-1.2 ± 0.1	7.9×	0.990	86.3 ± 0.3	$+1.0 \pm 0.4$	10.9×	0.800
Convex	80.3 ± 0.1	-0.6 ± 0.1	7.2×	0.999	57.6 ± 0.4	$+8.0 \pm 0.5$	23.0×	0.200
Rectangles	70.1 ± 0.1	$+1.5 \pm 0.2$	2.4×	0.990	95.8 ± 0.1	-2.3 ± 0.2	10.7×	0.990

The two Monolithic columns show the image retrieval precision (%) and the classification accuracy (%) of the monolithic method. The MBM columns show the corresponding change in performance due to using the MBM method for each task. The Speed columns show the corresponding speed ups ie. the test time for the Monolithic method divided by the test time for the MBM method.

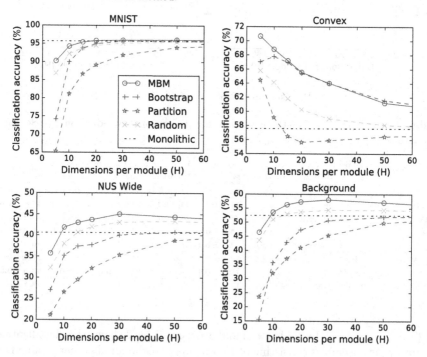

Fig. 4. Classification accuracy as a function of H (See Sect. 6.2).

The Diversity Parameter. The diversity parameter λ in the MBM algorithm controls the level of emphasis placed upon encouraging a diversity of representations. We found that the optimal performance (both in terms of information retrieval and classification) was typically attained with λ just below 1, with

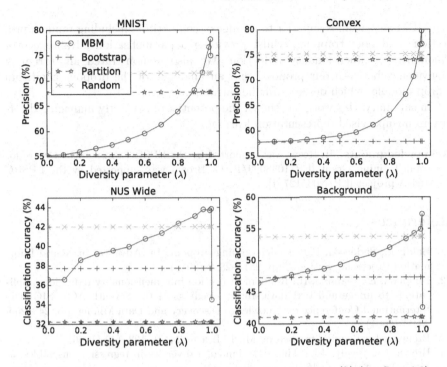

Fig. 5. Performance as a function of the diversity parameter (λ) (see Sect. 6.2).

performance declining sharply by taking $\lambda = 1$ (see Fig. 5, cols 1& 2). It is interesting to observe that the dropout algorithm often performs well with $p \approx 0.5$ and this corresponds a value of λ just below 1, when M is large (see Fig. 1). However, whilst this pattern was observed on all data sets for image retrieval (see Appendix 12.6) (see supplementary material), for some data sets the best classification performance was attained by taking much lower values of λ (see Fig. 5, col 3 and Appendix 12.6 (see supplementary material). Ultimately, the optimal value of λ is data dependent and must be set based on validation performance.

7   Discussion

We have investigated a method for *modular* unsupervised dimensionality reduction. Our method is based upon the *modular inner product loss* (Definition 3), an adaptation of concepts from both negative correlation learning [4,18] and kernel principal components analysis [21]. Whilst the modular loss could be optimised by gradient based methods we introduced a novel *module-by-module* algorithm, which converges at least twice as fast as a state of the art gradient based optimiser [14] without the need to tune the learning rate.

Modular representations have the potential to be applied on range of tasks. Empirical results on both image retrieval and classification tasks confirm that the MBM algorithm is superior to a range of competitors including random projections and bootstrapping, whilst providing a parallelisation advantage over "monolithic" dimensionality reduction. We also demonstrated an intriguing equivalency between our proposal and an analogue of the dropout algorithm - drop module, which deserves further attention.

In summary, this work has shown the potential of explicitly managing diversity in unsupervised representation learning.

Acknowledgments. H. Reeve was supported by the EPSRC through the Centre for Doctoral Training Grant [EP/1038099/1]. G. Brown was supported by the EPSRC LAMBDA project [EP/N035127/1].

References

1. Baldi, P., Sadowski, P.J.: Understanding dropout. In: Advances in Neural Information Processing Systems, pp. 2814–2822 (2013)
2. Bingham, E., Mannila, H.: Random projection in dimensionality reduction: applications to image and text data. In: Proceedings of the Seventh ACM SIGKDD International Conference on Knowledge Discovery and Data Mining, pp. 245–250. ACM (2001)
3. Breiman, L.: Bagging predictors. Mach. Learn. **24**(2), 123–140 (1996)
4. Brown, G., Wyatt, J.L., Tiňo, P.: Managing diversity in regression ensembles. J. Mach. Learn. Res. **6**, 1621–1650 (2005)
5. Chua, T.-S., Tang, J., Hong, R., Li, H., Luo, Z., Zheng, Y.: NUS-WIDE: a real-world web image database from National University of Singapore. In: Proceedings of the ACM International Conference on Image and Video Retrieval, p. 48. ACM (2009)
6. Cunningham, J.P., Ghahramani, Z.: Linear dimensionality reduction: survey, insights, and generalizations. J. Mach. Learn. Res. **16**, 2859–2900 (2015)
7. Dax, A.: Low-rank positive approximants of symmetric matrices. Adv. Linear Algebr. Matrix Theory **4**(3), 172 (2014)
8. Durrant, R.J., Kabán, A.: Random projections as regularizers: learning a linear discriminant ensemble from fewer observations than dimensions
9. Eckart, C., Young, G.: The approximation of one matrix by another of lower rank. Psychometrika **1**(3), 211–218 (1936)
10. Germain, P., Lacasse, A., Laviolette, F., Marchand, M., Roy, J.-F.: Risk bounds for the majority vote: from a Pac-Bayesian analysis to a learning algorithm. J. Mach. Learn. Res. **16**(1), 787–860 (2015)
11. Halko, N., Martinsson, P.-G., Tropp, J.A.: Finding structure with randomness: probabilistic algorithms for constructing approximate matrix decompositions. SIAM Rev. **53**(2), 217–288 (2011)
12. Ham, J., Lee, D.D., Mika, S., Schölkopf, B.: A kernel view of the dimensionality reduction of manifolds. In: Proceedings of the Twenty-First International Conference on Machine Learning, p. 47. ACM (2004)
13. Hinton, G.E., Srivastava, N., Krizhevsky, A., Sutskever, I., Salakhutdinov, R.R.: Improving neural networks by preventing co-adaptation of feature detectors (2012). arXiv preprint: arXiv:1207.0580
14. Kingma, D., Ba, J.: Adam: a method for stochastic optimization (2014). arXiv preprint: arXiv:1412.6980

15. Krogh, A., Vedelsby, J., et al.: Neural network ensembles, cross validation, and active learning. In: Advances in Neural Information Processing Systems, pp. 231–238 (1995)
16. Kwok, J.T.-Y., Tsang, I.W.-H.: The pre-image problem in kernel methods. IEEE Trans. Neural Netw. **15**(6), 1517–1525 (2004)
17. Larochelle, H., Erhan, D., Courville, A., Bergstra, J., Bengio, Y.: An empirical evaluation of deep architectures on problems with many factors of variation. In: Proceedings of the 24th International Conference on Machine Learning, pp. 473–480. ACM (2007)
18. Liu, Y., Yao, X.: Ensemble learning via negative correlation. Neural Netw. **12**(10), 1399–1404 (1999)
19. Lodhi, H., Saunders, C., Shawe-Taylor, J., Cristianini, N., Watkins, C.: Text classification using string kernels. J. Mach. Learn. Res. **2**, 419–444 (2002)
20. Schölkopf, B., Herbrich, R., Smola, A.J.: A generalized representer theorem. In: Helmbold, D., Williamson, B. (eds.) COLT 2001. LNCS (LNAI), vol. 2111, pp. 416–426. Springer, Heidelberg (2001). https://doi.org/10.1007/3-540-44581-1_27
21. Schölkopf, B., Smola, A., Müller, K.-R.: Kernel principal component analysis. In: Gerstner, W., Germond, A., Hasler, M., Nicoud, J.-D. (eds.) ICANN 1997. LNCS, vol. 1327, pp. 583–588. Springer, Heidelberg (1997). https://doi.org/10.1007/BFb0020217
22. Srivastava, N., Hinton, G., Krizhevsky, A., Sutskever, I., Salakhutdinov, R.: Dropout: a simple way to prevent neural networks from overfitting. J. Mach. Learn. Res. **15**(1), 1929–1958 (2014)
23. Storcheus, D., Rostamizadeh, A., Kumar, S.: A survey of modern questions and challenges in feature extraction. In: Proceedings of the 1st International Workshop on "Feature Extraction: Modern Questions and Challenges", NIPS, pp. 1–18 (2015)
24. Venna, J., Peltonen, J., Nybo, K., Aidos, H., Kaski, S.: Information retrieval perspective to nonlinear dimensionality reduction for data visualization. J. Mach. Learn. Res. **11**, 451–490 (2010)
25. Vishwanathan, S.V.N., Schraudolph, N.N., Kondor, R., Borgwardt, K.M.: Graph kernels. J. Mach. Learn. Res. **11**, 1201–1242 (2010)
26. Wang, S.I., Manning, C.D.: Fast dropout training
27. Williams, C., Seeger, M.: Using the Nyström method to speed up kernel machines. In: Proceedings of the 14th Annual Conference on Neural Information Processing Systems, number EPFL-CONF-161322, pp. 682–688 (2001)
28. Wu, X., Hauptmann, A.G., Ngo, C.-W.: Practical elimination of near-duplicates from web video search. In: Proceedings of the 15th ACM International Conference on Multimedia, pp. 218–227. ACM (2007)

Constructive Aggregation and Its Application to Forecasting with Dynamic Ensembles

Vitor Cerqueira[1,2]($\boxtimes$), Fabio Pinto[1], Luis Torgo[1,2,3], Carlos Soares[1,2], and Nuno Moniz[1,2]

[1] University of Porto, Porto, Portugal
vitor.cerqueira@fe.up.pt
[2] INESC TEC, Porto, Portugal
[3] Dalhousie University, Halifax, Canada

Abstract. While the predictive advantage of ensemble methods is nowadays widely accepted, the most appropriate way of estimating the weights of each individual model remains an open research question. Meanwhile, several studies report that combining different ensemble approaches leads to improvements in performance, due to a better trade-off between the diversity and the error of the individual models in the ensemble. We contribute to this research line by proposing an aggregation framework for a set of independently created forecasting models, i.e. heterogeneous ensembles. The general idea is to, instead of directly aggregating these models, first rearrange them into different subsets, creating a new set of combined models which is then aggregated into a final decision. We present this idea as constructive aggregation, and apply it to time series forecasting problems. Results from empirical experiments show that applying constructive aggregation to state of the art dynamic aggregation methods provides a consistent advantage. Constructive aggregation is publicly available in a software package. Data related to this paper are available at: https://github.com/vcerqueira/timeseriesdata. Code related to this paper is available at: https://github.com/vcerqueira/tsensembler.

Keywords: Ensemble learning · Forecasting
Constructive induction · Regression · Dynamic expert aggregation

1 Introduction

Supervised learning consists of searching for a model that represents an accurate hypothesis about some unknown function f we want to approximate. In particular, one way ensemble learning methods approach this problem is by constructing a set of M models $H = \{h^1, \ldots, h^M\}$, and aggregating them in some way to create a final decision $\hat{h}$:

$$\hat{h} = \sum_{i=1}^{M} w^i h^i \tag{1}$$

© Springer Nature Switzerland AG 2019
M. Berlingerio et al. (Eds.): ECML PKDD 2018, LNAI 11051, pp. 620–636, 2019.
https://doi.org/10.1007/978-3-030-10925-7_38

where w^i denotes the weight of model h^i, $\forall\ i \in \{1,\ldots,M\}$. The predictive advantage of combining different models over using a single one is nowadays widely accepted [3,20]. However, estimating the weighting factors for each model in an ensemble remains an open research question.

Previous work by Webb et al. [29,30] shows that combining ensemble learning techniques may improve predictive performance through a better trade-off between diversity and individual error of the ensemble members. These approaches typically combine different ensemble methods during the learning process. We hypothesise that similar effects can be obtained from a portfolio of heterogeneous models [5]. This is the main motivation for this work. This approach can be advantageous because models in a portfolio are typically independent and thus easily parallelised.

In this paper we propose an aggregation framework for a set of diverse and independently created models H, following the basic principles of constructive induction [31]. Constructive induction refers to procedures that modify a set of original attributes, where some of the attributes are removed, others are added, and some of the existing ones are aggregated [31]. This leads to a new set of attributes which hopefully provides an overall better description for approximating f relative to the original set. We follow a similar approach, but in our work the attributes refer to predictive models. To the best of our knowledge, there is no closely related approach for aggregating predictive models.

The aggregation framework proposed in this paper works by rearranging a set of diverse models H into different, overlapping subsets (denoted as $\mathcal{C}_H$). The elements in each of these subsets are then aggregated, leading to a new set of models H' (or sub-ensembles [30]) for approximating f. Similarly to constructive induction approaches, the search for $\mathcal{C}_H$ is done by analysing the individual predictive performance of the original models H in observations not used in the learning process (e.g. a validation set). Essentially, the new models $h' \in H'$ correspond to aggregated subsets of consistently top performing models h. We refer to this as **constructive aggregation** (CA). Our working hypothesis is that, similarly to approaches for combining different ensemble methods [30], CA leads to a decrease in the individual error of ensemble members, without overly decreasing the diversity among them.

To illustrate our idea, the workflow of CA is presented in Fig. 1 with $H = \{h^1, h^2, h^3\}$. After analysing the performance of each model in unseen observations, the set of committees $\mathcal{C}_H = \{\{h^1\}, \{h^1, h^2\}, \{h^2, h^3\}\}$ is created; then the models h^i within each committee are aggregated into models H'. Finally, the new set of combined models H' is aggregated into a final approximation $\hat{h}'$. Both of these aggregations are done according to a linear combination (Eq. 1). The construction of the committee set $\mathcal{C}_H$ is carried out by applying the concept of out-performance contiguity (OC) which is also formalised in this paper.

We apply CA to tackle time series forecasting tasks, where the goal is to predict future numeric values of time series. In experiments on 30 time series from several domains, aggregating a number of forecasting models using CA provides a consistent advantage in terms of predictive performance. That is,

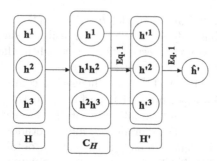

Fig. 1. Workflow of constructive aggregation: the set of available models H is rearranged into different subsets $\mathcal{C}_H$. Each subset is aggregated into H', and become hypotheses for approximating f. The models in H' are aggregated into the final decision $\hat{h}'$.

applying state of the art aggregation methods to H' leads to a better predictive performance relative to applying them to H. Moreover, we provide results that demonstrate that the constructive aggregation process entails a small execution time overhead. In summary, the contributions of this work are the following:

- Constructive Aggregation (CA), a new concept which consists in rearranging a portfolio of heterogeneous models H into different subsets $\mathcal{C}_H$, aggregating them, and using them as hypotheses H';
- Out-performance Contiguity (OC), an approach for building $\mathcal{C}_H$ in time-dependent forecasting problems;
- An extensive set of experiments including: paired comparisons that quantify the percentual difference in error between using CA and aggregating H directly (non-CA); execution time analysis of CA using OC; a sensitivity analysis of the main parameters behind the approach; and a study of CA in terms of bias-variance-covariance trade-off.

In the next section we overview the related work to this paper. In Sect. 3 we present CA, and formalise its application to forecasting problems via OC. In Sect. 4 we provide an extensive set of experiments to validate the proposal. Afterwards, in Sect. 5 we discuss the results and limitations of our work, outlining future directions. Finally, we summarise the paper in Sect. 6. The proposed method is publicly available as an R software package[1].

2   Related Work

In this section we briefly review the related work to this paper. We describe ensemble approaches for forecasting, focusing on dynamic aggregation of several experts designed for these problems. We then outline important approaches for combining different ensemble learning methods, which motivated this work.

---

[1] **tsensembler**: on CRAN or at https://github.com/vcerqueira/tsensembler.

2.1 Ensembles for Forecasting

An assumption of our work is that no forecasting model is universally applicable. Aiolfi and Timmermann [1] demonstrated this in empirical experiments, showing that different forecasters presented a varying relative performance over time. These findings can be regarded as a manifestation of the No Free Lunch theorem by Wolpert [33] (no learning algorithm is best suited for all tasks).

A successful approach to cope with this issue is to combine the opinion of a number of models, i.e. ensemble methods. While the simple average has been shown to be a robust combination method [28] (Simple), ensemble approaches for time series forecasting are typically dynamic: the weight of each model in the ensemble changes over time. This dynamic component is usually designed to cope with concept drift [13].

2.2 Dynamic Combination of Forecasting Experts

Windowing. One of the most common and successful approaches to combine predictive models in time dependent data is to weight them according to their recent performance [24,27] (WindowLoss). Essentially, these approaches are based on the assumption that the immediate future is likely to resemble the recent past.

Regret Minimisation. In the seminal work on online learning by Cesa-Bianchi and Lugosi [8], the authors describe several approaches for dynamically aggregating the opinion of a set of experts. These are typically designed for regret minimisation, having interesting theoretical properties. Essentially, regret is the average error suffered with respect to the best we could have obtained. In this paper we focus on the following three approaches: the exponentially weighted average [8, Sect. 2.1] (EWA), the polynomially weighted average [8, Sect. 2.1] (MLpol), and the fixed share aggregation [8, Sect. 5.2] (FixedShare). We will consider the application of these approaches using the gradient trick [8, Sect. 2.5]. We refer to the work by Cesa-Bianchi and Lugosi [8] for a comprehensive read on these approaches. Zinkevich [36] proposed an approach based on online convex programming and gradient descent that also minimises regret (OGD).

Combining by Learning. Another approach for dynamically combining forecasting models is to apply multiple regression to their output, similarly to stacked generalisation [32]. For example, Gaillard et al. [12] describe a setup in which Ridge regression is used to aggregate experts by minimising the L2-regularised least-squares (Ridge). Recently, Cerqueira et al. [7] proposed a metalearning approach named ADE (arbitrated dynamic ensemble), in which the weights of the experts are computed according to predictions of the loss that they will incur.

2.3 Combining Different Ensemble Approaches

Following the advantage shown by ensemble methods [19], further gains have been reported by approaches that combine different ensemble learning methods. Webb [29] developed `Multiboosting`, which combines `AdaBoost` [10] with `Wagging` [2]. In a posterior work [30], Webb and Zheng claim that `Multiboosting` and other similar approaches provide a better trade-off between the error of the individual members of the ensemble and the diversity among them. Yu et al. [35] presented an approach for combining a number of ensembles, dubbed `Cocktail Ensemble`, using the ambiguity decomposition. This kind of approaches have also been shown to be successful in popular machine learning competitions [25].

In this paper we follow a similar motivation for aggregating a number of forecasting models. Notwithstanding, the outlined approaches differ from our work in an important way: these build the different sub-ensembles during the learning process. Conversely, our approach settles on a portfolio of heterogeneous models [5] where the sub-ensembles are formed after the learning process, using a validation data set.

3 Methodology

This paper is focused on the problem of forecasting future values of uni-variate time series. A time series Y is a temporal sequence of values $Y = \{y_1, y_2, \ldots, y_t\}$, where $y_i \in \mathbb{R}, \forall\, y_i \in Y$ is the value of Y at time i.

As is standard in these problems, we construct a set of observations (the training cases) which are based on the past K lags of the time series. Each observation is composed of a feature vector $x_i \in \mathbb{X} \subset \mathbb{R}^K$, which denotes the previous K values, and a target value $y_i \in \mathbb{Y} \subset \mathbb{R}$, which represents the value we want to predict. The objective is to approximate an unknown function $f : \mathbb{X} \to \mathbb{Y}$, i.e. a function that maps the recently observed values of the series (as represented by the lagged feature vector), to the future value of the series.

3.1 Constructive Aggregation

Our approach is based on heterogeneous ensembles [5]: a set of models created in parallel and separately from each other. Diversity in the ensemble, a key ingredient in these methods [3], is introduced by varying the learning algorithm used to train each model $h \in H$.

We propose an aggregation approach for H based on constructive induction [31], and denote this idea as constructive aggregation (CA). CA works by rearranging the models $h \in H$ into different, possibly overlapping, subsets $\mathcal{C}_H$, and aggregating these into a new set of hypotheses H'. Given that forecasting models typically show a varying relative performance over time [1], our idea is that there are different subsets of models that work well in different time intervals. As such, aggregating these subsets into different combined opinions (H') may lead to better representations for approximating f.

Similarly to related approaches in the literature (c.f. Sect. 2.3) this approach may result in a better trade-off between diversity and the individual error of the ensemble members. Particularly, we hypothesise that aggregating H' decreases the individual error without overly decreasing diversity (relative to aggregating H).

CA can be split in three main steps (c.f. Fig. 1): (i) how to build C_H from H; (ii) how to aggregate the elements of C_H into a new set of hypotheses H'; and (iii) how to aggregate H' into a final decision $\hat{h}'$. In the next subsections we will address each of these steps in turn.

Algorithm 1. Out-performance contiguity for C_H

---

input : set of hypotheses H; validation set $\mathcal{D}$;
          smoothing window λ; contiguity window α

1 **foreach** *hypothesis h^i in H* **do**
2   **foreach** *observation (x_j, y_j) in $\mathcal{D}$* **do**
3           $e_j^i \leftarrow |\hat{y}_j^i - y_j|$ // absolute error of h^i in observation j
4       $e_j'^i \leftarrow moving\_average(e_j^i, \lambda)$
5 **foreach** *observation (x_j, y_j) in $\mathcal{D}$* **do**
6       $r_j \leftarrow rank(e_j')$ // rank of each hypothesis in observation j
7 $C_H \leftarrow \{\}$ // list of committees
8 **foreach** *size T from 1 to $(|H|$-1)* **do**
9       $TOP_T \leftarrow$ top T ranked hypothesis across $\mathcal{D}$
10   **foreach** *τ in TOP_T* **do**
11   **if** *τ is top ranked for at least α consecutive points* **then**
12              $C_H \leftarrow C_H \cup \tau$
13 **return** C_H

3.2   CA for Forecasting via Out-Performance Contiguity

We propose out-performance contiguity (OC) for building C_H from H. This approach is geared towards time series where observations are time-dependent.

Let $\mathcal{D}$ denote a set of validation observations. OC works by analysing the predictive performance of each model $h \in H$ in $\mathcal{D}$. More precisely, this procedure can be summarised as follows: a subset with size T of available models H is aggregated and used as an hypothesis for approximating f, if its elements are the top T performers (relative to H) for α contiguous observations. From the example in Fig. 1, the subset $\{h^1, h^2\}$ is aggregated into an hypothesis $h' \in H'$ because h^1 and h^2 outperform h^3 during a contiguous time interval of size α.

This idea is formalised in Algorithm 1. Initially (lines 1–4), for each model we compute its absolute error in observations of validation data ($\mathcal{D}$). To control for outliers, the loss is smoothed using a moving average of window size λ. Afterwards (lines 5–6), we compute the rank of each model $h \in H$ across $\mathcal{D}$, using the smoothed error. Then (line 8), for each possible size T of the subsets of H (except the size of the full set H), i.e. from 1 to $|H|$-1, we do as follows. All top ranked T models across $\mathcal{D}$ are retrieved (line 9). If the models composing this top is unchanged for α observations, the respective subset becomes an element of $\mathcal{C}_H$ (lines 10–12). In the extreme case in which $\mathcal{C}_H$ is empty, we revert to using the original set of hypotheses H.

Effectively, OC searches for groups of models that perform well in consecutive unseen observations to form $\mathcal{C}_H$. The goal of this search is to make a better exploration of the regions of the input space where these groups consistently perform better relative to other models in the available pool. Moreover, these groups may be relevant in the future due to the potential variance in relative performance shown by forecasters [1], or by other recurring concepts that typically characterise time series [13].

3.3   Aggregation Steps

$\mathcal{C}_H$ into H'. The preceding subsection presents an approach for retrieving the set of subsets $\mathcal{C}_H$ from H. Most of the elements in $\mathcal{C}_H$ are potentially comprised by several models $h \in H$. As such, they need to be aggregated, in order to form a single combined opinion $h' \in H'$.

In this work we accomplish this by using a windowing approach. Essentially, the elements $h \in H$ in each subset from $\mathcal{C}_H$ are aggregated according to their recent performance [24,27]. As mentioned in Sect. 2.2, the idea is that recent observations are more similar to the one we intend to predict, and thus they are considered more relevant. More formally, the weight of each model h^i in a committee $c \in \mathcal{C}_H$ is given by its relative loss in the last λ observations:

$$w^i = \frac{scale(-\overline{L}_\lambda^i)}{\sum_{i \in c} scale(-\overline{L}_\lambda^i)} \tag{2}$$

where $\overline{L}_\lambda^i$ denotes the average loss of model i in the last known λ observations, and *scale* represents the min–max scaling function.

H' into $\hat{h}'$. The objective of the CA process, from H to H', is to create a new representation that presents a better approximation to f. Therefore, our working idea is that applying state of the art aggregation methods to this new set of hypothesis H' is better than applying them directly to H.

4   Experimental Evaluation

We carried out several experiments to validate CA via OC for forecasting with dynamic ensembles. These address the following research questions:

Q1: How do state of the art approaches for dynamic combination of forecasters perform when applied using CA relative to non-CA?

Q2: How do state of the art aggregation methods applied with CA perform relative to state of the art methods for forecasting?

Q3: What are the implications of CA in terms of bias, variance, and covariance?
Q4: How sensitive is CA via OC to different values of α and λ?
Q5: How does CA scale in terms of execution time relative to non-CA?
Q6: Is CA simply pruning or avoiding poor models?

To address these questions we used 30 real-world time series from five different domains, briefly described in Table 1. The size of each time series was truncated to 3000 instances in order to speed up execution time.

Table 1. Datasets and respective summary

ID	Time series	Data source	Data characteristics	Size	K	$|\mathcal{C}_H|$
1	Rotunda AEP	Porto Water Consumption from different locations in the city of Porto [7]	Half-hourly values from Nov. 11, 2015 to Jan. 11, 2016	3000	16	41
2	Preciosa Mar			3000	8	32
3	Amial			3000	9	32
4	Global Horizontal Radiation	Solar Radiation Monitoring [7]	Hourly values from Apr. 25, 2016 to Aug. 25, 2016	3000	13	44
5	Direct Normal Radiation			3000	12	55
6	Diffuse Horizontal Radiation			3000	12	33
7	Average Wind Speed			3000	6	39
8	CO.GT	Air quality indicators in an Italian city [21]	Hourly values from Mar. 10, 2004 to Apr. 04 2005	3000	15	41
9	PT08.S1.CO			3000	6	39
10	C6H6.GT			3000	6	34
11	NMHC.GT			3000	10	33
12	PT08.S2.NMHC			3000	6	34
13	PT08.S4.NO2			3000	7	37
14	PT08.S5.O3			3000	6	32
15	NOx.GT			3000	10	35
16	NO2.GT			3000	15	36
17	PT08.S3.NOx			3000	6	32
18	Temperature			3000	8	25
19	RH			3000	6	34
20	Humidity			3000	7	41
21	Electricity Total Load	Hospital Energy Loads [7]	Hourly values from Jan. 1, 2016 to Mar. 25, 2016	3000	14	44
22	Equipment Load			3000	10	35
23	Gas Energy			3000	7	29
24	Gas Heat Energy			3000	8	26
25	Water heater Energy			3000	10	18
26	Foreign exchange rates	Data market [14]	Daily, from Dec. 31, 1979 to Dec. 31, 1998	3000	12	25
27	Rainfall in Melbourne		Daily, from from 1981 to 1990	3000	27	33
28	Mean flow Saugeen river		Daily, from from 1981 to 1990	3000	7	50
29	No. of Births in Quebec		Daily, from Jan. 1, 1977 to Dec. 31, 1990	3000	6	21
30	Mean Wave Height		Hourly, from Nov. 9, 2004 to Sep. 9, 2005	3000	17	39

4.1   Experimental Setup

The methods are evaluated using the root mean squared error (RMSE) on a repeated holdout procedure with 20 testing periods. For each repetition, a random point in time is chosen from the full time window available for each series, and the previous window consisting of 60% of the data set size is used for training the ensemble while the following window of size 10% is used for testing. This approach was evidenced to provide robust estimates relative to other estimation procedures [6]. The validation set $\mathcal{D}$ described in Algorithm 1 consists of the final 30% observations of the training set. Specifically, each model $h \in H$ is initially trained in 70% of the training set, and $\mathcal{C}_H$ is built using the following 30% observations. Then, all models are retrained using the complete training set.

We estimate the size K of the feature vectors $x \in \mathbb{X}$ using the false nearest neighbours approach [17]. The estimated value for each series is shown in Table 1. The parameters of OC, α and λ, are set to 30 and 100, respectively. This means that the elements of each subset in $\mathcal{C}_H$ show a better rank than the elements of other subsets of the same size for a contiguous window of 30 points. Each point denotes the average loss of each model in the last 100 observations. The number (averaged across the 20 testing periods) of subsets comprising $\mathcal{C}_H$ is also described in Table 1 ($|\mathcal{C}_H|$).

4.2   Set of Hypotheses H and Aggregation Approaches

The set of models H forming the ensemble are built using the following **15** learning algorithms: support vector regression with Gaussian, Laplace, and linear kernels (SVR_g, SVR_lp, SVR_l, respectively) [16]; Gaussian processes with Gaussian, Laplace, and linear kernels (GP_g, GP_lp, GP_l, respectively) [16]; LASSO regression [11], Ridge regression [11]; random forest (RF) [34]; multivariate adaptive regression splines (MARS) [23]; principal components regression (PCR) [22]; partial least squares (PLS) [22]; rule-based regression (RBR) [18]; and projection pursuit regression with SuperSmoother and splines smoothing methods (PPR_SS, and PPR_Spl, respectively) [26]. As a preliminary analysis, in Fig. 2 we show the distribution of the rank of each model across the 30 problems. A rank of 1 means that the respective model was the best performing one in a given dataset.

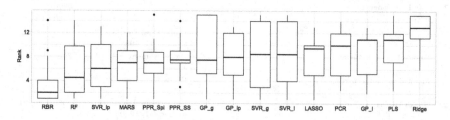

Fig. 2. Boxplot of the rank for each hypothesis in H across the 30 problems

These models are aggregated using the following approaches (c.f. Sect. 2.2 for an overview):

- WindowLoss: Weights computed according to the performance of the models in the last λ observations;
- Simple: Models are averaged using the arithmetic mean;
- FixedShare: The fixed share approach;
- Ridge: An approach that uses Ridge regression to aggregate models;
- ADE: A method for the arbitrage of forecasting models;
- OGD: An approach based on online gradient descent;
- MLpol: The polinomially weighted average forecast combination;
- EWA: Aggregation approach based on an exponentially weighted average.

When the methods are employed using CA, their name is denoted using the prefix "CA" (e.g. CA.Simple). We include the following baselines: LossTrain, in which the original hypotheses H are aggregated according to their RMSE in the training data; CA.SimpleWorst, a variant of CA.Simple in which C_H is built using the consistently worst performers, as opposed to using the best performers – this is accomplished by searching for the *bottom ranked hypotheses* (see line 9 in Algorithm 1); and CA.SimpleRandom, another variant of CA.Simple in which C_H is built randomly – the number of subsets and respective sizes are set according to OC, but these are filled with random models from the available pool. As a forecasting baseline, we include ARIMA [15], an online state of the art approach for time series forecasting.

4.3   Results

On Predictive Performance. Our first experiment is used to study the impact of CA on state of the art forecast combination approaches. In other words, we want to understand if, given an aggregation method, it is better to apply it to the set of hypotheses H' built using CA via OC, or directly to the set of original hypotheses H (non-CA).

Figure 3 addresses this question (**Q1**). The boxplots represent the log percentual difference in RMSE between CA and non-CA, for the different aggregation methods and across the 30 time series. Negative values denote better performance when the methods are applied using CA. These results show that, for almost all aggregation methods CA leads to a better predictive performance in the majority of problems. The exception is MLpol, which is relatively robust to CA.

These results are corroborated by the critical difference diagram in Fig. 4 [9]. In this, almost all state of the art aggregation methods show a better average rank when applied using CA via OC (again, the exception is MLpol). CA.ADE shows the best average rank of all methods. Almost all methods significantly outperform ARIMA, which shows that these are able predict future values of time series better than a state of the art approach (**Q2**).

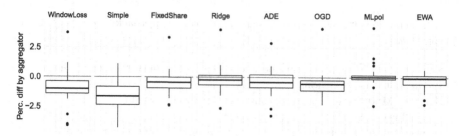

Fig. 3. Log percentual difference in RMSE between CA and non-CA for each aggregation method. Negative values denote a decrease in RMSE (better performance) when using CA.

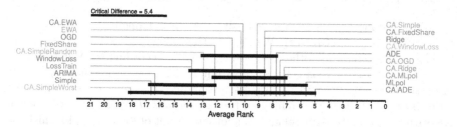

Fig. 4. Critical difference diagram for the post-hoc Nemenyi test [9]

Error Decomposition. From a regression perspective, and according to Brown et al. [4] we decomposed the squared error into bias, variance, and covariance terms as follows: $\overline{bias}^2 + \overline{var}\frac{1}{M} + (1 + \frac{1}{M})\overline{covar} + \sigma^2$, where $\overline{bias}^2$, $\overline{var}$, and $\overline{covar}$ represent the average bias, variance and covariance of the ensemble members, respectively. σ^2 is a constant irreducible term representing the variance of the noise. While a single estimator can be analysed using a bias-variance trade-off, the quadratic error generalisation of a regression ensemble depends on the bias, variance, and covariance of the individual models. The covariance term quantifies the diversity of the ensemble. Increasing values of average covariance denote less diversity. We refer to the work by Brown et al. [4] for a comprehensive read.

To analyse the impact of CA in this decomposition, we measured the percentual difference in each component relative to non-CA, across the 30 problems. Although the decomposition is valid for non-uniformly weighted ensembles [4], in the interest of brevity we focus on the simple average aggregation, i.e. difference between CA.Simple and Simple. This study is reported in Fig. 5. Negative values represent a percentual decrease in the respective term when applying CA. In the left part of the figure we present the decomposition following the proposed approach. For comparison, in the right side we present the same decomposition using the baseline CA.SimpleWorst.

According to the figure, CA.Simple shows an average decrease in the bias term relative to Simple. This outcome is reasonable since CA focus on searching consistently top performing subsets of models, i.e. regions where some multiple

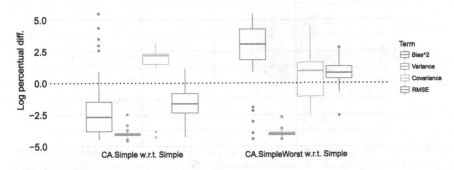

Fig. 5. Log percentual difference in $\overline{bias}^2$, $\overline{var}$, $\overline{covar}$, and RMSE between CA.Simple and Simple (left), and between CA.SimpleWorst and Simple (right)

individual models consistently show better rank than other equal-sized groups of models. There is also a considerable decrease in the average variance term. This is expected since most of the models in H' are combinations of models from H, which, when averaged, decrease variance. Oppositely, this leads to an average increase in the covariance term. This is also expected because each model from H can be part of multiple subsets that form the set of hypotheses H', leading to an increase of the ensembles' redundancy.

CA.SimpleWorst also leads to an average decrease in variance. Although less noticeable relative to CA.Simple, it also leads to a decrease in diversity (increase in covariance). The interesting part of this comparison is that for CA.SimpleWorst the bias term increases considerably, leading to a worse performance relative to Simple. This outcome suggests, as we hypothesised, that CA.Simple improves the performance through an improvement in the average bias of the members of the ensemble, even though it sacrifices some diversity to this effect (**Q3**).

Parameter Sensitivity and Execution Time. To address question **Q4** we analysed the sensitivity of parameters α and λ (c.f. Algorithm 1). This analysis is presented in Fig. 6a. This graphic shows an heatmap with the average rank of each combination of (α, λ) across the 30 problems. For simplicity we focused on the Simple aggregation method, and the set of values of each parameter were chosen arbitrarily. Overall, when α and λ are not set with too low values (from the searched grid) the average rank is relatively stable. In practice, the most appropriate set of values strongly depend on the data and the portfolio of models H.

Regarding question **Q5**, we studied the execution time of CA. Again, in the interest of conciseness we focused on the Simple aggregation method. To carry out this analysis we compute the time spent to train and aggregate the ensemble. Then, we measure the time difference between CA and non-CA for each time series. The results are reported in Fig. 6b, demonstrating that, on average, the

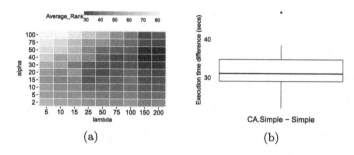

(a)   (b)

Fig. 6. (a) Heatmap illustrating the average rank CA for varying α and λ parameters with the `Simple` method. (b) distribution of difference in execution time (seconds) when using `Simple` aggregation with and without CA.

`Simple` method using CA takes around 30 s more than when not using CA. This overhead is caused by the creation of $\mathcal{C}_H$ via OC. Notwithstanding, we note that the difference in execution time also depends on the aggregation approach, i.e. how it scales with the number of predictors.

On Pruning. As we already mentioned, CA via OC builds the set of committees $\mathcal{C}_H$ focusing on consistently top performers. In this context, it might be argued that the improvements in performance are due to avoiding poor predictors, and it could be reached by simply pruning them from the aggregation rule.

To test this hypothesis we compared `CA.Simple` with an approach that quantifies the weight of each model according to the average rank in the last λ observations (denoted as `AvgRank`). We focus on the rank because it is the metric we use to build $\mathcal{C}_H$. Moreover, we apply `AvgRank` with a decreasing number of models, where the predictors are dynamically suspended (assigned weight 0) according to the `AvgRank`. For example, when using 10 out of the available 15 models (`AvgRank.10`) and for a given time step we do as follows. We compute the average rank of each of the 15 models in the last λ observations. Then, we drop the worst 5 models, weighing the remaining ones (w.r.t. `AvgRank`).

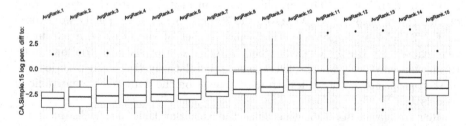

Fig. 7. Log percentual difference in RMSE of `AvgRank` with a decreasing number of predictors (denoted as sufix) relative to `CA.Simple`. Negative values denote lower RMSE by `CA.Simple`.

The results of this analysis are presented in Fig. 7. The suffix in the name of each approach denotes the number of members in each ensemble. The boxplots represent the log percentual difference in RMSE of each variant of `AvgRank` relative to `CA.Simple`, across the 30 time series. Results show that `CA.Simple` presents a better performance, which is increasing as `AvgRank` is applied with a decreasing number of models. This outcome suggests that CA is not simply pruning poor predictors (**Q6**) from the aggregation rule.

5 Discussion

5.1 On the Trade-Off Between Individual Error and Diversity

This paper follows evidence from previous work by Webb et al. [29,30], which showed that combining different ensemble approaches leads to a better trade-off between individual error of the ensemble members and diversity.

The original motivation of Webb with `Multiboosting` [29] was to increase diversity while maintaining a reasonable individual error. Notwithstanding, Webb and Zheng [30] later report different approaches where the inverse happens: cases where both diversity and individual error decrease, also leading to a lower ensemble error. The results from our experiments follow the second case. However interesting, these incite further investigation. Particularly, research into the mechanisms behind the success of this improved trade-off.

5.2 Other Limitations and Future Work

We presented OC for retrieving the set of committees C_H with a small execution time overhead. Despite the results showing a systematic improvement in predictive performance, our intuition is that this approach can be further improved. For example, as presented (c.f. Algorithm 1), OC searches for subsets of H of all sizes (except the full size of H), which can lead to an unnecessary redundancy. This can be particularly important if the portfolio of models H is larger than in our experimental design. We can potentially overcome this problem by introducing a *depth* parameter, which controls how large the subsets should be.

Contrary to other approaches, CA combines different sub-ensembles after the learning process, starting from a portfolio of heterogeneous models. This can be advantageous in terms of flexibility: sub-ensembles can be updated, new models can be added to the portfolio H, or obsolete ones removed.

This paper is focused on forecasting problems. In particular, the proposed algorithm OC capitalises on the time dependency among observations. Notwithstanding, our intuition is that the basic idea behind CA can be generalised to i.i.d. domains, e.g. standard regression tasks. We will also investigate this issue in future work.

6   Summary

Constructive aggregation (CA) rearranges a set of independently created hypotheses H into different subsets $\mathcal{C}_H$. This is achieved using out-performance contiguity (OC), which searches for groups of models that outperform other groups of same size contiguously during some time interval. These subsets are then aggregated into different combined hypotheses H'.

Applying state of the art aggregation approaches for forecasting to H' is demonstrated to provide better performance relative to applying them to H, on average. Moreover, this is accomplished with mild execution time overhead.

The results also suggest that the improvement in performance is mainly due to a decrease in individual error of the members of the ensemble. Although diversity also decreases, CA leads to a better trade-off between these two factors. The proposed method is publicly available in a software package.

Acknowledgements. This work is financed by Project "Coral - Sustainable Ocean Exploitation: Tools and Sensors/NORTE-01-0145-FEDER-000036", which is financed by the North Portugal Regional Operational Programme (NORTE 2020), under the PORTUGAL 2020 Partnership Agreement, and through the European Regional Development Fund (ERDF).

References

1. Aiolfi, M., Timmermann, A.: Persistence in forecasting performance and conditional combination strategies. J. Econ. **135**(1), 31–53 (2006)
2. Bauer, E., Kohavi, R.: An empirical comparison of voting classification algorithms: bagging, boosting, and variants. Mach. Learn. **36**(1–2), 105–139 (1999)
3. Brown, G.: Ensemble learning. In: Sammut, C., Webb, G.I. (eds.) Encyclopedia of Machine Learning, pp. 312–320. Springer, Boston (2010). https://doi.org/10.1007/978-0-387-30164-8_252
4. Brown, G., Wyatt, J.L., Tiňo, P.: Managing diversity in regression ensembles. J. Mach. Learn. Res. **6**, 1621–1650 (2005)
5. Caruana, R., Niculescu-Mizil, A., Crew, G., Ksikes, A.: Ensemble selection from libraries of models. In: Proceedings of the Twenty-First International Conference on Machine Learning, p. 18. ACM (2004)
6. Cerqueira, V., Torgo, L., Smailović, J., Mozetič, I.: A comparative study of performance estimation methods for time series forecasting, pp. 529–538. IEEE (2017)
7. Cerqueira, V., Torgo, L., Pinto, F., Soares, C.: Arbitrated ensemble for time series forecasting. In: Ceci, M., Hollmén, J., Todorovski, L., Vens, C., Džeroski, S. (eds.) ECML PKDD 2017. LNCS (LNAI), vol. 10535, pp. 478–494. Springer, Cham (2017). https://doi.org/10.1007/978-3-319-71246-8_29
8. Cesa-Bianchi, N., Lugosi, G.: Prediction, Learning, and Games. Cambridge University Press, New York (2006)
9. Demšar, J.: Statistical comparisons of classifiers over multiple data sets. J. Mach. Learn. Res. **7**, 1–30 (2006)
10. Freund, Y., Schapire, R.E.: A decision-theoretic generalization of on-line learning and an application to boosting. J. Comput. Syst. Sci. **55**(1), 119–139 (1997)

11. Friedman, J., Hastie, T., Tibshirani, R.: Regularization paths for generalized linear models via coordinate descent. J. Stat. Softw. **33**(1), 1–22 (2010)
12. Gaillard, P., Goude, Y.: Forecasting electricity consumption by aggregating experts; how to design a good set of experts. In: Antoniadis, A., Poggi, J.-M., Brossat, X. (eds.) Modeling and Stochastic Learning for Forecasting in High Dimensions. LNS, vol. 217, pp. 95–115. Springer, Cham (2015). https://doi.org/ 10.1007/978-3-319-18732-7_6
13. Gama, J., Žliobaitė, I., Bifet, A., Pechenizkiy, M., Bouchachia, A.: A survey on concept drift adaptation. ACM Comput. Surv. (CSUR) **46**(4), 44 (2014)
14. Hyndman, R.: Time series data library. http://data.is/TSDLdemo. Accessed 11 Dec 2017
15. Hyndman, R.J., et al.: forecast: Forecasting functions for time series and linear models, R package version 5.6 (2014)
16. Karatzoglou, A., Smola, A., Hornik, K., Zeileis, A.: kernlab - an S4 package for kernel methods in R. J. Stat. Softw. **11**(9), 1–20 (2004)
17. Kennel, M.B., Brown, R., Abarbanel, H.D.: Determining embedding dimension for phase-space reconstruction using a geometrical construction. Phys. Rev. A **45**(6), 3403 (1992)
18. Kuhn, M., Weston, S., Keefer, C., Coulter N.: C code for Cubist by Ross Quinlan. In: Cubist: Rule-and Instance-Based Regression Modeling, R package version 0.0.18 (2014)
19. Kuncheva, L.I.: A theoretical study on six classifier fusion strategies. IEEE Trans. Pattern Anal. Mach. Intell. **24**(2), 281–286 (2002)
20. Kuncheva, L.I.: Combining Pattern Classifiers: Methods and Algorithms. Wiley, New York (2004)
21. Lichman, M.: UCI machine learning repository (2013). https://archive.ics.uci.edu/ ml
22. Mevik, B.H., Wehrens, R., Liland, K.H.: pls: Partial Least Squares and Principal Component Regression, r package version 2.6-0 (2016)
23. Milborrow, S.: earth: Multivariate Adaptive Regression Spline Models. Derived from mda:mars by Trevor Hastie and Rob Tibshirani (2012)
24. Newbold, P., Granger, C.W.: Experience with forecasting univariate time series and the combination of forecasts. J. R. Stat. Society. Ser. (Gen.) **137**, 131–165 (1974)
25. Pfahringer, B.: Winning the KDD99 classification cup: bagged boosting. ACM SIGKDD Explor. Newsl. **1**(2), 65–66 (2000)
26. R Core Team: R: A Language and Environment for Statistical Computing. R Foundation for Statistical Computing, Austria, Vienna (2013)
27. van Rijn, J.N., Holmes, G., Pfahringer, B., Vanschoren, J.: The online performance estimation framework: heterogeneous ensemble learning for data streams. Mach. Learn. **107**, 1–28 (2018)
28. Timmermann, A.: Forecast combinations. In: Handbook of Economic Forecasting, vol. 1, pp. 135–196 (2006)
29. Webb, G.I.: Multiboosting: a technique for combining boosting and wagging. Mach. Learn. **40**(2), 159–196 (2000)
30. Webb, G.I., Zheng, Z.: Multistrategy ensemble learning: reducing error by combining ensemble learning techniques. IEEE Trans. Knowl. Data Eng. **16**(8), 980–991 (2004)
31. Wnek, J., Michalski, R.S.: Hypothesis-driven constructive induction in AQ17-HCI: a method and experiments. Mach. Learn. **14**(2), 139–168 (1994)

32. Wolpert, D.H.: Stacked generalization. Neural Netw. **5**(2), 241–259 (1992)
33. Wolpert, D.H.: The lack of a priori distinctions between learning algorithms. Neural Comput. **8**(7), 1341–1390 (1996)
34. Wright, M.N.: ranger: A Fast Implementation of Random Forests, R package (2015)
35. Yu, Y., Zhou, Z.H., Ting, K.M.: Cocktail ensemble for regression. In: 7th IEEE International Conference on Data Mining, ICDM 2007, pp. 721–726. IEEE (2007)
36. Zinkevich, M.: Online convex programming and generalized infinitesimal gradient ascent. In: Proceedings of the 20th International Conference on Machine Learning, ICML 2003, pp. 928–936 (2003)

MetaBags: Bagged Meta-Decision Trees for Regression

Jihed Khiari[1]([✉]), Luis Moreira-Matias[1], Ammar Shaker[1], Bernard Ženko[2], and Sašo Džeroski[2]

[1] NEC Laboratories Europe GmbH, Heidelberg, Germany
jihed.khiari@neclab.eu
[2] Jožef Stefan Institute, Ljubljana, Slovenia

Abstract. Methods for learning heterogeneous regression ensembles have not yet been proposed on a large scale. Hitherto, in classical ML literature, stacking, cascading and voting are mostly restricted to classification problems. Regression poses distinct learning challenges that may result in poor performance, even when using well established homogeneous ensemble schemas such as bagging or boosting. In this paper, we introduce MetaBags, a novel stacking framework for regression. MetaBags learns a set of meta-decision trees designed to select one base model (i.e. *expert*) for each query, and focuses on inductive bias reduction. Finally, these predictions are aggregated into a single prediction through a bagging procedure at meta-level. MetaBags is designed to learn a model with a fair bias-variance trade-off, and its improvement over base model performance is correlated with the prediction diversity of different experts on specific input space subregions. An exhaustive empirical testing of the method was performed, evaluating both generalization error and scalability of the approach on open, synthetic and real-world application datasets. The obtained results show that our method outperforms existing state-of-the-art approaches.

Keywords: Stacking · Regression · Meta-learning · Landmarking

1 Introduction

Ensemble refers to a collection of several models (i.e., experts) that are combined to address a given task (e.g. obtain a lower generalization error for supervised learning problems) [24]. Ensemble learning can be divided in three different stages [24]: (i) base model *generation*, where z multiple possible hypotheses $\hat{f}_i(x), i \in \{1..z\}$ to model a given phenomenon $f(x) = p(y|x)$ are generated; (ii) model *pruning*, where $c \leq z$ of those are kept and (iii) model *integration*, where these hypotheses are combined, i.e. $\hat{F}(\hat{f}_1(x), ..., \hat{f}_c(x))$. Naturally, the process may require large computational resources for (i) and/or large and representative training sets to avoid overfitting, since $\hat{F}$ is also learned on the training set, which was already used to train the base models $\hat{f}_i(x)$ in (i). Since the

© Springer Nature Switzerland AG 2019
M. Berlingerio et al. (Eds.): ECML PKDD 2018, LNAI 11051, pp. 637–652, 2019.
https://doi.org/10.1007/978-3-030-10925-7_39

pioneering Netflix competition in 2007 [1] and the introduction of cloud-based solutions for data storing and/or large-scale computations, ensembles have been increasingly used in industrial applications. For instance, *Kaggle*, the popular competition website, where, during the last five years, 50+% of the winning solutions involved at least one ensemble of multiple models [21].

Ensemble learning builds on the principles of committees, where there is typically never a single expert that outperforms all the others on each and every query. Instead, we may obtain a better overall performance by *combining* answers of multiple experts [28]. Despite the importance of the combining function $\hat{F}$ for the success of the ensemble, most of the recent research on ensemble learning is either focused on (i) model generation and/or (ii) pruning [24].

Model integration approaches are grouped in three clusters [30]: (a) voting (e.g. bagging [4]), (b) cascading [18] and (c) stacking [33]. In voting, the outputs of the ensemble is a (weighted) average of outputs of the base models. Cascading iteratively combines the outputs of the base experts by including them, one at a time, as another feature in the training set. Stacking learns a meta-model that combines the outputs of all the base models. Voting relies on base models to have complementary expertise, which is an assumption that is rarely true in practice (e.g. check Fig. 1(b,c)). On the other hand, cascading is typically too time-consuming to be put in practice, since it involves training of several models in a sequential fashion.

Stacking relies on the power of the meta-learning algorithm to approximate $\hat{F}$. Stacking approaches are of two types: parametric and non-parametric. The first (and most common [21]) assumes a (typically linear) functional form for $\hat{F}$, while its coefficients are either learned or estimated [7]. The second follows a strict meta-learning approach [3], where a meta-model for $\hat{F}$ is learned in a non-parametric fashion by relating the characteristics of problems (i.e. properties of the training data) with the performance of the experts. Notable approaches include instance-based learning [32] and decision trees [30]. However, novel approaches for model integration in ensemble learning are primarily designed for classification and, if at all, adapted later on for regression [24,30,32]. While such adaptation may be trivial in many cases, it is noteworthy that regression poses distinct challenges.

Formally, we formulate a regression problem as the problem of learning a function

$$\hat{f}_\theta : x_i \to \mathbb{R} \quad \text{such that} \quad \hat{f}(x_i; \theta) \simeq f(x_i) = y_i, \forall x_i \in X, y_i \in Y \qquad (1)$$

where $f(x_i)$ denotes the true unknown function which is generating the samples' target variable values, and $\hat{f}(x_i; \theta) = \hat{y}_i$ denotes an approximation dependent on the feature vector x_i and an unknown (hyper)parameter vector $\theta \in \mathbb{R}^n$. One of the key differences between regression and classification is that for regression the range of f is apriori undefined and potentially infinite. This issue raises practical hazards for applying many of the widely used supervised learning algorithms, since some of them cannot predict outside of the target range of their training set values (e.g. Generalized Additive Models (GAM) [20] or CART [6]).

Another major issue in regression problems are **outliers**. In classification, one can observe either *feature* or *concept* outliers (i.e. outliers in $p(x)$ and $p(y|x)$), while in regression one can also observe *target* outliers (in $p(y)$). Given that the true target domain is unknown, these outliers may be very difficult to handle with common preprocessing techniques (e.g. Tukey's boxplot or one-class SVM [9]). Figure 1 illustrates these issues in practice on a synthetic example with different regression algorithms. Although the idea of training different experts in parallel to subsequently combine them seems theoretically attractive, the above-mentioned issues make it difficult in practice, especially for regression. In this context, stacking is regarded as a complex *art* of finding the right combination of data preprocessing, model generation/pruning/integration and post-processing approaches for a given problem.

In this paper, we introduce MetaBags, a novel stacking framework for regression. MetaBags is a powerful meta-learning algorithm that learns a set of meta-decision trees designed to select one expert for each query thus reducing inductive bias. These trees are learned using different types of meta-features specially created for this purpose on data bootstrap samples, whereas the final meta-model output is the average of the outputs of the experts selected by each meta-decision tree for a given query. Our contributions are threefold:

1. A novel meta-learning algorithm to perform non-parametric stacking for regression problems with **minimum user expertise requirements**.
2. An approach for turning the traditional overfitting tendency of stacking into an advantage through the usage of **bagging at the meta-level**.
3. A novel set of **local landmarking meta-features** that characterize the learning process in feature subspaces and enable model integration for regression problems.

In the remainder of this paper, we describe the proposed approach, after discussing related work. We then present an exhaustive experimental evaluation of its efficiency and scalability in practice. This evaluation employs 17 regression datasets (including one real-world application) and compares our approach to existing ones.

2   Related Work

Since its first appearance, meta-learning has been defined in multiple ways that focus on different aspects such as collected experience, domain of application, interaction between learners and the knowledge about the learners [23]. Brazdil et al. [3] define meta-learning as the learning that deals with both types of bias, declarative and procedural. The declarative bias is imposed by the hypothesis space form which a base learner chooses a model, whereas the procedural bias defines how different hypotheses should be preferred. In a recent survey, Lemke et al. [23] characterize meta-learning as the learning that constitutes

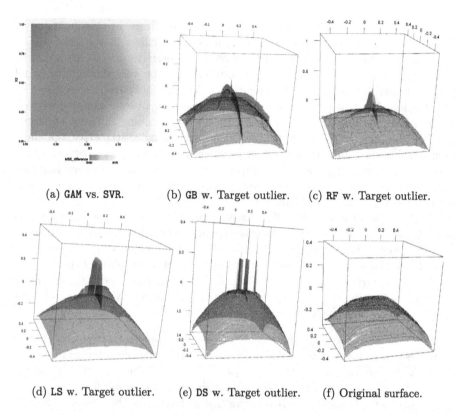

(a) GAM vs. SVR.   (b) GB w. Target outlier.   (c) RF w. Target outlier.

(d) LS w. Target outlier.   (e) DS w. Target outlier.   (f) Original surface.

Fig. 1. Illustration of distinctive issues in regression problems on a synthetic example. In all experiments, we generate 1k training examples for the function $y = (x_1^4 + x_2^4)^{\frac{1}{2}}$. In (a), x_1, x_2 training values are sampled from a uniform distribution constrained to $\in [0, 0.8]$, while the testing ones are $\in [0, 1]$. Panel (a) depicts the difference between RMSE between the two tested methods, GAM and SVR, where the hyperparameters were tuned using random search (60 points) and a 3-fold-CV procedure was used for error estimation. SVR's MSE is significantly larger than GAM's one, and still, there are several regions of the input space where GAM is outperformed (in light pink colors). Panels (b,c) depict the regression surface of two models learned using tree-based Gradient Boosting machines (GB) and Random Forests (RF), respectively, with 100 trees and default hyperparameter settings. To show their sensitivity to *target* outliers, we artificially imputed one extremely high value (in black) in the target of one single example (where the value is already expected to be maximum). In Panels (d, e), we analyze the same effects with two stacking approaches using the models fitted in (a, b, c) as base learners: Linear Stacking (LS) in (d) and Dynamic Selection (DS) with kNN in (e). Please note how deformed the regression surfaces (in gray) are in all settings (b–d). Panel (f) depicts the original surface. Best viewed in color. (Color figure online)

three essential aspects: (i) the adaptation with experience, (ii) the consideration of meta-knowledge of the data set (to be learned from) and (iii) the integration of meta-knowledge from various domains. Under this definition, both ensemble methods bagging [4] and boosting [14] do not qualify as meta-learners, since the base learners in bagging are trained independently of each other, and in boosting, no meta-knowledge from different domains is used when combining decisions from the base learners. Using the same argument, stacking [33] and cascading [18] cannot be definitely considered as meta-learners [23].

Algorithm recommendation, in the context of meta-learning, aims to propose the type of learner that best fits a specific problem. This recommendation can be performed after considering both the learner's performance and the characteristics of the problem [23]. Both aforementioned aspects qualify as meta-features that assist in deciding which learner could perform best on a specific problem. We note three classes of meta-features [3]: (i) meta-features of the dataset describing its statistical properties such as the number of classes and attributes, the ratio of target classes, the correlation between the attributes themselves, and between the attributes and the target concept, (ii) model-based meta-features that can be extracted from models learned on the target dataset, such as the number of support vectors when applying SVM, or the number of rules when learning a system of rules, and (iii) landmarkers, which constitute the generalization performance of diverse set of learners on the target dataset in order to gain insights into which type of learners fits best to which regions/subspaces of the studied problem. Traditionally, landmarkers have been mostly proposed in a classification context [3,27]. A notorious exception is proposed by Feurer et al. [12]. The authors use meta-learning to generate prior knowledge to feed a bayesian optimization procedure in order to find the best sequence of algorithms to address predefined tasks in either classification and regression pipelines. However, the original paper [13] focuses mainly on classification.

The dynamic approach of ensemble integration [24] postpones the integration step till prediction time so that the models used for prediction are chosen dynamically, depending on the query to be classified. Merz [25] applies dynamic selection (DS) locally by selecting models that have good performance in the neighborhood of the observed query. This can be seen as an integration approach that considers type-(iii) landmarkers. Tsymbal et al. [32] show how DS for random forests decreases the bias while keeping the variance unchanged.

In a classification setting, Todorovski and Džeroski [30] combine a set of base classifiers by using meta-decision trees which in a leaf node give a recommendation of a specific classifier to be used for instances reaching that leaf node. Meta-decision trees (MDT) are learned by stacking and use the confidence of the base classifiers as meta-features. These can be viewed as landmarks that characterizes the learner, the data used for learning and the example that needs to be classified. Most of the suggested meta-features MDT are applicable to classification problems only.

MetaBags can be seen as a generalization of DS [25,32] that uses meta-features instead. Moreover, we considerably reduce DS runtime complexity (generically, $\mathcal{O}(N)$ in test time, even with state-of-the-art search heuristics [2]), as well as the user-expertise requirements to develop a proper metric for each problem. Finally, the novel type of local landmarking meta-features characterize the local learning process - aiming to avoid overfitting.

3   Methodology

This Section introduces MetaBags and its three basic components: (1) First, we describe a novel algorithm to learn a decision tree that picks one expert among all available ones to address a particular query in a supervised learning context; (2) then, we depict the integration of base models at the meta-level with bagging to form the final predictor $\hat{F}$; (3) Finally, the meta-features used by MetaBags are detailed. An overview of the method is presented in Fig. 2.

3.1   Meta-Decision Tree for Regression

Problem Setting. In traditional stacking, $\hat{F}$ just depends on the base models $\hat{f}_i$. In practice, as stronger models may outperform weaker ones (c.f. Fig. 1(a)), they get assigned very high coefficients (assuming we combine base models with a linear meta-model). In turn, weaker models may obtain near-zero coefficients. This can easily lead to over-fitting if a careful model generation does not take place beforehand (c.f. Fig. 1(d,e)). However, even a model that is weak in the whole input space may be strong in some subregion. In our approach we rely on classic tree-based isothetic boundaries to identify *contexts* (e.g. subregions of the input space) where some models may outperform others, and by using only strong experts within each context, we improve the final model.

Let the dataset $\mathbb{D}$ be defined as $(x_i, y_i) \in \mathbb{D} \subset \mathbb{R}^n \times \mathbb{R} : i = \{1, \dots, N\}$ and generated by an unknown function $f(x) = y$, where n is the number of features of an instance x, and y denotes a numerical response. Let $\hat{f}_j(x) : j = \{1, .., M\}$ be a set of M base models (*experts*) learned using one or more base learning methods over $\mathbb{D}$. Let $\mathcal{L}$ denote a loss function of interest decomposable in independent bias/variance components (e.g. $L2$-loss). For each instance x_i, let $\{z_{i,1}, \dots, z_{i,Q}\}$ be the set of meta-features generated for that instance.

Starting from the definition of a decision tree for supervised learning introduced in CART [6], we aim to build a *classification* tree that, for a given instance x and its supporting meta-features $\{z_1, \dots, z_Q\}$, dynamically selects the expert that should be chosen for prediction, i.e., $\hat{F}(x, z_1, \dots, z_Q; \hat{f}_1, \dots, \hat{f}_M) = \hat{f}_j(x)$. As for the tree induction procedure, we aim, at each node, at finding the feature z_j and the splitting point z_j^t that leads to the maximum reduction of impurity. For the internal node p with the set of examples $\mathbb{D}_p \in \mathbb{D}$ that reaches p, the splitting point z_j^t splits the node p into the leaves p_l and p_r with the sets

$\mathbb{D}_{p_l} = \{x_i \in \mathbb{D}_p | z_{ij} \leq z_j^t\}$ and $\mathbb{D}_{p_r} = \{x_i \in \mathbb{D}_p | z_{ij} > z_j^t\}$, respectively. This can be formulated by the following optimization problem at each node:

$$\underset{z_j^t}{\arg\max} \quad \omega(z_j^t) \tag{2}$$

$$\text{s.t.} \quad \omega(z_j^t) = [I(p) - P_l I(p_l) - P_r I(p_r)] \tag{3}$$

where P_l, P_r denote the probability of each branch to be selected, while I denotes the so-called *impurity function*. In traditional classification problems, the functions applied here aim to minimize the entropy of the target variable. Hereby, we propose a new impurity function for this purpose denoted as Inductive Bias Reduction. It goes as follows:

$$I(p) = \text{IBR}(p) = \min_{j \in \{1...M\}} E\left[B\left(\mathcal{L}(p, \hat{f}_j)\right)^2\right] \tag{4}$$

where $B(\mathcal{L})$ denotes the inductive bias component of the loss $\mathcal{L}$.

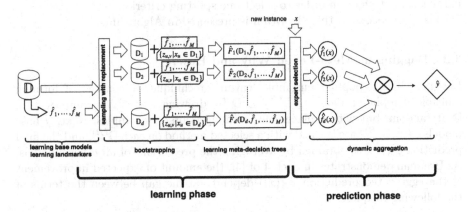

Fig. 2. MetaBags: The learning phase consists of (i) the learning of base models and the landmarkers, (ii) bootstrapping and finally (iii) the learning of the meta decision trees from each bootstrap. The prediction for an unseen example is achieved by consulting each meta decision tree and then aggregating their predictions.

Optimization. To solve the problem of Eq. (2), we address three issues: (i) splitting criterion/meta-feature, (ii) splitting point and (iii) stopping criterion. To select the splitting criterion, we start by constructing two auxiliary equally-sized matrices $a \in \mathbb{R}^{Q \times \phi}$ and $b : z_{i_{\min}} \leq b_{i,j} \leq z_{i_{\max}}, \forall i, j$, where $\phi \in \mathbb{N}, Q$ denote a user-defined hyperparameter and the number of meta-features used, respectively. Then, the matrices are populated with candidate values by elaborating over the Eq. (2, 3, 4) as

$$a_{i,j} = \omega(b_{i,j}), \quad b_{i,j} \sim U(z_{i_{\min}}, z_{i_{\max}}), \tag{5}$$

where $b_{i,j}$ is the jth splitting criterion for the ith meta feature.

First, we find the splitting criteria τ such that

$$\tau = \underset{i \in \{1..Q\}}{\arg\max}\ a_{i,j}, \quad \forall j \in \{1..\phi\}. \tag{6}$$

Secondly, we need to find the optimal splitting point according to the z_τ criteria. We can either take the splitting point already used to find τ or, alternatively, fine-tune the procedure by exploring further the domain of z_τ. For the latter problem, any scalable search heuristic can be applied (e.g.: Golden-section search algorithm [22]).

Thirdly, (iii) the stopping criteria to constraint Eq. (2). Here, like CART, we propose to create fully grown trees. Therefore, it goes as follows:

$$\omega\big(z_\tau^t\big) < \epsilon \ \lor \ |\mathbb{D}_p| < \upsilon : \epsilon \in \mathbb{R}^+, \upsilon \in \mathbb{N} \tag{7}$$

where ϵ, υ are user-defined hyperparameters. Intuitively, this procedure consists in randomly finding ϕ possible *partitioning* points on each meta-feature in a parallelizable fashion in order to select one splitting criterion.

The pseudocode of this algorithm is presented in Algorithm 1.

3.2 Bagging at Meta-Level: Why and How?

Bagging [4] is a popular ensemble learning technique. It consists of forming multiple d replicate datasets $\mathbb{D}^{(B)} \subset \mathbb{D}$ by drawing $s << N$ examples from $\mathbb{D}$ at random, but with replacement, forming bootstrap samples. Next, d base models $\varphi(x_i, \mathbb{D}^{(B)})$ are learned with a selected method on each $\mathbb{D}^{(B)}$, and the final prediction $\varphi_A(x_i)$ is obtained by averaging the predictions of all d base models. As Breiman demonstrates in Sect. 4 of [4], the amount of expected improvement of the aggregated prediction $\varphi_A(x_i)$ depends on the gap between the terms of the following inequality:

$$E\big[\mathcal{L}\big(\varphi(x_i, \mathbb{D}^{(B)})\big)\big]^2 \le E\big[\mathcal{L}\big(\varphi(x_i, \mathbb{D}^{(B)})\big)^2\big]. \tag{8}$$

In our case, $\varphi(x_i, \mathbb{D}^{(B)})$ is given by the $\hat{f}_j(x_i)$ selected by each meta-decision tree induced in each $\mathbb{D}^{(B)}$. By design, the procedure to learn this specific meta-decision tree is likely to overfit its training set, since all the decisions envisage reduction of inductive bias alone. However, when used in a bagging context, this turns to be an advantage because it causes instability of φ - as each tree may be selecting different predictors to each instance x_i.

3.3 Meta-Features

MetaBags is fed with three types of meta-features: (a) base, (b) performance-related and (c) local landmarking. These types are briefly explained below, as well as their connection with the state of the art.

(a) **Base Features.** Following [30], we propose to include all base features also as meta-features. This aims to stimulate a higher inequality in Eq. (8) due to the increase of inductive variance of each meta-predictor.

(b) **Performance-Related Features.** This type of meta-features describes the performance of specific learning algorithms in particular learning contexts on the same dataset. Besides the base learning algorithms, we also propose the usage of landmarkers. Landmarkers are ML algorithms that are computationally relatively *cheap* to run either in a train or test setting [27]. The resulting models aim to characterize the learning task (e.g. is the regression curve linear?). To the authors' best knowledge, so far, all proposed landmarkers and consequent meta-features have been primarily designed for classical meta-learning applications to classification problems [3,27], whereas we focus on model integration for regression. We use the following learning algorithms as landmarkers: LASSO [29], 1NN [10], MARS [15] and CART [6].

To generate the meta-features, we start by creating one landmarking model per method over the entire training set. Then, we design a *small* artificial neighborhood of size ψ of each training example x_i as $X'_i = \{x'_{i,1}, x'_{i,2}..x'_{i,\psi}\}$ by perturbing x_i with *gaussian noise* as follows:

$$x'_{i,j} = x_{i,j} + \xi : \xi \sim \mathcal{N}_n(0,1), \forall j \in \{1,..,\psi\} \qquad (9)$$

where ψ, is a user-defined hyperparameter. Then, we obtain outputs of each expert as well as of each landmarker given X'_i. The used meta-features are then descriptive statistics of the models' outputs: mean, stdev., 1st/3rd quantile. This procedure is applicable both to training and test examples, whereas the landmarkers are naturally obtained from the training set.

(c) **Local Landmarking Features.** In the original landmarking paper, Pfahringer *et al.* [27] highlight the importance on ensuring that our pool of landmarkers is diverse enough in terms of the different types of inductive bias that they employ, and the consequent relationship that this may have with the base learners performance. However, when observing performance on a neighborhood-level rather than on the task/dataset level, the low performance and/or high inductive bias may have different causes (e.g., inadequate data preprocessing techniques, low support/coverage of a particular subregion of the input space, etc.). These causes, may originate in different types of deficiencies of the model (e.g. low support of leaf nodes or high variance of the examples used to make the predictions in decision trees).

Hereby, we introduce a novel type of landmarking meta-features denoted **local landmarking**. Local landmarking meta-features are designed to characterize the landmarkers/models within the particular input subregion. More than finding a correspondence between the performance of landmarkers and base models, we aim to extract the knowledge that the landmarkers have learned about a particular input neighborhood. In addition to the prediction of each landmarker for a given test example, we compute the following characteristics:

- CART: depth of the leaf which makes the prediction; number of examples in that leaf and variance of these examples;
- MARS: width and mass of the interval in which a test example falls, as well as its distance to the nearest edge;
- 1NN: absolute distance to the nearest neighbor.

4   Experiments and Results

Empirical evaluation aims to answer the following four research questions:

(Q1) Does MetaBags systematically outperform its base models in practice?
(Q2) Does MetaBags outperform other model integration procedures?
(Q3) Do the local landmarking meta-features improve MetaBags performance?
(Q4) Does MetaBags scale on large-scale and/or high-dimensional data?

4.1   Regression Tasks

We used a total of 17 benchmarking datasets to evaluate MetaBags. They are summarized in Table 1. We include 4 proprietary datasets addressing a particular real-world application: public transportation. One of its most common research problems is travel time prediction (TTP). The work in [19] uses features such as scheduled departure time, vehicle type and/ or driver's meta-data. This type of data is known to be particularly noisy due to failures in the data collection, which in turn often lead to issues such as missing data, as well as several types of outliers [26].

Here, we evaluate MetaBags in a similar setting of [19], i.e. by using their four datasets and the original preprocessing. This case study is an undisclosed large urban bus operator in Sweden (BOS). We collected data on four high-frequency routes/datasets R11/R12/R21/R22. These datasets cover a time period of six months.

4.2   Evaluation Methodology

Hereby, we describe the empirical methodology designed to answer (Q1-Q4), including the hyperparameter settings of MetaBags and the algorithms selected for comparing the different experiments.

Hyperparameter Settings. Like other decision tree-based algorithms, MetaBags is expected to be robust to its hyperparameter settings. Table 2 presents the hyperparameters settings used in the empirical evaluation (a sensible default). If any, s and d can be regarded as more sensitive parameters. Their value ranges are recommended to be $0\% < s << 100\%$ and $100 \leq d << 2000$. Please note that these experiments did not included any hyperparameter sensitivity study neither a tuning procedure for MetaBags.

Testing Scenarios and Comparison Algorithms. We put in place two testing scenarios: A and B. In scenario A, we evaluate the generalization error of MetaBags with 5-fold cross validation (CV) with 3 repetitions. As base learners, we use four popular regression algorithms: Support Vector Regression (SVR) [11], Projection Pursuit Regression (PPR) [17], Random Forest RF [5] and Gradient Boosting GB [16]. The first two are popular methods in the chosen application domain [19], while the latter are popular voting-based ensemble methods for regression [21]. The base models had their hyperparameter values tuned with random search/3-fold CV (and 60 evaluation points). We used the implementations in the R package *caret* for both the landmarkers and the base learners. We compare our method to the following ensemble approaches: Linear Stacking LS [7], Dynamic Selection DS with kNN [25,32], and the best individual model. All methods used $l2$-loss as $\mathcal{L}$.

In scenario B, we extend the artificial dataset used in Fig. 1 to assess the computational runtime scalability of the decision tree induction process of MetaReg (using a CART-based implementation) in terms of number of examples and attributes. In this context, we compare our method's training stage to Linear Regression (used for LS) and kNN in terms of time to build k-d tree (DS). Additionally, we also benchmarked C4.5 (which was used in MDT [30]). For the latter, we discretized the target variable using the four quantiles.

4.3   Results

Table 3 presents the performance results of MetaBags against comparison algorithms: the base learners; SoA in model integration such as stacking with a linear model LS and kNN, i.e. DS, as well as the best base model selected using 3-CV i.e. Best; finally, we also included two variants of MetaBags: MetaReg – a singular decision tree, MBwLM – MetaBags without the novel landmarking features. Results are reported in terms of RMSE, as well as of statistical significance (using the using the two-sample t-test with the significance level $\alpha = 0.05$, with the null hypothesis that a given learner M wins against MetaBags after observing the results of all repetitions). Finally, Fig. 3 summarizes those results in terms of percentual improvements, while Fig. 4 depicts our empirical scalability study.

5   Discussion

The results, presented in Table 3, show that MetaBags outperforms existing SoA stacking methods. MetaBags is never statistically significantly worse than any of the other methods, which illustrates its generalization power.

Figure 3 summarizes well the contribution of introducing bagging at the meta-level as well as the novel local landmarking meta-features, with average relative percentages of improvement in performance across all datasets of 12.73% and 2.67%, respectively. The closest base method is GB, with an average percentage of improvement of 5.44%. However, if we weight this average by using the percentage of extreme target outliers of each dataset, the expected improvement goes up to 14.65% - illustrating well the issues of GB depicted earlier in Fig. 1(b).

Table 1. Datasets summary. Fields denote number of #ATTributes and #INStances, the Range of the Target variable, the number of #Target Outliers using Tukey's boxplot(ranges = 1.5,3), their ORIgin, as well as its TYPe (**P**roprietary/**O**pen) and Collection Process (**R**eal/**A**rtificial).

	Properties					Source and type		
	#ATT	#INS	RT	#TO(1.5)	#TO(3.0)	ORI	TYP	CP
R11	12	17953	[1306,10520]	66	9	BOS	P	R
R12	12	16353	[1507,9338]	154	6	BOS	P	R
R21	12	16280	[1434,6764]	341	27	BOS	P	R
R22	12	16353	[884,6917]	146	10	BOS	P	R
Cal. housing	8	20460	[14999,500001]	1071	0	StatLib	O	R
Concrete	8	1030	[2332,82599]	4	0	UCI	O	R
2Dplanes	10	40768	[−999.709,999.961]	4	0	dcc.fc.up.pt [31]	O	A
Delta Ailerons	6	7129	[−0.0021,0,0022]	107	12	dcc.fc.up.pt [31]	O	R
Elevators	18	16559	[0.012,0,078]	842	344	dcc.fc.up.pt [31]	O	R
Parkinsons Tele.	26	5875	[0.022,0732]	206	17	UCI	O	R
Physicochemical	9	45730	[15.228,55.3009]	0	0	UCI	O	R
Pole	48	15000	[0,100]	0	0	dcc.fc.up.pt [31]	O	R
Puma32H	32	8192	[−0.085173,0.088266]	56	0	DELVE	O	R
Red wine quality	12	1599	[3,8]	28	0	UCI	O	R
White wine quality	12	4898	[3,9]	200	0	UCI	O	R
Computer activity								
CPU-small	12	8192	[0,99]	430	294	DELVE	O	R
CPU-activity	21	8192	[0,99]	430	294	DELVE	O	R

Table 2. Hyperparameter settings used in MetaBags.

	Value	Description
ϕ	10	Number of random partitions performed on each meta-feature;
ϵ	$\|I(t_p) \cdot 10^{-2}\|$	Min. abs. bias reduction to perform split;
υ	$N \cdot 10^{-2}$	Min. examples in node to perform split;
ψ	100	Size of the artificial neighborhood generated to compute meta-features;
s	10%	Percentage of examples usage to generate the bootstraps;
d	300	Number of generated meta-decision trees;

Figure 4 also depicts how competitive MetaBags can be in terms of scalability. Although neither outperforming DS nor LS, we want to highlight that lazy learners have their cost in test time - while this study only covered the training stage. Moreover, many of its stages (learning of base learners, performance-based meta-features, local landmarking) as well as subroutines of the MetaBags are independent and thus, trivially parallelizable. Based in the above discussion, (Q1-Q4) can be answered affirmatively.

One possible drawback of MetaBags may be its space complexity - since it requires to train/maintain multiple decision trees and models in memory. Another possible issue when dealing with low latency data mining applications is that the computation of some of the meta-features is not trivial, which may slightly increase its runtimes in test stage. Both issues were out of the scope of the proposed empirical evaluation and represent open research questions.

Like any other stacking approach, MetaBags requires training of the base models apriori. This pool of models need to have some diversity on their responses. Hereby, we explore the different characteristics of different learning algorithms to stimulate that diversity. However, this may not be sufficient.

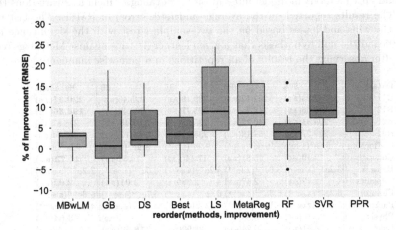

Fig. 3. Summary results of MetaBags using the percentage of improvement over its competitors. Note the consistently positive mean over all methods.

Algorithm 1. InduceMetaDecisionTreeRegression

Input: p: root (or current internal node) of the meta decision tree.
$\mathbb{D}_p \subset \mathbb{D}_i$: the subset of examples that reach the root (or internal node) p, $\mathbb{D}_i$ is a bootstrap sampled with replacement from $\mathbb{D}$.
$\{\hat{f}_j | j \in \{1..M\}\}$: the set of base models.
$\{z_{u,v} | x_u \in \mathbb{D}_p \wedge v \in \{1, \ldots, Q\}\}$: the set of meta features for each instance in $\mathbb{D}_p$.
/* check that the current node has the minimum number of supporting instances   */
1 **if** $|\mathbb{D}_p| \leq v$ **then**
2   ⌊ **return**
   /* create the matrices A and B */
3 $B = [b_{i,j}] \in \mathbb{R}^{Q \times \phi}$ s.t. $b_{i,j} \sim U(z_{i_{\min}}, z_{i_{\max}})$
4 $A = [a_{i,j}] \in \mathbb{R}^{Q \times \phi}$ s.t. $a_{i,j} = \omega(b_{i,j}) = [I(p) - P_l I(p_l) - P_r I(p_r)]$
   /* find the splitting criteria τ */
5 $\tau = \underset{i \in \{1..Q\}}{\arg\min} \ a_{i,j},$ for all $j \in \{1, \ldots, \phi\}$
6 **if** $\omega(z_\tau^t) \geq \epsilon$ **then**
   /* create the right and left leaf nodes, p_l and p_r */
7     $\mathbb{D}_{p_l} = \{x_i \in \mathbb{D}_p | z_{i\tau} \leq z_\tau^t\}$
8     $\mathbb{D}_{p_r} = \{x_i \in \mathbb{D}_p | z_{i\tau} > z_\tau^t\}$
9   InduceMetaDecisionTree$(p_l, \mathbb{D}_{p_l}, \{\hat{f}_j\}, \{z_{u,v}\})$
10   ⌊ InduceMetaDecisionTree$(p_r, \mathbb{D}_{p_r}, \{\hat{f}_j\}, \{z_{u,v}\})$
11 **return**

Formal approaches to strictly ensure diversity on model generation for ensemble learning in regression are scarce [8,24]. The best way to ensure such diversity within an advanced stacking framework like MetaBags is also an open research question.

Table 3. Detailed predictive performance results, comparing base learners vs. MetaBags (top) and SoA methods in model integration vs. MetaBags - including variations (bottom). The results reported on the average and (std. error) of RMSE. The last rows depict the wins and losses based on the two-sample t-test with the significance level $\alpha = 0.05$ and the null hypothesis that a given learner M wins against MetaBags (computed after observing the results of all repetitions in a pair-wise manner).

Dataset	SVR	PPR	RF	GB	MetaBags
R11	232.23(5.2)	242.12(8.1)	229.18(9.8)	225.45(6.5)	**220.37(4.6)**
R12	210.66(3.9)	217.34(5.4)	205.77(2.5)	200.49(6.7)	**194.36(4.2)**
R21	225.81(6.8)	240.12(5.3)	235.55(4.0)	**210.45(7.7)**	218.87(4.9)
R22	260.77(5.9)	269.19(6.5)	255.91(5.3)	**248.11(5.2)**	253.74(2.9)
C. Housing	$9.6e^4$(3.1e³)	$10.0e^4$(5.5e³)	$9.9e^4$(5.0e³)	**$8.9e^4$(3.9e³)**	$9.2e^4$(3.8e³)
Concrete	12.19(2.6)	15.81(2.4)	17.14(2.32)	14.54(1.7)	**11.73(0.3)**
Delta A.	$3.2e^{-4}$(1.2e⁻⁴)	$3.6e^{-4}$(2.4e⁻⁴)	$4.2e^{-4}$(2.3e⁻⁴)	$2.5e^{-4}$(1.8e⁻⁴)	**$2.2e^{-4}$(9.9e⁻⁵)**
2Dplanes	2.12(0.1)	2.57(0.2)	2.36(0.1)	**2.01(0.2)**	2.02(0.1)
Elevators	$6.2e^{-3}$(3.8e⁻⁴)	$6.3e^{-3}$(5.8e⁻⁴)	$6.4e^{-3}$(5.4e⁻⁴)	$6.7e^{-3}$(4.5e⁻⁴)	**$5.6e^{-3}$(5.8e⁻⁴)**
Parkinsons	$5.5e^{-2}$(7.0e⁻³)	$8.1e^{-2}$(7.0e⁻³)	$7.5e^{-2}$(6.0e⁻³)	$6.4e^{-2}$(6.0e⁻³)	**$4.9e^{-2}$(5.0e⁻³)**
Physic.	3.78(0.1)	3.90(0.2)	4.99(0.2)	3.70(0.2)	**3.64(0.2)**
Pole	24.11(7.2)	30.24(9.9)	28.90(5.9)	**18.54(9.3)**	20.12(3.1)
Puma32H	$3.0e^{-2}$(3.0e⁻³)	$2.9e^{-2}$(3.0e⁻³)	$2.8e^{-2}$(3.0e⁻³)	**$2.5e^{-2}$(6.0e⁻³)**	$2.7e^{-2}$(4.0e⁻³)
R. Wine	**0.69(0.0)**	0.91(0.0)	0.88(0.0)	0.79(0.0)	0.70(0.0)
W. Wine	0.70(0.0)	0.86(0.0)	0.76(0.0)	**0.62(0.0)**	**0.62(0.0)**
CPU_a.	**5.18(0.4)**	5.89(0.3)	6.87(0.2)	5.99(0.3)	5.45(0.2)
CPU_s.	6.07(0.1)	6.37(0.3)	8.71(0.4)	6.31(0.5)	**5.12(0.1)**
∅ Rank	2.82	4.47	4.12	2.12	**1.41**
Loss/Win	10/0	11/0	4/0	1/0	N/A

Dataset	LS	DS	Best	MetaReg	MBwLM	MetaBags
R11	230.25(9.2)	222.37(5.4)	223.20(8.9)	240.67(15.7)	228.32(7.0)	**220.37(4.6)**
R12	215.32(5.9)	**190.80(3.5)**	201.46(4.3)	219.76(19.2)	211.63(7.9)	194.37(4.2)
R21	234.37(7.1)	221.98(5.8)	226.27(4.2)	249.31(6.1)	220.01(7.5)	218.87(4.9)
R22	260.71(4.5)	**250.37(5.5)**	254.69(5.2)	273.12(5.4)	261.41(4.3)	253.744(2.9)
C. Housing	$9.8e^4$(4.3e³)	$9.3e^4$(3.2e³)	$9.5e^4$(4.6e³)	$9.8e^4$(4.1e³)	$9.5e^4$(3.7e³)	**$9.2e^4$(3.7e³)**
Concrete	12.14(0.3)	**11.70(0.3)**	12.00(0.3)	12.57(0.3)	11.91(0.3)	11.73(0.3)
Delta A.	$2.8e^{-4}$(2.3e⁻⁴)	$2.6e^{-4}$(1.8e⁻⁴)	$3.0e^{-4}$(1.8e⁻⁴)	$3.5e^{-4}$(1.7e⁻⁴)	$2.3e^{-4}$(1.8e⁻⁴)	**$2.2e^{-4}$(9.9e⁻⁵)**
2Dplanes	2.35(0.2)	2.26(0.1)	2.34(0.1)	2.39(0.1)	2.15(0.1)	**2.02(0.1)**
Elevators	$6.2e^{-3}$(5.5e⁻⁴)	$5.5e^{-3}$(7.0e⁻⁴)	$7.4e^{-3}$(6.1e⁻⁴)	$6.5e^{-3}$(4.6e⁻⁴)	$5.5e^{-3}$(6.8e⁻⁴)	**$5.7e^{-3}$(5.8e⁻⁴)**
Parkinsons	$5.9e^{-2}$(6.0e⁻³)	$6.5e^{-2}$(0.0e⁻³)	$6.2e^{-2}$(5.0e⁻³)	$5.4e^{-2}$(5.0e⁻³)	$5.1e^{-2}$(5.0e⁻³)	**$4.9e^{-2}$(5.0e⁻³)**
Physic.	**3.46(0.2)**	3.74(0.2)	3.79(0.2)	3.80(0.2)	3.71(0.2)	3.64(0.2)
Pole	26.22(3.3)	22.15(3.6)	21.61(3.3)	25.96(4.3)	20.91(3.6)	**20.12(3.1)**
Puma32H	$4.0e^{-2}$(5.0e⁻³)	$3.2e^{-2}$(6.0e⁻³)	$3.4e^{-2}$(4.0e⁻³)	$3.3e^{-2}$(6.0e⁻³)	$2.9e^{-2}$(4.0e⁻³)	**$2.7e^{-2}$(4.0e⁻³)**
R. Wine	0.87(0.0)	0.75(0.0)	**0.67(0.0)**	0.76(0.0)	0.70(0.0)	0.70(0.0)
W. Wine	0.82(0.0)	0.63(0.0)	0.67(0.0)	0.76(0.0)	**0.61(0.0)**	0.62(0.0)
CPU_a.	5.92(0.2)	5.93(5.5)	5.58(0.2)	5.54(0.3)	5.57(0.2)	**5.45(0.2)**
CPU_s.	5.92(0.2)	6.08(0.3)	5.90(0.2)	6.12(0.3)	5.30(0.2)	**5.12(0.2)**
∅ Rank	4.76	3.06	3.88	5.12	2.65	**1.47**
Loss/Win	4/0	2/0	4/0	6/0	0/0	N/A

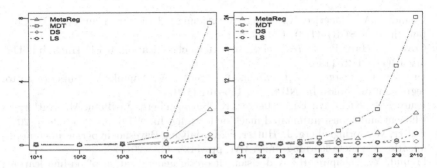

Fig. 4. Empirical runtime scalability analysis resorting to samples (left panel) and features (right panel) size. Times in seconds.

6 Final Remarks

This paper introduces MetaBags: a novel, practically useful stacking framework for regression. MetaBags uses meta-decision trees that perform on-demand selection of base learners at test time based on a series of innovative meta-features. These meta-decision trees are learned over data bootstrap samples, whereas the outputs of the selected models are combined by average. An exhaustive empirical evaluation, including 17 datasets and multiple comparison algorithms illustrates the ability of MetaBags to address model integration problems in regression. As future work, we aim to study which factors affect the performance of MetaBags, namely, at model generation level, as well as its time and spatial complexity in test time.

Acknowledgments. S.D. and B.Ž. are supported by The Slovenian Research Agency (grant P2-0103). B.Ž. is additionally supported by the European Commission (grant 769661 SAAM). S.D. further acknowledges support by the Slovenian Research Agency (via grants J4-7362, L2-7509, and N2-0056), the European Commission (projects HBP SGA2 and LANDMARK), ARVALIS (project BIODIV) and the INTERREG (ERDF) Italy-Slovenia project TRAIN.

References

1. Bell, R., Koren, Y.: Lessons from the netflix prize challenge. ACM SIGKDD Explor. Newsl. **9**(2), 75–79 (2007)
2. Beygelzimer, A., Kakade, S., Langford, J.: Cover trees for nearest neighbor. In: Proceedings of the 23rd ICML, pp. 97–104. ACM (2006)
3. Brazdil, P., Carrier, C., Soares, C., Vilalta, R.: Metalearning: Applications to Data Mining. Springer, Heidelberg (2008)
4. Breiman, L.: Bagging predictors. Mach. Learn. **24**(2), 123–140 (1996)
5. Breiman, L.: Random forests. Mach. Learn. **45**(1), 5–32 (2001)
6. Breiman, L., Friedman, J., Olshen, R., Stone, C.: Classification and regression trees (cart) wadsworth international group, CA, USA, Belmont (1984)
7. Breiman, L.: Stacked regressions. Mach. Learn. **24**(1), 49–64 (1996)
8. Brown, G., Wyatt, J.L., Tiňo, P.: Managing diversity in regression ensembles. J. Mach. Learn. Res. **6**, 1621–1650 (2005)

9. Chandola, V., Banerjee, A., Kumar, V.: Anomaly detection: a survey. ACM Comput. Surv. (CSUR) **41**(3), 15 (2009)
10. Cover, T., Hart, P.: Nearest neighbor pattern classification. IEEE Trans. Inf. Theory **13**(1), 21–27 (1967)
11. Drucker, H., Burges, C., Kaufman, L., Smola, A., Vapnik, V.: Support vector regression machines. In: NIPS, pp. 155–161 (1997)
12. Feurer, M., Klein, A., Eggensperger, K., Springenberg, J., Blum, M., Hutter, F.: Efficient and robust automated machine learning. In: NIPS, pp. 2962–2970 (2015)
13. Feurer, M., Springenberg, J., Hutter, F.: Initializing Bayesian hyperparameter optimization via meta-learning. In: AAAI, pp. 1128–1135(2015)
14. Freund, Y., Schapire, R.: A decision-theoretic generalization of on-line learning and an application to boosting. J. Comput. Syst. Sci. **55**(1), 119–139 (1997)
15. Friedman, J.: Multivariate adaptive regression splines. Ann. Stat. **19**, 1–67 (1991)
16. Friedman, J.: Greedy function approximation: a gradient boosting machine. Ann. Stat. **29**, 1189–1232 (2001)
17. Friedman, J., Stuetzle, W.: Projection pursuit regression. J. Am. Stat. Assoc. **76**(376), 817–823 (1981)
18. Gama, J., Brazdil, P.: Cascade generalization. Mach. Learn. **41**(3), 315–343 (2000)
19. Hassan, S.M., Moreira-Matias, L., Khiari, J., Cats, O.: Feature selection issues in long-term travel time prediction. In: Boström, H., Knobbe, A., Soares, C., Papapetrou, P. (eds.) IDA 2016. LNCS, vol. 9897, pp. 98–109. Springer, Cham (2016). https://doi.org/10.1007/978-3-319-46349-0_9
20. Hastie, T., Tibshirani, R.: Generalized additive models: some applications. J. Am. Stat. Assoc. **82**(398), 371–386 (1987)
21. Kaggle Inc.: https://www.kaggle.com/bigfatdata/what-algorithms-are-most-successful-on-kaggle. Technical report (Eletronic, Accessed in March 2018)
22. Kiefer, J.: Sequential minimax search for a maximum. Proc. Am. Math. Soc. **4**(3), 502–506 (1953)
23. Lemke, C., Budka, M., Gabrys, B.: Metalearning: a survey of trends and technologies. Artif. Intell. Rev. **44**(1), 117–130 (2015)
24. Mendes-Moreira, J., Soares, C., Jorge, A., Sousa, J.: Ensemble approaches for regression: a survey. ACM Comput. Surv. (CSUR) **45**(1), 10 (2012)
25. Merz, C.: Dynamical Selection of Learning Algorithms, pp. 281–290 (1996)
26. Moreira-Matias, L., Mendes-Moreira, J., Freire de Sousa, J., Gama, J.: On improving mass transit operations by using AVL-based systems: a survey. IEEE Trans. Intell. Transp. Syst. **16**(4), 1636–1653 (2015)
27. Pfahringer, B., Bensusan, H., Giraud-Carrier, C.: Meta-learning by landmarking various learning algorithms. In: ICML, pp. 743–750 (2000)
28. Schaffer, C.: A conservation law for generalization performance. In: Machine Learning Proceedings 1994, pp. 259–265. Elsevier (1994)
29. Tibshirani, R.: Regression shrinkage and selection via the lasso. J. R. Stat. Soc. Series B (Methodological) **58**(1), 267–288 (1996)
30. Todorovski, L., Dzeroski, S.: Combining classifiers with meta decision trees. Mach. Learn. **50**(3), 223–249 (2003)
31. Torgo, L.: Regression data sets. Eletronic (last access at 02/2018) (February 2018). http://www.dcc.fc.up.pt/~ltorgo/Regression/DataSets.html
32. Tsymbal, A., Pechenizkiy, M., Cunningham, P.: Dynamic integration with random forests. In: Fürnkranz, J., Scheffer, T., Spiliopoulou, M. (eds.) ECML 2006. LNCS (LNAI), vol. 4212, pp. 801–808. Springer, Heidelberg (2006). https://doi.org/10.1007/11871842_82
33. Wolpert, D.: Stacked generalization. Neural Netw. **5**(2), 241–259 (1992)

Evaluation

Visualizing the Feature Importance
for Black Box Models

Giuseppe Casalicchio$^{(\boxtimes)}$, Christoph Molnar, and Bernd Bischl

Department of Statistics, Ludwig-Maximilians-University Munich,
Ludwigstraße 33, 80539 Munich, Germany
giuseppe.casalicchio@stat.uni-muenchen.de

Abstract. In recent years, a large amount of model-agnostic methods to improve the transparency, trustability, and interpretability of machine learning models have been developed. Based on a recent method for model-agnostic global feature importance, we introduce a local feature importance measure for individual observations and propose two visual tools: partial importance (PI) and individual conditional importance (ICI) plots which visualize how changes in a feature affect the model performance on average, as well as for individual observations. Our proposed methods are related to partial dependence (PD) and individual conditional expectation (ICE) plots, but visualize the expected (conditional) feature importance instead of the expected (conditional) prediction. Furthermore, we show that averaging ICI curves across observations yields a PI curve, and integrating the PI curve with respect to the distribution of the considered feature results in the global feature importance. Another contribution of our paper is the Shapley feature importance, which fairly distributes the overall performance of a model among the features according to the marginal contributions and which can be used to compare the feature importance across different models. Code related to this paper is available at: https://github.com/giuseppec/featureImportance.

Keywords: Interpretable machine learning · Explainable AI
Feature importance · Variable importance · Feature effect
Partial dependence

1 Introduction and Related Work

Machine learning (ML) algorithms such as neural networks and support vector machines (SVM) are often considered to produce black box models because they do not provide any direct explanation for their predictions. However, these methods often outperform simple linear models or decision trees in predictive performance as they can model complex relationships in the data. Nevertheless, such simple models are still preferred in areas such as life sciences and social sciences due to their simplicity and interpretability [14]. Many researchers have therefore developed and implemented several model-agnostic interpretability tools, which quantify or visualize feature effects or feature importance [9,11,17].

© Springer Nature Switzerland AG 2019
M. Berlingerio et al. (Eds.): ECML PKDD 2018, LNAI 11051, pp. 655–670, 2019.
https://doi.org/10.1007/978-3-030-10925-7_40

In our context, the terms *feature effect*, feature contribution and feature attribution describe how or to what extent each feature contributes to the *prediction* of the model, either on a local or a global level. Methods for feature effects include partial dependence (PD) plots [10], individual conditional expectation (ICE) plots [11] and, more recently, SHAP values [15]. These methods visualize or quantify the relationship and contribution of each feature to the prediction of a model without requiring knowledge about the true values of the target variable. A method that measures feature effects based on the Shapley value [19] from coalitional game theory was first presented for classification in [21] and has been extended to regression and global analysis in [22]. Further developments, visualizations, and generalizations were introduced by [15,16]. Similar work proposing a general notion of a quantity of interest for the characteristic function of the Shapley value and focusing on the joint and marginal contributions of feature sets was introduced by [8].

In biomedical research, for example, measuring the effects of biomedical markers w.r.t. model prediction is as essential as measuring their added value regarding model performance [4]. We use the term *feature importance*[1] to describe how important the feature was for the *predictive performance* of the model, regardless of the shape (e.g., linear or nonlinear relationship) or direction of the feature effect. This implies that measures of feature importance require knowledge of the true values of the target variable. The most prominent approach is the permutation importance introduced by Breiman [3] for random forests. It computes the drop in out-of-bag performance after permuting the values of a feature. A model-agnostic global permutation-based feature importance (PFI) was recently introduced in [9].

Contributions: We review model-agnostic global PFI and propose an efficient approximation based on Monte-Carlo integration. We then introduce a local version of the global PFI, which measures the feature importance of individual observations. We provide visualizations for local and global PFI, which illustrate how changes in the considered feature affect model performance. We also relate our new visual tools to PD plots, ICE plots and show that the integral of our PI curve results in the global PFI measure. Furthermore, we propose a permutation-based Shapley feature importance (SFIMP) measure that fairly distributes the model performance among features and allows the comparison of feature importances across different models.

2   Preliminaries and Background on Feature Effects

In this section, we introduce the notation and describe methods focusing on feature effects, which we transfer to feature importance in Sects. 4 and 5.

---

[1] In the literature, the term feature importance is sometimes also used for methods that only work with model predictions. In our context, however, we would categorize them under feature effects as they do not take into account the model performance.

General Notation: Consider a p-dimensional feature space $\mathcal{X}_P = (\mathcal{X}_1 \times \ldots \times \mathcal{X}_p)$ with the feature index set $P = \{1, \ldots, p\}$ and a target space $\mathcal{Y}$. Suppose that there is an unknown functional relationship f between $\mathcal{X}_P$ and $\mathcal{Y}$. ML algorithms try to learn this relationship using training data with observations that have been drawn i.i.d. from an unknown probability distribution $\mathcal{P}$ on the joint space $\mathcal{X}_P \times \mathcal{Y}$. We consider an arbitrary prediction model $\hat{f}$, fitted on some training data to approximate f and analyze it with model-agnostic interpretability methods. Let $\mathcal{D} = \{(\mathbf{x}^{(i)}, y^{(i)})\}_{i=1}^n$ be a test data set sampled i.i.d. from $\mathcal{P}$ where n is the number of observations in the test set. We denote the corresponding random variables generated from the feature space by $X = (X_1, \ldots, X_p)$ and the random variable generated from the target space by Y. In our notation, the vector $\mathbf{x}^{(i)} = (x_1^{(i)}, \ldots, x_p^{(i)})^\top \in \mathcal{X}_P$ refers to the i-th observation, which is associated with the target variable $y^{(i)} \in \mathcal{Y}$, and $\mathbf{x}_j = (x_j^{(1)}, \ldots, x_j^{(n)})^\top$ denotes the realizations of the j-th feature. We denote the generalization error of a fitted model, which is measured by a loss function L on unseen test data from $\mathcal{P}$, by $GE(\hat{f}, \mathcal{P}) = \mathbb{E}(L(\hat{f}(X), Y))$. It can be estimated using the test data $\mathcal{D}$ by

$$\widehat{GE}(\hat{f}, \mathcal{D}) = \tfrac{1}{n} \sum_{i=1}^n L(\hat{f}(\mathbf{x}^{(i)}), y^{(i)}). \tag{1}$$

A better estimate for the generalization error of an ML algorithm can be obtained using resampling techniques such as cross-validation or bootstrap [1].

PD Plots [10]: They visualize the marginal relationship between features of interest and the expected prediction of a fitted model on a global level. Consider a subset of feature indices $S \subseteq P$ and its complement C. Each observation $\mathbf{x} \in \mathcal{X}_P$ can be partitioned into $\mathbf{x}_S \in \mathcal{X}_S$ and $\mathbf{x}_C \in \mathcal{X}_C$ containing only features from S and C, respectively. Let X_S and X_C be the corresponding random variables and let the prediction function using features in S, marginalized over features in C be the PD function defined by $f_S(\mathbf{x}_S) = \mathbb{E}_{X_C}(\hat{f}(\mathbf{x}_S, X_C))$. This definition also covers $f_\emptyset(\mathbf{x}_\emptyset)$ and results in a constant, the average prediction over $\mathcal{P}$. We can estimate the PD function using Monte-Carlo integration by averaging over feature values $\mathbf{x}_C^{(i)}$ in order to marginalize out features in C:

$$\hat{f}_S(\mathbf{x}_S) = \tfrac{1}{n} \sum_{i=1}^n \hat{f}_S^{(i)}(\mathbf{x}_S) = \tfrac{1}{n} \sum_{i=1}^n \hat{f}(\mathbf{x}_S, \mathbf{x}_C^{(i)}). \tag{2}$$

Here, $\hat{f}_S^{(i)}(\mathbf{x}_S) = \hat{f}(\mathbf{x}_S, \mathbf{x}_C^{(i)})$ can be read in two ways: (a) the prediction of the i-th observation with replaced feature values in S taken from $\mathbf{x}$ or (b) the prediction of $\mathbf{x}$ with replaced values in C taken from the i-th observation. Plotting the pairs $\{(\mathbf{x}_S^{*(k)}, \hat{f}_S(\mathbf{x}_S^{*(k)}))\}_{k=1}^m$ using (often $m < n$) grid points denoted by $\mathbf{x}_S^{*(1)}, \ldots, \mathbf{x}_S^{*(m)}$ yields a PD curve. Figure 1 illustrates the PD principle for a simple example.

ICE Plots [11]: The averaging in Eq. (2) of the PD function can obfuscate more complex relationships resulting from feature interactions, i.e. when the partial relationship of one or more observations depends on other features. ICE plots address this problem by visualizing to what extent the prediction of a

Fig. 1. PD plot for an example with $n = 2$, $p = 3$ and $S = \{1\}$ and $C = \{2,3\}$ (marginal effect of $\mathbf{x}_1$ on $\hat{f}$). We construct a grid using each observed value from $\mathbf{x}_1$, i.e., $x_1^{(1)} = 1$ and $x_1^{(2)} = 2$, and compute the PD function using these grid points.

single observation changes when the value of the considered feature changes. Instead of plotting the pairs $\{(\mathbf{x}_S^{*(k)}, \hat{f}_S(\mathbf{x}_S^{*(k)}))\}_{k=1}^m$, ICE plots visualize the pairs $\{(\mathbf{x}_S^{*(k)}, \hat{f}_S^{(i)}(\mathbf{x}_S^{*(k)}))\}_{k=1}^m$ for each observation indexed by $i \in \{1, \ldots, n\}$.

Shapley Value: A coalitional game is defined by a set of players P, which can form coalitions $S \subseteq P$. Each coalition S achieves a certain payout. The characteristic function $v : 2^P \to \mathbb{R}$ maps all 2^p possible coalitions to their payouts. The Shapley value [19] now fairly assigns a value to each player depending on their contribution in all possible coalitions. This concept was transferred to feature effect estimation in [21]. We could explain the prediction of a single, fixed observation $\mathbf{x}$ by regarding features as players, who form various coalitions (subsets) S to achieve the prediction $\hat{f}(\mathbf{x})$. For each coalition S, we are only allowed to access values of features from S. A natural definition of the payout is the PD value $f_S(\mathbf{x}_S)$, which we shift so that an empty set of no features is assigned a value of 0 – which is required by the general Shapley value definition:

$$v(\mathbf{x}_S) = \mathbb{E}_{X_C}(\hat{f}(\mathbf{x}_S, X_C)) - \mathbb{E}_X(\hat{f}(X)) = f_S(\mathbf{x}_S) - f_\emptyset(\mathbf{x}_\emptyset). \qquad (3)$$

The marginal contribution of feature j, joining a coalition S, is defined as

$$\Delta_j(\mathbf{x}_S) = v(\mathbf{x}_{S \cup \{j\}}) - v(\mathbf{x}_S) = f_{S \cup \{j\}}(\mathbf{x}_{S \cup \{j\}}) - f_S(\mathbf{x}_S).$$

Let Π be the set of all possible permutations over the index set P. For a permutation $\pi \in \Pi$, we denote the set of features that are in order *before* feature j as $B_j(\pi)$. For example, for $p = 4$, if we consider feature $j = 4$ and permutation $\pi = \{2,3,4,1\}$, then $B_4(\pi) = \{2,3\}$. For an observation $\mathbf{x}$ and its feature value for feature j, the Shapley value can be estimated by

$$\begin{aligned}
\hat{\phi}_j(\mathbf{x}) &= \tfrac{1}{p!} \sum_{\pi \in \Pi} \hat{\Delta}_j(\mathbf{x}_{B_j(\pi)}) \\
&= \tfrac{1}{p!} \sum_{\pi \in \Pi} \hat{f}_{B_j(\pi) \cup \{j\}}(\mathbf{x}_{B_j(\pi) \cup \{j\}}) - \hat{f}_{B_j(\pi)}(\mathbf{x}_{B_j(\pi)}) \\
&= \tfrac{1}{p! \cdot n} \sum_{\pi \in \Pi} \sum_{i=1}^n \hat{f}_{B_j(\pi) \cup \{j\}}^{(i)}(\mathbf{x}_{B_j(\pi) \cup \{j\}}) - \hat{f}_{B_j(\pi)}^{(i)}(\mathbf{x}_{B_j(\pi)}),
\end{aligned}$$

where $\hat{f}_{B_j(\pi)}$ and $\hat{f}_{B_j(\pi) \cup \{j\}}$ are estimated by Eq. (2). An efficient approximation based on Monte-Carlo integration using m rather than $p! \cdot n$ summands was proposed by [22]. Consider the following example to illustrate the Shapley value: The features enter a room in a random order specified by the permutation π. All features in the room participate in the game, i.e., they contribute to the model prediction. The Shapley value ϕ_j is the average additional contribution of feature j by joining whatever features already entered the room before.

3 Permutation-Based Feature Importance

Background. The permutation importance for random forests introduced in [3] measures the performance, e.g., the mean squared error (MSE), of each tree within a random forest using out-of-bag samples. The performance is measured once with and once without permuted values of the feature of interest. The difference between those two performance values is computed for each tree and averaged to yield the feature importance. Permuting the values of a feature breaks the association between the feature and the target variable and results in a large drop in performance if the considered feature is important. A model-agnostic global PFI for features included in S can be defined as

$$PFI_S = \mathbb{E}(L(\hat{f}(\tilde{X}_S, X_C), Y)) - \mathbb{E}(L(\hat{f}(X), Y)) \tag{4}$$

where $\tilde{X}_S$ refers to an independent replication of X_S, which is also independent of X_C and Y. This implies that $\tilde{X}_S$ is a new (multivariate) random variable, which is distributed as X_S, but independent of everything else. This definition is analogous to the permutation-based model reliance introduced by [9] and relates to the definition in [12] where the authors focus on random forests. The larger the value of PFI_S, the more substantial the increase in error when we permute feature values in S, and the more important we deem the feature set S. According to [9], the use of the ratio $PFI_S = \mathbb{E}(L(\hat{f}(\tilde{X}_S, X_C), Y))/\mathbb{E}(L(\hat{f}(X), Y))$ instead of the difference might be more comparable across different models, as it always refers to the relative drop in performance with respect to the standard generalization error. However, using the ratio can result in numerically unstable estimations if the denominator is close or equal to zero. Thus, both definitions have drawbacks that we try to are address in Sect. 5.

Estimating and Approximating the PFI. The first term of Eq. (4) encodes the expected generalization error under perturbation of features in feature set S, which can be formulated as:

$$
\begin{aligned}
\mathbb{E}(L(\hat{f}(\tilde{X}_S, X_C), Y)) &= \mathbb{E}_{(X_C, Y)}(\mathbb{E}_{\tilde{X}_S \mid (X_C, Y)}(L(\hat{f}(\tilde{X}_S, X_C), Y))) \\
&= \mathbb{E}_{(X_C, Y)}(\mathbb{E}_{\tilde{X}_S}(L(\hat{f}(\tilde{X}_S, X_C), Y))) \\
&= \mathbb{E}_{(X_C, Y)}(\mathbb{E}_{X_S}(L(\hat{f}(X_S, X_C), Y)))
\end{aligned}
$$

In the derivation above, the first equality follows from the "law of total expectation", the second from the independence of $\tilde{X}_S$ from (X_C, Y), and the third because $\tilde{X}_S$ is distributed as X_S. We can plug in an estimator for the inner expected value and denote the estimate of this quantity by

$$\widehat{GE}_C(\hat{f}, \mathcal{D}) = \frac{1}{n}\sum_{i=1}^{n}\frac{1}{n}\sum_{k=1}^{n}L(\hat{f}(\mathbf{x}_S^{(k)}, \mathbf{x}_C^{(i)}), y^{(i)}). \tag{5}$$

The index C in GE_C emphasizes that the set of features in C were not replaced with a perturbed random variable and can thus be seen as the model performance using features in C (and ignoring those in S). The above estimator is analogous to the V-statistic [18] and may also be replaced by the unbiased U-statistic using

$\frac{1}{n}\sum_{i=1}^{n}\frac{1}{n-1}\sum_{k\neq i}L(\hat{f}(\mathbf{x}_S^{(k)},\mathbf{x}_C^{(i)}),y^{(i)})$ as proposed by [9].[2] The estimator scales with $O(n^2)$ (for a given set C, and assuming $\hat{f}$ can be computed in constant time), which can be expensive when n is large. However, we can use a different formulation to motivate an approximation for Eq. (5): Let $\{\boldsymbol{\tau}_1,\ldots,\boldsymbol{\tau}_{n!}\}$ be the set of all possible permutation vectors over the observation index set $\{1,\ldots,n\}$. As shown by [9], we can replace Eq. (5) by the equivalent formulation

$$\widehat{GE}_{C,\mathrm{perm}}(\hat{f},\mathcal{D}) = \frac{1}{n}\sum_{i=1}^{n}\frac{1}{n!}\sum_{k=1}^{n!}L(\hat{f}(\mathbf{x}_S^{(\tau_k^{(i)})},\mathbf{x}_C^{(i)}),y^{(i)}).$$

If we approximate $\widehat{GE}_{C,\mathrm{perm}}$ by Monte-Carlo integration using only m randomly selected permutations rather than all $n!$ permutations, we obtain

$$\widehat{GE}_{C,\mathrm{approx}}(\hat{f},\mathcal{D}) = \frac{1}{n}\sum_{i=1}^{n}\frac{1}{m}\sum_{k=1}^{m}L(\hat{f}(\mathbf{x}_S^{(\tau_k^{(i)})},\mathbf{x}_C^{(i)}),y^{(i)}). \quad (6)$$

The approximation refers to permuting features in S repeatedly (i.e., m times) and averaging the resulting model performances.[3] The PFI from Eq. (4) can be estimated using Eq. (5) for the first term and using Eq. (1) for the last term. Including the summands into an iterated sum yields the estimate

$$\widehat{PFI}_S = \frac{1}{n^2}\sum_{i=1}^{n}\sum_{k=1}^{n}\left(L(\hat{f}(\mathbf{x}_S^{(k)},\mathbf{x}_C^{(i)}),y^{(i)}) - L(\hat{f}(\mathbf{x}^{(i)}),y^{(i)})\right). \quad (7)$$

If we use Eq. (6) rather than Eq. (5), we obtain the approximation

$$\widehat{PFI}_{S,\mathrm{approx}} = \frac{1}{n\cdot m}\sum_{i=1}^{n}\sum_{k=1}^{m}\left(L(\hat{f}(\mathbf{x}_S^{(\tau_k^{(i)})},\mathbf{x}_C^{(i)}),y^{(i)}) - L(\hat{f}(\mathbf{x}^{(i)}),y^{(i)})\right). \quad (8)$$

Equation (8) is identical to the permutation importance of random forests formalized in [12] if we consider m as the number of trees, replace n with the number of out-of-bag samples per tree and replace the model $\hat{f}$ with the individual trees fitted within a random forest, i.e., $\hat{f}_k$.

4   Visualizing Global and Local Feature Importance

Consider the summands in Eq. (7) and denote them by

$$\Delta L^{(i)}(\mathbf{x}_S) = L(\hat{f}(\mathbf{x}_S,\mathbf{x}_C^{(i)}),y^{(i)}) - L(\hat{f}(\mathbf{x}^{(i)}),y^{(i)}).$$

This quantity refers to the change in performance between the i-th observation with and without replaced feature values $\mathbf{x}_S$. Inspired by ICE plots, we

---

[2] For the sake of simplicity, we consider the V-statistic throughout the article. However, all calculations and approximations based on Eq. (5) still apply – with slight modifications – when using the U-statistic.

[3] By the same logic, we could also directly approximate Eq. (5) by summing over m randomly selected feature values for features in S instead of using all of them. We here opted for Eq. (6), due to the in our opinion interesting relation to the random forest permutation importance explained at the end of this section.

introduce *individual conditional importance* (ICI) plots which visualize the pairs $\{(\mathbf{x}_S^{(k)}, \Delta L^{(i)}(\mathbf{x}_S^{(k)}))\}_{k=1}^n$ for all observations $i = 1, \ldots, n$. We define the local feature importance of the i-th observation (regarding features in S) as the integral of the corresponding ICI curve with respect to the distribution of X_S. It is estimated by $\widehat{PFI}_S^{(i)} = \frac{1}{n} \sum_{k=1}^n \Delta L^{(i)}(\mathbf{x}_S^{(k)})$ and can be interpreted as the expected change in performance of the i-th observation after marginalizing its features in S. It also refers to the contribution of the i-th observation to the global PFI (see later in Eq. (9)). To the best of our knowledge, a similar definition for local feature importance only exists in the context of random forests, e.g., in [7].

Analogous to the PD function from Eq. (2), we introduce the *partial importance* (PI) function as the expected change in performance at a specific value $\mathbf{x}_S$, which can be estimated by $\widehat{PI}_S(\mathbf{x}_S) = \frac{1}{n} \sum_{i=1}^n \Delta L^{(i)}(\mathbf{x}_S)$. Consequently, a PI plot visualizes the pairs $\{(\mathbf{x}_S^{(k)}, \widehat{PI}_S(\mathbf{x}_S^{(k)}))\}_{k=1}^n$ and refers to the pointwise average of all ICI curves across all observations at fixed grid points $\mathbf{x}_S$.

Figure 2 illustrates the computation of ICI and PI curves for the first feature. It also shows the n grid points for which $\Delta L^{(i)}(\mathbf{x}_S^{(i)}) = 0 \;\forall i$. We can omit these points by plotting the pairs $\{(\mathbf{x}_S^{(k)}, \Delta L^{(i)}(\mathbf{x}_S^{(k)}))\}_{k \in \{1,\ldots,n\} \setminus \{i\}}$ to visualize the unbiased estimation of the feature importance proposed by [9]. Visualizing the ICI curves for the approximation in Eq. (8) implies that some grid points are randomly skipped because the feature values used as grid points are implicitly determined by the randomly selected permutations in Eq. (8). The ICI curves, the PI curve, and the global PFI are related: Averaging all ICI curves pointwise yields a PI curve. Integrating the PI curve (as well as averaging the integral of all ICI curves) using Monte-Carlo integration over all points $\{\mathbf{x}_S^{(k)}\}_{k=1}^n$ yields an equivalent estimate of the global PFI from Eq. (7):

$$\widehat{PFI}_S = \frac{1}{n} \sum_{i=1}^n \widehat{PFI}_S^{(i)} = \frac{1}{n} \sum_{k=1}^n \widehat{PI}_S(\mathbf{x}_S^{(k)}). \tag{9}$$

We propose to additionally inspect the PI and ICI curves instead of focusing on a single PFI value. PI curves enable the user to identify regions in which the feature importance is higher or lower than its global PFI. ICI curves additionally enable the user to identify (suspicious) observations that impact the global PFI strongly and can reveal heterogeneity in the feature importance among the observations, which remain hidden in the PI plots (see also Sect. 6).

Algorithm 1 describes a procedure for obtaining PI and PD plots, which also allows to return ICI and ICE plots by visualizing $\{(\mathbf{x}_S^{*(k)}, \Delta L^{(i)}(\mathbf{x}_S^{*(k)}))\}_{k=1}^m$ and $\{(\mathbf{x}_S^{*(k)}, \hat{f}_S^{(i)}(\mathbf{x}_S^{*(k)}))\}_{k=1}^m$ for all observations i. Similar to PD and ICE plots, we can use all $k = 1, \ldots, n$ or a random sample (of size $m < n$) of feature values from S as grid points for PI and ICI plots.

5   Shapley Feature Importance

In this section, we introduce the *S*hapley *F*eature *IMP*ortance (SFIMP) measure, which allows to easily visualize and interpret the contribution of each feature to the model performance. Our goal is to fairly distribute the performance

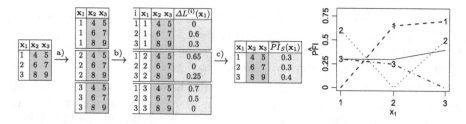

Fig. 2. The tables on the left side illustrate the required steps to create ICI curves and PI plots as described in Algorithm 1. The right plot visualizes the ICI curves of individual observations for $i = 1, 2, 3$ (dotted and dashed lines) and the PI curve (solid line) which is the average of ICI curves at each point of the abscissa. All points belonging to the same observation are connected by a line to produce the ICE curves.

Algorithm 1. PD plot and PI plot

---

1. Choose m grid points $x_S^{*(1)}, \dots, x_S^{*(m)}$.
2. Repeat the following steps for the k-th grid point:
   (a) Modify the data by replacing all observed values in $\mathbf{x}_S$ with the constant values from the k-th grid point $x_S^{*(k)}$.
   (b) Use the modified data from a), the prediction function $\hat{f}$ and the loss function L and calculate for all individual observations:
       (i) $\hat{f}_S^{(i)}(\mathbf{x}_S^{*(k)}) = \hat{f}(\mathbf{x}_S^{*(k)}, \mathbf{x}_C^{(i)})$
       (ii) $\Delta L^{(i)}(\mathbf{x}_S^{*(k)}) = L(\hat{f}(\mathbf{x}_S^{*(k)}, \mathbf{x}_C^{(i)}), y^{(i)}) - L(\hat{f}(\mathbf{x}^{(i)}), y^{(i)})$
   (c) Aggregate the individual values:
       (i) $\hat{f}_S(\mathbf{x}_S^{*(k)}) = \frac{1}{n} \sum_{i=1}^{n} \hat{f}_S^{(i)}(\mathbf{x}_S^{*(k)})$
       (ii) $\widehat{PI}_S(\mathbf{x}_S^{*(k)}) = \frac{1}{n} \sum_{i=1}^{n} \Delta L^{(i)}(\mathbf{x}_S^{(k)})$
3. Plot the pairs $\{(\mathbf{x}_S^{*(k)}, \hat{f}_S(\mathbf{x}_S^{*(k)}))\}_{k=1}^{m}$ and $\{(\mathbf{x}_S^{*(k)}, \widehat{PI}_S(\mathbf{x}_S^{*(k)}))\}_{k=1}^{m}$.

---

difference among the individual features between the scenario when all features are used and when all features are ignored, which is illustrated in Fig. 3.

The Shapley value was used in [6] for a fair attribution of the difference in model performance. However, the authors focused on feature selection which requires refitting the model by leaving out or including features. This can lead to different results of the learning algorithm since different relationships can be learned due to the absence of features. This is reasonable in the context of feature selection. However, as we measure the feature importance of an already fitted model, we prefer marginalizing over features rather than omitting them completely. Inspired by Eq. (3), we define the characteristic function of the coalition of features in $S \subseteq P$ based on Eq. (5) as:

$$v_{GE}(S) = \widehat{GE}_S(\hat{f}, \mathcal{D}) - \widehat{GE}_\emptyset(\hat{f}, \mathcal{D}). \tag{10}$$

The characteristic function measures the change in performance between using features in S (i.e., ignoring features in its complement C by marginalizing over

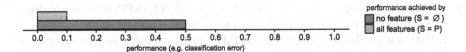

Fig. 3. Illustration of the difference in model performance that we want to fairly distribute among the features. The model performance (e.g., classification error) is 0.1 when using all features (green bar) and 0.5 when ignoring all features (red bar). Our goal is to fairly distribute the resulting performance difference of 0.4 among all involved features based on their marginal contribution.

them) and ignoring all features. This is similar to Eq. (7) which, in contrast, measures the change in performance between ignoring features in S and using all features. Since the error $\widehat{GE}_\emptyset(\hat{f}, \mathcal{D})$ (no features are considered, i.e., all features are marginalized out) is usually greater than $\widehat{GE}_S(\hat{f}, \mathcal{D})$, $v_{GE}(S)$ will have negative values.[4] The marginal contribution of a feature j to a coalition of features in S is given by

$$\Delta_j(S) = v_{GE}(S \cup \{j\}) - v_{GE}(S) = \widehat{GE}_{S \cup \{j\}}(\hat{f}, \mathcal{D}) - \widehat{GE}_S(\hat{f}, \mathcal{D}).$$

If we consider a permuted order $\pi \in \Pi$ of the features, where $B_j(\pi)$ is the set of features occurring before feature j, we obtain the Shapley value estimation

$$\begin{aligned}\hat{\phi}_j(v_{GE}) &= \tfrac{1}{p!} \sum_{\pi \in \Pi} \Delta_j(B_j(\pi)) \\ &= \tfrac{1}{p!} \sum_{\pi \in \Pi} \widehat{GE}_{B_j(\pi) \cup \{j\}}(\hat{f}, \mathcal{D}) - \widehat{GE}_{B_j(\pi)}(\hat{f}, \mathcal{D}),\end{aligned} \tag{11}$$

which refers to the SFIMP measure of feature j. Computing Eq. (11) is computationally expensive when the number of features p is large, even if we use the approximation of the model performance from Eq. (6). We therefore suggest an efficient procedure in Algorithm 2. The Shapley value satisfies the following four desirable properties as already worked out in [6]:

1. Efficiency: $\sum_{j=1}^p \phi_j = v_{GE}(P)$. All SFIMP values add up to $v_{GE}(P)$, i.e., the difference in performance between the scenario when all features are used and when all features are ignored. This allows us to calculate the proportion of explained importance for each feature j using $\frac{\phi_j}{\sum_{j=1}^p \phi_j}$.
2. Symmetry: If $v_{GE}(S \cup \{j\}) = v_{GE}(S \cup \{k\})$ for all $S \subseteq \{1, \ldots, p\} \setminus \{j, k\}$, then $\phi_j = \phi_k$. Two features j and k have the same SFIMP values if their marginal contribution to all possible coalitions is the same.
3. Dummy property: If $v_{GE}(S \cup \{j\}) = v_{GE}(S)$ for all $S \subseteq P$, then $\phi_j = 0$. The SFIMP value of a feature j is zero if its marginal contribution does not change no matter to which coalition S the feature is added.

---

[4] We prefer the definition in Eq. (10) as it directly shows the relation to Eq. (3), however, we could also swap the sign as discussed at the end of this section.

4. Additivity: $\phi_j(v_{GE}+w_{GE}) = \phi_j(v_{GE})+\phi_j(w_{GE})$. The SFIMP value resulting from a single game with two combined performance measures $\phi_j(v_{GE}+w_{GE})$ is the same as the sum of the two SFIMP values resulting from two separate games with corresponding characteristic functions, i.e., $\phi_j(v_{GE}) + \phi_j(w_{GE})$. Linearity: $\phi_j(c \cdot v_{GE}) = c \cdot \phi_j(v_{GE})$. Any multiplication of the performance measure with a constant c does not affect the feature ranking.

Algorithm 2. Approximation of SFIMP values: Contribution of j-th feature towards the model performance.

Input: m_{feat}, m_{obs}, $\hat{f}$, L, $\mathcal{D} = \{(\mathbf{x}^{(i)}, y^{(i)})\}_{i=1}^n$

1 **forall** $k \in \{1, \ldots, m_{\text{feat}}\}$ **do**

2   choose a random permutation of the feature indices $\pi \in \Pi$.

3   set $S = B_j(\pi)$ containing features that won't be permuted.

4   set $\widehat{GE}_{S,\text{perm}} = 0$ and $\widehat{GE}_{S\cup\{j\},\text{perm}} = 0$.

5   **forall** $l \in \{1, \ldots, m_{\text{obs}}\}$ **do**

6   choose a random permutation of observation indices $\tau \in \{\tau_1, \ldots, \tau_{n!}\}$.

7   measure performance by permuting features w.r.t. $\tau = (\tau^{(1)}, \ldots, \tau^{(n)})$:

$$\widehat{GE}_{S,\text{perm}} = \widehat{GE}_{S,\text{perm}} + \tfrac{1}{n} \sum_{i=1}^n L(\hat{f}(\mathbf{x}_S^{(i)}, \mathbf{x}_C^{(\tau^{(i)})}), y^{(i)}))$$

$$\widehat{GE}_{S\cup\{j\},\text{perm}} = \widehat{GE}_{S\cup\{j\},\text{perm}} + \tfrac{1}{n} \sum_{i=1}^n L(\hat{f}(\mathbf{x}_{S\cup\{j\}}^{(i)}, \mathbf{x}_{C\setminus\{j\}}^{(\tau^{(i)})}), y^{(i)}))$$

8   compute marginal contribution for feature j in iteration k:

$$\Delta_j^{(k)}(S) = \tfrac{1}{m_{\text{obs}}} \cdot (\widehat{GE}_{S\cup\{j\},\text{perm}} - \widehat{GE}_{S,\text{perm}})$$

9 **return** $\hat{\phi}_j = \tfrac{1}{m_{\text{feat}}} \sum_{k=1}^{m_{\text{feat}}} \Delta_j^{(k)}(S)$

The properties above imply that fairly distributing the drop in performance using $v_{PFI}(S) = \widehat{PFI}_S = \widehat{GE}_C(\hat{f}, \mathcal{D}) - \widehat{GE}_P(\hat{f}, \mathcal{D})$ results in the same Shapley values (except for the sign) and is equivalent to using $-v_{GE}(P)$. The SFIMP measure can thus be seen as an extension of the PFI measure in the sense that it additionally fairly distributes the importance values among features. The PFI measure ignores features in S by permuting or marginalizing over them, which destroys any correlation and interaction of features in C with features in S. Consequently, the PFI of a feature also includes the importance of any interaction with that feature and features in C and therefore an interaction will be fully attributed to all involved features. The SFIMP measure solves this issue as it considers the marginal contribution of a feature and equally distributes the importance of interactions among the interacting features. This allows comparing feature importances across different models.

6 Simulations and Application

For full reproducibility, all our proposed methods are available in the R package featureImportance[5]. The repository also contains the R code, which is partly based on batchtools [13], for the application and simulation in this section.

---

[5] https://github.com/giuseppec/featureImportance.

6.1   Simulations

PI and ICI Plots. Consider the following data-generating model:

$$Y = X_1 + X_2 + 10X_1 \cdot \mathbb{1}_{X_3=0} + 10X_2 \cdot \mathbb{1}_{X_3=1} + \epsilon,$$

$$X_1, X_2 \overset{i.i.d}{\sim} \mathcal{N}(0,1), X_3 \sim \mathcal{B}(1, 0.5), \epsilon \sim \mathcal{N}(0, 0.5).$$

We simulate a training data set with 10000 observations, train a random forest and compute the global PFI on 100 test sets of size $n = 100$ sampled from the same distribution. We demonstrate that, by merely inspecting the global PFI, the features X_1 and X_2 would be considered equally important. However, due to the interactions, it is clear that feature X_1 should be considered more important than X_2 when $X_3 = 0$ and vice-versa when $X_3 = 1$.

According to Eq. (9), averaging the local feature importances (i.e., the integral of all ICI curves) results in the global PFI. Having at hand the local feature importance of each observation allows calculating the PFI conditional on other features. This does not require additional time-consuming calculations, as we only have to average the already computed local feature importances according to the condition considered in the conditional PFI. The relevance of conditional feature importance in the case of random forests with correlated features was discussed in [20]. In Fig. 4, we illustrate the usefulness of a model-agnostic conditional PFI in case of interactions by showing the PI curves of X_1 and X_2 conditional on the binary feature X_3. The integral of these conditional PI curves refers to the PFI conditional on X_3. Its value differs depending on the two groups introduced by feature X_3, which suggests that there is an interaction between the features X_1 and X_3 as well as X_2 and X_3.

Table 1 shows that feature X_1 and X_2 are almost equally important if we consider the unconditional global PFI. However, a different ranking of features is obtained when we compute the PFI conditional on X_3. Thus, inspecting PI and ICI curves conditional on other feature values may help in detecting interactions.

Table 1. The mean and the standard deviation (numbers in brackets) of the PFI values estimated using the 100 simulated test data sets.

	X_1	X_2
global PFI	77.976 (14.15)	76.764 (13.89)
PFI for $X_3 = 0$	152.49 (26.06)	1.428 (1.32)
PFI for $X_3 = 1$	1.261 (1.03)	151.489 (24.69)

Shapley Feature Importance. We illustrate how the SFIMP measure can be used to compare the feature importance across different models and present the results of a small simulation study to compare the SFIMP measure introduced in Sect. 5 with the difference-based and the ratio-based PFI discussed in Sect. 3. Consider the following data-generating linear model with a simple interaction:

$$Y = X_1 + X_2 + X_3 + X_1 \cdot X_2 + \epsilon, \quad X_1, X_2, X_3 \overset{i.i.d}{\sim} \mathcal{N}(0,1), \epsilon \sim \mathcal{N}(0, 0.5).$$

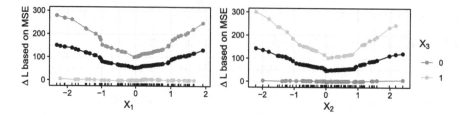

Fig. 4. PI curves of X_1 and X_2 calculated using all observations (black line) and conditional on $X_3 = 0$ (red line) and $X_3 = 1$ (green line). The points plotted on the lines refer to the observed feature values that were used as grid points to produce the corresponding PI curves as described in Algorithm 1.

All three features and the interaction of X_1 and X_2 have the same linear effect on the target Y. We simulate training data with 10000 observations and train four learning algorithms using the `mlr` R package [2] in their defaults: An SVM with Gaussian kernel (`ksvm`), a random forest (`randomForest`), a simple linear model (`lm`) and another one that considers 2-way interaction effects (`rsm`). We use a test set with $n = 100$ observations sampled from the same distribution and compute the SFIMP values according to Eq. (11). Panel (a) of Fig. 5 displays how the SFIMP measure distributes the total explainable performance among all features and shows the proportion of explained importance for each feature. We repeat the experiment 500 times on different test sets of equal size and additionally compute the difference-based and ratio-based PFI. The results are shown in panel (b) of Fig. 5. For the linear model without interaction effects, the calculated importance of all three features is equal (median ratio of 1). For all other models, we obtained a higher importance for the interacting features, indicating that these models were able to grasp the interaction effect. However, as permuting a feature destroys any interaction with that feature, the PFI values of a feature will also include the importance of any interaction with that feature. Thus, the importance of the interaction between X_1 and X_2 is contained in the PFI value for feature X_1 as well as in the PFI value for feature X_2. This will overestimate the importance of X_1 and X_2 with respect to X_3 since X_1 and X_2 share the same interaction. In panel (b), we thus show the ratio of the importance values with respect to X_3. The results suggest that the difference-based PFI considers X_1 and X_2 twice as important as X_3 as the median ratio is around 2. In contrast, the median ratio of SFIMP is around 1.5 as the importance of the interaction is fairly distributed among X_1 and X_2.

6.2 Application on Real Data

We demonstrate our graphical tools on the Boston housing data, which is publicly available on OpenML [23] with data set ID 531. The data set contains 13 features that may affect the median home price of 506 metropolitan areas of Boston. We used the `OpenML` R package [5] and created the OpenML task with

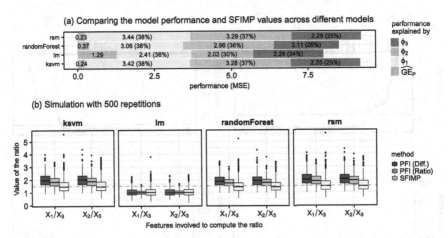

Fig. 5. Panel (a) shows the results of a single run, consisting of sampling test data and computing the importance on the previously fitted models. The first numbers on the left refer to the model performance (MSE) using all features. The other numbers are the SFIMP values which sum up to the total explainable performance $v_{GE}(P)$ from Eq. (10). The percentages refer to the proportion of explained importance. Panel (b) shows the results of 500 repetitions of the experiment. The plots display the distribution of ratios of the importance values for X_1 and X_2 with respect to X_3 computed by SFIMP, by the difference-based PFI, and by the ratio-based PFI.

Table 2. PFI values calculated for a random forest trained on the Boston housing training set and using the MSE on the test data. The PFI values in row (1) are based on all observations from the test set, in row (2) on a subset where LSTAT < 10 and in row (3) after removing observations having a negative ICI integral.

	LSTAT	RM	NOX	DIS	CRIM	PTRATIO	AGE	INDUS	TAX	RAD	B	ZN	CHAS
(1)	32.0	15.6	3.9	2.7	2.6	2.2	1.2	1.0	1.0	0.8	0.8	0.1	0.1
(2)	10.4	29.6	1.5	3.3	0.8	2.3	0.8	0.5	1.2	1.1	0.6	0.2	0.2
(3)	35.3	17.0	4.3	2.4	2.5	2.5	1.1	1.2	0.8	0.9	0.8	0.1	0.1

ID 167147 containing a holdout split ($\frac{2}{3}$ vs. $\frac{1}{3}$) for training a random forest and producing the PI and ICI plots on the test set.

Row (1) of Table 2 shows the global PFI values of all features. They are estimated using Eq. (7) by taking into account all $166 \cdot 166$ points of the test data. Figure 6 shows the corresponding PI and ICI curves for the two most important features (LSTAT and RM). They visualize which regions of each feature and which observations have a high impact on the computed PFI values on a global and local level, which follows from the relation in Eq. (9).

PI plots visualize the expected change in performance at each position of the abscissa. An expected change close to zero across the whole range of the feature values suggests an unimportant feature. The PI plot of LSTAT in Fig. 6 suggests that the feature is more important if LSTAT < 10. For illustration purposes, we omit all observations for which LSTAT $\geq$ 10 and recompute the conditional PFI

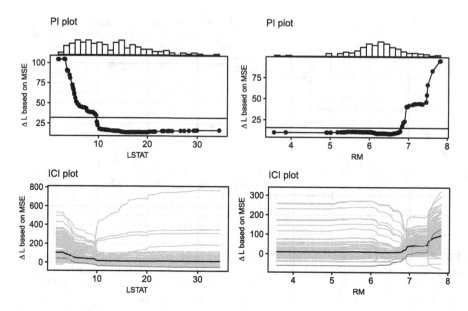

Fig. 6. PI and ICI plots for a random forest and the two most important features of the Boston housing data (LSTAT and RM). The horizontal lines in the PI plots represent the value of the global PFI (i.e., the integral of the PI curve). Marginal distribution histograms for features are added to the PI margins. The ICI curve with the largest integral is highlighted in green and the curve with the smallest integral in red.

values, which are displayed in Row (2) of Table 2. The resulting conditional PFI values are smaller, i.e., excluding observations for which LSTAT $\geq$ 10 makes the LSTAT feature less important. Note that omitting observations change the empirical distribution of the features and thus also affects the importance of other features when the PFI values are recomputed.

ICI curves additionally reveal the most (and the least) influential observations for the feature importance by considering their integral (see highlighted lines in Fig. 6). We can, for example, omit observations with a negative ICI curve integral. In our test set, we observe 18 of 166 ICI curves with a negative integral for the LSTAT feature. These observations have a negative impact on the global PFI according to the relation in Eq. (9). We omit them and recompute the PFI values. The results are listed in row (3) of Table 2 and show an increased PFI value for LSTAT.

7   Conclusion and Future Work

It is essential for practitioners to peek inside black box models to get a better understanding of how features contribute to model predictions or how they affect the model performance. Model-agnostic visualization methods can simplify this task tremendously. Regarding the feature importance, the PI and ICI curves are a convenient choice for visualizing how features affect model performance. We demonstrated how to disaggregate the global PFI into its individual local PFI components, which enabled us to visualize the feature importance on a local and global level. It also allows practitioners to analyze and compare the feature importance across different groups of observations in the data, e.g., by subsetting the data according to other feature values and computing a conditional feature importance similar to [20] on the subsetted data which may reveal interactions. Another interesting aspect, which we leave for future work, is aggregating the local feature importances of individual observations (i.e., the integral of ICI curves) across different features to obtain a measure for the importance of individual observations. This could be used to find clusters of observations in the data that were important for the model performance similar to [15], but based on feature importance rather than feature effects. Furthermore, it is also possible to disaggregate the Shapley feature importance introduced in Sect. 5 and produce plots similar to Shapley dependency plots that were recently introduced in [15], but we leave this for future work. Our proposed methods serve as an evaluation tool that is applied to a data set *after* a model has been fitted. As a consequence, our methods can be used to either assess the feature importance based on the "in-sample performance" or based on the "out-of-sample performance" of a fitted model. In the former case, the same data could be used to fit the model and to calculate the quantities involved in the definition of our methods. We focused on the latter case with independent test data. However, we could also investigate the variability introduced by the estimation of the model itself via resampling and plot or aggregate the resulting set of quantities.

Acknowledgments. This work is funded by the Bavarian State Ministry of Education, Science and the Arts in the framework of the Centre Digitisation. Bavaria (ZD.B).

References

1. Bischl, B., Mersmann, O., Trautmann, H., Weihs, C.: Resampling methods for meta-model validation with recommendations for evolutionary computation. Evol. Comput. **20**(2), 249–275 (2012)
2. Bischl, B., Lang, M., Kotthoff, L., Schiffner, J., Richter, J., Studerus, E., Casalicchio, G., Jones, Z.M.: mlr: machine learning in R. J. Mach. Learn. Res. **17**(170), 1–5 (2016)
3. Breiman, L.: Random forests. Mach. Learn. **45**(1), 5–32 (2001)
4. Casalicchio, G., Bischl, B., Boulesteix, A.L., Schmid, M.: The residual-based predictiveness curve: a visual tool to assess the performance of prediction models. Biometrics **72**(2), 392–401 (2016)
5. Casalicchio, G., et al.: OpenML: an R package to connect to the machine learning platform OpenML. Comput. Stat. 1–15 (2017). https://doi.org/10.1007/s00180-017-0742-2
6. Cohen, S., Dror, G., Ruppin, E.: Feature selection via coalitional game theory. Neural Comput. **19**(7), 1939–1961 (2007)
7. Cutler, A., Cutler, D.R., Stevens, J.R.: Random forests. Ensemble Machine Learning, pp. 157–175. Springer, Boston (2012). https://doi.org/10.1007/978-1-4419-9326-7_5

8. Datta, A., Sen, S., Zick, Y.: Algorithmic transparency via quantitative input influence: theory and experiments with learning systems. In: Proceedings of IEEE Symposium on Security and Privacy, SP, pp. 598–617 (2016)

9. Fisher, A., Rudin, C., Dominici, F.: Model class reliance: variable importance measures for any machine learning model class, from the "Rashomon" perspective (2018). arXiv preprint arXiv:1801.01489

10. Friedman, J.H.: Greedy function approximation: a gradient boosting machine. Annals of Statistics, pp. 1189–1232 (2001)

11. Goldstein, A., Kapelner, A., Bleich, J., Pitkin, E.: Peeking inside the black box: visualizing statistical learning with plots of individual conditional expectation. J. Comput. Graph. Stat. **24**(1), 44–65 (2015)

12. Gregorutti, B., Michel, B., Saint-Pierre, P.: Correlation and variable importance in random forests. Stat. Comput. **27**(3), 659–678 (2017)

13. Lang, M., Bischl, B., Surmann, D.: batchtools: tools for R to work on batch systems. J. Open Source Softw. **2**(10) (2017)

14. Lipton, Z.C.: The mythos of model interpretability. In: ICML WHI 2016 (2016)

15. Lundberg, S.M., Erion, G.G., Lee, S.I.: Consistent individualized feature attribution for tree ensembles (2018). arXiv preprint arXiv:1802.03888

16. Lundberg, S.M., Lee, S.I.: A unified approach to interpreting model predictions. In: NIPS, vol. 30, pp. 4765–4774. Curran Associates, Inc., Red Hook (2017)

17. Molnar, C., Casalicchio, G., Bischl, B.: iml: an R package for interpretable machine learning. J. Open Source Softw. **3**(27), 786 (2018). https://doi.org/10.21105/joss.00786

18. Serfling, R.J.: Approximation Theorems of Mathematical Statistics, vol. 162. Wiley, New York (2009)

19. Shapley, L.S.: A value for n-person games. Contrib. Theory Games **2**(28), 307–317 (1953)

20. Strobl, C., Boulesteix, A.L., Kneib, T., Augustin, T., Zeileis, A.: Conditional variable importance for random forests. BMC Bioinf. **9**, 307 (2008)

21. Štrumbelj, E., Kononenko, I., Wrobel, S.: An efficient explanation of individual classifications using game theory. J. Mach. Learn. Res. **11**(Jan), 1–18 (2010)

22. Štrumbelj, E., Kononenko, I.: A general method for visualizing and explaining black-box regression models. In: Dobnikar, A., Lotrič, U., Šter, B. (eds.) ICANNGA 2011, Part II. LNCS, vol. 6594, pp. 21–30. Springer, Heidelberg (2011). https://doi.org/10.1007/978-3-642-20267-4_3

23. Vanschoren, J., Van Rijn, J.N., Bischl, B., Torgo, L.: OpenML: networked science in machine learning. ACM SIGKDD Explor. Newsl. **15**(2), 49–60 (2014)

Efficient Estimation of AUC
in a Sliding Window

Nikolaj Tatti[(✉)]

F-Secure, Helsinki, Finland
nikolaj.tatti@gmail.com

Abstract. In many applications, monitoring area under the ROC curve (AUC) in a sliding window over a data stream is a natural way of detecting changes in the system. The drawback is that computing AUC in a sliding window is expensive, especially if the window size is large and the data flow is significant.

In this paper we propose a scheme for maintaining an approximate AUC in a sliding window of length k. More specifically, we propose an algorithm that, given ϵ, estimates AUC within $\epsilon/2$, and can maintain this estimate in $\mathcal{O}((\log k)/\epsilon)$ time, per update, as the window slides. This provides a speed-up over the exact computation of AUC, which requires $\mathcal{O}(k)$ time, per update. The speed-up becomes more significant as the size of the window increases. Our estimate is based on grouping the data points together, and using these groups to calculate AUC. The grouping is designed carefully such that (i) the groups are small enough, so that the error stays small, (ii) the number of groups is small, so that enumerating them is not expensive, and (iii) the definition is flexible enough so that we can maintain the groups efficiently.

Our experimental evaluation demonstrates that the average approximation error in practice is much smaller than the approximation guarantee $\epsilon/2$, and that we can achieve significant speed-ups with only a modest sacrifice in accuracy. Code related to this paper is available at: https://bitbucket.org/orlyanalytics/streamauc.

Keywords: AUC · Approximation guarantee · Sliding window

1 Introduction

Consider monitoring prediction performance in a stream of data points. That is, we first receive a data point d without the label, and we predict the missing label with a score of s, after the prediction we receive the true label ℓ. We are interested in monitoring how well s predicts ℓ as the stream evolves over time.

A good example of such a task is a monitoring system for corporate computers that detects abnormal behavior based on event logs. Here the positive label represents an abnormal event that requires a closer inspection, and such a label can be given, for example, by an expert or triggered automatically. The produced

© Springer Nature Switzerland AG 2019
M. Berlingerio et al. (Eds.): ECML PKDD 2018, LNAI 11051, pp. 671–686, 2019.
https://doi.org/10.1007/978-3-030-10925-7_41

score can be used for decision making, and can be a specific feature or a simple statistic, or the result of some classifier, such as logistic regression. It is vital to monitor such a system continuously to notice breakdowns early. Possible causes may be changes in the underlying distribution or a system failure, due to the software update.

A natural choice to monitor the predictive power of a real-valued score is the area under the ROC curve (AUC) in a sliding window over the stream of events as proposed by Brzezinski and Stefanowski [5]. Unfortunately, maintaining the exact AUC requires $\mathcal{O}(k)$ time, per new event, where k is the size of the window. This may be too expensive if k is large and the rate of the events is significant.

In this paper we propose a technique for estimating AUC efficiently in a sliding window. Namely, we propose an approximation scheme that has $\epsilon/2$ approximation error guarantee while having $\mathcal{O}((\log k)/\epsilon)$ update time. That is, the scheme provides a trade-off between the accuracy and computational complexity.

Our approach is straightforward. Computing AUC exactly requires sorting data points and summing over all data points (see Eq. 1 for the exact formula). Maintaining points sorted can be done using binary search trees. However, estimating the sum requires additional tricks. We approach the problem by grouping neighboring data points together, that is, treating them as if the classifier given them the same score.

The key step is to design a grouping such that 3 properties hold at the same time: (*i*) the groups are small enough so that the relative error is small, more specifically, $|a\tilde{u}c - auc|/auc \le \epsilon/2$, (*ii*) the number of groups is small enough, more specifically, it should be in $\mathcal{O}((\log k)/\epsilon)$, and (*iii*) the definition should be flexible enough so that we can do quick updates whenever points arrive or leave the sliding window.

Roughly speaking, in order to accommodate all 3 demands, we will maintain the groups with the two following properties: (*i*) the number of positive labels in a group is less than or equal to $(1 + \epsilon)$ than the total number of positive labels in all the previous groups, (*ii*) the number of positive labels in a group, *and the next group*, is larger than $(1 + \epsilon)$ than the total number of positive labels in all the previous groups. The first property will yield the approximation guarantee, while the second property guarantees that the number of groups remains small. Moreover, these properties are flexible enough so we can perform update procedures quickly.

The rest of the paper is organized as follows. We begin by reminding ourselves the definition of AUC in Sect. 2. Updating the groups of data points quickly requires several auxiliary structures, which we introduce in Sect. 3. We then proceed describing AUC estimation in Sect. 4. The related work is given in Sect. 5. In Sect. 6, we demonstrate that the relative error in practice is much smaller than the guaranteed bound, as well as, study the trade-off between the error and the computational cost. Finally, we conclude the paper with discussion in Sect. 7.

2   Preliminaries

We start with the definition of AUC, and provide a formula for computing it.

Assume that we are given a set of k pairs $W = (s_i, \ell_i)_i^k$, where ℓ_i is the true label of the ith instance, $\ell_i = 0, 1$, and s_i is score produced by the classification algorithm. The larger s_i, the more we believe that ℓ_i should be 0.[1]

In order to predict a label, we need a threshold σ, and predict that $\ell_i = 0$ if $s_i \geq \sigma$, and $\ell_i = 1$ otherwise. The ROC curve is obtained by varying σ and plotting true positive rate as a function of false positive rate. AUC is the area under the ROC curve. To compute AUC, we can use the following formula. Let

$$n(s) = |\{i \mid s_i = s, \ell_i = 0\}| \quad \text{and} \quad p(s) = |\{i \mid s_i = s, \ell_i = 1\}|$$

be the counts of labels with a score of s. Define also $hp(s) = \sum_{t<s} p(t)$. Then,

$$auc = \frac{1}{A} \sum_s (hp(s) + \frac{1}{2}p(s))n(s), \tag{1}$$

where $A = |\{i \mid \ell_i = 0\}||\{i \mid \ell_i = 1\}|$ is the normalization factor. Equation 1 can be computed in $\mathcal{O}(k \log k + k)$ time by first sorting W, computing hp, and enumerating over the sum of Eq. 1.

In a streaming setting, W is a sliding window, and our goal is to compute AUC as W slides over a stream of predictions and labels.

3   Supporting Data Structures for Estimating AUC

In this section we introduce supporting data structures that are needed to compute AUC in a streaming setting. Additional structures and the actual logic for computing AUC are given in the next section. We begin by describing the data structures, then follow with introducing the needed query operations, and finally finish with explaining the update procedures.

3.1   Data Structures

Assume that we have a sequence of pairs $W = (s_i, \ell_i)_{i=1}^k$, where s_i is the score produced by the classifier, and $\ell_i \in \{0, 1\}$ is the true label.

We store W in a red-black tree T sorted by the scores s_i. Let $v \in T$ be a node in T. We will denote the corresponding score of v by $s(v)$. We store and maintain the following information:

- Counter $p(v) = |\{i \mid s_i = s(v), \ell_i = 1\}|$, number of pairs in W with a score $s(v)$ and a positive label.
- Counter $n(v) = |\{i \mid s_i = s(v), \ell_i = 0\}|$, number of pairs in W with a score $s(v)$ and a negative label.

---

[1] We chose this direction due to the notational convenience.

- Counter $accpos(v)$, the total sum of $p(w)$, where w ranges over all descendant nodes of v in T, including v itself.
- Counter $accneg(v)$, the total sum of $n(w)$, where w ranges over all descendant nodes of v in T, including v itself.

For simplicity, we will add two sentinel nodes to T. The first node will have a score of $-\infty$ and the second node has a score ∞. We will assume that the actual entries will never achieve these values. Both sentinel nodes have 0 positive labels and 0 negative labels.

Note that if the scores s_i are unique, then we have either $p(v) = 1$, $n(v) = 0$, or $p(v) = 0$, $n(v) = 1$. However, if there are duplicate scores, then we may have any integer combinations.

In addition to red-black trees, we need to maintain several linked lists, for which we will now introduce the notation. Assume that we are given a subset U of nodes in T. We would like to maintain U in a linked list L, sorted by the score. For that we will need two pointers for each node $u \in U$, namely, $next(u; L)$ indicating the next node in L, and $prev(u; L)$ indicating the previous node in L. Let $u \in U$ and assume that $v = next(u; L)$ exists. Let

$$B = \{w \in T \mid s(u) \leq s(w) < s(v)\}$$

be the set of nodes in T between u and v. We define

$$gp(u; L) = \sum_{w \in B} p(w) \quad \text{and} \quad gn(u; L) = \sum_{w \in B} n(w)$$

to be the total sums of the labels in the gap B. We will refer to L as *weighted linked list*. Note that deleting an element from L and maintaining the gap counters can be done in constant time. We will refer to the deletion algorithm by REMOVE(L, v). Moreover, adding a new element, say v, to L after u can be also done in constant time, if we already know the total sums of labels, say p and n, between u and v. We will refer to the insertion algorithm by ADD(L, u, v, p, n).

We say that the node $v \in T$ is *positive*, if $p(v) > 0$. Similarly, we say that the node v is *negative*, if $n(v) > 0$. Note that v can be both negative and positive.

We maintain all positive nodes in a weighted linked list, which we will refer as P. Finally, we also store all positive nodes in its own dedicated red-black tree, denoted by TP. For simplicity, we also store the sentinel nodes of T in P and TP as the first and the last nodes.

3.2 Query Procedures

The first query that we need is MAXPOS(s), returning the *positive* node v with the largest score such that $s(v) \leq s$. This can be done in $\mathcal{O}(\log k)$ time using TP, where k is the number of elements in the window.

Maintaining $accpos(v)$ and $accneg(v)$ allows us to query a cumulative sums of counts. Specifically, given a score s, we are interested in

$$hp(v) = \sum_{v \in T \mid s(v) < s} p(v) \quad \text{and} \quad hn(v) = \sum_{v \in T \mid s(v) < s} n(v). \tag{2}$$

Algorithm 1. HEADSTATS(s), computes the cumulative counts of labels, $hp(v)$ and $hn(v)$. Assumes that a node in T with a score s exists.

---

1   $hp \leftarrow 0;\ hn \leftarrow 0;$
2   $v \leftarrow$ root of $T;$
3   **while true do**
4   **if** $s(v) < s$ **then**
5         $v \leftarrow left(v);$
6   **else**
7   **if** $left(v)$ **then**
8            $hp \leftarrow hp + accpos(left(v));$
9            $hn \leftarrow hn + accneg(left(v));$
10   **if** $s(v) = s$ **then**
11   **return** $hp, hn;$
12   **else**
13           $hp \leftarrow hp + p(v);$
14           $hn \leftarrow hn + n(v);$
15           $v \leftarrow right(v);$

We can compute both of these sums with HEADSTATS(s), given in Algorithm 1.

The algorithm assumes that there is a node in T containing s, and proceeds to find it; during the search whenever we go the right branch we add the accumulative sums from the left branch. We omit the trivial proof of correctness. Since the tree is balanced, the running time of HEADSTATS(s) is $\mathcal{O}(\log k)$, where k is the number of entries in the window.

3.3 Update Procedures

We now continue to the maintenance procedures as we slide the window. This comes down to two procedures: (i) removing an entry from the window and (ii) adding an entry to the window.

We will first describe removing an entry with a positive label and a score s. First we will find the node, say v, with the score s, and reduce the counter $p(v)$ by 1. We will need to update the *accpos* counters. However, we only need to do it for the ancestors of v, and there are only $\mathcal{O}(\log k)$ of them, where k is the number of entries in the window, since T is balanced. We also reduce $gp(v; P)$ by 1. In the process, v may become non-positive, and we need to delete it from TP as well as from P.

Finally, if $p(v) = n(v) = 0$, we need to delete the node from T. This may result in rebalancing of the tree, and during the balancing we need to make sure that the counters *accpos* and *accneg* are properly updated. Luckily, the red-black tree balancing is based on left and right rotations. During these rotations it is easy to maintain the counters without additional costs.

We will refer to this procedure as REMOVETREEPOS(s) and the pseudo-code is given in Algorithm 2. REMOVETREEPOS(s) runs in $\mathcal{O}(\log k)$ time.

Algorithm 2. REMOVETREEPOS(s, T, TP, P), removes an entry to T with a positive label and a score s.

---

1 $v \leftarrow$ node with score s in T;
2 update $p(v)$, $gp(v; P)$, and *accpos* counters of the ancestors of v;
3 **if** $p(v) = 0$ **then** remove v from the linked list P and the search tree TP;
4 **if** $p(v) = n(v) = 0$ **then** remove v from T;

Deleting an entry with a negative label and a score s is simpler. First, we find the node, say v, with the score s, and reduce the $n(v)$ counter by 1. If needed, we delete v from T. Finally we use MAXPOS(s) to find u, the largest positive node with $s(u) \leq u$, and reduce $gn(u; P)$ by 1. The procedure, referred as REMOVETREENEG, runs in $\mathcal{O}(\log k)$ time.

Next, we will describe the addition of a positive entry with a score s. First, we will add the entry s to T, possibly creating a new node in the process. Let v be the node in T with the score s.

If v is a new node, then we need to add it to the weighted linked list P. First, we find the node, say $w =$ MAXPOS(s), after which v is supposed to be added. We need to compute the new gap counter $gn(v; P)$. By definition, this value is equal to the total count of negative labels of nodes between w and v, including w. Thus, this new gap counter is equal to $hn(w) - hn(v)$. Both counters can be obtained using HEADSTATS in $\mathcal{O}(\log k)$ time.

We will refer to this procedure as ADDTREEPOS(s), and the pseudo-code is given in Algorithm 3. ADDTREEPOS(s) runs in $\mathcal{O}(\log k)$ time.

Algorithm 3. ADDTREEPOS(s), adds an entry to T with a positive label and a score s.

---

1 $w \leftarrow$ MAXPOS(s);
2 add s to T (possibly creating new node), and update *accpos* and p counters;
3 $v \leftarrow$ node with score s in T;
4 **if** $w \neq v$ **then**
5 $\quad$ add v to TP;
6 $\quad$ $p_1, n_1 \leftarrow$ HEADSTATS($s(w)$);
7 $\quad$ $p_2, n_2 \leftarrow$ HEADSTATS($s(v)$);
8 $\quad$ ADD($P, w, v, 1, n_2 - n_1$) ;
9 **return** v;

Adding an entry with negative label and a score s is simpler. First, we will add the entry s to T, possibly creating a new node in the process. Let v be the node in T with a score s. Then, we use MAXPOS(s) to find u, the largest positive node with $s(u) \leq u$, and increase $gn(u)$ by 1. The procedure, referred as ADDTREENEG, runs in $\mathcal{O}(\log k)$ time.

4   Estimating AUC Efficiently

In order to approximate AUC, we will use Eq. 1 as a basis. However, instead of enumerating over every node we will enumerate only over some selected nodes. The key is how to select the nodes such that we will obtain the approximation guarantee while keeping the number of nodes small.

We will maintain a weighted linked list C. Given $\alpha > 1$, we say that C is α-compressed, if for every two consecutive nodes in C, say v and w, it holds that

$$hp(w) \leq \alpha(hp(v) + p(v)), \tag{3}$$

and if $u = next(w; C)$ exists, then

$$hp(u) > \alpha(hp(v) + p(v)). \tag{4}$$

Equation 3 will yield the approximation guarantee, while the Eq. 4 will guarantee the running time.

4.1   Computing Approximate AUC

Our next step is to show how we can approximate AUC using a compressed list L in $\mathcal{O}(L)$ time. The idea is as follows. Let B be the set of nodes between two consecutive nodes v and w in L. Normally, we would have to go over each individual node in B when computing AUC. Instead, we will group B to a *single* node. We will use the total number of positive labels in B, that is, $gp(v; L) - p(v)$, for the number of positive labels for this node. Similarly, we will use $gn(v; L) - n(v)$ for the negative labels. The pseudo-code for the algorithm is given in Algorithm 4.

Algorithm 4. APPROXAUC(L) computes approximate AUC using a weighted linked list.

---

1 $hp \leftarrow 0; a \leftarrow 0$;
2 **while** $v \in L$ **do**
3      $p \leftarrow p(v); n \leftarrow n(v)$;
4      $a \leftarrow a + (hp + p/2)n$;
5      $hp \leftarrow hp + p$;
6      $p \leftarrow gp(v; L) - p(v); n \leftarrow gn(v; L) - n(v)$;
7      $a \leftarrow a + (hp + p/2)n$;
8      $hp \leftarrow hp + p$;

9 $A \leftarrow$ (total number of positive labels) × (total number of negative labels);
10 **return** a/A;

Let us first establish that APPROXAUC produces an accurate estimate.

Proposition 1. *Let L be $(1 + \epsilon)$-compressed list constructed from the search tree T. Let $a\tilde{u}c=$ APPROXAUC(L) be an approximate AUC, and let auc be the correct AUC. Then $|a\tilde{u}c - auc| \leq \epsilon auc/2$.*

Proof. Let A be as defined in APPROXAUC. Let $v \in T$ be a node, and let u be the node in L with the largest score such that $s(u) < s(v)$. Let $w = next(u; L)$ be the next node. Define

$$c_v = \frac{1}{2}(hp(u) + p(u) + hp(w)) \quad .$$

Then, APPROXAUC returns

$$a\tilde{u}c = \frac{1}{A} \sum_{v \in L} (hp(v) + \frac{1}{2}p(v))n(v) + \sum_{v \in T \setminus L} c_v n(v). \tag{5}$$

We will argue the approximation guarantee by comparing the terms in Eqs. 1 and 5. Let v be a node in L. Then the corresponding term can be found in sums of both equations.

Let $v \in T \setminus L$, and write $b = hp(v) + \frac{1}{2}p(v)$. Let u be the node in L with the largest score such that $s(u) \leq s(v)$. Let $w = next(u; L)$ be the next node. By definition, we have $hp(u) + p(u) \leq b \leq hp(w)$. Since c_v is the average of the lower bound and the upper bound, we have

$$|b - c_v| \leq \frac{1}{2}(hp(w) - hp(u) - p(u)) \leq \frac{\epsilon}{2}(hp(u) + p(u)) \leq \frac{\epsilon b}{2},$$

where the second inequality follows since L is $(1 + \epsilon)$-compressed.

We have shown that the approximation holds for individual terms. Consequently, it holds for the summands $a\tilde{u}c$ and auc, completing the proof. □

Two remarks are in order. First, since AUC is always smaller than 1, Proposition 1 implies that the approximation is also absolute, $|a\tilde{u}c - auc| \leq \epsilon/2$. The relative approximation is more accurate if AUC is small. However, if AUC is close to 1, it may make sense to reverse the approximation guarantee, that is, modify the algorithm such that we have a guarantee of $|a\tilde{u}c - auc| \leq (1 - auc)\epsilon/2$. This can be done by flipping the labels, and using $1 - $ APPROXAUC(C) as the estimate.

APPROXAUC runs in $\mathcal{O}(|L|)$ time. Next we establish that $|L|$ is small.

Proposition 2. *Let L be $(1 + \epsilon)$-compressed list. Then $|L| \in \mathcal{O}\left(\frac{\log k}{\epsilon}\right)$, where k is the number of entries in the sliding window.*

Proof. Write $L = u_0, \ldots, u_m$. Since L is $(1 + \epsilon)$-compressed, $hp(u_2) \geq 1$ and $hp(u_{i+2}) > (1 + \epsilon)hp(u_i)$. Since $hp(u_m) \leq k$, we have $(1 + \epsilon)^{\lfloor m/2 \rfloor - 1} \leq k$. Solving for m leads to $m \in \mathcal{O}\left(\frac{\log k}{\log 1 + \epsilon}\right) \subseteq \mathcal{O}\left(\frac{\log k}{\epsilon}\right)$. □

4.2 Updating the Data Structures

Our final step is to describe procedures for maintaining C as the data window slides. In the previous section, we already described how to update the search trees T and TP as well as the weighed linked list P. Our next step is to make sure that the weighted linked list C stays α-compressed.

We will need two utility routines. The first routine, ADDNEXT, given in Algorithm 5, takes as input a node included in both P and C, and adds to C the next node in P. This procedure will be used extensively to add extra nodes to C so that Eq. 3 is satisfied.

Algorithm 5. ADDNEXT(v, L, P), adds the following node of v in P to L. Here P is the weighted linked list of all positive labels, and v is a node in P and L.

---

1   $w \leftarrow next(v, P)$;
2   $p \leftarrow gp(v, P)$; $n \leftarrow gn(v, P)$;
3   **if** $w \notin L$ **then** ADD(L, v, w, p, n);

Next, we demonstrate how ADDNEXT enforces Eq. 3.

Lemma 1. *Assume that a linked list L satisfies Eq. 3 for consecutive positive nodes v and w. Add or remove a single positive entry with a score s, and assume that v and w are still positive. Let u be the next positive node from v in P, and let L' be the list obtained from L by adding a positive node u. Then Eq. 3 holds for L' for the nodes v and u as well as for the nodes u and w.*

Proof. Let us write $c_x = hp(x)$ before modifying T, and $c'_x = hp(x)$ after the modification. Similarly, write $b_x = p(x)$ before the modification, and $b'_x = p(x)$ after the modification.

Since u is the next positive node of v, we have $c'_u = c'_v + b'_v \leq \alpha(c'_v + b'_v)$, proving the case of v and u.

If $s \geq s(w)$, then $c'_w = c_w \leq \alpha(c_v + b_v) = \alpha c_u = \alpha c'_u \leq \alpha(c'_u + b'_u)$.

If we are adding s and $s < s(w)$, then

$$c'_w = c_w + 1 \leq \alpha(c_v + b_v + 1) \leq \alpha(c'_v + b'_v + 1) = \alpha(c'_u + 1) \leq \alpha(c'_u + b'_u),$$

where the last inequality holds since u is a positive node.

If we are removing s and $s < s(w)$, then $c_v + b_v - 1 \leq c'_v + b'_v$, and so

$$c'_w \leq c_w \leq \alpha(c_v + b_v) \leq \alpha(c'_v + b'_v + 1) = \alpha(c'_u + 1) \leq \alpha(c'_u + b'_u).$$

This proves the case for u and w, and completes the proof.   □

Note that the execution of ADDNEXT is done in constant time, the key step for this being able to obtain $gp(v, P) = p(v)$ and $gn(v, P)$ in constant time. This is the main reason why we maintain P.

While the first utility algorithm adds new entries to C, our second utility algorithm, COMPRESS, given in Algorithm 6 tries to delete as many entries as possible. It assumes that the input list C already satisfies Eq. 3, and searches for violations of Eq. 4. Whenever such violation is found, the algorithm proceeds deleting the middle node. Note that deleting this node will not violate Eq. 3. Consequently, upon termination, the resulted linked list will be α-compressed. The computational complexity of COMPRESS(C, α) is $\mathcal{O}(|C|)$.

Algorithm 6. COMPRESS(L, α), forces a weighted linked list L that satisfies Equation 3 to also satisfy Equation 4, making L α-compressed.

---

1 $v \leftarrow$ first element in L;
2 $c \leftarrow 0$;
3 **while** $next(next(v; L) ; L)$ exists **do**
4      $w \leftarrow next(v; L)$;
5   **if** $c + gp(v; L) + gp(w; L) \leq \alpha(c + p(v))$ **then**
6   delete w from L;
7   **else**
8          $c \leftarrow c + gp(v; L)$;
9          $v \leftarrow w$;

Next, we describe the update steps. We will start with the easier ones:

Adding negative entry: Given a negative entry with a score s, we first invoke ADDTREENEG. Then we search $u \in C$ with the largest score such that $s(u) \leq s$. Once this entry is found, we increase $gn(u; C)$ by 1.

Removing negative entry: Given a negative entry with a score s, we first invoke REMOVETREENEG. Then we search $u \in C$ with the largest score such that $s(u) \leq s$. Once this entry is found, we decrease $gn(u; C)$ by 1.

Since the positive labels are not modified, C remains α-compressed, so there is no need for modifying C. The running time for both routines is $\mathcal{O}\left(\log k + \frac{\log k}{\epsilon}\right)$.

Let us now consider more complex cases:

Adding positive entry: Given a positive entry with a score s, we first invoke ADDTREEPOS. Then we search $u \in C$ with the largest score such that $s(u) \leq s$. Once this entry is found, we increase $gp(u; C)$ by 1. By doing so, we may have violated Eq. 3 for u. Lemma 1 states that we can correct the problem by adding the next positive node for each violation. However, a closer inspection of the proof shows that there can be only one violation, namely u. Consequently, we check if Eq. 3 holds for u, and if it fails, we add the next positive node by invoking ADDNEXT(u, C, P). Finally, we call COMPRESS(C, α) to force Eq. 4; ensuring that C is α-compressed. The pseudo-code for ADDPOS is given in Algorithm 7.

Algorithm 7. ADDPOS($s, \alpha; T, TP, P, C$), adds an entry with a positive label and a score s, updates the tree structures T and TP and the weighted linked lists P and C.

---

1   $v \leftarrow$ ADDTREEPOS(s, T, TP, P);
2   $u \leftarrow \arg\max \{s(w) \mid w \in C, s(w) \leq s\}$;
3   $gp(u; C) \leftarrow gp(u; C) + 1$;
4   $c \leftarrow \sum_{w \in C \mid s(w) < s(u)} gp(w; C)$; $\{c = hp(u)\}$
5   **if** $c + gp(u; C) > \alpha(c + p(v))$ **then** ADDNEXT(u, C, P) ;
6   COMPRESS(C, α);

Removing positive entry: Assume that we are given a positive entry with a score s. First we search $u \in C$ with the largest score such that $s(u) \leq s$. Once this entry is found, we decrease $gp(u; C)$ by 1. If u is no longer positive, we add the next positive entry to C and delete u from C. The reason for this is explained later. We proceed by deleting the entry from the search trees with REMOVETREEPOS.

Next we make sure that Eq. 3 holds for every consecutive nodes v and w. There are two possible cases: (i) v and w were consecutive nodes in C before the deletion, or (ii) u was deleted from C, and w was the next positive node before the deletion. In the first case, Lemma 1 guarantees that using ADDNEXT forces Eq. 3. In the second case, note that $hp(w)$ *after* the deletion is equal to $hp(u)$ *before* the deletion of u. This implies that since Eq. 3 held for v and u before the deletion, Eq. 3 holds for v and w after the deletion. Finally, we enforce Eq. 4 with COMPRESS. The pseudo-code for REMOVEPOS is given in Algorithm 8.

Algorithm 8. REMOVEPOS($s, \alpha; T, TP, P, C$), removes an entry with a positive label and a score s, updates the tree structures T and TP and the weighted linked lists P and C.

---

1   $u \leftarrow \arg\max \{s(w) \mid w \in C, s(w) \leq s\}$;
2   $gp(u) \leftarrow gp(u) - 1$;
3   **if** $u \in C$ **and** $p(u) = 1$ **then**
4 | ADDNEXT(u, C, P);
5 | REMOVE(C, u);
6   REMOVETREEPOS(s, T, TP, P);
7   $v \leftarrow$ first element in C;
8   $c \leftarrow 0$;
9   **while** $next(v; C)$ exists **do**
10 | $w \leftarrow next(v; C)$;
11 | $x \leftarrow gp(v; C)$;
12 | **if** $c + x > \alpha(c + p(v))$ **then** ADDNEXT(v, C, P);
13 | $c \leftarrow c + x$;
14 | $v \leftarrow w$;
15   COMPRESS(C, α);

In both routines, modifying the search trees is done in $\mathcal{O}(\log k)$ time, while modifying C is done in $\mathcal{O}(|C|) \subseteq \mathcal{O}\left(\frac{\log k}{\epsilon}\right)$ time.

5   Related Work

The closest related work is a study by Bouckaert [3], where the author divided the ROC curve area into bins, allowing only to maintain the counters for individual bins. However, the number of the bins as well as the bins were static, and no direct approximation guarantees were provided.

Using AUC in a streaming setting was proposed in a paper by Brzezinski and Stefanowski [5]. Here the authors use red-black tree, similar to T, to maintain the order of the data points in a sliding window, but they recompute the AUC from scratch every time, leading to a update time of $\mathcal{O}(k + \log k)$. In fact, our approach is essentially equivalent to their approach if we set $\epsilon = 0$.

Note that using AUC is useful if we do not have a threshold to binarize the score. If we do have such a threshold, then we can easily maintain a confusion matrix, which allows us to compute many metrics, such as, accuracy, recall, $F1$-measure [8,9], and Kappa-statistic [2,13]. However, determining such a threshold may be extremely difficult since it depends on the misclassification costs. Selecting such costs may come down to a(n educated) guess.

We based our AUC calculation on a sliding window, that is, we abruptly forget the data points after certain period of time. The other option is to gradually forget the data points, for example using an exponential decay (see a survey by Gama et al. [10] for such examples). There are currently no methodology for efficiently estimating AUC under exponential decay, and this is a promising future line of work.

In a related line of work, training a classifier by optimizing AUC in a static setting has been proposed by Ataman et al. [1], Brefeld and Scheffer [4], Ferri et al. [7], Herschtal and Raskutti [12]. Here, AUC is used as an optimization criterion, and needs to be recomputed from scratch in $\mathcal{O}(|D| \log |D|)$ time. Naturally, this may be too expensive for large databases. Calders and Jaroszewicz [6] estimated AUC as a continuous function. This allowed to view AUC as a smooth function, and optimize the parameters of the underlying classifier efficiently using gradient descent techniques. While the underlying problem is the same as ours, that is, computing AUC from scratch is expensive, the maintenance procedures make problems orthogonal: in our settings we are required to do updates when a single data point leaves or enters to our window, whereas here AUC needs to be recomputed since the scores (and the order) for all existing data points have changed. However, it may be possible and fruitful to use similar tricks in order to speed-up the AUC calculation when optimizing classifiers. We leave this as a future line of work.

Hand [11] proposed a fascinating alternative for AUC. Namely, the author views AUC as the optimal classification loss averaged (with weights) over misclassification cost ratio. He then argues that AUC evaluates incoherently, namely the cost ratio weights depend on the ROC curve, and then he proposes a different coherent alternative. The computation of proposed metric, though more complex, shares some similarity with AUC, and it may be possible to use similar techniques as in this paper to approximate this measure efficiently in a stream.

6   Experimental Evaluation

In this section we present our experimental evaluation. We have two goals: to demonstrate the relative error in practice as a function of the guaranteed error, and to demonstrate the trade-off between the computational cost and the error.

We implemented calculation of AUC using C++, and conducted the experiments using Macbook Air (1.6 GHz Intel Core i5 / 8 GB Memory).[2] As a classifier we used Python's scikit implementation of logistic regression. Computing AUC was done in a separate job from training the classifier as well as scoring new data points; the reported running times measure only the computation of AUC over the whole test data.

We used 3 UCI datasets[3] for our experiments, see Table 1: (*i*) *Hepmass*, a dataset containing features from simulated particle collisions, split in training and test datasets. We used the *Hepmass*-1000 variant. Due to the memory restrictions of Python, we only used a sample of 500 000 data points from training data. We used the whole test dataset. (*ii*) *Miniboone*: a data used to distinguish electron neutrinos from muon neutrinos. Since the original data has data points ordered by label, we permuted the dataset and split it to training and test data. (*iii*) *Tvads*: a data containing features for identifying commercials from TV news channels. We used BBC and CNN channels as training data, and the remaining channels as test data.

Table 1. Basic characteristics of the benchmark datasets.

Dataset	Size of training dataset	Size of test dataset
Hepmass	500 000	3 500 000
Miniboone	30 064	100 000
Tvads	40 265	89 420

Actual Error vs. Guarantee: Proposition 1 states that the error cannot be more than $\epsilon/2$. First, we test the actual relative error, that is, $|a\tilde{u}c - auc|/auc$ as a function of ϵ. Here we set the sliding window size to be 1000.

The top row of Fig. 1 shows the relative error, averaged over all sliding windows, and the bottom row of Fig. 1 shows the relative error, maximized over all sliding windows. From the results we see that both maximum and average error

---

[2] See https://bitbucket.org/orlyanalytics/streamauc for the implementation.
[3] https://archive.ics.uci.edu/.

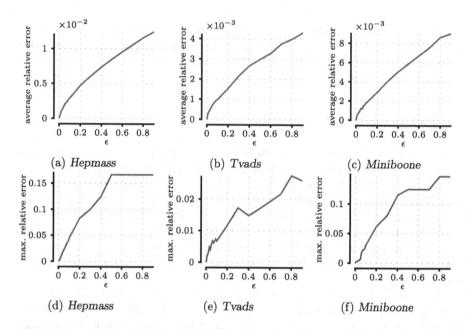

Fig. 1. Actual relative error as a function of ϵ. Top row: average error, bottom row: maximum error. Proposition 1 states that error cannot be larger than $\epsilon/2$.

are smaller than the guaranteed. Especially, the average error is typically smaller of several orders than the theoretical guarantee. As expected, both errors tend to increase as ϵ increases.

Computational Cost vs. Error: Next, we test the trade-off between the computational cost and the relative error. The top row of Fig. 2 shows the running time as a function of the average error, while the bottom row of Fig. 2 shows the size of $(1 + \epsilon)$-compressed list as a function of the average error. Here, we used a window size of 1000.

From the results, we see the trade-off between the error and the running time: as the error increases, the running time drops. This is mainly due to the fewer elements in the compressed list as demonstrated in the bottom row. The running stabilizes for larger errors; this is due to the operations that do not depend on ϵ, such as maintaining binary tree T.

Computational Cost vs. Window Size: Computing exact AUC requires $\mathcal{O}(k)$ time while estimating AUC is $\mathcal{O}(\log k/\epsilon)$. Consequently, the speed-up should increase as the size of the sliding window increases. We demonstrate this effect in Fig. 3 using the *Miniboone* dataset. We see that the speed-up increases as a function of window size: computing estimates using $\epsilon = 0.1$ is 17 times faster for a window size of 10 000.

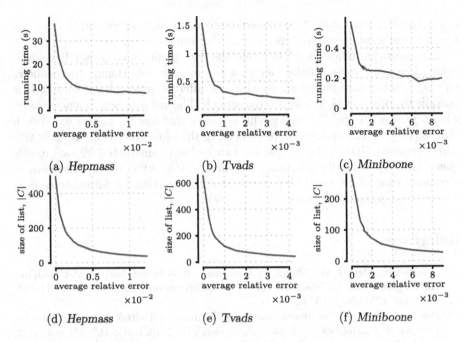

Fig. 2. Top row: running time as a function of average relative error. Bottom row: size of the compressed list $|C|$ as a function of average relative error.

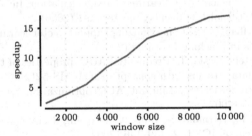

Fig. 3. A speed-up of estimating AUC with $\epsilon = 0.1$ against computing AUC exactly, as a function of sliding window size. The dataset is *Miniboone*.

7   Concluding Remarks

In this paper we introduced an approximation scheme that allows to maintain an estimate AUC in a sliding window within the guaranteed relative error of $\epsilon/2$ in $\mathcal{O}((\log k)/\epsilon)$ time. The key idea behind the estimator is to group the data points. The grouping has to be done cleverly so that the error stays small, the number of groups stay small, and the list can be updated quickly. We achieve this by maintaining groups, where the number of positive labels can only increase relatively by $(1 + \epsilon)$ within one group, and must increase by at least $(1 + \epsilon)$ within two groups. Our experimental evaluation suggests that the average error in practice is much smaller than the guaranteed approximation, and that we can achieve significant speed-up, especially as the window size grows.

Our algorithm relies on the fact that the data points have no weights, specifically, Lemma 1 relies on the fact that the update may change the counters only by 1. If the data points are weighted, a different approach is required: It is possible to construct $(1+\epsilon)$-list from a scratch. The key idea here is a new query, where, given a threshold σ, we look for a node v that has the largest $hp(v)$ such that $hp(v) \leq \sigma$. This query can be done using the same trick as in HEADSTATS, and it requires $\mathcal{O}(\log k)$ time. The list can be then constructed by calling this query with exponentially increasing thresholds $\mathcal{O}((\log k)/\epsilon)$ times. This leads to a running time of $\mathcal{O}((\log^2 k)/\epsilon)$. An interesting direction for future work is to improve this complexity to, say, $\mathcal{O}((\log k)/\epsilon)$.

References

1. Ataman, K., Streetr, W., Zhang, Y.: Learning to rank by maximizing AUC with linear programming. In: International Joint Conference on Neural Networks, IJCNN 2006, pp. 123–129. IEEE (2006)
2. Bifet, A., Frank, E.: Sentiment knowledge discovery in Twitter streaming data. In: Pfahringer, B., Holmes, G., Hoffmann, A. (eds.) DS 2010. LNCS (LNAI), vol. 6332, pp. 1–15. Springer, Heidelberg (2010). https://doi.org/10.1007/978-3-642-16184-1_1
3. Bouckaert, R.R.: Efficient AUC learning curve calculation. In: Australasian Joint Conference on Artificial Intelligence, pp. 181–191 (2006)
4. Brefeld, U., Scheffer, T.: AUC maximizing support vector learning. In: Workshop on ROC Analysis in Machine Learning (2005)
5. Brzezinski, D., Stefanowski, J.: Prequential AUC: properties of the area under the ROC curve for data streams with concept drift. KAIS 52(2), 531–562 (2017)
6. Calders, T., Jaroszewicz, S.: Efficient AUC optimization for classification. In: PKDD, pp. 42–53 (2007)
7. Ferri, C., Flach, P., Hernández-Orallo, J.: Learning decision trees using the area under the ROC curve. ICML 2, 139–146 (2002)
8. Gama, J.: Knowledge Discovery from Data Streams. CRC Press, Boca Raton (2010)
9. Gama, J., Sebastião, R., Rodrigues, P.P.: On evaluating stream learning algorithms. Mach. Learn. 90(3), 317–346 (2013)
10. Gama, J., Žliobaitė, I., Bifet, A., Pechenizkiy, M., Bouchachia, A.: A survey on concept drift adaptation. ACM Comput. Surv. 46(4), 44 (2014)
11. Hand, D.J.: Measuring classifier performance: a coherent alternative to the area under the ROC curve. Mach. Learn. 77(1), 103–123 (2009)
12. Herschtal, A., Raskutti, B.: Optimising area under the roc curve using gradient descent. In: Proceedings of the twenty-first international conference on Machine learning, p. 49. ACM (2004)
13. Žliobaitė, I., Bifet, A., Read, J., Pfahringer, B., Holmes, G.: Evaluation methods and decision theory for classification of streaming data with temporal dependence. Mach. Learn. 98(3), 455–482 (2015)

Controlling and Visualizing the Precision-Recall Tradeoff for External Performance Indices

Blaise Hanczar[1]([✉]) and Mohamed Nadif[2]

[1] IBISC, University of Paris-Saclay, Univ. Evry, Evry, France
`blaise.hanczar@ibisc.univ-evry.fr`
[2] LIPADE, University of Paris Descartes, Paris, France
`mohamed.nadif@parisdescartes.fr`

Abstract. In many machine learning problems, the performance of the results is measured by indices that often combine precision and recall. In this paper, we study the behavior of such indices in function of the tradeoff precision-recall. We present a new tool of performance visualization and analysis referred to the tradeoff space, which plots the performance index in function of the precision-recall tradeoff. We analyse the properties of this new space and show its advantages over the precision-recall space. Code related to this paper is available at: https://sites.google.com/site/bhanczarhomepage/prerec.

Keywords: Evaluation · Precision-recall

1 Introduction

In machine learning, precision and recall are usual measures to assess the performances of the results. These measures are particularly used in supervised learning [18], information retrieval [16], clustering [13] and recently in biclustering contexts [12]. In supervised learning, the classifier performances are assessed by comparing the predicted classes to the actual classes of a test set. These comparisons can be measured by using the precision and recall of the positive class. The precision-recall is generally used in the problems which present very unbalanced classes where the couple sensitivity-specificity is not relevant. In information retrieval, the performance of a search algorithm is assessed by analysing from the similarity between the set of target documents and the set returned by the algorithm. This similarity is generally based on the precision and recall values. In clustering or biclustering, the algorithms identify the clusters or biclusters in the data matrix which are then compared to the clusters or biclusters of reference. It is very common to combine precision and recall in order to construct a performance index such as the F-measure or Jaccard indices; see for instance [1].

By default, the performance indices give the same weight to the precision and recall measures. However, in many contexts, one of these two measures is more

M. Berlingerio et al. (Eds.): ECML PKDD 2018, LNAI 11051, pp. 687–702, 2019.
https://doi.org/10.1007/978-3-030-10925-7_42

important than the other. For example, in genomics, we use clustering algorithms in order to identify clusters of genes with similar expression profiles [6]. These clusters are compared to a gene clustering constructed from genomics databases in order to evaluate their biological relevance. The objective of these analyses is to identify as much biological information as possible in the clusters of genes. In this context the recall is more important than the precision, it is therefore more convenient to use an index that favors the recall. In some performance indices, a parameter has been introduced in order to control the precision-recall tradeoff, for example, the parameter β in F-measure.

In this paper, we analyze the performance indices in function of the precision-recall tradeoff and point out their characteristics. For the analysis and visualization of the performances, we also propose a new tool called the tradeoff space. This new space has many advantages compared to the classic precision-recall space.

The paper is organized as follows. In Sect. 2, we present the performance indices and their variants which are sensitive to the precision-recall tradeoff. In Sect. 3, we give the properties of the precision-recall space and analyze the performance indices in this space. In Sect. 4, we introduce the tradeoff space and show how to represent the performances with the tradeoff curves. Section 5 is devoted to applications in unsupervised and supervised contexts, we point out the advantages of the tradeoff space to model selection and comparison of algorithms. Finally, we present our conclusions and give some recommendations on the choice of the performance index.

2   Performance Indices Based on the Precision and Recall Measures

In this section, the definitions are given in the context of unsupervised learning, however all these methods can also be used in the context of supervised learning. In Sect. 5 all the indices and methods are applied to both contexts.

2.1   Definitions

Let D be a dataset containing N elements. Let $T \subset D$ be a target cluster that we want to find and let X be the cluster returned by an algorithm referred as A whose objective is to find the target cluster. The goodness of X is estimated by using a performance index $I(T, X)$ measuring the similarity between T and X. Some performance indices rely on two basic measures of precision and recall given by

$$\begin{cases} pre = precision(T, X) = \frac{|T \cap X|}{|X|}, \\ rec = recall(T, X) = \frac{|T \cap X|}{|T|} \end{cases}$$

where $|.|$ denotes the cardinality. The main performance indices are a combination of the precision and recall measures. These indices give the same importance to precision and recall, however we can define weighted version that may favor the precision or the recall. We introduce in each index a parameter $\lambda \in [0, 1]$ that controls the tradeoff; λ gives the importance of recall and $1 - \lambda$ the importance of precision. The weighted indices have to respect the following conditions. For $\lambda = 0$ (resp. $\lambda = 1$) only precision (resp. recall) matters. The index must return 0 when the intersection $|T \cap X|$ is null and 1 when $T = X$. For $\lambda = 0.5$, the same importance is given to precision and recall. Formally the conditions are as follows:

$$\begin{cases} I_{weighted}(T, X, \lambda) \in [0, 1] \\ I_{weighted}(T, X, 0) = pre \\ I_{weighted}(T, X, 1) = rec \\ I_{weighted}(T, X, 0.5) = I_{non-weighted}(T, X) \\ I_{weighted}(T, X, \lambda) = 0 \Rightarrow |T \cap X| = 0 \\ I_{weighted}(T, T, \lambda) = 1. \end{cases}$$

In this paper, we study the four most popular indices: Kulczynski, F-measure, Folke and Jaccard. However, our work can easily be extended to other indices.

2.2   Kulczynski Index

The Kulczynski index is the arithmetic mean between precision and recall.

$$I_{Kul}(T, X) = \frac{1}{2}(pre + rec).$$

The weighted version introduces parameter $\rho \in [0, +\infty[$ that controls the precision-recall tradeoff. The importance of precision increases with the value of ρ, the pivotal point is at $\rho = 1$. In order to respect the conditions on the weighted indices, we rewrite this index in setting: $\lambda = \frac{\rho}{\rho+1}$.

$$\begin{cases} I_{Kul}(T, X, \rho) = \frac{1}{\rho+1}(\rho.pre + rec) \\ I_{Kul}(T, X, \lambda) = \lambda rec + (1 - \lambda)pre. \end{cases}$$

2.3   F-Measure

The F1-measure, also called the Dice index, is the ratio between the intersection and the sum of the sizes of cluster X and target cluster T. It is the harmonic mean between precision and recall.

$$I_{Fmes}(T, X) = \frac{2|T \cap X|}{|T| + |X|} = \frac{2}{\frac{1}{rec} + \frac{1}{pre}} = \frac{2pre.rec}{pre + rec}.$$

The F-measure is a weighted version of the F1-measure. The parameter $\beta \in [0, +\infty]$ controls the precision-recall tradeoff. The importance of precision increases with the value of β, the pivotal point is at $\beta = 1$. In order to respect the conditions on the weighted indices, we rewrite this index in setting: $\lambda = \frac{\beta^2}{1+\beta^2}$ and we obtain

$$I_{Fmes}(T, X, \beta) = \frac{1 + \beta^2}{\frac{\beta^2}{rec} + \frac{1}{pre}} = (1 + \beta^2)\frac{pre.rec}{\beta^2 pre + rec}$$

$$I_{Fmes}(T, X, \lambda) = \frac{1}{\frac{\lambda}{rec} + \frac{1-\lambda}{pre}} = \frac{pre.rec}{\lambda pre + (1 - \lambda)rec}.$$

2.4   Folke Index

The Folke index is the geometric mean of precision and recall

$$I_{Fk}(T, X) = \frac{|T \cap X|}{\sqrt{|T||X|}} = \sqrt{pre.rec}.$$

To obtain the weighted version of the Folke index, we introduce the parameter λ such that:

$$I_{Fk}(T, X, \lambda) = \frac{|T \cap X|}{|X|^{1-\frac{\lambda}{2}}|T|^{\frac{\lambda}{2}}} = rec^{\frac{\lambda}{2}}pre^{1-\frac{\lambda}{2}}.$$

2.5   Jaccard Index

The Jaccard index is the ratio between the intersection and the union of cluster X and target cluster T.

$$I_{Jac}(T, X) = \frac{|T \cap X|}{|T| + |X| - |T \cap X|} = \frac{pre.rec}{pre + rec - pre.rec}.$$

It is not easy to define a weighted version of the Jaccard index because of the term $pre.rec$ in the denominator. In order to respect the conditions on the weighted indices, we introduce the two weight functions $w(\lambda) = min\{2\lambda, 1\}$ and $v(\lambda) = 1 - |1 - 2\lambda|$ where λ controls the precision-recall tradeoff. The weighted Jaccard index is defined by:

$$I_{Jac}(T, X, \lambda) = \frac{pre.rec}{w(\lambda).pre + w(1 - \lambda)rec - v(\lambda).pre.rec}.$$

3   The Precision-Recall Space

A common analysis and visualization tool of the performances is the precision-recall space. It is a 2D space which represents the precision on the y-axis and recall on the x-axis. The performance of a cluster is represented by a point in this space (Fig. 1) [3].

The precision-recall space is close to the ROC space that is defined by the false and true positive rates. Some relationships between precision-recall and ROC spaces have been identified [7]. A point on the precision-recall space represents the performance of all clusters with the same size $|X| = |T|\frac{rec}{pre}$ and the same intersection $|T \cap X| = |T|rec$. The $\bullet$ symbol at $(1, 1)$ which maximizes

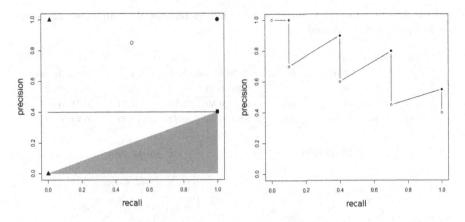

Fig. 1. The precision-recall space (left). The precision-recall curve (right).

both precision and recall, represents the perfect cluster, i.e. that equal to the target cluster $(T = X)$. The ■ symbol at $(1, \frac{|T|}{|D|})$ represents the case where the returned cluster is equal to the whole data matrix $X = D$. The horizontal bold line corresponds to the expected performances of a random cluster, i.e. a cluster whose elements are randomly selected. Since it depends on the size of the target cluster $|T|$, the expected precision of a random cluster is constant and $\mathbb{E}[pre] = \frac{|T|}{|D|}$. The expected recall of a random cluster depends on the size of the cluster $|X|$, it is equal to $\mathbb{E}[rec] = \frac{|X|}{|D|}$. The ▲ symbol at $(\frac{1}{|T|}, 1)$ represents the clusters with a unique element belonging to the target cluster. The ▲ symbol at $(0, 0)$ represents the clusters whose intersection with the target cluster is null. The gray area represents performances that cannot be reached by a cluster. Since we have $pre \geq \frac{|T|}{|D|}rec$ and $|X| \leq |D|$, then all the clusters whose performance are represented by a point on the $pre = \frac{|T|}{|D|}rec$ line are the clusters with the minimal intersection possible $|T \cap X|$ for a given size $|X|$. In clustering and biclustering, the algorithms may have a parameter controlling the size of the returned clusters $|X|$. In varying this parameter, an algorithm produces different clusters having different values of precision and recall. The performance of an algorithm is therefore represented by a set of points that can be approximated by a curve. Figure 1 (right) gives an example of this precision-recall curve. Some information can be drawn from this curve. If a point A dominates another point B i.e. $pre(T, A) > pre(T, B)$ and $rec(T, A) > rec(T, B)$, then the performances of cluster A are better than the performances of cluster B, whatever the performance index used to compare the two clusters. In Fig. 1 (right) the black points are the dominant points, they represent the performance of the best clusters. However, there is no domination relation between these points, we can not compare them from the precision-recall space, the use of a performance index is needed.

Table 1. The values of the four performance indices on several examples with different values of precision, recall and λ.

Cluster	Precision	Recall	$\lambda = 0.5$				$\lambda = 0.2$			
			Kulczynski	Fmeasure	Folke	Jaccard	Kulczynski	Fmeasure	Folke	Jaccard
C_1	0.70	0.70	0.70	0.70	0.70	0.54	0.70	0.70	0.70	0.62
C_2	0.75	0.60	0.67	0.67	0.67	0.50	0.72	0.71	0.71	0.62
C_3	0.80	0.50	0.65	0.61	0.63	0.44	0.74	0.71	0.73	0.60

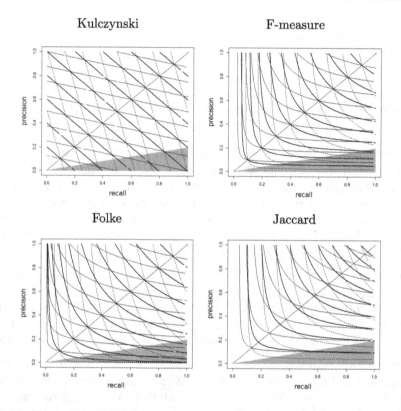

Fig. 2. The isolines of the four performance indices in the precision-recall space.

The behavior of the performances indices can be visualized in plotting their isolines in the precision-recall space. An isoline is a set of points in the precision-recall space having the same value of the performance index [9,12]. Figure 2 shows the isolines of the Kulczynski, F-measure, Folke and Jaccard indices. The bold lines represent the isolines when $\lambda = 0.5$ while the dotted lines and full lines represent the isolines for respectively $\lambda = 0.2$ and $\lambda = 0.8$. For the four indices, we observe a symmetry of the isoline around the axis $pre = rec$, this means that precision and recall have the same importance. Nevertheless, the different indices do not record the difference between precision and recall ($pre - rec$) in the same way. This difference is not taken into account in the Kulczynski index, whereas the other indices penalize it. The Folke index penalizes less than the F-measure and Jaccard indices. Note that the F-measure and Jaccard indices are equivalent because they are compatible i.e. $I_{Fmes}(T, X_1) \geq I_{Fmes}(T, X_2) \Leftrightarrow I_{Jac}(T, X_1) \geq I_{Jac}(T, X_2)$.

A modification of λ value changes the shape of the isolines and gives more importance to precision or recall. Note that for $pre = rec$ the Kulczynski, F-measure and Folke indices return the same value, whatever λ (the bold, dotted and full lines cross the line $pre = rec$). The Jaccard index is different, it penalizes the fact that λ is close to 0.5. Table 1 gives some examples illustrating the consequences of these characteristics. We see that the recorded value and rank of each point depends on the index. For $\lambda = 0.5$ all indices consider C_3 as the best cluster. For $\lambda = 0.2$ the indices do not agree anymore, and the best cluster depends on the performance index.

As we have seen, the choice of the performance index has a high impact on the analysis of the results and especially when the precision-recall tradeoff is far from 0.5. It is a crucial step that must depend on the context. We discuss this point in the next part.

4   The Tradeoff Space

4.1   Definitions

We propose a new tool, called the tradeoff space, in order to visualize the performance of the algorithms in function on the precision-recall tradeoff. The x-axis and y-axis represent respectively λ and the performance index. This method is inspired by the cost curves used in supervised classification [8]. The performance of a result is represented on this space by a curve: $I(T, X, \lambda)$, this curve depends only on λ because X and T are fixed. There is a bijection between the points on the precision-recall space and the curves on the tradeoff space. Figure 3 gives an example of these curves for a result whose performances are $pre = 0.85$ and $rec = 0.5$. The bold curve represents the performance index. The extremities of the curves give the precision and recall of the cluster, we have $I(T, X, 0) = pre$ and $I(T, X, 1) = rec$. The full line shows the performances of the maximal cluster, i.e. the cluster containing all the elements, this corresponds to the $(1, \frac{|X|}{|D|})$ point in the precision-recall space. This curve defines the domain of application of the performance index for a given dataset, it is illustrated by the white area in Fig. 3. A point in the gray area means that the corresponding cluster has worse performances than the maximal cluster and may be considered irrelevant. The application domain of the Kulczynski index is much smaller than the application domain of the other indices because this index does not penalize the difference between precision and recall. The other extreme case is the empty cluster containing no element. However the precision of the empty cluster is not defined because its size is null $|X| = 0$. We therefore consider the cluster containing a unique element belonging to the target cluster as the minimal cluster whose

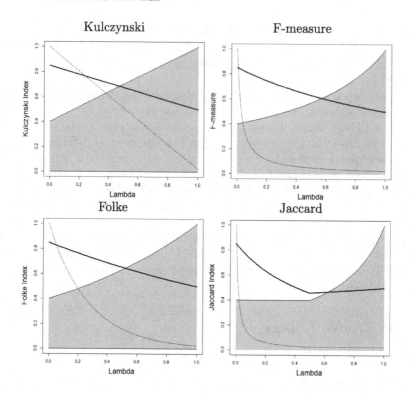

Fig. 3. The tradeoff spaces for the four performance indices.

performances are $pre = 1$ and $rec = 1/|T|$. The dotted line represents the performance of the minimal cluster. The clusters below this line are worse than the trivial minimal cluster. Note that this line is relevant only for the Kulczynski index, for the other indices the line falls sharply when the tradeoff value is not close to zero. The perfect cluster is represented by the $I(X, T, \lambda) = 1$ line while the clusters with a null intersection with the target cluster are represented by the $I(X, T, \lambda) = 0$ line. All curves representing the expected performances of a random cluster pass through point $(0, \frac{|T|}{|D|})$.

4.2 The Optimal Curve of Tradeoff

The performance of an algorithm can be represented by a curve in the precision-recall space as illustrated in Fig. 1 (right). Each point of this curve corresponds to a curve in the tradeoff space. The precision-recall curve is therefore represented by a set of curves in the tradeoff space. Figure 4 shows the representation of the precision-recall curve of Fig. 1 in the tradeoff space for the four performance indices. We call optimal tradeoff curve the upper envelope of the set of curves (in bold in Fig. 4). The optimal tradeoff curve of the algorithm $\mathbb{A}$, noted $I^*(T, \mathbb{A}, \lambda)$, is a piece-wise curve obtained by keeping the best tradeoff curve for each value of

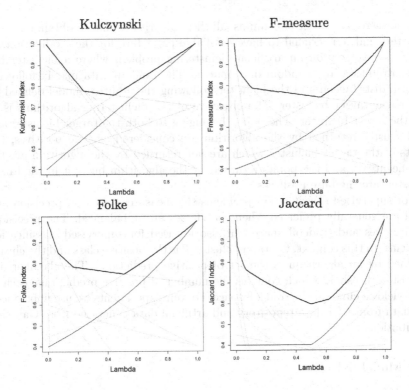

Fig. 4. Representation of the precision-recall curve of Fig. 1 in the tradeoff spaces for the four performance indices. The gray curves are the tradeoff curves, the bold curves are the optimal tradeoff curves.

λ. It represents the best performances of the algorithm for any value of tradeoff λ. Note that the curves forming the upper envelope correspond to dominant points of the precision-recall curve. The curve of the dominated points is always below the optimal tradeoff curve. For the Kulczynski index, the curves forming the upper envelope correspond to the points of the convex hull of the precision-recall curve, which is our next point.

5   Application of the Tradeoff Curves

We show in this section that the tradeoff curves are a better visualization tool and easier to interpret than the precision-recall curves. We focus especially on the application of tradeoff curves, and more precisely, on two problems: the model selection and the comparison of algorithms. To this end, we study these problems in the context of biclustering and supervised binary classification.

Biclustering, referred also to as co-clustering, is a major tool of data science in many domains and many algorithms have emerged in recent years [4,10,11, 14,15,17]. Knowing that a bicluster is a subset of rows which exhibit a similar behavior across a subset of columns, all these algorithms aim to obtain coherent biclusters and it is crucial to have a reliable procedure for their validation. In this paper we rely on an artificial biclustering problem where a bicluster has been introduced in a random data matrix. The random data matrix follows a uniform distribution and the bicluster, following the additive model defined by [15], is the target bicluster. The objective of the biclustering algorithms is to find the target bicluster. The points belonging to both the target bicluster and the bicluster returned by the algorithm are considered as true positives, the points of the target bicluster which are not returned by the algorithm are the false negatives, and the points returned by the algorithm but not in the target bicluster, are the false positives.

For supervised classification problems the measures based on precision and recall are generally preferred when the classes are unbalanced. The precision-recall curves and tradeoff space can also be used for supervised classification problems. In this context, the target cluster T is the positive class and the cluster returned by an algorithm is the set of positive predictions. The classifier has a decision threshold which controls the number of positive predictions, each of them yields a curve in the tradeoff space. For our experiments we use unbalanced real data from UCI data repository and artificial data generated from Gaussian distributions.

5.1   Model Selection

To deal with the biclustering aim, we used the well known CC (Cheng & Church) algorithm to find the bicluster [5]. The similarity between the bicluster returned by the CC algorithm and the target bicluster is computed with the four performance indices. The CC algorithm has a hyper-parameter controlling the size of the returned bicluster. The performance of the algorithm can, therefore, be represented by a precision-recall curve (Fig. 5 1st graphics), each bicluster returned by the CC algorithm is identified by its size. From this curve, it is not easy to define the best bicluster because there are several dominant points. Even if we plot the isolines on the graph, the comparison of the different biclusters is not intuitive. The graphics 2–5 in the Fig. 5 represent the optimal tradeoff curve for the Kulczynski, F-measure, Folke and Jaccard indices. From the tradeoff curve, we can immediately identify the best bicluster for a given tradeoff value. There is a decomposition of the value of λ in a set of intervals, represented in Fig. 5 by the vertical dotted lines. For each interval, the best bicluster is identified. In our example, there are seven intervals for the F-measure, only the seven corresponding biclusters are therefore relevant. For the last interval ($\lambda > 0.74$) the best bicluster is the maximal bicluster, the optimal tradeoff curve is the curve of the maximal bicluster. We can do the same analysis with the tradeoff curves of the Kulczynski, Folke and Jaccard indices, containing respectively eight, seven and six intervals and relevant biclusters. Note that the identification of the best bicluster depends on the chosen performance index. It is not possible to identify the best biclusters in the precision-recall space because they do correspond neither to the set of the dominant points nor to the convex hull of the precision-recall curve (except for the Kulczynski index). Furthermore, it is also easy to use constraints on the precision and recall in the tradeoff space. The precision and recall can be read at the extremity of the tradeoff curve. If a minimal precision pre_{min} is required, we simply have to select the curves that start above the minimal precision, i.e. $I(X, T, 0) > pre_{min}$. In the same way, with a required minimal recall rec_{min}, we keep only the curves that finish above the minimal recall i.e. $T(X, T, 1) > rec_{min}$.

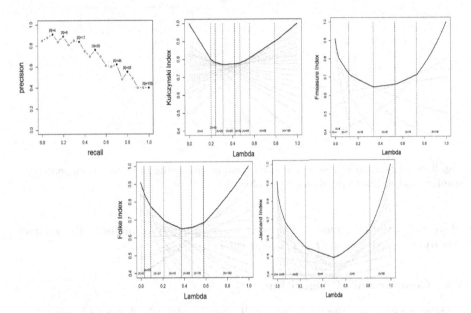

Fig. 5. Identification of the best biclusters in the precision-recall space (1st graphic) and in the tradeoff space (graphics 2–5).

To deal with the supervised classification problems, we use the linear discriminant analysis (LDA) to find the positive class. For each test example the classifier estimates a probability to belong to the positive class, this probability is then compared to a decision threshold t, in order to assign the positive or negative class to the example. By default this threshold is 0.5 but it can be changed to favor positive or negative classes in the context of unbalanced classes. The decision threshold is a hyper-parameter to optimize in order to maximize the performance index. The first graphic in the Fig. 6 shows the precision-recall curve of the LDA classifier, each point represents a different value of decision threshold. As with the biclustering problem, it is difficult to identify the best thresholds from this curve for a given performance index. The tradeoff curves, represented in the graphics 2–5 of the Fig. 6, are much more useful to find the models maximizing the F-measure or the Jaccard index whatever the precision-recall tradeoff. To each interval corresponds a value of the decision threshold

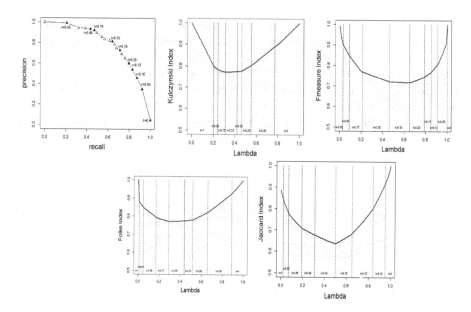

Fig. 6. Identification of the best classification model in the precision-recall space (1st graphic) and in the tradeoff space (graphics 2–5).

which yields the best model. The models assigning all examples to the positive $(t = 0)$ or negative class $(t = 1)$ are at the extremities.

5.2 Comparison of Algorithms

Here we consider the precision-recall curves and tradeoff curves as visualization tools to compare the algorithms. We keep the same illustrative problems used in the previous section. Concerning the situation of biclustering, a second algorithm, ISA [2], is used to find the target bicluster. The objective is to compare the performances of the two algorithms and identify the best one. The first graphic of the Fig. 7 shows the performance of the CC algorithm (in black) and ISA (in gray) in the precision-recall space. In the precision-recall space, the two curves cross each other several times, no algorithm is strictly better than the other. It is hard to identify the conditions in which CC is better than ISA and *vice versa*. The graphics 2–5 of the Fig. 7 represents the tradeoff curves of the algorithms for the Kulczynski, F-measure, Folke and Jaccard indices. In the tradeoff space, we immediately visualize which is the best algorithm whatever the tradeoff value. According to F-measure, CC is the best for $\lambda < 0.28$, ISA is the best for $0.28 < \lambda < 0.83$, for $\lambda > 0.83$ both algorithms return the maximal bicluster and have the same performances. According to the Folke index, CC is the best for $\lambda < 0.30$, ISA for $0.30 < \lambda < 0.69$ and for $\lambda > 0.69$ both algorithms return the maximal bicluster. According to the Jaccard index, CC is the best for $\lambda < 0.2$, ISA for $0.2 < \lambda < 0.87$ and for $\lambda > 0.87$ both algorithms

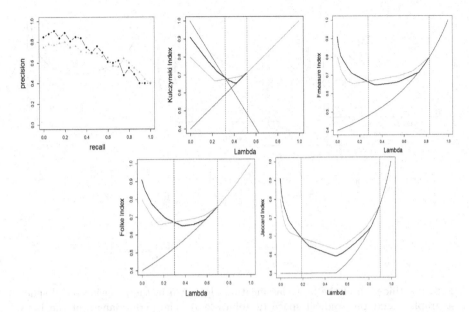

Fig. 7. Identification of the best biclustering algorithm in the precision-recall space (1st graphic) and the tradeoff space (graphics 2–5). The performances of the CC algorithm are in black and ISA in gray.

return the maximal bicluster. In the Kulczynski figure, we add the line representing the extreme case where the algorithm return the minimum bicluster i.e. a bicluster containing only one true positive. For $\lambda < 0.33$, CC is better than ISA but both algorithms are worse than the minimal bicluster. ISA is the best for $0.33 < \lambda < 0.53$, for $\lambda > 0.53$ both algorithms return the maximal bicluster. The interval $[0.33, 0.53]$ is the tradeoff range in which the algorithms are useful, outside this interval trivial solutions are better than algorithm's solutions. We can, therefore, conclude that ISA is strictly better than CC for the Kulczynski index. Defining the conditions where CC is better than ISA from the precision-recall curve is much more difficult. In the precision-recall space, the two curves cross each other three times which implies that the interval of λ, where CC is better than ISA, is not continued. Actually, the tradeoff curves show that the best algorithm changes only once. In the precision-recall space, CC has a better precision than ISA, 14 times out of 20. We can therefore conclude that CC is better than ISA for a large range of λ values. The tradeoff space shows that the opposite is true.

For the supervised classification problems, we compare LDA with the linear support vector machine (SVM). The Fig. 8 shows the performances of LDA (in black) and SVM (in gray) in the precision-recall space (1st graphic) and the tradeoff space (graphics 2–5). As observed with the biclustering situation, we can easily show which algorithm is the best for any tradeoff values and define the range of tradeoff for which the algorithms are better than a trivial solution.

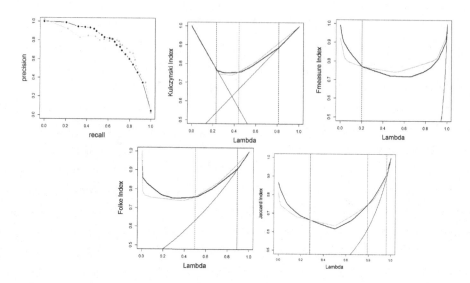

Fig. 8. Identification of the best classification algorithm in the precision-recall space (1st graphic) and the tradeoff space (graphics 2–5). The performances of the LDA algorithm are in black and SVM in gray.

According to the Kulczynski index, LDA is the best for $0.23 < \lambda < 0.44$ and SVM for $0.44 < \lambda < 0.81$. According to F-measure, LDA is the best for $\lambda < 0.23$, SVM for $0.23 < \lambda$. According to the Folke index, LDA is the best for $\lambda < 0.50$ and SVM for $0.50 < \lambda < 0.89$. According to the Jaccard index, LDA is the best for $\lambda < 0.29$ and $0.79 < \lambda < 0.96$, and SVM for $0.29 < \lambda < 0.79$.

6 Conclusion

In this paper, we have presented new methods to deal with the precision-recall tradeoff for the different performance indices. The analysis of these indices in the precision-recall space shows several properties depending on the difference between the precision and recall measures, and the λ tradeoff. These characteristics should guide the choice of the performance index in order to select the most suitable one to the dataset and context.

The tradeoff space is a new tool to visualize the performance in function of the tradeoff. In this space, the model selection and comparison of algorithms are much easier and more intuitive than with the precision-recall space. We have also proposed new performance indices weighted by a probability density function representing the knowledge about the tradeoff precision-recall. This work focuses on four indices (Kulczynsky, F-measure, Folke and Jaccard) but it can easily be extended to any other index that relies on the precision and recall measures.

To conclude this paper, we give some recommendations to the user who wants to use a performance index adapted to a given problem. First, choose the type of index (Jaccard, Folke, F-measure,...) in function of its characteristics; the visualization of its isolines can be helpful. Secondly, define the precision-recall tradeoff of the problem. If the exact value of the tradeoff is known, use the index given in Sect. 2 in setting λ to this value. If there is no knowledge about the precision-recall tradeoff, draw the optimal tradeoff curve and compute its AUC for computing a numerical value. If some vague information about the precision-recall tradeoff is known, draw the optimal tradeoff curve on the range of the possible values of λ. These recommendations should produce evaluation procedures which will be more suitable to the problem and would, therefore, improve the robustness and accuracy of the experimental studies.

References

1. Albatineh, A.N., Niewiadomska-Bugaj, M.: Correcting Jaccard and other similarity indices for chance agreement in cluster analysis. Adv. Data Anal. Classif. **5**(3), 179–200 (2011)
2. Bergmann, S., Ihmels, J., Barkai, N.: Iterative signature algorithm for the analysis of large-scale gene expression data. Phys. Rev. E Stat. Nonlin. Soft Matter. Phys. **67**, 031902 (2003)
3. Buckland, M., Gey, F.: The relationship between recall and precision. J. Am. Soc. Inform. Sci. **45**, 12–19 (1994)
4. Busygin, S., Prokopyev, O., Pardalos, P.: Biclustering in data mining. Comput. Oper. Res. **35**(9), 2964–2987 (2008)
5. Cheng, Y., Church, G.M.: Biclustering of expression data. Proc. Int. Conf. Intell. Syst. Mol. Biol. **8**, 93–103 (2000)
6. Datta, S., Datta, S.: Methods for evaluating clustering algorithms for gene expression data using a reference set of functional classes. BMC Bioinform. **7**, 397 (2006)
7. Davis, J., Goadrich, M.: The relationship between precision-recall and ROC curves. In: Proceedings of the 23rd International Conference on Machine Learning, ICML 2006, pp. 233–240 (2006)
8. Drummond, C., Holte, R.C.: Cost curves: an improved method for visualizing classifier performance. Mach. Learn. **65**, 95–130 (2006)
9. Flach, P.A.: The geometry of ROC space: understanding machine learning metrics through ROC isometrics. In: ICML, pp. 194–201 (2003)
10. Govaert, G., Nadif, M.: Co-clustering: Models, Algorithms and Applications. Wiley, Hoboken (2013)
11. Hanczar, B., Nadif, M.: Ensemble methods for biclustering tasks. Pattern Recogn. **45**(11), 3938–3949 (2012)
12. Hanczar, B., Nadif, M.: Precision recall space to correct external indices for biclustering. In: International Conference on Machine Learning ICML, vol. 2, pp. 136–144 (2013)
13. Jain, A.K.: Data clustering: 50 years beyond k-means. Pattern Recogn. Lett. **31**(8), 651–666 (2010)
14. Lazzeroni, L., Owen, A.: Plaid models for gene expression data. Technical report, Stanford University (2000)
15. Madeira, S.C., Oliveira, A.L.: Biclustering algorithms for biological data analysis: a survey. IEEE/ACM Trans. Comput. Biol. Bioinform. **1**(1), 24–45 (2004)
16. Manning, C.D., Raghavan, P., Schtze, H.: Introduction to Information Retrieval. Cambridge University Press, Cambridge (2008)

17. Salah, A., Nadif, M.: Directional co-clustering. Adv. Data Anal. Classif, 1–30 (2018)
18. Sokolova, M., Japkowicz, N., Szpakowicz, S.: Beyond accuracy, F-score and ROC: a family of discriminant measures for performance evaluation. In: Proceedings of the 19th Australian Joint Conference on Artificial Intelligence: Advances in Artificial Intelligence, AI 2006, pp. 1015–1021 (2006)

Evaluation Procedures for Forecasting with Spatio-Temporal Data

Mariana Oliveira[1,3]([✉]), Luís Torgo[1,2,3], and Vítor Santos Costa[1,3]

[1] University of Porto, Porto, Portugal
[2] Dalhousie University, Halifax, Nova Scotia, Canada
[3] INESC TEC, Porto, Portugal
mariana.r.oliveira@inesctec.pt

Abstract. The amount of available spatio-temporal data has been increasing as large-scale data collection (e.g., from geosensor networks) becomes more prevalent. This has led to an increase in spatio-temporal forecasting applications using geo-referenced time series data motivated by important domains such as environmental monitoring (e.g., air pollution index, forest fire risk prediction). Being able to properly assess the performance of new forecasting approaches is fundamental to achieve progress. However, the dependence between observations that the spatio-temporal context implies, besides being challenging in the modelling step, also raises issues for performance estimation as indicated by previous work. In this paper, we empirically compare several variants of cross-validation (CV) and out-of-sample (OOS) performance estimation procedures that respect data ordering, using both artificially generated and real-world spatio-temporal data sets. Our results show both CV and OOS reporting useful estimates. Further, they suggest that blocking may be useful in addressing CV's bias to underestimate error. OOS can be very sensitive to test size, as expected, but estimates can be improved by careful management of the temporal dimension in training. Code related to this paper is available at: https://github.com/mrfoliveira/Evaluation-procedures-for-forecasting-with-spatio-temporal-data.

Keywords: Evaluation methods · Performance estimation
Cross-validation · Spatio-temporal data · Geo-referenced time series
Reproducible research

1 Introduction

The problem of identifying whether a machine learning solution will perform well on unseen data is at the core of predictive analytics. Two questions must be addressed: **(i)** are the evaluation metrics an appropriate fit to the application

Electronic supplementary material The online version of this chapter (https://doi.org/10.1007/978-3-030-10925-7_43) contains supplementary material, which is available to authorized users.

M. Berlingerio et al. (Eds.): ECML PKDD 2018, LNAI 11051, pp. 703–718, 2019.
https://doi.org/10.1007/978-3-030-10925-7_43

domain; and **(ii)** does the evaluation procedure make the best use of valuable data to obtain accurate estimates of these metrics. This paper focuses on the second question, in the context of forecasting with geo-referenced time series data. The answer is not always obvious as spatio-temporal dependencies are present in the data.

Performance estimation procedures can be classified into two classes of methods, both widely used: *out-of-sample OOS* estimation and *cross-validation (CV)* strategies.

Hold-out validation is the simplest of OOS estimators. It operates by splitting the data into a training set – used to learn a model–, and a test set – used to estimate the loss of the learned model in "unseen" data [12]. In the context of this study, only OOS procedures that respect an underlying order of the data are considered (e.g., in time series, the test set is always comprised of the more recent observations). These can also be called "last-block" procedures [3].

In CV, the total data is split several times into different training sets and test sets. Estimates of performance are obtained by averaging the losses over the several splits [30]. The use of different splits allows the whole data set to be used in the test set at least once. The data may be split in an exhaustive or partial manner, with partial splitting often being more computationally viable. The classical example of exhaustive splitting is leave-one-out cross-validation (LOOCV) where each observation plays the role of test set once. A common way to partially split the data is to divide it into K subsets of approximately the same size, and then having each subset successively used as test set – this strategy is referred to as K-fold CV [15]. However, standard CV procedures such as this assume that each test set is independent from the training set, which does not hold for many types of data sets, such as time series [2]. Several variations of CV procedures that do not require independence between sets have been proposed, with most of them being geared toward a time series setting [5,11,27]. Some of these methods have been proposed for spatio-temporal settings [18].

Our study aims at: **(i)** providing a review of validation strategies in the presence of spatio-temporal dependencies; and **(ii)** investigating the predictive ability of different cross-validation and out-of-sample strategies in a geo-referenced time series forecasting setting. To accomplish this goal we compare the loss estimated by different procedures against the loss incurred in previously withheld data. We consider artificial as well as real-world experimental settings.

2   Performance Estimation with Spatio-Temporal Dependence Structures

Observations that have been made at different times and/or at neighbouring locations may be related through internal dependence structures within data sets, as there is a tendency for values of close observations to be more similar (or otherwise related) than distant ones.

Dependence between training and test sets may lead to overly optimistic estimates of the loss a model will incur when presented with previously unseen, independent data, and may also lead to structural overfitting and poor generalization ability [28]. In fact, more than one study has proven that CV overfits for choosing the bandwidth of a kernel estimator in regression [13,22].

2.1 Temporal Dependence

Several performance estimation methods specifically designed to deal with temporal dependency have been proposed in the past.

In terms of OOS procedures in time series settings, decisions must be made regarding the split point between training/test sets, and how long a time-interval to include in the training set, that is, the window settings (sliding/growing). Two approaches are worth mentioning: (a) For *repeated time-wise holdout*, it is advised in [31] that holdout procedures should be repeated over different periods of time so that loss estimates are more robust. The selection of split points for each repetition of holdout may be randomized, with a window of preceding observations used for training and a fraction of the following instances used for testing. Training and test sets may potentially overlap across repetitions, similarly to random sub-sampling. These are also referred to as Monte Carlo experiments [32]; (b) *Prequential evaluation* or interleaved-test-then-train evaluation is often used in data stream mining. Each observation (or block of non-overlapping observations) is first used to test and then to train the model [19] in a sequential manner. The term prequential usually refers to the case where the training window is growing, i.e., a block of observations that is used for testing in one iteration will be merged with all previous blocks and used for training in the next iteration.

Four alternatives to standard CV proposed for time series should be highlighted: (a) *Modified CV* is similar to K-fold CV, except that l observations preceding and following the observation(s) in the test set are discarded from the training set after shuffling and fold assignment [11]. Also referred to as non-dependent cross-validation in [3]; (b) *Block CV* is a procedure similar to K-fold CV where, instead of the observations being randomly assigned to folds, each fold is a sequential, non-interrupted time series [29]; (c) *h-block CV* is based on LOOCV, except h observations preceding and following the observation in the test set are removed from the training set [5], and (d) *hv-block CV* is a modification of h-block CV where, instead of having single observations as test sets, a block of v observations preceding and following each observation is used for testing (causing test sets to overlap), with h observations before and after each block being removed from the training set [27].

Note that while in all types of block-CV, each test set is composed of a sequential non-interrupted time series (or a single observation), a fold in modified CV will almost certainly have non-sequential observations. If K is set to the number of observations in modified CV, then it works the same as h-block CV. Moreover, note that only hv-block CV allows test sets to overlap.

A number of empirical studies compare estimation methods for time series. Bergmeir *et al.* [3,4] suggest that cross-validation (in particular, hv-block CV) may have advantage over OOS approaches, especially when samples are small and the series stationary. Cerqueira *et al.* [9] indicate that, although this might be valid for synthetic time series, the same might not apply in real-world scenarios where methods preserving the order of the series (such as OOS Monte Carlo) seem to better estimate loss in withheld data. Mozetic *et al.* [20] reinforce the notion that blocking is important for time-ordered data.

2.2 Spatial Dependence

A major change when switching from temporal dependence to spatial dependence is that there is not a clear unidirectional ordering of data in 2D- or 3D- space as there is in time. This precludes using prequential evaluation strategies in the spatial domain. However, other strategies can be adapted quite straightforwardly to deal with spatial dependence.

Cross-validation approaches seem to be most commonly used in spatial settings. To avoid the problems arising from spatial dependence, block CV is often adopted. As in the temporal case, blocks can be designed to include neighbouring geographic points, forcing testing on more spatially distant records, and thus decreasing spatial dependence and reducing optimism in error estimates [33]. Methods that would correspond to h-block or hv-block CV are usually referred to as "buffered" CV in the spatial domain as a geographic vicinity of the testing block is removed from the training set.

The validity of these procedures was empirically tested by Roberts *et al.* in [28]. The authors find that block CV (with a block size substantially larger than residual autocorrelation) and "buffered" LOOCV (a spatial version of h-block CV, with h equivalent to the distance at which residual autocorrelation is zero) better approximate the error obtained when predicting onto independent simulations of species abundances data depending on spatially autocorrelated "environmental" variables.

2.3 Spatio-Temporal Dependence

When both spatial and temporal structures are present in the data, authors often resort to one of the procedures described in previous sections, effectively treating the data as if it was spatial-only (e.g., [16]) or temporal-only (e.g., [1,8]) for evaluation purposes. Others, while treating the problem mostly from a temporal perspective, then make an effort towards breaking down the results across space (e.g., [21]), or vice-versa (e.g., [7]), without the evaluation procedure itself being specifically designed to accommodate this.

In [28], no experimental results are presented specifically for spatio-temporal data, but there is a mention of data often being structured in both space and time in the context of avoiding extrapolation in cross-validation. When models are only meant to interpolate, the provided intuitions are that blocks should be no larger than necessary, models should be trained with as much data as possible, and predictors should be equally represented across blocks or folds. While conservatively large blocks can help avoid overly optimistic error estimates, the potential for introducing extrapolation is also increased. It is suggested that this effect may be mitigated by using "optimised random" or systematic (patterned) assignment of blocks to folds. Roberts *et al.* [28] also provide a general guide on blocking for CV, proposing the following five steps: assess dependence structures in the data, determine prediction objectives, block according to objectives and structure, perform cross-validation, and make "final" predictions.

Recent work by Meyer *et al.* [18] highlights how, for spatio-temporal interpolation problems, the results of conventional CV differ from the results of what they call "target-oriented" CV (versions of CV that address each and/or both dimensions, namely, "leave-location-out","leave-time-out" and "leave-location-and-time-out"). The authors attribute the lower error estimated by conventional CV to spatio-temporal over-fitting of the models and propose a forward feature selection procedure to improve interpolation results.

The applicability of solutions that consider the temporal and/or spatial autocorrelation is worth exploring, but the optimal strategy will depend on the modeling goal. It is important to make the distinction, as previous works have, between interpolation and forecasting problems. Unlike previous work on spatio-temporal data, the focus of this study is on forecasting, meaning that the aim is to make predictions about the future/new locations. Even after that is established, it may still be the case that the best evaluation procedure when the goal is to make predictions about unseen locations might differ from the best strategy when the aim is to make predictions in known sites.

3   Experiments

The different estimation procedures being compared are presented in Sect. 3.1. We first investigate their performance on data sets of randomly-generated artificial spatio-temporal data, as they provide a foundation for understanding the real-world case studies presented in Sect. 3.2. Section 3.3 describes the experimental design. Code for replication of these experiments is freely available.

3.1   Estimation Procedures

The estimators tested here included time-wise holdout methods (one-time, H, Monte Carlo, MC), cross-validation (CV), and prequential evaluation (P).

Train/Test Allocation Strategies. Table 1 summarises the different train/test assignment procedures used for CV and prequential evaluation methods.

Methods to assign observations into cross-validation folds that were tested include: standard CV, where instances are randomly assigned to folds, ($tRsR$), ignoring both dependency dimensions; time-sliced CV, where the spatial dimension is ignored and time-slices are assigned to folds randomly ($tRsA$); spatial block CV (also referred to as "leave-location-out" CV), where the temporal dimension is ignored and spatial blocks are assigned to folds either randomly

($tAsR$), in contiguous blocks ($tAsC$), or in a systematic, checkered pattern ($tAsS$).

When time is divided into blocks, prequential evaluation can also be applied. In this scenario, ($tBsA$), also referred to as "leave-time-out" CV, fold assignment ignores the spatial dimension. If space is also divided into blocks, then different types of spatio-temporal CV can be achieved by having the spatial assignment of folds be either random ($tBsR$), in contiguous blocks ($tBsC$), or in a (systematic) checkered pattern ($tBsS$).

Note that in what we call prequential evaluation, temporal order is always respected even when dividing data into spatio-temporal blocks, i.e., if a block in space-time is used for testing, then only blocks with previous time-stamps are used for training. Whether the spatial region in the test set is included in the training set is optional (rmS indicates that spatio-temporal data from the past but in the spatial region of the test set are not used in training). Moreover, the number of previous blocks in time used for training can be either fixed – sliding window (slW), or increase at each blocked time step – growing window (grW).

Buffered CV. Methods that remove a block of observations in the neighbourhood of the test set (in the temporal and/or spatial dimensions) from the training set have also been considered.

In the case of standard CV, for each instance in the test set, a number of past and future observations at that location are removed and/or past observations within a certain distance from the location are removed (CV-T, CV-S or CV-ST). This is akin to modified CV mentioned earlier in a time series context. The same process can be applied to spatio-temporal CV. In that scenario, if the buffer is set to the maximum distances between any two points in space/time (CV-STM), the result is what is called "leave-location-and-time-out" CV.

When time block CV is used, then a number of previous and future observations are removed around the test set (CV-T). This is similar to hv-block CV. However, while hv-block CV is repeated for each instance of the whole set (therefore including overlapping test sets), the procedure is only repeated here for each non-overlapping block of sequential time.

In spatial random or contiguous block CV, a spatial buffer can be applied, so that locations within a pre-defined spatial distance of the test set are removed from the training set (CV-S). This is, again, similar to hv-block CV in space.

3.2 Data Sets

As previously mentioned, both artificially generated and real-world data sets were used for this study.

Artificial Data Sets. Artificial data was generated by stationary spatio-temporal autoregressive moving average (STARMA) models as proposed in [24] and implemented in R package *starma* [10].

Table 1. Cross-validation and prequential evaluation fold assignment procedures

		Time	Space	
Cross-validation	Standard	random	random	tRsR • † ‡
	Time-sliced		all	tRsA
	Spatial block		random block	tAsR •
	Checkered spatial block	all	systematic	tAsS
	Contiguous spatial block		contiguous	tAsC •
Cross-validation + Prequential evaluation	Time block	block	all	tBsA †
	Spatio-temporal block		random block	tBsR ‡
	Spatio-temporal checkered block		systematic	tBsS
	Spatio-temporal contiguous block		contiguous	tBsC

† Time-buffered CV variation included
• Space-buffered CV variation included
‡ Space-time buffered CV variation included

The models are denoted by $STARMA(p_{\lambda_1\lambda_2...\lambda_p}, q_{m_1m_2...p})$ where p is the autoregressive order, q is the moving average order, λ_l is the spatial order of the kth autoregressive term, m_k is the spatial order of the kth moving average term. If $q = 0$, then $STAR(p_{\lambda_1...\lambda_p})$ will suffice; if $p = 0$, then it may be denoted by $STMA(q_{m_1...p})$. Non-linear versions of STAR models, $NLSTAR(p_{\lambda_1...\lambda_p})$, are generated by applying a non-linear function at each autoregressive step (similar to what is done in [3] to obtain non-linear AR models).

In data sets generated by a $STAR(2_{10})$ model, a value measured at location i and time t will be directly influenced by the values of location i and of its first-degree neighbours at time $t-1$, and by the values of location i at time $t-2$. Note that neighbours of lower order must be considered "closer" than neighbours of higher order (according to some metric of distance).

In this study, for each model of type STARMA (with $p = q$), STMA, STAR, and NLSTAR, two sets of coefficients of each order 2_{10}, 2_{01} and 2_{11} are generated randomly (within intervals likely to respect stationarity conditions) until the resulting STARMA models are stationary. In the case of NLSTAR, a non-linear function is also randomly selected from a pre-defined set. Then, using grids of 10×10 and 22×22 equally spaced locations, data is generated with time series lengths of 250 and 400. However, after this step, the first 100 observations at each location are discarded in an effort to avoid dependence on initial conditions; outer locations are ignored so each used location has information for its four first order neighbours – top, bottom, left and right. Thus, 150 and 300 observations on 8×8 and 20×20 grids are kept for forecasting performance analysis. For details on the data generation process, consult the supplementary material.

Spatio-Temporal Embedding. In order to apply standard regression techniques to the spatio-temporal forecasting problem, the generated data sets have to be transformed in some way so each instance has a set of predictors. A simple way to do this is by spatio-temporal embedding, i.e., by using previous values measured at the given location and its neighbours as predictors. The order of spatio-temporal embedding can be denoted in the same way as the $STARMA$ order. All artificially generated data sets were embedded with order 3_{110}. In total, 96 artificial data sets were generated and embedded.

Real-World Data Sets. Seventeen variables from seven different real-world data sources were used as independent univariate data sets for experimental validation of the performance evaluation procedures. The measured variables describe environmental monitoring, from air pollution to climate and soil characteristics. A summary of the characteristics of each data set can be found in Table 2. The size of the data sets varies from small networks of 20 sensors to larger networks of 900 geolocations. Though most sensor networks are irregularly distributed in space, one of them forms a regular grid of $0.5 \times 0.5°$ of longitude/latitude. The data sets also vary in terms of time series size (from 280 time points to over 11 k) and sampling frequency (from hourly to monthly). About half of the variables were measured at every point in time and space, with no missing values. However, for others, only a percentage of location and time-stamp pairs (from 39% to 74%) have available values, due to, for instance, some sensors only being installed later in the measurement period.

Table 2. Real-world data sets

Data set	#	Variables	Time #IDs	frequency	Locations #IDs	distribution	Total #	% available	Source
MESA Air Pollution	1	NO$_X$ concentration	280	bi-weekly	20	irregular	5.6 k	100	[25][a]
NCDC Air Climate	2	precipitation, solar energy	105	monthly	72	irregular	7.6 k	100	[25][a]
TCE Air Climate	3	ozone concentration, air temperature, wind speed	330	hourly	26	irregular	8.6–9.4 k	100	[25][a]
COOK Agronomy Farm	3	water content, temperature, conductivity	729	daily	42	irregular	22–23 k	73–74	[14,17][b]
SAC Air Climate	1	air temperature	144	monthly	900	regular	130 k	100	[25][a]
RURAL airBase	1	PM10 concentration	4382	daily	70	irregular	149 k	49	[23][b]
BEIJ Beijing UrbanAir	6	PM25, PM10 & NO$_X$ concentration, air temperature, humidity	11357	hourly	36	irregular	404–409 k	39–41	[35][c]

[a] Downloaded at: http://www.di.uniba.it/~appice/software/COSTK/index.htm
[b] Loaded from *R* packages *GSIF* (0.5-4) and *spacetime* (1.2-1).
[c] Downloaded at: https://www.microsoft.com/en-us/research/publication/u-air-when-urban-air-quality-inference-meets-big-data/

Spatio-Temporal Indicators. In order to compare performance, a learning approach had to be selected that would work with the different data set characteristics. Unlike the artificial data sets, most real-world sensor networks are not distributed in a regular grid, so the simple spatio-temporal embedding used for the artificial data sets seemed over-simplistic. The approach adopted instead was the one proposed in [21], using as predictors a temporal embed of values measured at the location, spatio-temporal indicators built by calculating summary statistics from the neighbouring observations within 3 data set specific boundaries of spatio-temporal distance, and ratios between the indicators of spatio-temporal neighbourhoods of increasing radius. The temporal embed size was set to 7, resulting in a total of 20 predictors.

Missing Data. Some of the data sets have missing data, either due to failures in data acquisition or due to sensors being set up at later times. After calculating the predictors but before any experiments are carried out, all columns that have 20% or more of their data missing from the first 80% of time-points, are discarded as they should not be very useful predictors. The remaining missing data is dealt with as follows: first, any rows that have too many predictors missing (set at 20% of columns) are discarded from the training set; then, missing values for both the training and test sets are imputed as the median of that column in the respective set.

3.3 Experimental Design

For each data set: **(1)** The data is divided into an in-set and out-set. This is performed time-wise, so that the out-set consists of a percentage of the most recent observations; **(2)** A regression model is trained in the in-set and tested on the out-set. The error on the out-set is considered to be the "gold standard" error that estimation methods should be able to estimate accurately; **(3)** Several error estimation methods (cross-validation, prequential and out-of-sample methods), applied exclusively on data from the in-set, are used to approximate the "gold standard". The differences between the "gold standard" error and the error estimated by each estimation methodology can be compared over all data sets and learning model pairs.

Train/Test Sizing. The in-set was set to be 80% of the time-points. When using cross-validation or prequential evaluation on the in-set, 16 folds were used for artificial data and 9 folds for real data. When using OOS procedures on the in-set, the splits are always made time-wise. For holdout, estimations were made with test sizes of 20% (same proportion as the out-set) and 6%/9% for artificial/real in-set data (the proportions used in the last block of time-block CV).

Note that the data set is divided into the same number (16 or 9) of equally-sized folds across all variations of CV. In the interest of fairness, the test size of time-wise holdout was defined to correspond to the size of one fold in CV. All of these methodologies use the whole given in-set to make estimates. However, time-wise Monte Carlo estimations, by definition, use only a fraction of the data set for each iteration – meaning the sizing of these competing procedures can never be made entirely "fair". The option taken was to keep the proportion between train and test sizes the same as that used in CV, i.e., the percentages used for training and testing in Monte Carlo correspond to the estimation on the last block of a 16-fold or 9-fold time block CV performed on 50% or 60% of the in-set. Thus, Monte Carlo estimations were averaged over 16 repetitions with training (testing) performed on 47% (3%) and 56% (4%) of the in-set for artificial data, and averaged over 9 repetitions of training (testing) on 44% (6%)

and 53% (7%) of real data. Buffer sizes are set to the highest embed size or spatio-temporal neighbourhood radius.

Learning Models. The process is repeated over each data set using two different learning algorithms: a linear regression model, *LM* (*R* package *stats* [26]) and a random forest, *RF* (*R* package *ranger* [34]).

Error Metrics. The error of learning algorithms is measured by Normalized Mean Absolute Error (NMAE), defined by Eq. 1 where z is the observed value, $\hat{z}$ is the prediction, and $\bar{z}$ is the mean of Z. By opting for a normalized metric instead of the more widely used MAE, comparisons between error estimation methods across data sets can be made more easily.

$$NMAE = \frac{\sum_{i=0}^{n} |\hat{z} - z|}{\sum_{i=0}^{n} |z - \bar{z}|} \tag{1}$$

4 Results

The estimation error is defined as the difference between the error estimated by a procedure using the in-set, Est, and the "gold standard" error incurred on the out-set, $Gold$, $Err = Est - Gold$. Note that experiments with methods that rely on non-random spatial blocking were not carried out using real-world data sets due to issues arising from irregular spatial distributions. Time-buffering without time-blocking in real-world scenarios caused issues related with buffer size/neighbourhood radius. Results for variations of prequential evaluation using sliding window and/or removing locations in the test set from the training set are not reported as they were consistently out-performed by their growing window counterparts (though the difference was not statistically significant).

4.1 Median Errors

Figures 1 and 2 show the distribution of estimation errors for artificial and real-world data sets. The sign of the median error indicates whether the procedure tends to underestimate the error meaning it is overly optimistic (negative median error), or overestimate it (positive median error).

In Fig. 1, all procedures appear centered around zero. However, most cross-validation procedures under-estimate the error in more than half the cases, even when using some form of block CV. This effect is mitigated when a type of buffering is applied (either temporal, spatial or spatio-temporal). Most OOS procedures overestimate the error in more than half the cases, with the exception of holdout at 80%.

Figure 2 shows significant differences between procedures. Is is important to note that standard CV (*CVtRsR*) under-estimates the error in over 75% of cases. We observe this problem even after applying a spatial buffer. Note that spatial-buffered CV estimates were not obtained for a fraction of real data sets due to problems associated with the irregularity of sensor network locations.

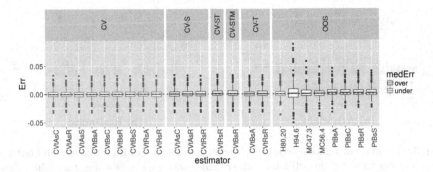

Fig. 1. Box plots of estimation errors incurred by cross-validation and out-of-sample procedures on 96 artificial data sets using 2 learning algorithms

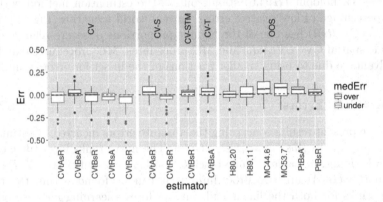

Fig. 2. Box plots of estimation errors incurred by cross-validation and out-of-sample procedures on 17 real world data sets using 2 learning algorithms

Spatial block CV ($CVtAsR$) and time-sliced CV ($CVtRsA$) are also overly optimistic in their error estimates. However, OOS procedures and variations of CV using time-blocks and/or time-buffers seem to be less prone to underestimate the error.

4.2   Relative Errors

Another useful metric to analyse is the relative error as defined by $RelErr = |Est - Gold|/Gold$. Figure 3 shows the distribution of low, moderate and high errors. The binning is somewhat arbitrary but chosen so that comparisons might be useful. In the real-world case, relative errors are generally higher so bins were chosen accordingly. Possible explanations for the lower relative errors found for artificial data sets when compared to the real-world case include the absence of missing data, the regularity of the grids and stationarity of the underlying data generation process.

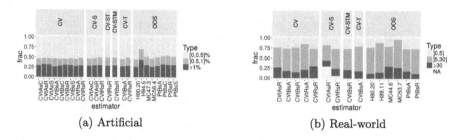

(a) Artificial   (b) Real-world

Fig. 3. Bar plots of relative estimation errors incurred by cross-validation and out-of-sample procedures on 96 artificial and 17 real-world data sets using 2 learning algorithms. Note the different legends

In Fig. 3a, holdout (H94.6) stands out as the estimation method with the lowest percentage of low relative errors. In real-world scenarios (Fig. 3b), standard CV ($CVtRsR$) has one of the highest proportions of severe relative error, alongside spatial CV ($CVtAsR$) and MC procedures. MC procedures may be at a disadvantage due to using smaller fractions of the in-set for error estimation.

4.3   Absolute Errors

Finally, we present results concerning the absolute errors incurred by estimation procedures, that is, $AbsErr = |Est - Gold|$. The mean ranks for artificial data sets can be found in Table 3. Time-block CV either time-buffered ($CVtBsA\_T$) or plain ($CVtBsA$) are two procedures that can be found within the top 5 average ranks for both the linear and random forests learning models. Within OOS procedures, holdout (H80.20) and Monte Carlo (MC56.4) can be found in the top 3 average ranks for both learning models.

Table 4 shows average ranks for real-world data sets. Spatio-temporal block CV ($CVtBsR$) is within the top 5 average rank of both learning models, alongside space-buffered standard CV ($CVtRsR\_S$). The top 3 OOS procedures are consistently spatio-temporal block prequential evaluation ($PtBsR$), holdout (H80.20) and time-block prequential evaluation ($PtBsA$).

(a) LM   (b) RF

Fig. 4. Critical difference diagram according to Friedman-Nemenyi test (at 5% confidence level) for a subset of estimation procedures using 24 artificial data sets (64×64 grid; 150 time-points each)

Table 3. Average ranks of absolute errors incurred by cross-validation and out-of-sample procedures when estimating performance on 96 artificial data sets. Best results in bold

	CVtAsC	CVtAsC_S	CVtAsR	CVtAsR_S	CVtAsS	CVtBsA	CVtBsA_T	CVtBsC
LM	8.35	8.56	8.07	**7.57**	8.64	8.16	8.02	8.41
RF	8.47	8.57	9.01	8.44	8.54	7.95	**7.55**	8.67
	CVtBsR	CVtBsR_STM	CVtBsS	CVtRsA	CVtRsR	CVtRsR_S	CVtRsR_ST	CVtRsR_T
LM	8.67	8.11	8.73	8.29	9.21	9.07	8.86	9.27
RF	8.50	8.96	8.30	7.93	9.07	8.75	8.88	8.42
	H80.20	H94.6	MC47.3	MC56.4	PtBsA	PtBsC	PtBsR	PtBsS
LM	4.23	6.21	4.24	**3.99**	4.09	4.33	4.54	4.36
RF	**4.02**	5.95	4.34	4.20	4.51	4.43	4.14	4.42

Table 4. Average ranks of absolute errors incurred by cross-validation and out-of-sample procedures when estimating performance on 17 real-world data sets. Best results in bold

	CVtAsR	CVtAsR_S	CVtBsA	CVtBsA_T	CVtBsR	CVtBsR_STM	CVtRsA	CVtRsR	CVtRsR_S
LM	4.53	7.06	5.29	5.35	**3.94**	5.12	4.82	4.35	4.53
RF	5.00	5.88	**4.00**	4.41	4.41	4.41	5.47	6.59	4.82
	H80.20	H89.11	MC44.6	MC53.7	PtBsA	PtBsR			
LM	**2.47**	3.71	4.47	4.35	3.47	2.53			
RF	3.12	3.35	4.12	4.41	3.18	**2.82**			

Only the aforementioned procedures, along with any other method that appears as the best for a certain learning model, and standard CV, are considered for statistical significance testing. The Friedman-Nemenyi test is applied, with estimation procedures used as the "classifiers" or "treatments" (using R package *scmamp* [6]). Since there is an assumption that the data sets should be independent, separate Friedman tests were carried out for the results obtained by linear and random forest learning models. Moreover, a test is performed for each subset of 24 artificial data sets with the same grid and time series size, since the same STARMA coefficients were re-used across different data sizes. Figures 4 and 5 show critical difference diagrams for a subset of artificial data sets and

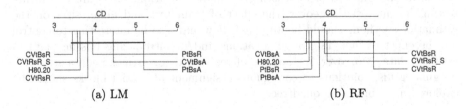

(a) LM (b) RF

Fig. 5. Critical difference diagram according to Friedman-Nemenyi test (at 5% confidence level) for a subset of estimation procedures using real data sets

all the real-world data sets. In most cases, no significant difference between estimation procedures was found at a 5% confidence level. However, a significant difference was found for the smaller artificial data sets as seen in Fig. 4b.

5   Conclusion

The problem of how to properly evaluate spatio-temporal forecasting methods is still an open one. Previous studies have empirically shown that dependence between observations negatively impacts performance estimation using standard error estimation methods like cross-validation for time series [3, 4, 9], time-ordered Twitter data [20], spatial and phylogenetic data [28], and spatio-temporal interpolation [18].

In this paper, an extensive empirical study of performance estimation for forecasting problems using both artificially generated and real-world spatio-temporal data sets is provided. First, we observe that most often error estimates are reasonably accurate. Standard CV does have problems: it underfits and it exhibits a number of outliers of severe error underestimation. Moreover, though the best estimator in terms of absolute difference to the "gold standard" is not always the same, most top-performers block the data set in time. This is in line with previous research on time-ordered data [3, 20]. Indeed, for artificial data sets, time-buffered time-block CV is one of the best in terms of approximating the "gold standard" error while also avoiding being overly optimistic in the estimates. For real-world data sets, spatio-temporal block CV and, when using random forests, time-block CV (this time without the buffer) not only approximate the error better than other methods, they also mostly avoid being overly optimistic about errors. Note that the fact that time-buffered time block CV did not perform as well in real-world data sets might have to do with buffer size parametrization. Out-of-sample procedures, in general, did not do as well in terms of absolute difference to the gold-standard, but they did tend to avoid underestimation of the error in almost all cases which might still be seen as an advantage over cross-validation. These results seem to point to the temporal dimension being more important to respect when evaluating spatio-temporal forecasting methods.

There is some bias in the experimental design, but results are still fairly consistent and some issues can be addressed in future work. Varying the in-set/out-set ratio, and setting the "gold standard" as forecasting future observations in new locations (instead of forecasting for known locations only) are two future settings of interest. Moreover, the effect of train/test and buffer sizes on the estimation methods should be analysed. It would also be interesting to control for the effect of including outer locations and/or introducing missing data in artificial data. Moreover, in the case of real-world (or artificial) data sets with irregular grids, solutions to contiguous assignment of spatial blocks should be explored, possibly using quadtrees.

Acknowledgments. This work is partially funded by the ERDF through the COMPETE 2020 Programme within project POCI-01-0145-FEDER-006961, and by

National Funds through the FCT as part of project UID/EEA/50014/2013. Mariana Oliveira is supported by a FCT/MAPi PhD research grant (PD/BD/128166/2016). Vítor Santos Costa is supported by the project POCI-01-0145-FEDER-016844.

References

1. Appice, A., Pravilovic, S., Malerba, D., Lanza, A.: Enhancing regression models with spatio-temporal indicator additions. In: Baldoni, M., Baroglio, C., Boella, G., Micalizio, R. (eds.) AI*IA 2013. LNCS (LNAI), vol. 8249, pp. 433–444. Springer, Cham (2013). https://doi.org/10.1007/978-3-319-03524-6_37
2. Arlot, S., Celisse, A.: A survey of cross-validation procedures for model selection. Stat. Surv. 4, 40–79 (2010). https://doi.org/10.1214/09-SS054
3. Bergmeir, C., Benítez, J.M.: On the use of cross-validation for time series predictor evaluation. Inf. Sci. (Ny) 191, 192–213 (2012). https://doi.org/10.1016/j.ins.2011.12.028
4. Bergmeir, C., Costantini, M., Benítez, J.M.: On the usefulness of cross-validation for directional forecast evaluation. Comput. Stat. Data Anal. 76, 132–143 (2014). https://doi.org/10.1016/j.csda.2014.02.001
5. Burman, P., Chow, E., Nolan, D.: A cross-validatory method for dependent data. Biometrika 81(2), 351–358 (1994). https://doi.org/10.1093/biomet/81.2.351
6. Calvo, B., Santafé Rodrigo, G.: scmamp: statistical comparison of multiple algorithms in multiple problems. R J. 8(1), August 2016
7. Carroll, S.S., Cressie, N.: Spatial modeling of snow water equivalent using covariances estimated from spatial and geomorphic attributes. J. Hydrol. 190(1–2), 42–59 (1997). https://doi.org/10.1016/S0022-1694(96)03062-4
8. Ceci, M., Corizzo, R., Fumarola, F., Malerba, D., Rashkovska, A.: Predictive modeling of PV energy production: How to set up the learning task for a better prediction? IEEE T. Ind. Inform. 13(3), 956–966 (2017)
9. Cerqueira, V., Torgo, L., Smailovi, J., Mozeti, I.: A comparative study of performance estimation methods for time series forecasting. In: International Conference on Data Science and Advanced Analytics (DSAA), pp. 529–538 (2017). https://doi.org/10.1109/DSAA.2017.7
10. Cheysson, F.: starma: Modelling Space Time AutoRegressive Moving Average. In: (STARMA) Processes (2016)
11. Chu, C.K., Marron, J.S.: Comparison of two bandwidth selectors with dependent errors. Ann. Stat. 19(4), 1906–1918 (1991)
12. Devroye, L., Wagner, T.: Distribution-free performance bounds for potential function rules. IEEE Trans. Inf. Theory 25(5), 601–604 (1979)
13. Diggle, P.: Analysis of Longitudinal Data. Oxford University Press, Oxford (2002)
14. Gasch, C.K., Hengl, T., Gräler, B., Meyer, H., Magney, T.S., Brown, D.J.: Spatio-temporal interpolation of soil water, temperature, and electrical conductivity in 3D+ T: the cook agronomy farm data set. Spat. Stat. 14, 70–90 (2015)
15. Geisser, S.: The predictive sample reuse method with applications. J. Am. Stat. Assoc. 70(350), 320–328 (1975)
16. Haberlandt, U.: Geostatistical interpolation of hourly precipitation from rain gauges and radar for a large-scale extreme rainfall event. J. Hydrol. 332(1–2), 144–157 (2007). https://doi.org/10.1016/j.jhydrol.2006.06.028
17. Hengl, T.: GSIF: Global Soil Information Facilities (2017). R package version 0.5-4

18. Meyer, H., Reudenbach, C., Hengl, T., Katurji, M., Nauss, T.: Improving performance of spatio-temporal machine learning models using forward feature selection and target-oriented validation. Environ. Model. Softw. **101**, 1–9 (2018). https://doi.org/10.1016/j.envsoft.2017.12.001
19. Modha, D.S., Masry, E.: Prequential and cross-validated regression estimation. Mach. Learn. **33**(1), 5–39 (1998). https://doi.org/10.1109/ISIT.1998.708964
20. Mozetič, I., Torgo, L., Cerqueira, V., Smailović, J.: How to evaluate sentiment classifiers for Twitter time-ordered data? PLoS One **13**(3), 1–20 (2018). https://doi.org/10.1371/journal.pone.0194317
21. Ohashi, O., Torgo, L.: Wind speed forecasting using spatio-temporal indicators. In: Proceedings of the 20th European Conference on Artificial Intelligence, pp. 975–980. IOS Press (2012)
22. Opsomer, J., Wang, Y., Yang, Y.: Nonparametric regression with correlated errors. Stat. Sci. **16**(2), 134–153 (2001). https://doi.org/10.1214/ss/1009213287
23. Pebesma, E.: spacetime: Spatio-temporal data in R. J. Stat. Softw. **51**(7), 1–30 (2012). http://www.jstatsoft.org/v51/i07/
24. Pfeifer, P.E., Deutsch, S.J.: A three-stage iterative procedure for space-time modeling. Technometrics **22**(1), 35–47 (1980)
25. Pravilovic, S., Appice, A., Malerba, D.: Leveraging correlation across space and time to interpolate geophysical data via CoKriging. Int. J. Geogr. Inf. Sci. **32**(1), 191–212 (2018). https://doi.org/10.1080/13658816.2017.1381338
26. R Core Team: R: a language and environment for statistical computing. In: R Foundation for Statistical Computing, Austria, Vienna (2017)
27. Racine, J.: Consistent cross-validatory model-selection for dependent data: hv-block cross-validation. J. Econom. **99**(1), 39–61 (2000)
28. Roberts, D.R., et al.: Cross-validation strategies for data with temporal, spatial, hierarchical, or phylogenetic structure. Ecography **40**(8), 913–929 (2017)
29. Snijders, T.A.B.: On cross-validation for predictor evaluation in time series. In: Dijkstra, T.K. (ed.) On Model Uncertainty and its Statistical Implications. LNE, pp. 56–69. Springer, Berlin, Heidelberg (1988). https://doi.org/10.1007/978-3-642-61564-1_4
30. Stone, M.: Cross-validatory choice and assessment of statistical predictions. J. R. Stat. Soc. B 111–147 (1974)
31. Tashman, L.J.: Out-of-sample tests of forecasting accuracy : an analysis and review. Int. J. Forecast. **16**(4), 437–450 (2000)
32. Torgo, L.: Data Mining with R: Learning with Case Studies. CRC Press, Boca Raton (2016)
33. Trachsel, M., Telford, R.J.: Estimating unbiased transfer-function performances in spatially structured environments. Clim. Past **12**(5), 1215–1223 (2016)
34. Wright, M.N., Ziegler, A.: Ranger: a fast implementation of random forests for high dimensional data in C++ and R. J. Stat. Softw. **77**(1), 1–17 (2017). https://doi.org/10.18637/jss.v077.i01
35. Zheng, Y., Liu, F., Hsieh, H.P.: U-Air: when urban air quality inference meets big data. In: Proceedings of the 19th ACM SIGKDD International Conference on Knowledge Discovery Data Mining, pp. 1436–1444. ACM (2013). https://doi.org/10.1145/2487575.2488188

A Blended Metric for Multi-label Optimisation and Evaluation

Laurence A. F. Park[1](✉) and Jesse Read[2]

[1] School of Computing, Engineering and Mathematics,
Western Sydney University, Sydney, Australia
lapark@scem.westernsydney.edu.au
[2] DaSciM team, LIX Laboratory, École Polytechnique, 91120 Palaiseau, France
jesse.read@polytechnique.edu

Abstract. In multi-label classification, a large number of evaluation metrics exist, for example Hamming loss, exact match, and Jaccard similarity – but there are many more. In fact, there remains an apparent uncertainty in the multi-label literature about which metrics should be considered and when and how to optimise them. This has given rise to a proliferation of metrics, with some papers carrying out empirical evaluations under 10 or more different metrics in order to analyse method performance. We argue that further understanding of underlying mechanisms is necessary. In this paper we tackle the challenge of having a clearer view of evaluation strategies. We present a blended loss function. This function allows us to evaluate under the properties of several major loss functions with a single parameterisation. Furthermore we demonstrate the successful use of this metric as a surrogate loss for other metrics. We offer experimental investigation and theoretical backing to demonstrate that optimising this surrogate loss offers best results for several different metrics than optimising the metrics directly. It simplifies and provides insight to the task of evaluating multi-label prediction methodologies. Data related to this paper are available at: http://mulan.sourceforge.net/datasets-mlc.html, https://sourceforge.net/projects/meka/files/Datasets/, http://www.ces.clemson.edu/~ahoover/stare/.

1 Introduction

The major challenge in multi-label classification is dealing with multiple output labels simultaneously, which has important ramifications on building models, and also evaluating them. There have been an impressive number of new methods proposed in recent years, proposing different ways of modelling labels together, but relatively little investigation into the study of *which* loss functions methods optimise, which functions they *should* optimise, and how well they can be expected to achieve this for a given problem. As a result, empirical studies look at up to a dozen evaluation metrics. Our study is targeted at bringing new clarity and insight to this situation.

© Springer Nature Switzerland AG 2019
M. Berlingerio et al. (Eds.): ECML PKDD 2018, LNAI 11051, pp. 719–734, 2019.
https://doi.org/10.1007/978-3-030-10925-7_44

In multi-label classification, optimisation as part of a predictive model inherently involves multiple dimensions; one for each label. A review of multi-label classification is given in [11]. A number of common benchmark algorithms are reviewed in [5,15]. Often, a vector is used to represent the labelling, e.g., $\mathbf{y}_k = [y_{k1}, y_{k2}, \ldots, y_{kL}]$ where $y_{kj} = 1$ iff the j-th of L labels is relevant to the k-th example $\mathbf{x}_k$ (and $y_{kj} = 0$ otherwise). To evaluate the performance of a multi-label classifier, typically predicted vectors $\mathbf{y}_k$ must be compared the vector of true labels $\mathbf{t}_k$ over all examples $k = 1, \ldots, K$ (for a test set of K examples).

Three of the most common similarity functions used in multi-label learning and evaluation are the Jaccard index, Hamming loss, and 0/1 loss. Jaccard index is known as accuracy in some publications, e.g., [3,9], Hamming loss and 0/1 loss are known often as Hamming score and exact match in their payoff-form (higher is better), respectively [7]. However the basic principal of all multi-label metrics is the same for any metric: provide a single number indicating the *similarity* of the set (or vector[1]) of *predicted* labels compared to the set of *true* labels, i.e., a score that may be normalised to between 0 and 1. We henceforth refer to each of these mentioned metrics as a *similarity* or *loss* function, interchangeably. There is a large number of multi-label criteria, including rankings and micro and macro evaluation; a recent survey of multi-label metrics and unified view of them is given by [14].

Consider the examples in Table 1. The similarity functions are defined, for the k-th instance, as

$$\text{Hamming} := \frac{1}{L} \sum_{j=1}^{L} \mathbb{I}[y_{kj} = t_{kj}] \tag{1}$$

$$\text{Exact} := \mathbb{I}[\mathbf{y}_k = \mathbf{t}_k] \tag{2}$$

$$\text{Jacard} := \frac{|\mathbf{y}_k \wedge \mathbf{t}_k|}{|\mathbf{y}_k \vee \mathbf{t}_k|} \tag{3}$$

for L possible labels, where $\vee$ and $\wedge$ are the logical OR and AND operations, applied vector-wise, and $\mathbb{I}[\cdot]$ is an indicator function returning 1 if the inner condition holds (0 otherwise).

Each of these similarity functions measure different accuracy qualities of a multi-label classification system. Hamming similarity provides the proportion of labels predicted correctly, Exact similarity provides the proportion of label sets predicted correctly, while Jaccard similarity only examines the proportion of correctly predicted positive labels out of the potential positive set (predicted positive and actually positive). Therefore, a multi-label classification system should be optimised according to the desired similarity function.

The problem of optimisation in multi-label classification is complicated by the label dimension and the interdependence of labels. Resulting search spaces are usually non-convex and non-differentiable, and all the difficulties of such a search

---

[1] Notation alternates among papers, since given a set of labels $\mathcal{L}$, a (sub)set $Y \subseteq \mathcal{L}$ can be represented as vector $\mathbf{y} = [y_1, \ldots, y_L]$ where $y_j = 1 \Leftrightarrow y_j \in Y$.

Table 1. An example of multi-label evaluation. The average Hamming, Exact, and Jaccard similarity is 0.80, 0.40, and 0.67, respectively. Note that real-world multi-label data sets are typically much sparser with respect to labelling.

	$\mathbf{t}_k$	$\mathbf{y}_k$	Hamming	Exact	Jaccard
$\mathbf{x}_1$	[1 0 1 0]	[1 0 0 1]	0.5	0	0.33
$\mathbf{x}_2$	[0 1 0 1]	[0 1 0 1]	1	1	1
$\mathbf{x}_3$	[1 0 0 1]	[1 0 0 1]	1	1	1
$\mathbf{x}_4$	[0 1 1 0]	[0 1 0 0]	0.75	0	0.5
$\mathbf{x}_5$	[1 0 0 0]	[1 0 0 1]	0.75	0	0.5

are inherited. A solution may fall into a local maximum/minimum and provide a non-optimal solution. We can transform the problem into simpler problems which provide convex search spaces, but even in these case, optimisation is typically much more difficult than a traditional single-label problem.

One solution is to optimise each label individually and independently. This approach is known as Hamming similarity, and defines the so-called *binary relevance* method, widely known across the multi-label literature as a baseline approach. Indeed we see that Hamming similarity is proportional to the sum of individual label similarities, therefore, when optimising for Hamming similarity, we can simply optimise the accuracy of each label. An example of this is to model each label with a logistic regression; resulting in L tasks of convex-optimisation.

However, it is typical to motivate methods that model labels together (e.g., [3,8,9,12,15] and many references therein). As shown thoroughly in those papers and many others, it is desirable to model an explicit or implicit dependence among labels. And this, in turn, motivates the *evaluation* of labels together also. Exact similarity does not decompose across labels (i.e., it requires optimisation of all labels jointly) and therefore encourages modelling labels together. In fact, for this metric it is theoretically optimal to model label combinations as single class values in a large multi-*class* problem [2]. This transformation is often called the *label powerset* method. Hence each vector/set $\mathbf{y}_k$ is treated as a *single* value that may be assigned to each instance. This can also be formulated as a differentiable and convex optimization problem, such as multi-class logistic regression, since the multiple classes can be modelled under a single softmax function. Nevertheless, optimisation is still not expected to be easy or necessarily effective in practice (see, e.g., [10,12]): most label combination (class) are likely to be very sparse or not present in the training data and the metric is particularly sensitive to noise in the data, since even 1 of the L labels being incorrectly predicted will lead to a 0 score for that particular instance. Other methods such as classifier chains [1,9] provide an alternative, which divides the optimisation across labels. However, unlike the binary relevance method, the labels are linked together in a chain cascade and thus can no longer be optimized separately (at least, not in terms of optimising exact similarity), and thus becomes thus a non-convex problem and remains a difficult task, hence paving the way for several extensions (e.g., [1,2,8]) with approximate inference, and alternatives (e.g., [10,12]) with exact inference on small sub-problems.

Jaccard similarity is also used for evaluating labels together. It is often preferred since it can be seen as a midpoint between Hamming and exact similarity as two extremes. However, fewer theoretical results exist for its optimisation, although relationships with F-measure has been outlined [13]. It remains popular, and it is typically assumed that good results in both Hamming and exact similarity will reflect good results also in Jaccard similarity. Numerous empirical studies also show this [5,8].

Therefore, to summarise so far: Hamming similarity is easy to optimise, but may not correspond to labellings that we would find desirable in real-world problems. On the other hand, optimising a multi-label classification system with respect to Jaccard or Exact similarity is difficult, since it requires optimisation of all labels at once, leading to a large optimisation search space. Solutions involve:

- Problem transformation (i.e., binary relevance, or label powerset)
- Non-convex optimisation such as local search and hill-climbing methods
- Using a surrogate loss
- Smoothing and regularisation of the loss to improve results.

In this paper we propose a blended metric, combining Jaccard and Exact similarity functions with the Hamming function as a surrogate metric for both Jaccard and Exact similarity, providing a simpler search space and thus more efficient and effective optimisation. Under a series of investigations we show the benefit of this blended metric. In particular, we obtain a very important result: best results for exact match can be obtained using our surrogate blended metric, rather than the exact match minimizer (label powerset) itself. In general, we are able to conclude making the recommendation of exploring different surrogate metrics, rather than optimising separately across an ever-growing set of metrics.

Our novel contribution is covered as follows:

- In Sect. 2 we formulate our surrogate blended metric: a single function that is a blend of the three similarity functions (Hamming, Exact, and Jaccard).
- In Sect. 3 we derive the gradient for optimisation of this function.
- In Sects. 4 and 5 we examine how optimisation over the blended function affects the accuracy with regard to the maximisation of the three similarity functions, on a number of real-world multi-label datasets; and we identify the spectrum where each of the three similarity functions are maximised and we discuss in detail and draw conclusions and recommendations.

2   Multi-label Evaluation and Undetermined Predictions

To begin our investigation, we formulate the blended Hamming, Jaccard, Exact similarity function and show that it behaves well when using undetermined label values (such as votes or probabilistic predictions $\in [0, 1]$ for each label).

2.1 A Blended Metric

Hamming and Jaccard similarity can be represented in terms of true/false positive/negative counts[2]:

$$\text{Hamming} = \frac{TP + TN}{L} \qquad \text{Jaccard} = \begin{cases} \frac{TP}{TP+FP+FN} & \text{if } TN \neq L \\ 1 & \text{otherwise} \end{cases}$$

where $TP + TN + FP + FN = L$ (the total number of labels). We also note that exact similarity can also be represented in terms of Hamming and Jaccard similarity.

$$\text{Exact} = \lim_{\beta \to \infty} \text{Hamming}^{\beta} = \lim_{\beta \to \infty} \text{Jaccard}^{\beta}$$

Of the three similarity functions, Hamming similarity is the simplest to optimise over. Hamming similarity is the mean of the accuracy of each label prediction and therefore, each label can be optimised individually (details in [2]). Both Jaccard and Exact similarity depend on the collection of labels, so the optimisation is non-decomposable.

We reason that a blending of Hamming similarity with either Jaccard or Exact similarity provides a greater optimisation of Jaccard or Exact similarity. The insight to this is that the Hamming optimisation search space is likely to be less chaotic than the Jaccard and Exact optimisation search spaces, providing a simpler optimisation path. On the other hand, this may lead to non-optimal solutions. Therefore we will investigate this effect throughout the rest of this paper.

We propose the blended Hamming, Jaccard similarity function:

$$s(\mathbf{y}, \mathbf{t}; \alpha) = \frac{TP + \alpha TN}{TP + FN + FP + \alpha TN} \qquad (4)$$

$$= \frac{TP + \alpha TN}{L - (1 - \alpha)TN} \qquad (5)$$

given the vector of true labels $\mathbf{t}$, the vector of determined or undetermined predictions $\mathbf{y}$, and a parameter α which controls the blending. This metric provides Hamming similarity when $\alpha = 1$ and Jaccard when $\alpha = 0$. We additionally note that this form avoids the divide-by-zero problem of the Jaccard similarity. By taking the limit as $\alpha \to 0$, the equation results in $s(\mathbf{y}, \mathbf{t}; \alpha \to 0) = TN/TN = 1$, when TP, FN and FP are zero.

2.2 Behaviour Using Undetermined Labels

The blended similarity function is a function of true/false positive/negative counts. Optimising over these functions is difficult due to the discrete nature

---

[2] Throughout this work we refer to F-measure and Jaccard in an instance-wise evaluation context (noting that these metrics can also be used in a micro- or macro-averaging context).

of these counts, which leads to a discontinuous optimisation function and a gradient containing values of either zero of infinity.

Of course, many multi-label classifiers compute undetermined label values z_{kj} in the $[0, 1]$ range, that are used to obtain the label predictions $y_{kj} \in \{0, 1\}$, such that

$$y_{kj} = \mathbb{I}\left[z_{kj} \geq \tau_j\right] \tag{6}$$

is the class prediction for the j-th label ($\mathbb{I}[\cdot] = 1$ if the inner condition holds), where τ_j is some threshold, typically set to 0.5 but may also be tuned (see, e.g., [4,9]) either per label or the same for all labels $j = 1, \ldots, L$. The vector $\mathbf{z}_k = [z_{k1}, \ldots, z_{kL}]$ contains the undetermined label values and in many cases $z_{kj} \approx P(y_{kj} = 1 | \mathbf{x}_k)$ is related to probabilistic models or approximations[3], or is a normalised sum of ensemble votes (e.g., [1,10,12]).

Using these undetermined label values z_{kj} to obtain undetermined true/false positive/negative values, provides a continuous optimisation space, and a gradient for gradient-based optimisation. For example, given the three labels $\mathbf{y}_k = [1\ 0\ 1]$ and the predicted undetermined label values $\mathbf{z}_k = [0.6\ 0.4\ 0.2]$, the undetermined values would be rounded to obtain determined the values $\mathbf{y}_k = [1\ 0\ 0]$, giving TP = 1, FP = 0, FN = 1, TN = 1. Using the undetermined values instead, we get TP = 0.8, FP = 0.4, FN = 1.2, TN = 0.6. For both cases TP + TN + FP + FN = 3.

The undetermined values can also be used for evaluation, such as in the case of the log loss metric (applied to multi-label evaluation in, e.g., [9]). Nevertheless, such metrics are relatively less popular in the multi-label literature, perhaps because not all classifiers can provide them. Nevertheless, they help considerably in the optimisation to smooth out the search space.

2.3   Blending the Exact Similarity

TP and TN are dependent only on the correctness of the predictions (the correctly predicted 1 and 0 values), so we instead use the vector of correctness values $\mathbf{p}_k = [p_{k1}, \ldots, p_{kL}]$, where p_{ki} (each element of $\mathbf{p}_k$) is equal to

$$\begin{cases} z_{ki} & \text{when } t_{ki} = 1 \\ 1 - z_{ki} & \text{when } t_{ki} = 0 \end{cases} \tag{7}$$

where t_{ki} is the true label. The combined Hamming-Jaccard score using undetermined labels can now be represented as

$$s(\mathbf{z}_k, \mathbf{t}_k; \alpha) = \frac{\mathbf{p}_k^\top \mathbf{a}_k}{L + \mathbf{p}_k^\top \mathbf{b}_k}$$

where $\mathbf{a}_k$ contains 1 when true label $t_{ki} = 1$ and α when true label is 0; and $\mathbf{b}_k$ contains 0 when the true label $t_{ki} = 1$ and $\alpha - 1$ when the true label is 0. The set of vectors $\mathbf{a}_k$ and $\mathbf{b}_k$ are constant for the optimisation.

---

[3] Note, however, that metrics like F-measure and Jaccard measure cannot be optimised only under consideration of marginal probabilities.

To include the Exact similarity into this spectrum, we add the parameter β

$$s(\mathbf{z}_k, \mathbf{t}_k; \alpha, \beta) = \left[\frac{\mathbf{p}_k^\mathsf{T} \mathbf{a}_k}{L + \mathbf{p}_k^\mathsf{T} \mathbf{b}_k} \right]^\beta$$

which is equivalent to Hamming similarity when $\alpha = 1, \beta = 1$, and equivalent to Jaccard similarity when $\alpha \to 0$, $\beta = 1$ and to Exact when $\beta \to \infty$. A continuous spectrum exists between the three similarity functions for $\alpha \in (0, 1]$ and $\beta \in [1, \infty)$.

Given that we obtain the Exact similarity as $\beta \to \infty$, let us look at how α effects the rate at which the limit is approached, and if so which value of α provides the fastest rate of convergence. Namely, if α is adjusted to $\alpha + \delta$, where $\delta > 0$, the score changes to:

$$s(\mathbf{z}_k, \mathbf{t}_k; \alpha + \delta, \beta) = \left[\frac{\mathbf{p}_k^\mathsf{T} \mathbf{a}_k + \delta(1 - \mathbf{t}_k)}{L + \mathbf{p}_k^\mathsf{T} \mathbf{b}_k + \delta(1 - \mathbf{t}_k)} \right]^\beta \tag{8}$$

This value is bound between 0 and 1, so this addition of $\delta(1 - \mathbf{t}_k)$ to the numerator and denominator will increase the score. And, thus,

$$s(\mathbf{z}_k, \mathbf{t}_k; \alpha, \beta) \leq s(\mathbf{y}_k, \mathbf{t}_k; \alpha + \delta, \beta)$$

for $0 \leq \alpha \leq [\alpha + \delta] \leq 1$. The limit as $\beta \to \infty$ will approach the correct Exact score faster when any score that is not 1, is closer to zero (since Exact requires that all fractional scores should be mapped to zero). Since decreasing α decreases the score, the limit will approach the correct score faster for smaller α. Therefore, we expect that the best estimate of Exact will occur when $\alpha \to 0$ and $\beta \to \infty$.

3   Optimising the Combined Metric

In this section, we present the linear model, the optimisation function used to fit the linear model, a derivation of the optimisation gradient and inclusion of a penalty to deter over-fitting.

3.1   Optimisation Function

Suppose the linear model

$$\log \left(\frac{\mathbf{z}_k}{1 - \mathbf{z}_k} \right) = W \mathbf{x}_k$$

with unknown matrix W, undetermined prediction vector $\mathbf{z}_k \in (0, 1)^L$ and feature vector $\mathbf{x}_k \in \mathbb{R}^M$. This model resembles a set of L parallel logistic regressions, but unlike logistic regression, we are not determining the W that maximises the likelihood of the data. We compute W that maximises the chosen evaluation function score $s(\mathbf{z}_k, \mathbf{t}_k; \alpha, \beta)$ for all k. To optimise the average score, the optimisation problem becomes:

$$\max_{W} \frac{1}{K} \sum_{k=1}^{K} s(\mathbf{z}_k, \mathbf{t}_k; \alpha, \beta) \tag{9}$$

where W is the weight matrix containing the elements w_{ij}, $s_k \in [0, 1)$ is the score for each of the K objects.

$$s(\mathbf{z}_k, \mathbf{t}_k; \alpha, \beta) = s_k = \left[\frac{\sum_{i=1}^{L} p_{ki} a_{ki}}{L + \sum_{i=1}^{L} p_{ki} b_{ki}} \right]^{\beta}$$

L is the number of labels, $p_{ki} \in (0, 1)$ is the correctness of the undetermined label prediction z_{ki} (recall Eq. (7))

$$p_{ki} = z_{ki}^{t_{ki}} (1 - z_{ki})^{(1-t_{ki})}$$

and $z_{ki} \in (0, 1)$ is the sigmoid of the mapped feature vector $\mathbf{x}_k$

$$z_{ki} = \frac{1}{1 + \exp\left(-\sum_j w_{ij} x_{kj}\right)}$$

where x_{kj} is an element of the feature vector $\mathbf{x}_k$, and the element 1 is appended to $\mathbf{x}_k$ as a bias term. The constants $a_{ki} \in \{\alpha, 1\}$ and $b_{ki} \in \{0, 1 - \alpha\}$ depend only on the optimisation parameter α and the true label values t_{ki}.

$$a_{ki} = t_{ki} + \alpha(1 - t_{ki}) \qquad b_{ki} = (\alpha - 1)(1 - t_{ki})$$

$\alpha \in (0, 1]$ and $\beta \in [1, \infty)$ are parameters to set the desired optimisation (e.g. $\alpha = 1$, $\beta = 1$ for Hamming similarity). Also note that $a_{ki} - b_{ki} = 1$.

The optimisation of Eq. (9) provides us with W which we can use to make determined predictions by

$$\mathbf{y} = \text{sign}(W\mathbf{x}_k)$$

where sign returns the L signs of the L elements in $W\mathbf{x}_k$, making sure that 1 is appended to $\mathbf{x}$ if it was used to compute a bias term during optimisation. Note that the bias term has the same role as the threshold described in Sect. 2.2.

3.2   Metric Gradient

We can optimise the similarity function using gradient descent or stochastic gradient descent for large problems. To do so, we need the gradient of the function with respect to each elements of the matrix W. The derivative of the sigmoid function is

$$\frac{dz_{ki}}{dw_{ij}} = x_{kj} z_{ki} (1 - z_{ki})$$

The derivative of the correctness function is

$$\frac{dp_{ki}}{dz_{ki}} = p_{ki} \left[\frac{t_{ki}}{z_{ki}} - \frac{1 - t_{ki}}{1 - z_{ki}} \right]$$

and the derivative of the score function with respect to the correctness is

$$\frac{ds_k}{dp_{ki}} = \beta s_k \left[\frac{a_{ki}}{\sum_{i=1}^{L} p_{ki} a_{ki}} - \frac{b_{ki}}{L + \sum_{i=1}^{L} p_{ki} b_{ki}} \right]$$

Combining the partial derivatives provides the derivative of the score function with respect to the unknown weights;

$$\frac{ds_k}{dw_{ij}} = \frac{ds_k}{dp_{ki}} \frac{dp_{ki}}{dz_{ki}} \frac{dz_{ki}}{dw_{ij}} = \beta s_k p_{ki}(t_{ki} - z_{ki}) \left[\frac{a_{ki}}{\sum_i^{L} p_{ki} a_{ki}} - \frac{b_{ki}}{L + \sum_i^{L} p_{ki} b_{ki}} \right] x_{kj} \tag{10}$$

Using gradient descent, we update weights W until the optimisation function stops increasing,

$$w_{ij} \leftarrow w_{ij} + \lambda \frac{ds_k}{dw_{ij}}$$

where λ controls the rate of convergence.

Note that if $\alpha = 1$ and $\beta = 1$ (giving Hamming similarity), we obtain $a_{ki} = 1$, $b_{ki} = 0$ and $s_k = \sum_i^{L} p_{ki}/L$ giving the gradient

$$\frac{ds_k}{dw_{ij}} = \frac{p_{ki}(t_{ki} - z_{ki})x_{kj}}{L}$$

$$= \begin{cases} \frac{z_{ki}(1-z_{ki})x_{kj}}{L} & \text{if } t_{ki} = 1 \\ \frac{-z_{ki}(1-z_{ki})x_{kj}}{L} & \text{if } t_{ki} = 0 \end{cases} = \frac{(-1)^{1-t_{ki}}}{L} \frac{dz_{ki}}{dw_{ij}}$$

Thus we see that the gradient that optimises Hamming similarity with respect to w_{ij} is independent of the weights for other values of i, meaning that each label can be fitted independently.

3.3 Optimisation Penalty

To reduce the chance of over-fitting the training data, we include a penalisation term. Thus Eq. (9) becomes

$$\max_W \frac{1}{K} \sum_{k=1}^{K} s(\mathbf{z}_k, \mathbf{t}_k; \alpha, \beta) - \underbrace{\frac{\gamma}{2} \sum_{i=1} \sum_{j=1} w_{ij}^2}_{\text{penalty}}$$

for some positive γ, typically chosen via cross-validation. We make a corresponding minor adjustment to the gradient, hence Eq. (10) becomes

$$\frac{ds_k}{dw_{ij}} = \beta s_k p_{ki}(t_{ki} - z_{ki}) \left[\frac{a_{ki}}{\sum_i^{L} p_{ki} a_{ki}} - \frac{b_{ki}}{L + \sum_i^{L} p_{ki} b_{ki}} \right] x_{kj} - \gamma w_{ij}$$

4   Parameter Investigation

Having proposed a parametrisable blended metric, and derived its optimiser, we now turn to explore the question: Does blending the Hamming similarity with either Jaccard or Exact similarity provide a greater optimisation of Jaccard or Exact similarity itself?

We answer this question by optimising the model over different data sets, using different α, β parameter combinations and examining the results using the three similarity functions.

Namely, we use the Emotions, Enron, Scene, Slashdot, Stare and Yeast multi-label data sets[4]. Each of these data sets are commonly used for multi-label machine learning except for Stare, which is a data set for detecting cardiovascular disease from retinal features [6]. A 50/50 train/test split is used and the models are optimised over the training portion. We experiment with values of α from 0.1 to 1 in increments of 0.1 and also $\alpha = 0.00001$ to approximate the limit of α approaching zero ($\alpha = 0$ was not used to avoid the problems associated to the Jaccard metric). We also experiment with eight values $\beta \in \{2^0, \ldots, 2^7\}$ (higher values lead to approximately zero gradients), giving 88 $\alpha \times \beta$ combinations. The penalty parameter γ was computed using cross validation. Each of the 88 optimised models for each data set was evaluated using Hamming, Jaccard and Exact similarity to examine the effect of the parameters on the evaluation scores. The optimisation was performed using undetermined scores, but the evaluation scores (shown in the figures) are computed using the final determined labels. The results for the training data are shown in Fig. 1, and the results for the testing data are shown in Fig. 2. The top six plots in each figure show the Hamming similarity on each of the six data sets, the middle six show the Jaccard similarity, and the bottom six show the Exact accuracy. The shade of each block in a given plot shows the accuracy of the model optimised using the set α, β parameters. For example, the first plot in Fig. 1 shows the Hamming similarity on the training portion of the Emotions data; the bottom row of the plot shows the grey level transitioning from light to dark grey, meaning that when $\beta = 1$ ($\log_2 \beta = 0$) the Hamming accuracy increases as α changes from 0 to 1.

5   Results and Discussion

The top sections of Figs. 1 and 2 show the mean Hamming similarity between the predicted and true label sets for the six data sets on the training and testing portions. Each block in the figures show the effect of changing α and β on the Hamming similarity. We would expect that the optimal configuration for Hamming similarity is $\alpha = 1$, $\beta = 1$ (the lower right corner of each block plot) since it optimises Hamming similarity. We can see that this is the case for most of the data sets for both training and testing. For the remaining cases, we find that

---

[4] All available from http://mulan.sourceforge.net/datasets-mlc.html, https://sourceforge.net/projects/meka/files/Datasets/ (Slashdot), and http://www.ces.clemson.edu/~ahoover/stare/ (Stare).

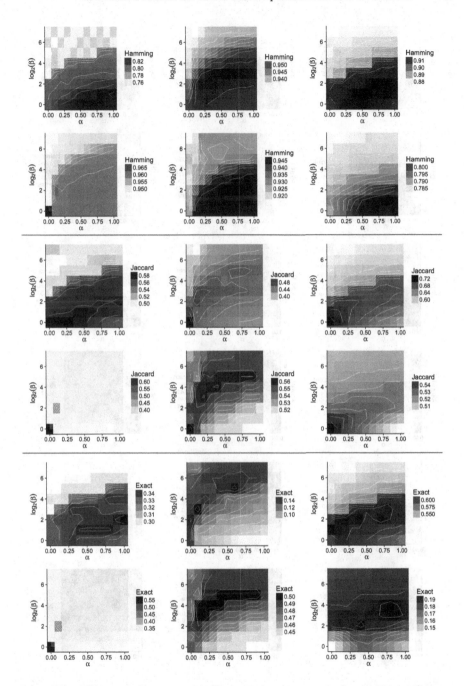

Fig. 1. The Hamming (top section), Jaccard (middle section) and Exact (bottom section) *training* similarity of the (top row from left to right in each section) Emotions, Enron, Scene (bottom row from left to right in each section) Slashdot, Stare and Yeast data, optimised using the shown values of α and β.

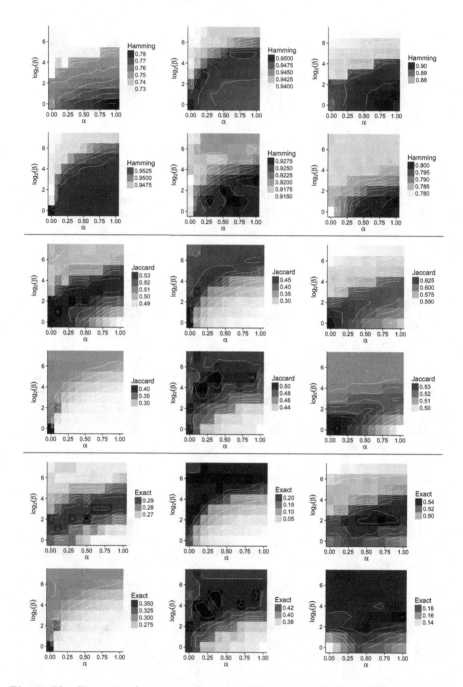

Fig. 2. The Hamming (top section), Jaccard (middle section) and Exact (bottom section) *testing* similarity of the (top row from left to right in each section) Emotions, Enron, Scene (bottom row from left to right in each section) Slashdot, Stare and Yeast data, optimised using the shown values of α and β.

Table 2. The location (α, β) of the optimal Jaccard scores, and their Base score ($\alpha = 0$, $\beta = 1$). A dash means that the optimal score was located at the Base score position.

Data	Type	α	β	Jaccard	Base
Emotions	Train	0.3	1	0.580	0.580
	Test	0.1	2	0.535	0.535
Stare	Train	1e−05	1	-	0.561
	Test	0.1	16	0.505	0.501
Scene	Train	1e−05	1	-	0.720
	Test	1e−05	1	-	0.643
Yeast	Train	1e−05	1	-	0.551
	Test	1e−05	2	0.534	0.533
Slashdot	Train	1e−05	1	-	0.628
	Test	1e−05	1	-	0.407
Enron	Train	1e−05	1	-	0.514
	Test	1e−05	2	0.489	0.488

the accuracy difference between the $\alpha = 1$ and $\beta = 1$ and optimal configuration is small (of the order of 0.01 or less).

The middle sections of Figs. 1 and 2 show the Jaccard scores for each of the training and testing data sets. The model optimises over Jaccard similarity when $\alpha \to 0$ and $\beta = 1$. Therefore it is expected that the Jaccard scores are optimal or close to optimal for precisely these values (bottom left corners of the block plots).

The plots show that the Jaccard score is high in the lower left corner, but it also remains high as both α and β increase (moving diagonally along the block plot). Slashdot training is a notable exception, where the lower left corner seems to be a "sweet spot", obtaining an approximate 0.2 increase in Jaccard similarity over the rest of the plot. We note that this is not as dramatic for the Slashdot test data, but the lower left corner still provides a high Jaccard score relative to the remaining configurations.

The lower sections of Figs. 1 and 2 show the Exact match scores for each of the training and testing sets. Exact similarity is being optimised when $\beta \to \infty$ (the top of the plots). However, these plots show interesting and surprising behaviour, with many of them being similar to the Jaccard plots. In other cases there does not seem to be consistency in the optimal regions, which seems to be located at about $\alpha = 0.5$, with fluctuating values of β.

We now proceed to examine the difference in accuracy of the optimal α, β configuration and expected α, β configuration. Tables 2, 3 and 4 contain these details for the Jaccard, Hamming and Exact similarity functions.

Table 2 shows the Jaccard scores, where Base is the score for the expected optimal configuration ($\alpha \to 0$, $\beta = 1$). We find that seven of the twelve data sets provide the optimal score at the expected configuration, while the remaining

Table 3. The location (α, β) of the optimal Hamming scores, and their Base score ($\alpha = 1$, $\beta = 1$). A dash means that the optimal score was located at the Base score position.

Data	Type	α	β	Hamming	Base
Emotions	Train	1	2	0.827	0.823
	Test	1	1	-	0.783
Stare	Train	0.6	4	0.946	0.944
	Test	0.3	2	0.927	0.927
Scene	Train	0.7	4	0.916	0.912
	Test	0.8	1	0.901	0.899
Yeast	Train	0.9	1	0.804	0.804
	Test	1	2	0.799	0.798
Slashdot	Train	1e−05	1	0.968	0.959
	Test	1e−05	1	0.954	0.953
Enron	Train	0.6	8	0.954	0.952
	Test	1e−05	1	0.950	0.947

five provide a very slight increase (at most 0.004) over the Base score. Thus we confirm that when evaluating using Jaccard similarity, we should optimise with respect to Jaccard similarity.

Table 3 shows the Hamming scores, where Base is the score for the expected optimal configuration ($\alpha = 1$, $\beta = 1$). Similar conclusions can me made for Hamming similarity: when evaluating using this metric, we should likewise optimise using Hamming similarity; as expected.

Table 4 shows the Exact scores, where BaseH and BaseJ give the scores for the expected optimal configuration ($\alpha = 1$ or $\alpha \to 0$, $\beta = 2^7$). Only one of the data sets provides the optimal score at the expected $\beta = 2^7$. We also find that there are large differences between the optimal scores and the BaseH and BaseJ scores. There are many values of small α (nine values ≤ 0.2). But it is clearly seen that the selection of α and β for optimal Exact similarity is dependent on the data.

Therefore, our findings may be summarised as the following recommendations: if optimising Hamming: use $\alpha = 1$, $\beta = 1$; if Jaccard: use $\alpha \to 0$, $\beta = 1$; and – we emphasise – for Exact similarity: we should in fact explore the α and β space. This has important and far-reaching implications, since Exact similarity is a widely used metric, and often used to promote novel classifiers because classifiers that model labels together are more likely to outperform the independent baseline under this metric. With a more careful and exploratory optimisation scheme as we define, we have showed how it is possible to achieve even higher predictive performance.

Table 4. The location (α, β) of the optimal Exact scores, and their Jaccard Base score BaseJ $(\alpha = 0, \beta = 128)$ and Hamming Base score BaseH $(\alpha = 1, \beta = 128)$.

Data	Type	α	β	Exact	BaseJ	BaseH
Emotions	Train	1	4	0.348	0.296	0.296
	Test	0.5	4	0.297	0.265	0.265
Stare	Train	0.1	8	0.500	0.470	0.490
	Test	1e−05	4	0.439	0.427	0.427
Scene	Train	1e−05	2	0.624	0.526	0.532
	Test	0.7	4	0.545	0.480	0.489
Yeast	Train	0.1	8	0.192	0.185	0.186
	Test	0.2	16	0.194	0.191	0.190
Slashdot	Train	1e−05	1	0.553	0.351	0.351
	Test	1e−05	1	0.360	0.300	0.301
Enron	Train	0.1	8	0.150	0.115	0.137
	Test	0.2	128	0.239	0.205	0.229

5.1 Conclusions and Future Work

We have analysed multi-label evaluation, and outlined the difficulties in optimising several of the well-known loss metrics. To tackle the issues that arise in multi-label optimisation, we proposed a surrogate loss, in the form of a blended metric. This blended metric forms a smooth spectrum between the Hamming and exact match metrics, where it falls depending on its parameterisation. Using particular parameterisations of this function, we show that optimisation is more effective, on account of its smoothness. Indeed, for example we demonstrated that one can obtain better results under exact match by optimising the blended metric, than by optimising exact match directly (which is difficult to do). This is important because exact match is a common metric in the literature and many empirical evaluations are based around it. We also made a series of other recommendations to multi-label researchers, in reflection of our findings.

To fully evaluate the potential of our proposal under a complete range of contexts, further experimental comparison will be necessary, for example with structured prediction like probabilistic classifier chains under different inference algorithms, and structural SVMs. We cannot claim that our proposal optimises in a Bayes optimal way, due to the approximation used; Further theoretical analysis is needed on this front, particularly under exact similarity, Jaccard, and F-measure. Finally, we can point out that with some modification, the blended metric could be used to maximize F-measure – a promising line of future investigation.

References

1. Dembczyński, K., Cheng, W., Hüllermeier, E.: Bayes optimal multilabel classification via probabilistic classifier chains. In: 27th International Conference on Machine Learning, ICML 2010, Haifa, Israel, pp. 279–286. Omnipress, June 2010
2. Dembczyński, K., Waegeman, W., Cheng, W., Hüllermeier, E.: On label dependence and loss minimization in multi-label classification. Mach. Learn. **88**(1–2), 5–45 (2012)
3. Godbole, S., Sarawagi, S.: Discriminative methods for multi-labeled classification. In: Dai, H., Srikant, R., Zhang, C. (eds.) PAKDD 2004. LNCS (LNAI), vol. 3056, pp. 22–30. Springer, Heidelberg (2004). https://doi.org/10.1007/978-3-540-24775-3_5
4. Largeron, C., Moulin, C., Géry, M.: MCut: a thresholding strategy for multi-label classification. In: Hollmén, J., Klawonn, F., Tucker, A. (eds.) IDA 2012. LNCS, vol. 7619, pp. 172–183. Springer, Heidelberg (2012). https://doi.org/10.1007/978-3-642-34156-4_17
5. Madjarov, G., Kocev, D., Gjorgjevikj, D., Džeroski, S.: An extensive experimental comparison of methods for multi-label learning. Pattern Recognit. **45**(9), 3084–3104 (2012)
6. Nguyen, U.T., et al.: An automated method for retinal arteriovenous nicking quantification from color fundus images. IEEE Trans. Biomed. Eng. **60**(11), 3194–3203 (2013)
7. Park, L.A.F., Simoff, S.: Using entropy as a measure of acceptance for multi-label classification. In: Fromont, E., De Bie, T., van Leeuwen, M. (eds.) IDA 2015. LNCS, vol. 9385, pp. 217–228. Springer, Cham (2015). https://doi.org/10.1007/978-3-319-24465-5_19
8. Read, J., Martino, L., Luengo, D.: Efficient Monte Carlo methods for multidimensional learning with classifier chains. Pattern Recognit. **47**(3), 1535–1546 (2014)
9. Read, J., Pfahringer, B., Holmes, G., Frank, E.: Classifier chains for multi-label classification. Mach. Learn. **85**(3), 333–359 (2011)
10. Read, J., Puurula, A., Bifet, A.: Multi-label classification with meta labels, In: IEEE International Conference on Data Mining (ICDM 2014), pp. 941–946. IEEE, December 2014
11. Tsoumakas, G., Katakis, I., Vlahavas, I.: Mining multi-label data. In: Maimon, O., Rokach, L. (eds.) Data Mining and Knowledge Discovery Handbook, pp. 667–685. Springer, Boston (2009). https://doi.org/10.1007/978-0-387-09823-4_34
12. Tsoumakas, G., Katakis, I., Vlahavas, I.: Random k-labelsets for multi-label classification. IEEE Trans. Knowl. Data Eng. **23**(7), 1079–1089 (2011)
13. Waegeman, W., Dembczyńki, K., Jachnik, A., Cheng, W., Hüllermeier, E.: On the bayes-optimality of f-measure maximizers. J. Mach. Learn. Res. **15**(1), 3333–3388 (2014)
14. Wu, X., Zhou, Z.: A unified view of multi-label performance measures. In: ICML, vol. 70, pp. 3780–3788. PMLR (2017)
15. Zhang, M.-L., Zhou, Z.-H.: A review on multi-label learning algorithms. IEEE Trans. Knowl. Data Eng. **26**(8), 1819–1837 (2014)

Author Index

Printed in the United States
By Bookmasters